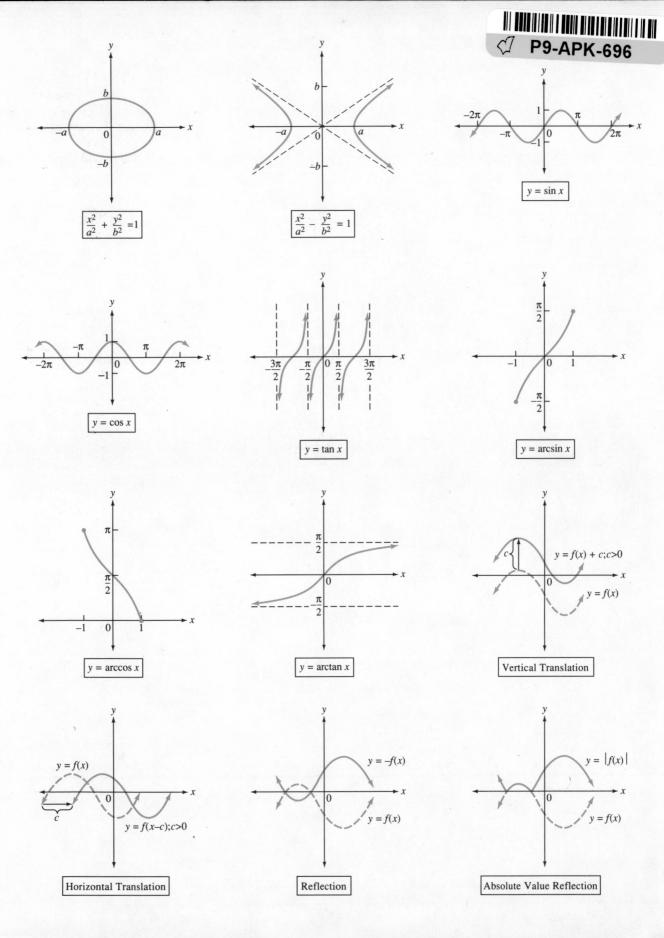

$$\frac{x^2}{a^2} + \frac{y^2}{b^2} = 1$$

$$\frac{x^2}{a^2} - \frac{y^2}{b^2} = 1$$

$y = \sin x$

$y = \cos x$

$y = \tan x$

$y = \arcsin x$

$y = \arccos x$

$y = \arctan x$

Vertical Translation

$y = f(x) + c; c > 0$

$y = f(x)$

Horizontal Translation

$y = f(x)$

$y = f(x-c); c > 0$

Reflection

$y = -f(x)$

$y = f(x)$

Absolute Value Reflection

$y = |f(x)|$

$y = f(x)$

ALGEBRA
AND
TRIGONOMETRY

Fourth Edition

ALGEBRA
AND
TRIGONOMETRY

Max A. Sobel
Montclair State College

Norbert Lerner
State University of New York at Cortland

PRENTICE HALL, Englewood Cliffs, NJ 07632

Library of Congress Cataloging-in-Publication Data

Sobel, Max A.
    Algebra and trigonometry / Max A. Sobel,
  Norbert Lerner. – 4th ed.
     p.   cm.
   ISBN 0-13-025818-0
   1. Algebra. 2. Trigonometry.   I. Lerner, Norbert.   II. Title.
QA154.2.S59 1991
512'.13–dc20
                                       90-35447
                                         CIP

Editorial/production supervision: Rachel J. Witty, Letter Perfect, Inc.
Acquisitions editor: Priscilla McGeehon
Interior design: Meryl Poweski
Cover design: Meryl Poweski
Prepress buyer: Paula Massenaro
Manufacturing buyer: Lori Bulwin

© 1991 by Prentice-Hall, Inc.
A Division of Simon & Schuster
Englewood Cliffs, New Jersey 07632

Printed in the United States of America
10 9 8 7 6 5 4 3 2 1

ISBN 0-13-025818-0

Prentice-Hall International (UK) Limited, *London*
Prentice-Hall of Australia Pty. Limited, *Sydney*
Prentice-Hall Canada Inc., *Toronto*
Prentice-Hall Hispanoamericana, S.A., *Mexico*
Prentice-Hall of India Private Limited, *New Delhi*
Prentice-Hall of Japan, Inc., *Tokyo*
Simon & Schuster Asia Pte. Ltd., *Singapore*
Editora Prentice-Hall do Brasil, Ltda., *Rio de Janeiro*

CONTENTS

CHAPTER 10
LINEAR SYSTEMS, MATRICES, AND DETERMINANTS   503

CHAPTER 11
SEQUENCES AND SERIES   558

CHAPTER 12
PERMUTATIONS, COMBINATIONS, PROBABILITY   601

PREFACE

Algebra and Trigonometry has been written to provide the essential concepts and skills of algebra, trigonometry and the study of functions, which are needed for further study in mathematics. Since calculus is a subject numerous students study after this course, a special emphasis is given to the preparation for the study of calculus. Thus, one of the *major objectives* of the book is *to help you make a comfortable transition from elementary mathematics to calculus*. (However, the objectives of your particular course may not require this special pre-calculus emphasis. You will find out from your instructor or from the course syllabus.)

A major difficulty that students have in future mathematics courses, especially in calculus, involves the lack of adequate algebraic skills. To help you overcome such deficiencies, an extensive review of the fundamentals of algebra has been included in Chapter 1. You are encouraged to refer to this chapter throughout the course if you encounter algebraic difficulties.

Of special interest are the numerous features in this book that have been designed to assist you in learning the subject matter of the course. These features are listed below, with descriptions of their purpose and suggested use.

Margin Notes

Throughout the text, notes appear in the margin to enhance the exposition, raise questions, point out interesting facts, explain why some things are done as they are, show alternate procedures, give references and reminders, and caution you to avoid errors.

Test Your
Understanding

These are short sets of exercises (in addition to the end-of-section exercises) that are found within most sections of the text. These encourage you to *think carefully* and to test your knowledge of new material just developed, prior to attempting to solve the exercises at the end of each section. Answers to all of these are given at the end of each chapter and thus provide an excellent means of self-study. For example, note the following set of exercises, which appears on page 37:

TEST YOUR	*Write each of the following using fractional exponents.*				
UNDERSTANDING					
Think Carefully	**1.** $\sqrt{7}$	**2.** $\sqrt[3]{-10}$	**3.** $\sqrt[4]{7}$	**4.** $\sqrt[3]{7^2}$	**5.** $(\sqrt[4]{5})^3$
	Evaluate.				
	6. $25^{1/2}$	**7.** $64^{1/3}$	**8.** $(\frac{1}{36})^{1/2}$	**9.** $49^{-1/2}$	**10.** $(-\frac{1}{27})^{-1/3}$
(Answers: Page 75)	**11.** $4^{3/2}$	**12.** $4^{-3/2}$	**13.** $(\frac{81}{16})^{3/4}$	**14.** $(-8)^{2/3}$	**15.** $(-8)^{-2/3}$

Where appropriate, these items appear in the margin notes or in the text and alert you to the typical kinds of errors that you should avoid. Frequently the caution will show the errors and the corrections side by side. For example, note the following set of caution items, which appears on page 60:

CAUTION: Learn to Avoid These Mistakes	
WRONG	**RIGHT**
$\dfrac{2}{3} + \dfrac{x}{5} = \dfrac{2 + x}{3 + 5}$	$\dfrac{2}{3} + \dfrac{x}{5} = \dfrac{2 \cdot 5 + 3 \cdot x}{3 \cdot 5} = \dfrac{10 + 3x}{15}$
$\dfrac{1}{a} + \dfrac{1}{b} = \dfrac{1}{a + b}$	$\dfrac{1}{a} + \dfrac{1}{b} = \dfrac{b + a}{ab}$
$\dfrac{2x + 5}{4} = \dfrac{x + 5}{2}$	$\dfrac{2x + 5}{4} = \dfrac{2x}{4} + \dfrac{5}{4} = \dfrac{x}{2} + \dfrac{5}{4}$
$2 + \dfrac{x}{y} = \dfrac{2 + x}{y}$	$2 + \dfrac{x}{y} = \dfrac{2y + x}{y}$
$3\left(\dfrac{x + 1}{x - 1}\right) \neq \dfrac{3(x + 1)}{3(x + 1)}$	$3\left(\dfrac{x + 1}{x - 1}\right) = \dfrac{3(x + 1)}{x - 1}$
$a \div \dfrac{b}{c} = \dfrac{1}{a} \cdot \dfrac{b}{c}$	$a \div \dfrac{b}{c} = a \cdot \dfrac{c}{b} = \dfrac{ac}{b}$
$\dfrac{1}{a^{-1} + b^{-1}} = a + b$	$\dfrac{1}{a^{-1} + b^{-1}} = \dfrac{1}{\dfrac{1}{a} + \dfrac{1}{b}} = \dfrac{ab}{b + a}$
$\dfrac{x^2 + 4x + 6}{x + 2} = \dfrac{x^2 + 4x + \overset{3}{\cancel{6}}}{x + \underset{1}{\cancel{2}}}$ $= \dfrac{x^2 + 4x + 3}{x + 1}$	$\dfrac{x^2 + 4x + 6}{x + 2}$ is in simplest form.

The text contains numerous illustrative examples with detailed solutions, designed to help you to understand new concepts and learn new skills. You should study these examples, and carefully follow each step with paper and pencil in hand. If this is done, you will be well prepared for the exercises at the end of each section.

It is generally agreed that students need to be able to practice writing skills in their mathematics courses. Therefore, throughout the book, written assignments will be found that ask you to write an explanation or description rather than just to solve a problem. These are designated in the text by this symbol: ▐▭▷

At several places in each chapter you will be challenged to *think creatively* and attempt to solve a problem that is more difficult than others, or that has an unusual twist to it. For example, consider the challenge that appears on page 13:

Two boats begin their journeys back and forth across a river at the same time, but from opposite sides of the river. The first time that they pass each other, they are 700 feet from one of the shores of the river. After they each make one turn, they pass each other once again at a distance of 400 feet from the other shore. Assuming that each boat goes at a constant speed and that there is no loss of time making the turn, how wide is the river?

To solve this problem, you might consider the problem-solving strategy of drawing a diagram and then using it to find the solution as follows:

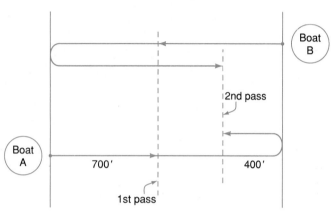

Explorations

Within each chapter there are sets of exercises that encourage you to *think critically*. These demand higher-order thinking skills and do not depend alone on the routine application of skills. For example, consider this set of Explorations, which appear on page 63:

EXPLORATIONS
Think Critically

1. Explain why for any real number b, $b \neq 0$, b^0 is defined to be equal to 1.
2. Many students claim that $\sqrt{25} = \pm 5$. Explain why this is incorrect.

3. Provide a convincing argument to explain why $3\left(\dfrac{x+1}{x-1}\right) \neq \dfrac{3x+1}{3x-1}$.

4. Note that $\dfrac{12}{\sqrt{150}} = \dfrac{2}{5}\sqrt{6}$. Is either form preferable to the other? Explain your answer.

5. When factoring polynomials, it is useful first to factor out common factors if possible. What is the advantage in doing this?

6. Translate the following statement into symbolic form:

   The square of the sum of two numbers is *at least as large* as four times the product.

   Prove the statement is true. *Hint:* Assume the inequality is true and work backward.

7. Three numbers can be arranged vertically in fraction form to produce these cases:

$$\frac{\dfrac{a}{b}}{c} = \frac{a}{bc} \qquad \frac{a}{\dfrac{b}{c}} = \frac{ac}{b}$$

   Find all possible distinct cases using the four numbers a, b, c, and d arranged vertically in fractional form.

Review Exercises

Each chapter has a set of review exercises that are exactly the same as the illustrative examples developed in the text. Thus you can use this as a self-study review of the chapter by comparing your results with the worked-out solutions that can be found in the body of the chapter.

Key Terms

At the end of each chapter a list of key terms will be found. These allow you to test your knowledge of new items found in that particular chapter and provide for yet another means of self-study.

Chapter Tests

Each chapter concludes with two forms of a chapter test: standard answer and multiple choice. You should use these to test your knowledge of the work of the chapter, and check your answers with those provided at the back of the book.

Boxed Displays

Boxed displays for important results, definitions, formulas, and summaries appear throughout the text and serve to alert you to major concepts and results.

Inside Covers

The inside of the covers contain summaries of useful information. The front cover contains a collection of basic graphs; the back contains useful algebraic, geometric, and trigonometric formulas.

Solutions Manual

There is a solutions manual accompanying this text that includes worked-out solutions to all the odd numbered section exercises, and for all of the questions in the chapter tests. You may wish to check with your instructor about this.

There are two other important features that are identified by special symbols:

 This symbol indicates that an exercise or set of exercises would best be solved by use of a scientific calculator. It would be advantageous to have such a calculator, and the instructor of the course will probably indicate any limitations for its use. Although calculator displays appear in the text as part of the solution to some problems, the variety of different calculators in common usage is such that you are encouraged to refer to the manual that accompanies your calculator for specific instruction in its use.

 This symbol identifies exercises that are *directly supportive* to topics in calculus. Also, when this symbol appears next to a section heading, it means that the entire section and its exercises fall into this support category.

How does one succeed in a mathematics course? Unfortunately, there is no universal prescription guaranteed to work. However, our experience with many students suggests that the most important thing to do is to *get involved in the mathematical process*. Don't use this text only as a source for exercises. Rather, read the book, attend class regularly, study your class notes, and make use of the special features described earlier. Furthermore, keep up to date. Don't let yourself fall behind; it often leads to poor results. If difficulties arise and you begin to fall behind, then don't hesitate to get additional assistance from your instructor or from a classmate. Working together with a friend can often be beneficial, as long as individual efforts are also made.

Be positive! Don't give up! We are convinced that even if you have an initial negative attitude toward mathematics, an honest attempt to learn it properly will result not only in greater success, but will lead to a self-awareness that you have more ability and talent than you ever gave yourself credit for!

We hope that you will find this book enjoyable, and that you will learn the skills and concepts for future study of mathematics. We encourage you to write to us; your comments, criticisms, and suggestions might be useful for future editions. Also, despite all of our efforts, errors occasionally creep into a book. We would appreciate it if you would call any of these to our attention. We promise to respond with our letter of thanks.

Max A. Sobel
Norbert Lerner

CHAPTER

1

FUNDAMENTALS OF ALGEBRA

1.1 REAL NUMBERS AND THEIR PROPERTIES

Throughout this course we will be working with the **set of real numbers.** Here are some examples of such numbers:

$$6 \qquad -5 \qquad 0 \qquad \frac{3}{4} \qquad \sqrt{2} \qquad -3.25 \qquad \pi$$

We can show these numbers on a **number line,** where each real number is the **coordinate** of some point on the line.

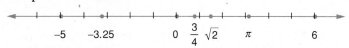

WATCH THE MARGINS!
We will use these for special notes, added explanations, and hints throughout the book.

At times we will use a **subset,** or part, of the real numbers in a discussion. For example, we will deal with such sets of numbers as the following:

The set N of **natural** or **counting numbers**: $\qquad \{1, 2, 3, \ldots\}$

The set W of **whole numbers**: $\qquad \{0, 1, 2, 3, \ldots\}$

The set I of **integers**: $\qquad \{\ldots, -3, -2, -1, 0, 1, 2, 3, \ldots\}$

*The set of natural numbers is also called the **set of positive integers,** and the set of whole numbers is often referred to as the **set of nonnegative integers.***

Note that every integer can be written in fractional form. For example:

$$3 = \frac{3}{1} \qquad -2 = \frac{-2}{1} \qquad 0 = \frac{0}{1}$$

1

However, there are fractions that cannot be written as integers, such as $\frac{2}{3}$ and $-\frac{3}{4}$. The collection of all such integers and fractions is called the set of **rational numbers.**

For every integer a we have $a = \frac{a}{1}$. Therefore, every integer is a rational number.

> A **rational number** is one that can be written in the form $\frac{a}{b}$, where a and b are integers, $b \neq 0$.

In Exercises 57 and 58 you will learn how to convert repeating decimals into the form a/b.

Every rational number a/b can be converted into decimal form by dividing b into a. The decimal form will either *terminate* as in $11/4 = 2.75$, or it will *repeat endlessly* as in

$$\frac{2}{3} = .666 \ldots \qquad \frac{4}{11} = .363636 \ldots$$

Decimals that neither terminate nor repeat are called **irrational numbers** such as

The irrational number Pi (π) is the ratio of the circumference of a circle to its diameter. Using a supercomputer, the decimal form of π has recently been calculated to more than one billion decimal places.

$$\sqrt{3} = 1.73205 \ldots \qquad \pi = 3.14159 \ldots$$

Irrational numbers *cannot* be expressed as the ratio of integers. Some other examples of irrational numbers are

$$\sqrt{5} \qquad \sqrt{12} \qquad \sqrt[3]{4} \qquad -\sqrt{17} \qquad \sqrt[4]{\frac{16}{9}}$$

The collection of rational numbers and irrational numbers comprises the set of **real numbers.** The relationships between these various sets of numbers can be shown by means of this *tree diagram*:

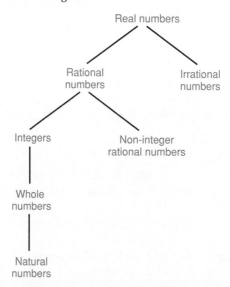

Reading upward, each set of numbers is a subset of the set of numbers listed above it.

EXAMPLE 1 To which subsets of the real numbers do each of the following numbers belong?
(a) 5   **(b)** $\frac{2}{3}$ **(c)** $\sqrt{7}$ **(d)** -14

Solution

(a) 5 is a natural number, a whole number, an integer, a rational number, and a real number.

(b) $\frac{2}{3}$ is a rational number and a real number.

(c) $\sqrt{7}$ is an irrational number and a real number.

(d) -14 is an integer, a rational number, and a real number. ∎

EXAMPLE 2 Classify as true or false: Every whole number is a natural number.

Solution In order for a statement to be true, it must be true for all possible cases; otherwise it is false. Since 0 is a whole number but is *not* a natural number, the statement is false. ∎

0 is the identity element for addition; 1 is the identity element for multiplication.

Throughout this course we shall be using various properties of the set of real numbers, most of which you have encountered before. Here is a summary of some of these important properties.

PROPERTIES OF THE REAL NUMBERS

For all real numbers a, b, and c:	Addition	Multiplication
Closure Properties	$a + b$ is a real number	$a \cdot b$ is a real number
Commutative Properties	$a + b = b + a$	$a \cdot b = b \cdot a$
Associative Properties	$(a + b) + c = a + (b + c)$	$(a \cdot b) \cdot c = a \cdot (b \cdot c)$
Distributive Property	$a(b + c) = ab + ac$	
Identity Properties	$0 + a = a + 0 = a$	$1 \cdot a = a \cdot 1 = a$
Inverse Properties	$a + (-a) = 0$	$a \cdot \dfrac{1}{a} = \dfrac{1}{a} \cdot a = 1, a \neq 0$
Multiplication Property		$0 \cdot a = a \cdot 0 = 0$
Zero-product Property	If $ab = 0$, then $a = 0$ or $b = 0$, or both $a = 0$ and $b = 0$.	

TEST YOUR UNDERSTANDING
Think Carefully

Throughout this book we shall occasionally pause for you to test your understanding of the ideas just presented. If you have difficulty with these brief sets of exercises, you should reread the material in the section before going ahead. Answers are given at the end of the chapter.

Name the property of real numbers being illustrated.

1. $3 + (\frac{1}{2} + 5) = (3 + \frac{1}{2}) + 5$
2. $3 + (\frac{1}{2} + 5) = (\frac{1}{2} + 5) + 3$
3. $6 + 4(2) = 4(2) + 6$
4. $8(-6 + \frac{2}{3}) = 8(-6) + 8(\frac{2}{3})$
5. $(17 \cdot 23)59 = (23 \cdot 17)59$
6. $(17 \cdot 23)59 = 17(23 \cdot 59)$
7. Does $3 - 5 = 5 - 3$? Is there a commutative property for subtraction?
8. Given a *counterexample* to show that the set of real numbers is not commutative with respect to division. (That is, use a specific example to show that $a \div b \neq b \div a$.)
9. Does $(8 - 5) - 2 = 8 - (5 - 2)$? Is there an associative property for subtraction?
10. Give a counterexample to show that the set of real numbers is not associative with respect to division.
11. Are $3 + \sqrt{7}$ and $3\sqrt{7}$ real numbers? Explain.

(Answers: Page 74)

Subtraction

Use this fact to show that multiplication distributes over subtraction:
$a(b - c) = ab - ac.$
Begin with $a[b + (-c)]$.

The preceding properties have been described primarily in terms of addition and multiplication. The basic operations of subtraction and division can now be defined in terms of addition and multiplication, respectively.

The difference $a - b$ of two real numbers a and b is defined as

$$a - b = a + (-b)$$

For example, $5 - 8 = 5 + (-8) = -3$. Alternatively, we say

$$a - b = c \text{ if and only if } c + b = a$$

Thus $5 - 8 = -3$ because $-3 + (8) = 5$.

"p if and only if q" is a form often used in mathematics.

When we say "statement p *if and only if* statement q" it means that if either statement is true, then so is the other. Here we say that $a - b = c$ if and only if $c + b = a$ and thus imply these two results:

Whenever $a - b = c$ is true, then so is $c + b = a$.

Whenever $c + b = a$ is true, then so is $a - b = c$.

These two statements may also be written in the following briefer "if–then" form.

If $a - b = c$, then $c + b = a$.

If $c + b = a$, then $a - b = c$.

Division

The quotient $a \div b$ of two real numbers a and b is defined as

$$a \div b = a \cdot \frac{1}{b} \qquad b \neq 0$$

For example, $8 \div 2 = 8 \times \frac{1}{2} = 4$. Alternatively, we say

$$a \div b = c \text{ if and only if } c \times b = a \qquad b \neq 0$$

Thus $8 \div 2 = 4$ because $4 \times 2 = 8$.

The statement $a \div b = c$ *if and only if* $c \times b = a$ means:

If $a \div b = c$, then $c \times b = a$ and if $c \times b = a$, then $a \div b = c$.

*This is very important: DIVISION BY ZERO IS NOT POSSIBLE. We demonstrate this here by using an **indirect proof**, that is, by assuming the opposite to be true and arriving at a false or contradictory result.*

Using this definition of division we can see why division by zero is not possible. Suppose we assume that division by zero is possible. Assume, for example, that $2 \div 0 = x$, where x is some real number. Then, by the definition of division, $0 \cdot x = 2$. But $0 \cdot x = 0$, leading to the false statement $2 = 0$. This argument can be duplicated where 2 is replaced by any nonzero number. Can you explain why $0 \div 0$ is indeterminate? (See Exercise 59.)

EXERCISES 1.1

List the elements in each set.

1. The set of natural numbers less than 5.
2. The set of whole numbers greater than 100.
3. The set of whole numbers between 2 and 7.
4. The set of integers greater than -3.
5. The set of negative integers greater than -3.
6. The set of positive integers less than 5.
7. The set of integers less than 1.
8. The set of integers that are not whole numbers.
9. The set of whole numbers that are not integers.
10. The set of integers that are also rational numbers.

Answer true or false to each statement. If false, give a specific counterexample to justify your answer. (Recall that a true statement must be true for all possible cases; otherwise it is false.)

11. The set of whole numbers is closed with respect to multiplication.
12. The set of natural numbers is closed with respect to subtraction.
13. The set of integers is closed with respect to division.
14. Except for 0, the set of rational numbers is closed with respect to division.
15. The set of integers is commutative with respect to subtraction.
16. The set of rational numbers is associative with respect to multiplication.
17. The set of rational numbers contains the additive inverse for each of its members.
18. Except for 0, the set of rational numbers contains the multiplicative inverse for each of its members.
19. The product of any two real numbers is a real number.
20. The quotient of any two real numbers is a real number.

Classify each number as a member of one or more of these sets: (a) natural numbers; (b) whole numbers; (c) integers; (d) rational numbers; (e) irrational numbers; (f) real numbers.

21. -15 22. 72 23. $\sqrt{51}$ 24. $-\frac{3}{4}$ 25. $\frac{16}{2}$
26. 0.01 27. 0 28. 1000 29. $\sqrt{12}$ 30. $-\sqrt{2}$

Name the property illustrated by each of the following.

31. $5 + 7$ is a real number.
32. $8 + \sqrt{7} = \sqrt{7} + 8$
33. $(-5) + 5 = 0$
34. $9 + (7 + 6) = (9 + 7) + 6$
35. $(5 \times 7) \times 8 = (7 \times 5) \times 8$
36. $(5 \times 7) \times 8 = 5 \times (7 \times 8)$
37. $\frac{1}{4} + \frac{1}{2} = \frac{1}{2} + \frac{1}{4}$
38. $(4 \times 5) + (4 \times 8) = 4(5 + 8)$
39. $-13 + 0 = -13$
40. $1 \times \frac{1}{9} = \frac{1}{9}$
41. $\frac{1}{2} + (-\frac{1}{2}) = 0$
42. $3 - 7 = 3 + (-7)$
43. $0(\sqrt{2} + \sqrt{3}) = 0$
44. $\sqrt{2} \times \pi$ is a real number.
45. $(3 + 9)(7) = (3)(7) + (9)(7)$
46. $\dfrac{1}{\sqrt{2}} \cdot \sqrt{2} = 1$

Replace the variable n by a real number to make each statement true.

47. $7 + n = 3 + 7$
48. $\sqrt{5} \times 6 = 6 \times n$
49. $(3 + 7) + n = 3 + (7 + 5)$
50. $6 \times (5 \times 4) = (6 \times n) \times 4$
51. $5(8 + n) = (5 \times 8) + (5 \times 7)$
52. $(3 \times 7) + (3 \times n) = 3(7 + 5)$

53. If $5(x - 2) = 0$, then $x = 2$. Explain this statement using an appropriate property of zero.

54. Give an example showing that addition does *not* distribute over multiplication; your example should show that $a + (b \cdot c) \neq (a + b) \cdot (a + c)$.

55. Note that $2^4 = 4^2$. Is the operation of raising a number to a power a commutative operation? Justify your answer.

56. Explain why the real numbers have the closure property for both subtraction and division, excluding division by 0.

57. Every repeating decimal can be expressed as a rational number in the form a/b. Consider, for example, the decimal $0.727272\ldots$ and the following process:

$$\text{Let } n = 0.727272\ldots \text{. Then } 100n = 72.727272\ldots$$

$$100n = 72.727272\ldots$$

$$\text{Subtract:} \qquad \underline{n = 0.727272\ldots}$$

$$99n = 72$$

$$\text{Solve for } n: \qquad n = \frac{72}{99} = \frac{8}{11}$$

Use this method to express each decimal as the quotient of integers and check your answer by division.

(a) $0.454545\ldots$ **(b)** $0.373737\ldots$ **(c)** $0.234234\ldots$
(*Hint:* In part (c) let $n = 0.234234\ldots$; multiply by 1000.)

58. Study the illustration given below for $n = 0.2737373\ldots$ and then convert each repeating decimal into a quotient of integers. Often a bar is placed over the set of digits that repeat. Thus we may write $0.2737373\ldots$ as $0.2\overline{73}$.

Multiply n by 1000: $1000n = 273.\overline{73}$ The decimal point is *behind* the first cycle.

Multiply n by 10: $\underline{10n = 2.\overline{73}}$ The decimal point is in *front* of the first cycle.

Subtract: $990n = 271$

Divide: $n = \dfrac{271}{990}$

(a) $0.4585858\ldots$ **(b)** $3.21444\ldots$ **(c)** $2.0\overline{146}$ **(d)** $0.00\overline{123}$

59. Explain why $\frac{0}{0}$ is indeterminate. (*Hint:* If $\frac{0}{0}$ is to be some value, then it must be a unique value.)

60. Take a circular object (like a tin can) and let the diameter be 1 unit. Use this as a unit on a number line. Mark a point on the circular object with a dot and place the circle so that the dot coincides with zero on the number line. Roll the circle to the right and mark the point where the dot coincides with the number line after one revolution. What number corresponds to this position? Why?

Written Assignment: From a mathematical point of view, it is common practice to introduce the integers after the counting numbers have been discussed and then develop the rational numbers. Historically, however, the positive rational numbers were developed before the negative integers.

(a) Think of some everyday situations where the negative integers are not usually used but could easily be introduced.

(b) Speculate on the historical necessity for the development of the positive rational numbers.

A statement such as $2x - 3 = 7$ is said to be a **conditional equation.** It is true for some replacements of the variable x, but not true for others. For example, $2x - 3 = 7$ is a true statement for $x = 5$ but is false for $x = 9$. On the other hand, an equation such as $3(x + 2) = 3x + 6$ is called an **identity** because it is true for *all* real numbers x.

To *solve* an equation means to find the real numbers x for which the given equation is true; these are called the **solutions** or **roots** of the given equation. Let us solve the equation $2x - 3 = 7$, showing the important steps.

$$2x - 3 = 7 \qquad \text{Add 3 to each side of the equation.}$$
$$(2x - 3) + 3 = 7 + 3$$
$$2x = 10 \qquad \text{Now multiply each side by } \tfrac{1}{2}.$$
$$\tfrac{1}{2}(2x) = \tfrac{1}{2}(10)$$
$$x = 5$$

We can check by substituting 5 for x in the original equation.

$$2(5) - 3 = 10 - 3 = 7$$

Note: We cannot be certain that a solution is correct until it has been checked.

Thus the solution is $x = 5$.

In the preceding solution, we made use of the following two basic *properties of equality*.

*The strength of these two properties is that they produce **equivalent equations,** equations having the same roots. Thus the addition property converts $2x - 3 = 7$ into the equivalent form $2x = 10$.*

ADDITION PROPERTY OF EQUALITY:

For all real numbers a, b, c, if $a = b$, then $a + c = b + c$.

MULTIPLICATION PROPERTY OF EQUALITY:

For all real numbers a, b, c, if $a = b$, then $ac = bc$.

The properties of equality can also be used to solve more complicated equations. The procedure is to *collect all terms with the variable on one side of the equation and all constants on the other side.* Although most of the steps shown can be done mentally, the following example will include the essential details that comprise a formal solution.

EXAMPLE 1 Solve for x: $3(x + 3) = x + 5$

Solution The first step is to eliminate the parentheses by applying the distributive property.

$$3(x + 3) = x + 5$$
$$3x + 9 = x + 5$$
$$3x + 9 + (-9) = x + 5 + (-9) \qquad \text{(addition property)}$$
$$3x = x - 4$$

*This is an example of a **linear equation** because the variable x appears only to the first power.*

$$3x + (-x) = x - 4 + (-x) \qquad \text{(addition property)}$$

$$2x = -4$$

$$\tfrac{1}{2}(2x) = \tfrac{1}{2}(-4) \qquad \text{(multiplication property)}$$

$$x = -2$$

Show that $x = -2$ is the solution by checking. Does $3[(-2) + 3] = (-2) + 5$? ■

TEST YOUR UNDERSTANDING
Think Carefully

(Answers: Page 74)

Solve each linear equation for x.

1. $x + 3 = 9$
2. $x - 5 = 12$
3. $x - 3 = -7$
4. $2x + 5 = x + 11$
5. $3x - 7 = 2x + 6$
6. $3(x - 1) = 2x + 7$
7. $3x + 2 = 5$
8. $5x - 3 = 3x + 1$
9. $x + 3 = 13 - x$
10. $2(x + 2) = x - 5$
11. $2x - 7 = 5x + 2$
12. $4(x + 2) = 3(x - 1)$

The properties of equality can be used to solve equations having more than one variable. The next example shows the use of properties of equality to solve a *formula* for one of the variables in terms of the others.

EXAMPLE 2 The formula relating degrees Fahrenheit and degrees Celsius is $\dfrac{5F - 160}{9} = C$. Solve for F in terms of C.

Solution Try to explain each step shown.

$$\frac{5F - 160}{9} = C$$

$$5F - 160 = 9C$$

$$5F = 9C + 160$$

$$F = (\tfrac{1}{5})(9C + 160)$$

$$F = \tfrac{9}{5}C + 32 \qquad ■$$

Here is an example of the use of this formula: When $C = 20$, $F = \frac{9}{5}(20) + 32 = 36 + 32 = 68$. Thus $20°$ Celsius $= 68°$ Fahrenheit.

Let us now explore the solution of problems that are expressed in words. Our task will be to translate the English sentences of a problem into suitable mathematical language and to develop an equation that we can solve.

In the following problem, guidelines are suggested that are important for critical thinking. Study the solution carefully, as well as those in the examples that follow.

The look of a strategy for solving word problems often creates difficulties for many students. To be a good problem solver you need some patience and much practice.

Problem: The length of a rectangle is 1 centimeter less than twice the width. The perimeter is 28 centimeters. Find the dimensions of the rectangle.

1. *Reread the problem and try to picture the situation given. Make note of all the information stated in the problem.*

The length is one less than twice the width.
The perimeter is 28.

2. *Determine what it is you are asked to find. Introduce a suitable variable, usually to represent the quantity to be found. When appropriate, draw a figure.*

Let w represent the width.
Then $2w - 1$ represents the length.

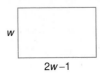

w

$2w-1$

3. *Use the available information to compose an equation that involves the variable.*

The perimeter is the distance around the rectangle. This provides the necessary information to write an equation.

$$w + (2w - 1) + w + (2w - 1) = 28$$

4. *Solve the equation.*

$$w + (2w - 1) + w + (2w - 1) = 28$$
$$6w - 2 = 28$$
$$6w = 30$$
$$w = 5$$

5. *Return to the original problem to see whether the answer obtained makes sense. Does it appear to be a reasonable solution? Have you answered the question posed in the problem?*

The original problem asked for both dimensions. If the width w is 5 centimeters, then the length $2w - 1$ must be 9 centimeters.

6. *Check the solution by direct substitution of the answer into the original statement of the problem.*

As a check, note that the length of the rectangle, 9 centimeters, is 1 centimeter less than twice the width, 5, as given in the problem. Also, the perimeter is 28 centimeters.

7. *Finally, state the solution in terms of the appropriate units of measure.*

The dimensions are 5 centimeters by 9 centimeters.

EXAMPLE 3   A car leaves a certain town at noon, traveling due east at 40 miles per hour. At 1:00 P.M. a second car leaves the town traveling in the same direction at a rate of 50 miles per hour. In how many hours will the second car overtake the first car?

Solution   Problems of motion of this type often prove difficult to students of algebra, but need not be. The basic relationship to remember is that *rate multiplied by time equals distance* ($r \times t = d$). For example, a car traveling at a rate of 60 miles per hour for 5 hours will travel 60×5 or 300 miles.

| Read the problem. |
| List the given information. |

↓

| What is to be found? |
| Introduce a variable and |
| state what the variable |
| represents. Draw a figure |
| or use a table if appropriate. |

↓

| Write an equation. |

↓

| Solve the equation. |

↓

| Does the answer seem |
| reasonable? Have you |
| answered the question |
| stated in the problem? |

↓

| Check your answer to |
| the equation in the |
| original problem. |

↓

| State the solution |
| to the problem. |

Guidelines such as the preceding are useful in organizing your work, which, in turn, will assist you to think more critically and ultimately lead to more success in problem solving.

Now we need to reread the problem to see what part of the information given will help form an equation. The two cars travel at different rates, and for different amounts of time, but both travel the same distance from the point of departure until they meet. This is the clue: *Represent the distance each travels and equate these quantities.*

Let us use x to represent the number of hours it will take the second car to overtake the first. Then the first car, having started an hour earlier, travels $x + 1$ hours before they meet. You may find it helpful to summarize this information in tabular form.

	Rate	Time	Distance
First car	40	$x + 1$	$40(x + 1)$
Second car	50	x	$50x$

Equating the distances, we have an equation that can be solved for x:

$$50x = 40(x + 1)$$
$$50x = 40x + 40$$
$$10x = 40$$
$$x = 4$$

The second car overtakes the first in 4 hours. Does this solution seem reasonable? Let us check the solution. The first car travels 5 hours at 40 miles per hour for a total of 200 miles. The second car travels 4 hours at 50 miles per hour for the same total of 200 miles.

Solution: The second car takes 4 hours to overtake the first. ∎

EXAMPLE 4   David has a total of $4.10 in nickels, dimes, and quarters. He has twice as many nickels as dimes, and two more quarters than dimes. How many dimes does he have?

Solution   Begin by letting x represent the number of dimes since the number of nickels and quarters are both related to the number of dimes. Then from the statement of the problem we can let $2x$ represent the number of nickels, and $x + 2$ the number of quarters.

Next, observe that whenever you have a certain number of a particular coin, then *the total value is the number of coins times the value of that coin.* For example, if you have 8 dimes, then you have 10×8 or 80¢. If you have 9 nickels, then you have 5×9 or 45¢. For this problem:

$10x$ represents the value of the dimes, in cents.
$5(2x)$ represents the value of the nickels, in cents.
$25(x + 2)$ represents the value of the quarters, in cents.

This information is summarized in the following table.

Although some of these problems may not seem to be very practical to you, they will help you develop the basic skills of problem solving that will be helpful later when you encounter more realistic applications.

	Value of each coin (¢)	Number of coins	Value of coins (¢)
Nickels	5	$2x$	$5(2x)$
Dimes	10	x	$10x$
Quarters	25	$x + 2$	$25(x + 2)$

Since the total value of these coins is \$4.10, or 410¢, we write and solve the following equation:

$$\underset{\substack{\downarrow}}{\underset{\text{nickels}}{\text{of}}} \text{Value} + \underset{\substack{\downarrow}}{\underset{\text{dimes}}{\text{of}}} \text{Value} + \underset{\substack{\downarrow}}{\underset{\text{quarters}}{\text{of}}} \text{Value} = \underset{\substack{\downarrow}}{\underset{\text{value}}{\text{Total}}}$$

$$5(2x) + 10x + 25(x + 2) = 410$$

$$10x + 10x + 25x + 50 = 410$$

$$45x + 50 = 410$$

$$45x = 360$$

$$x = 8$$

Always check by returning to the original statement of the problem. If you just check in the constructed equation, and the equation you formed was incorrect, you will not detect the error.

Check: If David has 8 dimes, then he must have $2x$ or 16 nickels, and $x + 2$ or 10 quarters. The total amount is

$$8(10) + 16(5) + 10(25) = 80 + 80 + 250 = 410, \quad \text{that is,} \quad \$4.10$$

Solution: David has 8 dimes. ∎

EXAMPLE 5 Marcella has a base salary of \$250 per week. In addition, she receives a commission of 12% of her sales. Last week her total earnings were \$520. What were her total sales for the week?

Solution Let x represent her total sales for the week. Then we can make use of this relationship:

$$\text{Salary} + \text{Commission} = \text{Total earnings}$$

$$\underset{\downarrow}{250} + \underset{\downarrow}{0.12x} = \underset{\downarrow}{520} \qquad \text{(Recall: } 12\% = 0.12\text{)}$$

$$0.12x = 270$$

$$x = \frac{270}{0.12}$$

$$x = 2250$$

Solution: Marcella's total sales for the week were \$2250.
Check: Show that $250 + 0.12(2250) = 520$. ∎

Now it's your turn! Use the guidelines suggested earlier in this section and try to solve as many of the problems as you can. Don't become discouraged if you have difficulty; be assured that most mathematics students have trouble with word problems. Time and practice will most certainly help you develop your critical thinking skills.

EXERCISES 1.2

Solve for x and check each result.

1. $3x - 2 = 10$
2. $5x + 1 = 21$
3. $-2x + 1 = 9$
4. $-3x - 2 = 10$
5. $-3x - 5 = 7$
6. $3x + 2 = -13$
7. $2x - 1 = -17$
8. $-2x + 3 = -12$
9. $2(x + 1) = 11$
10. $3(x - 2) = 15$
11. $3x + 7 = 2x - 2$
12. $2.5x - 8 = x + 3$
13. $\frac{1}{2}x + 7 = 2x - 3$
14. $\frac{5}{2}x - 5 = 3x + 7$
15. $\frac{4}{3}x - 7 = \frac{1}{3}x + 8$
16. $5x - 1 = 5x + 1$
17. $\frac{3}{5}(x - 5) = x + 1$
18. $5(x + 4) = \frac{5}{2}x - 5$
19. $\frac{7}{2}x + 5 + \frac{1}{2}x = \frac{5}{2}x - 6$
20. $2(x + 3) - x = 2x + 8$
21. $-3(x + 2) + 1 = x - 25$
22. $\frac{4}{3}(x + 8) = \frac{3}{4}(2x + 12)$
23. $1 - 12x = 7(1 - 2x)$
24. $2(3x - 7) - 4x = -2$
25. $x + 2(\frac{1}{6}x + 2) = \frac{6}{5}x + 16$

Solve for the indicated variable.

26. $P = 4s$ for s
27. $P = 2l + 2w$ for w
28. $F = \frac{9}{5}C + 32$ for C
29. $N = 10t + u$ for t
30. $7a - 3b = c$ for b
31. $C = 2\pi r$ for r
32. $2(r - 3s) = 6t$ for s
33. $6 + 4v = w - 1$ for v

34. Find a number such that two-thirds of the number increased by one is 13.

35. Find the dimensions of a rectangle whose perimeter is 56 inches if the length is three times the width.

36. Each of the two equal sides of an isosceles triangle is 3 inches longer than the base of the triangle. The perimeter is 21 inches. Find the length of each side.

37. Carlos spent $4.85 on stamps, in denominations of 10¢, 20¢, and 25¢. He bought one-half as many 25¢ stamps as 10¢ stamps, and three   more 20¢ stamps than 10¢ stamps. How many of each type did he buy?

38. Maria has $169 in ones, fives, and tens. She has twice as many one-dollar bills as she has five-dollar bills, and five more ten-dollar bills than five-dollar bills. How many of each type bill does she have?

39. Two cars leave a town at the same time and travel in opposite directions. One car travels at the rate of 45 miles per hour, and the other at 55 miles per hour. In how many hours will the two cars be 350 miles apart?

40. Robert goes for a walk at a speed of 3 miles per hour. Two hours later Roger attempts to overtake him by jogging at the rate of 7 miles per hour. How long will it take him to reach Robert?

41. Prove that the measures of the angles of a triangle cannot be represented by consecutive odd integers. (*Hint:* The sum of the measures is 180.)

42. The width of a painting is 4 inches less than the length. The frame that surrounds the painting is 2 inches wide and has an area of 240 square inches. What are the dimensions of the painting? (*Hint:* The total area minus the area of the painting alone is equal to the area of the frame.)

43. The length of a rectangle is 1 inch less than three times the width. If the length is increased by 6 inches and the width is increased by 5 inches, then the length will be twice the width. Find the dimensions of the rectangle.

44. The units' digit of a two-digit number is three more than the tens' digit. The number is equal to four times the sum of the digits. Find the number. (*Hint:* We can represent a two-digit number as $10t + u$.)

45. Find three consecutive odd integers such that their sum is 237. (*Hint:* The three integers can be represented as x, $x + 2$, and $x + 4$.)

46. The length of a rectangle is 1 inch less than twice the width. If the length is increased by

11 inches and the width is increased by 5 inches, then the length will be twice the width. What can you conclude about the data for this problem?

47. Amy travels 27.5 miles to get to work by car. The first part of her trip is along a country road on which she averages 35 miles per hour, and the second part is on a highway where she averages 48 miles per hour. If the time she travels on the highway is 5 times the amount of time she travels on the country road, what is the total time for the trip?

48. A taxi charges 80¢ for the first $\frac{1}{6}$ mile and 20¢ for each additional $\frac{1}{6}$ mile. If a passenger paid $6.00, how far did the taxi travel?

49. A financial advisor invested an amount of money at an annual rate of interest of 9%. She invested $2700 more than this amount at 12% annually. The total yearly income from these investments was $1794. How much did she invest at each rate? (*Note:* Use the formula $I = Prt$, where I is the interest earned on the principal of P dollars invested at the rate r (in decimal form) per year. In this case the time, t, is 1 year.)

50. Luis earns a monthly salary of $1225, plus a commission of 8% of his total sales for the month. Last month his total earnings were $1750. What were his total sales?

51. Leslie paid $9010 for a used car, which included a 6% sales tax on the base cost of the car. What was the cost of the car, without the sales tax?

52. The total cost of two certificates of deposit is $12,800. The annual interest rates of these certificates are 8% and 9%. The yearly interest on the 9% certificate is $217 more than that on the 8% certificate. What is the cost of each certificate?

53. Following is a set of directions for a mathematical trick. First try it. Then use algebraic representations for each phrase (direction) to show why the trick works.

> Think of a number.
> Add 2.
> Multiply by 3.
> Add 9.
> Multiply by 2.
> Divide by 6.
> Subtract the number with which you started.
> The result is 5.

**CHALLENGE
Think Creatively**

Two boats begin their journeys back and forth across a river at the same time, but from opposite sides of the river. The first time that they pass each other they are 700 feet from one of the shores of the river. After they each make one turn, they pass each other once again at a distance of 400 feet from the other shore. How wide is the river? Assume each boat travels at a constant speed and that there is no loss of time in making a turn.

**1.3
STATEMENTS
OF INEQUALITY
AND THEIR
GRAPHS**

As you continue your study of mathematics you will find a great deal of attention given to *inequalities*. We begin our discussion of this topic by considering the ordering of the real numbers on the number line. In the following figure we say that *a is less than b* because *a* lies to the left of *b*. In symbols, we write $a < b$.

Also note that *b* lies to the right of *a*. That is, $b > a$; this is read "*b is greater than a*." Two inequalities, one using the symbol $<$ and the other $>$, are said to have the *opposite sense*.

Here are two examples of the use of these *symbols of inequality*.

It may help to note that the inequality symbol "points" to the smaller of the two numbers and "opens wide" to the larger.

$$3 < 7 \qquad 3 \text{ is less than } 7$$
$$5 > -2 \qquad 5 \text{ is greater than } -2$$

Since positive numbers are to the right of the origin on a number line, $a > 0$ means that a is positive. Similarly, $a < 0$ means that a is negative.

Algebraically, we define $a < b$ as follows:

For any two real numbers a and b, $a < b$ (or $b > a$) if and only if $b - a$ is a positive number; that is, if and only if $b - a > 0$.

For example, $3 < 7$ because $7 - 3 = 4$, a positive number; that is, $7 - 3 > 0$. Also, since $5 - (-2) > 0$, then $5 > -2$.

A fundamental property of inequalities states that for any two real numbers a and b, either a is less than b, or a equals b, or a is greater than b. In symbols, we have the following property.

TRICHOTOMY PROPERTY

For any real numbers a and b, exactly one of the following is true:

$$a < b \qquad a = b \qquad a > b$$

The same number may be added to, or subtracted from, each side of an inequality and the *sense* of the new inequality will be the same. For example:

Note: $5 < 10$ and $8 < 13$ have the same sense.

Since $5 < 10$, then $5 + 3 < 10 + 3$; that is, $8 < 13$.
Since $9 > 5$, then $9 - 2 > 5 - 2$; that is, $7 > 3$.

This gives rise to the following important property of inequalities:

ADDITION PROPERTY OF ORDER

For all real numbers, a, b, and c:

If $a < b$, then $a + c < b + c$.

If $a > b$, then $a + c > b + c$.

To solve an inequality in the variable x means that *all* values of x that make the inequality true need to be found.

This property can now be used to solve an inequality, such as $x + 2 < 7$.

$$x + 2 < 7$$
$$x + 2 + (-2) < 7 + (-2) \qquad \text{(addition property of order)}$$
$$x < 5$$

When nothing is said to the contrary, it will always be assumed that we are using the set of real numbers. Therefore, the *solution set* here consists of all real numbers that are less than 5. Rather than use a verbal description, we can use braces and write this solution set in **set-builder notation** as

$$\{x \,|\, x < 5\}$$

This is read as "the set of all x such that x is less than 5."

EXAMPLE 1   Solve the inequality: $-4n - (3 - 5n) > 8$

Solution   First simplify the left side.

$$-4n - (3 - 5n) > 8$$
$$-4n - 3 + 5n > 8$$
$$n - 3 > 8$$
$$n - 3 + 3 > 8 + 3$$
$$n > 11$$

The solution set is $\{n \,|\, n > 11\}$. ∎

The addition property of order also applies for inequalities using the symbols $\leq$ and $\geq$. For example, if $a \leq b$, then $a + c \leq b + c$.

These additional symbols of inequality are also used quite often:

$a \leq b$ means *a is less than or is equal to b;* that is, $a < b$ or $a = b$.
$a \geq b$ means *a is greater than or is equal to b;* that is, $a > b$ or $a = b$.

EXAMPLE 2   Find the solution set: $3x + 7 \leq 2x - 1$

Solution   Apply the addition property of order twice.

$$3x + 7 \leq 2x - 1$$
$$3x + 7 + (-7) \leq 2x - 1 + (-7)$$
$$3x \leq 2x - 8$$
$$3x + (-2x) \leq 2x - 8 + (-2x)$$
$$x \leq -8$$

The solution set consists of all real numbers that are less than or equal to -8, that is, $\{x \,|\, x \leq -8\}$. ∎

When the addition property of order is applied, **equivalent inequalities** are produced. That is, the new inequality has the same solution set as the original one. Now let us see what happens when we multiply (or divide) each side of an inequality by the same number. Here are several illustrations:

$$\left.\begin{array}{l} 8 < 12 \rightarrow 2(8) < 2(12); \text{ that is, } 16 < 24 \\ 20 > -15 \rightarrow \tfrac{1}{5}(20) > \tfrac{1}{5}(-15); \text{ that is, } 4 > -3 \end{array}\right\} \quad \text{The sense is preserved.}$$

$$5 < 6 \rightarrow -2(5) > -2(6); \text{ that is, } -10 > -12$$
$$6 > -4 \rightarrow -\tfrac{1}{2}(6) < -\tfrac{1}{2}(-4); \text{ that is, } -3 < 2$$

The sense is reversed.

These illustrations support the following property:

MULTIPLICATION PROPERTY OF ORDER

Note: Similar properties hold for the inequality when $a > b$. Also, note that when $c = 0$, then $ac = bc$; that is, both sides will be equal to 0.

For all real numbers a, b, and c:

If $a < b$ and c is positive, then $ac < bc$.

If $a < b$ and c is negative, then $ac > bc$.

EXAMPLE 3 Solve for x: $5(3 - 2x) \geq 10$

Solution Multiply each side by $\tfrac{1}{5}$ (or divide by 5).

$$\tfrac{1}{5} \cdot 5(3 - 2x) \geq \tfrac{1}{5}(10) \qquad \text{(multiplication property of order)}$$
$$3 - 2x \geq 2$$

Add -3 to each side.

$$3 - 2x + (-3) \geq 2 + (-3) \qquad \text{(addition property of order)}$$
$$-2x \geq -1$$

Note the change in the sense of the resulting inequality because we are multiplying (or dividing) by a negative number.

Multiply by $-\tfrac{1}{2}$ (or divide by -2).

$$-\tfrac{1}{2}(-2x) \leq -\tfrac{1}{2}(-1) \qquad \text{(multiplication property of order)}$$
$$x \leq \tfrac{1}{2}$$

The solution set is $\{x \mid x \leq \tfrac{1}{2}\}$. ∎

At times a statement of inequality can be solved by our knowledge of the properties of numbers. Thus, in the following example, the solution is obtained because we know that the given fraction will be negative if the numerator and denominator are of opposite signs.

EXAMPLE 4 Solve: $\dfrac{2}{x + 4} < 0$

Solution Since the numerator of the fraction is positive, the fraction will be less than zero if and only if the denominator is negative. Thus

$$x + 4 < 0$$
$$x < -4$$

The solution set is $\{x \mid x < -4\}$. ∎

Find the solution set.

1. $x + 3 < 12$ **2.** $x - 5 < 13$ **3.** $x - 1 > 8$
4. $x + 7 > 2$ **5.** $x + (-5) \leq 9$ **6.** $x + (-3) \geq -5$
7. $3x + 8 < 2x + 12$ **8.** $3x - 6 \geq x + 8$ **9.** $5(x + 7) \leq 3x - 7$
10. $2(x - 1) \leq 5x + 1$ **11.** $\dfrac{5}{3 - x} < 0$ **12.** $\frac{1}{2}x + 3 < \frac{1}{3}x - 2$

The two inequalities $a < b$ *and* $b < c$ may be written as $a < b < c$. Similar forms apply when one or both of the inequality symbols $<$ are replaced by $\leq$. Thus, $a \leq b < c$ means $a \leq b$ *and* $b < c$. Also, forms such as $a \geq b \geq c$ are obtained when the sense is reversed. Such inequalities can be used to define *bounded intervals* on the number line. This is shown in the following figure together with a specific example of each type. Observe, for example, that the set of all real numbers x where $a < x < b$ is the interval denoted by the symbol (a, b).

Note that the parenthesis in $(a, b]$ means that the number a is not included in the interval, and that the bracket means that b is included—that is:

$$(a, b] = \{x \mid a < x \leq b\}$$

The solid dot means that the boundary point is included in the interval, and the open circle means it is not included.

Also, regardless of the parentheses or brackets, the numbers a and b are boundaries for all of the values x in the interval.

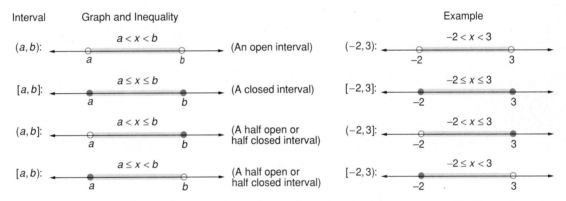

There are also *unbounded intervals*. For example, the set of all $x > 5$ is denoted by $(5, \infty)$. Similarly, $(-\infty, 5]$ represents all $x \leq 5$. The symbols ∞ and $-\infty$ are read "plus infinity" and "minus infinity" but do *not* represent numbers. They are symbolic devices used to indicate that *all* x in a given direction, without end, are included, as in the following figure.

Note, for example, that $(-\infty, b)$ means $\{x \mid x < b\}$.

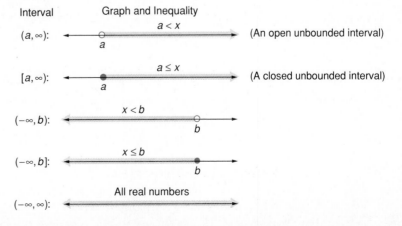

17

The number line can be used to show the graphs of solution sets of inequalities, as in the following examples.

EXAMPLE 5   Graph the solution set: $3x - 4 > 11$

Solution

Because we are multiplying (or dividing) by a positive number, the sense is preserved.

$$3x - 4 > 11$$
$$3x > 15$$
$$x > 5$$

The solution set is $\{x \mid x > 5\}$ and may be graphed as follows:

$(5, \infty)$:

The heavily shaded arrow is used to show that all points in the indicated direction are included. The open circle indicates that 5 is *not* included in the solution. Why not? When the open circle is replaced by a solid dot we have the following graph, which represents the solution of the inequality $3x - 4 \geq 11$. In this case 5 is included in the solution of the inequality.

$[5, \infty)$:

EXAMPLE 6   Graph the solution set: $-2(x - 1) \geq 4$

Solution

$$-2(x - 1) \geq 4$$
$$x - 1 \leq -2$$
$$x \leq -1$$

The solution set is $\{x \mid x \leq -1\}$ and has the following graph:

$(-\infty, -1]$:

EXAMPLE 7   If $x < 2$, is $x - 5$ positive or negative?

Solution

$$x < 2$$
$$x - 5 < 2 - 5 \qquad \text{(Subtract 5)}$$
$$x - 5 < -3$$

Therefore, $x - 5$ is negative.

Inequalities can also be used to solve applied problems. This is illustrated in Example 8, which makes use of inequalities of the form $a \leq b \leq c$.

EXAMPLE 8 A small family business employs two part-time workers per week. The total amount of wages they pay to these employees ranges from $128 to $146 per week. If one employee will earn $18 more than the other, what are the possible amounts earned by each per week?

Solution Let x = wages paid to the employee who earns the smaller amount. Then $x + 18$ = wages for the other employee. Since the sum of the wages is at least $128 but no more than $146, the sum $x + (x + 18)$ satisfies this *compound inequality:*

$$128 \leq x + (x + 18) \leq 146$$

Now simplify to get the possibilities for x.

Note: -18 is added to each part of the compound inequality, and then each part is divided by 2.

$$128 \leq 2x + 18 \leq 146$$
$$110 \leq 2x \leq 128$$
$$55 \leq x \leq 64$$

To get the result for the other employee, add 18 to the preceding inequality and simplify.

$$55 + 18 \leq x + 18 \leq 64 + 18$$
$$73 \leq x + 18 \leq 82$$

One part-time employee earns from $55 to $64 per week and the other earns from $73 to $82 per week. ■

There are a number of additional properties of order that are fundamental for later work. We list them in terms of the following rules.

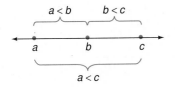

These rules can also be restated by reversing the sense. For instance, Rule 2 would then read as: If $a > b$ and $c > d$, then $a + c > b + d$.

RULE 1. **If $a < b$ and $b < c$, then $a < c$.**
This is known as the **transitive property of order.** Geometrically, it says that if a is to the left of b, and b is to the left of c, then a must be to the left of c on a number line.

RULE 2. **If $a < b$ and $c < d$, then $a + c < b + d$.**
Since $5 < 10$ and $-15 < -4$, then $5 + (-15) < 10 + (-4)$; that is, $-10 < 6$.

RULE 3. **If $0 < a < b$ and $0 < c < d$, then $ac < bd$.**
Since $3 < 7$ and $5 < 9$, then $(3)(5) < (7)(9)$; that is $15 < 63$.

RULE 4. **If $a < b$ and $ab > 0$, then $\dfrac{1}{a} > \dfrac{1}{b}$.**
Since $5 < 10$, then $\frac{1}{5} > \frac{1}{10}$.
Since $-3 < -2$, then $-\frac{1}{3} > -\frac{1}{2}$.

EXERCISES 1.3

Classify each statement as true or false. If it is false, give a specific counterexample to explain your answer.

1. If $x < 2$, then x is negative.
2. If $x > 1$ and $y > 2$, then $x + y > 3$.
3. If $0 < x$, then $-x < 0$.
4. If $x < 5$ and $y < 6$, then $xy < 30$.
5. If $0 < x$, then $x < x^2$.
6. If $x < y < -2$, then $\dfrac{1}{x} > \dfrac{1}{y}$.
7. If $x \le -5$, then $x - 2 \le -7$.
8. If $x \le y$ and $y < z$, then $x < z$.

Find the solution set.

9. $x + 5 > 17$
10. $x - 8 < 5$
11. $x - 7 \ge -3$
12. $x + 6 \le -7$
13. $5x - 4 < 6 + 4x$
14. $3x + 12 > 2x - 5$
15. $3x > -21$
16. $9x \le -45$
17. $-5x < 50$
18. $3x + 5 \ge 17$
19. $5x - 3 \le 22$
20. $-2x + 1 \le 19$
21. $2x + 7 \le 5 - 6x$
22. $3x - 2 > x + 5$
23. $-5x + 5 < -3x + 1$
24. $3(x - 1) \ge 2(x - 1)$
25. $2(x + 1) < x - 1$
26. $3x + 5 + x > 2(x - 1)$
27. $\frac{1}{2}x - 5 > \frac{1}{4}x + 3$
28. $\frac{3}{4}x + 2 < \frac{5}{8}x - 3$
29. $-\frac{3}{5}x - 6 < -\frac{2}{5}x + 7$
30. $\dfrac{2 - x}{5} \ge 0$
31. $\dfrac{1}{x} < 0$
32. $-\dfrac{2}{x + 1} > 0$

Solve for x.

33. $3x + 5 \ne 8$ (The symbol $\ne$ is read "is not equal to.")
34. $2x + 1 \not> 5$ (The symbol $\not>$ is read "is not greater than.")
35. $3x - 2 \not< 1$ (The symbol $\not<$ is read "is not less than.")
36. $12x + 9 \ne 15(x - 2)$
37. $3(x - 1) \not> 5(x + 2)$
38. $2(x + 3) \not< 3(x + 1)$
39. $\frac{5}{3}x \not< 2x - 1$
40. $\frac{2}{9}(3x + 7) \not\ge 1 - \frac{4}{3}x$

Show each of the following intervals as a graph on a number line.

41. $(-3, -1)$
42. $(-3, -1]$
43. $[-3, -1)$
44. $[-3, -1]$
45. $[0, 5]$
46. $(-1, 3)$
47. $(-\infty, 0]$
48. $[2, \infty)$

Express each inequality in interval notation.

49. $-5 \le x \le 2$
50. $0 < x < 7$
51. $-6 \le x < 0$
52. $-2 < x \le 4$
53. $-10 < x < 10$
54. $3 \le x \le 7$
55. $x < 5$
56. $x \le -2$
57. $-2 \le x$
58. $2 < x$
59. $x \le -1$
60. $x < 3$

Write an inequality for each of the following graphs. Also express each as an interval of real numbers.

For example: 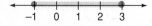 $-1 \le x \le 3; [-1, 3]$

61.
62.
63.
64.
65.
66.

Solve and graph each inequality.

67. $2x + 3 < 11$ **68.** $-3x + 1 > -2$ **69.** $\frac{1}{2}x + 2 \leq 1$

70. $-3(x + 1) \geq 3$ **71.** $2(x + 1) < 3(x + 2)$ **72.** $-2(x - 2) > 3x - 3$

73. The sum of an integer and 5 less than three times this integer is between 34 and 54. Find all possible pairs of such integers.

74. In order to get a grade of B$^+$ in an algebra course, a student must have a test average of at least 86% but less than 90%. If the student's grades on the first three tests were 85%, 86%, and 93%, what grades on the fourth test would guarantee a grade of B$^+$?

75. In Exercise 74, what grades on the fourth test would guarantee a B$^+$ if the fourth test counts twice as much as each of the other tests?

76. If x satisfies $\frac{7}{4} < x < \frac{9}{4}$, then what are the possible values of y where $y = 4x - 8$? (*Hint:* Apply the multiplication and addition properties of inequality to the given inequality to obtain $4x - 8$ as the middle part.)

77. For a given time period, the temperature in degrees Celsius varied between 25° and 30°. What was the range in degrees Fahrenheit for this time period? (*Hint:* Begin with $25 < C < 30$ and apply the idea in Exercise 76 using $F = \frac{9}{5}C + 32$.)

78. Suppose that a machine is programmed to produce rectangular metal plates such that the length of the plate will be 1 more than twice the width w. When the entry $w = 2$ cm is made, the design of the machine will guarantee only that the width is within a one-tenth tolerance of 2. That is, $2 - 0.1 < w < 2 + 0.1$.

   (a) Within what tolerance of 5 centimeters is the length?

   (b) Find the range of values for the area.

79. A delivery service will accept a package only if the sum of its length ℓ and its girth g is no more than 110 inches. It also requires that each of the three dimensions be at least 2 inches.

   (a) If $\ell = 42$ inches, what are the allowable values for the girth g?

   (b) If $\ell = 42$ inches and $w = 18$ inches, what are the allowable values of h?

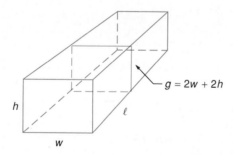

80. A store has two part-time employees who together are paid a weekly total from $150 to $180. If one of the employees earns $15 more than the other, what are the possible amounts earned by each per week?

81. A store has three part-time employees who together are paid a weekly total from $210 to $252. Two of them earn the same amount and the third earns $12 less than the others. Find the possible amounts earned by each per week.

82. A supermarket has twenty part-time employees who together are paid a weekly total from $1544 to $1984. Twelve of them earn the same amount and the remaining eight each earn $22 less. Find the possible amounts earned by each employee per week.

➡ **Written Assignment:** In your own words, state the addition and multiplication properties of equality and of order.

What do the numbers -5 and $+5$ have in common? Obviously, they are different numbers and are the coordinates of two distinct points on the number line. However, they are both the same distance from 0, the **origin,** on the number line.

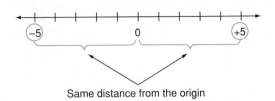

Same distance from the origin

In other words, -5 is as far to the left of 0 as $+5$ is to the right of 0. We show this fact by using **absolute value notation** as follows:

$$|-5| = 5 \text{ read as "The } absolute\ value \text{ of } -5 \text{ is 5."}$$

$$|+5| = 5 \text{ read as "The } absolute\ value \text{ of } +5 \text{ is 5."}$$

Geometrically, for any real number x, $|x|$ is the distance (without regard to direction) that x is from the origin. Note that for a positive number, $|x| = x$; $|+5| = 5$. For a negative number, $|x| = -x$; that is, $|-5| = -(-5) = 5$. Also, since 0 is the origin, it is natural to have $|0| = 0$.

We can summarize this with the following definition.

In words, if x is positive or zero, the absolute value of x is x. If x is negative, the absolute value is the opposite of x. For example:

$$|+3| = 3$$
$$|-3| = -(-3) = 3$$
$$|0| = 0$$

DEFINITION OF ABSOLUTE VALUE

For any real number x,

$$|x| = \begin{cases} x & \text{if } x \geq 0 \\ -x & \text{if } x < 0 \end{cases}$$

EXAMPLE 1 Solve for x: $\dfrac{x - 3}{|x - 3|} = 1$

The solution set for Example 1 is $\{x \mid x > 3\}$. However, for the sake of simplicity, we will often omit the set-builder notation. Thus, when we write $x > 3$ as the solution, it is understood that the solution consists of all x such that $x > 3$.

Solution This fraction will be equal to 1 if the numerator and denominator have the same value. By definition, $|x - 3| = x - 3$ only if $x - 3 \geq 0$, or $x \geq 3$. However, *to avoid division by 0,* we note that $x \neq 3$. Thus the solution is $x > 3$. Test several values of x for $x < 3$ and $x > 3$ to verify this solution. (You may find it easier to think of this problem in the form $\dfrac{a}{|a|} = 1$, where $a = x - 3$.) ∎

Some useful properties of absolute value follow, presented with illustrations but without formal proof.

PROPERTY 1.   For $k > 0$, $|a| = k$ if and only if $a = k$ or $-a = k$, that is, $a = k$ or $a = -k$; also, $|0| = 0$.

This property follows immediately from the definition of absolute value. For example, an equation such as $|x| = 2$ is just another way of saying that $x = 2$ or

$x = -2$. The graph consists of the two points with coordinates 2 and -2; each of these points is 2 units from the origin.

EXAMPLE 2 Solve: $|5 - x| = 7$

We use Property 1 for this solution, noting that $5 - x$ plays the role of a.

Solution

$$5 - x = 7 \quad \text{or} \quad -(5 - x) = 7$$

$$-x = 2 \qquad\qquad -5 + x = 7$$

$$x = -2 \qquad\qquad x = 12$$

These two solutions, -2 and 12, can be checked by subsitution into the original equation.
Check: $|5 - (-2)| = |7| = 7; |5 - 12| = |-7| = 7.$ ■

Now consider the inequality $|x| < 2$. Here we are considering all the real numbers whose absolute value *is less than* 2. On the number line, these are the points whose distance from the origin is less than 2 units, that is, all of the real numbers between -2 and 2. This inequality can be written in either of these two forms:

$$x > -2 \quad \text{and} \quad x < 2 \qquad -2 < x < 2$$

The graph of $|x| < 2$ is the interval $(-2, 2)$.

$$|x| < 2: \qquad \underset{\substack{-2 \ -1 \ \ 0 \ \ 1 \ \ 2}}{\longleftrightarrow}$$

Conversely, the graph also shows that if $-2 < x < 2$, then $|x| < 2$. If the endpoints are included, then we would have $-2 \le x \le 2$; that is, $|x| \le 2$.

Note: $-k < a < k$ is the same as saying that $a < k$ and $a > -k$. Also, $|a| \le k$ if and only if $-k \le a \le k$.

We can now generalize by means of the following property.

PROPERTY 2. For $k > 0$, $|a| < k$ if and only if $-k < a < k$.

EXAMPLE 3 Solve: $|x - 2| \le 3$.

Note that $|x - 2| \le 3$ means $x - 2 \ge -3$ and $x - 2 \le 3$.

Solution Let $x - 2$ play the role of a in Property 2. Consequently, $|x - 2| \le 3$ is equivalent to $-3 \le x - 2 \le 3$. Now add 2 to each part to isolate x in the middle.

$$-3 \le x - 2 \le \ \ 3$$
$$\underline{+2 \qquad +2 \quad +2}$$
$$-1 \le x + 0 \le \ \ 5 \qquad \text{Thus } -1 \le x \le 5.$$ ■

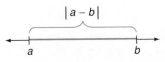

The quantity $|a - b|$ also represents the distance between the points a and b on the number line. For example, the distance between 8 and 3 on the number line is $5 = |8 - 3| = |3 - 8|$. The idea of distance between points on a number line can be used to give an alternative way to solve Example 3. Think of the expression $|x - 2|$ as the distance between x and 2 on the number line, and then consider all the points x whose distance from 2 is less than or equal to 3 units.

Think: $|x - 2| < 3$ *means that x is within 3 units of 2. For $|x - 2| \le 3$ include the endpoints.*

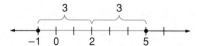

Observe that the *midpoint* or *center* of the interval in the preceding discussion has coordinate 2. This value can be found by taking the average of the numbers -1 and 5; $\dfrac{-1 + 5}{2} = 2$. In general, we have the following result.

MIDPOINT FORMULA FOR A LINE SEGMENT

If x_1 and x_2 are the endpoints of an interval, then the coordinate of the midpoint is

$$\frac{x_1 + x_2}{2}$$

EXAMPLE 4 Find the coordinate of the midpoint M of each line segment AB.

(a)
$$\begin{array}{ccc} -4 & 0 & 6 \\ A & M & B \end{array}$$

(b)
$$\begin{array}{ccc} -10 & -2 & 0 \\ A & M & B \end{array}$$

Solution **(a)** $\dfrac{-4 + 6}{2} = 1$ **(b)** $\dfrac{-10 + (-2)}{2} = -6$ ∎

Throughout the text you will find sets of CAUTION items. These illustrate errors that are often made by students. Study these carefully so that you understand the errors shown and can avoid such mistakes.

CAUTION: Learn to Avoid These Mistakes	
WRONG	RIGHT
$\left|\frac{3}{4} - 2\right| = \frac{3}{4} - 2 = -\frac{5}{4}$	$\left|\frac{3}{4} - 2\right| = \left|-\frac{5}{4}\right| = \frac{5}{4}$
$|5 - 7| = |5| - |7| = -2$	$|5 - 7| = |-2| = 2$
$|x| = -2$ has the solution $x = -2$.	There is no solution; the absolute value of a number can never be negative.
$|x - 1| < 3$ if and only if $x < 4$.	$|x - 1| < 3$ if and only if $-3 < x - 1 < 3$; that is, $-2 < x < 4$.

Next, consider the inequality $|x| > 2$. We know that $|x| < 2$ implies that $-2 < x < 2$. Therefore, $|x| > 2$ consists of those values of x for which $x < -2$ or $x > 2$. The graph of this set may be drawn as follows.

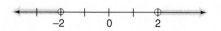

This graph shows the set of points whose distance from 0 is more than 2 units. The property involved is summarized in this way.

Also $|a| \geq k$ if and only if $a \leq -k$ or $a \geq k$.

PROPERTY 3. For $k \geq 0$, $|a| > k$ if and only if $a < -k$ or $a > k$.

EXAMPLE 5 Graph: $|x + 1| > 2$

Solution Think of $|x + 1|$ as $|x - (-1)|$ so as to have it in the form $|a - b|$. This represents the distance between x and -1, and we wish to find all the points that are more than 2 units away from the point -1.

CAUTION
*A common error that students make is to write the solution for Example 5 as $1 < x < -3$. Explain why this is **not** correct.*

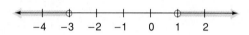

The solution consists of all $x < -3$ or all $x > 1$. ∎

In Example 5 we could have solved the inequality by using Property 3.

$$|x + 1| > 2 \quad \text{means} \quad x + 1 < -2 \quad \text{or} \quad x + 1 > 2$$

Thus $x < -3$ or $x > 1$. The word "or" indicates that we are to consider the values for x that satisfy either one of these two conditions; that is, x may be less than -3 or x may be greater than 1.

Other useful properties of absolute value are

*The statement $|x + y| \leq |x| + |y|$ is called the **triangle inequality**.*

$$|xy| = |x||y| \qquad \left|\frac{x}{y}\right| = \frac{|x|}{|y|} \qquad |x + y| \leq |x| + |y|$$

(See Exercises 46 and 47.)

EXERCISES 1.4

Classify each statement as true or false.

1. $-\left|-\frac{1}{3}\right| = \frac{1}{3}$

2. $|-1000| < 0$

3. $\left|-\frac{1}{2}\right| = 2$

4. $|\sqrt{2} - 5| = 5 - \sqrt{2}$

5. $\left|\frac{x}{y}\right| = |x| \cdot \frac{1}{|y|}$

6. $2 \cdot |0| = 0$

7. $\|x\| = |x|$

8. $|-(-1)| = -1$

9. $|a| - |b| = a - b$

10. $|a - b| = b - a$

11. Find the coordinate of the midpoint of a line segment with endpoints as given:
    (a) 3 and 9   (b) -8 and -2 (c) -12 and 0   (d) -5 and 8

12. One endpoint of a line segment is located at -7. The midpoint is located at 3. What is the coordinate of the other endpoint of the line segment?

Solve for x.

13. $|x| = \frac{1}{2}$

14. $|3x| = 3$

15. $|x - 1| = 3$

16. $|3x - 4| = 0$

17. $|2x - 3| = 7$

18. $|6 - 2x| = 4$

19. $|4 - x| = 3$

20. $|3x - 2| = 1$

21. $|3x + 4| = 16$

22. $\left|\frac{1}{x - 1}\right| = 2$

23. $\frac{|x|}{x} = 1$

24. $\frac{|x|}{x} = -1$

Solve for x and graph.

25. $|x + 1| = 3$ 26. $|x - 1| \leq 3$ 27. $|x - 1| \geq 3$

28. $|x + 2| = 3$ 29. $|x + 2| \leq 3$ 30. $|x + 2| \geq 3$

31. $|-x| = 5$ 32. $|x| \leq 5$ 33. $|x| \geq 5$

34. $|x - 5| \neq 3$ 35. $|x - 5| \leq 3$ 36. $|x - 5| \geq 3$

37. $|x - 3| < 0.1$ 38. $|2 - x| < 3$ 39. $|2x - 1| < 7$

40. $|3x - 6| < 9$ 41. $|4 - x| < 2$ 42. $|1 + 5x| < 1$

43. $|x - 4| \not> 1$ (*Note:* $\not>$ means $\leq$.)   44. $|x - 2| \not\geq 0$ (*Note:* $\not\geq$ means $<$.)   45. $\dfrac{1}{|x - 3|} > 0$

46. (a) Prove the product rule $|xy| = |x| \cdot |y|$ for the case $x < 0$ and $y > 0$.

     (b) Prove the quotient rule $\left|\dfrac{x}{y}\right| = \dfrac{|x|}{|y|}$ for the case $x < 0$ and $y < 0$.

47. Cite four different examples to confirm each inequality, using the cases
     (i) $x > 0$, $y > 0$; (ii) $x > 0$, $y < 0$; (iii) $x < 0$, $y > 0$; (iv) $x < 0$, $y < 0$.
     (a) $|x - y| \geq ||x| - |y||$ (b) $|x + y| \leq |x| + |y|$

Written Assignment: Explain why $x^2 = |x^2| = |x|^2$ for any real number x.

EXPLORATIONS
Think Critically

1. How can you show that the set of real numbers is *not* associative with respect to division?

2. Show that the set of real numbers does not contain the multiplicative inverse for each of its members.

3. Explain the use of a specific counterexample to disprove a statement claimed to be true in general. Give several examples.

4. What are the possible values for the quotient $\dfrac{|2x|}{2x}$?

5. What inequality represents the set of real numbers that are less than three units from the point -2 on the number line?

6. In what sense does the concept of subtraction depend on addition?

7. If $a < b$ and $c < d$, is it true that $a - c < b - d$? Explain your answer.

1.5
INTEGRAL
EXPONENTS

Much of mathematical notation can be viewed as efficient abbreviations of lengthier statements. For example:

$$4^9 = 4 \times 4 \times 4 \times 4 \times 4 \times 4 \times 4 \times 4 \times 4$$

This illustration makes use of a *positive integral exponent*. In this section we shall explore the use of integers as exponents.

DEFINITION OF POSITIVE INTEGRAL EXPONENT

If n is a positive integer and b is any real number, then

$$\underbrace{b^n = b \cdot b \cdot \cdots \cdot b}_{n \text{ factors}}$$

The most common ways of referring to b^n are "b to the nth power," "b to the nth," or "the nth power of b."

The number b is called the **base** and n is called the **exponent**.

Here are some illustrations of the definition:

$$b^1 = b$$

$$(a + b)^2 = (a + b)(a + b)$$

$$(-2)^3 = (-2)(-2)(-2) = -8$$

$$10^6 = 10 \cdot 10 \cdot 10 \cdot 10 \cdot 10 \cdot 10 = 1,000,000$$

CAUTION
Be careful when you work with parentheses. Note that
$(-3)^2 \neq -3^2$.
$$(-3)^2 = (-3)(-3) = 9$$
$$-3^2 = -(3 \times 3) = -9$$

A number of important rules concerning positive integral exponents are easily established on the basis of the preceding definition. Here is a list of these rules in which m and n are positive integers, a and b are any real numbers, and with the usual understanding that denominators cannot be zero.

When you multiply powers with a common base you add the powers and use the same base.

RULE 1. $b^m b^n = b^{m+n}$

Illustrations:

$$2^3 \cdot 2^4 = 2^{3+4} = 2^7 \qquad x^3 \cdot x^4 = x^7$$

RULE 2. $\dfrac{b^m}{b^n} = \begin{cases} b^{m-n} & \text{if } m > n \\ 1 & \text{if } m = n \\ \dfrac{1}{b^{n-m}} & \text{if } m < n \end{cases}$

Illustrations:

$$(m > n) \qquad\qquad (m = n) \qquad\qquad (m < n)$$

$$\frac{2^5}{2^2} = 2^{5-2} = 2^3 \qquad \frac{5^2}{5^2} = 1 \qquad \frac{2^2}{2^5} = \frac{1}{2^{5-2}} = \frac{1}{2^3}$$

$$\frac{x^5}{x^2} = x^{5-2} = x^3 \qquad \frac{x^2}{x^2} = 1 \qquad \frac{x^2}{x^5} = \frac{1}{x^{5-2}} = \frac{1}{x^3}$$

The power of a power is the product of the powers with the same base.

RULE 3. $(b^m)^n = b^{mn}$

Illustrations:

$$(2^3)^2 = 2^{3 \cdot 2} = 2^6 \qquad (x^3)^2 = x^{3 \cdot 2} = x^6$$

A power of a product is the product of the powers.

RULE 4. $(ab)^m = a^m b^m$

Illustrations:

$$(2 \cdot 3)^5 = 2^5 \cdot 3^5 \qquad (xy)^5 = x^5 y^5$$

The power of a quotient is the quotient of the powers.

RULE 5. $\left(\dfrac{a}{b}\right)^m = \dfrac{a^m}{b^m}$

Illustrations:

$$\left(\frac{3}{2}\right)^5 = \frac{3^5}{2^5} \qquad \left(\frac{x}{y}\right)^5 = \frac{x^5}{y^5}$$

Here is a proof of Rule 4. You can try proving the other rules by using similar arguments.

$$(ab)^m = \underbrace{(ab)(ab) \cdots (ab)}_{m \text{ times}} \qquad \text{(by definition)}$$

$$= \underbrace{(a \cdot a \cdots \cdot a)}_{m \text{ times}} \underbrace{(b \cdot b \cdots \cdot b)}_{m \text{ times}} \qquad \left\{ \begin{array}{l} \text{(by repeated use of the} \\ \text{commutative and associative} \\ \text{laws for multiplication)} \end{array} \right.$$

$$= a^m b^m \qquad \text{(by definition)}$$

The proper use of these rules can simplify computations, as in the following example.

EXAMPLE 1 Evaluate in two ways: $12^3(\frac{1}{6})^3$

Solution

(a) $12^3(\frac{1}{6})^3 = 1728(\frac{1}{216})$ **(b)** $12^3(\frac{1}{6})^3 = (12 \cdot \frac{1}{6})^3$
$= \dfrac{1728}{216}$ $= 2^3$
$= 8$ $= 8$

EXAMPLE 2 Simplify: **(a)** $2x^3 \cdot x^4$ **(b)** $\dfrac{(x^3y)^2y^3}{x^4y^6}$

Solution

(a) $2x^3 \cdot x^4 = 2x^{3+4} = 2x^7$ **(b)** $\dfrac{(x^3y)^2y^3}{x^4y^6} = \dfrac{(x^3)^2y^2y^3}{x^4y^6} = \dfrac{x^6y^5}{x^4y^6} = \dfrac{x^2}{y}$

Care must be taken with both the multiplication and the division rules when the bases are not the same, as in the following example.

EXAMPLE 3 Simplify: $\dfrac{4^5}{8^3}$

Solution The bases are not the same, so that Rule 2 does not apply. However, rather than finding 4^5 and 8^3, the problem can be simplified in this way:

$$\frac{4^5}{8^3} = \frac{(2^2)^5}{(2^3)^3} = \frac{2^{10}}{2^9} = 2$$

TEST YOUR UNDERSTANDING
Think Carefully

Evaluate each of the following.

1. 5^3 **2.** $(-\frac{1}{2})^5$ **3.** $(-\frac{2}{3})^3 + \frac{8}{27}$ **4.** $(10^3)^2$

5. $2^3(-2)^3$ **6.** $(\frac{1}{2})^3 8^3$ **7.** $\dfrac{17^8}{17^9}$ **8.** $\dfrac{(-2)^3 + 3^2}{3^3 - 2^2}$

9. $\dfrac{(-12)^4}{4^4}$ **10.** $(ab^2)^3(a^2b)^4$ **11.** $\dfrac{2^2 \cdot 16^3}{(-2)^8}$ **12.** $\dfrac{(2x^3)^2(3x)^2}{6x^4}$

(Answers: Page 74)

This discussion provides meaning for the use of 0 as an exponent. That is, an expression like 5^0 will now be defined.

Our discussion of exponents has been restricted to the use of positive integers only. Now let us consider the meaning of 0 as an exponent. In particular, what is the meaning of 5^0? We know that 5^3 means that we are to use 5 three times as a factor. But certainly it makes no sense to use 5 zero times. The rules of exponents will help to resolve this dilemma.

We would like these laws for exponents to hold even if one of the exponents happens to be zero. That is, we would *like* Rule 2 to give

$$\frac{5^2}{5^2} = 5^{2-2} = 5^0$$

But it is already known that

$$\frac{5^2}{5^2} = \frac{25}{25} = 1$$

Thus 5^0 ought to be assigned the value 1. Consequently, in order to *preserve* the rules of exponents, we decide to let $5^0 = 1$. That is, from now on, we agree to the following:

Notice that the definition calls for b to be a real number different from 0. That is, we do not define an expression such as 0^0; this is said to be undefined (see Exercise 74).

DEFINITION OF ZERO EXPONENT

If b is a real number different from 0, then

$$b^0 = 1$$

Our next objective is to give meaning to negative integer exponents. For example, we want to decide the meaning of x^{-3}. Our guideline in making this decision will be that *the preceding rules of exponents are to apply for all kinds of integer exponents*. That is, we want to *preserve* the structure of the basic rules. With this in mind, observe the effect of Rule 1 when a negative exponent is involved.

$$(x^3)(x^{-3}) = x^{3+(-3)} = x^0 = 1$$

Dividing both sides of $(x^3)(x^{-3}) = 1$ first by x^3 and then by x^{-3} produces these two statements

$$x^{-3} = \frac{1}{x^3} \quad \text{and} \quad x^3 = \frac{1}{x^{-3}}$$

We are now ready to make the following definition:

Note that the exponents $-n$ and n are opposites, and that $-n$ may be negative or positive.

DEFINITION OF b^{-n}

If n is an integer and $b \neq 0$, then

$$b^{-n} = \frac{1}{b^n}$$

Can you show that $\left(\dfrac{a}{b}\right)^{-2} = \left(\dfrac{b}{a}\right)^{2}$?

It follows from this definition that $\left(\dfrac{a}{b}\right)^{-1} = \dfrac{b}{a}$, since $\left(\dfrac{a}{b}\right)^{-1} = \dfrac{1}{\dfrac{a}{b}} = \dfrac{b}{a}$. In other words, a fraction to the -1 power is the reciprocal of the fraction.

As in Illustration (a), more than one correct procedure is often possible. Finding the most efficient procedure depends largely on experience.

Illustrations:

(a) $\left(\dfrac{1}{7}\right)^{-2} = \dfrac{1}{\left(\frac{1}{7}\right)^{2}} = \dfrac{1}{\frac{1}{49}} = 49$ or $\left(\dfrac{1}{7}\right)^{-2} = (7^{-1})^{-2} = 7^{2} = 49$

(b) $\dfrac{5^{-2}}{15^{-2}} = \left(\dfrac{5}{15}\right)^{-2} = \left(\dfrac{1}{3}\right)^{-2} = (3^{-1})^{-2} = 3^{2} = 9$

(c) $\left(\dfrac{400 - 10^{4}}{80^{2}}\right)^{0} = 1$

It should be noted that the three cases of Rule 2 can now be condensed into this single form.

CAUTION
Be careful when applying Rule 2, especially when an exponent is negative. Thus
$\dfrac{b^{3}}{b^{-2}} = b^{3-(-2)} = b^{5};$
$\dfrac{b^{3}}{b^{-2}} \neq b^{3-2}.$

RULE 2.   (revised): $\dfrac{b^{m}}{b^{n}} = b^{m-n}$

Illustrations:

$$\dfrac{3^{4}}{3^{2}} = 3^{2} \qquad \dfrac{3^{2}}{3^{4}} = 3^{-2} \qquad \dfrac{3^{2}}{3^{2}} = 3^{0} = 1$$

$$\dfrac{x^{8}}{x^{2}} = x^{6} \qquad \dfrac{x^{2}}{x^{8}} = x^{-6} \qquad \dfrac{x^{2}}{x^{2}} = x^{0} = 1$$

EXAMPLE 5   Simplify $\left(\dfrac{a^{-2}b^{3}}{a^{3}b^{-2}}\right)^{5}$ and express the answer using only positive exponents.

Solution   There are several ways to proceed; here are two.

(i) $\left(\dfrac{a^{-2}b^{3}}{a^{3}b^{-2}}\right)^{5} = (a^{-5}b^{5})^{5} = (a^{-5})^{5}(b^{5})^{5} = a^{-25}b^{25} = \dfrac{b^{25}}{a^{25}}$

(ii) $\left(\dfrac{a^{-2}b^{3}}{a^{3}b^{-2}}\right)^{5} = \left(\dfrac{b^{5}}{a^{5}}\right)^{5} = \dfrac{(b^{5})^{5}}{(a^{5})^{5}} = \dfrac{b^{25}}{a^{25}}$ ∎

Exponential notation is used in a variety of situations. Example 6 shows the use of exponents to analyze a situation in which a substance is decreasing exponentially.

EXAMPLE 6   Suppose that a radioactive substance decays so that $\frac{1}{2}$ the amount remains after each hour. If at a certain time there are 320 grams of the substance, how much will remain after 8 hours? How much after n hours?

Solution   Since the amount remaining after each hour is $\frac{1}{2}$ of the grams at the end of the preceding hour, we find the remaining amount by multiplying the preceding number of grams by $\frac{1}{2}$.

	Grams Remaining
Start: 0 hours	$320(\frac{1}{2})^0 = 320$
After 1 hour	$320(\frac{1}{2})^1 = 160$
After 2 hours	$320(\frac{1}{2})^2 = 80$
After 3 hours	$320(\frac{1}{2})^3 = 40$
$\vdots$	$\vdots$
After 8 hours	$320(\frac{1}{2})^8 = \frac{5}{4}$

Observe that the power of $\frac{1}{2}$ is the same as the number of hours that the substance has been decaying. Assuming that the same pattern will continue, we conclude that there are $320(\frac{1}{2})^n = \dfrac{320}{2^n}$ grams remaining after n hours.   ∎

Many errors are made when working with exponents because of misuses of the basic rules and definitions. This list shows some of the common errors that you should try to avoid.

CAUTION: Learn to Avoid These Mistakes	
WRONG	RIGHT
$5^2 \cdot 5^4 = 5^8$ (Do not multiply exponents.) $5^2 \cdot 5^4 = 25^6$ (Do not multiply the base numbers.)	$5^2 \cdot 5^4 = 5^6$ (Rule 1)
$\dfrac{5^6}{5^2} = 5^3$ (Do not divide the exponents.) $\dfrac{5^6}{5^2} = 1^4$ (Do not divide the base numbers.)	$\dfrac{5^6}{5^2} = 5^4$ (Rule 2)
$(5^2)^6 = 5^8$ (Do not add the exponents.)	$(5^2)^6 = 5^{12}$ (Rule 3)
$(-2)^4 = -2^4$ (Misreading the parentheses)	$(-2)^4 = (-1)^4 2^4 = 2^4$ (Rule 4)
$(-5)^0 = -1$ (Misreading definition of b^0)	$(-5)^0 = 1$ (Definition of b^0)
$2^{-3} = -\dfrac{1}{2^3}$ (Misreading definition of b^{-n})	$2^{-3} = \dfrac{1}{2^3}$ (Definition of b^{-n})

EXERCISES 1.5

Classify each statement as true or false. If it is false, correct the right side of the equality to obtain a true statement.

1. $3^4 \cdot 3^2 = 3^8$

2. $(2^2)^3 = 2^8$

3. $2^5 \cdot 2^2 = 4^7$

4. $\dfrac{9^3}{9^3} = 1$

5. $\dfrac{10^4}{5^4} = 2^4$

6. $\left(\dfrac{2}{3}\right)^4 = \dfrac{2^4}{3}$

7. $(-27)^0 = 1$

8. $(2^0)^3 = 2^3$

9. $3^4 + 3^4 = 3^8$

10. $(a^2 b)^3 = a^2 b^3$

11. $(a + b)^0 = a + 1$

12. $a^2 + a^2 = 2a^2$

13. $\dfrac{1}{2^{-3}} = -2^3$

14. $(2 + \pi)^{-2} = \dfrac{1}{4} + \dfrac{1}{\pi^2}$

15. $\dfrac{2^{-5}}{2^3} = 2^{-2}$

Evaluate.

16. -10^5

17. $2^0 + 2^1 + 2^2$

18. $(-3)^2(-2)^3$

19. $\left(\frac{2}{3}\right)^0 + \left(\frac{2}{3}\right)^1$

20. $[(\frac{1}{2})^3]^2$

21. $(\frac{1}{2})^4(-2)^4$

22. $\dfrac{3^2}{3^0}$

23. $\dfrac{(-2)^5}{(-2)^3}$

24. $(-\frac{3}{4})^3$

25. $\dfrac{2^3 \cdot 3^4 \cdot 4^5}{2^2 \cdot 3^3 \cdot 4^4}$

26. $\dfrac{8^3}{16^2}$

27. $\dfrac{2^{10}}{4^3}$

28. $\dfrac{3^{-3}}{4^{-3}}$

29. $(\frac{2}{3})^{-2} + (\frac{2}{3})^{-1}$

30. $[(-7)^2(-3)^2]^{-1}$

Simplify and express each answer using positive exponents only.

31. $(x^{-3})^2$

32. $(x^3)^{-2}$

33. $x^3 \cdot x^9$

34. $\dfrac{x^9}{x^3}$

35. $(2a)^3(3a)^2$

36. $(-2x^3 y)^2(-3x^2 y^2)^3$

37. $(-2a^2 b^0)^4$

38. $(2x^3 y^2)^0$

39. $\dfrac{(x^2 y)^4}{(xy)^2}$

40. $\left(\dfrac{3a^2}{b^3}\right)^2\left(\dfrac{-2a}{3b}\right)^2$

41. $\left(\dfrac{x^3}{y^2}\right)^4\left(\dfrac{-y}{x^2}\right)^2$

42. $\dfrac{(x - 2y)^6}{(x - 2y)^2}$

43. $\dfrac{x^{-2} y^3}{x^3 y^{-4}}$

44. $\dfrac{(x^{-2} y^2)^3}{(x^3 y^{-2})^2}$

45. $\dfrac{5x^0 y^{-2}}{x^{-1} y^{-2}}$

46. $\dfrac{(2x^3 y^{-2})^2}{8x^{-3} y^2}$

47. $\dfrac{(-3a)^{-2}}{a^{-2} b^{-2}}$

48. $\dfrac{3a^{-3} b^2}{2^{-1} c^2 d^{-4}}$

49. $\dfrac{(a + b)^{-2}}{(a + b)^{-8}}$

50. $\dfrac{8x^{-8} y^{-12}}{2x^{-2} y^{-6}}$

51. $\dfrac{-12x^{-9} y^{10}}{4x^{-12} y^7}$

52. $\dfrac{(2x^2 y^{-1})^6}{(4x^{-6} y^{-5})^2}$

53. $\dfrac{(a + 3b)^{-12}}{(a + 3b)^{10}}$

54. $\dfrac{(-a^{-5} b^6)^3}{(a^8 b^4)^2}$

55. $x^{-2} + y^{-2}$

56. $(a^{-2} b^3)^{-1}$

57. $\left(\dfrac{x^{-2}}{y^3}\right)^{-1}$

△ **58.** $-2(1 + x^2)^{-3}(2x)$

△ **59.** $-2(4 - 5x)^{-3}(-5)$ △ **60.** $-7(x^2 - 3x)^{-8}(2x - 3)$

Find a value of x to make each statement true.

61. $2^x \cdot 2^3 = 2^{12}$

62. $2^{-3} \cdot 2^x = 2^6$

63. $2^x \cdot 2^x = 2^{16}$

64. $2^x \cdot 2^{x-1} = 2^7$

65. $\dfrac{2^x}{2^2} = 2^{-5}$

66. $\dfrac{2^{-3}}{2^x} = 2^4$

67. Assume that a substance decays such that $\frac{1}{2}$ the amount remains after each hour. If there were 640 grams at the start, how much remains after 7 hours? How much after n hours?

68. If a rope is 243 feet long and you successively cut off $\frac{2}{3}$ of the rope, how much remains after five cuts? How much after n cuts?

△ Throughout this text, this symbol will be used to identify exercises or groups of exercises that are of special value with respect to the future study of calculus. See the preface for further discussion.

69. For the rope in Exercise 68, how much remains after five cuts if each time you cut off $\frac{1}{3}$? How much is left after n cuts?

70. A company has a 4-year plan to increase its work force by $\frac{1}{4}$ for each of the 4 years. If the current work force is 2560, how big will it be at the end of the 4-year plan? Write an exponential expression that gives the work force after n years.

71. When an investment of P dollars earns $i\%$ interest per year and the interest is compounded annually, the formula for the final amount A after n years is $A = P(1 + i)^n$, where i is in decimal form. Find the amount A if \$1000 is invested at 10% compounded annually for 3 years.

72. Use the formula in Exercise 71 to approximate the number of years it would take for a \$1000 investment to double when it is invested at 10% interest compounded annually.

73. "Raising to a power" is neither a commutative nor an associative property. Verify this by showing a counterexample for each property.

74. We have said that 0^0 is undefined. The following shows why we have not defined this to be equal to 1.

Suppose that $0^0 = 1$. Then $1 = \dfrac{1}{1} = \dfrac{1^0}{0^0} = \left(\dfrac{1}{0}\right)^0$.

   (a) What rule for exponents is being used in the last step?

   (b) What went wrong?

75. Use the definition of a^{-n} to prove that $\dfrac{1}{a^{-n}} = a^n$.

**CHALLENGE
Think Creatively**

Suppose you snap your fingers and then wait 1 minute before you snap them again. Next, wait 2 minutes and snap your fingers, then 4 minutes, 8 minutes, 16 minutes, etc. That is, double the interval of time between successive snaps. If you were to keep this up for 1 year, how many times would you snap your fingers in that interval? First guess before doing any computation.

1.6
RADICALS
AND RATIONAL
EXPONENTS

You may recall using radicals in the past as you worked with square roots. For example, $\sqrt{25}$ is called a *radical* and denotes the positive number whose square is 25. Now it is true that $5^2 = 25$ and $(-5)^2 = 25$, but the **radical sign** $\sqrt{}$ implies the positive or *principal square root of a number*. Thus we say that $\sqrt{25} = 5$. We denote the negative square root as $-\sqrt{25} = -5$.

In general, the **principal nth root** of a real number a is denoted by $\sqrt[n]{a}$, but the expression $\sqrt[n]{a}$ does not always have meaning. For example, let us try to evaluate $\sqrt[4]{-16}$:

It is a fundamental property of real numbers that every positive real number a has exactly one positive nth root. Furthermore, every negative real number has a negative nth root provided that n is odd.

$$2^4 = 16 \qquad (-2)^4 = 16$$

It appears that there is no real number x such that $x^4 = -16$. In general, *there is no real number that is the even root of a negative number*.

Throughout this text, we will use the calculator symbol to indicate that it is advantageous or essential to use a calculator to complete an exercise.

DEFINITION OF $\sqrt[n]{a}$; THE PRINCIPAL nTH ROOT OF a

Let a be a real number and n a positive integer, $n \geq 2$.

Note, for example:
$\sqrt[3]{64} = 4$ *because* $4^3 = 64$
$\sqrt[3]{-8} = -2$ *because*
$\qquad (-2)^3 = 8$

1. If $a > 0$, then $\sqrt[n]{a}$ is the positive number x such that $x^n = a$.
2. $\sqrt[n]{0} = 0$.
3. If $a < 0$ and n is odd, then $\sqrt[n]{a}$ is the negative number x such that $x^n = a$.
4. If $a < 0$ and n is even, then $\sqrt[n]{a}$ is not a real number.

The symbol $\sqrt[n]{a}$ is also said to be a **radical;** $\sqrt{}$ is the radical sign, n is the **index** or **root,** and a is called the **radicand.**

EXAMPLE 1 Evaluate those radicals that are real numbers and check. If an expression is not a real number, give a reason.

(a) $\sqrt[3]{-125}$ (b) $\sqrt{-9}$ (c) $\sqrt[5]{-\dfrac{x^{10}}{32}}$ (d) $(\sqrt[3]{8})^3$

Solution

(a) $\sqrt[3]{-125} = -5$ *Check:* $(-5)^3 = -125$

(b) $\sqrt{-9}$ is not a real number since it is the even root of a negative number.

In general, $(\sqrt[n]{a})^n = a$

(c) $\sqrt[5]{-\dfrac{x^{10}}{32}} = -\dfrac{x^2}{2}$ *Check:* $\left(-\dfrac{x^2}{2}\right)^5 = -\dfrac{x^{10}}{32}$

(d) $(\sqrt[3]{8})^3 = 8$ *Check:* $\sqrt[3]{8} = 2$ and $2^3 = 8$ ∎

In order to multiply or divide radicals, the index must be the same. Here are illustrations that motivate Rules 1 and 2 stated below.

$$\left.\begin{array}{l} \sqrt[3]{8} \cdot \sqrt[3]{-27} = (2)(-3) = -6 \\ \sqrt[3]{(8)(-27)} = \sqrt[3]{-216} = -6 \end{array}\right\} \quad \sqrt[3]{8} \cdot \sqrt[3]{-27} = \sqrt[3]{(8)(-27)}$$

$$\left.\begin{array}{l} \dfrac{\sqrt{36}}{\sqrt{4}} = \dfrac{6}{2} = 3 \\[2mm] \sqrt{\dfrac{36}{4}} = \sqrt{9} = 3 \end{array}\right\} \quad \dfrac{\sqrt{36}}{\sqrt{4}} = \sqrt{\dfrac{36}{4}}$$

In the following rules it is assumed that all radicals exist according to the definition of $\sqrt[n]{a}$ and, as usual, denominators are not zero.

RULES FOR RADICALS

The proofs of these rules are called for in Exercises 72–74.

If all the indicated radicals are real numbers, then

1. $\sqrt[n]{a} \cdot \sqrt[n]{b} = \sqrt[n]{ab}$ (Multiplication of radicals)

2. $\dfrac{\sqrt[n]{a}}{\sqrt[n]{b}} = \sqrt[n]{\dfrac{a}{b}}$ (Division of radicals $b \neq 0$)

3. $\sqrt[m]{\sqrt[n]{a}} = \sqrt[mn]{a}$

Illustrations: (Assume that all variables represent positive numbers.)

$$\sqrt{6x} \cdot \sqrt{7y} = \sqrt{6x \cdot 7y} = \sqrt{42xy}$$

$$\frac{\sqrt[3]{81x^7}}{\sqrt[3]{-3x}} = \sqrt[3]{\frac{81x^7}{-3x}} = \sqrt[3]{-27x^6} = -3x^2 \qquad (\textit{Note: } (-3x^2)^3 = -27x^6.)$$

$$\sqrt[2]{\sqrt[3]{64}} = \sqrt[6]{64} = 2 \qquad (\textit{Note: } 2^6 = 64.)$$

$$\sqrt[5]{16x} \cdot \sqrt[5]{-2x^4} = \sqrt[5]{(16x)(-2x^4)} = \sqrt[5]{-32x^5} = -2x$$

EXAMPLE 2   Evaluate the product $\sqrt{5} \cdot \sqrt{14}$ to three decimal places.

Solution   By Rule 2, $\sqrt{5} \cdot \sqrt{14} = \sqrt{70}$. Using Table I at the back of the book, we have $\sqrt{70} = 8.367$. Using a calculator you should find that to six places, $\sqrt{70} = 8.366600$; rounded to three decimal places, this gives the square root as 8.367.   ∎

In applying the rules for radicals, care must be taken to avoid the type of error that results from the incorrect assumption of the existence of nth root. For example, $\sqrt{-4}$ is *not* a real number, but if this is not noticed, then *false* results such as the following occur:

CAUTION
Here is a common error you must avoid. Explain why the step $\sqrt{(-4)(-4)} = \sqrt{-4}\sqrt{-4}$ is not permissible.

$$4 = \sqrt{16} = \sqrt{(-4)(-4)} = \sqrt{-4} \cdot \sqrt{-4} = (\sqrt{-4})^2 = -4$$

We are now ready to make the extension of the exponential concept to include fractional exponents. Once again our guideline will be to *preserve* the earlier rules for integer exponents. First, consider the expression $b^{1/5}$. If Rule 3 for exponents is to apply, then $(b^{1/5})^5 = b^{(1/5)(5)} = b$. Thus $b^{1/5}$ is the fifth root of b; $b^{1/5} = \sqrt[5]{b}$. This motivates the following definition for $b^{1/n}$.

Since $\sqrt{-1}$ is not a real number, $(-1)^{1/2}$ is not defined. In general, $b^{1/n}$ is not defined within the set of real numbers when $b < 0$ and n is even.

DEFINITION OF $b^{1/n}$

For a real number b and a positive integer n ($n \geq 2$),

$$b^{1/n} = \sqrt[n]{b}$$

provided that $\sqrt[n]{b}$ exists.

Illustrations:

$$(-27)^{1/3} = \sqrt[3]{-27} = -3 \qquad 9^{-1/2} = \frac{1}{9^{1/2}} = \frac{1}{\sqrt{9}} = \frac{1}{3}$$

$$(-16)^{1/4} \text{ is not defined since } \sqrt[4]{-16} \text{ is not a real number}$$

Now that $b^{1/n}$ has been defined, we are able to define $b^{m/n}$, where $\dfrac{m}{n}$ is any rational number. Once again we want the earlier rules of exponents to apply. Observe, for example, the two ways that $8^{2/3}$ can be evaluated on the *assumption* that Rule 3 applies.

$$8^{2/3} = 8^{(1/3)\cdot 2} = (8^{1/3})^2 = (\sqrt[3]{8})^2 = 2^2 = 4$$

$$\uparrow$$
$$\text{Rule 3}$$
$$\downarrow$$

$$8^{2/3} = 8^{2\cdot(1/3)} = (8^2)^{1/3} = \sqrt[3]{8^2} = \sqrt[3]{64} = 4$$

These observations lead to this definition:

DEFINITION OF $b^{m/n}$

Note that a rational number can always be expressed with a positive denominator; for example, $\dfrac{2}{-3} = \dfrac{-2}{3}$.

Let $\dfrac{m}{n}$ be a rational number with $n \geq 2$. If b is a real number such that $\sqrt[n]{b}$ is defined, then

$$b^{m/n} = (\sqrt[n]{b})^m = \sqrt[n]{b^m} \quad \text{or} \quad b^{m/n} = (b^{1/n})^m = (b^m)^{1/n}$$

Illustrations:

For most such problems it is easier to take the nth root first and then raise to the mth power, rather than the reverse.

$$(-64)^{2/3} = (\sqrt[3]{-64})^2 = (-4)^2 = 16 \qquad (\text{using } b^{m/n} = (\sqrt[n]{b})^m)$$

or

$$(-64)^{2/3} = \sqrt[3]{(-64)^2} = \sqrt[3]{4096} = 16 \qquad (\text{using } b^{m/n} = \sqrt[n]{b^m})$$

Observe that the earlier definition $\quad b^{-n} = \dfrac{1}{b^n} \quad$ extends to the case $b^{-(m/n)}$ as follows:

$$b^{-(m/n)} = b^{-m/n} = (b^{1/n})^{-m} = \frac{1}{(b^{1/n})^m} = \frac{1}{b^{m/n}}$$

EXAMPLE 3 Evaluate: $8^{-2/3} + (-32)^{-2/5}$

Solution First rewrite each part using positive exponents. Then apply the definition and add.

$$8^{-2/3} + (-32)^{-2/5} = \frac{1}{8^{2/3}} + \frac{1}{(-32)^{2/5}}$$

$$= \frac{1}{(\sqrt[3]{8})^2} + \frac{1}{(\sqrt[5]{-32})^2}$$

$$= \frac{1}{(2)^2} + \frac{1}{(-2)^2}$$

$$= \frac{1}{4} + \frac{1}{4}$$

$$= \frac{1}{2}$$

Since the definition of rational exponents was made on the basis of preserving the basic rules for integer exponents, it can be shown that these rules also apply to rational exponents as in the following example.

EXAMPLE 4   Simplify: $\left(\dfrac{-8a^3}{b^{-6}}\right)^{2/3}$ ←when will this ever be useful

Solution

$$\left(\frac{-8a^3}{b^{-6}}\right)^{2/3} = \frac{(-8a^3)^{2/3}}{(b^{-6})^{2/3}} \qquad \text{(Rule 5)}$$

$$= \frac{(-8)^{2/3}(a^3)^{2/3}}{(b^{-6})^{2/3}} \qquad \text{(Rule 4)}$$

$$= \frac{\sqrt[3]{(-8)^2}\,a^2}{b^{-4}}$$

$$= 4a^2b^4 \qquad (\sqrt[3]{(-8)^2} = \sqrt[3]{64} = 4) \qquad ■$$

When the index of a radical is n, and the radicand is a perfect nth power, then there is usually no difficulty in computing with the radical. For example:

$$\sqrt{36} + \sqrt[3]{-27} = 6 + (-3) = 3$$

When the radicand is not a perfect nth power, as in $\sqrt{24}$, we simplify the radical so that no perfect square appears as a factor under the radical sign. Thus

$$\sqrt{24} = \sqrt{4 \cdot 6} = \sqrt{4} \cdot \sqrt{6} = 2\sqrt{6}$$

Note: The fundamental idea used here is that $\sqrt[n]{ab} = \sqrt[n]{a} \cdot \sqrt[n]{b}$ where $a \geq 0$ and $b \geq 0$.

We say that $2\sqrt{6}$ is the *simplified form* of $\sqrt{24}$.

EXAMPLE 5 Simplify: **(a)** $\sqrt{50}$ **(b)** $\sqrt[3]{16}$ **(c)** $\sqrt[3]{-54}$

Solution

(a) $\sqrt{50} = \sqrt{25 \cdot 2} = \sqrt{25} \cdot \sqrt{2} = 5\sqrt{2}$

(b) $\sqrt[3]{16} = \sqrt[3]{8 \cdot 2} = \sqrt[3]{8} \cdot \sqrt[3]{2} = 2\sqrt[3]{2}$

In parts (b) and (c) we search for a perfect cube as a factor under the radical sign.

(c) $\sqrt[3]{-54} = \sqrt[3]{(-27)(2)} = \sqrt[3]{-27} \cdot \sqrt[3]{2} = -3\sqrt[3]{2}$ ■

Recall: In order to be able to multiply two radicals, they must have the same index.

The rule for multiplication of radicals provides a way to find the product of any two radicals having the same index. Does a similar pattern work for the addition of radicals? That is, is the sum of the square roots of two numbers equal to the square root of their sum? Does $\sqrt{4} + \sqrt{9} = \sqrt{4 + 9}$? This can easily be checked as follows:

$$\sqrt{4} + \sqrt{9} = 2 + 3 = 5$$

But $\sqrt{4 + 9} = \sqrt{13}$. Therefore $\sqrt{4} + \sqrt{9} \neq \sqrt{4 + 9}$.

In general, $\sqrt{a} \pm \sqrt{b} \neq \sqrt{a \pm b}$.

In order to be able to add or subtract radicals, they must have the same index and the same radicand.

For example, we may use the distributive property to add radicals as follows:

$$3\sqrt{5} + 4\sqrt{5} = (3 + 4)\sqrt{5} = 7\sqrt{5}$$

You may find the following list of squares and cubes helpful when simplifying radicals.

Integer	Perfect squares	Perfect cubes
1	1	1
2	4	8
3	9	27
4	16	64
5	25	125
6	36	216
7	49	343
8	64	512
9	81	729
10	100	1000

EXAMPLE 6 Simplify: $\sqrt[3]{-24x^3} + 2\sqrt[3]{375x^3} - \sqrt[4]{162x^4}$ $(x \geq 0)$

Solution: Simplify each term by searching for perfect powers as factors.

$$\sqrt[3]{-24x^3} = \sqrt[3]{(-8x^3)(3)} = -2x\sqrt[3]{3} \qquad [(-2x)^3 = -8x^3]$$

$$2\sqrt[3]{375x^3} = 2\sqrt[3]{(125x^3)(3)} = 2(5x)\sqrt[3]{3} = 10x\sqrt[3]{3} \qquad [(5x)^3 = 125x^3]$$

$$\sqrt[4]{162x^4} = \sqrt[4]{(81x^4)(2)} = 3x\sqrt[4]{2} \qquad [(3x)^4 = 81x^4]$$

Now combine the terms that have the same index and radicand.

$$-2x\sqrt[3]{3} + 10x\sqrt[3]{3} - 3x\sqrt[4]{2} = 8x\sqrt[3]{3} - 3x\sqrt[4]{2}$$ ■

At times a fraction can be simplified by a process known as **rationalizing the denominator.** This consists of eliminating a radical from the denominator of a fraction. For example, consider the fraction $4/\sqrt{2}$. To rationalize the denominator, multiply the numerator and denominator by $\sqrt{2}$.

$$\frac{4}{\sqrt{2}} = \frac{4 \cdot \sqrt{2}}{\sqrt{2} \cdot \sqrt{2}} = \frac{4\sqrt{2}}{2} = 2\sqrt{2}$$

If a calculator is to be used, $\dfrac{4}{\sqrt{2}}$ is just as easy to evaluate as $2\sqrt{2}$ and would therefore be an acceptable form.

One reason for rationalizing denominators is to make computations easier. For example, suppose that we wish to evaluate $\dfrac{4}{\sqrt{2}}$ to three decimal places. It certainly is easier to multiply $\sqrt{2} = 1.414$ by 2 than to divide 4 by 1.414. Another reason for rationalizing denominators is to obtain a form for radical expressions in which they are more easily combined or compared.

EXAMPLE 7   Rationalize the denominator: **(a)** $\dfrac{6}{\sqrt{8}}$ **(b)** $\dfrac{5}{\sqrt[3]{2}}$

Solution

Here is another way to simplify part (a):
$$\dfrac{6}{\sqrt{8}} = \dfrac{6}{2\sqrt{2}} = \dfrac{3}{\sqrt{2}} = \dfrac{3\sqrt{2}}{2}$$

(a) $\dfrac{6}{\sqrt{8}} = \dfrac{6 \cdot \sqrt{2}}{\sqrt{8} \cdot \sqrt{2}} = \dfrac{6\sqrt{2}}{\sqrt{16}} = \dfrac{6\sqrt{2}}{4} = \dfrac{3\sqrt{2}}{2}$

(b) Multiply numerator and denominator by $\sqrt[3]{4}$ in order to have a perfect cube in the denominator.

$$\dfrac{5}{\sqrt[3]{2}} = \dfrac{5 \cdot \sqrt[3]{4}}{\sqrt[3]{2} \cdot \sqrt[3]{4}} = \dfrac{5\sqrt[3]{4}}{\sqrt[3]{8}} = \dfrac{5\sqrt[3]{4}}{2}$$ ∎

EXAMPLE 8   Combine: $\dfrac{6}{\sqrt{3}} + 2\sqrt{75} - \sqrt{3}$

Solution   Rationalize the denominator in the first term, simplify the second term, and combine:

$$\dfrac{6}{\sqrt{3}} + 2\sqrt{75} - \sqrt{3} = \dfrac{6 \cdot \sqrt{3}}{\sqrt{3} \cdot \sqrt{3}} + 2\sqrt{25 \cdot 3} - \sqrt{3}$$
$$= 2\sqrt{3} + 10\sqrt{3} - 1\sqrt{3} = 11\sqrt{3}$$ ∎

TEST YOUR UNDERSTANDING
Think Carefully

(Answers: Page 75)

Simplify each expression, if possible.

1. $\sqrt{8} + \sqrt{32}$
2. $\sqrt{12} + \sqrt{48}$
3. $\sqrt{45} - \sqrt{20}$
4. $\sqrt[3]{-16} + \sqrt[3]{54}$
5. $\sqrt[3]{128} + \sqrt[3]{125}$
6. $\sqrt[3]{-81} - \sqrt[3]{-24}$
7. $\dfrac{8}{\sqrt{2}} + \sqrt{98}$
8. $\dfrac{9}{\sqrt{3}} + \sqrt{300}$
9. $2\sqrt{20} - \dfrac{5}{\sqrt{5}}$
10. $3\sqrt{63} - \dfrac{14}{\sqrt{7}}$
11. $\dfrac{8}{\sqrt[3]{4}} + \sqrt[3]{16}$
12. $\sqrt[3]{81} - \dfrac{3}{\sqrt[3]{9}}$

Do you think that $\sqrt{x^2} = x$? If this were true, then for $x = -5$, we would have $\sqrt{(-5)^2} = -5$. However, as stated on page 33, the radical sign, $\sqrt{}$, means the positive square root. Therefore, $\sqrt{(-5)^2} = \sqrt{5^2} = 5$. This leads to the following important result:

> For all real numbers a, $\sqrt{a^2} = |a|$

Also recall that $(\sqrt[n]{a})^n = a$ for n even or odd as long as $\sqrt[n]{a}$ is a real number.

$$\sqrt[n]{a^n} = |a|, \qquad \text{if } n \text{ is even.}$$
$$\sqrt[n]{a^n} = a, \qquad \text{if } n \text{ is odd.}$$

Illustrations:

$$\sqrt{75x^2} = \sqrt{25 \cdot 3}\sqrt{x^2} = 5\sqrt{3}\,|x|$$
$$\sqrt[7]{(-3)^7} = -3$$
$$\sqrt[8]{(-\tfrac{1}{2})^8} = |-\tfrac{1}{2}| = \tfrac{1}{2}$$

EXAMPLE 9   Simplify: $2\sqrt{8x^3} + 3x\sqrt{32x} - x\sqrt{18x}$

Note: The expressions under the radicals would be negative for $x < 0$. Since the index is even, we must assume that $x \geq 0$ in order for these to have meaning.

Solution   For this problem, $x \geq 0$. Thus we need not make use of absolute-value notation.

$$2\sqrt{8x^3} = 2\sqrt{4 \cdot 2 \cdot x^2 \cdot x} = 4x\sqrt{2x}$$
$$3x\sqrt{32x} = 3x\sqrt{16 \cdot 2x} = 12x\sqrt{2x}$$
$$x\sqrt{18x} = x\sqrt{9 \cdot 2x} = 3x\sqrt{2x}$$

In each case the radicand and the index are the same, so the distributive property can be used to simplify.

$$4x\sqrt{2x} + 12x\sqrt{2x} - 3x\sqrt{2x} = (4x + 12x - 3x)\sqrt{2x} = 13x\sqrt{2x} \qquad \blacksquare$$

CAUTION: Learn to Avoid These Mistakes	
WRONG	**RIGHT**
$\sqrt{9 + 16} = \sqrt{9} + \sqrt{16}$	$\sqrt{9 + 16} = \sqrt{25}$
$(a + b)^{1/3} = a^{1/3} + b^{1/3}$	$(a + b)^{1/3} = \sqrt[3]{a + b}$
$\sqrt[3]{8} \cdot \sqrt[2]{8} = \sqrt[6]{64}$	$\sqrt[3]{8} \cdot \sqrt[2]{8} = 2 \cdot 2\sqrt{2} = 4\sqrt{2}$
$2\sqrt{x + 1} = \sqrt{2x + 1}$	$2\sqrt{x + 1} = \sqrt{4(x + 1)}$ $= \sqrt{4x + 4}$
$2 - \dfrac{1}{\sqrt{2}} = \dfrac{2 - 1}{\sqrt{2}}$	$2 - \dfrac{1}{\sqrt{2}} = 2 - \dfrac{\sqrt{2}}{2} = \dfrac{4 - \sqrt{2}}{2}$
$\sqrt{(x - 1)^2} = x - 1$	$\sqrt{(x - 1)^2} = |x - 1|$
$\sqrt{x^9} = x^3$	$\sqrt{x^9} = \sqrt{x^8 \cdot x} = x^4\sqrt{x}$
$a^{-1/2} + b^{-1/2} = \dfrac{1}{\sqrt{a + b}}$	$a^{-1/2} + b^{-1/2} = \dfrac{1}{\sqrt{a}} + \dfrac{1}{\sqrt{b}}$

EXERCISES 1.6

Evaluate.

1. $81^{-1/2}$
2. $\sqrt[3]{-64}$
3. $(64)^{-2/3}$
4. $(-64)^{1/3}$
5. $(-125)^{2/3}$
6. $(-125)^{-2/3}$
7. $\sqrt[3]{9} \cdot \sqrt[3]{-3}$
8. $\sqrt{5} \cdot \sqrt{20}$
9. $\dfrac{\sqrt[3]{-3}}{\sqrt[3]{-24}}$
10. $\dfrac{\sqrt{75}}{\sqrt{3}}$
11. $\dfrac{\sqrt{9}}{27^{-1/3}}$
12. $\dfrac{9^{1/2}}{\sqrt[3]{27}}$
13. $\sqrt[3]{(-125)(-1000)}$
14. $\sqrt{\sqrt{625}}$
15. $\sqrt[3]{\sqrt[3]{-512}}$
16. $\sqrt{\sqrt[3]{729}}$
17. $\sqrt{144 + 25}$
18. $\sqrt[5]{(-243)^2} \cdot (49)^{-1/2}$
19. $\left(\dfrac{1}{8} + \dfrac{1}{27}\right)^{1/3}$
20. $\sqrt{144} + \sqrt{25}$
21. $\left(\dfrac{16}{81}\right)^{3/4} + \left(\dfrac{256}{625}\right)^{1/4}$
22. $\left(\dfrac{1}{8}\right)^{1/3} + \left(\dfrac{1}{27}\right)^{-1/3}$
23. $\left(-\dfrac{125}{8}\right)^{1/3} - \left(\dfrac{1}{64}\right)^{1/3}$
24. $\left(\dfrac{8}{27}\right)^{-2/3} + \left(-\dfrac{32}{243}\right)^{2/5}$

Simplify.

25. $\sqrt{2} + \sqrt{18}$
26. $\sqrt{32} + \sqrt{72}$
27. $\sqrt{6} \cdot \sqrt{12}$
28. $\sqrt[3]{4} \cdot \sqrt[3]{12}$
29. $2\sqrt{5} + 3\sqrt{125}$
30. $-5\sqrt{24} - 2\sqrt{54}$
31. $2\sqrt{200} - 5\sqrt{8}$
32. $3\sqrt{45} - 2\sqrt{20}$
33. $\sqrt[3]{128} + \sqrt[3]{16}$
34. $\sqrt[3]{-24} + \sqrt[3]{81}$
35. $\dfrac{8}{\sqrt{2}} + 2\sqrt{50}$
36. $\sqrt[3]{-54x} + \sqrt[3]{250x}$
37. $3\sqrt{8x^2} - \sqrt{50x^2}$
38. $5\sqrt{75x^2} - 2\sqrt{12x^2}$
39. $3\sqrt{10} + 4\sqrt{90} - 5\sqrt{40}$
40. $\dfrac{2}{\sqrt{3}} + 10\sqrt{3} - 2\sqrt{12}$
41. $\dfrac{10}{\sqrt{5}} + 3\sqrt{45} - 2\sqrt{20}$
42. $10\sqrt{3x} - 2\sqrt{75x} + 3\sqrt{243x}$
43. $3\sqrt{9x^2} + 2\sqrt{16x^2} - \sqrt{25x^2}$
44. $\sqrt{2x^2} + 5\sqrt{32x^2} - 2\sqrt{98x^2}$
45. $\sqrt{x^2y} + \sqrt{8x^2y} + \sqrt{200x^2y}$

Rationalize the denominators and simplify.

46. $\dfrac{24}{\sqrt{6}}$
47. $\dfrac{8x}{\sqrt{2}}$
48. $\dfrac{10}{\sqrt{5}} + \dfrac{8}{\sqrt{4}}$
49. $\dfrac{1}{\sqrt{18}}$
50. $\dfrac{1}{\sqrt{27}}$
51. $\dfrac{24}{\sqrt{3x^2}}$
52. $\dfrac{20x}{\sqrt{5x^3}}$
53. $\dfrac{8}{\sqrt[3]{2}}$

Simplify, and express all answers with positive exponents. (Assume that all letters represent positive numbers.)

54. $(a^{-4}b^{-8})^{3/4}$
55. $(a^{-1/2}b^{1/3})(a^{1/2}b^{-1/3})$
56. $\dfrac{a^2b^{-1/2}c^{1/3}}{a^{-3}b^0c^{-1/3}}$
57. $\left(\dfrac{64a^6}{b^{-9}}\right)^{2/3}$
58. $\dfrac{(49a^{-4})^{-1/2}}{(81b^6)^{-1/2}}$
59. $\left(\dfrac{a^{-2}b^3}{a^4b^{-3}}\right)^{-1/2}\left(\dfrac{a^4b^{-5}}{ab}\right)^{-1/3}$
60. $\dfrac{2}{3}(3x - 1)^{-1/3} \cdot 3$
61. $\dfrac{1}{2}(3x^2 + 2)^{-1/2} \cdot 6x$
62. $\dfrac{1}{3}(x^3 + 2)^{-2/3} \cdot 3x^2$
63. $\dfrac{1}{2}(x^2 + 4x)^{-1/2}(2x + 4)$
64. $\dfrac{2}{3}(x^3 - 6x^2)^{-1/3}(3x^2 - 12x)$

65. The diagonal d of a rectangle is given by the formula $d = \sqrt{\ell^2 + w^2}$, where ℓ is the length and w is the width.

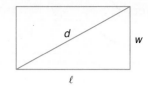

(a) Find d if $\ell = 20$ centimeters and $w = 15$ centimeters.

(b) Find d if $\ell = 16$ centimeters and $w = 10$ centimeters. Use a calculator and give the answer to one decimal place.

66. The formula $A = \sqrt{s(s - a)(s - b)(s - c)}$ is known as **Heron's formula.** It gives the area A of a triangle with sides of length a, b, c and semiperimeter $s = \frac{1}{2}(a + b + c)$. Show that for an equilateral triangle each of whose sides is of length a, Heron's formula gives $A = \dfrac{\sqrt{3}}{4}a^2$.

Combine and simplify.

67. $\sqrt[3]{\dfrac{32}{x^2}} - \dfrac{2\sqrt[3]{x}}{\sqrt[3]{2x^3}}$ **68.** $\dfrac{\sqrt{72a^3}}{3b} - \dfrac{a\sqrt{50a}}{2b} + \dfrac{12a^2}{b\sqrt{2a}}$

Simplify, and express the answers without radicals, using only positive exponents. (Assume that n is a positive integer and that all other letters represent positive numbers.)

69. $\sqrt[3]{\dfrac{x^{3n+1}y^n}{x^{3n+4}y^{4n}}}$ **70.** $\left(\dfrac{x^n}{x^{n-2}}\right)^{-1/2}$ **71.** $\sqrt{\dfrac{x^n}{x^{n-2}}}$

Prove the following assuming all radicals are real numbers.

72. $\sqrt[n]{a} \cdot \sqrt[n]{b} = \sqrt[n]{ab}$ (*Hint:* Let $\sqrt[n]{a} = x$ and $\sqrt[n]{b} = y$. Then $x^n = a$ and $y^n = b$.)

73. $\dfrac{\sqrt[n]{a}}{\sqrt[n]{b}} = \sqrt[n]{\dfrac{a}{b}}$

74. $\sqrt[m]{\sqrt[n]{a}} = \sqrt[mn]{a}$ (*Hint:* Let $x = \sqrt[mn]{a}$. Then $a = x^{mn}$.)

75. Use the result $\sqrt{a^2} = |a|$ and the rules for radicals to prove that $|xy| = |x| \cdot |y|$, where x and y are real numbers.

 Written Assignment: Use the definitions of $\sqrt{a}$ and $|a|$ to explain why $\sqrt{x^2} = |x|$ for all real numbers x.

1.7 FUNDAMENTAL OPERATIONS WITH POLYNOMIALS

The expression $5x^3 - 7x^2 + 4x - 12$ is called a **polynomial in the variable x.** Its *degree* is 3 because 3 is the largest power of the variable x. The *terms* of this polynomial are $5x^3$, $-7x^2$, $4x$, and -12. The *coefficients* are 5, -7, 4, and -12.

A nonzero constant, like 7, is classified as a polynomial of degree zero, since $7 = 7x^0$. The number zero is also referred to as a constant polynomial, but it is not assigned any degree.

A polynomial is in *standard form* if its terms are arranged so that the powers of the variable are in descending or ascending order. Here are some illustrations.

Note: All the exponents of the variable of a polynomial must be nonnegative integers. Therefore, $x^3 + x^{1/2}$ and $x^{-2} + 3x + 1$ are not polynomials because of the fractional and negative exponents.

Some of these polynomials have "missing" terms. For example, $x^3 - 3x + 12$ has no x^2 term, but it is still a third-degree polynomial.

Polynomial	Degree	Standard form
$x^3 - 3x + 12$	3	Yes
$\frac{2}{3}x^{10} - 4x^2 + \sqrt{2}\,x^4$	10	No
$32 - y^5 + 2y^2$	5	No
$6 + 2x - x^2 + x^3$	3	Yes

Polynomials having one, two, or three terms have special names:

Number of terms	Name of polynomial	Illustration
one	monomial	$17x^4y$
two	binomial	$\frac{1}{2}x^3 - 6x$
three	trinomial	$x^5 - x^2 + 2$

In general, an nth-degree polynomial in the variable x may be written in either one of these standard forms:

$$a_n x^n + a_{n-1}x^{n-1} + \cdots + a_2 x^2 + a_1 x + a_0$$

$$a_0 + a_1 x + a_2 x^2 + \cdots + a_{n-1}x^{n-1} + a_n x^n$$

You should become familiar with this notation, as you will encounter it frequently in future mathematics courses.

The coefficients $a_n, a_{n-1}, \ldots, a_0$ are real numbers and the exponents are nonnegative integers. The *leading coefficient* is $a_n \neq 0$, and a_0 is called the *constant term*. (a_0 may also be considered as the coefficient of the term $a_0 x^0$.)

Note: Polynomials represent real numbers regardless of the specific choice of the variable. Thus computations with polynomials are based on the fundamental properties for real numbers.

In a polynomial like $3x^2 - x + 4$ the variable x represents a real number. Therefore, when a specific real value is substituted for x, the result will be a real number. For instance, using $x = -3$ in this polynomial gives

$$3(-3)^2 - (-3) + 4 = 34$$

Adding or subtracting polynomials involves the combining of **like terms** (those having the same exponent on the variable). This can be accomplished by first rearranging and regrouping the terms (associative and commutative properties) and then combining by using the distributive property.

EXAMPLE 1   Add: $(4x^3 - 10x^2 + 5x + 8) + (12x^2 - 9x - 1)$

Solution

(a) $(4x^3 - 10x^2 + 5x + 8) + (12x^2 - 9x - 1)$
$= 4x^3 + (12x^2 - 10x^2) + (5x - 9x) + (8 - 1)$
$= 4x^3 + (12 - 10)x^2 + (5 - 9)x + 7$
$= 4x^3 + 2x^2 - 4x + 7$

In method (b) we list the polynomials in column form, putting like terms in the same column.

(b) $\quad\;\; 4x^3 - 10x^2 + 5x + 8$
$(+) \quad\underline{\qquad\quad 12x^2 - 9x - 1}$
$\quad\;\; 4x^3 + \;\; 2x^2 - 4x + 7$ ∎

Polynomials may contain more than one variable, as in Example 2. Again, the subtraction can be completed in two different ways, as shown.

Example 2 is of the form $a - b$. Think of this as $a - 1 \cdot b$ and use the distributive property to simplify in method (a).

EXAMPLE 2   Subtract: $(4a^3 - 10a^2b + 5b + 8) - (12a^2b - 9b - 1)$

Solution

(a) $(4a^3 - 10a^2b + 5b + 8) - (12a^2b - 9b - 1)$
$= 4a^3 - 10a^2b + 5b + 8 - 12a^2b + 9b + 1$
$= 4a^3 - 10a^2b - 12a^2b + 5b + 9b + 8 + 1$
$= 4a^3 - 22a^2b + 14b + 9$

(b)

$$\begin{array}{r} 4a^3 - 10a^2b + 5b + 8 \\ (-)\quad \underline{12a^2b - 9b - 1} \\ 4a^3 - 22a^2b + 14b + 9 \end{array}$$

■

The use of the distributive property is fundamental when multiplying polynomials. Perhaps the simplest situation calls for the product of a **monomial** (a polynomial having only one term) times a polynomial of two or more terms, as follows:

$$3x^2(4x^7 - 3x^4 - x^2 + 15) = 3x^2(4x^7) - 3x^2(3x^4) - 3x^2(x^2) + 3x^2(15)$$
$$= 12x^9 - 9x^6 - 3x^4 + 45x^2$$

In the first line we used an extended version of the distributive property, namely,

$$a(b - c - d + e) = ab - ac - ad + ae$$

Next, observe how the distributive property is used to multiply two **binomials** (polynomials having two terms).

$$(2x + 3)(4x + 5) = (2x + 3)4x + (2x + 3)5$$
$$= (2x)(4x) + (3)(4x) + (2x)(5) + (3)(5)$$
$$= 8x^2 + 12x + 10x + 15$$
$$= 8x^2 + 22x + 15$$

Here is a shortcut that can be used to multiply two binomials.

$(2x + 3)(4x + 5)$:

$$(2x + 3)(4x + 5) = 8x^2 + 22x + 15$$

with $8x^2$, 15, $12x$, $10x$ labeled.

$8x^2$ is the product of the *first* terms in the binomials.
$10x$ and $12x$ are the products of the *outer* and *inner* terms.
15 is the product of the *last* terms in the binomials.

In general, we may write the product $(a + b)(c + d)$ in this way:

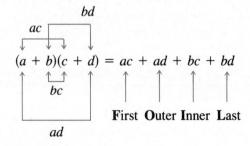

$$(a + b)(c + d) = ac + ad + bc + bd$$

First Outer Inner Last

with bd, ac, bc, ad labeled.

Keep this diagram in mind as an aid to finding the product of two binomials mentally. Some students find it helpful to remember the first letter of each step, FOIL.

EXAMPLE 3 Find the product of $ax + b$ and $cx + d$ by:
(a) Making use of the distributive property.
(b) Using the visual inspection method (FOIL).

Solution:
(a) $(ax + b)(cx + d) = (ax + b)cx + (ax + b)d$
$$= acx^2 + bcx + adx + bd$$
$$= acx^2 + (bc + ad)x + bd$$

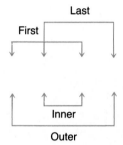

First
Last
Inner
Outer

(b)

$$(ax + b)(cx + d) = acx^2 + (bc + ad)x + bd$$

acx^2 bd
bcx
adx

∎

The distributive property can be extended to multiply polynomials with three terms, **trinomials,** as in Example 4.

EXAMPLE 4 Multiply $3x^3 - 8x + 4$ by $2x^2 + 5x - 1$

Note: The column method is a convenient way to organize your work. Be certain to keep like terms in the same column. Let $x = 2$ and check the solution.

Solution

$$3x^3 - 8x + 4$$
$$2x^2 + 5x - 1$$

(add) $\begin{cases} - 3x^3 + 8x - 4 & (-1 \text{ times } 3x^3 - 8x + 4) \\ 15x^4 - 40x^2 + 20x & (5x \text{ times } 3x^3 - 8x + 4) \\ 6x^5 - 16x^3 + 8x^2 & (2x^2 \text{ times } 3x^3 - 8x + 4) \end{cases}$

$$6x^5 + 15x^4 - 19x^3 - 32x^2 + 28x - 4$$

∎

TEST YOUR
UNDERSTANDING
Think Carefully

Combine.

1. $(x^2 + 2x - 6) + (-2x + 7)$ **2.** $(x^2 + 2x - 6) - (-2x + 7)$
3. $(5x^4 - 4x^3 + 3x^2 - 2x + 1) + (-5x^4 + 6x^3 + 10x)$
4. $(x^2y + 3xy + xy^2) + (3x^2y - 6xy) - (2xy - 5xy^2)$

Find the products.

5. $(-3x)(x^3 + 2x^2 - 1)$ **6.** $(2x + 3)(3x + 2)$
7. $(6x - 1)(2x + 5)$ **8.** $(4x + 7)(4x - 7)$
9. $(2x - 3y)(2x - 3y)$ **10.** $(x^2 - 3x + 5)(2x^3 + x^2 - 3x)$

(Answers: Page 75)

Sometimes more than one operation is involved, as shown in the next example.

EXAMPLE 5 Simplify by performing the indicated operations:

$$(x^2 - 5x)(3x^2) + (x^3 - 1)(2x - 5)$$

Solution Multiply first and then combine like terms.

$$(x^2 - 5x)(3x^2) + (x^3 - 1)(2x - 5) = (3x^4 - 15x^3) + (2x^4 - 5x^3 - 2x + 5)$$
$$= 5x^4 - 20x^3 - 2x + 5 \qquad \blacksquare$$

The product of $a + b$ times itself is given by

$$(a + b)(a + b) = a^2 + 2ab + b^2$$

Using exponents, we may write the **expanded form** of $(a + b)^2$:

$$(a + b)^2 = a^2 + 2ab + b^2$$

Explain how these figures provide a geometric interpretation for the expansion of $(a + b)^2$.

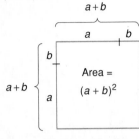

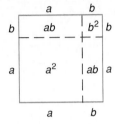

EXAMPLE 6 Expand: $(a + b)^3$

Solution First write $(a + b)^3 = (a + b)(a + b)^2$. Then use the expansion of $(a + b)^2$:

$$(a + b)^3 = (a + b)(a^2 + 2ab + b^2)$$
$$= a^3 + 2a^2b + ab^2 + a^2b + 2ab^2 + b^3$$
$$= a^3 + 3a^2b + 3ab^2 + b^3 \qquad \blacksquare$$

The result of Example 6 provides a formula for the cube of a binomial, that is, for all expansions of the form $(a + b)^3$.
CAUTION
$(a + b)^2 \neq a^2 + b^2$.

The following example makes use of the product $(a - b)(a + b) = a^2 - b^2$ to rationalize denominators. Each of the two factors $a - b$ and $a + b$ is called the *conjugate* of the other.

EXAMPLE 7 Rationalize the denominator: $\dfrac{5}{\sqrt{10} - \sqrt{3}}$

Solution

$\sqrt{10} + \sqrt{3}$ is the conjugate of $\sqrt{10} - \sqrt{3}$.

$$\frac{5}{\sqrt{10} - \sqrt{3}} = \frac{5}{\sqrt{10} - \sqrt{3}} \cdot \frac{\sqrt{10} + \sqrt{3}}{\sqrt{10} + \sqrt{3}}$$

$$= \frac{5(\sqrt{10} + \sqrt{3})}{10 - 3}$$

$$= \frac{5(\sqrt{10} + \sqrt{3})}{7} \qquad \blacksquare$$

Add.

1. $3x^2 + 5x - 2$
   $5x^2 - 7x + 9$

2. $5x^2 - 9x - 1$
   $2x^2 + 2x + 7$

3. $3x^3 - 7x^2 - 8x + 12$
   $x^3 - 2x^2 + 8x - 9$

4. $x^3 - 3x^2 + 2x - 5$
   $5x^2 - x + 9$

5. $4x^2 + 9x - 17$
   $2x^3 - 3x^2 + 2x - 11$

6. $2x^3 + x^2 - 7x + 1$
   $- 2x^2 - x + 8$

Subtract.

7. $3x^3 - 2x^2 - 8x + 9$
   $2x^3 + 5x^2 + 2x + 1$

8. $x^3 - 2x^2 + 6x + 1$
   $- x^2 - 6x - 1$

9. $4x^3 + x^2 - 2x - 13$
   $2x^2 + 3x + 9$

Simplify by performing the indicated operations.

10. $(3x + 5) + (3x - 2)$
11. $5x + (1 - 2x)$
12. $(7x + 5) - (2x + 3)$
13. $(y + 2) + (2y + 1) + (3y + 3)$
14. $h - (h + 2)$
15. $(x^3 + 3x^2 + 3x + 1) - (x^2 + 2x + 1)$
16. $7x - (3 - x) - 2x$
17. $5y - [y - (3y + 8)]$
18. $(2x^3y^2 - 5xy + x^2y^3) + (3xy - x^2y^3) - (x^3y^2 + 2xy)$
19. $(5x - 2xy + x^2y^2) - (3x + 7xy) - (2x + xy - x^2y^2)$
20. $x(3x^2 - 2x + 5)$
21. $2x^2(2x + 1 - 10x^2)$
22. $-4t(t^4 - \frac{1}{4}t^3 + 4t^2 - \frac{1}{16}t + 1)$
23. $(x + 1)(x + 1)$
24. $(2x + 1)(2x - 1)$
25. $(4x - 2)(x + 7)$
26. $(12x - 8)(7x + 4)$
27. $(-2x + 3)(3x + 6)$
28. $(-2x - 3)(3x + 6)$
29. $(-2x - 3)(3x - 6)$
30. $(\frac{1}{2}x + 4)(\frac{1}{2}x - 4)$
31. $(\frac{2}{3}x + 6)(\frac{2}{3}x + 6)$
32. $(7 + 3x)(9 - 4x)$
33. $(7 - 3x)(4x - 9)$
34. $(x + \frac{3}{4})(x + \frac{3}{4})$
35. $(\frac{1}{5}x - \frac{1}{4})(\frac{1}{5}x - \frac{1}{4})$
36. $(x - \sqrt{3})(x + \sqrt{3})$
37. $(\sqrt{x} - 10)(\sqrt{x} + 10)$
38. $(\frac{1}{10}x - \frac{1}{100})(\frac{1}{10}x - \frac{1}{100})$
39. $(\sqrt{x} + \sqrt{2})(\sqrt{x} - \sqrt{2})$
40. $(x^2 - 3)(x^2 + 3)$
41. $(x^2 + x + 9)(x^2 - 3x - 4)$
42. $(x^{1/3} - 2)(x^{2/3} + 2x^{1/3} + 4)$
43. $(x - 2)(x^2 + 2x + 4)$
44. $(x - 2)(x^3 + 2x^2 + 4x + 8)$
45. $(x - 2)(x^4 + 2x^3 + 4x^2 + 8x + 16)$
46. $(x^n - 4)(x^n + 4)$
47. $(x^{2n} + 1)(x^{2n} - 2)$
48. $5(x + 5)(x - 5)$
49. $3x(1 - x)(1 - x)$
50. $(x + 3)(x + 1)(x - 4)$
51. $(2x + 1)(3x - 2)(3 - x)$
52. $(2x^2 + 3)(9x^2) + (3x^3 - 2)(4x)$
53. $(x^3 - 2x + 1)(2x) + (x^2 - 2)(3x^2 - 2)$
54. $(2x^3 - x^2)(6x - 5) + (3x^2 - 5x)(6x^2 - 2x)$
55. $(x^4 - 3x^2 + 5)(2x + 3) + (x^2 + 3x)(4x^3 - 6x)$

Expand each of the following and combine like terms.

56. $(a - b)^2$
57. $(x - 1)^3$
58. $(x + 1)^4$
59. $(a + b)^4$
60. $(a - b)^4$
61. $(2x + 3)^3$
62. $(\frac{1}{2}x - 4)^2$
63. $(\frac{1}{3}x + 3)^3$
64. $(\frac{1}{2}x - 1)^3$

Rationalize the denominator and simplify.

65. $\dfrac{12}{\sqrt{5} - \sqrt{3}}$ **66.** $\dfrac{20}{3 - \sqrt{2}}$ **67.** $\dfrac{14}{\sqrt{2} - 3}$

68. $\dfrac{\sqrt{x}}{\sqrt{x} + \sqrt{y}}$ **69.** $\dfrac{\sqrt{x} + \sqrt{y}}{\sqrt{x} - \sqrt{y}}$ **70.** $\dfrac{1}{\sqrt{x} + 2}$

Rationalize the numerator.

71. $\dfrac{\sqrt{5} + 3}{\sqrt{5}}$ **72.** $\dfrac{\sqrt{5} + \sqrt{3}}{\sqrt{2}}$ **73.** $\dfrac{\sqrt{x} + \sqrt{y}}{\sqrt{x} - \sqrt{y}}$

◤ *Show how to convert the first fraction given into the form of the second one.*

74. $\dfrac{\sqrt{x} - 2}{x - 4}$; $\dfrac{1}{\sqrt{x} + 2}$ **75.** $\dfrac{\sqrt{4 + h} - 2}{h}$; $\dfrac{1}{\sqrt{4 + h} + 2}$

76. Find a geometric interpretation for the formula $(a - b)^2 = a^2 - 2ab + b^2$.

1.8 FACTORING POLYNOMIALS

In the preceding section, we multiplied polynomials in this way:

$$(x + 2)(x - 2)(x - 3) = (x^2 - 4)(x - 3)$$
$$= x^3 - 3x^2 - 4x + 12$$

In a sense, factoring is "unmultiplying."

Now we want to *begin* with $x^3 - 3x^2 - 4x + 12$ and **factor** (*unmultiply*) it into the form $(x + 2)(x - 2)(x - 3)$.

One of the basic methods of factoring is the reverse of multiplying by a monomial. Consider this multiplication problem:

$$6x^2(4x^7 - 3x^4 - x^2 - 15) = 24x^9 - 18x^6 - 6x^4 + 90x^2$$

As you read this equation from left to right it involves multiplication by $6x^2$. Reading from right to left, it involves "factoring out" the *common monomial factor* $6x^2$. These are the details of this factoring process:

Note the use of the distributive property in this illustration.

$$24x^9 - 18x^6 - 6x^4 + 90x^2 = 6x^2(4x^7) - 6x^2(3x^4) - 6x^2(x^2) + 6x^2(15)$$
$$= 6x^2(4x^7 - 3x^4 - x^2 + 15)$$

Suppose, instead, we were to use $3x$ as the common factor:

$$24x^9 - 18x^6 - 6x^4 + 90x^2 = 3x(8x^8 - 6x^5 - 2x^3 + 30x)$$

The instruction to factor will always mean that we are to find the complete factored form for the given expression.

This is a correct factorization but it is not considered the *complete factored form* because $2x$ can still be factored out of the expression in the parentheses.

We can extend factoring techniques to polynomials with more than one variable as in the example that follows.

EXAMPLE 1 Factor:
(a) $21x^4y - 14x^5y^2 - 63x^8y^3 + 91x^{11}y^4$
(b) $8x^2(x - 1) + 4x(x - 1) + 2(x - 1)$

Solution

(a) $21x^4y - 14x^5y^2 - 63x^8y^3 + 91x^{11}y^4 = 7x^4y(3 - 2xy - 9x^4y^2 + 13x^7y^3)$

(b) $8x^2(x - 1) + 4x(x - 1) + 2(x - 1) = 2(x - 1)(4x^2 + 2x + 1)$ ∎

We will now consider several basic procedures for factoring polynomials. You should check each of these by multiplication. These forms are very useful and you need to learn to recognize and apply each one.

THE DIFFERENCE OF TWO SQUARES

$$a^2 - b^2 = (a - b)(a + b)$$

Illustrations:

$$x^2 - 9 = x^2 - 3^2 = (x - 3)(x + 3)$$

$$(x + 1)^2 - 4 = (x + 1)^2 - 2^2 = (x + 1 - 2)(x + 1 + 2)$$

$$= (x - 1)(x + 3)$$

The trick is to look for a common factor first in each of the given terms. This step not only will save you work in some problems but also may well be the difference between success and failure.

Can $3x^2 - 75$ be factored by using the difference of two squares? It becomes possible if we first *factor out the common factor* 3.

$$3x^2 - 75 = 3(x^2 - 25) = 3(x - 5)(x + 5)$$

THE DIFFERENCE (SUM) OF TWO CUBES

$$a^3 - b^3 = (a - b)(a^2 + ab + b^2)$$

$$a^3 + b^3 = (a + b)(a^2 - ab + b^2)$$

Be sure to check each of these facts by multiplication.

Illustrations:

$$8x^3 - 27 = (2x)^3 - 3^3 = (2x - 3)(4x^2 + 6x + 9)$$

$$2x^3 + 128y^3 = 2(x^3 + 64y^3) = 2[x^3 + (4y)^3] = 2(x + 4y)(x^2 - 4xy + 16y^2)$$

TEST YOUR UNDERSTANDING
Think Carefully

Factor out the common monomial.

1. $3x - 9$ **2.** $-5x + 15$ **3.** $5xy + 25y^2 + 10y^5$

Factor as the difference of squares.

4. $x^2 - 36$ **5.** $4x^2 - 49$ **6.** $(a + 2)^2 - 25b^2$

Factor as the difference (sum) of two cubes.

7. $x^3 - 27$ **8.** $x^3 + 27$ **9.** $8a^3 - 125$

Factor the following by first considering common monomial factors.

(Answers: Page 75) **10.** $3x^2 - 48$ **11.** $ax^3 + ay^3$ **12.** $2hx^2 - 8h^3$

It is not possible to factor $x^2 - 5$ or $x - 8$ as the difference of two squares by using polynomial factors with integer coefficients. There are times, however, when it is desirable to allow other factorizations, such as the following:

$$x^2 - 5 = x^2 - (\sqrt{5})^2 = (x + \sqrt{5})(x - \sqrt{5})$$
$$x - 8 = (\sqrt{x})^2 - (\sqrt{8})^2 = (\sqrt{x} + \sqrt{8})(\sqrt{x} - \sqrt{8})$$
$$x - 8 = (x^{1/3})^3 - 2^3 = (x^{1/3} - 2)(x^{2/3} + 2x^{1/3} + 4)$$

In general, we follow this rule when factoring:

> *Unless otherwise indicated, use the same type of numerical coefficients and exponents in the factors as appear in the given unfactored form.*

Just as we learned how to factor the difference of two squares or cubes, we can also learn how to factor the difference of two fourth powers, two fifth powers, and so on. All these situations can be collected in the following single general form that gives the factorization of the difference of two nth powers, where n is an integer greater than 1.

The factorization of $a^n - b^n$ is one of the most useful in mathematics; it will be needed in calculus. Study it carefully here so that in later work you will be able to recall it with minimal effort.

THE DIFFERENCE OF TWO nTH POWERS

$$a^n - b^n = (a - b)(a^{n-1} + a^{n-2}b + a^{n-3}b^2 + \cdots + ab^{n-2} + b^{n-1})$$

To help describe the second factor, we may rewrite it like this:

$$a^{n-1}b^0 + a^{n-2}b^1 + a^{n-3}b^2 + \cdots + a^1b^{n-2} + a^0b^{n-1}$$

You can see that the exponents for a begin with $n - 1$ and decrease to 0, whereas for b they begin with 0 and increase to $n - 1$. Note also that the sum of the exponents for a and b, for each term, is $n - 1$.

EXAMPLE 2   Factor: $a^5 - b^5$

Solution   Use the general form for $a^n - b^n$ with $n = 5$.

$$a^5 - b^5 = (a - b)(a^4 + a^3b + a^2b^2 + ab^3 + b^4)$$ ∎

EXAMPLE 3   Use the result of Example 2 to factor $3y^5 - 96$.

Solution

$$3y^5 - 96 = 3(y^5 - 32) = 3(y^5 - 2^5)$$
$$= 3(y - 2)(y^4 + 2y^3 + 4y^2 + 8y + 16)$$ ∎

EXAMPLE 4   Factor: $x^6 - 1$

Solution   Several different approaches are possible.
(a) Using the formula for the difference of two nth powers, we have

$$x^6 - 1 = (x - 1)(x^5 + x^4 + x^3 + x^2 + x + 1)$$

(b) Using the formula for the difference of two cubes, we have

$$x^6 - 1 = (x^2)^3 - 1^3 = (x^2 - 1)(x^4 + x^2 + 1)$$
$$= (x - 1)(x + 1)(x^4 + x^2 + 1)$$

Notice that in (c) we continued to factor the difference and sum of two cubes. This gives the complete factorization for $x^6 - 1$ since neither of the quadratic factors is factorable.

(c) Using the formula for the difference of two squares, we have

$$x^6 - 1 = (x^3)^2 - 1^2 = (x^3 - 1)(x^3 + 1)$$
$$= (x - 1)(x^2 + x + 1)(x + 1)(x^2 - x + 1) \quad \blacksquare$$

Some polynomials of four terms that do not appear to be factorable, because there is no common monomial factor in all the terms, can be factored by a method known as **grouping.** In the following, we group the given polynomial into two binomials, each of which is factorable.

CAUTION
$x^2(x + 3) + 9(x + 3)$ is not a factored form of the given expression.

$$x^3 + 3x^2 + 9x + 27 = (x^3 + 3x^2) + (9x + 27)$$
$$= x^2(x + 3) + 9(x + 3)$$
$$= (x^2 + 9)(x + 3)$$

Here is an alternative grouping that leads to the same answer.

$$x^3 + 3x^2 + 9x + 27 = (x^3 + 27) + (3x^2 + 9x)$$
$$= (x + 3)(x^2 - 3x + 9) + (x + 3)(3x)$$
$$= (x + 3)(x^2 - 3x + 9 + 3x)$$
$$= (x + 3)(x^2 + 9)$$

Not all groupings are productive. Thus in Example 5 the grouping $(ax^2 + 15) + (-3x - 5ax)$ does not lead to a solution.

EXAMPLE 5   Factor $ax^2 + 15 - 5ax - 3x$ by grouping.

Solution

$$ax^2 + 15 - 5ax - 3x = (ax^2 - 5ax) + (15 - 3x)$$
$$= ax(x - 5) + 3(5 - x)$$
$$= ax(x - 5) - 3(x - 5)$$
$$= (ax - 3)(x - 5) \quad \blacksquare$$

By multiplying $a + b$ by $a + b$ and $a - b$ by $a - b$, the following formulas can be established:

PERFECT SQUARE TRINOMIALS

$$a^2 + 2ab + b^2 = (a + b)^2$$

$$a^2 - 2ab + b^2 = (a - b)^2$$

EXAMPLE 6 Factor: **(a)** $x^2 + 10x + 25$ **(b)** $9x^3 - 42x^2 + 49x$

Solution

(a) $x^2 + 10x + 25 = x^2 + 2(x \cdot 5) + 5^2$

$\qquad\qquad\qquad\quad = (x + 5)^2$

(b) $9x^3 - 42x^2 + 49x = x(9x^2 - 42x + 49)$

$\qquad\qquad\qquad\qquad = x[(3x)^2 - 2(3x)(7) + 7^2]$

$\qquad\qquad\qquad\qquad = x(3x - 7)^2$ ∎

Another factoring technique that we will consider deals with trinomials that are not necessarily perfect squares. Let us factor $x^2 + 7x + 12$.

From our experience with multiplying binomials we can anticipate that the factors of $x^2 + 7x + 12$ will be of this form:

$$(x + \underline{\ ?\ })(x + \underline{\ ?\ })$$

We need to fill in the blanks with two integers whose product is 12. Furthermore, the middle term of the product must be $+7x$. The possible choices for the two integers are

12 and 1   6 and 2   4 and 3

To find the correct pair is now a matter of trial and error. These are the possible factorizations:

$(x + 12)(x + 1)$ $(x + 6)(x + 2)$ $(x + 4)(x + 3)$

Only the last form gives the correct middle term of $7x$. Therefore, we conclude that $x^2 + 7x + 12 = (x + 4)(x + 3)$.

EXAMPLE 7 Factor: $x^2 - 10x + 24$

Solution The final term, $+24$, must be the product of two positive numbers or two negative numbers. (Why?) Since the middle term, $-10x$, has a minus sign, the factorization must be of this form:

$$(x - \underline{\ ?\ })(x - \underline{\ ?\ })$$

Now try all the pairs of integers whose product is 24: 24 and 1, 12 and 2, 8 and 3, 6 and 4. Only the last of these gives $-10x$ as the middle term. Thus

$$x^2 - 10x + 24 = (x - 6)(x - 4)$$ ∎

Let us now consider a more complicated factoring problem. If we wish to factor the trinomial $15x^2 + 43x + 8$, we need to consider possible factors both of 15 and of 8. Because of the "+" signs in the trinomial, the factorization will be of this form:

$$(\underline{?}\,x + \underline{?})(\underline{?}\,x + \underline{?})$$

Here are the different possibilities for factoring 15 and 8:

$$15 = 15 \cdot 1 \qquad 15 = 5 \cdot 3$$
$$8 = 8 \cdot 1 \qquad 8 = 4 \cdot 2$$

Using $15 \cdot 1$, write the form

$$(15x + \underline{?})(x + \underline{?})$$

*With a little luck, and much more experience, you can often avoid exhausting **all** the possibilities before finding the correct factors. You will then find that such work can often be shortened significantly.*

Try 8 and 1 in the blanks, *both ways,* namely:

$$(15x + 8)(x + 1) \qquad (15x + 1)(x + 8)$$

Neither gives a middle term of $43x$; so now try 4 and 2 in the blanks both ways. Again you see that neither case works. Next consider the form

$$(5x + \underline{?})(3x + \underline{?})$$

Once again try 4 and 2 both ways, and 8 with 1 both ways. The correct answer is:

$$15x^2 + 43x + 8 = (5x + 1)(3x + 8)$$

Example 8 shows us that not every trinominal can be factored.

EXAMPLE 8 Factor: $12x^2 - 9x + 2$

Solution Consider the forms

$$(12x - \underline{?})(x - \underline{?}) \qquad (6x - \underline{?})(2x - \underline{?}) \qquad (4x - \underline{?})(3x - \underline{?})$$

In each form try 2 with 1 both ways. None of these produces a middle term of $-9x$; hence we say that $12x^2 - 9x + 2$ is *not factorable* with integral coefficients. ∎

TEST YOUR UNDERSTANDING
Think Carefully

Factor the following trinomial squares.

1. $a^2 + 6a + 9$ **2.** $x^2 - 10x + 25$ **3.** $4x^2 + 12xy + 9y^2$

Factor each trinomial if possible.

4. $x^2 + 8x + 15$ **5.** $a^2 - 12a + 20$ **6.** $x^2 + 3x + 4$

7. $10x^2 - 39x + 14$ **8.** $6x^2 - 11x - 10$ **9.** $6x^2 + 6x - 5$

(Answers: Page 75)

EXAMPLE 9   Factor $15x^2 + 7x - 8$

You may find it easier to leave out the sign in the two binomial forms as you try various cases. Because of the term -8, the signs must be opposites.

Solution   Try these forms:

$$(5x \ \underline{?})(3x \ \underline{?}) \qquad (15x \ \underline{?})(x \ \underline{?})$$

In each form try the factors of 8 (2 and 4, 1 and 8) and search for a middle term of $7x$. Then insert the appropriate signs to obtain $(15x - 8)(x + 1)$.   ∎

All the trinomials we have considered were of degree 2. The same methods can be modified to factor certain trinomials of higher degree, as in Example 10.

EXAMPLE 10   Factor: **(a)** $x^4 - x^2 - 12$ **(b)** $a^6 - 3a^3b - 18b^2$

Solution

(a) Note that $x^4 = (x^2)^2$ and let $u = x^2$. Then

The substitution step in Example 10(a) changes the factoring of a fourth degree trinomial into the factoring of a second degree trinomial. Letting $u = a^3$ in part (b) will have the same effect.

$$x^4 - x^2 - 12 = u^2 - u - 12$$
$$= (u - 4)(u + 3)$$
$$= (x^2 - 4)(x^2 + 3)$$
$$= (x - 2)(x + 2)(x^2 + 3)$$

(b) $a^6 - 3a^3b - 18b^2 = (a^3 - 6b)(a^3 + 3b)$ ∎

CAUTION: Learn to Avoid These Mistakes	
WRONG	RIGHT
$(x + 2)3 + (x + 2)y = (x + 2)3y$	$(x + 2)3 + (x + 2)y$ $= (x + 2)(3 + y)$
$3x + 1 = 3(x + 1)$	$3x + 1$ is not factorable by using integers.
$x^3 - y^3 = (x - y)(x^2 + y^2)$	$x^3 - y^3 = (x - y)(x^2 + xy + y^2)$
$x^3 + 8$ is not factorable.	$x^3 + 8 = (x + 2)(x^2 - 2x + 4)$
$x^2 + y^2 = (x + y)(x + y)$	$x^2 + y^2$ is not factorable by using real numbers.
$4x^2 - 6xy + 9y^2 = (2x - 3y)^2$	$4x^2 - 6xy + 9y^2$ is not a perfect trinomial square and cannot be factored using integers.

EXERCISES 1.8

Factor as the difference of two squares.

1. $4x^2 - 9$ 2. $25x^2 - 144y^2$ 3. $a^2 - 121b^2$

Factor as the difference (sum) of two cubes.

4. $x^3 - 8$ 5. $x^3 + 64$ 6. $8x^3 + 1$
7. $125x^3 - 64$ 8. $8 - 27a^3$ 9. $8x^3 + 343y^3$

Factor as the difference of two squares, allowing irrational numbers as well as radical expressions. (All letters represent positive numbers.)

10. $a^2 - 15$ 11. $3 - 4x^2$ 12. $x - 1$
13. $x - 36$ 14. $2x - 9$ 15. $8 - 3x$

Factor as the difference (sum) of two cubes, allowing irrational numbers as well as radical expressions.

16. $x^3 - 2$ 17. $7 + a^3$ 18. $1 - h$
19. $27x + 1$ 20. $27x - 64$ 21. $3x - 4$

Factor by grouping.

22. $a^2 - 2b + 2a - ab$ 23. $x^2 - y - x + xy$ 24. $x + 1 + y + xy$
25. $-y - x + 1 + xy$ 26. $ax + by + ay + bx$ 27. $2 - y^2 + 2x - xy^2$

Factor completely.

28. $8a^2 - 2b^2$ 29. $7x^3 + 7h^3$ 30. $81x^4 - 256y^4$
31. $a^8 - b^8$ 32. $40ab^3 - 5a^4$ 33. $a^5 - 32$
34. $x^6 + x^2y^4 - x^4y^2 - y^6$ 35. $a^3x - b^3y + b^3x - a^3y$ 36. $7a^2 - 35b + 35a - 7ab$
37. $x^5 - 16xy^4 - 2x^4y + 32y^5$ look in nb.

Factor each trinomial.

38. $12a^2 - 13a + 1$ 39. $20a^2 - 9a + 1$ 40. $4 - 5b + b^2$ 41. $9x^2 + 6x + 1$
42. $5x^2 + 31x + 6$ 43. $14x^2 + 37x + 5$ 44. $9x^2 - 18x + 5$ 45. $8x^2 - 9x + 1$
46. $6x^2 + 12x + 6$ 47. $8x^2 - 16x + 6$ 48. $12x^2 + 92x + 15$ 49. $12a^2 - 25a + 12$
50. $6a^2 - 5a - 21$ 51. $4x^2 + 4x - 3$ 52. $15x^2 + 19x - 56$ 53. $24a^2 + 25ab + 6b^2$

Factor each trinomial when possible. When appropriate, first factor out the common monomial.

54. $x^2 + x + 1$ 55. $a^2 - 2a + 2$ 56. $49r^2s - 42rs + 9s$
57. $6x^2 + 2x - 20$ 58. $15 + 5y - 10y^2$ 59. $2b^2 + 12b + 16$
60. $4a^2x^2 - 4abx^2 + b^2x^2$ 61. $a^3b - 2a^2b^2 + ab^3$ 62. $12x^2y + 22xy^2 - 60y^3$
63. $16x^2 - 24x + 8$ 64. $16x^2 - 24x - 8$ 65. $25a^2 + 50ab + 25b^2$
66. $x^4 - 2x^2 + 1$ 67. $a^6 - 2a^3 + 1$ 68. $2x^4 + 8x^2 + 42$
69. $6x^5y - 3x^3y^2 - 30xy^3$

△ *Simplify by factoring.*

70. $(1 + x)^2(-1) + (1 - x)(2)(1 + x)$ 71. $(x + 2)^3(2) + (2x + 1)(3)(x + 2)^2$
72. $(x^2 + 2)^2(3) + (3x - 1)(2)(x^2 + 2)(2x)$ 73. $(x^3 + 1)^3(2x) + (x^2 - 1)(3)(x^3 + 1)^2(3x^2)$

74. Find these products
(a) $(a + b)(a^2 - ab + b^2)$
(b) $(a + b)(a^4 - a^3b + a^2b^2 - ab^3 + b^4)$
(c) $(a + b)(a^6 - a^5b + a^4b^2 - a^3b^3 + a^2b^4 - ab^5 + b^6)$

75. Use the results of Exercise 74 to factor the following:
(a) $x^5 + 32$ (b) $128x^8 + xy^7$

76. Write the factored form of $a^n + b^n$, where n is an odd positive integer greater than 1.

77. Compare the results in parts (b) and (c) of Example 4, page 00, to obtain the factorization of $x^4 + x^2 + 1$.

78. The result in Exercise 76 can also be found using this procedure:

$$x^4 + x^2 + 1 = (x^4 + 2x^2 + 1) - x^2$$
$$= (x^2 + 1)^2 - x^2$$
$$= (x^2 + 1 - x)(x^2 + 1 + x)$$
$$= (x^2 - x + 1)(x^2 + x + 1)$$

Use this procedure to factor the following:
(a) $x^4 + 3x^2 + 4$ (b) $9x^4 + 3x^2 + 4$ (c) $a^4 + 2a^2b^2 + 9b^4$ (d) $a^4 - 3a^2 + 1$

△ 79. Factor completely: $5(4x^2 + 4x + 1)^4(8x + 4)$

△ 80. Factor completely: $3(x^4 - 2x^2 + 1)^2(4x^3 - 4x)$

△ 81. If $y = x\sqrt{x^2 + 2}$, show that $\sqrt{1 + y^2} = x^2 + 1$

△ 82. If $y = 2x(x^2 + 1)^{1/2}$, show that $\sqrt{1 + y^2} = 2x^2 + 1$

| 1.9 FUNDAMENTAL OPERATIONS WITH RATIONAL EXPRESSIONS | A **rational expression** is a ratio of polynomials. Rational expressions are the "algebraic extensions" of rational numbers, so the fundamental rules for operating with rational numbers extend to rational expressions. |

The important rules for operating with rational expressions, also referred to as *algebraic fractions*, will now be considered. In each case we shall give an example of the rule under discussion in terms of arithmetic fractions so that you can compare the procedures being used. Also, in each case, we exclude values of the variable for which the denominator is equal to zero.

Negative of a Fraction

RULE 1. $-\dfrac{a}{b} = \dfrac{-a}{b} = \dfrac{a}{-b}$ $\left[-\dfrac{2}{3} = \dfrac{-2}{3} = \dfrac{2}{-3}\right]$

The negative of a fraction is the same as the fraction obtained by taking the negative of the numerator or of the denominator.

Illustration:

Note that since $-(2 - x) = x - 2$, each of these fractions is also equal to $\dfrac{x - 2}{x^2 - 5}$.

$$-\frac{2 - x}{x^2 - 5} = \frac{-(2 - x)}{x^2 - 5} = \frac{2 - x}{-(x^2 - 5)}$$

Reducing Fractions

This rule says that a fraction can be simplified by dividing both the numerator and denominator by the same nonzero quantity.

RULE 2. $\quad \dfrac{ac}{bc} = \dfrac{a}{b} \qquad \left[\dfrac{2 \cdot 3}{5 \cdot 3} = \dfrac{2}{5} \right]$

Reading this formula from left to right shows how to reduce a fraction to *lowest terms* so that the numerator and denominator of the resulting fraction have no common factors.

Illustrations:

(a) $\qquad \dfrac{x^2 + 5x - 6}{x^2 + 6x} = \dfrac{(x - 1)(x + 6)}{x(x + 6)} = \dfrac{x - 1}{x} \quad$ (Rule 2)

(b) $\qquad \dfrac{5a - 3b}{3b - 5a} = \dfrac{(-1)(-5a + 3b)}{(1)(3b - 5a)} = \dfrac{(-1)(3b - 5a)}{(1)(3b - 5a)} = -1$

The work in the preceding illustration can be shortened by dividing the numerator and denominator by $3b - 5a$:

Any nonzero number divided by its opposite is equal to -1:
$\dfrac{x - y}{y - x} = -1.$

$$\dfrac{5a - 3b}{3b - 5a} = \dfrac{\overset{-1}{\cancel{5a - 3b}}}{\underset{1}{\cancel{3b - 5a}}} = \dfrac{-1}{1} = -1$$

Multiplication of Fractions

RULE 3. $\quad \dfrac{a}{b} \cdot \dfrac{c}{d} = \dfrac{ac}{bd} \qquad \left[\dfrac{2}{3} \cdot \dfrac{4}{5} = \dfrac{2 \cdot 4}{3 \cdot 5} = \dfrac{8}{15} \right]$

Illustration:

$$\dfrac{x + 1}{x - 1} \cdot \dfrac{2 - x - x^2}{5x} = \dfrac{(x + 1)(2 - x - x^2)}{(x - 1)5x} \quad \text{(Rule 3)}$$

Whenever we are working with rational expressions it is taken for granted that final answers are reduced to lowest terms.

$$= \dfrac{(x + 1)\overset{-1}{\cancel{(1 - x)}}(2 + x)}{\underset{1}{\cancel{(x - 1)}}5x}$$

$$= -\dfrac{(x + 1)(x + 2)}{5x} \quad \text{or} \quad -\dfrac{x^2 + 3x + 2}{5x}$$

Division of Fractions

RULE 4. $\quad \dfrac{a}{b} \div \dfrac{c}{d} = \dfrac{a}{b} \cdot \dfrac{d}{c} = \dfrac{ad}{bc} \qquad \left[\dfrac{3}{5} \div \dfrac{2}{3} = \dfrac{3}{5} \cdot \dfrac{3}{2} = \dfrac{3 \cdot 3}{5 \cdot 2} = \dfrac{9}{10} \right]$

Here is an alternative form for Rule 4:
$$\dfrac{\dfrac{a}{b}}{\dfrac{c}{d}} = \dfrac{a}{b} \div \dfrac{c}{d} = \dfrac{ad}{bc}$$

Illustration:

(a) $\qquad \dfrac{(x + 1)^2}{x^2 - 6x + 9} \div \dfrac{3x + 3}{x - 3} = \dfrac{(x + 1)^2}{(x - 3)} \cdot \dfrac{x - 3}{3(x + 1)} \quad \text{(Rule 4)}$

$$= \dfrac{x + 1}{3(x - 3)}$$

Simplify each expression by reducing to lowest terms.

1. $\dfrac{x^2 + 2x}{x}$　　　2. $\dfrac{x^2}{x^2 + 2x}$　　　3. $\dfrac{4b^2 - 4ab}{3a^2 - 3ab}$

4. $\dfrac{x^2 + 6x + 5}{x^2 - x - 2}$　　　5. $\dfrac{x^2 - 4}{x^4 - 16}$　　　6. $\dfrac{3x^2 + x - 10}{5x - 3x^2}$

Perform the indicated operation and simplify.

7. $\dfrac{4 - 2x}{2} \cdot \dfrac{x + 2}{x^2 - 4}$　　　8. $\dfrac{x - 2}{3(x + 1)} \div \dfrac{x^2 + 2x}{x + 2}$

(Answers: Page 75)

9. $\dfrac{a}{b} \div \dfrac{a^2 - ab}{ab + b^2}$　　　10. $\dfrac{x + y}{x - y} \cdot \dfrac{x^2 - 2xy + y^2}{x^2 - y^2}$

**Addition
and Subtraction
of Fractions—
Same Denonimators**

RULE 5.　$\dfrac{a}{d} + \dfrac{c}{d} = \dfrac{a + c}{d}$　　$\left[\dfrac{3}{7} + \dfrac{2}{7} = \dfrac{3 + 2}{7} = \dfrac{5}{7}\right]$

RULE 6.　$\dfrac{a}{d} - \dfrac{c}{d} = \dfrac{a - c}{d}$　　$\left[\dfrac{7}{9} - \dfrac{2}{9} = \dfrac{7 - 2}{9} = \dfrac{5}{9}\right]$

Illustration:

$$\dfrac{6x^2}{2x^2 - x - 10} - \dfrac{15x}{2x^2 - x - 10} = \dfrac{6x^2 - 15x}{2x^2 - x - 10} \qquad \text{(Rule 6)}$$

$$= \dfrac{3x\,\cancel{(2x - 5)}}{(x + 2)\,\cancel{(2x - 5)}} \qquad \text{(factoring)}$$

$$= \dfrac{3x}{x + 2} \qquad \text{(Rule 2)}$$

**Addition
and Subtraction
of Fractions—
Different
Denominators**

RULE 7.　$\dfrac{a}{b} + \dfrac{c}{d} = \dfrac{ad + bc}{bd}$　　$\left[\dfrac{2}{3} + \dfrac{3}{4} = \dfrac{2 \cdot 4 + 3 \cdot 3}{3 \cdot 4} = \dfrac{8 + 9}{12} = \dfrac{17}{12}\right]$

RULE 8.　$\dfrac{a}{b} - \dfrac{c}{d} = \dfrac{ad - bc}{bd}$　　$\left[\dfrac{4}{5} - \dfrac{2}{3} = \dfrac{4 \cdot 3 - 5 \cdot 2}{5 \cdot 3} = \dfrac{12 - 10}{15} = \dfrac{2}{15}\right]$

Illustration:

$$\dfrac{3}{x^2 + x} + \dfrac{2}{x^2 - 1} = \dfrac{3(x^2 - 1) + 2(x^2 + x)}{(x^2 + x)(x^2 - 1)} \qquad \text{(Rule 7)}$$

$$= \dfrac{5x^2 + 2x - 3}{(x^2 + x)(x^2 - 1)} \qquad \text{(combining terms)}$$

$$= \dfrac{(5x - 3)(x + 1)}{x(x + 1)(x^2 - 1)} \qquad \text{(factoring)}$$

$$= \dfrac{5x - 3}{x(x^2 - 1)} \qquad \text{(Rule 2)}$$

Here is an alternative method for the preceding illustration. It makes use of the **least common denominator (LCD)** of the two fractions.

The LCD of the two fractions is $x(x + 1)(x - 1)$. We express each fraction using this common denominator and then add the numerators.

$$\frac{3}{x^2 + x} + \frac{2}{x^2 - 1} = \frac{3}{x(x + 1)} + \frac{2}{(x + 1)(x - 1)}$$

$$= \frac{3(x - 1)}{x(x + 1)(x - 1)} + \frac{2x}{(x + 1)(x - 1)x} \qquad \text{(Rule 2)}$$

$$= \frac{3(x - 1) + 2x}{x(x + 1)(x - 1)} \qquad \text{(Rule 5)}$$

$$= \frac{5x - 3}{x(x^2 - 1)}$$

EXAMPLE 1 Combine and simplify: $\dfrac{3}{2} - \dfrac{4}{3x(x + 1)} - \dfrac{x - 5}{3x^2}$

Solution The least common denominator of the fractions is $6x^2(x + 1)$.

$$\frac{3}{2} - \frac{4}{3x(x + 1)} - \frac{x - 5}{3x^2} = \frac{3 \cdot 3x^2(x + 1)}{2 \cdot 3x^2(x + 1)} - \frac{4 \cdot 2x}{3x(x + 1) \cdot 2x} - \frac{(x - 5) \cdot 2(x + 1)}{3x^2 \cdot 2(x + 1)}$$

To find the LCD, use each factor that appears in the factored forms of the denominators the maximum number of times it appears in any one of the factored forms.

$$= \frac{(9x^3 + 9x^2) - 8x - (2x^2 - 8x - 10)}{6x^2(x + 1)}$$

$$= \frac{9x^3 + 7x^2 + 10}{6x^2(x + 1)} \qquad \blacksquare$$

TEST YOUR UNDERSTANDING
Think Carefully

Combine and simplify.

1. $\dfrac{x}{5} + \dfrac{2x}{3}$

2. $\dfrac{4x}{3} - \dfrac{x}{2}$

3. $\dfrac{2}{3x^2} - \dfrac{1}{2x}$

4. $\dfrac{3}{2x} + \dfrac{5}{3x} + \dfrac{1}{x}$

5. $\dfrac{7}{x - 2} + \dfrac{3}{x + 2}$

6. $\dfrac{9}{x - 3} + \dfrac{7}{x^2 - 9}$

7. $\dfrac{5}{(x - 1)(x + 2)} - \dfrac{8}{4 - x^2}$

8. $\dfrac{1 - 4x}{2x + 5} + \dfrac{8x^2 - 16x}{4x^2 - 25} - \dfrac{1}{2x - 5}$

(Answers: Page 75)

The fundamental properties of fractions can be used to simplify rational expressions whose numerators and denominators may themselves also contain fractions.

The expression in Example 2 is a type that you will encounter in calculus. Be certain that you understand each step in the solution.

EXAMPLE 2 Simplify: $\dfrac{\dfrac{1}{5 + h} - \dfrac{1}{5}}{h}$

Solution Combine the fractions in the numerator and then divide.

$$\frac{\dfrac{1}{5 + h} - \dfrac{1}{5}}{h} = \frac{\dfrac{5 - (5 + h)}{5(5 + h)}}{h} = \frac{\dfrac{-h}{5(5 + h)}}{h}$$

$$= \frac{-h}{5(5 + h)} \cdot \frac{1}{h} = \frac{-h}{5h(5 + h)} = -\frac{1}{5(5 + h)} \qquad \blacksquare$$

EXAMPLE 3 Simplify: $\dfrac{x^{-2} - y^{-2}}{\dfrac{1}{x} - \dfrac{1}{y}}$

In this method both the numerator and denominator are multiplied by x^2y^2 to simplify.

Solution

$$\frac{x^{-2} - y^{-2}}{\dfrac{1}{x} - \dfrac{1}{y}} = \frac{\dfrac{1}{x^2} - \dfrac{1}{y^2}}{\dfrac{1}{x} - \dfrac{1}{y}} = \frac{\left(\dfrac{1}{x^2} - \dfrac{1}{y^2}\right)(x^2y^2)}{\left(\dfrac{1}{x} - \dfrac{1}{y}\right)(x^2y^2)} \quad \text{(Rule 2)}$$

This problem can also be solved by using the procedure shown in Example 2.

$$= \frac{y^2 - x^2}{xy^2 - x^2y}$$

$$= \frac{(y - x)(y + x)}{xy(y - x)}$$

$$= \frac{y + x}{xy}$$

Working with fractions often creates difficulties for many students. Study this list; it may help you avoid some common pitfalls.

CAUTION: Learn to Avoid These Mistakes	
WRONG	**RIGHT**
$\dfrac{2}{3} + \dfrac{x}{5} = \dfrac{2 + x}{3 + 5}$	$\dfrac{2}{3} + \dfrac{x}{5} = \dfrac{2 \cdot 5 + 3 \cdot x}{3 \cdot 5} = \dfrac{10 + 3x}{15}$
$\dfrac{1}{a} + \dfrac{1}{b} = \dfrac{1}{a + b}$	$\dfrac{1}{a} + \dfrac{1}{b} = \dfrac{b + a}{ab}$
$\dfrac{2x + 5}{4} = \dfrac{x + 5}{2}$	$\dfrac{2x + 5}{4} = \dfrac{2x}{4} + \dfrac{5}{4} = \dfrac{x}{2} + \dfrac{5}{4}$
$2 + \dfrac{x}{y} = \dfrac{2 + x}{y}$	$2 + \dfrac{x}{y} = \dfrac{2y + x}{y}$
$3\left(\dfrac{x + 1}{x - 1}\right) = \dfrac{3(x + 1)}{3(x + 1)}$	$3\left(\dfrac{x + 1}{x - 1}\right) = \dfrac{3(x + 1)}{x - 1}$
$a \div \dfrac{b}{c} = \dfrac{1}{a} \cdot \dfrac{b}{c}$	$a \div \dfrac{b}{c} = a \cdot \dfrac{c}{b} = \dfrac{ac}{b}$
$\dfrac{1}{a^{-1} + b^{-1}} = a + b$	$\dfrac{1}{a^{-1} + b^{-1}} = \dfrac{1}{\dfrac{1}{a} + \dfrac{1}{b}} = \dfrac{ab}{b + a}$
$\dfrac{x^2 + 4x + 6}{x + 2} = \dfrac{x^2 + 4x + \overset{3}{\cancel{6}}}{x + \underset{1}{\cancel{2}}}$ $= \dfrac{x^2 + 4x + 3}{x + 1}$	$\dfrac{x^2 + 4x + 6}{x + 2}$ is in simplest form.

EXERCISES 1.9

Classify each statement as true or false. If it is false, correct the right side to get a correct equality.

1. $\dfrac{5}{7} - \dfrac{2}{3} = \dfrac{3}{4}$

2. $\dfrac{2x + y}{y - 2x} = -2\left(\dfrac{x + y}{x - y}\right)$

3. $\dfrac{3ax - 5b}{6} = \dfrac{ax - 5b}{2}$

4. $\dfrac{x + x^{-1}}{xy} = \dfrac{x + 1}{x^2 y}$

5. $x^{-1} + y^{-1} = \dfrac{y + x}{xy}$

6. $\dfrac{2}{\frac{3}{4}} = \dfrac{8}{3}$

Simplify, if possible.

7. $\dfrac{8xy}{12yz}$

8. $\dfrac{24abc^2}{36bc^2 d}$

9. $\dfrac{45x^3 + 15x^2}{15x^2}$

10. $\dfrac{9y^2 + 12y^8 - 15y^6}{3y^2}$

11. $\dfrac{12x^3 + 8x^2 + 4x}{4x}$

12. $\dfrac{5a^2 - 10a^3 + 15a^4}{5a^2}$

13. $\dfrac{a^2 b^2 + ab^2 - a^2 b^3}{ab^2}$

14. $\dfrac{-6a^3 + 9a^6 - 12a^9}{-3a^3}$

15. $\dfrac{6a^2 x^2 - 8a^4 x^6}{2a^2 x^2}$

16. $\dfrac{-8a^3 x^3 + 4ax^3 - 12a^2 x^6}{-4ax^3}$

17. $\dfrac{x^2 - 5x}{5 - x}$

18. $\dfrac{n - 1}{n^2 - 1}$

19. $\dfrac{n + 1}{n^2 + 1}$

20. $\dfrac{(x + 1)^2}{1 - x^2}$

21. $\dfrac{3x^2 + 3x - 6}{2x^2 + 6x + 4}$

22. $\dfrac{x^3 - x}{x^3 - 2x^2 + x}$

23. $\dfrac{4x^2 + 12x + 9}{4x^2 - 9}$

24. $\dfrac{x^2 + 2x + xy + 2y}{x^2 + 4x + 4}$

25. $\dfrac{a^2 - 16b^2}{a^3 + 64b^3}$

26. $\dfrac{a^2 - b^2}{a^2 - 6b - ab + 6a}$

Perform the indicated operations and simplify.

27. $\dfrac{2x^2}{y} \cdot \dfrac{y^2}{x^3}$

28. $\dfrac{3x^2}{2y^2} \div \dfrac{3x^3}{y}$

29. $\dfrac{2a}{3} \cdot \dfrac{3}{a^2} \cdot \dfrac{1}{a}$

30. $\left(\dfrac{a^2}{b^2} \cdot \dfrac{b}{c^2}\right) \div a$

31. $\dfrac{3x}{2y} - \dfrac{x}{2y}$

32. $\dfrac{a + 2b}{a} + \dfrac{3a + b}{a}$

33. $\dfrac{a - 2b}{2} - \dfrac{3a + b}{3}$

34. $\dfrac{7}{5x} - \dfrac{2}{x} + \dfrac{1}{2x}$

35. $\dfrac{x - 1}{3} \cdot \dfrac{x^2 + 1}{x^2 - 1}$

36. $\dfrac{x^2 - x - 6}{x^2 - 3x} \cdot \dfrac{x^3 + x^2}{x + 2}$

37. $\dfrac{1 - x}{2 + x} \div \dfrac{x^2 - x}{x^2 + 2x}$

38. $\dfrac{x^2 + 3x}{x^2 + 4x + 3} \div \dfrac{x^2 - 2x}{x + 1}$

39. $\dfrac{2}{x} - y$

40. $\dfrac{x^2}{x - 1} - \dfrac{1}{1 - x}$

41. $\dfrac{3y}{y + 1} + \dfrac{2y}{y - 1}$

42. $\dfrac{2a}{a^2 - 1} - \dfrac{a}{a + 1}$

43. $\dfrac{2x^2}{x^2 + x} + \dfrac{x}{x + 1}$

44. $\dfrac{3x + 3}{2x^2 - x - 1} + \dfrac{1}{2x + 1}$

45. $\dfrac{5}{x^2 - 4} - \dfrac{3 - x}{4 - x^2}$

46. $\dfrac{1}{a^2 - 4} + \dfrac{3}{a - 2} - \dfrac{2}{a + 2}$

47. $\dfrac{2x}{x^2 - 9} + \dfrac{x}{x^2 + 6x + 9} - \dfrac{3}{x + 3}$

48. $\dfrac{x}{x - 1} + \dfrac{x + 7}{x^2 - 1} - \dfrac{x - 2}{x + 1}$

49. $\dfrac{x + 3}{5 - x} - \dfrac{x - 5}{x + 5} + \dfrac{2x^2 + 30}{x^2 - 25}$

50. $\dfrac{a^2 + 2ab + b^2}{a^2 - b^2} \div \dfrac{a^2 + 3ab + 2b^2}{a^2 - 3ab + 2b^2}$

51. $\dfrac{x^3 + x^2 - 12x}{x^2 - 3x} \cdot \dfrac{3x^2 - 10x + 3}{3x^2 + 11x - 4}$

52. $\dfrac{n^2 + n}{2n^2 + 7n - 4} \cdot \dfrac{4n^2 - 4n + 1}{2n^2 - n - 3} \cdot \dfrac{2n^2 + 5n - 12}{2n^3 - n^2}$

53. $\dfrac{n^3 - 8}{n + 2} \cdot \dfrac{2n^2 + 8}{n^3 - 4n} \cdot \dfrac{n^3 + 2n^2}{n^3 + 2n^2 + 4n}$

54. $\dfrac{a^3 - 27}{a^2 - 9} \div \left(\dfrac{a^2 + 2ab + b^2}{a^3 + b^3} \cdot \dfrac{a^3 - a^2 b + ab^2}{a^2 + ab}\right)$

Simplify.

55. $\dfrac{\dfrac{5}{x^2-4}}{\dfrac{10}{x-2}}$

56. $\dfrac{\dfrac{1}{x}-\dfrac{1}{4}}{x-4}$

57. $\dfrac{\dfrac{1}{4+h}-\dfrac{1}{4}}{h}$

58. $\dfrac{\dfrac{1}{x^2}-\dfrac{1}{9}}{x-3}$

59. $\dfrac{\dfrac{1}{x+3}-\dfrac{1}{3}}{x}$

60. $\dfrac{\dfrac{1}{4}-\dfrac{1}{x^2}}{x-2}$

61. $\dfrac{\dfrac{1}{x^2}-\dfrac{1}{16}}{x+4}$

62. $\dfrac{x^{-2}-\dfrac{1}{4}}{\dfrac{1}{x}-\dfrac{1}{2}}$

63. $\dfrac{x^{-1}-y^{-1}}{\dfrac{1}{x^2}-\dfrac{1}{y^2}}$

64. $\dfrac{\dfrac{4}{x^2}-\dfrac{1}{y^2}}{\dfrac{2}{x}-\dfrac{1}{y}}$

△ 65. $\dfrac{(1+x^2)(-2x)-(1-x^2)(2x)}{(1+x^2)^2}$

△ 66. $\dfrac{(x^2-9)(2x)-x^2(2x)}{(x^2-9)^2}$

△ 67. $\dfrac{x^2(4-2x)-(4x-x^2)(2x)}{x^4}$

△ 68. $\dfrac{(x+1)^2(2x)-(x^2-1)(2)(x+1)}{(x+1)^4}$

Simplify, and express as a single fraction without negative exponents.

69. $\dfrac{a^{-1}-b^{-1}}{a-b}$

70. $\dfrac{(a+b)^{-1}}{a^{-1}+b^{-1}}$

71. $\dfrac{x^{-2}-y^{-2}}{xy}$

72. There are three tests and a final examination given in a mathematics course. Let a, b, and c be the numerical grades of the tests, and let d represent the examination grade.

(a) If the final grade is computed by allowing the exam to count the same as the average of the three tests, show that the final average is given by the expression $\dfrac{a+b+c+3d}{6}$.

(b) Assume that the average of the three tests accounts for 60% of the final grade and that the examination accounts for 40%. Show that the final average is given by the expression $\dfrac{a+b+c+2d}{5}$.

73. Some calculators require that certain calculations be performed in a different manner to accommodate the machine. Show that in each case the expression on the left can be computed by using the equivalent expression on the right.

(a) $\dfrac{A}{B}+\dfrac{C}{D}=\dfrac{\dfrac{A\cdot D}{B}+C}{D}$

(b) $A\cdot B+C\cdot D+E\cdot F=\left[\dfrac{\left(\dfrac{A\cdot B}{D}+C\right)\cdot D}{F}+E\right]\cdot F$

△ 74. If $x^2+y^2=4$, show that $-\dfrac{y-x\left(-\dfrac{x}{y}\right)}{y^2}=-\dfrac{4}{y^3}$.

△ 75. If $y^3-x^3=8$, show that $\dfrac{2xy^2-2x^2y\left(\dfrac{x^2}{y^2}\right)}{y^4}=\dfrac{16x}{y^5}$.

△ 76. If $y=x^2-\dfrac{1}{4x^2}$, show that $\sqrt{1+y^2}=x^2+\dfrac{1}{4x^2}$.

△ 77. If $y=\dfrac{x^2}{8}-\dfrac{2}{x^2}$, show that $\sqrt{1+y^2}=\dfrac{x^2}{8}+\dfrac{2}{x^2}$.

CHALLENGE
Think Creatively

A man left 17 horses to his three children. He left one-half to the oldest, one-third to the middle child, and one-ninth to the youngest. Since 17 is not divisible by 2, 3, or 9, the children borrowed a horse from a neighbor in order to have a total of 18 horses. Then the oldest child received $\frac{1}{2}\times18=9$ horses, the middle child received $\frac{1}{3}\times18=6$ horses, the youngest child received $\frac{1}{9}\times18=2$ horses. Since $9+6+2=17$, the number of horses left to the three children, it was possible to return the extra horse to the neighbor! What is wrong with this story?

1. Explain why for any real number b, $b \neq 0$, b^0 is defined to be equal to 1.
2. Many students claim that $\sqrt{25} = \pm 5$. Explain why this is incorrect.
3. Provide a convincing argument to explain why $3\left(\dfrac{x+1}{x-1}\right) \neq \dfrac{3x+1}{3x-1}$.

4. Note that $\dfrac{12}{\sqrt{150}} = \dfrac{2}{5}\sqrt{6}$. Is either form preferable to the other? Explain your answer.

5. When factoring polynomials, it is useful first to factor out common factors if possible. What is the advantage in doing this?

6. Translate the following statement into symbolic form:

   The square of the sum of two numbers is *at least as large* as four times the product.

   Prove the statement is true. *Hint:* Assume the inequality is true and work backward.

7. Three numbers can be arranged vertically in fraction form to produce these cases:

$$\frac{\dfrac{a}{b}}{c} = \frac{a}{bc} \qquad \frac{a}{\dfrac{b}{c}} = \frac{ac}{b}$$

   Find all possible distinct cases using the four numbers a, b, c, and d arranged vertically in fractional form.

1.10 INTRODUCTION TO COMPLEX NUMBERS

In the definition of a radical, care was taken to avoid the even root of a negative number, such as $\sqrt{-4}$. This was necessary because there is no real number x whose square is -4. Consequently, there can be no real number that satisfies the equation $x^2 + 4 = 0$. Suppose, for the moment, that we could solve $x^2 + 4 = 0$ using our algebraic methods. Then we might write the following:

$$x^2 + 4 = 0$$
$$x^2 = -4$$
$$x = \pm\sqrt{-4}$$
$$= \pm\sqrt{4(-1)}$$
$$= \pm\sqrt{4}\sqrt{-1}$$
$$= \pm 2\sqrt{-1}$$

Although it could be claimed that $2\sqrt{-1}$ is a solution of $x^2 + 4 = 0$, it is certainly not a *real number* solution. Therefore we introduce $\sqrt{-1}$ as a new kind of number; it will be the unit for a new set of numbers, the *imaginary numbers*. The symbol i is used to stand for this number and is defined as follows.

DEFINITION OF i

$$i = \sqrt{-1} \quad \text{and} \quad i^2 = -1$$

Using i, the square root of a negative real number is now defined:

$$\text{For } x > 0, \quad \sqrt{-x} = \sqrt{-1}\sqrt{x} = i\sqrt{x}$$

EXAMPLE 1 Simplify: **(a)** $\sqrt{-16} + \sqrt{-25}$ **(b)** $\sqrt{-16} \cdot \sqrt{-25}$

Solution

In the example, 4i and 5i are combined by using the usual rules of algebra. You will see later that such procedures apply for this new kind of number.

(a) $\sqrt{-16} + \sqrt{-25} = \sqrt{-1} \cdot \sqrt{16} + \sqrt{-1} \cdot \sqrt{25}$

$$= i \cdot 4 + i \cdot 5 = 4i + 5i = 9i$$

(b) $\sqrt{-16} \cdot \sqrt{-25} = (4i)(5i) = 20i^2 = 20(-1) = -20$ ∎

TEST YOUR UNDERSTANDING
Think Carefully

Express as the product of a real number and i.

1. $\sqrt{-9}$ **2.** $\sqrt{-49}$ **3.** $\sqrt{-5}$ **4.** $-2\sqrt{-1}$ **5.** $\sqrt{-\dfrac{4}{9}}$

Simplify.

6. $\sqrt{-64} \cdot \sqrt{-225}$ **7.** $\sqrt{9} \cdot \sqrt{-49}$

(Answers: Page 75)

8. $\sqrt{-50} + \sqrt{-32} - \sqrt{-8}$ **9.** $3\sqrt{-20} + 2\sqrt{-45}$

An indicated product of a real number times the imaginary unit i, such as $7i$ or $\sqrt{2}\,i$, is called a **pure imaginary number.** The sum of a real number and a pure imaginary number is called a **complex number.**

> A complex number has the form $a + bi$, where a and b are real numbers and $i = \sqrt{-1}$.

We say that the real number a is the **real part** of $a + bi$ and the real number b is called the **imaginary part** of $a + bi$. In general, two complex numbers are equal only when both their real parts and their imaginary parts are equal. Thus

$$a + bi = c + di \quad \text{if and only if} \quad a = c \text{ and } b = d$$

The collection of complex numbers contains all the real numbers, since any real number a can also be written as $a = a + 0i$. Similarly, if b is real, $bi = 0 + bi$, so that the complex numbers also contain the pure imaginaries.

Note, for example, that $\sqrt{\frac{4}{9}}$ is complex, real, and rational; $\sqrt{-10}$ is complex and pure imaginary.

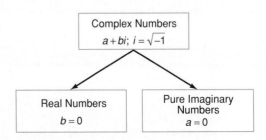

How should complex numbers be added, subtracted, multiplied, or divided? In answering this question, it must be kept in mind that the real numbers are included in the collection of complex numbers. Thus the definitions we construct for the complex numbers must preserve the established operations for the reals.

We add and subtract complex numbers by combining their real and their imaginary parts separately, according to these definitions.

SUM AND DIFFERENCE OF COMPLEX NUMBERS

$$(a + bi) + (c + di) = (a + c) + (b + d)i$$
$$(a + bi) - (c + di) = (a - c) + (b - d)i$$

These procedures are quite similar to those used for combining polynomials. For example:

$$(2 + 3i) + (5 + 7i) = (2 + 5) + (3 + 7)i = 7 + 10i$$
$$(8 + 5i) - (3 + 2i) = (8 - 3) + (5 - 2)i = 5 + 3i$$

Multiplication of two complex numbers is similar to the multiplication of two binomials. For example:

$$(3 + 2i)(5 + 3i) = 15 + 10i + 9i + 6i^2$$

This can be simplified by noting that $10i + 9i = 19i$ and $6i^2 = 6(-1) = -6$. The result is $9 + 19i$.

In general:

$$(a + bi)(c + di) = ac + adi + bci + bdi^2$$
$$= ac + (ad + bc)i + bd(-1)$$
$$= (ac - bd) + (ad + bc)i$$

Note: In practice, it is easier to find the product by using the procedure for multiplying binomials rather than by memorizing the formal definition.

PRODUCT OF COMPLEX NUMBERS

$$(a + bi)(c + di) = (ac - bd) + (ad + bc)i$$

Now consider the quotient of two complex numbers such as

$$\frac{2 + 3i}{3 + i}$$

Our objective is to express this quotient in the form $a + bi$. To do so, we use a method similar to rationalizing the denominator. Note what happens when $3 + i$ is multiplied by its **conjugate**, $3 - i$:

$$(3 + i)(3 - i) = 9 + 3i - 3i - i^2 = 9 - i^2 = 9 + 1 = 10$$

In general, $(a + bi)(a - bi) = a^2 - b^2i^2 = a^2 + b^2$.

We are now ready to complete the division problem.

The term "imaginary" is said to have been first applied to these numbers precisely because they seem mysteriously to vanish under certain multiplications.

$$\frac{2 + 3i}{3 + i} = \frac{2 + 3i}{3 + i} \cdot \frac{3 - i}{3 - i}$$

$$= \frac{6 + 9i - 2i - 3i^2}{9 - i^2}$$

$$= \frac{9 + 7i}{10} \qquad (-3i^2 = -3(-1) = 3)$$

$$= \frac{9}{10} + \frac{7}{10}i \qquad \leftarrow \text{This is the quotient in the form } a + bi.$$

In general, multiplying the numerator and denominator of $\dfrac{a + bi}{c + di}$ by the conjugate of $c + di$ leads to the following definition for division (see Exercise 59).

Rather than memorize this definition, simply find quotients as in the preceding illustration.

QUOTIENT OF COMPLEX NUMBERS

$$\frac{a + bi}{c + di} = \frac{ac + bd}{c^2 + d^2} + \frac{bc - ad}{c^2 + d^2}i \qquad c + di \neq 0$$

Although we will not go into the details here, it can be shown that some of the basic rules for real numbers apply for the complex numbers. For example, the commutative, associative, and distributive laws hold, whereas the rules of order do not apply.

It is also true that the rules for integer exponents apply for complex numbers. For example, $(2 - 3i)^0 = 1$ and $(2 - 3i)^{-1} = \dfrac{1}{2 - 3i}$. In particular, the positive integral powers of i are easily evaluated. For example:

$$i = \sqrt{-1} \qquad\qquad i^5 = i^4 \cdot i = 1 \cdot i = i$$

$$i^2 = -1 \qquad\qquad i^6 = i^4 \cdot i^2 = 1 \cdot i^2 = -1$$

$$i^3 = i^2 \cdot i = -1 \cdot i = -i \qquad i^7 = i^4 \cdot i^3 = 1 \cdot i^3 = -i$$

$$i^4 = i^2 \cdot i^2 = (-1)(-1) = 1 \qquad i^8 = i^4 \cdot i^4 = 1 \cdot i^4 = 1$$

Observe that the first four powers of i at the left repeat for the next four powers shown at the right. This cycle of i, -1, $-i$, and 1 continues endlessly. Therefore, to simplify i^n when $n > 4$, we search for the largest multiple of 4 in the integer n as in the next example.

EXAMPLE 2 Simplify: **(a)** i^{22} **(b)** i^{39}

Since 20 = 5(4), 5 is the largest multiple of 4 in 22. Since 36 = 9(4), 9 is the largest multiple of 4 in 39.

Solution

(a) $i^{22} = i^{20} \cdot i^2 = (i^4)^5 \cdot i^2 = 1^5 \cdot i^2 = i^2 = -1$

(b) $i^{39} = i^{36} \cdot i^3 = (i^4)^9 \cdot i^3 = 1^9 \cdot i^3 = i^3 = -i$ ∎

The following example illustrates how to operate with a negative integral power of i.

EXAMPLE 3 Express $2i^{-3}$ as the indicated product of a real number and i.

As an alternative solution, write $\frac{2}{i^3} = \frac{2}{-i}$. Then multiply numerator and denominator by i.

Solution First note that $2i^{-3} = \frac{2}{i^3}$. Next multiply numerator and denominator by i to obtain a real number in the denominator.

$$2i^{-3} = \frac{2}{i^3} \cdot \frac{i}{i} = \frac{2i}{i^4} = \frac{2i}{1} = 2i$$ ∎

EXERCISES 1.10

Classify each statement as true or false.

1. Every real number is a complex number.
2. Every complex number is a real number.
3. Every irrational number is a complex number.
4. Every integer can be written in the form $a + bi$.
5. Every complex number may be expressed as an irrational number.
6. Every negative integer may be written as a pure imaginary number.

Express each of the following numbers in the form $a + bi$.

7. $5 + \sqrt{-4}$ 8. $7 - \sqrt{-7}$ 9. -5 10. $\sqrt{25}$

Express in the form bi.

11. $\sqrt{-16}$ 12. $\sqrt{-81}$ 13. $\sqrt{-144}$ 14. $-\sqrt{-9}$
15. $\sqrt{-\frac{9}{16}}$ 16. $\sqrt{-3}$ 17. $-\sqrt{-5}$ 18. $-\sqrt{-8}$

Simplify.

19. $\sqrt{-9} \cdot \sqrt{-81}$ 20. $\sqrt{4} \cdot \sqrt{-25}$ 21. $\sqrt{-3} \cdot \sqrt{-2}$ 22. $(2i)(3i)$
23. $(-3i^2)(5i)$ 24. $(i^2)(i^2)$ 25. $\sqrt{-9} + \sqrt{-81}$ 26. $\sqrt{-12} + \sqrt{-75}$
27. $\sqrt{-8} + \sqrt{-18}$ 28. $2\sqrt{-72} - 3\sqrt{-32}$ 29. $\sqrt{-9} - \sqrt{-3}$ 30. $3\sqrt{-80} - 2\sqrt{-20}$

Complete the indicated operation. Express all answers in the form $a + bi$.

31. $(7 + 5i) + (3 + 2i)$ 32. $(8 + 7i) + (9 - i)$ 33. $(8 + 2i) - (3 + 5i)$
34. $(7 + 2i) - (4 - 3i)$ 35. $(7 + \sqrt{-16}) + (3 - \sqrt{-4})$ 36. $(8 + \sqrt{-49}) - (2 - \sqrt{-25})$
37. $2i(3 + 5i)$ 38. $3i(5i - 2)$ 39. $(3 + 2i)(2 + 3i)$
40. $(\sqrt{5} + 3i)(\sqrt{5} - 3i)$ 41. $(5 - 2i)(3 + 4i)$ 42. $(\sqrt{3} + 2i)^2$

43. $\dfrac{3 + 5i}{i}$ **44.** $\dfrac{5 - i}{i}$ **45.** $\dfrac{5 + 3i}{2 + i}$

46. $\dfrac{7 - 2i}{2 - i}$ **47.** $\dfrac{3 - i}{3 + i}$ **48.** $\dfrac{8 + 3i}{3 - 2i}$

Simplify

49. $3i^3$ **50.** $-5i^5$ **51.** $2i^7$ **52.** $3i^{-3}$ **53.** $-4i^{18}$ **54.** i^{-32}

Simplify and express each answer in the form $a + bi$.

55. $(3 + 2i)^{-1}$ **56.** $(3 + 2i)^{-2}$

57. One of the basic rules for operating with radicals is that $\sqrt{ab} = \sqrt{a} \cdot \sqrt{b}$, where a and b are nonnegative real numbers. Prove that this rule does not work when both a and b are negative by showing that $\sqrt{(-4)(-9)} \neq \sqrt{-4} \cdot \sqrt{-9}$.

58. Use the definition $\sqrt{-x} = i\sqrt{x}$ $(x > 0)$ to prove that $\sqrt{ab} = \sqrt{a} \cdot \sqrt{b}$ when $a < 0$ and $b \geq 0$.

59. Write $\dfrac{a + bi}{c + di}$ in the form $x + yi$. (*Hint:* Multiply the numerator and denominator by the conjugate of $c + di$.)

60. The set of complex numbers satisfies the associative property for addition. Verify this by completing this problem in two different ways.

$$(3 + 5i) + (2 + 3i) + (7 + 4i)$$

61. Repeat Exercise 60 for multiplication, using $(3 + i)(3 - i)(4 + 3i)$.

62. We say that $0 = 0 + 0i$ is the additive identity for the complex numbers since $0 + z = z$ for any $z = a + bi$. Find the additive inverse (negative) of z.

Perform the indicated operations and express the answers in the form $a + bi$.

63. $(5 + 4i) + 2(2 - 3i) - i(1 - 5i)$ **64.** $2i(3 - 4i)(3 - 6i) - 7i$

65. $\dfrac{(2 + i)^2(3 - i)}{2 + 3i}$ **66.** $\dfrac{1 - 2i}{3 + 4i} - \dfrac{2i - 3}{4 - 2i}$

Use complex numbers to factor each polynomial.
    Example: $x^2 + 9 = x^2 - (-9) = x^2 - (3i)^2 = (x + 3i)(x - 3i)$

67. $x^2 + 1$ **68.** $9x^2 + 4$ **69.** $3x^2 + 75$ **70.** $x^2 + 4ix - 3$

KEY TERMS

*Review these **key terms** so that you are able to define or describe them. A clear understanding of these terms will be very helpful when reviewing the developments of this chapter.*

Real numbers: irrational, rational, integers, whole, natural • Real number properties: closure, commutative, associative, distributive, identity, inverse, zero-product • Properties of equality: addition, multiplication • Properties of inequalities: trichotomy, transitive, addition, multiplication • Intervals: bounded, unbounded • Absolute value • Rules for exponents • Definitions of b^0 and b^{-n} • Definition of $\sqrt[n]{a}$ • Rules for radicals • Rationalizing denominators • Combining radicals • Definitions of $b^{\frac{1}{n}}$ and $b^{\frac{m}{n}}$ • Monomial • Binomial • Trinomial • Polynomial • Fundamental operations with polynomials • Factoring methods: common monomial, difference of squares, difference (sum) of cubes, difference of nth powers, grouping, trinomials • Rational expression • Fundamental operations with rational expressions • Definition of i • Powers of i • Complex number • Pure imaginary number • Operations with complex numbers

REVIEW EXERCISES

The solutions to the following exercises can be found within the text of Chapter 1. Try to answer each question before referring to the text.

Section 1.1

1. What is meant by the "set of whole numbers"?
2. State the definition of a rational number.
3. Give an example of a repeating decimal.
4. To which subsets of the real numbers do each of the following belong?
   (a) 5   (b) $\frac{2}{3}$ (c) $\sqrt{7}$ (d) -14
5. Classify as true or false, and explain your answer: Every whole number is a natural number.
6. Given two real numbers a and b, write in symbols the commutative property for addition and for multiplication.
7. Repeat Exercise 6 for the associative property.
8. If $ab = 0$, what do you know about the numbers a and b?
9. State the definition of subtraction in terms of addition.
10. State the definition of division in terms of multiplication.
11. Explain what the phrase "if and only if" means.
12. Why is division by zero not possible?

Section 1.2

13. State the addition property of equality.
14. State the multiplication property of equality.
15. Solve for x: $3(x + 3) = x + 5$
16. The formula relating degrees Fahrenheit and degrees Celsius is $\dfrac{5F - 160}{9} = C$. Solve for F in terms of C.
17. The length of a rectangle is 1 centimeter less than twice the width. The perimeter is 28 centimeters. Find the dimensions of the rectangle.
18. A car leaves a certain town at noon, traveling due east at 40 miles per hour. At 1:00 P.M. a second car leaves the town traveling in the same direction at a rate of 50 miles per hour. In how many hours will the second car overtake the first car?
19. David has a total of $4.10 in nickels, dimes, and quarters. He has twice as many nickels as dimes, and two more quarters than dimes. How many dimes does he have?
20. Marcella has a base salary of $250 per week. In addition, she receives a commission of 12% of her sales. Last month her total earnings were $520. What were her total sales for the week?

Section 1.3

21. State the algebraic definition for the inequality $a < b$.
22. State the addition property of order.
23. State the multiplication property of order.
24. Solve the inequality: $-4n - (3 - 5n) > 8$
25. Find the solution set: $3x + 7 \leq 2x - 1$
26. Solve for x: (a) $5(3 - 2x) \geq 10$ (b) $\dfrac{2}{x + 4} < 0$
27. State the trichotomy law for any two real numbers a and b.

28. State the transitive property of order for real numbers a, b, and c.

29. Show each of the following intervals as a graph on the number line.
   (a) $(-2, 3)$ (b) $[-2, 3]$ (c) $(-2, 3]$ (d) $[-2, 3)$

30. Graph the solution set: $3x - 4 > 11$

31. Graph the solution set: $-2(x - 1) \geq 4$

32. If $x < 2$, is $x - 5$ positive or negative?

33. A small family business employs two part-time workers per week. The total amount of wages they pay to these employees ranges from \$128 to \$146 per week. If one employee will earn \$18 more than the other, what are the possible amounts earned by each per week?

Section 1.4

34. State the definition of the absolute value of a number.

35. Solve for x: $\dfrac{x - 3}{|x - 3|} = 1$

36. Solve: (a) $|5 - x| = 7$ (b) $|x - 2| \leq 3$

37. Graph: (a) $|x| < 2$ (b) $|x + 1| > 2$

38. What does $|a - b|$ represent with respect to points a and b on a number line? Illustrate with two specific values.

39. What is the coordinate of the midpoint of a line segment with endpoints x_1 and x_2?

Section 1.5

40. Simplify: (a) $12^3(\frac{1}{6})^3$ (b) $2x^3 \cdot x^4$ (c) $\dfrac{(x^3 y)^2 y^3}{x^4 y^6}$ (d) $\dfrac{4^5}{8^3}$

41. Simplify and write without negative exponents: $\left(\dfrac{a^{-2}b^3}{a^3 b^{-2}}\right)^5$

42. Suppose that a radioactive substance decays so that $\frac{1}{2}$ the amount remains after each hour. If at a certain time there were 320 grams of the substance, how much will remain after 8 hours? How much after n hours?

43. Explain the motivation behind the definition $b^0 = 1$.

Section 1.6

Evaluate those radicals that are real numbers and check. If an expression is not a real number, give a reason.

44. $\sqrt[3]{-125}$ 45. $\sqrt{-9}$ 46. $(\sqrt[3]{x})^3$ 47. $\sqrt[5]{-\dfrac{x^{10}}{32}}$

Simplify. Assume that all variables represent positive numbers.

48. $\sqrt{6x} \cdot \sqrt{7y}$ 49. $\dfrac{\sqrt[3]{81x^7}}{\sqrt[3]{-3x}}$ 50. $\sqrt[2]{\sqrt[3]{64}}$ 51. $\sqrt[5]{16x} \cdot \sqrt[5]{-2x^4}$

52. Evaluate: (a) $(-64)^{2/3}$ (b) $8^{-2/3} + (-32)^{-2/5}$

53. Simplify: $\left(\dfrac{-8a^3}{b^{-6}}\right)^{2/3}$

54. Simplify: (a) $\sqrt{50}$ (b) $\sqrt[3]{16}$ (c) $\sqrt[3]{-54}$ (d) $3\sqrt{5} + 4\sqrt{5}$

55. Simplify $\sqrt[3]{-24x^3} + 2\sqrt[3]{375x^3} - \sqrt[4]{162x^4}$

56. Rationalize the denominator: (a) $\dfrac{6}{\sqrt{8}}$ (b) $\dfrac{5}{\sqrt[3]{2}}$

57. Combine: $\dfrac{6}{\sqrt{3}} + 2\sqrt{75} - \sqrt{3}$

58. Simplify.

   (a) $\sqrt{75x^2}$ **(b)** $\sqrt{(-5)^2}$ **(c)** $\sqrt[7]{(-3)^7}$ **(d)** $\sqrt[8]{(-\frac{1}{2})^8}$

59. Simplify: $2\sqrt{8x^3} + 3x\sqrt{32x} - x\sqrt{18x}$

Section 1.7

60. Add: $(4x^3 - 10x^2 + 5x + 8) + (12x^2 - 9x - 1)$

61. Subtract: $(4a^3 - 10a^2b + 5b + 8) - (12a^2b - 9b - 1)$

Multiply.

62. $3x^2(4x^7 - 3x^4 - x^2 + 15)$ **63.** $(2x + 3)(4x + 5)$ **64.** $(ax + b)(cx + d)$

65. Multiply $3x^3 - 8x + 4$ by $2x^2 + 5x - 1$.

66. Simplify by performing the indicated operations: $(x^2 - 5x)(3x^2) + (x^3 - 1)(2x - 5)$

67. Expand: **(a)** $(a + b)^2$ **(b)** $(a + b)^3$

68. Rationalize the denominator: $\dfrac{5}{\sqrt{10} - \sqrt{3}}$

Section 1.8

Factor completely.

69. $24x^9 - 18x^6 - 6x^4 + 90x^2$ **70.** $21x^4y - 14x^5y^2 + 91x^{11}y^4$ **71.** $8x^2(x - 1) + 4x(x - 1) + 2(x - 1)$

72. $3x^2 - 75$ **73.** $8x^3 - 27$ **74.** $(x + 1)^2 - 4$

75. $2x^3 + 128y^3$ **76.** $a^5 - b^5$ **77.** $3y^5 - 96$

78. $x^6 - 1$ **79.** $ax^2 + 15 - 5ax - 3x$ **80.** $x^3 + 3x^2 + 9x + 27$

81. Factor as the difference of squares using irrational numbers: $x^2 - 5$.

82. Factor as the difference of cubes using radical expressions: $x - 8$.

Factor if possible.

83. $a^2 + 2ab + b^2$ **84.** $a^2 - 2ab + b^2$ **85.** $x^2 + 10x + 25$ **86.** $9x^3 - 42x^2 + 49x$

87. $x^2 - 10x + 24$ **88.** $15x^2 + 43x + 8$ **89.** $15x^2 + 7x - 8$ **90.** $12x^2 - 9x + 2$

91. $x^4 - x^2 - 12$ **92.** $a^6 - 3a^3b - 18b^2$

Section 1.9

Reduce to lowest terms.

93. $\dfrac{x^2 + 5x - 6}{x^2 + 6x}$ **94.** $\dfrac{5a - 3b}{3b - 5a}$

95. Multiply: $\dfrac{x + 1}{x - 1} \cdot \dfrac{2 - x - x^2}{5x}$ **96.** Divide: $\dfrac{(x + 1)^2}{x^2 - 6x + 9} \div \dfrac{3x + 3}{x - 3}$

97. Subtract: $\dfrac{6x^2}{2x^2 - x - 10} - \dfrac{15x}{2x^2 - x - 10}$ **98.** Add: $\dfrac{3}{x^2 + x} + \dfrac{2}{x^2 - 1}$

99. Combine and simplify: $\dfrac{3}{2} - \dfrac{4}{3x(x + 1)} - \dfrac{x - 5}{3x^2}$

100. Simplify: $\dfrac{\dfrac{1}{5 + h} - \dfrac{1}{5}}{h}$ **101.** Simplify: $\dfrac{x^{-2} - y^{-2}}{\dfrac{1}{x} - \dfrac{1}{y}}$

102. Simplify: (a) $\sqrt{-16} + \sqrt{-25}$ (b) $\sqrt{-16} \cdot \sqrt{-25}$
103. Perform the indicated operations:
    (a) $(2 + 3i) + (5 + 7i)$ (b) $(8 + 5i) - (3 + 2i)$
104. Multiply: $(3 + 2i)(5 + 3i)$ 105. Divide: $\dfrac{2 + 3i}{3 + i}$
106. Simplify: (a) i^{22} (b) i^{39}
107. Express $2i^{-3}$ as the indicated product of a real number and i.

CHAPTER 1 TEST: STANDARD ANSWER

Use these questions to test your knowledge of the basic skills and concepts of Chapter 1. Then check your answers with those given at the back of the book.

1. Classify each statement as true or false.
    (a) Negative irrational numbers are not real numbers.
    (b) Every integer is a rational number.
    (c) Some irrational numbers are integers.
    (d) Zero is a rational number.
    (e) If $x < y$, then $x - 5 > y - 5$.
    (f) The absolute value of a sum equals the sum of the absolute values.
    (g) If $-5x < -5y$, then $x > y$.
    (h) $|x - 2| < 3$ means that x is within 2 units of 3 on the number line.

2. One endpoint of a line segment is located at -7. The midpoint is located at -2. What is the coordinate of the other endpoint of the line segment?
3. Solve for x: $\frac{2}{3}(x - 3) + 1 = 2x + 3$
4. Find the dimensions of a rectangle whose perimeter is 52 inches if the length is 5 inches more than twice the width.
5. A car leaves from point B at noon traveling at the rate of 55 miles an hour. One hour later a second car leaves from the same point, traveling in the opposite direction at 45 miles per hour. At what time will they be 200 miles apart?
6. Show each interval on a number line: (a) $(-2, 2]$ (b) $[-1, 1]$

Solve for x.

7. $\dfrac{|x + 2|}{x + 2} = -1$ 8. $|x + 2| < 1$

Solve and graph each inequality.

9. $2(5x - 1) < x$ 10. $|2x - 1| \geq 3$

11. Classify each statement as true or false:
    (a) $\dfrac{x^3(-x)^2}{x^5} = x$ (b) $\left(\dfrac{3}{2 + a}\right)^{-1} = \dfrac{2}{3} + \dfrac{a}{3}$ (c) $(-27)^{-1/3} = 3$
    (d) $(x + y)^{3/5} = (\sqrt[3]{x + y})^5$ (e) $\sqrt{9x^2} = 3|x|$ (f) $(8 + a^3)^{1/3} = 2 + a$

12. Simplify. Express the answers using positive exponents: (a) $\dfrac{(2x^3y^{-2})^2}{x^{-2}y^3}$ (b) $\dfrac{(3x^2y^{-3})^{-1}}{(2x^{-2}y^2)^{-2}}$

Perform the indicated operations and simplify.

13. (a) $\sqrt{8} \cdot \sqrt{6}$ (b) $\dfrac{\sqrt{360}}{2\sqrt{2}}$ (c) $\dfrac{\sqrt[3]{-243x^8}}{\sqrt[3]{3x^2}}$

14. (a) $\sqrt{50} + 3\sqrt{18} - 2\sqrt{8}$ **(b)** $\dfrac{12}{\sqrt{3}} + 2\sqrt{3}$

15. $5\sqrt{3x^2} + 2\sqrt{27x^2} - 3\sqrt{48x^2}$ **16.** $(x^2 + 3x)(3x^2) + (x^3 - 1)(2x + 3)$

Factor completely.

17. $64 - 27b^3$ **18.** $6x^2 - 7x - 3$ **19.** $2x^2 - 6xy - 3y^3 + xy^2$

Perform the indicated operations and simplify.

20. $\dfrac{x^2 - 9}{x^3 + 4x^2 + 4x} \cdot \dfrac{2x^2 + 4x}{x^2 + 2x - 15}$ **21.** $\dfrac{x^3 + 8}{x^2 - 4x - 12} \div \dfrac{x^3 - 2x^2 + 4x}{x^3 - 6x^2}$

22. $\dfrac{\dfrac{1}{x^2} - \dfrac{1}{49}}{x - 7}$ **23.** $\dfrac{1}{x + 3} - \dfrac{2}{x^2 - 9} + \dfrac{x}{2x^2 + x - 15}$

24. Multiply the complex numbers $3 + 7i$ and $5 - 4i$ and express the result in the form $a + bi$.

25. Divide $3 + 7i$ by $5 - 4i$ and express the result in the form $a + bi$.

CHAPTER 1 TEST: MULTIPLE CHOICE

1. Which of these statements are true?

   I. Every integer is the coordinate of some point on the number line.

   II. Every rational number is a real number.

   III. Every point on the number line can be named by a rational number.

   (a) Only I **(b)** Only II **(c)** Only I and II **(d)** I, II, and III **(e)** None of the preceding

2. Express the inequality $-5 \le x$ in interval notation.

   (a) $(\infty, -5)$ **(b)** $(-\infty, -5)$ **(c)** $[-5, \infty)$ **(d)** $(\infty, -5]$ **(e)** None of the preceding

3. If $a + b$ is negative, then $|a + b| =$

   (a) 0 **(b)** $a + b$ **(c)** $-a + b$ **(d)** $-(a + b)$ **(e)** None of the preceding

4. Which of the following is *false?*

   (a) $|8 - 2| = 8 - 2$ **(b)** $|2 - 7| = 2 - 7$ **(c)** $|6 + 8| = |6| + |8|$ **(d)** $|3 - 5| = -(3 - 5)$

   (e) None of the preceding

5. The inequality $|x| \ge k$ $(k \ge 0)$ is true if and only if

   (a) $x \le -k$ or $x \ge k$ **(b)** $x \le k$ or $x \ge -k$ **(c)** $-k \le x \le k$ **(d)** $x \ge k$ **(e)** None of the preceding

6. Which of these statements are true for all real numbers a, b, and c?

   I. If $a < b$, then $a + c < b + c$.

   II. If $a < b$, then $ac < bc$ for $c > 0$.

   III. If $a < b$, then $ac > bc$ for $c < 0$.

   (a) Only I **(b)** Only II **(c)** Only III **(d)** I, II, and III **(e)** None of the preceding

7. In interval notation, the solution of the inequality $-3(x + 1) < 2x + 2$ is

   (a) $(-\infty, -1)$ **(b)** $(-1, \infty)$ **(c)** $(1, \infty)$ **(d)** $(-\infty, 1)$ **(e)** None of the preceding

8. The length of a rectangle is 2 inches less than three times the width. If the length is increased by 5 inches and the width decreased by 1 inch, the length will be five times the width. Using x to denote the original width, which of these equations can be used to solve for x?

   (a) $x - 1 = 5(3x + 3)$ **(b)** $3x - 2 = 5x - 1$ **(c)** $3x + 3 = 5(x - 1)$ **(d)** $3x + 3 = 5x - 1$

   (e) None of the preceding

9. Which of the following is equivalent to $(a + b)^{-1}$?

   (a) $a^{-1} + b^{-1}$ **(b)** $\dfrac{1}{a + b}$ **(c)** $(-a) + (-b)$ **(d)** $\dfrac{1}{a} + \dfrac{1}{b}$ **(e)** None of the preceding

10. Which of the following are true?

    I. $a^{-1/2} + b^{-1/2} = \dfrac{1}{\sqrt{a}} + \dfrac{1}{\sqrt{b}}$ **II.** $(-x)^{-1/3} = x^{1/3}$ **III.** $x^{3/4} = (\sqrt[3]{x})^4$

    (a) Only I   **(b)** Only II   **(c)** Only III   **(d)** I, II, and III   **(e)** None of the preceding

11. Rationalize the denominator: $\dfrac{8}{\sqrt{5} - 1}$.

    (a) $8(\sqrt{5} + 1)$ **(b)** $2(\sqrt{5} - 1)$ **(c)** $2(\sqrt{5} + 1)$ **(d)** $2\sqrt{5} + 1$ **(e)** None of the preceding

12. Which of the following are true?

    I. $2\sqrt{x + 1} = \sqrt{2x + 1}$ **II.** $\sqrt{(x - 1)^2} = |x - 1|$ **III.** $(x + y)^{1/3} = x^{1/3} + y^{1/3}$

    (a) Only I   **(b)** Only II   **(c)** Only III   **(d)** I, II, and III   **(e)** None of the preceding

13. What is the complete factored form for the expression $x^2(x - 1) - 2x(x - 1) + (x - 1)$?

    (a) $x(x - 2)(x - 1)$ **(b)** $(x - 1)(x^2 - 2x + 1)$ **(c)** $(x - 2x)(3x - 3)$ **(d)** $(x - 1)^3$

    (e) None of the preceding

14. Which of the following are true?

    I. $x^3 - y^3 = (x - y)(x^2 + y^2)$ **II.** $x^2 + y^2 = (x + y)(x + y)$ **III.** $4x^2 - 6xy + 9y^2 = (2x - 3y)^2$

    (a) Only I   **(b)** Only II   **(c)** Only III   **(d)** Only II and III   **(e)** None of the preceding

15. Which of the following are true?

    I. Every integer can be written in the form $a + bi$, where $i = \sqrt{-1}$.

    II. Every real number is a complex number.

    III. The sum, difference, and product of two complex numbers are complex numbers.

    (a) Only I   **(b)** Only II   **(c)** Only III   **(d)** I, II, and III   **(e)** None of the preceding

ANSWERS TO THE TEST YOUR UNDERSTANDING EXERCISES

Page 3

1. Associative property for addition.   2. Commutative property for addition.
3. Commutative property for addition.   4. Distributive property.
5. Commutative property for multiplication.   6. Associative property for multiplication.
7. No; no.   8. $12 \div 3 \neq 3 \div 12$ 9. No; no.   10. $(8 \div 4) \div 2 \neq 8 \div (4 \div 2)$
11. Yes; the closure properties for addition and multiplication of real numbers, respectively.

Page 8

1. $x = 6$ 2. $x = 17$ 3. $x = -4$ 4. $x = 6$ 5. $x = 13$ 6. $x = 10$
7. $x = 1$ 8. $x = 2$ 9. $x = 5$ 10. $x = -9$ 11. $x = -3$ 12. $x = -11$

Page 17

1. $\{x \mid x < 9\}$ 2. $\{x \mid x < 18\}$ 3. $\{x \mid x > 9\}$ 4. $\{x \mid x > -5\}$ 5. $\{x \mid x \leq 14\}$
6. $\{x \mid x \geq -2\}$ 7. $\{x \mid x < 4\}$ 8. $\{x \mid x \geq 7\}$ 9. $\{x \mid x \leq -21\}$ 10. $\{x \mid x \geq -1\}$
11. $\{x \mid x > 3\}$ 12. $\{x \mid x < -30\}$

Page 28

1. 125 2. $-\frac{1}{32}$ 3. 0 4. $1{,}000{,}000$ 5. -64 6. 64 7. $\frac{1}{17}$ 8. $\frac{1}{23}$
9. 81 10. $a^{11}b^{10}$ 11. 64 12. $6x^4$

Page 37

1. $7^{1/2}$ 2. $(-10)^{1/3}$ 3. $7^{1/4}$ 4. $7^{2/3}$ 5. $5^{3/4}$ 6. 5   7. 4   8. $\frac{1}{6}$
9. $\frac{1}{7}$ 10. -3 11. 8   12. $\frac{1}{8}$ 13. $\frac{27}{8}$ 14. 4   15. $\frac{1}{4}$ 16. $6x$
17. $-\dfrac{1}{2x}$ 18. $-\frac{3}{2}x^2$

Page 39

1. $6\sqrt{2}$ 2. $6\sqrt{3}$ 3. $\sqrt{5}$ 4. $\sqrt[3]{2}$ 5. $4\sqrt[3]{2}+5$ 6. $-\sqrt[3]{3}$ 7. $11\sqrt{2}$
8. $13\sqrt{3}$ 9. $3\sqrt{5}$ 10. $7\sqrt{7}$ 11. $6\sqrt[3]{2}$ 12. $2\sqrt[3]{3}$

Page 45

1. x^2+1 2. $x^2+4x-13$ 3. $2x^3+3x^2+8x+1$ 4. $4x^2y-5xy+6xy^2$
5. $-3x^4-6x^3+3x$ 6. $6x^2+13x+6$ 7. $12x^2+28x-5$ 8. $16x^2-49$
9. $4x^2-12xy+9y^2$ 10. $2x^5-5x^4+4x^3+14x^2-15x$

Page 49

1. $3(x-3)$ 2. $-5(x-3)$ 3. $5y(x+5y+2y^4)$ 4. $(x+6)(x-6)$
5. $(2x+7)(2x-7)$ 6. $(a+2+5b)(a+2-5b)$ 7. $(x-3)(x^2+3x+9)$
8. $(x+3)(x^2-3x+9)$ 9. $(2a-5)(4a^2+10a+25)$ 10. $3(x+4)(x-4)$
11. $a(x+y)(x^2-xy+y^2)$ 12. $2h(x+2h)(x-2h)$

Page 53

1. $(a+3)^2$ 2. $(x-5)^2$ 3. $(2x+3y)^2$ 4. $(x+3)(x+5)$ 5. $(a-10)(a-2)$
6. Not factorable   7. $(5x-2)(2x-7)$ 8. $(3x+2)(2x-5)$ 9. Not factorable

Page 58

1. $x+2$ 2. $\dfrac{x}{x+2}$ 3. $-\dfrac{4b}{3a}$ 4. $\dfrac{x+5}{x-2}$ 5. $\dfrac{1}{x^2+4}$ 6. $-\dfrac{x+2}{x}$ 7. -1
8. $\dfrac{x-2}{3x(x+1)}$ 9. $\dfrac{a+b}{a-b}$ 10. 1

Page 59

1. $\dfrac{13x}{15}$ 2. $\dfrac{5x}{6}$ 3. $\dfrac{4-3x}{6x^2}$ 4. $\dfrac{25}{6x}$ 5. $\dfrac{2(5x+4)}{(x-2)(x+2)}$ 6. $\dfrac{9x+34}{x^2-9}$
7. $\dfrac{13x-18}{(x-1)(x^2-4)}$ 8. $\dfrac{2}{2x+5}$

Page 64

1. $3i$ 2. $7i$ 3. $i\sqrt{5}$ 4. $-2i$ 5. $\frac{2}{3}i$ 6. -120 7. $21i$ 8. $7i\sqrt{2}$ 9. $12i\sqrt{5}$

2 LINEAR FUNCTIONS AND INEQUALITIES

2.1 INTRODUCTION TO THE FUNCTION CONCEPT

Suppose that you are riding in a car that is averaging 40 miles per hour. Then the distance traveled is determined by the time traveled.

$$\text{distance} = \text{rate} \times \text{time}$$

Symbolically, this relationship can be expressed by the equation

$$s = 40t$$

where s is the distance traveled in time t (measured in hours). For $t = 2$ hours, the distance traveled is

$$s = 40(2) = 80 \text{ miles}$$

*Note, for example, that the equation $y^2 = 12x$ does **not** define y as a function of x; for a given value of $x > 0$ there is **more than one** corresponding value for y. If $x = 3$, for example, then $y^2 = 36$ and $y = 6$ or $y = -6$.*

Similarly, for each specific value of $t \geq 0$ the equation produces *exactly one* value for s. This correspondence between the distance s and the time t is an example of a *functional relationship*. More specifically, we say that the equation $s = 40t$ defines s as a *function* of t because for *each* choice of t there corresponds *exactly one* value for s. We first choose a value of t. Then there is a corresponding value of s that depends on t; s is the *dependent variable* and t is the *independent variable* of the function defined by $s = 40t$.

Because the variable t represents time in the equation $s = 40t$, it is reasonable to say that $t \geq 0$. This set of allowable values for the independent variable is called the **domain** of the function. The set of corresponding values for the dependent variable is called the **range** of the function.

DEFINITION OF FUNCTION

This is an important definition. Much of the work in calculus deals with the study of functions.

A **function** is a correspondence between two sets, the domain and the range, such that for each value in the domain there corresponds exactly one value in the range.

The specific letters used for the independent and dependent variables are of no consequence. Usually, we will use x for the independent variable and y for the dependent variable.

Most of the expressions encountered earlier in this text can be used to define functions. Here are some illustrations. Note that in each case *the domain of the function is taken as the largest set of real numbers for which the defining expression in x leads to a real value*. Unless otherwise indicated, this will be our policy throughout this text.

Note that we must restrict the domain of a function so that the denominator of a fraction is not zero, and so that there is not an even root of a negative number. These two cases produce results that are not real numbers.

Function given by	Domain
$y = 6x^4 - 3x^2 + 7x + 1$	All real numbers
$y = \dfrac{2x}{x^2 - 4}$	All reals except 2 and -2
$y = \lvert x \rvert$	All real numbers
$y = \sqrt{x}$	All real $x \geq 0$

EXAMPLE 1 Explain why the following equation defines y as a function of x, and find the domain: $y = \dfrac{1}{\sqrt{x - 1}}$.

Solution For each allowable x the expression $\dfrac{1}{\sqrt{x - 1}}$ produces just one y-value. Therefore, the given equation defines a function. To find the domain, note that $\sqrt{x - 1}$ is only defined if $x - 1 \geq 0$, or $x \geq 1$. However, since $\sqrt{x - 1}$ is in the denominator, $x = 1$ produces division by 0 and so it must be excluded. Thus the domain consists of all $x > 1$. ∎

TEST YOUR UNDERSTANDING
Think Carefully

Decide whether the given equation defines y to be a function of x. For each function, find the domain.

1. $y = (x + 2)^2$ **2.** $y = \dfrac{1}{(x + 2)^2}$ **3.** $y = \dfrac{1}{x^2 + 2}$

4. $y = \pm 3x$ **5.** $y = \dfrac{1}{\sqrt{x^2 + 2x + 1}}$ **6.** $y^2 = x^2$

7. $y = \dfrac{x}{\lvert x \rvert}$ **8.** $y = \sqrt{2 - x}$

(Answers: Page 132)

Sometimes we say that an equation, such as $y = 40x$, is a function. Such informal language is commonly used and should not cause difficulty.

Thus far only equations have been used to define functions. One could gain the impression that equations are the only way to state functions. This impression is *not* correct. Since a function defined by an equation is the *correspondence* between the variables, such correspondences can be stated in many ways. Here, for example, is a *table of values*. The table defines y to be a function of x because for each domain value x there corresponds *exactly one* value for y.

In the table, note that for each x there is exactly one value for y. However, it is permissible to have a value in the range (such as 6) associated with more than one value in the domain.

x	1	2	5	-7	23	$\sqrt{2}$
y	6	-6	6	-4	0	5

Instead of using a single equation to define a function, there will be times when a function is defined in terms of more than one equation. For instance, the following three equations define a function whose domain is the set of all real numbers.

$$y = \begin{cases} 1 & \text{if } x \leq -6 \\ x^2 & \text{if } -6 < x < 0 \\ 2x + 1 & \text{if } x \geq 0 \end{cases}$$

Here are several illustrations of y-values found for specific replacements of x in this function:

For $x = -7$, $y = 1$. $(x \leq -6)$
For $x = -5$, $y = x^2 = (-5)^2 = 25$. $(-6 < x < 0)$
For $x = 5$, $y = 2x + 1 = 11$. $(x \geq 0)$

EXAMPLE 2 Decide whether these two equations define y to be a function of x.

$$y = \begin{cases} 3x - 1 & \text{if } x \leq 1 \\ 2x + 1 & \text{if } x \geq 1 \end{cases}$$

Solution If $x = 1$, the first equation gives $y = 3(1) - 1 = 2$. However, for $x = 1$ the second equation gives $y = 2(1) + 1 = 3$. Since we have two different y-values for the same x-value, the two equations do *not* define a function. ∎

EXAMPLE 3 Use the definition of absolute value (Section 1.4) to explain why the equation $y = |x|$ defines a function whose domain consists of all the real numbers x.

Solution The definition of absolute value says that $|x| = x$ for $x \geq 0$ and $|x| = -x < 0$. Thus $y = |x|$ defines a function because for each real value x there is just one corresponding y-value. ∎

A useful way to refer to a function is to name it by using a specific letter, such as f, g, F, and the like. For example, the function defined by $y = \dfrac{1}{x - 3}$ may be referred to as f. The domain of f is the set of all real numbers not equal to 3; that is, $x \neq 3$. We write the range values as follows:

$$f(x) = \frac{1}{x - 3}$$

to mean the value of the function f at x is $\frac{1}{x-3}$. For example:

$$f(4) = 1 \text{ means that when } x = 4,\ y = 1$$
$$f(4) = 1 \text{ is read as “} f \text{ of 4 is 1” or “} f \text{ at 4 is 1”}$$

We use $f(x)$ to represent the range value for the specific value of x given in the parentheses. In this situation f stands for the function that is given by $y = \frac{1}{x-3}$.

CAUTION
Note that $f(x)$ does **not** mean that we are to multiply f by x; f does not stand for a number.

For $x \neq 3$, $f(x) = \frac{1}{x-3}$. Then

$$f(0) = \frac{1}{0-3} = -\frac{1}{3}$$

$$\text{and}\quad f(9) = \frac{1}{9-3} = \frac{1}{6} \leftarrow \begin{cases} 9 \text{ is the } input \\ \frac{1}{6} \text{ is the } output \end{cases}$$

Note that $f(3)$ is undefined. Can you explain why?

The preceding illustration can be demonstrated by means of this diagram:

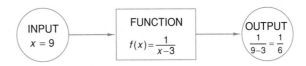

INPUT $x = 9$ → FUNCTION $f(x) = \frac{1}{x-3}$ → OUTPUT $\frac{1}{9-3} = \frac{1}{6}$

Let us explore the function notation with another example. If g is the function defined by $y = g(x) = x^2$, then

$$g(1) = 1^2 = 1 \qquad g(2) = 2^2 = 4 \qquad g(3) = 3^2 = 9$$

Note that $g(1) + g(2) \neq g(3)$. To write $g(1) + g(2) = g(1 + 2)$ would be to assume, *incorrectly,* that the distributive property holds for the functional notation. This is not true in general, which comes as no great surprise since g is not a number.

Keep in mind that the variable x in $g(x) = x^2$ is only a placeholder. Any letter could serve the same purpose. For example, $g(t) = t^2$ and $g(z) = z^2$ both define the same function with domain all real numbers.

EXAMPLE 4 For the function g defined by $g(x) = \frac{1}{x}$, find:

(a) $3g(x)$ **(b)** $g(3x)$ **(c)** $3 + g(x)$ **(d)** $g(3) + g(x)$ **(e)** $g(3 + x)$

Solution **(a)** $3g(x) = 3 \cdot \frac{1}{x} = \frac{3}{x}$ **(b)** $g(3x) = \frac{1}{3x}$

(c) $3 + g(x) = 3 + \frac{1}{x}$ **(d)** $g(3) + g(x) = \frac{1}{3} + \frac{1}{x}$

(e) $g(3 + x) = \frac{1}{3 + x}$

EXAMPLE 5 Let $f(x) = -x^2 + 3x$ and find $f(x - 2)$.

Solution Wherever there is an x in the given function, replace it by $x - 2$. Thus

$$f(x - 2) = -(x - 2)^2 + 3(x - 2)$$
$$= -(x^2 - 4x + 4) + 3(x - 2)$$
$$= -x^2 + 4x - 4 + 3x - 6$$
$$= -x^2 + 7x - 10$$

∎

**TEST YOUR UNDERSTANDING
Think Carefully**

(Answers: Page 132)

Let $f(x) = x^2 - 3x$ and find each of the following.

1. $f(-3)$
2. $f(5)$
3. $f(0)$
4. $f(\frac{1}{2})$
5. $f(-\frac{1}{2})$
6. $f(a)$
7. $f(2x)$
8. $2f(x)$
9. $f(x - 3)$
10. $f(3 - x)$
11. $f\left(\dfrac{1}{x}\right)$
12. $\dfrac{1}{f(x)}$

△ **EXAMPLE 6** Let $g(x) = x^2$. Evaluate and simplify the *difference quotient:*

Difference quotients will be used in the study of calculus.

$$\frac{g(x) - g(4)}{x - 4}, \qquad x \neq 4$$

Solution $g(x) = x^2$ and $g(4) = 16$

$$\frac{g(x) - g(4)}{x - 4} = \frac{x^2 - 16}{x - 4}$$
$$= \frac{(x - 4)(x + 4)}{x - 4}$$
$$= x + 4$$

∎

△ **EXAMPLE 7** Let $g(x) = \dfrac{1}{x}$. Evaluate and simplify the *difference quotient:*

$$\frac{g(4 + h) - g(4)}{h}, \qquad h \neq 0$$

Solution Three steps are involved in finding this difference quotient:

1. Find $g(4 + h)$: $\qquad g(4 + h) = \dfrac{1}{4 + h}$

2. Subtract $g(4)$: $\qquad g(4 + h) - g(4) = \dfrac{1}{4 + h} - \dfrac{1}{4}$

3. Divide by h and simplify:

$$\frac{g(4+h)-g(4)}{h} = \frac{\dfrac{1}{4+h}-\dfrac{1}{4}}{h} = \frac{\dfrac{4-(4+h)}{4(4+h)}}{h}$$

$$= \frac{-h}{4h(4+h)}$$

$$= -\frac{1}{4(4+h)} \qquad \blacksquare$$

CAUTION: Learn to Avoid Mistakes Like These	
In each of the following, the function f is defined by $f(x) = 3x^2 - 4$.	
WRONG	**RIGHT**
$f(0) = 0$	$f(0) = 3(0)^2 - 4 = -4$
$f(-2) = -f(2)$	$f(-2) = 3(-2)^2 - 4 = 8$ $-f(2) = -[3(2)^2 - 4] = -8$
$f\left(\dfrac{1}{2}\right) = \dfrac{1}{f(2)}$	$f\left(\dfrac{1}{2}\right) = 3\left(\dfrac{1}{2}\right)^2 - 4 = -\dfrac{13}{4}$ $\dfrac{1}{f(2)} = \dfrac{1}{8}$
$[f(2)]^2 = f(4)$	$[f(2)]^2 = 8^2 = 64$ $f(4) = 3(4)^2 - 4 = 44$
$2 \cdot f(5) = f(10)$	$2 \cdot f(5) = 2[3(5)^2 - 4] = 142$ $f(10) = 3(10)^2 - 4 = 296$
$f(5) + f(2) = f(7)$	$f(5) + f(2) =$ $3(5)^2 - 4 + 3(2)^2 - 4 = 79$ $f(7) = 3(7)^2 - 4 = 143$

i wanna be a Toys R Us Kid!

EXERCISES 2.1

Decide whether the given equation defines y to be a function of x. For each function, find the domain.

1. $y = x^3$

2. $y = \sqrt[3]{x}$

3. $y = \dfrac{1}{\sqrt{x}}$

4. $y = |2x|$

5. $y^2 = 2x$

6. $y = x \pm 3$

7. $y = \dfrac{1}{x+1}$

8. $y = \dfrac{x-2}{x^2+1}$

9. $y = \dfrac{1}{1 \pm x}$

10. $y = \dfrac{1}{\sqrt[3]{x^2-4}}$

Classify each statement as true or false. If it is false, correct the right side to get a correct equation. For each of these statements, use $f(x) = -x^2 + 3$.

11. $f(3) = -6$ **12.** $f(2)f(3) = -33$ **13.** $3f(2) = -33$

14. $f(3) + f(-2) = 2$ **15.** $f(3) - f(2) = -5$ **16.** $f(2) - f(3) = 11$

17. $f(x) - f(4) = -(x - 4)^2 + 3$ **18.** $f(x) - f(4) = x^2 + 19$ **19.** $f(4 + h) = -h^2 - 8h - 13$

20. $f(4 + h) = -h^2 - 10$

Find (a) $f(-1)$, (b) $f(0)$, and (c) $f(\frac{1}{2})$, if they exist.

21. $f(x) = 2x - 1$ **22.** $f(x) = -5x + 6$ **23.** $f(x) = x^2$

24. $f(x) = x^2 - 5x + 6$ **25.** $f(x) = x^3 - 1$ **26.** $f(x) = (x - 1)^2$

27. $f(x) = x^4 + x^2$ **28.** $f(x) = -3x^3 + \frac{1}{2}x^2 - 4x$ **29.** $f(x) = \dfrac{1}{x - 1}$

30. $f(x) = \sqrt{x}$ **31.** $f(x) = \dfrac{1}{\sqrt[3]{x}}$ **32.** $f(x) = \dfrac{1}{3|x|}$

33. For $g(x) = x^2 - 2x + 1$, find:

    (a) $g(10)$ **(b)** $5g(2)$ **(c)** $g(\frac{1}{2}) + g(\frac{1}{3})$ **(d)** $g(\frac{1}{2} + \frac{1}{3})$

34. Let h be given by $h(x) = x^2 + 2x$. Find $f(3)$ and $h(1) + h(2)$ and compare.

35. Let h be given by $h(x) = x^2 + 2x$. Find $3h(2)$ and $h(6)$ and compare.

36. Let h be given by $h(x) = x^2 + 2x$. Find:

    (a) $h(2x)$ **(b)** $h(2 + x)$ **(c)** $h\left(\dfrac{1}{x}\right)$ **(d)** $h(x^2)$

Find the value for y for these values of x: (a) -5 (b) -2 (c) 0 (d) 2 (e) 5

37. $y = \begin{cases} 2x - 1 & \text{if } x \leq -2 \\ 1 - 2x & \text{if } x > 2 \end{cases}$ **38.** $y = \begin{cases} |1 - x| & \text{if } x < -2 \\ 2x - 3 & \text{if } -2 \leq x \leq 2 \\ x^2 - 2 & \text{if } x > 2 \end{cases}$

△ *Find the difference quotient $\dfrac{f(x) - f(3)}{x - 3}$ and simplify for the given function f.*

39. $f(x) = x^2$ **40.** $f(x) = x^2 - 1$ **41.** $f(x) = \dfrac{1}{x}$

42. $f(x) = \sqrt{x}$ **43.** $f(x) = 2x + 1$ **44.** $f(x) = -x^3 + 1$

△ *Find the difference quotient $\dfrac{f(2 + h) - f(2)}{h}$ and simplify for the given function f.*

45. $f(x) = x$ **46.** $f(x) = -x + 3$ **47.** $f(x) = -x^2$

48. $f(x) = \sqrt{x + 2}$ **49.** $f(x) = \dfrac{1}{x^2}$ **50.** $f(x) = \dfrac{1}{x - 1}$

Written Assignment: Explain, in your own words, the meaning of a function as well as the domain and range of a function. Illustrate with a specific function of your own.

2.2
GRAPHING
LINES IN THE
RECTANGULAR
COORDINATE
SYSTEM

A great deal of information can be learned about a functional relationship by studying its graph. A fundamental objective of this course is to acquaint you with the graphs of some important functions, as well as to develop basic graphing procedures. First we need to review the structure of a **rectangular coordinate system.**

In a plane take any two lines that intersect at right angles and call their point of intersection the **origin.** Let each of these lines be a number line with the origin corresponding to zero for each line. Unless otherwise specified, the unit length is the same on both lines. On the horizontal line the positive direction is taken to be to the right of the origin, and on the vertical line it is taken to be above the origin. Each of these two lines will be referred to as an **axis** of the system (plural: **axes**).

The union of algebra and geometry, credited to French mathematician René Descartes (1596–1650), led to the development of analytic geometry. In his honor, we often refer to the rectangular coordinate system as the **Cartesian coordinate system** or simply the **Cartesian plane.**

The horizontal line is usually called the **x-axis,** and the vertical line the **y-axis.** The axes divide the plane into four regions called **quadrants.** The quadrants are numbered in a counterclockwise direction as shown in the following figure.

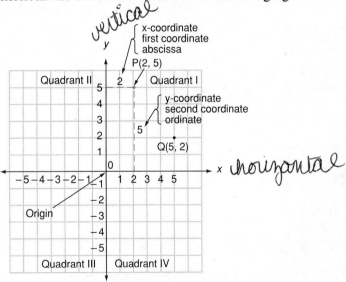

Note that the ordered pair (2, 5) is not the same as the pair (5, 2). Each gives the coordinates of a different point on the plane.

The points in the plane (denoted by the capital letters) are matched with pairs of numbers, referred to as the **coordinates** of these points. For example, starting at the origin, P can be reached by moving 2 units to the right, along the x-axis; then 5 units up, parallel to the y-axis. Thus the first coordinate, 2, of P is called the **x-coordinate** (another name is **abscissa**) and the second coordinate, 5, is the **y-coordinate** (also called **ordinate**). We say that the *ordered pair of numbers* (2, 5) are the coordinates of P.

All points in the first quadrant are to the right and above the origin and therefore have positive coordinates. Any point in quadrant II is to the left and above the origin and therefore has a negative x-coordinate and a positive y-coordinate. In quadrant III both coordinates are negative, and in the fourth quadrant they are positive and negative, respectively.

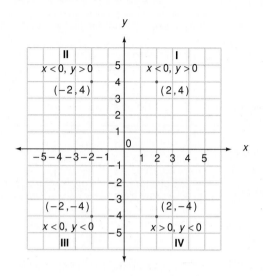

EXAMPLE 1   Find the coordinates of the given points. Also state the quadrant or axis in which each point is located.

Usually, the points on the axes are labeled with just a single number. In that case it is understood that the missing coordinate is 0. Thus in Example 1, B has coordinates (0, 2) and the coordinates of point F are (3, 0). The coordinates of the origin are (0, 0).

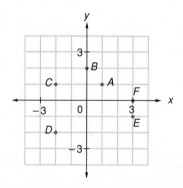

Solution   Point A is located at $(1, 1)$ in the first quadrant. Point B is at $(0, 2)$, on the y-axis. Point C is at $(-2, 1)$, in the second quadrant. Point D is at $(-2, -2)$, in the third quadrant. Point E is at $(3, -1)$, in the fourth quadrant. Point F is at $(3, 0)$, on the x-axis.   ■

The equality $y = x + 2$ is an equation in two variables. When a specific value for x is substituted into this equation, we get a corresponding y-value. For example, substituting 3 for x gives $y = 3 + 2 = 5$. We therefore say that the ordered pair $(3, 5)$ *satisfies* the equation $y = x + 2$.

> *To graph an equation in the variables x and y means to locate all the points in a rectangular system whose coordinates satisfy the given equation.*

There is an infinite number of ordered pairs that satisfy $y = x + 2$, and all are located on the same straight line. The following table of values shows some ordered pairs of numbers that satisfy the equation $y = x + 2$. These have been *plotted* (located) in a rectangular system and connected by a straight line. The arrowheads in the figure suggest that the line continues endlessly in both directions.

The straight line contains exactly those points whose ordered pairs (x, y) satisfy the equation $y = x + 2$. Any point not on the line has an ordered pair (x, y) where $y \neq x + 2$. Thus $(1, 5)$ is not on the line, since $5 \neq 1 + 2$.

x	y = x + 2
−3	−1
−2	0
−1	1
0	2
1	3
2	4

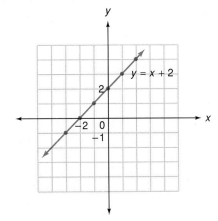

The graph of $y = x + 2$ is a straight line and the equation is called a **linear equation.** Since two points determine a line, a convenient way to graph a line is to locate its two **intercepts.** The **x-intercept** for $y = x + 2$ is −2, the abscissa of the point where the line crosses the x-axis. The **y-intercept** is 2, the ordinate of the point where the line crosses the y-axis.

EXAMPLE 2   Graph the linear equation $y = 2x - 1$ using the intercepts.

Solution   To find the x-intercept, let $y = 0$.

When the line crosses the x-axis, the y-value is 0.

$$2x - 1 = 0$$
$$2x = 1$$
$$x = \tfrac{1}{2} \quad \longleftarrow \text{ the x-intercept}$$

When the line crosses the y-axis, the x-value is 0.

To find the y-intercept, let $x = 0$.

It is generally wise to locate a third point to verify your work. Thus for $x = 2$, $y = 3$, and the line passes through the point $(2, 3)$.

$$y = 2(0) - 1$$
$$y = -1 \quad \longleftarrow \text{ the y-intercept}$$

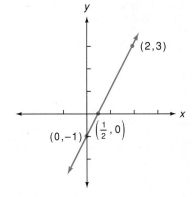

Plot the points $(\tfrac{1}{2}, 0)$ and $(0, -1)$ and draw the line through them to determine the graph.   ∎

An equation such as $y = 2x - 1$ defines y as a *linear function* of x, and is often written in the form $f(x) = 2x - 1$. Usually the domain of such a function is the set of all real numbers. However, at times we may wish to limit the domain of a function. For example, the graph of $y = 2x - 1$ for $-2 \leq x \leq 1$ is a line segment with endpoints at $(-2, -5)$ and $(1, 1)$ as in the following figure.

$$y = 2x - 1$$

Domain: $-2 \le x \le 1$

Range: $-5 \le y \le 1$

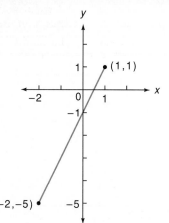

For convenience, the change in y may be referred to as "Δy" (read "delta y"); the change in x is denoted as "Δx" (read "delta x").

$$\frac{\Delta y}{\Delta x} = \frac{2 - 0}{3 - 2} = \frac{2}{1} = 2$$

Show that the same ratio is obtained by using any other two points on the line, such as $(1, -2)$ and $(0, -4)$.

The following figure shows the graph of the linear equation $y = 2x - 4$, including the coordinates of four specific points. From the diagram you will see that the y-value increases 2 units each time that the x-value increases by 1 unit. The ratio of this change in y compared to the corresponding change in x is $\frac{2}{1} = 2$. Using the coordinates of the points $(3, 2)$ and $(2, 0)$, we have the following:

$$\frac{\text{change in } y\text{-values}}{\text{change in } x\text{-values}} = \frac{2 - 0}{3 - 2} = \frac{2}{1} = 2$$

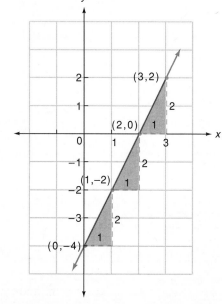

We call this ratio the **slope** of the line, defined as follows.

DEFINITION OF SLOPE

If two points (x_1, y_1) and (x_2, y_2) are on a line ℓ, then the slope m of line ℓ is defined by

$$m = \frac{y_2 - y_1}{x_2 - x_1}, \qquad x_2 \ne x_1$$

Notice that in the definition $x_2 - x_1$ cannot be zero; that is, $x_2 \ne x_1$. The only time that $x_2 = x_1$ is when the line is vertical.

In the following figure the coordinates of two points A and B have been labeled (x_1, y_1) and (x_2, y_2). The change in the y direction from A to B is given by the difference $y_2 - y_1$; the change in the x direction is $x_2 - x_1$. If a different pair of points is chosen, such as P and Q, then the ratio of these differences is still the same because the resulting triangles (ABC and PQR) are similar. Thus, since corresponding sides of similar triangles are proportional, we have

This discussion shows that there can be only one slope for a given line.

$$m = \frac{y_2 - y_1}{x_2 - x_1} = \frac{AC}{CB} = \frac{PR}{RQ}$$

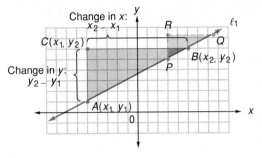

It may be helpful to think of the slope of a line in any of these ways:

Another descriptive language for slope is $m = \dfrac{rise}{run}$, where
rise is the vertical change and
run is the horizontal change.

$$m = \frac{y_2 - y_1}{x_2 - x_1} = \frac{\text{change in } y}{\text{change in } x} = \frac{\text{vertical change}}{\text{horizontal change}}$$

EXAMPLE 3   Find the slope of line ℓ determined by the points $(-3, 4)$ and $(1, -6)$.

Solution   Use $(x_1, y_1) = (-3, 4)$; $(x_2, y_2) = (1, -6)$. Then

$$m = \frac{-6 - 4}{1 - (-3)} = -\frac{10}{4} = -\frac{5}{2}$$

CAUTION
Do not mix up the coordinates like this:
$\dfrac{y_2 - y_1}{x_1 - x_2} = \dfrac{4 - (-6)}{1 - (-3)} = \dfrac{5}{2}$
This is the negative of the slope.

Note: It makes no difference which of the two points is called (x_1, y_1) or (x_2, y_2) since the ratio will still be the same. If $(x_1, y_1) = (1, -6)$ and $(x_2, y_2) = (-3, 4)$, for example, then $\dfrac{y_2 - y_1}{x_2 - x_1} = \dfrac{4 - (-6)}{-3 - 1} = -\dfrac{5}{2} = m.$ ∎

Reading from left to right, a rising line has a positive slope and a falling line has a negative slope.

EXAMPLE 4   Graph the line with slope $\frac{3}{2}$ that passes through the point $(-2, -2)$.

Solution   Think of $\dfrac{3}{2}$ as $\dfrac{\text{change in } y}{\text{change in } x}$. Now start at $(-2, -2)$ and move 3 units up and 2 units to the right. This locates the point $(0, 1)$. Draw the straight line through these two points.

Alternately, start at $(-2, -2)$ and move 3 units down and 2 units to the left to locate $(-4, -5)$.

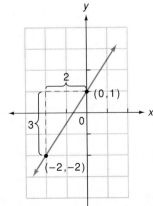

Next consider lines that are parallel to the axes, that is, horizontal and vertical lines.

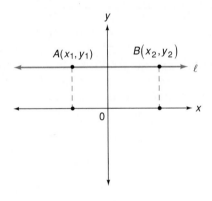 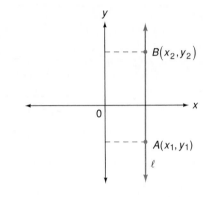

CAUTION
Do not confuse slope 0 for horizontal lines with undefined slope for vertical lines.

Since ℓ is parallel to the x-axis, $y_1 = y_2$ and $y_2 - y_1 = 0$. Thus *the slope of a horizontal line is* 0.

$$m = \frac{y_2 - y_1}{x_2 - x_1} = \frac{0}{x_2 - x_1} = 0$$

Since ℓ is parallel to the y-axis, $x_1 = x_2$ and $x_2 - x_1 = 0$. Since division by 0 is undefined, we say that *the slope of a vertical line is undefined;* that is, vertical lines do not have slope.

Two nonvertical lines are parallel if and only if they have the same slope. The slope property for perpendicular lines is not as obvious. The figure below suggests the following (see Exercise 45).

Two lines not parallel to the coordinate axes are perpendicular if and only if their slopes are negative reciprocals of one another.

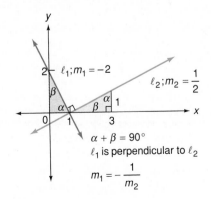

EXAMPLE 5 In the figure, line ℓ_1 has slope $\frac{2}{3}$ and is perpendicular to ℓ_2. If the lines intersect at $P(-1, 4)$, use the slope of ℓ_2 to find the coordinates of another point on ℓ_2.

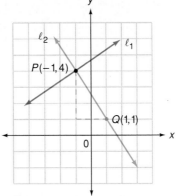

Solution Since the lines are perpendicular, the slope of ℓ_2 is $-\frac{3}{2}$. Now start at P and count 3 units downward and 2 units to the right to reach point $(1, 1)$ on ℓ_2. Other solutions are possible. Can you locate a point on ℓ_2 that lies in the second quadrant? ∎

In summary, we have the following concerning slopes of lines.

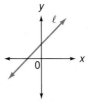

Line ℓ has a positive slope; ℓ is rising from left to right.

Line ℓ has a negative slope; ℓ is falling from left to right.

Line ℓ has a slope of 0; ℓ is horizontal.

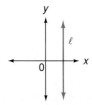

The slope of line ℓ is undefined; ℓ is vertical.

In business it is often noted that *demand* is related to the *selling price* of a commodity. That is, usually the higher the price, the less is the demand. Sometimes this relationship can be expressed as a linear function having a *negative* slope. For example, suppose that for a certain calculator it is determined that the weekly demand, D, is related to the price x (in dollars) by the function below:

$$y = D = 500 - 20x$$

where $0 \le x \le 25$

From the graph we see that the demand is greatest when the price is zero dollars! The demand is zero when the price is $25.

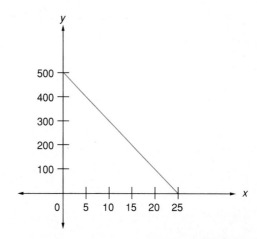

Note that the graph uses a different scale on each axis in order to accommodate the data shown.

Now suppose that the weekly supply of calculators, S, is also a linear function of the price, x, and is given by $y = S = 10x + 200$. When the graph of this supply function is drawn in the same coordinate system as the demand function, then the **equilibrium point** is the point where the two lines intersect. At this point, the demand = supply for the same price x. To find this x-value, set $D = S$ and solve for the common value x.

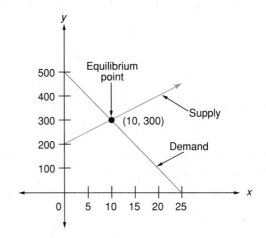

$$500 - 20x = 10x + 200$$
$$300 = 30x$$
$$10 = x$$

Note that the point of intersection is (10, 300), the equilibrium point. At this point the coordinates are the same for both lines.

At $10 per calculator, 300 calculators will be supplied and sold per week. This result, $10, represents the price at which supply equals demand; the demand will be met, and no surplus will exist.

EXERCISES 2.2

Copy and complete each table of values. Then graph the line given by the equation.

1. $y = x - 2$

x	-3	-2	-1	0	1	2
y						

2. $y = -x + 1$

x	-3	-2	-1	0	1	2	3
y							

3. $y = 2x - 4$

x	-2	-1	0	1	2
y					

4. $y = -2x + 3$

x	-2	-1	0	1	2
y					

Find the x- and y-intercepts and use these to graph each of the following lines.

5. $x + 2y = 4$ 6. $2x + y = 4$ 7. $x - 2y = 4$ 8. $2x - y = 4$
9. $2x - 3y = 6$ 10. $3x + y = 6$ 11. $y = -3x - 9$ 12. $y + 2x = -5$
13. $y = 2x - 1$

14. Sketch the following on the same set of axes.

   (a) $y = x$

   (b) By adding 1 to each y-value (ordinate) in part (a), graph $y = x + 1$. In other words, *shift* each point of $y = x$ one unit up.

   (c) By subtracting 1 from each y-value in part (a), graph $y = x - 1$. That is, shift each point of $y = x$ one unit down.

15. Repeat Exercise 14 for:

   (a) $y = -x$ (b) $y = -x + 1$ (c) $y = -x - 1$

16. Sketch the following on the same set of axes.

   (a) $y = x$

   (b) By multiplying each y-value in part (a) by 2, graph $y = 2x$. In other words, *stretch* each y-value of $y = x$ to twice its size.

   (c) Graph $y = 2x + 3$ by shifting $y = 2x$ three units upward.

Graph the points that satisfy each equation for the given values of x.

17. $y = \frac{1}{2}x$; $-6 \le x \le 6$

18. $y = -2x + 1$; $-2 \le x \le 2$

19. $y = 3x - 5$; $1 \le x \le 4$

20. $y = -\frac{1}{2}x + 2$; $-2 \le x \le 2$

21. Use the coordinates of each of the following pairs of points to find the slope of ℓ.

   (a) A, C (b) B, D (c) C, D (d) A, E (e) B, E (f) C, E

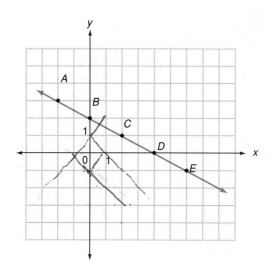

Compute the slope, if it exists, for the line determined by the given pair of points.

22. $(3, 4)$; $(2, -5)$

23. $(4, 3)$; $(-5, 2)$

24. $(-7, 6)$; $(-7, 106)$

25. $(6, -7)$; $(106, -7)$

26. $(-9, \frac{1}{2})$; $(2, \frac{1}{2})$

27. $(2, -\frac{3}{4})$; $(-\frac{1}{3}, \frac{2}{3})$

Draw the line through the given point having slope m.

28. $(-\frac{1}{2}, 0)$; $m = -1$

29. $(0, 2)$; $m = \frac{3}{4}$

30. $(3, -4)$; $m = 4$

31. $(-3, 4)$; $m = -\frac{1}{4}$

32. $(1, 1)$; $m = 2$

33. $(-2, \frac{3}{2})$; $m = 0$

34. Graph each of the lines with the following slopes through the point $(5, -3)$:

   $m = -2$ $m = -1$ $m = 0$ $m = 1$ $m = 2$

35. In the same coordinate system, draw the line:

   (a) Through $(1, 0)$ with $m = -1$

   (b) Through $(0, 1)$ with $m = 1$

   (c) Through $(-1, 0)$ with $m = -1$

   (d) Through $(0, -1)$ with $m = 1$

36. Line ℓ passes through $(-4, 5)$ and $(8, -2)$.

   (a) Draw the line through $(0, 0)$ perpendicular to ℓ.

   (b) What is the slope of any line perpendicular to line ℓ?

37. Why is the line determined by the points $(6, -5)$ and $(8, -8)$ parallel to the line through $(-3, 12)$ and $(1, 6)$?

38. Verify that the points $A(1, 2)$, $B(4, -1)$, $C(2, -2)$, and $D(-1, 1)$ are the vertices of a parallelogram. Sketch the figure.

39. Consider the four points $P(5, 11)$, $Q(-7, 16)$, $R(-12, 4)$, and $S(0, -1)$. Show that the four angles of the quadrilateral $PQRS$ are right angles. Also show that the diagonals are perpendicular.

40. Lines ℓ_1 and ℓ_2 are perpendicular and intersect at point $(-2, -6)$. ℓ_1 has slope $-\frac{2}{5}$. Use the slope of ℓ_2 to find the y-intercept of ℓ_2.

41. Any horizontal line is perpendicular to any vertical line. Why were such lines excluded from the result, which states that lines are perpendicular if and only if their slopes are negative reciprocals?

42. Find t if the line through $(-1, 1)$ and $(3, 2)$ is parallel to the line through $(0, 6)$ and $(-8, t)$.

43. Find t if the line through $(-1, 1)$ and $(1, \frac{1}{2})$ is perpendicular to the line through $(1, \frac{1}{2})$ and $(7, t)$. Use the fact that two perpendicular lines have slopes that are negative reciprocals of one another.

44. (a) Prove that nonvertical parallel lines have equal slopes by considering two parallel lines ℓ_1, ℓ_2 as in the figure. On ℓ_1 select points A and B, and choose A' and B' on ℓ_2. Now form the appropriate right triangles ABC and $A'B'C'$ using points C and C' on the x-axis. Prove they are similar and write a proportion to show the slopes of ℓ_1 and ℓ_2 are equal.

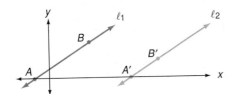

   (b) Why is the converse of the fact given in part (a) also true?

45. This exercise gives a proof that if two lines are perpendicular, then they have slopes that are negative reciprocals of one another.

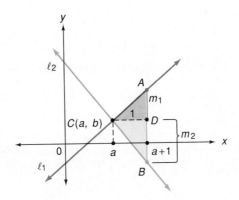

   Let line ℓ_1 be perpendicular to line ℓ_2 at the point $C(a, b)$. Use m_1 for the slope of ℓ_1, and m_2 for ℓ_2. We want to show the following:

$$m_1 m_2 = -1 \quad \text{or} \quad m_1 = -\frac{1}{m_2}$$

Add 1 to the x-coordinate a of point C and draw the vertical line through $a + 1$ on the x-axis. This vertical line will meet ℓ_1 at some point A and ℓ_2 at some point B, forming right triangle ABC with right angle at C. Draw the perpendicular from C to AB meeting AB at D. Then CD has length 1.

(a) Using the right triangle CDA, show that $m_1 = DA$.

(b) Show that $m_2 = DB$. Is m_2 positive or negative?

(c) For right triangle ABC, CD is the mean proportional between segments BD and DA on the hypotenuse. Use this fact to conclude that $\dfrac{m_1}{1} = \dfrac{1}{-m_2}$, or $m_1 m_2 = -1$.

46. For a certain type of toy, the weekly demand, D, is related to the price x, in dollars, by the function $y = D = 1000 - 40x$. The weekly supply is given by the function $y = S = 15x + 340$. Find the price at which the supply equals the demand. Graph both equations on the same set of axes and show the equilibrium point.

47. The monthly demand, D, for a certain portable radio is related to the price x by the function $D = 5000 - 25x$. The monthly supply is given by the function $S = 10x + 275$. At what price will the supply equal the demand? How many radios will be sold at this price?

2.3 ALGEBRAIC FORMS OF LINEAR FUNCTIONS

Pictured below is a line with slope m and y-intercept b. To find the equation of ℓ we begin by considering any point $P(x, y)$ on ℓ other than $(0, b)$.

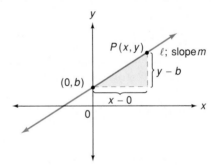

Since the slope of ℓ is given by any two of its points, we may use $(0, b)$ and (x, y) to write

$$m = \frac{y - b}{x - 0} = \frac{y - b}{x}$$

If both sides are multiplied by x, then

$$y - b = mx \quad \text{or} \quad y = mx + b$$

This leads to the following *y-form of the equation of a line*.

SLOPE-INTERCEPT FORM OF A LINE

$$y = mx + b$$

A point (x, y) is on this line if and only if the coordinates satisfy this equation.

where m is the slope and b is the y-intercept.

The equation $y = mx + b$ also defines a function; thus we may think of $y = f(x) = mx + b$ as a linear function with the domain consisting of all the real numbers.

EXAMPLE 1 Graph the linear function f defined by $y = f(x) = 2x - 1$ by using the slope and y-intercept. Also indicate the domain and range of f, and display $f(2) = 3$ geometrically; that is, show the point corresponding to $f(2) = 3$.

Both the domain and range of f consist of all real numbers.

Solution The y-intercept is -1. Locate $(0, -1)$ and use $m = 2 = \frac{2}{1}$ to reach $(1, 1)$, another point on the line.

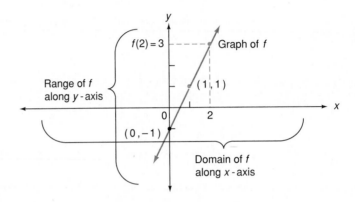

EXAMPLE 2 Write, in slope intercept form, the equation of the line with slope $m = \frac{2}{3}$ passing through the point $(0, -5)$.

Solution Since $m = \frac{2}{3}$ and $b = -5$, the slope-intercept form $y = mx + b$ gives

$$y = \frac{2}{3}x + (-5)$$

$$= \frac{2}{3}x - 5$$

A special case of $y = f(x) = mx + b$ is obtained when $m = 0$. Then

$$y = 0(x) + b \quad \text{or} \quad y = b$$

This says that for *each* input x, the output $f(x)$ is always the same value b.

Domain: all reals
Range: the single
         value b

Can a vertical line be the graph of a function where y depends on x? Explain.

Since $f(x) = b$ is constant for all x, this linear function is also referred to as a **constant function.** Its graph is a horizontal line.

EXAMPLE 3 Write and graph the equation of the line through the point $(2, 3)$ that is:

(a) Parallel to the x-axis   **(b)** Parallel to the y-axis

Solution

(a) The y-intercept is 3 and the slope is 0. Thus the equation is

$$y = 0(x) + 3 \text{ or } y = 3.$$

(b) The line has no y-intercept. Also, the slope is undefined. From the figure we note that y can be any value but that x is always 2. Thus the equation of the line is $x = 2$.

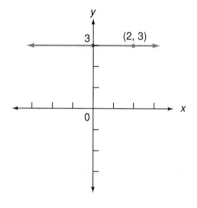

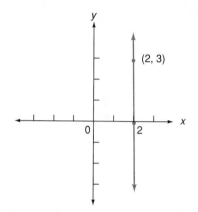

In general, a horizontal line through the point (h, k) is of the form $y = k$.

In general, a vertical line through the point (h, k) is of the form $x = h$.   ∎

Now let ℓ be a line with slope m that passes through a specific point (x_1, y_1). We wish to determine the conditions on the coordinates of any point $P(x, y)$ that is on the line ℓ.

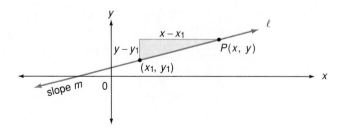

From the figure you can see that $P(x, y)$ will be on ℓ if and only if the ratio $\dfrac{y - y_1}{x - x_1}$ is the same as m. That is, P is on ℓ if and only if

$$m = \frac{y - y_1}{x - x_1}$$

Multiply both sides of this equation by $x - x_1$:

$$m(x - x_1) = y - y_1$$

This leads to another form for the equation of a straight line.

<table>
<tr><td>

POINT-SLOPE FORM OF A LINE

$$y - y_1 = m(x - x_1)$$

where m is the slope and (x_1, y_1) is a point on the line.

</td></tr>
</table>

This is the form of a line that is most frequently used in calculus. A point (x, y) is on this line if and only if the coordinates satisfy this equation.

EXAMPLE 4 Write the point-slope form of the line ℓ with slope $m = 3$ that passes through the point $(-1, 1)$. Verify that $(-2, -2)$ is on the line.

Solution Since $m = 3$, any (x, y) on ℓ satisfies this equation:

$$y - 1 = 3[x - (-1)]$$
$$y - 1 = 3(x + 1)$$

CAUTION
Pay attention to the minus signs on the coordinates when used in the point-slope form. Note the substitution of $x_1 = -1$ in this example.

Let $x = -2$:

$$y - 1 = 3(-2 + 1) = -3$$
$$y = -2$$

Thus $(-2, -2)$ is on the line. ∎

<table>
<tr><td>

TEST YOUR UNDERSTANDING
Think Carefully

(Answers: Page 132)

</td><td>

Write the slope-intercept form of the line with the given slope and y-intercept.

1. $m = 2$; $b = -2$ **2.** $m = -\frac{1}{2}$; $b = 0$ **3.** $m = \sqrt{2}$; $b = 1$

Write the point-slope form of the line through the given point with slope m.

4. $(2, 6)$; $m = -3$ **5.** $(-1, 4)$; $m = \frac{1}{2}$ **6.** $(5, -\frac{2}{3})$; $m = 1$
7. $(0, 0)$; $m = -\frac{1}{4}$ **8.** $(-3, -5)$; $m = 0$ **9.** $(1, -1)$; $m = -1$
10. Which, if any, of the preceding produce a constant linear function? State its range.

</td></tr>
</table>

EXAMPLE 5 Write the slope-intercept form of the line through the two points $(6, -4)$ and $(-3, 8)$.

Solution First compute the slope.

$$m = \frac{-4 - 8}{6 - (-3)} = \frac{-12}{9} = -\frac{4}{3}$$

Use either point to write the point-slope form, and then convert to the slope-intercept form. Thus, using the point (6, −4),

$$y - (-4) = -\tfrac{4}{3}(x - 6)$$

$$y + 4 = -\tfrac{4}{3}x + 8$$

$$y = -\tfrac{4}{3}x + 4$$

Show that the same final form is obtained using the point (−3, 8). ∎

EXAMPLE 6 Write an equation of the line that is perpendicular to the line $5x - 2y = 2$ and that passes through the point (−2, −6).

Solution First find the slope of the given line by writing it in slope-intercept form.

$$5x - 2y = 2$$

$$-2y = -5x + 2$$

$$y = \frac{5}{2}x - 1 \qquad \text{The slope is } \frac{5}{2}.$$

The perpendicular line has slope $= -\dfrac{2}{5}$. Since this line also goes through $P(-2, -6)$, the point-slope form gives

$$y + 6 = -\tfrac{2}{5}(x + 2) \quad \text{or} \quad y = -\tfrac{2}{5}x - \tfrac{34}{5} \qquad ∎$$

A linear equation such as $y = -\tfrac{2}{3}x + 4$ can be converted to other equivalent forms. In particular, when this equation is multiplied by 3, we obtain the form $2x + 3y = 12$. This is an illustration of the *general linear equation*.

GENERAL LINEAR EQUATION

$$Ax + By = C$$

where A, B, C are constants and A, B are not both 0.

The general linear equation $Ax + By = C$ is said to define y as a function of x *implicitly* if $B \neq 0$. In other words, we have these equivalent forms:

$$Ax + By = C \qquad \longleftarrow \text{\textit{implicit} form of the linear function}$$

$$By = -Ax + C$$

$$y = -\frac{A}{B}x + \frac{C}{B} \qquad \longleftarrow \text{\textit{explicit} form of the linear function; slope-intercept form}$$

EXAMPLE 7   Find the slope and y-intercept of the line given by the equation $-6x + 2y - 5 = 0$.

Solution   Convert to the explicit form of the linear function.

$$-6x + 2y - 5 = 0$$
$$2y = 6x + 5$$
$$y = 3x + \tfrac{5}{2} \quad \longleftarrow \text{slope-intercept form}$$

Thus $m = 3$ and $b = \tfrac{5}{2}$.   ∎

EXAMPLE 8   Find the equation of the line through the points $(2, -3)$ and $(3, -1)$. Write the equation:
(a) In point-slope form   **(b)** In slope-intercept form   **(c)** In the general form $Ax + By = C$

Solution

(a)   The slope m is $\dfrac{-1 - (-3)}{3 - 2} = 2$. Use this slope and either point, such as $(2, -3)$.

$y - y_1 = m(x - x_1):$ $y - (-3) = 2)(x - 2)$ point-slope form
$$y + 3 = 2(x - 2)$$

(b)   Use the solution for part (a) and solve for y.

$y = mx + b:$ $y + 3 = 2(x - 2)$
$$y + 3 = 2x - 4$$
$$y = 2x - 7 \qquad \text{slope-intercept form}$$

(c)   Rewrite the solution for part (b).

$Ax + By = C:$ $y = 2x - 7$
$$-2x + y = -7 \qquad \text{the form } Ax + By = C$$
$$\text{or} \quad 2x - y = 7$$

Note that all three forms are different ways of expressing the same equation for the given line through $(2, -3)$ and $(3, -1)$. Show that $x = 3$ and $y = -1$ satisfies each form.   ∎

Here is a summary of the algebraic forms of a line that we have explored.

Slope-intercept form	Point-slope form	General linear equation
$y = mx + b$	$y - y_1 = m(x - x_1)$	$Ax + By = C$
Line with slope m and y-intercept b.	Line with slope m and through the point (x_1, y_1).	Line with slope $-\dfrac{A}{B}$ and y-intercept $\dfrac{C}{B}$, if $B \neq 0$.

For the general linear equation $Ax + By = C$, note the following:

If $B = 0$ and $A \neq 0$, then $x = \dfrac{C}{A}$, the equation of a vertical line.

If $A = 0$ and $B \neq 0$, then $y = \dfrac{C}{B}$, the equation of a horizontal line.

CAUTION: Learn to Avoid Mistakes Like These	
WRONG	RIGHT
The slope of the line through $(2, 3)$ and $(5, 7)$ is $$m = \frac{7-3}{2-5} = -\frac{4}{3}$$	The slope of the line through $(2, 3)$ and $(5, 7)$ is $$m = \frac{7-3}{5-2} = \frac{4}{3}$$
The slope of the line $2x - 3y = 7$ is $-\frac{2}{3}$.	The slope is $\frac{2}{3}$.
The line through $(-4, -3)$ with slope 2 is $y - 3 = 2(x - 4)$.	The equation is $y - (-3) = 2[x - (-4)]$ or $y + 3 = 2(x + 4)$.

EXERCISES 2.3

Write the equation of the line with the given slope m and y-intercept b.

1. $m = 2, b = 3$ 2. $m = -2, b = 1$ 3. $m = 1, b = 1$ 4. $m = -1, b = 2$
5. $m = 0, b = 5$ 6. $m = 0, b = -5$ 7. $m = \frac{1}{2}, b = 3$ 8. $m = -\frac{1}{2}, b = 2$
9. $m = \frac{1}{4}, b = -2$

Write the point-slope form of the line through the given point with the indicated slope.

10. $(3, 4); m = 2$ 11. $(2, 3); m = 1$ 12. $(1, -2); m = 0$
13. $(-2, 3); m = 4$ 14. $(-3, 5); m = -2$ 15. $(-3, 5); m = 0$
16. $(8, 0); m = -\frac{2}{3}$ 17. $(2, 1); m = \frac{1}{2}$ 18. $(-6, -3); m = \frac{4}{3}$
19. $(0, 0); m = 5$ 20. $(-\frac{3}{4}, \frac{2}{5}); m = 1$ 21. $(\sqrt{2}, -\sqrt{2}); m = 10$

22. **(a)** Find the slope of the line determined by the points $A(-3, 5)$ and $B(1, 7)$, and write its equation in point-slope form, using the coordinates of A.
   (b) Do the same in part (a) using the coordinates of B.
   (c) Verify that the equations obtained in parts (a) and (b) give the same slope-intercept form.

Write each equation in slope-intercept form; give the slope and y-intercept.

23. $3x + y = 4$ 24. $2x - y = 5$ 25. $6x - 3y = 1$ 26. $4x + 2y = 10$
27. $3y - 5 = 0$ 28. $x = \frac{3}{2}y + 3$ 29. $4x - 3y - 7 = 0$ 30. $5x - 2y + 10 = 0$
31. $\frac{1}{4}x - \frac{1}{2}y = 1$

Write the equation of the line through the two given points in the form $Ax + By = C$.

32. $(-1, 2), (2, -1)$ 33. $(2, 3), (3, 2)$ 34. $(1, 1), (-1, -1)$ 35. $(3, 0), (0, -3)$

36. $(3, -4)$, $(0, 0)$ **37.** $(-1, -13)$, $(-8, 1)$ **38.** $(\frac{1}{2}, 7)$, $(-4, -\frac{3}{2})$ **39.** $(10, 27)$, $(12, 27)$
40. $(\sqrt{2}, 4\sqrt{2})$, $(-3\sqrt{2}, -10\sqrt{2})$

41. Two lines, parallel to the coordinate axes, intersect at the point $(5, -7)$. What are their equations?

42. Write the equation of the line that is parallel to $y = -3x - 6$ and with y-intercept 6.

43. Write the equation of the line parallel to $2x + 3y = 6$ that passes through the point $(1, -1)$.

Write the equation of the line that is perpendicular to the given line and passes through the indicated point.

44. $y = -10x$; $(0, 0)$ **45.** $y = 3x - 1$; $(4, 7)$ **46.** $3x + 2y = 6$; $(6, 7)$ **47.** $y - 2x = 5$; $(-5, 1)$

48. The vertices of a triangle are located at $(-1, -1)$, $(1, 3)$, and $(4, 2)$. Write the equations for the sides of the triangle.

49. In Exercise 48, write the equations of the three altitudes of the triangle.

50. The vertices of a rectangle are located at $(2, 2)$, $(6, 2)$, $(6, -3)$, and $(2, -3)$. What is the relationship between the slopes of the diagonals?

51. The vertices of a square are located at $(2, 2)$, $(5, 2)$, $(5, -1)$, and $(2, -1)$. What is the relationship between the slopes of the diagonals?

Graph the linear function f by using the slope and y-intercept. Display the point corresponding to $y = f(-2)$ on the graph.

52. $y = f(x) = -2x + 1$ **53.** $y = f(x) = x + 3$ **54.** $y = f(x) = 3x - \frac{1}{2}$ **55.** $y = f(x) = \frac{1}{2}x - 3$

The domain is given for the function defined by the equation $y = 3x - 7$.
Graph each function.

56. All $x \le 4$.   **57.** All $x \ge 0$.   **58.** All x where $-1 \le x \le 3$.   **59.** $x = -1, 0, 1, 2, 3$

Write the equation for each graph. State the domain and range if it is the graph of a function.

60.   **61.**   **62.**

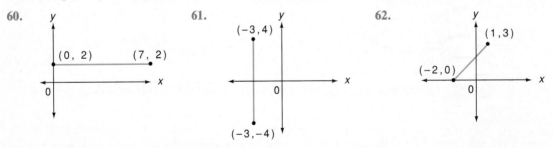

63.   **64.**   **65.**

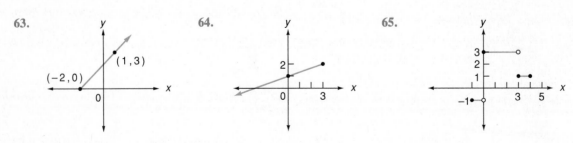

66. The sides of the parallelogram with vertices $(-1, 1)$, $(0, 3)$, $(2, 1)$, and $(3, 3)$ are the graphs of four different functions. In each case find the equation that defines the function and state the domain.

67. Any line having a nonzero slope that does not pass through the origin always has both an x- and y-intercept. Let ℓ be such a line having equation $ax + by = c$.

(a) Why is $c \neq 0$?

(b) What are the x- and y-intercepts?

(c) Derive the equation $\dfrac{x}{q} + \dfrac{y}{p} = 1$ where q and p are the x- and y-intercepts, respectively. This is known as the *intercept form* of a line.

(d) Use the intercept form to write the equation of the line passing through $(\frac{3}{2}, 0)$ and $(0, -5)$.

(e) Use the two points in part (d) to find the slope, write the slope-intercept form, and compare with the result in part (d).

68. Replace m in the point-slope form of a line through points (x_1, y_1) and (x_2, y_2) by $\dfrac{y_2 - y_1}{x_2 - x_1}$. Show that this gives the *two-point form* for the equation of a line:

$$\frac{y - y_1}{y_2 - y_1} = \frac{x - x_1}{x_2 - x_1}$$

69. Use the result of Exercise 68 to find the equation of the line through $(-2, 3)$ and $(5, -2)$. Write the equation in point-slope form.

Written Assignment: Select a point and a nonzero slope of your choice. Then describe the various algebraic forms that can be used to identify the line through your given point and with your given slope.

EXPLORATIONS
Think Critically

1. We often say that an equation such as $y = x^2 + 2x$ is a function. Explain why technically this is not correct.

2. Suppose $f(x) = 2x + 3$. Find a meaning for the expression $f(f(x))$.

3. How can you prove that a point $P(x, y)$ is *not* located on a line with a given equation?

4. Consider the line defined by the equation $y = mx + b$. Under what conditions on m and b will the line pass through quadrants I, III, and IV?

5. Consider the graphs of the two equations $y = ax + c$ and $y = bx + d$. Under what conditions are the x-intercepts the same? Under what conditions are the two lines parallel but not vertical?

2.4
SPECIAL
FUNCTIONS
AND LINEAR
INEQUALITIES

We have seen that the graph of a linear function is a straight line. Linear functions can also be used to define other functions which, in themselves, are not linear but may be described as being "partly" or "piecewise" linear. An important example is the **absolute-value function.**

$$y = f(x) = |x| = \begin{cases} x & \text{if } x \geq 0 \\ -x & \text{if } x < 0 \end{cases}$$

See Section 1.4 to review the meaning of absolute value.

To graph this function, first draw the line $y = -x$ and eliminate all those points on it for which x is positive. Then draw the the line $y = x$ and eliminate the part for which x is negative. Now join these two parts to get the graph of $y = |x|$ as shown on the following page.

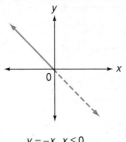

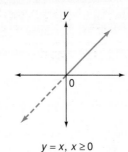

$y = -x,\ x \le 0$ $y = x,\ x \ge 0$ $y = |x|$

The graph of $y = |x|$ *consists of two perpendicular rays intersecting at the origin.* Now $y = |x|$ is not a linear function, but it is linear in parts; the two halves $y = x,\ y = -x$ are linear.

Note that the graph is symmetric about the y-axis. (If the paper were folded along the y-axis, the two parts would coincide.) This symmetry can be observed by noting that the y-values for x and $-x$ are the same. That is,

$$|x| = |-x| \qquad \text{for all } x$$

EXAMPLE 1   Graph: $y = |x - 2|$

Solution   For $x \ge 2$, we find that $x - 2 \ge 0$, which implies that $y = |x - 2| = x - 2$. This is the ray through (and to the right of) the point $(2, 0)$, with slope 1. For $x < 2$, we get $x - 2 < 0$, which implies that $y = |x - 2| = -(x - 2) = -x + 2$. This gives the left half of the graph shown.

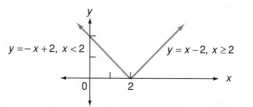

See Exercise 31 for an alternative way to draw this graph.

Just as $y = f(x) = |x|$ was defined in two parts, so can other functions be defined in several parts, as in Example 2.

EXAMPLE 2   Graph: $y = f(x) = \begin{cases} 2x & \text{if } 0 \le x \le 1 \\ -x + 2 & \text{if } 1 < x \end{cases}$

What are the domain and range of f?

Solution   The domain of f is all $x \ge 0$. From the graph we see that the range consists of all $y \le 2$.

Note: The open dot at $(1, 1)$ *means that this point is not part of the graph; but the point* $(1, 2)$ *is on the graph since* $f(1) = 2 \cdot 1 = 2$.

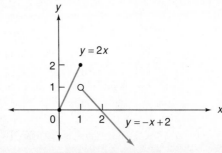

EXAMPLE 3 Graph the function f given by $y = f(x) = \dfrac{|x|}{x}$. What is the domain of f?

Solution The domain consists of all $x \neq 0$. When $x > 0$, $|x| = x$ and

$$f(x) = \frac{|x|}{x} = \frac{x}{x} = 1$$

Thus $f(x)$ is the constant 1, for all positive x. Similarly, for $x < 0$, $|x| = -x$ and

$$f(x) = \frac{|x|}{x} = \frac{-x}{x} = -1.$$

Example 3 is an illustration of a **step function.** Such a function may be described as a function whose graph consists of parts of horizontal lines. Here is another step function defined for $-2 \leq x < 4$. Note that each step is the graph of one of the six equations used to define f.

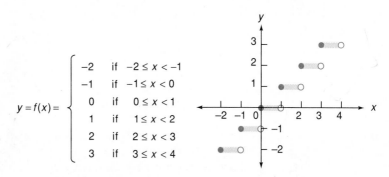

$$y = f(x) = \begin{cases} -2 & \text{if} \ -2 \leq x < -1 \\ -1 & \text{if} \ -1 \leq x < 0 \\ 0 & \text{if} \ \ \ 0 \leq x < 1 \\ 1 & \text{if} \ \ \ 1 \leq x < 2 \\ 2 & \text{if} \ \ \ 2 \leq x < 3 \\ 3 & \text{if} \ \ \ 3 \leq x < 4 \end{cases}$$

Observe that in each case, say $2 \leq x < 3$, the integer 2 at the left of the inequality is also the corresponding y-value for each x within this inequality. Putting it another way, we say that the y-value 2 is *the greatest integer less than or equal to x*.

For each number x there is an integer n such that $n \leq x < n + 1$. Therefore, the greatest integer less than or equal to x equals n. In other words, the preceding step function may be extended to a step function with domain of *all* real numbers; its graph would consist of an infinite number of steps.

[x] is the greatest integer not exceeding x itself.

We use the symbol **[x]** to mean *the greatest integer less than or equal to x.* Thus, the **greatest integer function** is given by

$$y = [x] = \text{greatest integer} \leq x$$

Here are a few illustrations:

$$[2\tfrac{1}{2}] = \ \ 2 \ \text{because 2 is the greatest integer} \leq 2\tfrac{1}{2}$$
$$[0.64] = \ \ 0 \ \text{because 0 is the greatest integer} \leq 0.64$$
$$[-2.8] = -3 \ \text{because} -3 \ \text{is the greatest integer} \leq -2.8$$
$$[-5] = -5 \ \text{because} -5 \ \text{is the greatest integer} \leq -5$$

TEST YOUR UNDERSTANDING
Think Carefully

(Answers: Page 132)

Evaluate:

1. [12.3] 2. [12.5] 3. [12.9] 4. [13]
5. $[-3\frac{3}{4}]$ 6. $[-3\frac{1}{4}]$ 7. [−3] 8. [0]
9. [−0.25] 10. [0.25] 11. [−0.75] 12. [0.75]

A linear equation such as $y = 2x - 1$ can also be used to identify two-dimensional regions in the plane. First observe that the graph of $y = 2x - 1$ divides the plane into two *half-planes*. These two half-planes represent the graphs for the two *linear inequalities*, $y < 2x - 1$ (below the line) and $y > 2x - 1$ (above the line). To show these graphs we would use a dashed line for $y = 2x - 1$ and shade the appropriate half-plane, as in the following figures.

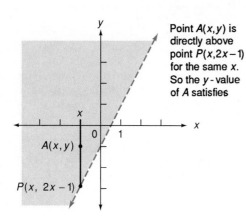

Point $A(x,y)$ is directly above point $P(x, 2x-1)$ for the same x. So the y-value of A satisfies

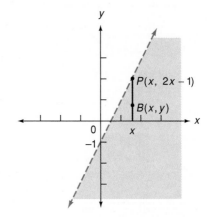

Point $B(x,y)$ is directly below point $P(x, 2x-1)$ for the same x. So the y-value of B satisfies $y < 2x - 1$.

In general, the graph of $y > mx + b$ consists of the points *above* the line $y = mx + b$. The points *below* the line show the graph of $y < mx + b$.

In order to identify the region satisfying an inequality like $3x - y < -2$, it is useful first to graph the corresponding line $3x - y = -2$. Then there are a number of ways to proceed. Here are two methods:

1. Solve the given inequality for y; that is, $y > 3x + 2$. Therefore, the region is above the line ℓ.

2. After graphing the line, pick any convenient point, such as $(-2, 2)$, that is *not* on the line and substitute into the given inequality $3x - y < -2$. This gives the *true statement* $-8 < -2$. Therefore, $(-2, 2)$ must be on the correct side of ℓ, which may now be indicated by the shading in the following graph.

To show the graph of $3x - y \le -2$ draw the straight line as a solid rather than as a dashed line. The graph would consist of the same shaded half-plane together with the solid line.

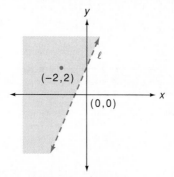

As a check, use a point on the other side, say $(0, 0)$. Substituting into the inequality $3x - y < -2$ gives the *false statement* $0 < -2$. Consequently, $(0, 0)$ is on the wrong side, and you may safely.conclude that $3x - y < -2$ works for the points on the other side.

EXAMPLE 4 Locate (shade) the region that satisfies the linear inequality $2x + 3y < 1$.

Solution   Here are two methods:

1.   Solve for y: $y < -\frac{2}{3}x + \frac{1}{3}$. Thus the required region is the half-plane below the line.

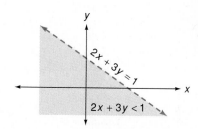

2.   Substitute $(0, 0)$ into $2x + 3y < 1$ to get $0 < 1$. Hence $(0, 0)$ is on the correct side, namely below the line $2x + 3y = 1$.   ■

EXERCISES 2.4

Evaluate.

1. $[99.1]$ 2. $[100.1]$ 3. $[-99.1]$ 4. $[-100.1]$
5. $[-\frac{7}{2}]$ 6. $[10^3]$ 7. $[\sqrt{2}]$ 8. $[\pi]$

Graph each function and state the domain and range.

9. $y = |x - 1|$ 10. $y = |x + 3|$ 11. $y = |2x|$
12. $y = |2x - 1|$ 13. $y = |3 - 2x|$ 14. $y = |\frac{1}{2}x + 4|$
15. $y = \begin{cases} 3x & \text{if } -1 \le x \le 1 \\ -x & \text{if } 1 < x \end{cases}$

16. $y = \begin{cases} -2x + 3 & \text{if } x < 2 \\ x + 1 & \text{if } x > 2 \end{cases}$

17. $y = \begin{cases} x & \text{if } -2 < x \le 0 \\ 2x & \text{if } 0 < x \le 2 \\ -x + 3 & \text{if } 2 < x \le 3 \end{cases}$

18. $y = \begin{cases} x & \text{if } 0 \le x < 1 \\ x - 1 & \text{if } 1 \le x < 2 \\ x - 2 & \text{if } 2 \le x < 3 \\ x - 3 & \text{if } 3 \le x \le 4 \end{cases}$

Graph each step function for its given domain.

19. $y = \dfrac{x}{|x|}$; all $x \ne 0$ 20. $y = \dfrac{|x - 2|}{x - 2}$; all $x \ne 2$ 21. $y = \dfrac{x + 3}{|x + 3|}$; all $x \ne -3$

22. $y = \begin{cases} -1 & \text{if } -1 \le x \le 0 \\ 0 & \text{if } 0 < x \le 1 \\ 1 & \text{if } 1 < x \le 3 \\ 2 & \text{if } 2 < x \le 3 \end{cases}$ **23.** $y = [x]; \; -3 \le x \le 3$ **24.** $y = [2x]; \; 0 \le x \le 2$

25. $y = 2[x]; \; 0 \le x < 2$ **26.** $y = [3x]; \; -1 \le x \le 1$ **27.** $y = [x - 1]; \; -2 \le x \le 3$

28. The postage for mailing packages depends on the weight and destination. Let the rates for a certain destination be as follows:

x = weight (pounds)	y = postage (cost)
Under 1	$0.80
1 or more but under 2	$0.90
2 or more but under 3	$1.00
3 or more but under 4	$1.10
4 or more but under 5	$1.20
5 or more but under 6	$1.30
6 or more but under 7	$1.40
7 or more but under 8	$1.50
8 or more but under 9	$1.60
9 or more but under 10	$1.70

This table defines y to be a function of x. If we use P (for postage) we may write $y = P(x)$ for the cost in dollars for x pounds. The table above gives $P(x) = 1.20$ for $4 \le x < 5$. Graph this function on its domain $0 < x < 10$. To achieve clarity, you may want to use a larger unit along the vertical axis than on the x-axis.

29. A formula for $P(x)$ in Exercise 28 can be given in terms of the greatest integer function. Find such a formula.

30. (a) The following information shows the first class letter rates in a recent year for pieces not exceeding 6 ounces.

For pieces not exceeding x ounces	The rate $R(x)$ is
1	0.25
2	0.45
3	0.65
4	0.85
5	1.05
6	1.25

Draw the graph of $y = R(x)$.

(b) Find a formula for the rate $R(x)$ that makes use of the greatest integer function. (*Hint:* For $0 < x \le 6$, consider $[-x]$.)

31. The graph of $y = |x - 2|$ was found in Example 1. Here is an alternative procedure. First graph $y = x - 2$ using a dashed line for the part below the x-axis. Now reflect the negative part (the dashed part) through the x-axis to get the final graph of $y = |x - 2|$.

32. Follow the procedure in Exercise 31 to graph $y = |2x + 5|$.

Write the linear inequality whose graph is the shaded region.

33.

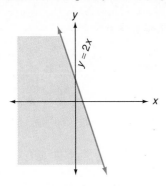

34.

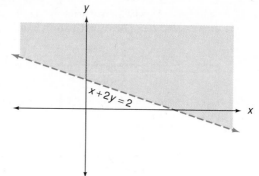

35.

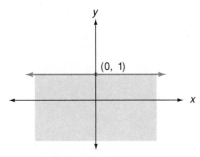

36.

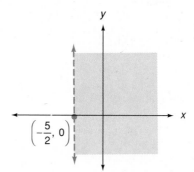

Graph each linear inequality.

37. $y > x + 2$ **38.** $y \leq x + 2$ **39.** $y \geq x - 1$ **40.** $y < x - 1$ **41.** $y > x$

42. $y - 2x < 4$ **43.** $2x + y > 4$ **44.** $x - 2y + 2 \leq 0$ **45.** $2y + x - 4 < 0$ **46.** $3y + x + 6 \geq 0$

 Written Assignment: Describe at least one example, other than the postage function, of a step function that can be found in everyday life. Explain your answer.

CHALLENGE
Think Creatively

Graph $y = |[x]|$ and $y = [|x|]$. For what values of x are the two equal to each other?

2.5
SYSTEMS
OF LINEAR
EQUATIONS

Any two nonparallel lines in a plane intersect in exactly one point. One of our objectives will be to find the coordinates of this point using the equations of the lines. Here, for example, is a *system* of two linear equations in two variables and their graphs, drawn in the same coordinate system.

The coordinates of the point of intersection P could be estimated by careful inspection of the graph. However, to find the exact coordinates requires algebraic procedures. These coordinates are called the solution of the system of equations.

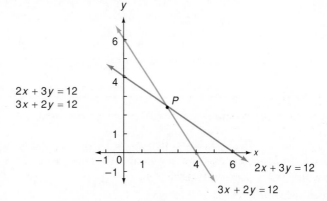

Two procedures will now be described to solve such systems. The underlying idea for each of them is that *the coordinates of the point of intersection must satisfy each equation*. The **substitution method** will be taken up first.

EXAMPLE 1   Solve the system by substitution.

$$2x + 3y = 12$$

$$3x + 2y = 12$$

Solution   We begin by letting (x, y) be the coordinates of the point of intersection of the preceding system. Then these x- and y-values fit both equations. Hence either equation may be solved for x or y and then subsituted into the other equation. For example, solve the second equation for y:

$$y = -\tfrac{3}{2}x + 6$$

Substitute this expression into the first equation and solve for x.

$$2x + 3y = 12$$

$$2x + 3(-\tfrac{3}{2}x + 6) = 12$$

$$2x - \tfrac{9}{2}x + 18 = 12$$

$$-\tfrac{5}{2}x = -6$$

$$x = (-\tfrac{2}{5})(-6) = \tfrac{12}{5}$$

To find the y-value, substitute $x = \tfrac{12}{5}$ into either of the given equations. It is easiest to use the equation that was written in y-form:

Check the ordered pair by substituting these values into each of the given equations for this system. Recall that we really cannot call this a solution until it has been checked.

$$y = -\tfrac{3}{2}x + 6$$

$$y = -\tfrac{3}{2}(\tfrac{12}{5}) + 6 = \tfrac{12}{5}$$

The solution is $(\tfrac{12}{5}, \tfrac{12}{5})$.   ∎

Note that all these methods have as their objective the elimination of one variable.

Now we will solve the same system as before, this time using the **multiplication-addition** (or **multiplication-subtraction**) **method**. The idea here is to alter the equations so that the coefficients of one of the variables are either negatives of one another or equal to each other. We may, for example, multiply the first equation by 3 and multiply the second by -2:

Original system	Resulting system
$2x + 3y = 12$	$6x + 9y = 36$
$3x + 2y = 12$	$-6x - 4y = -24$

Keep in mind that we are looking for the pair (x, y) that fits both equations. Thus, for these x and y values, we may *add equals to equals* in the resulting system to eliminate the variable x.

$$5y = 12$$

$$y = \tfrac{12}{5}$$

As before, x is now found by substituting $y = \tfrac{12}{5}$ into one of the given equations.

If in the preceding solution the second equation is multiplied by 2 instead of -2, the system becomes

$$6x + 9y = 36$$

$$6x + 4y = 24$$

Now x can be eliminated by *subtracting* the equations.

As illustrated in the next example, this method can be condensed into a compact procedure.

EXAMPLE 2   Solve the system by the multiplication-addition method.

$$\tfrac{1}{3}x - \tfrac{2}{5}y = 4$$

$$7x + 3y = 27$$

Solution

Multiply both sides of the first equation by 15, and both sides of the second equation by 2. Observe that $2(7x + 3y = 27)$ is an abbreviation for $2(7x + 3y) = 2(27)$.

$$15(\tfrac{1}{3}x - \tfrac{2}{5}y = 4) \Rightarrow 5x - 6y = 60$$

$$2(7x + 3y = 27) \Rightarrow \underline{14x + 6y = 54}$$

$$\text{Add: } 19x \quad\quad = 114$$

$$x \quad\quad = 6$$

Substitute to solve for y.

$$7x + 3y = 27 \Rightarrow 7(6) + 3y = 27$$

$$3y = -15$$

$$y = -5$$

Check: $\tfrac{1}{3}(6) - \tfrac{2}{5}(-5) = 4$; $7(6) + 3(-5) = 27$

The solution is the ordered pair $(6, -5)$.

■

TEST YOUR UNDERSTANDING
Think Carefully

(Answers: Page 132)

Use the substitution method to solve each linear system.

1. $y = 3x - 1$
  $y = -5x + 7$

2. $y = 4x + 16$
  $y = -\tfrac{2}{5}x + \tfrac{14}{5}$

3. $4x - 3y = 11$
  $y = 6x - 13$

4. $2x + 2y = \tfrac{4}{5}$
  $-7x + 2y = -1$

5. $x + 7y = 3$
  $5x + 12y = -8$

6. $4x - 2y = 40$
  $-3x + 3y = 45$

Use the multiplication-addition (or subtraction) method to solve each linear system.

7. $3x + 4y = 5$
  $5x + 6y = 7$

8. $-8x + 5y = -19$
  $4x + 2y = -4$

9. $\tfrac{1}{7}x + \tfrac{5}{2}y = 2$
  $\tfrac{1}{2}x - 7y = -\tfrac{17}{4}$

When a linear system has a unique solution, as in the preceding examples, we say that the system is **consistent.** Graphically, this means that the lines intersect. There are two other possibilities, as demonstrated next.

An **inconsistent** system:

$$39x - 91y = -28$$
$$6x - 14y = 7$$

A **dependent** system:

$$y = -\tfrac{2}{3}x + 5$$
$$2x + 3y = 15$$

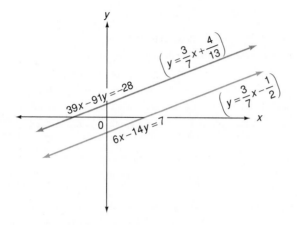

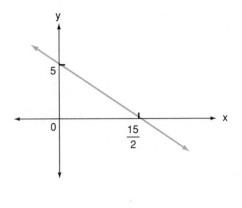

The two lines have the same slope, $\tfrac{3}{7}$, but different y-intercepts. There is no point of intersection. *An inconsistent system has no solution.*

The graph is a single line. Each ordered pair (x, y) that satisfies one equation must also satisfy the other. *A dependent system has an infinite number of solutions, that is, all points on the line.*

A dependent system can be identified algebraically when a statement that is always true (an identity) is obtained when trying to solve the system.

An inconsistent system can also be identified algebraically by obtaining a *false result* when trying to solve such a system. Thus for the preceding system, if the second equation is solved for y, we obtain

$$y = \tfrac{3}{7}x - \tfrac{1}{2}$$

Now substitute into the first equation.

$$39x - 91(\tfrac{3}{7}x - \tfrac{1}{2}) = -28$$
$$39x - 39x + \tfrac{91}{2} = -28$$
$$\tfrac{91}{2} = -28 \qquad \text{False}$$

*Arriving at a false result, as shown here, is not a wasted effort. As long as there are no computational errors, such a false conclusion tells us that the given system has no solution. That is, **the system has been solved by learning that it has no solution.***

This false result tells us that the system is inconsistent.

There is an important feature that the substitution method and the multiplication-addition (subtraction) method have in common. They all begin by *eliminating* one of the two variables. Thus you soon reach one equation in one unknown. This basic strategy of reducing the number of unknowns can be applied to "larger" linear systems, as demonstrated in Examples 3 and 4 for a system of three linear equations in three variables.

Geometrically, an equation of the form $ax + by + cz = d$ can be interpreted as a plane in space. For the following example, there are three planes in space that intersect at the unique common point $(-1, 3, 7)$.

EXAMPLE 3 Solve the system.

$$
\begin{aligned}
(1) \quad 2x - 5y + z &= -10 \\
(2) \quad x + 2y + 3z &= 26 \\
(3) \quad -3x - 4y + 2z &= 5
\end{aligned}
$$

Solution To solve this system we may begin by eliminating the variable x from the first two equations.

$$
\left.\begin{aligned}
2x - 5y + z &= -10 \\
-2(x + 2y + 3z &= 26)
\end{aligned}\right\} \Rightarrow
\begin{aligned}
2x - 5y + z &= -10 \\
-2x - 4y - 6z &= -52
\end{aligned}
$$

Add: $\quad -9y - 5z = -62$

or $\quad\quad 9y + 5z = 62$

Another equation in y and z can be obtained from Equations (2) and (3) by eliminating x:

$$
\left.\begin{aligned}
3(x + 2y + 3z &= 26) \\
-3x - 4y + 2z &= 5
\end{aligned}\right\} \Rightarrow
\begin{aligned}
3x + 6y + 9z &= 78 \\
-3x - 4y + 2z &= 5
\end{aligned}
$$

Three planes intersecting at a common point

Add: $\quad\quad 2y + 11z = 83$

Now we have this system in two variables:

$$
\begin{aligned}
9y + 5z &= 62 \\
2y + 11z &= 83
\end{aligned}
$$

You can solve this system as before to find $y = 3$ and $z = 7$. Then substitute these values into an earlier equation, say (2).

$$
\begin{aligned}
x + 2y + 3z &= 26 \\
x + 2(3) + 3(7) &= 26 \\
x &= -1
\end{aligned}
$$

The remaining equations can be used for checking:

$$
\begin{aligned}
(1) \quad 2(\quad) - 5(3) + 7 &= -10 \\
(3) \quad -3(\quad) - 4(3) + 2(7) &= 5
\end{aligned}
$$

The solution is $x = -1$, $y = 3$, $z = 7$, which may also be written as the *ordered triple* $(-1, 3, 7)$. ∎

EXAMPLE 4   Solve the system.

$$3x - 2y + z = 1$$
$$x - y - z = 2$$
$$6x - 4y + 2z = 3$$

Solution   Add the first two equations to eliminate z.

$$3x - 2y + z = 1$$
$$\underline{x - y - z = 2}$$
$$4x - 3y \quad\quad = 3$$

Now eliminate z from the second and third equations. To do so, multiply the second equation by 2 and add.

$$2x - 2y - 2z = 4$$
$$\underline{6x - 4y + 2z = 3}$$
$$8x - 6y \quad\quad = 7$$

We now have the following system of two equations in two variables to solve:

$$4x - 3y = 3$$
$$8x - 6y = 7$$

However, note what happens if you multiply the first of these equations by 2 and subtract:

$$8x - 6y = \quad 6$$
$$\underline{8x - 6y = \quad 7}$$
$$0 = -1$$

This conclusion, which is false, indicate that the original system of equations is inconsistent and has no solution. Geometrically, in this example, two of the three planes represented by the given equations are parallel. ■

EXERCISES 2.5

Solve each system by the substitution method.

1. $2x + y = -10$
   $6x - 3y = 6$

2. $-3x + 6y = 0$
   $4x + y = 9$

3. $v - w = 14$
   $3v + w = 2$

4. $4x - y = 6$
   $2x + 3y = 10$

5. $x + 5y = -9$
   $4x - 3y = -13$

6. $s + 2t = 5$
   $-3s + 10t = -7$

Solve each system by the multiplication-addition or subtraction method.

7. $-3x + y = 16$
   $2x - y = 10$

8. $2x + 4y = 24$
   $-3x + 5y = -25$

9. $3y - 9x = 30$
   $8x - 4y = 24$

10. $2u - 6v = -16$
    $5u - 3v = 8$

11. $4x - 5y = 3$
    $16x + 2y = 3$

12. $\frac{1}{2}x + 3y = 6$
    $-x - 8y = 18$

Solve each system.

13. $x - 2y = 3$
    $y - 3x = -14$

14. $2x + y = 6$
    $3x - 4y = 12$

15. $-3x + 8y = 16$
    $16x - 5y = 103$

16. $3x - 8y = -16$
    $7x + 19y = -188$

17. $16x - 5y = 103$
    $7x + 19y = -188$

18. $4x = 7y - 6$
    $9y = -12x + 12$

19. $3s + t - 3 = 0$
    $2s - 3t - 2 = 0$

20. $\frac{1}{2}x - \frac{1}{3}y = 2$
    $\frac{3}{4}x + \frac{2}{5}y = -1$

21. $\frac{1}{4}x + \frac{1}{3}y = \frac{5}{12}$
    $\frac{1}{2}x + y = 1$

22. $0.1x + 0.2y = 0.7$
    $0.01x - 0.01y = 0.04$

23. $\dfrac{x}{2} + \dfrac{y}{6} = \dfrac{1}{2}$
    $0.2x - 0.3y = 0.2$

24. $2(x + y) = 4 - 3y$
    $\frac{1}{2}x + y = \frac{1}{2}$

25. $2(x - y - 1) = 1 - 2x$
    $6(x - y) = 4 - 3(3y - x)$

26. $\dfrac{x - 2}{5} + \dfrac{y + 1}{10} = 1$
    $\dfrac{x + 2}{3} - \dfrac{y + 3}{2} = 4$

Decide whether the given systems are consistent, inconsistent, or dependent.

27. $x + y = 2$
    $x + y = 3$

28. $3x - y = 7$
    $-9x + 3y = -21$

29. $2x - 3y = 8$
    $-8x + 12y = 33$

30. $\frac{1}{2}x + 5y = -4$
    $7x - 3y = 17$

31. $-6x + 3y = 9$
    $10x + 5y = -1$

32. $20x + 36y = -27$
    $\frac{5}{27}x + \frac{1}{3}y = -\frac{1}{4}$

33. $4x - 12y = 3$
    $x + \frac{1}{3}y = 3$

34. $-7x + y = 2$
    $28x - 4y = 2$

35. $2x + 5y = -20$
    $x = -\frac{5}{2}y - 10$

36. $x - y = 3$
    $-\frac{1}{3}x + \frac{1}{3}y = 1$

37. $x - 5y = 15$
    $0.01x - 0.05y = 0.5$

38. $4y = 3x + 2$
    $2x = 3y - 3$

Solve each linear system.

39. $x + 2y + 3z = 5$
    $-4x + z = 6$
    $3x - y = -3$

40. $x + y + z = 2$
    $x - y + 3z = 12$
    $2x + 5y + 2z = -2$

41. $x + 2y + 3z = -4$
    $4x + 5y + 6z = -4$
    $7x - 15y - 9z = 4$

42. $-3x + 3y + z = -10$
    $4x + y + 5z = 2$
    $x - 8y - 2z = 12$

43. $x + 2y - z = 3$
    $2x - 3y + 3z = 0$
    $y - 2z = 6$

44. $2x + y = 5$
    $-3x + 2z = 7$
    $3y - 8z = 5$

45. $4x - 2y - z = 1$
    $2x + y + 2z = 9$
    $x - 3y - z = \frac{3}{2}$

46. $5x - y - 2z = 2$
    $3y + 2z = 5$
    $-5x + 4y + 4z = -3$

47. $x + y + z = 4$
    $3x - 2y + z = 6$
    $2x + 5y + 3z = 11$

48. $2x - 2y + z = -7$
    $3x + y + 2z = -2$
    $5x + 3y - 3z = -7$

49. $2x + 3y - z = 8$
    $x + y + z = 5$
    $-4x - 6y + 2z = 3$

50. $2x + 2y - 2z = 9$
    $3x - y + 4z = 0$
    $6x + 4y + 6z = 13$

51. $0.1x + 0.2y + 0.2z = 0.2$
    $0.3x + 0.5y + 0.1z = -0.1$
    $0.2x - 0.3y - 0.5z = 0.7$

52. $\frac{1}{2}x - \frac{1}{2}y - z = 11$
    $\frac{3}{4}x + \frac{2}{3}y - \frac{1}{4}z = -2$
    $\frac{3}{2}x + \frac{1}{3}y + 2z = 2$

Solve the systems for the common pair (x, y) in terms of c.

53. $-x + y = 6c$
$x + y = 3c$

54. $2x - 10c = -y$
$7x - 2y = 2c$

55. Points $(-8, -16)$, $(0, 10)$, and $(12, 14)$ are three vertices of a parallelogram. Find the coordinates of the fourth vertex if it is located in the third quadrant.

56. Find the point of intersection for the diagonals of the parallelogram given in Exercise 55.

57. A line with equation $ax + by = 3$ passes through $(6, 3)$ and $(-1, -1)$. Find a and b without finding the slope.

58. A line with equation $y = mx + b$ passes through the points $(-\frac{1}{3}, -6)$ and $(2, 1)$. Find m and b by substituting the coordinates into the equation and solving the resulting system.

59. Solve the system for the common pair (x, y).

$$ax + by = c$$
$$dx + ey = f$$

Assume a, b, c, d, e, and f are constants and $ae - bd \neq 0$.

2.6 APPLICATIONS OF LINEAR SYSTEMS

There are many word problems that can be solved by using linear systems. When solving such a problem you must first use the given statement to obtain a system of equations that can then be used to find the answer. You will find that the translation of the verbal form into the mathematical form is usually the most difficult part of the solution. Unfortunately, because of the variety of problems, as well as the numerous ways in which a problem can be stated, there are no fixed methods of translation that apply to all situations. However, the general guidelines given in Section 1.2 for solving word problems in one variable can be adjusted and used here.

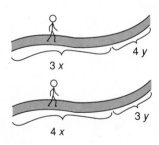

EXAMPLE 1   For her participation in a recent "walk-for-poverty," Ellen collected \$2 per mile for a total of \$52. She recorded that for a certain time she walked at the rate of 3 miles per hour (mph) and the rest at 4 mph. Afterward she mentioned that it was too bad that she did not have the energy to reverse the rates. For if she could have walked 4 mph for the same time that she actually walked 3 mph, and vice versa, she would have collected a total of \$60. How long did her walk take?

Solution   The problem asks for the total time of the walk. This time is broken into two parts: the time she walked at 3 mph and the time at 4 mph.

Let x = time at 3 mph

and y = time at 4 mph

We want to find $x + y$.

Since *distance = rate × time*, $3x$ is the distance at the 3-mph rate and $4y$ is the distance at 4 mph; the total distance is the sum $3x + 4y$. This total must equal 26 because she earned \$52 at \$2 per mile. Thus

$$3x + 4y = 26$$

The $60 she could have earned would have required walking 30 miles. And this 30 miles, she said, would have been possible by reversing the rates for the actual times that she did walk. This says that $4x + 3y = 30$. We now have the system

$$3x + 4y = 26$$
$$4x + 3y = 30$$

Notice that the answer in Example 1 was checked by returning to the language of the original problem rather than to any of the equations. It is essential to do this for word problems because the equations themselves may be incorrect translations of the problem.

The common solution is (6, 2), and therefore the total time for the walk is $6 + 2$, or 8 hours.

Check: She walked $3 \times 6 = 18$ miles at 3 mph, and $4 \times 2 = 8$ miles at 4 mph, for a total of 26 miles. At $2 per mile, she collected $52. Reversing the rates gives $(4 \times 6) + (3 \times 2) = 30$ miles, for which she would have earned $60. ∎

EXAMPLE 2 A grocer sells Brazilian coffee at $5.00 per pound and Colombian coffee at $8.50 per pound. How many pounds of each should he mix in order to have a blend of 50 pounds that he can sell at $7.10 per pound?

Solution

Let x = the number of pounds of Brazilian coffee to be used.

Let y = the number of pounds of Colombian coffee to be used.

It is frequently helpful, although not necessary, to summarize the given information in a table such as the following.

	Number of pounds	Cost per pound	Total cost
Brazilian	x	5.00	$5.00x$
Colombian	y	8.50	$8.50y$
Blend	50	7.10	50(7.10)

Now use the information in the table to obtain this system of equations:

$$x + \quad y = 50$$
$$5.00x + 8.50y = 355 \qquad 50(7.10) = 355$$

To check this solution show that 20 pounds of the $5.00 coffee and 30 pounds of the $8.50 coffee will give the same income as 50 pounds of a $7.10 blend.

Show that the solution for the system is $x = 20$ and $y = 30$. Thus the grocer should mix 20 pounds of the Brazilian coffee with 30 pounds of the Colombian coffee. This will give him 50 pounds of a blend to sell at $7.10 per pound. ∎

The next example demonstrates how a company can find the **break-even point** for a product that they produce and sell. This is the point at which the company's expenses equal its revenues. Beyond this point the company makes a profit, and below it the company has a loss.

EXAMPLE 3 It costs the Roller King Company $8 to produce one skateboard. In addition, there is a $200 daily fixed cost for building maintenance.

(a) Find the total daily cost for producing x skateboards per day.

(b) Find the total daily revenue if the company sells x skateboards per day for $16 each.

(c) Find the daily break-even point. That is, find the coordinates of the point at which the cost equals the revenue.

Solution

(a)
$$\begin{pmatrix} \text{Total daily} \\ \text{cost} \end{pmatrix} = \begin{pmatrix} \text{Cost per} \\ \text{item} \end{pmatrix} \cdot \begin{pmatrix} \text{Number} \\ \text{of items} \end{pmatrix} + \begin{matrix} \text{Fixed} \\ \text{cost} \end{matrix}$$

$$y \quad = \quad 8 \cdot x \quad + \ 200$$

(b)
$$\begin{pmatrix} \text{Total daily} \\ \text{revenue} \end{pmatrix} = \begin{pmatrix} \text{Sale price} \\ \text{per item} \end{pmatrix} \cdot \begin{pmatrix} \text{Number} \\ \text{of items} \end{pmatrix}$$

$$y \quad = \quad 16 \cdot x$$

(c) The daily profit or loss can be observed from the graphs of the cost and revenue lines drawn in the same coordinate system. The point where the lines intersect is the break-even point. At this point the cost for producing x skateboards is the same as the revenue for selling x skateboards on a daily basis. To find the break-even point, we solve this system.

$$\text{Cost:} \qquad y = 8x + 200$$

$$\text{Revenue:} \quad y = 16x$$

Thus, $16x = 8x + 200$, which gives $x = 25$ and $y = 16(25) = 400$. The break-even point is $(25, 400)$. When the company produces and sells 25 skateboards per day, their cost and revenue are $400 each.

For $x > 25$, the revenue is greater than the cost, resulting in a profit. For $x < 25$, the cost is more than the revenue, resulting in a loss.

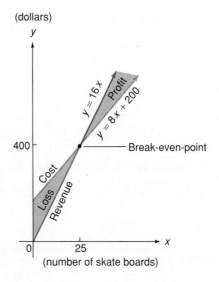

The final example is an application that calls for the solution of a system of three equations in three variables.

EXAMPLE 4   A veterinarian wants to control the diet of an animal so that on a monthly basis the animal consumes (besides hay, grass, and water) 60 pounds of oats, 75 pounds of corn, and 55 pounds of soybeans. The veterinarian has three feeds available, each consisting of oats, corn, and soybeans as shown in the table. How many pounds of each feed should be used to obtain the desired mix?

	Oats	Corn	Soybeans
1 lb of feed A	6 oz	5 oz	5 oz
1 lb of feed B	6 oz	6 oz	4 oz
1 lb of feed C	4 oz	7 oz	5 oz

Solution

$$\text{Let} \quad x = \text{pounds of feed } A$$
$$y = \text{pounds of feed } B$$
$$z = \text{pounds of feed } C$$

Then

$$6x = \text{ounces of oats in } x \text{ pounds of feed } A$$
$$6y = \text{ounces of oats in } y \text{ pounds of feed } B$$
$$4z = \text{ounces of oats in } z \text{ pounds of feed } C$$

The total number of pounds of oats required is 60. In ounces this is $60 \times 16 = 960$. Thus:

$$6x + 6y + 4z = 960$$

Analysis for the total ounces of corn and soybeans leads to this linear system:

$$
\begin{aligned}
6x + 6y + 4z &= 960 && \text{(total ounces of oats)} \\
5x + 6y + 7z &= 1200 && \text{(total ounces of corn)} \\
5x + 4y + 5z &= 880 && \text{(total ounces of soybeans)}
\end{aligned}
$$

To solve this system we may begin by eliminating x from the second and third equations, by subtraction, and obtain

$$2y + 2z = 320$$

or

$$y + z = 160$$

Eliminating x from the first two equations leads to this system in y and z:

$$
\begin{aligned}
y + z &= 160 \\
3y + 11z &= 1200
\end{aligned}
$$

Solving for y and z, we have $y = 70$ and $z = 90$. Then using the first equation, we get

$$6x + 6(70) + 4(90) = 960$$
$$6x = 180$$
$$x = 30$$

Therefore, 30 pounds of feed A, 70 pounds of feed B, and 90 pounds of feed C should be combined to obtain the desired mix. ∎

EXERCISES 2.6

Use a system of two linear equations in two variables to answer the following questions.

1. The total points that a basketball team scored was 96. If there were two-and-a-half times as many field goals as free throws, how many of each were there? (Field goals count 2 points; free throws count 1 point. There are no 3-point field goals.)

2. During a game, a golfer scored only fours and fives per hole. If he played 18 holes and his total score was 80, how many holes did he play in four strokes and how many in five?

3. The perimeter of a rectangle is 60 centimeters. If the length is three more than twice the width, find the dimensions.

4. Three times the larger of two numbers is 10 more than twice the smaller. Five times the smaller is 11 less than four times the larger. What are the numbers?

5. A shopper pays \$3.82 for $9\frac{1}{2}$ pounds of vegetables consisting of potatoes and string beans. If potatoes cost 35¢ per pound and string beans cost 68¢ per pound, how much of each was purchased?

6. A football team scored 54 points. They scored touchdowns at 6 points each, some points-after-touchdown at 1 point each, and some field goals at 3 points each. If there were two-and-one-third as many touchdowns as field goals, and just as many points-after-touchdown as field goals, how many of each were there?

7. The tuition fee at a college plus the room and board comes to \$8400 per year. The room and board is \$600 more than half the tuition. How much is the tuition, and what does the room and board cost?

8. A college student had a work-study scholarship that paid \$4.20 per hour. He also made \$2 per hour babysitting. His income one week was \$80 for a total of $23\frac{1}{2}$ hours of employment. How many hours did he spend on each of the two jobs?

9. There are only nickels and quarters in a child's bank. There are 22 coins in the bank having a total value of \$3.90. How many nickels and how many quarters are there?

10. The perimeter of a rectangle is 72 inches. The length is three-and-one-half times as large as the width. Find the dimensions.

11. The treasurer of the student body reported that the receipts for the last concert totaled \$916 and that 560 people attended. If students paid \$1.25 per ticket and nonstudents paid \$2.25 per ticket, how many students attended the concert?

12. The cost of 10 pounds of potatoes and 4 pounds of apples is \$6.16, while 4 pounds of potatoes and 8 pounds of apples cost \$6.88. What is the cost per pound of potatoes and of apples?

13. David walked 19 kilometers. The first part of the walk was at 5 kph (kilometers per hour) and the rest at 3 kph. He would have covered 2 kilometers less if he had reversed the rates, that is, if he had walked at 3 kph and at 5 kph for the same times that he actually walked at 5 kph and 3 kph, respectively. How long did it take him to walk the 19 kilometers?

14. Karin walked at a steady rate for a half-hour and then rode a bicycle for another half-hour, also at a constant rate. The total distance traveled was 7 miles. The next day, going at the same rates, she covered 6 miles, only she walked for two-thirds of an hour and rode for a third of an hour. What were her speeds walking and riding?

15. An airplane, flying with a tail wind, takes 2 hours and 40 minutes to travel 1120 miles. The return trip, against the wind, takes 2 hours and 48 minutes. What is the wind velocity and what is the speed of the plane in still air? (Assume that both velocities are constant; add the velocities for the downwind trip, and subtract them for the return trip.)

16. A swimmer going downstream takes 1 hour, 20 minutes to travel a certain distance. It takes the swimmer 4 hours to make the return trip against the current. If the river flows at the rate of $1\frac{1}{2}$ miles per hour, find the rate of the swimmer in still water and the distance traveled one way.

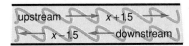

17. A store paid $299.50 for a recent mailing. Some of the letters cost 25¢ postage and the rest needed 45¢ postage. How many letters at each rate were mailed if the total number sent out was 910?

18. Ellen and Robert went to the store to buy some presents. They had a total of $22.80 to spend and came home with $6.20. If Ellen spent two-thirds of her money and Robert spent four-fifths of his money, how much did they each have to begin with?

19. The annual return on two investments totals $464. One investment gives 8% interest and the other $7\frac{1}{2}\%$. How much is invested at each rate if the total investment is $6000?

20. A wholesaler has two grades of oil that ordinarily sell for $1.14 per quart and 94¢ per quart. She wants a blend of the two oils to sell at $1.05 per quart. If she anticipates selling 400 quarts of the new blend, how much of each grade should she use? (One of the equations makes use of the fact that the total income will be 400(1.05), or $420.)

21. A student in a chemistry laboratory wants to form a 32-milliliter mixture of two solutions to contain 30% acid. Solution A contains 42% acid and solution B contains 18% acid. How many milliliters of each solution must be used? (Hint: Use the fact that the final mixture will have 0.30(32) = 9.6 milliliters of acid.)

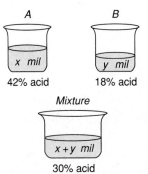

22. It costs $16.25 to send a 16-word telegram from city A to city B. A 21-word telegram costs $19. These charges consist of a flat rate for the first 10 words plus a fixed amount for each additional word. What is the cost for an additional word and how much is the flat rate?

23. A salesperson said that it did not matter whether one pair of shoes was sold for $31 or two pairs for $49, because the profit was the same for each sale. How much does one pair of shoes cost the salesperson and what is the profit?

24. A purchase of 6 dozen oranges and 10 pounds of peaches costs $10.60. If the price per dozen oranges increases 10%, and if peaches increase 5% per pound, the same order would cost $11.46. What is the initial cost of a pound of peaches and of a dozen oranges?

25. To go to work Juanita first averages 36 miles per hour driving her car to the train station and then rides the train, which averages 60 miles per hour. The entire trip takes 1 hour and 22 minutes. It costs her 15¢ per mile to drive the car and 6¢ per mile to ride the train. If the total cost is $5.22, find the distances traveled by car and by train.

26. Suppose that someone asked you to find the two numbers in the following puzzle:
   The larger of two numbers is 16 more than twice the smaller. The difference between $\frac{1}{4}$ of the larger and $\frac{1}{2}$ the smaller is 2.
   Why can you say that no answer is possible for this puzzle?

27. How many answers are there for the following puzzle?
   The difference between two numbers is 3. The larger number decreased by 1 is the same as $\frac{1}{3}$ the sum of the smaller plus twice the larger.

28. How many answers are there to the puzzle in Exercise 27 if the difference between the two numbers is 2?

29. A bag containing a mixture of 6 oranges and 12 tangerines sold for $2.34. A smaller bag containing 2 oranges and 4 tangerines sold for 77¢. An alert shopper asked the salesclerk if it was a better buy to purchase the larger bag. The clerk was not sure, but said that it really made no difference because the price of each package was based on the same unit price for each kind of fruit. Why was the clerk wrong?

30. If Sue gave Sam one of her dollars, then Sam would have half as many dollars as Sue. If Sam gave Sue one of his dollars, then Sue would have five times as many dollars as Sam. How many dollars did each of them have?

31. Refer to Example 3 of this section and determine the daily profit when 40 skateboards are produced and sold per day.

32. Refer to Example 3 of this section and determine the loss when 20 skateboards are produced and sold per day.

33. Refer to Example 3 of this section and determine how many skateboards must be made and sold per day to obtain a $520 profit.

34. Find the break-even point for each pair of cost and revenue equations. (Assume the equations represent dollars.)

Cost	Revenue
(a) $y = 15x + 450$	$y = 30x$
(b) $y = 240x + 1600$	$y = 400x$

35. The Electro Calculator Company can produce a calculator at a cost of $12. Their daily fixed cost is $720, and they plan to sell each calculator for $20. Graph the cost and revenue equations in the same coordinate system and find the break-even point.

36. What is the profit or loss for 200 calculators? For 300 calculators? (Refer to Exercise 35.)

37. How many calculators must be sold for there to be a $200 daily profit? (See Exercise 35.)

38. A bakery makes extra large chocolate chip cookies at a cost of $0.12 per cookie and sells them for $0.60 apiece. If the daily overhead expenses are $60, find the break-even point.

Write a system of three equations in three variables for each problem, and solve.

39. Daniel has $575 in one-dollar, five-dollar, and ten-dollar bills. Altogether he has 95 bills. The number of one-dollar bills plus the the number of ten-dollar bills is five more than twice the number of five-dollar bills. How many of each type of bill does he have?

40. The sum of three numbers is 33. The largest number is one less than twice the smallest number. Three times the smallest number is one less than the sum of the other two numbers. Find the three numbers.

41. The sum of the angles of a triangle is 180°. The largest angle is equal to the sum of the other two angles. Twice the smallest angle is 10° less than the largest angle. Find the measure for each angle.

42. The treasurer of a club invested $5000 of their savings into three different accounts, at annual yields of 8%, 9%, and 10%. The total interest earned for the year was $460. The amount earned by the 10% account was $20 more than that earned by the 9% account. How much was invested at each rate?

43. A grocer sells peanuts at $2.80 per pound, pecans at $4.50 per pound, and Brazil nuts at $5.40 per pound. He wants to make a mixture of 50 pounds of mixed nuts to sell at $4.44 per pound. The mixture is to contain as many pounds of Brazil nuts as the other two types combined. How many pounds of each type must he use in this mixture?

44. Answer the question in Example 4 of this section after making the following changes. The monthly consumption of oats, corn and soybeans is 45, 60, and 45 pounds, respectively. Also, the contents of the feeds are given in this table:

	Oats	Corn	Soybeans
1 lb of feed A	4 oz	6 oz	6 oz
1 lb of feed B	8 oz	4 oz	4 oz
1 lb of feed C	3 oz	8 oz	5 oz

45. A dietician wants to combine three foods so that the resulting mixture contains 900 units of vitamins, 750 units of minerals, and 350 units of fat. The units of vitamins, minerals, and fat contained in each gram of the three foods are shown in the table. How many grams of each food should be combined to obtain the required mixture?

	Vitamins	Minerals	Fat
1 gram of food A	35 units	15 units	10 units
1 gram of food B	10 units	20 units	10 units
1 gram of food C	20 units	15 units	5 units

EXPLORATIONS
Think Critically

1. Explain why the current cost of first-class postage is an example of a step function.
2. In terms of both their graphs and their solutions, describe the difference between an inconsistent system and a dependent system of two linear equations in two variables.
3. Describe the different ways in which three planes can intersect in space.
4. Under what conditions does the following represent an inconsistent system?

$$ax + by = c$$
$$dx + ey = f$$

5. An infinite set of points (x, y) are on the line $-3x + 7y = 6$, and another infinite set of points (x, y) are on the line $11x - 7y = 10$. Why then, when solving the system of the two equations, may we add to obtain $8x = 16$?

2.7 EXTRACTING FUNCTIONS FROM GEOMETRIC FIGURES

In this chapter you have learned how linear functions correspond to straight lines. This connection and many other interrelationships between algebra and geometry are vital to the study of mathematics. The purpose of this section is to learn how to "extract" algebraic functions from a variety of geometric situations, many of which make use of straight lines and linear functions. This process will prove to be of great value to you when you study applications of calculus.

A summary of some useful geometric formulas is given on the inside back cover of the text.

EXAMPLE 1 In the figure, right triangle ABE is similar to triangle ACD; $CD = 8$ and $BC = 10$; h and x are the measures of the altitude and base of triangle ABE. Express h as a function of x.

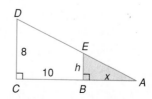

Solution Since corresponding sides of similar triangles are proportional, $\dfrac{BE}{AB} = \dfrac{CD}{AC}$. Substitute as follows:

$$\frac{h}{x} = \frac{8}{x + 10} \qquad AC = AB + BC = x + 10$$

$$h = \frac{8x}{x + 10} \qquad \text{Multiply by } x.$$

To emphasize that h is a function of x, use the functional notation to write the answer in this form:

$$h(x) = \frac{8x}{x + 10} \qquad\qquad \blacksquare$$

EXAMPLE 2 A water tank is in the shape of a right circular cone with altitude 30 feet and radius 8 feet. The tank is filled to a depth of h feet. Let x be the radius of the circle at the top of the water level. Solve for h in terms of x and use this to express the volume of water as a function of x.

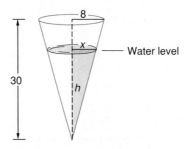

Solution The shaded right triangle is similar to the larger right triangle having base 8 and altitude 30. Therefore

$$\frac{h}{x} = \frac{30}{8}$$

$$h = \frac{15}{4}x \qquad \text{Solve for } h.$$

Now substitute for h in the formula for the volume of a right circular cone.

$$V = \tfrac{1}{3}\pi r^2 h \qquad\qquad \text{volume of a right circular cone}$$
$$V(x) = \tfrac{1}{3}\pi x^2\left(\tfrac{15}{4}x\right) \qquad \text{Substitute for } h \text{ and } r.$$
$$= \tfrac{5}{4}\pi x^3 \qquad\qquad\qquad \blacksquare$$

EXAMPLE 3 A window is in the shape of a rectangle with a semicircular top as shown. The perimeter of the window is 15 feet. Use r as the radius of the semicircle and express the area of the window as a function of r.

Solution The length of the semicircle is $\frac{1}{2}(2\pi r) = \pi r$. Now subtract this from 15 to get

$$15 - \pi r \quad \text{total length of the three straight sides}$$

The base of rectangle has length $2r$. Then

$$\frac{(15 - \pi r) - 2r}{2} \quad \begin{array}{l} \text{length of each} \\ \text{vertical side} \end{array}$$

Using A for the area of the window, we have

$$A(r) = \text{area of semicircle} + \text{area of rectangle}$$

$$= \tfrac{1}{2}\pi r^2 + 2r\left(\frac{15 - \pi r - 2r}{2}\right)$$

$$= \tfrac{1}{2}\pi r^2 + 15r - \pi r^2 - 2r^2$$

$$= 15r - \tfrac{1}{2}\pi r^2 - 2r^2 \qquad \blacksquare$$

EXAMPLE 4 A 50-inch piece of wire is to be cut into two parts AP and PB as shown. If part AP is to be used to form a square and PB is used to form a circle, express the total area enclosed by these figures as a function of x.

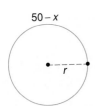

Solution Since the perimeter of the square is x, one of its sides is $\dfrac{x}{4}$. Then

$$\left(\frac{x}{4}\right)\left(\frac{x}{4}\right) = \frac{x^2}{16} \quad \text{area of square}$$

The circumference of the circle is $50 - x$, which can be used to find the radius of the circle.

$$2\pi r = 50 - x$$

$$r = \frac{50 - x}{2\pi}$$

Then

$$\pi r^2 = \pi\left(\frac{50 - x}{2\pi}\right)^2 \qquad \text{area of circle}$$

$$= \frac{(50 - x)^2}{4\pi}$$

Using F to represent the sum of the areas, we have

$$F(x) = \frac{x^2}{16} + \frac{(50 - x)^2}{4\pi} \qquad\qquad \blacksquare$$

EXERCISES 2.7

(Note: Geometric formulas needed for these exercises can be found on the inside back cover of the text.)

1. In the figure, the shaded right triangle, with altitude x, is similar to the larger triangle that has altitude h. Express h as a function of x.

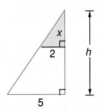

2. Use the result in Exercise 1 to express the area A of the larger triangle as a function of x.

3. In the figure, the shaded triangle is similar to triangle ABC. If $BC = 20$ and the altitude of triangle $ABC = 9$, express w as a function of the altitude h of the shaded triangle.

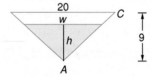

4. In Exercise 3, express the area of the shaded triangle as a function of w.

5. In the figure, s is the length of the shadow cast by a 6-foot person standing x feet from a light source that is 24 feet above the level ground. Express s as a function of x.

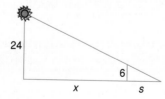

6. Triangle ABC is an isosceles right triangle with right angle at C. h is the measure of the perpendicular from C to side AB. Express the area of the triangle ABC as a function of h.

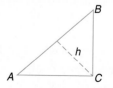

7. Express the area of rectangle *PQRS* as a function of $x = OP$.

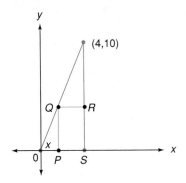

8. Express the area of triangle *OPQ* in Exercise 7 as a function of x.

9. A square piece of tin is 50 centimeters on a side. Congruent squares are cut from the four corners of the square so that when the sides are folded up a rectangular box (without a top) is formed. If the four congruent squares are x centimeters on a side, what is the volume of the box?

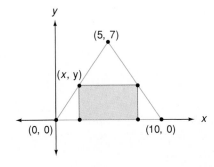

10. Replace the square piece of tin in Exercise 9 by a rectangular piece of tin with dimensions 30 centimeters by 60 centimeters and find the volume of the resulting box in terms of x.

11. An athletic field is semicircular at each end as shown. If the radius of each semicircle is r, and if the total perimeter of the field is 400 meters, express the area of the field in terms of r.

12. A closed tin can with height h and radius r has volume 5 cubic centimeters. Solve for h in terms of r, and express the surface area of the tin can as a function of r.

13. A closed rectangular-shaped box is x units wide and it is twice as long. Let h be the altitude of the box. If the total surface area of the box is 120 square units, express the volume of the box as a function of x. (*Hint:* First solve for h in terms of x.)

14. The vertices of the right triangle are $(0, y)$, $(0, 0)$, and $(x, 0)$. The hypotenuse passes through $(3, 1)$. Express the area of the triangle as a function of x.

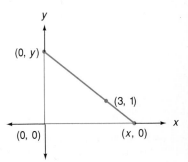

15. A shaded rectangle is inscribed in an isosceles triangle as shown. Express the area of the rectangle as a function of x.

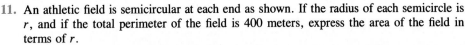

16. The figure represents a solid with a circular base having radius 2. A plane cutting the solid perpendicular to the xy-plane and to the x-axis, between -2 and 2, forms a cross section in the shape of a square. Express the area of the cross section in terms of the variable x. (*Hint:* The length of a side of the square is $2y$.)

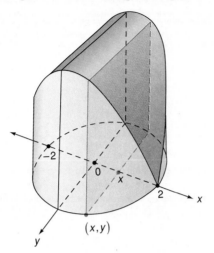

17. A rectangle is inscribed in a semicircle of radius 12 as shown. Express the area of the rectangle as a function of x.

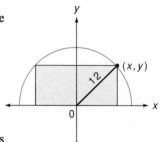

18. Triangle ABC is inscribed in a semicircle of radius 8 so that one of its sides coincides with a diameter. Express the area of the triangle as a function of $x = AC$.

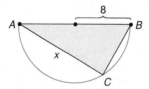

19. $ABCD$ is an isosceles trapezoid in which sides AB and DC are parallel. Express the area of the trapezoid as a function of altitude h.

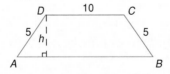

20. The figure shows a right circular cone in which r is the radius of the base, and the slant height is 10. Express the volume of the cone as a function of r.

REVIEW EXERCISES

The solutions to the following exercises can be found within the text of Chapter 2. Try to answer each question before referring to the text.

Section 2.1

1. State the definition of a function.

2. Explain why the equation $y = \dfrac{1}{\sqrt{x-1}}$ defines y to be a function of x.

3. y is a function of x defined by:

$$y = \begin{cases} 1 & \text{if } x \le -6 \\ x^2 & \text{if } -6 < x < 0 \\ 2x + 1 & \text{if } x \ge 0 \end{cases}$$

   Find the values of y for $x = -7$, for $x = -5$, and for $x = 5$.

4. Decide whether these two equations define y to be a function of x.

$$y = \begin{cases} 3x - 1 & \text{if } x \le 1 \\ 2x + 1 & \text{if } x \ge 1 \end{cases}$$

Find the domain.

5. $y = \dfrac{2x}{x^2 - 4}$ 6. $y = |x|$ 7. $y = \sqrt{x}$

8. $y = 6x^4 - 3x^2 + 7x + 1$ 9. $y = \dfrac{1}{x - 1}$

10. For $g(x) = \dfrac{1}{x}$, find:

    (a) $3g(x)$ (b) $g(3x)$ (c) $3 + g(x)$ (d) $g(3) + g(x)$ (e) $g(3 + x)$

11. Let $f(x) = -x^2 + 3x$ and find $f(x - 2)$.

12. For $g(x) = x^2$, evaluate and simplify the difference quotient $\dfrac{g(x) - g(4)}{x - 4}$.

13. For $g(x) = \dfrac{1}{x}$, evaluate and simplify the difference quotient $\dfrac{g(4 + h) - g(4)}{h}$.

Section 2.2

14. Graph the equation $y = x + 2$.
15. Find the x- and y-intercepts for $y = x + 2$.
16. Graph the linear equation $y = 2x - 1$ using the intercepts.
17. Graph $y = 2x - 1$ for $-2 \leq x \leq 1$.
18. State the definition of the slope of a line.
19. What is the slope of the line through the points (x_1, y_1) and (x_2, y_2)?
20. Find the slope of the line determined by the points $(-3, 4)$ and $(1, -6)$.
21. Graph the line with slope $\frac{3}{2}$ that passes through the point $(-2, -2)$.
22. What is the slope of a horizontal line? Of a vertical line?
23. What is the relationship between the slopes of two perpendicular lines?
24. Line ℓ_1 has slope $\frac{2}{3}$ and is perpendicular to line ℓ_2. If the lines intersect at $P(-1, 4)$, find the second coordinate of the point on ℓ_2 whose first coordinate is 1.
25. Suppose that for a certain calculator the weekly demand, D, is given by $y = D = 500 - 2x \; (0 \leq x \leq 25)$, where x is the price per calculator. Also, the weekly supply can be given by the formula $y = S = 10x + 200$. Graph both equations on the same set of axes and find the equilibrium point.

Section 2.3

26. Write the slope-intercept form of a line where m is the slope and b is the y-intercept.
27. Write the point-slope form of a line where m is the slope and (x_1, y_1) is a point on the line.
28. Graph the linear function f defined by $y = f(x) = 2x - 1$ by using the slope and y-intercept.
29. Find the equation of the line with slope $\frac{2}{3}$ passing through the point $(0, -5)$.
30. Describe the graph of a constant function.
31. Write the equation of the line through the point $(2, 3)$ that is:
    (a) Parallel to the x-axis   (b) Parallel to the y-axis
32. Write the point-slope form of a line with $m = 3$ that passes through the point $(-1, 1)$. Verify that $(-2, -2)$ is on the line.
33. Write the slope-intercept form of the line through the points $(6, -4)$ and $(-3, 8)$.
34. Write the equation of the line that is perpendicular to the line $5x - 2y = 2$ and that passes through the point $(-2, -6)$.
35. Find the slope and y-intercept of the line $-6x + 2y - 5 = 0$.
36. Find the equation of the line through the points $(2, -3)$ and $(3, -1)$. Write the equation:
    (a) In point-slope form   (b) In slope-intercept form   (c) In the form $Ax + By = C$

Section 2.4

37. Graph $y = f(x) = |x|$.
38. Graph $y = |x - 2|$.
39. Graph the function f defined by the following two equations.

$$y = f(x) = \begin{cases} 2x & \text{if } 0 \leq x \leq 1 \\ -x + 2 & \text{if } 1 < x \end{cases}$$

40. Graph the function f given by $y = f(x) = \dfrac{|x|}{x}$. What is the domain of f?
41. Draw the graph of the step function $y = [x]$ for $-2 \leq x < 4$.
42. Graph the inequality $2x + 3y < 1$.

Section 2.5

43. Solve and graph this system: $2x + 3y = 12$
$$3x + 2y = 12$$

Solve each system, and state whether it is consistent, inconsistent, or dependent.

44. $\frac{1}{3}x - \frac{2}{3}y = 4$
$7x + 3y = 27$

45. $39x - 91y = -28$
$6x - 14y = 7$

46. $y = -\frac{2}{3}x + 5$
$2x + 3y = 15$

Solve each system.

47. $2x - 5y + z = -10$
$x + 2y + 3z = 26$
$-3x - 4y + 2z = 5$

48. $3x - 2y + z = 1$
$x - y - z = 2$
$6x - 4y + 2z = 3$

Section 2.6

49. For her participation in a recent "walk-for-poverty," Ellen collected $2 per mile for a total of $52. She recorded that for a certain time she walked at the rate of 3 miles per hour (mph) and the rest at 4 mph. Afterward she mentioned that it was too bad that she did not have the energy to reverse the rates. For if she could have walked 4 mph for the same time that she actually walked 3 mph, and vice versa, she would have collected a total of $60. How long did her walk take?

50. A grocer sells Brazilian coffee at $5.00 per pound and Colombian coffee at $8.50 per pound. How many pounds of each should he mix in order to have a blend of 50 pounds that he can sell at $7.10 per pound?

51. It costs the Roller King Company $8 to produce one skateboard. In addition, there is a $200 daily fixed cost for building maintenance.

(a) Find the total daily cost for producing x skateboards per day.

(b) Find the total daily revenue if the company sells x skateboards per day for $16 each.

(c) Find the daily break-even point. That is, find the coordinates of the point at which the cost equals the revenue.

52. A veterinarian wants to control the diet of an animal so that on a monthly basis the animal consumes (besides hay, grass, and water) 60 pounds of oats, 75 pounds of corn, and 55 pounds of soybeans. The veterinarian has three feeds available, each consisting of oats, corn, and soybeans, as shown in the table. How many pounds of each feed should be used to obtain the desired mix?

	Oats	Corn	Soybeans
1 lb of feed A	6 oz	5 oz	5 oz
1 lb of feed B	6 oz	6 oz	4 oz
1 lb of feed C	4 oz	7 oz	5 oz

CHAPTER 2 TEST: STANDARD ANSWER

Use these equations to test your knowledge of the basic skills and concepts of Chapter 2. Then check your answers with those given at the back of the book.

1. Find the domain of the function given by $y = \dfrac{1}{\sqrt[3]{x^3 + 8}}$.

2. Let $g(x) = \dfrac{3}{x}$. Find **(a)** $g(2 + x)$ **(b)** $g(2) + g(x)$.

3. Let $g(x) = x^2 - x$. Evaluate and simplify: $\dfrac{g(x) - g(9)}{x - 9}$.

4. Find the x- and y-intercepts and use these to graph $3x - 2y = 6$.

5. Find the slope of the line determined by these points:
   (a) $(2, -3)$ and $(-1, 4)$ **(b)** $(-3, 2)$ and $(4, 2)$

6. Graph. State the domain and range: $y = 2x + 1$

7. Write the equation of the line through the point $(3, -2)$ that is:
   (a) Parallel to the y-axis   **(b)** Parallel to the x-axis

8. Write the equation of the line with slope $m = \frac{1}{2}$ and y-intercept $b = -3$.

9. Write the equation $2x - 3y = 5$ in slope-intercept form. Give the slope and y-intercept.

10. Write the slope-intercept form of the line through the points $(3, -5)$ and $(-2, 4)$.

11. Write the point-slope form of the line through the point $(2, 8)$ that is perpendicular to the line $y = -\frac{2}{5}x + 3$.

12. Graph: $y = |x + 2|$ 13. Graph the step function $y = \dfrac{|x + 1|}{x + 1}$.

14. Graph $y = 2 - x$ for $-1 \le x \le 2$.

Graph each inequality.

15. $y \ge x - 2$ 16. $2x + y > 6$

17. A square piece of tin is 40 centimeters on each side. Congruent squares of length x centimeters are cut from the four corners, and the sides are then folded up to form a rectangular box without a top. What is the volume of the box in terms of x?

Solve each system.

18. $\begin{aligned} 2x - y &= 7 \\ -3x + 2y &= -11 \end{aligned}$ 19. $\begin{aligned} 2x + 3y - z &= -1 \\ 3x - y + 2z &= 5 \\ -3x + 4y - 2z &= 1 \end{aligned}$

Decide whether each system is consistent, inconsistent, or dependent. If it is consistent, find the solution.

20. $\begin{aligned} 4x - 5y &= 12 \\ -2x + \tfrac{5}{2}y &= -6 \end{aligned}$ 21. $\begin{aligned} 7x + 4y &= 1 \\ -3x + 2y &= -19 \end{aligned}$ 22. $\begin{aligned} -6x + 4y &= -24 \\ 9x - 6y &= 14 \end{aligned}$

23. A Porsche and a Mercedes pass each other and continue traveling in opposite directions on a straight autobahn (there is no speed limit on the autobahn). The rate of the Mercedes is four-fifths the rate of the Porsche. If the cars are 60 kilometers apart after $\frac{1}{6}$ hour, find the rate of the Porsche in kilometers per hour.

24. It costs a company \$2.50 to produce a certain item that it sells for \$3.10. In addition to the production cost per item, the company has a daily overhead cost of \$450. Find the daily break-even point.

25. A nutritionist wants to combine three foods so that the resulting mixture will contain 550 units of ingrediant A, 300 units of ingredient B, and 350 units of ingredient C. The units of each ingredient per ounce of food are given in the table. How many ounces of each food should be combined to obtain the required mixture?

	A	B	C
1 oz of food I	25 units	20 units	15 units
1 oz of food II	35 units	15 units	25 units
1 oz of food III	40 units	20 units	20 units

1. The line parallel to the line $2x - 3y = 5$ that passes through the point $(-8, 4)$ has equation:

   (a) $y - 4 = -\frac{2}{3}(x + 8)$ **(b)** $y - 4 = \frac{2}{3}(x + 8)$ **(c)** $y - 4 = \frac{3}{2}(x + 8)$

   (d) $y - 4 = -\frac{3}{2}(x + 8)$ **(e)** None of the preceding

2. Which of the following statements are correct for the inequality $2x - 3y > 6$?

   I. The graph consists of all points below the line $y = \frac{2}{3}x - 2$.

   II. The graph consists of all points above the line $2x - 3y = 6$.

   III. The graph consists of all points on or below the line $3y - 2x = -6$.

   (a) Only I   **(b)** Only II   **(c)** Only III   **(d)** Only I and III   **(e)** None of the preceding

3. Let $g(x) = (x - 2)^2$, Then $\dfrac{g(x) - g(7)}{x - 7} =$

   (a) $x - 7$ **(b)** $x + 7$ **(c)** $x - 3$ **(d)** $x + 3$ **(e)** None of the preceding

4. Which of the following is the equation of the line that passes through the point $(2, -3)$ and is parallel to the y-axis?

   (a) $x = 2$ **(b)** $x = -2$ **(c)** $y = 3$ **(d)** $y = -3$ **(e)** None of the preceding

5. Which of the following statements are correct?

   I. The slope of a horizontal line is undefined?

   II. The slope of a vertical line is 0.

   III. The slopes of two perpendicular lines (not parallel to the coordinate axes) are negative reciprocals of one another.

   (a) Only I   **(b)** Only II   **(c)** Only III   **(d)** Only I and II   **(e)** None of the preceding

6. Which of the following is the equation of a line perpendicular to $2x - 3y = 6$?

   (a) $2x + 3y = 6$ **(b)** $3x + 2y = 6$ **(c)** $3x - 2y = 6$ **(d)** $3y - 2x = 6$ **(e)** None of the preceding

7. What is the slope-intercept form of the line through $(2, -3)$ and $(-1, 6)$?

   (a) $y = 3x + 3$ **(b)** $y = -3x + 3$ **(c)** $y = -\frac{1}{3}x + \frac{11}{3}$ **(d)** $y = -3x - 3$ **(e)** None of the preceding

8. The vertices of a right triangle are at $(0, 0)$, $(x, 0)$, and (x, y), with (x, y) in the first quadrant. If $(5, 2)$ is on the hypotenuse of the triangle, then the area of the triangle as a function of x is given by:

   (a) $A(x) = \frac{5}{2}x^2$ **(b)** $A(x) = \frac{5}{4}x^2$ **(c)** $A(x) = \frac{2}{5}x^2$ **(d)** $A(x) = \frac{1}{5}x^2$ **(e)** None of the preceding

9. Which of the following is the equation of a line with intercepts at $(0, -4)$ and $(2, 0)$?

   (a) $2x + 4y = 8$ **(b)** $4x + 2y = 8$ **(c)** $4x + 2y = -8$ **(d)** $2x - 4y = 8$ **(e)** None of the preceding

10. Consider the function defined below and find $f(-3) + f(1)$.

$$f(x) = \begin{cases} x - 2 & \text{if } -3 \le x \le 1 \\ -x + 1 & \text{if } 1 < x \end{cases}$$

   (a) -6 **(b)** 5   **(c)** 4   **(d)** -5 **(e)** None of the preceding

11. Consider the function defined as $y = f(x) = \dfrac{|x - 1|}{x - 1}$. What are the domain and range of this function?

   (a) Domain: all x; range: all y

   (b) Domain: all $x \neq 1$; range: $y = 1$ or $y = -1$

   (c) Domain: all $x \neq 1$; range: all y

   (d) Domain: all x; range: $y = 1$ or $y = -1$

   (e) None of the preceding

12. Given: $y = [|x|]$. For $-2 < x < -1$, $y =$

   (a) -2 **(b)** 2   **(c)** -1 **(d)** 1   **(e)** None of the preceding

13. Which of the following statements is true for the system shown at the right? $\quad 2x - \ y = 4$
$$x + 3y = 7$$

(a) The graph consists of two parallel lines.　　(b) The graph consists of a single line.

(c) The graph consists of two lines that intersect in the first quadrant.

(d) The graph consists of two lines that intersect in the fourth quadrant.

(e) None of the preceding.

14. Which of the following is one of the three numbers in the solution of this system? $\quad 2x - 3y + \ z = \ 11$
$$3x + \ y - \ z = \ 9$$
$$-x + 3y - 2z = -1$$

(a) -3　　(b) -2　　(c) 5　　(d) -8　　(e) None of the preceding

15. A toy manufacturer can produce a rag doll at a cost of $3 and can sell it for $9. The daily fixed cost for the production of these dolls is $150. What is the daily break-even point?

(a) (12.5, 112.5)　　(b) (12.5, 187.5)　　(c) (25, 125)　　(d) (25, 225)　　(e) None of the preceding

ANSWERS TO THE TEST YOUR UNDERSTANDING EXERCISES

Page 77

1. Function; all reals.　2. Function; all real $x \neq -2$.　3. Function; all reals.　4. Not a function.

5. Function; all real $x \neq -1$.　6. Not a function.　7. Function; all $x \neq 0$.　8. Function; $x \le 2$

Page 80

1. 18　2. 10　3. 0　4. $-\frac{5}{4}$　5. $\frac{7}{4}$　6. $a^2 - 3a$　7. $4x^2 - 6x$

8. $2x^2 - 6x$　9. $x^2 - 9x + 18$　10. $x^2 - 3x$　11. $\frac{1}{x^2} - \frac{3}{x}$　12. $\frac{1}{x^2 - 3x}$

Page 88

1. $\frac{6-5}{4-1} = \frac{1}{3}$　2. $\frac{3-(-5)}{-3-3} = \frac{8}{-6} = -\frac{4}{3}$　3. $\frac{1-(-3)}{-1-(-2)} = \frac{4}{1} = 4$　4. $\frac{1-0}{0-(-1)} = \frac{1}{1} = 1$

5.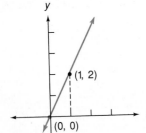

6.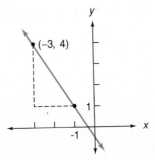

Page 96

1. $y = 2x - 2$　2. $y = -\frac{1}{2}x$　3. $y = \sqrt{2}x + 1$　4. $y - 6 = -3(x - 2)$

5. $y - 4 = \frac{1}{2}(x + 1)$　6. $y + \frac{2}{3} = 1(x - 5)$　7. $y = -\frac{1}{4}x$　8. $y + 5 = 0(x + 3)$

9. $y + 1 = -(x - 1)$　10. Exercise 8; $y = -5$ and the range is -5.

Page 104

1. 12　2. 12　3. 12　4. 13　5. -4　6. -4　7. -3　8. 0　9. -1

10. 0　11. -1　12. 0

Page 109

1. (1, 2)　2. $(-3, 4)$　3. $(2, -1)$　4. $(\frac{1}{5}, \frac{1}{5})$　5. $(-4, 1)$　6. (35, 50)

7. $(-1, 2)$　8. $(\frac{1}{2}, -3)$　9. $(\frac{3}{2}, \frac{5}{7})$

CHAPTER 3

QUADRATIC FUNCTIONS AND THE CONIC SECTIONS, WITH APPLICATIONS

3.1 GRAPHING QUADRATIC FUNCTIONS

A function defined by a polynomial expression of degree 2 is refered to as a **quadratic function** in x. Thus the following are all examples of quadratic functions in x:

$$f(x) = -3x^2 + 4x + 1 \qquad g(x) = 7x^2 - 4 \qquad h(x) = x^2$$

The most general form of such a quadratic function is

$$f(x) = ax^2 + bx + c$$

If $a = 0$, then the resulting polynomial no longer represents a quadratic function; $f(x) = bx + c$ is a linear function.

where a, b, and c represent constants, with $a \neq 0$.

The simplest quadratic function is given by $f(x) = x^2$. The graph of this quadratic function will serve as the basis for drawing the graph of any quadratic function $f(x) = ax^2 + bx + c$. We can save some labor by noting the *symmetry* that exists. For example, note the following:

$$f(-3) = f(3) = 9 \qquad f(-1) = f(1) = 1$$

In general, for this function,

$$f(-x) = (-x)^2 = x^2 = f(x)$$

Note: When $f(-x) = f(x)$, the graph is said to be *symmetric with respect to the y-axis*.

Greater accuracy can be obtained by using more points. But since we can never locate an infinite number of points, we must admit that there is a certain amount of faith involved in connecting the points as we did.

The following table of values gives several ordered pairs of numbers that are coordinates of points on the graph of $y = x^2$. When these points are located on a rectangular system and connected by a smooth curve, the graph of $y = f(x) = x^2$ is obtained. The curve is called a **parabola,** and every quadratic function $y = ax^2 + bx + c$ has a parabola as its graph. The domain of the function is the set of all real numbers. The range of the function depends on the constants a, b, and c. For the function $y = f(x) = x^2$, the range consists of all $y \geq 0$.

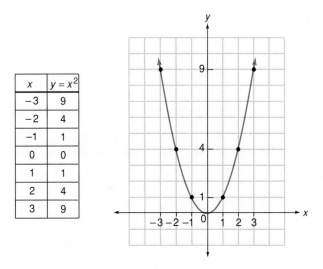

x	$y = x^2$
-3	9
-2	4
-1	1
0	0
1	1
2	4
3	9

A function symmetric about the y-axis is also called an **even function.**

An important feature of such a parabola is that it is symmetric about a vertical line called its **axis of symmetry.** The graph of $y = x^2$ is symmetric with respect to the y-axis. This symmetry is due to the fact that $(-x)^2 = x^2$.

The parabola has a *turning point,* called the **vertex,** which is located at the intersection of the parabola with its axis of symmetry. For the preceding graph the coordinates of the vertex are $(0, 0)$, and 0 is the *minimum value* of the function.

From the graph you can see that, reading from left to right, the curve is "falling" down to the origin and then is "rising." These features are technically described as f **decreasing** and f **increasing.**

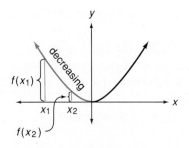

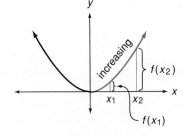

$f(x) = x^2$ is decreasing on $(-\infty, 0]$ because for *each* pair x_1, x_2 in this interval, if $x_1 < x_2$ then $f(x_1) > f(x_2)$

$f(x) = x^2$ is increasing on $[0, \infty)$ because for *each* pair x_1, x_2 in this interval, if $x_1 < x_2$ then $f(x_1) < f(x_2)$

The graph of $y = x^2$ can be used as a guide to draw the graphs of other quadratic functions. In the following illustrations, the graph of $y = x^2$ is shown as a dashed curve.

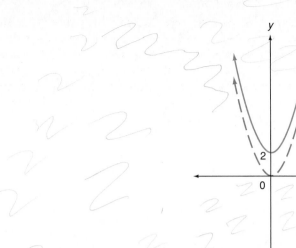

$$y = x^2 + 2$$

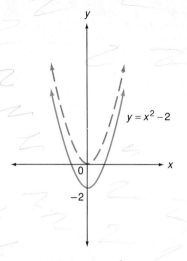

$$y = x^2 - 2$$

The graph of $y = x^2 + 2$ is *congruent* to that of $y = x^2$, but is shifted 2 units up.

The graph of $y = x^2 - 2$ is congruent to that of $y = x^2$, but is shifted 2 units down.

Next consider the graph of $y = (x + 2)^2$. In this case first add 2 to x, and then square. For $x = -2$, $y = 0$ and the vertex of the curve is at $(-2, 0)$. The graph is congruent to that for $y = x^2$, but is shifted 2 units to the left. In a similar fashion, note that the graph of $y = (x - 2)^2$ is shifted 2 units to the right. Verify the entries in the table of values given beside the following graphs.

Both of these parabolas are congruent to the basic parabola $y = x^2$. Each may be graphed by translating (shifting) the parabola $y = x^2$ by 2 units, to the right for $y = (x - 2)^2$ and to the left for $y = (x + 2)^2$.

$y = (x + 2)^2$

x	−5	−4	−3	−2	−1	0	1
y	9	4	1	0	1	4	9

The axis of symmetry is the line $x = -2$

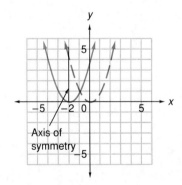

Axis of symmetry

$y = (x - 2)^2$

x	−1	0	1	2	3	4	5
y	9	4	1	0	1	4	9

The axis of symmetry is the line $x = 2$

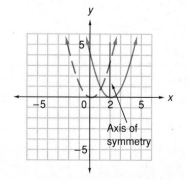

Axis of symmetry

The graph of $y = -x^2$ may also be obtained by multiplying each of the ordinates of $y = x^2$ by -1. This step has the effect of "flipping" the parabola $y = x^2$ downward, a reflection in the x-axis.

In each of the graphs drawn thus far, the coefficient of x^2 has been 1. If the coefficient is -1, it has the effect of reflecting the graph through the x-axis. The domain of the function is still the set of real numbers, but the range is the set of nonpositive real numbers. The vertex is at $(0, 0)$ and 0 is the *maximum value* of the function. Again, the graph of $y = x^2$ is shown as a dashed curve. Since the graph of $y = x^2$ bends "upward," we say that the curve is **concave up.** Also, since $y = -x^2$ bends "downward," we say that the curve is **concave down.**

$y = -x^2$

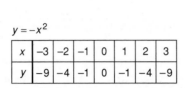

x	-3	-2	-1	0	1	2	3
y	-9	-4	-1	0	-1	-4	-9

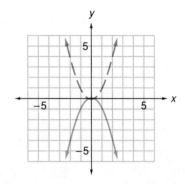

When the coefficient a in $y = ax^2$ is other than 1, then the graph of $y = ax^2$ can be obtained by multiplying each ordinate of $y = x^2$ by the number a, as in the following example.

EXAMPLE 1   Graph: **(a)** $y = \frac{1}{2}x^2$ **(b)** $y = 2x^2$

Solution

(a) $y = \frac{1}{2}x^2$ **(b)** $y = 2x^2$

Note that the graph of $y = 2x^2$ is "steeper" than that of $y = x^2$; the graph of $y = \frac{1}{2}x^2$ is not as steep as that of $y = x^2$.

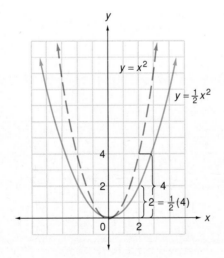

Each ordinate (*y*-value) is one-half that of the graph for $y = x^2$, shown as a dashed curve.

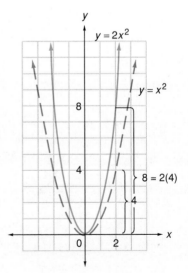

Each ordinate is twice that of the graph for $y = x^2$, shown as a dashed curve.

■

EXAMPLE 2 Graph $y = f(x) = -x^2 + 2$. State where the function is increasing or decreasing. What is the concavity?

Observe that to say "f is increasing for all $x \leq 0$" means the same as saying "f is increasing on $(-\infty, 0]$. Similarly, we say that f is decreasing on $[0, \infty)$.

Solution   Consider the graph of $y = -x^2$ and shift it up 2 units. The function f is increasing for all $x \leq 0$, and it is decreasing for all $x \geq 0$. The curve is concave down.

$y = -x^2 + 2$

x	-3	-2	-1	0	1	2	3
y	-7	-2	1	2	1	-2	-7

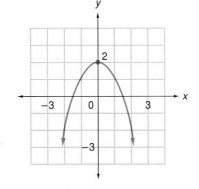

TEST YOUR UNDERSTANDING
Think Carefully

Match each graph with one of the given quadratic equations.

1.

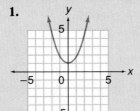

2.

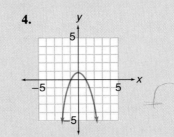

(a) $y = (x + 1)^2$
(b) $y = x^2 + 1$
(c) $y = (x - 1)^2$
(d) $y = x^2 - 1$
(e) $y = -(x - 1)^2$
(f) $y = -x^2 + 1$

3.

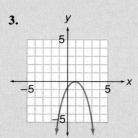

4.

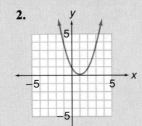

5.

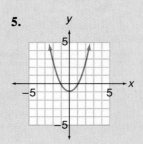

6.

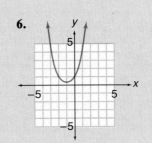

(Answers: Page 201)

Let us now put several ideas together and draw the graph of this function:

$$y = f(x) = (x + 2)^2 - 2$$

An effective way to do this is to begin with the graph of $y = x^2$ and shift the graph 2 units to the left for $y = (x + 2)^2$ and then 2 units down for the graph of $f(x) = (x + 2)^2 - 2$, as in the following figures.

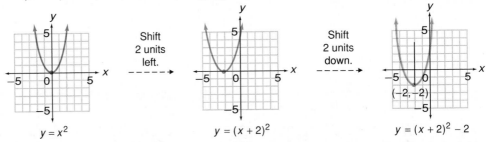

$y = x^2$ Shift 2 units left. $y = (x + 2)^2$ Shift 2 units down. $y = (x + 2)^2 - 2$

Note that the graph of $y = (x + 2)^2 - 2$ is congruent to the graph of $y = x^2$. The vertex of the curve is at $(-2, -2)$, and the axis of symmetry is the line $x = -2$. The minimum value of the function, -2, occurs at the vertex. Also observe that the domain consists of all numbers x, and the range consists of all numbers $y \geq -2$.

EXAMPLE 3   Graph $y = f(x) = -(x - 2)^2 + 1$, give the coordinates of the vertex, the equation of the axis of symmetry, and state the domain and range of f.

Solution   Consider the graph of $y = -x^2$, and shift this 2 units to the right and 1 unit up. The vertex is at $(2, 1)$, the highest point of the curve, and the axis of symmetry is the line $x = 2$. Also, the domain of f is the set of all real numbers and the range consists of all $y \leq 1$.

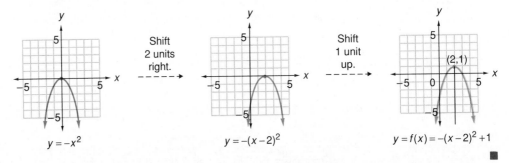

$y = -x^2$ Shift 2 units right. $y = -(x - 2)^2$ Shift 1 unit up. $y = f(x) = -(x - 2)^2 + 1$

We may summarize our results thus far as follows:

All such parabolas may be described as being vertical since their axis of symmetry is vertical. Vertical parabolas open either upward or downward.

The graph of $y = a(x - h)^2 + k$ is congruent to the graph of $y = ax^2$ but is shifted h units horizontally and k units vertically.
   (a) The horizontal shift is to the right if $h > 0$ and to the left if $h < 0$.
   (b) The vertical shift is upward if $k > 0$ and downard if $k < 0$.

The axis of symmetry is $x = h$, the vertex is (h, k), and k is either a maximum or a minimum value.
   (a) If $a < 0$, the parabola opens downward, and k is the maximum value.
   (b) If $a > 0$, the parabola opens upward, and k is the minimum value.

The function is increasing on $(-\infty, 3]$, *decreasing on* $[3, \infty)$, *and the curve is concave down.*

EXAMPLE 4 Graph the parabola $y = f(x) = -2(x - 3)^2 + 4$.

Solution The graph will be a parabola congruent to $y = -2x^2$, with vertex at $(3, 4)$ and with $x = 3$ as axis of symmetry. A brief table of values, together with the graph, is shown.

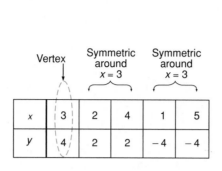

	Vertex	Symmetric around $x = 3$		Symmetric around $x = 3$	
x	3	2	4	1	5
y	4	2	2	−4	−4

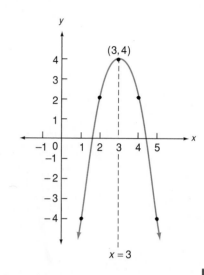

Interchanging the variables x and y in the equation $y = x^2$ produces $x = y^2$, whose graph is a horizontal parabola that opens to the right.

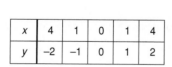

x	4	1	0	1	4
y	−2	−1	0	1	2

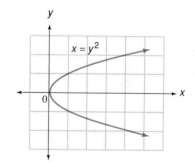

The graph of any quadratric equation of the form $x = ay^2 + by + c$ is a horizontal parabola that opens to the right when $a > 0$, or the left when $a < 0$. These parabolas can be graphed using procedures similar to those used for the vertical parabolas. For example, the parabola $x = y^2 + 3$ can be obtained by shifting the basic horizontal parabola 3 units to the right. The axis of symmetry is the x-axis.

EXAMPLE 5 Graph $x = (y + 2)^2 - 4$ and identify the vertex and axis of symmetry.

Solution Begin with the parabola $x = y^2$; shift 2 units downward and 4 units to the left as shown on the following page. The vertex is $(-4, -2)$ and $y = -2$ is the axis of symmetry.

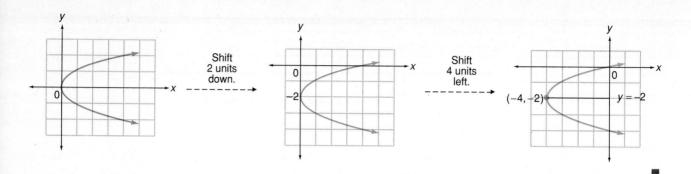

Shift 2 units down.

Shift 4 units left.

$y = -2$

$(-4, -2)$

Vertical Line Test for Functions

Observe that a horizontal parabola is *not* the graph of a function of x, since there are two y-values for a given x. This can also be observed using the following geometric test.

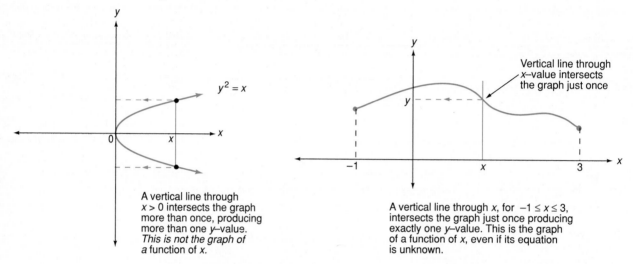

$y^2 = x$

A vertical line through $x > 0$ intersects the graph more than once, producing more than one y–value. *This is not the graph of a function of x.*

Vertical line through x–value intersects the graph just once

A vertical line through x, for $-1 \leq x \leq 3$, intersects the graph just once producing exactly one y–value. This is the graph of a function of x, even if its equation is unknown.

The vertical line test applied to a horizontal parabola shows that such a parabola cannot be the graph of a function of x. Thus the equations $x = y^2$ and $x = (y + 2)^2 - 4$ do *not* define functions of x. However, any equation in two variables is said to define a **relation.** Thus all functions are relations, but many relations are not functions.

Note that a vertical parabola divides the plane into two regions, one above the parabola and one below. For example, the graph of the **quadratic inequality in two variables** $y \geq x^2$ is shown below. It consists of all the points where $y = x^2$, as well as the points above the curve where $y > x^2$. (The unshaded region of the plane represents the graph of $y < x^2$.)

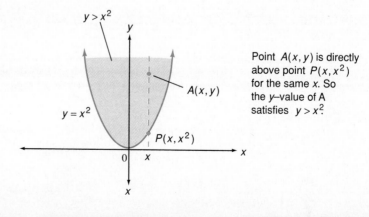

$y > x^2$

$y = x^2$

$A(x, y)$

$P(x, x^2)$

Point $A(x, y)$ is directly above point $P(x, x^2)$ for the same x. So the y–value of A satisfies $y > x^2$.

EXAMPLE 6 Graph the quadratic inequality $y + (x - 2)^2 \le 5$ by shading the appropriate region.

Solution First isolate y on one side of the inequality symbol.

$$y + (x - 2)^2 \le 5$$

$$y \le -(x - 2)^2 + 5$$

Then draw the related parabola $y = -(x - 2)^2 + 5$ and shade the region below the curve, as shown.

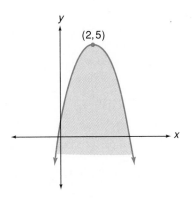

$y + (x - 2)^2 \le 5$

The graph consists of all points on or below the parabola. A solid curve is used to indicate that the points on the parabola are included.

EXERCISES 3.1

Draw each set of graphs on the same axes.

Plot pts.

1. **(a)** $y = x^2$ **(b)** $y = (x - 1)^2$ **(c)** $y = (x - 1)^2 + 3$
2. **(a)** $y = x^2$ **(b)** $y = (x + 1)^2$ **(c)** $y = (x + 1)^2 - 3$
3. **(a)** $y = -x^2$ **(b)** $y = -(x - 1)^2$ **(c)** $y = -(x - 1)^2 + 3$
4. **(a)** $y = -x^2$ **(b)** $y = -(x + 1)^2$ **(c)** $y = -(x + 1)^2 - 3$
5. **(a)** $y = x^2$ **(b)** $y = 2x^2$ **(c)** $y = 3x^2$
6. **(a)** $y = -x^2$ **(b)** $y = -\frac{1}{2}x^2$ **(c)** $y = -\frac{1}{2}x^2 + 1$

Draw the graph of each function.

7. $y = x^2 + 3$ 8. $y = (x + 3)^2$ 9. $y = -x^2 + 3$ 10. $y = -(x + 3)^2$
11. $y = 3x^2$ 12. $y = 3x^2 + 1$ 13. $y = \frac{1}{4}x^2$ 14. $y = \frac{1}{4}x^2 - 1$
15. $y = \frac{1}{4}x^2 + 1$ 16. $y = -2x^2$ 17. $y = -2x^2 + 2$ 18. $y = -2x^2 - 2$

Graph each of the following functions. Where is the function increasing and decreasing? What is the concavity?

19. $f(x) = (x - 1)^2 + 2$ 20. $f(x) = (x + 1)^2 - 2$ 21. $f(x) = -(x + 1)^2 + 2$
22. $f(x) = -(x + 1)^2 - 2$ 23. $f(x) = 2(x - 3)^2 - 1$ 24. $f(x) = 2(x + \frac{5}{4})^2 + \frac{5}{4}$

State (a) the coordinates of the vertex, (b) the equation of the axis of symmetry, (c) the domain, and (d) the range for each of the following functions.

25. $y = f(x) = (x - 3)^2 + 5$ 26. $y = f(x) = (x + 3)^2 - 5$ 27. $y = f(x) = -(x - 3)^2 + 5$
28. $y = f(x) = -(x + 3)^2 - 5$ 29. $y = f(x) = 2(x + 1)^2 - 3$ 30. $y = f(x) = \frac{1}{2}(x - 4)^2 + 1$

31. $y = f(x) = -2(x - 1)^2 + 2$ **32.** $y = f(x) = -\frac{1}{2}(x + 2)^2 - 3$ **33.** $y = f(x) = \frac{1}{4}(x + 2)^2 - 4$
34. $y = f(x) = 3(x - \frac{3}{4})^2 + \frac{4}{5}$

35. Graph the function f where

$$f(x) = \begin{cases} x^2 & \text{if } -2 \le x < 1 \\ x & \text{if } 1 < x \le 3 \end{cases}$$

36. Graph f where

$$f(x) = \begin{cases} x^2 - 9 & \text{if } -2 \le x < 4 \\ -3x + 15 & \text{if } 4 \le x < 6 \\ 3 & \text{if } x = 6 \end{cases}$$

37. Consider the equation $x = y^2$. Why is y not a function of x?

38. Compare the graphs of $y = x^2$ and $y = |x|$. In what ways are they alike?

39. What is the relationship between the graph of $y = x^2 - 4$ on a plane and of $x^2 - 4 > 0$ on a line?

40. Repeat Exercise 39 for $y = x^2 - 9$ and for $x^2 - 9 < 0$.

41. The graph of $y = ax^2$ passes through the point $(1, -2)$. Find a.

42. The graph of $y = ax^2 + c$ has its vertex at $(0, 4)$ and passes through the point $(3, -5)$. Find the values for a and c.

43. Find the value for k so that the graph of $y = (x - 2)^2 + k$ will pass through the point $(5, 12)$.

44. Find the value for h so that the graph of $y = (x - h)^2 + 5$ will pass through the point $(3, 6)$.

Graph each horizontal parabola.

45. $x = y^2 - 4$ **46.** $x = -y^2 + 3$ **47.** $x = (y - 3)^2$
48. $x = (y + 1)^2$ **49.** $x = 2y^2$ **50.** $x = -\frac{1}{2}y^2$

Draw each set of parabolas on the same axes.

51. (a) $x = y^2$ **(b)** $x = (y - 1)^2$ **(c)** $x = (y - 1)^2 - 5$
52. (a) $x = -y^2$ **(b)** $x = -(y + 3)^2$ **(c)** $x = -(y + 3)^2 + 4$

Write the equation of the parabola labeled P, which is obtained from the dashed curve by shifting it horizontally and vertically.

53.

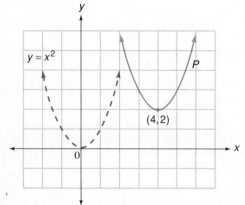

54.

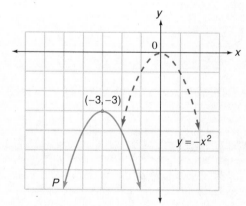

55.

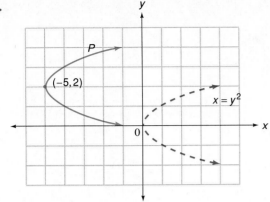

56.

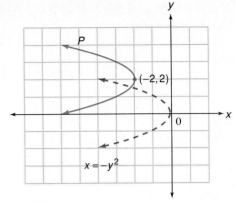

57.

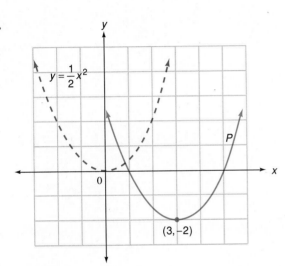

58.

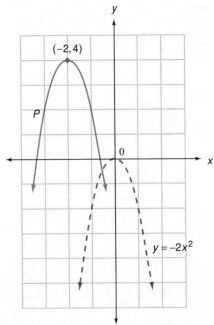

Graph each of the following inequalities.

59. $y \geq (x + 2)^2$

60. $y \leq (x + 2)^2$

61. $y \leq -x^2$

62. $y \leq -(x + 2)^2$

63. $y \geq x^2 - 4$

64. $y \geq -x^2 + 4$

65. $y < (x + 3)^2 + 4$

64. $y - (x - 4)^2 \geq -1$

67. $y + (x - 1)^2 > 3$

Use the vertical line test to decide if the given graph can be the graph of a function having the indicated set A as the domain.

68.

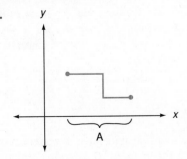

69.

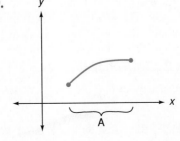

70.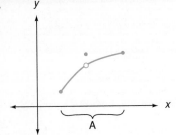

71.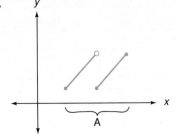

72.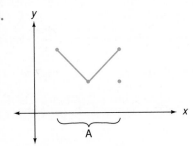

73.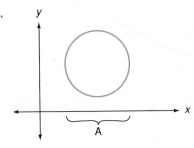

74.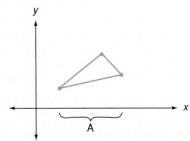

75.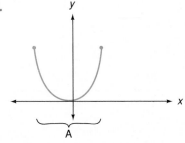

 Written Assignment: Use a specific example of your own and describe in words how to draw the graph of $y = (x - h)^2 + k$ from the graph of $y = x^2$.

Written Assignment: Explain, with specific examples, the distinction between a function and a relation.

**3.2
COMPLETING
THE SQUARE**

When a quadratic function is given in the form $f(x) = ax^2 + bx + c$, the properties of the graph are not evident. However, if this function is converted to the **standard form** $f(x) = a(x - h)^2 + k$, then the methods of Section 3.1 can be used to sketch the parabola. Our objective here is to learn how to make this algebraic conversion.

Let us begin with the quadratic function given by $y = x^2 + 4x + 3$. First rewrite the equation in this way:

$$y = (x^2 + 4x + \underline{\quad ? \quad}) + 3$$

Note that if the question mark is replaced by 4, then we will have a *perfect square trinomial* within the parentheses. However, since this changes the original equation, we must also subtract 4. The completed work looks like this:

$$y = x^2 + 4x + 3$$
$$= (x^2 + 4x + 4) + 3 - 4$$
$$= (x + 2)^2 - 1$$

From this last form you should recognize the graph to be a parabola with vertex at $(-2, -1)$, and $x = -2$ as axis of symmetry.

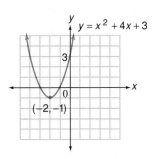

The technique that we have just used is called **completing the square.** Study these illustrations of perfect squares that have been completed. Note that in each case the coefficient of the x^2-term is 1.

The process of completing the square makes use of one of these two identities:

$$(x + h)^2 = x^2 + 2hx + h^2$$
$$(x - h)^2 = x^2 - 2hx + h^2$$

In the trinomials, h^2 is the square of one-half the coefficient of the x-term (without regard to sign). That is, the third term $= [\frac{1}{2}(2h)]^2 = h^2$.

$$x^2 + 8x + \ ? \longrightarrow x^2 + 8x + 16 = (x + 4)^2$$
$$\left[\frac{1}{2}(8)\right]^2 = 4^2 = 16$$

$$x^2 - 3x + \ ? \longrightarrow x^2 - 3x + \frac{9}{4} = \left(x - \frac{3}{2}\right)^2$$
$$\left[\frac{1}{2}(-3)\right]^2 = \left(-\frac{3}{2}\right)^2 = \frac{9}{4}$$

$$x^2 + \frac{b}{a}x + \ ? \longrightarrow x^2 + \frac{b}{a}x + \frac{b^2}{4a^2} = \left(x + \frac{b}{2a}\right)^2$$
$$\left[\frac{1}{2}\left(\frac{b}{a}\right)\right]^2 = \left(\frac{b}{2a}\right)^2 = \frac{b^2}{4a^2}$$

The process of completing the square can be extended to the case where the coefficient of x^2 is a number different from 1, such as $y = 2x^2 - 12x + 11$. The first step is to factor the coefficient of x^2 from the two variable terms only.

$$y = 2x^2 - 12x + 11 = 2(x^2 - 6x + \underline{\ \ ? \ \ }) + 11$$

Next, add 9 within the parentheses to form the perfect square $x^2 - 6x + 9 = (x - 3)^2$. However, because of the coefficient in front of the parentheses, we are really adding $2 \times 9 = 18$; thus 18 must also be subtracted.

$$y = 2(x^2 - 6x + 9) + 11 - 18 = 2(x - 3)^2 - 7$$

This is a parabola that opens upward with vertex as $(3, -7)$; $x = 3$ is the equation of the axis of symmetry.

Example 1 illustrates the procedure for completing the square when the coefficient of x^2 is a negative number or a fraction.

EXAMPLE 1   Convert the equation $y = -\frac{1}{3}x^2 - 2x + 1$ into the standard form $y = a(x - h)^2 + k$.

Solution   First factor $-\frac{1}{3}$ from the first two terms only.

$$y = -\tfrac{1}{3}x^2 - 2x + 1 = -\tfrac{1}{3}(x^2 + 6x) + 1$$

Next add 9 inside the parentheses to form the perfect square $x^2 + 6x + 9 = (x + 3)^2$. Because of the coefficient in front of the parentheses, however, we will really be adding $-\frac{1}{3}(9) = -3$. Thus 3 must also be added outside the parentheses.

$$y = -\tfrac{1}{3}x^2 - 2x + 1 = -\tfrac{1}{3}(x^2 + 6x + 9) + 1 + 3$$
$$= -\tfrac{1}{3}(x + 3)^2 + 4$$

The graph of the function $y = -\frac{1}{3}x^2 - 2x + 1$ is a parabola that opens downward with vertex at $(-3, 4)$ and with $x = -3$ as axis of symmetry.

To match the general form $y = a(x - h)^2 + k$, the answer may be written as

$$y = -\tfrac{1}{3}[x - (-3)]^2 + 4$$

TEST YOUR UNDERSTANDING

Think Carefully

Complete so as to express y as a perfect square trinomial.

1. $y = x^2 + 8x + \underline{}$ **2.** $y = x^2 + 10x + \underline{}$ **3.** $y = x^2 - 6x + \underline{}$

4. $y = x^2 - 12x + \underline{}$ **5.** $y = x^2 + 3x + \underline{}$ **6.** $y = x^2 - 5x + \underline{}$

Write in standard form: $y = a(x - h)^2 + k$.

7. $y = x^2 + 4x - 3$ **8.** $y = x^2 - 6x + 7$ **9.** $y = x^2 - 2x + 9$

(Answers: Page 201)   **10.** $y = 2x^2 + 8x - 1$ **11.** $y = -x^2 + x - 2$ **12.** $y = \frac{1}{2}x^2 - 3x + 2$

We have seen that any quadratic equation $y = ax^2 + bx + c$ may be written in the form $y = a(x - h)^2 + k$. From this form we can identify the vertex, (h, k), the axis of symmetry, $x = h$, and other information to help us graph the parabola, as illustrated in the following examples.

EXAMPLE 2   Graph the function $y = x^2 - 4x + 4$ and find all its intercepts, the vertex, and the axis of symmetry.

Solution

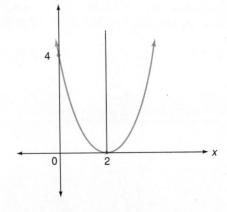

The x–axis is *tangent* to the graph at x = 2, the x–intercept. When x = 0, y = 4, the y–intercept.

The standard form of this equation is $y = (x - 2)^2$. This is the equation of a parabola that opens upward with vertex at $(2, 0)$, and $x = 2$ is the equation of the axis of symmetry. ∎

△ **EXAMPLE 3**   Write $y = -2 - 4x + 3x^2$ in the form $y = a(x - h)^2 + k$. Find the vertex, axis of symmetry, and graph. On which interval is the function increasing or decreasing? What is the concavity?

Solution   First write the given equation in the form $y = 3x^2 - 4x - 2$. Then complete the square.

*Note: Since the original equation is in the form $y = ax^2 + bx + c$, it is very easy to find the y-intercept by letting $x = 0$. This gives the point $(0, -2)$. Since this point is $\frac{2}{3}$ unit to the left of the axis of symmetry $x = \frac{2}{3}$, we quickly find the **symmetric point** $(\frac{4}{3}, -2)$.*

$$y = 3x^2 - 4x - 2$$
$$= 3(x^2 - \tfrac{4}{3}x) - 2$$
$$= 3(x^2 - \tfrac{4}{3}x + \tfrac{4}{9}) - 2 - \tfrac{4}{3}$$
$$= 3(x - \tfrac{2}{3})^2 - \tfrac{10}{3}$$

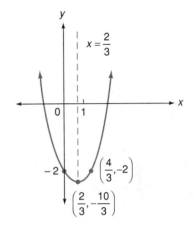

Vertex: $(\frac{2}{3}, -\frac{10}{3})$
Axis of symmetry: $x = \frac{2}{3}$

The function is decreasing on $(-\infty, \frac{2}{3}]$ and increasing on $[\frac{2}{3}, \infty)$, and the curve is concave up. ∎

EXAMPLE 4   Graph the function $y = f(x) = |x^2 - 4|$. Find the range.

Solution   First graph the parabola $y = x^2 - 4$. Then take the part of this curve that is below the x-axis (these are the points for which $x^2 - 4$ is negative) and reflect it through the x-axis.

The curve is decreasing on $(-\infty, -2]$ and on $[0, 2]$. It is increasing on $[-2, 0]$ and on $[2, \infty)$. The curve is concave up on $(-\infty, -2)$ and on $(2, \infty)$, and concave down on $(-2, 2)$.

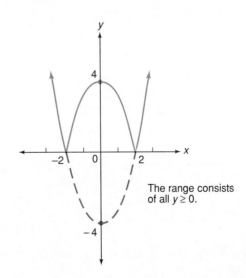

The range consists of all $y \geq 0$.

∎

EXAMPLE 5 State the conditions on the values a and k so that the parabola $y = a(x - h)^2 + k$ opens downward and intersects the x-axis in two points. What are the domain and range of this function?

Solution In order for the parabola to open downward we must have $a < 0$. If $k > 0$, the parabola will intersect the x-axis in two distinct points. The domain is the set of all real numbers and the range consists of $y \le k$. ∎

EXERCISES 3.2

Complete so as to express y as a perfect square trinomial.

1. $y = x^2 + 6x +$ _____
2. $y = x^2 - 8x +$ _____
3. $y = x^2 + 4x +$ _____
4. $y = x^2 - 20x +$ _____
5. $y = x^2 - 3x +$ _____
6. $y = x^2 + x +$ _____

Write in the standard form $y = a(x - h)^2 + k$.

7. $y = x^2 + 2x - 5$
8. $y = x^2 - 2x + 5$
9. $y = -x^2 - 6x + 2$
10. $y = x^2 - 3x + 4$
11. $y = -x^2 + 3x - 4$
12. $y = x^2 - 5x - 2$
13. $y = x^2 + 5x - 2$
14. $y = 2x^2 - 4x + 3$
15. $y = 2x^2 + 4x - 3$
16. $y = 5 - 6x + 3x^2$
17. $y = -5 + 6x + 3x^2$
18. $y = -3x^2 - 6x + 5$
19. $y = x^2 - \frac{1}{2}x + 1$
20. $y = -x^2 - \frac{1}{2}x + 1$
21. $y = \frac{3}{4}x^2 - x - \frac{1}{3}$
22. $y = -\frac{3}{4}x^2 + x - \frac{1}{3}$
23. $y = -5x^2 - 2x + \frac{4}{5}$
24. $y = ax^2 + bx + c$
25. $y = (x + 1)^2 - 3(x + 1) - \frac{3}{4}$
26. $y = ax^2 - 2ahx + ah^2 + k$
27. Compare Exercises 24 and 26 and express h and k in terms of a, b, and c.

Write each of the following in standard form. Identify the coordinates of the vertex, the equation of the axis of symmetry, the y-intercept, and check by graphing.

28. $y = x^2 + 2x - 1$
29. $y = x^2 - 4x + 7$
30. $y = -x^2 + 4x - 1$
31. $y = x^2 - 6x + 5$
32. $y = 3x^2 + 6x - 3$
33. $y = 2x^2 - 4x - 4$

34. By Exercise 27, $h = -\dfrac{b}{2a}$ is the first coordinate of the vertex of the parabola $y = ax^2 + bx + c$. Show that $(2h, c)$ is on the parabola and explain why this point is the reflection of $(0, c)$ through the axis of symmetry $x = h$.

Graph the indicated parabola using the three points shown.
(Note: $y = ax^2 + bx + c = a(x - h)^2 + k$.)

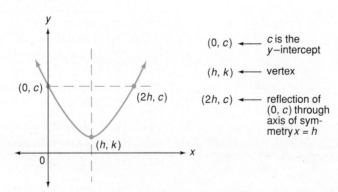

35. $y = x^2 - 3x + \frac{9}{4}$
36. $y = -x^2 + 2x$
37. $y = \frac{1}{2}x^2 + \frac{5}{2}x + 5$
38. $y = 3x^2 - 12x + \frac{29}{2}$
39. $y = -2x^2 - 6x - \frac{9}{2}$
40. $y = 3x^2 + 3x + \frac{3}{4}$

Graph the function f and decide where f is increasing or decreasing. Where is the curve concave up or down?

41. $f(x) = |9 - x^2|$

42. $f(x) = |x^2 - 1|$

43. $f(x) = |x^2 - x - 6|$

44. Graph the function f where

$$f(x) = \begin{cases} -x^2 + 4 & \text{if } -2 \leq x < 2 \\ x^2 - 10x + 21 & \text{if } \ 2 \leq x \leq 7 \end{cases}$$

45. Graph the function f where

$$f(x) = \begin{cases} 1 & \text{if } -3 \leq x < 0 \\ x^2 - 4x + 1 & \text{if } \ 0 \leq x < 5 \\ -2x + 16 & \text{if } \ 5 \leq x < 9 \end{cases}$$

 State the conditions on the values a and k so that the parabola $y = a(x - h)^2 + k$ has the properties indicated.

46. Concave down and has range $y \leq 0$

47. Concave up and has vertex at $(h, 0)$.

48. Concave up and does not intersect the x-axis.

49. Concave up and has range $y \geq 2$.

3.3 THE QUADRATIC FORMULA

The graph of a quadratic function may or may not intersect the x-axis. Here are some typical cases:

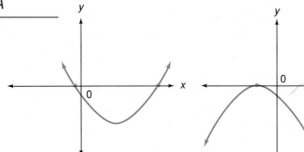

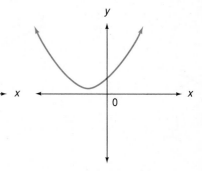

Two x–intercepts;
two solutions for
$y = ax^2 + bx + c = 0$

One x–intercept;
one solution for
$y = ax^2 + bx + c = 0$

No x–intercepts;
no solutions for
$y = ax^2 + bx + c = 0$

It is clear from these figures that if there are x-intercepts, then these values of x are the solutions to the equation $ax^2 + bx + c = 0$. If there are no x-intercepts, then the equation will not have any real solutions. In this section we learn procedures to handle all cases.

As a first example let us find the x-intercepts of the parabola $y = f(x) = x^2 - x - 6$. This calls for those values of x for which $y = 0$. That is, we need to solve the equation $y = f(x) = 0$ for x. This can be done by factoring:

$$x^2 - x - 6 = 0$$

$$(x + 2)(x - 3) = 0$$

To solve a quadratic equation by factoring, we make use of this fact: If $A \cdot B = 0$, then $A = 0$ or $B = 0$ or both $A = 0$ and $B = 0$.

Since the product of two factors is zero only when one or both of them are zero, it follows that

$$x + 2 = 0 \quad \text{or} \quad x - 3 = 0$$

$$x = -2 \quad \text{or} \quad x = 3$$

The x-intercepts are -2 and 3. The x-intercepts of the parabola are also called the *roots* of the equation $f(x) = 0$.

Observe that the x-intercepts are symmetrically spaced about the axis of symmetry, $x = \frac{1}{2}$. That is, -2 and 3 are each $2\frac{1}{2}$ units from the axis.

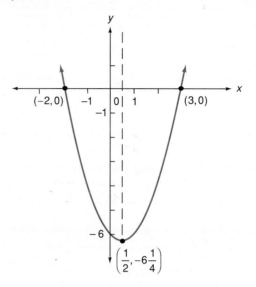

EXAMPLE 1   Find the x-intercepts of $f(x) = 25x^2 + 30x + 9$.

Solution   Let $f(x) = 0$:

$$25x^2 + 30x + 9 = 0$$
$$(5x + 3)(5x + 3) = 0$$
$$5x + 3 = 0$$
$$x = -\tfrac{3}{5}$$

Check: $25(\frac{9}{25}) + 30(-\frac{3}{5}) + 9 = 9 - 18 + 9 = 0$.

*The number $-\frac{3}{5}$ is referred to as a **double root** of $25x^2 + 30x + 9 = 0$.*

Note: Since there is only *one* answer to this quadratic equation, it follows that the parabola $y = 25x^2 + 30x + 9$ has only one x-intercept; the x-axis is *tangent* to the parabola at its vertex $(-\frac{3}{5}, 0)$. ∎

The x-intercepts of a function, if there are any, can be found by completing the square, as in the following example.

EXAMPLE 2   Find the x-intercepts of $f(x) = 2x^2 - 9x - 18$.

Try to explain each step in this solution.

Solution   Set $f(x) = 0$ and solve for x.

$$2x^2 - 9x - 18 = 0$$
$$2x^2 - 9x = 18$$
$$x^2 - \tfrac{9}{2}x = 9$$
$$x^2 - \tfrac{9}{2}x + \tfrac{81}{16} = 9 + \tfrac{81}{16}$$

$$\left(x - \tfrac{9}{4}\right)^2 = \tfrac{225}{16}$$

$$x - \tfrac{9}{4} = \pm \tfrac{15}{4}$$

$$x = \tfrac{9}{4} \pm \tfrac{15}{4}$$

$$x = 6 \quad \text{or} \quad x = -\tfrac{3}{2}$$

As another check, solve this quadratic equation by factoring.

Check: $2(6)^2 - 9(6) - 18 = 72 - 54 - 18 = 0$
$2(-\tfrac{3}{2})^2 - 9(-\tfrac{3}{2}) - 18 = \tfrac{9}{2} + \tfrac{27}{2} - 18 = 0$ ∎

The method of completing the square can always be used to solve quadratic equations of the form $ax^2 + bx + c = 0$. However, as you have seen in Example 2, this method soon becomes tedious. To avoid some of this labor we will now develop an efficient formula that can be used in all cases. This will be done by solving the general quadratic equation $ax^2 + bx + c = 0$ for x in terms of the constants a, b, and c.

We begin with the general quadratic equation.

$$ax^2 + bx + c = 0 \qquad a \neq 0$$

Add $-c$ to each side:

$$ax^2 + bx = -c$$

Divide each side by a $(a \neq 0)$.

$$x^2 + \frac{b}{a}x = -\frac{c}{a}$$

Add $\left[\frac{1}{2}\left(\frac{b}{a}\right)\right]^2 = \frac{b^2}{4a^2}$ to each side:

$$x^2 + \frac{b}{a}x + \frac{b^2}{4a^2} = \frac{b^2}{4a^2} - \frac{c}{a}$$

Factor on the left and combine on the right:

$$\left(x + \frac{b}{2a}\right)^2 = \frac{b^2 - 4ac}{4a^2}$$

Take the square root of each side and solve for x.

$$x + \frac{b}{2a} = \pm\sqrt{\frac{b^2 - 4ac}{4a^2}} \qquad \text{(If } t^2 = k, \text{ then } t = \pm\sqrt{k}.)$$

$$x + \frac{b}{2a} = \pm\frac{\sqrt{b^2 - 4ac}}{2a}$$

$$x = -\frac{b}{2a} \pm \frac{\sqrt{b^2 - 4ac}}{2a}$$

Combine terms to obtain the **quadratic formula.**

SECTION 3.3 The Quadratic Formula   151

$$\boxed{\begin{array}{c} \text{QUADRATIC FORMULA} \\[4pt] \text{If}\quad ax^2 + bx + c = 0,\ a \neq 0, \\[8pt] \text{then}\quad x = \dfrac{-b \pm \sqrt{b^2 - 4ac}}{2a} \end{array}}$$

The values $x = \dfrac{-b \pm \sqrt{b^2 - 4ac}}{2a}$ and $x = \dfrac{-b - \sqrt{b^2 - 4ac}}{2a}$ are the **roots** of the quadratic equation. These values are also the x-intercepts of the parabola $y = ax^2 + bx + c$, provided they are real numbers.

This formula now allows you to solve any quadratic equation in terms of the constants used. Let us apply it to the equation $2x^2 - 5x + 1 = 0$:

You can use $\sqrt{17} = 4.12$ from the square root table (Table 1) or from a calculator to obtain rational approximations for the solutions. To two decimal places, $x = 2.28$ or $x = 0.22$.

$$2x^2 - 5x + 1 = 0 \qquad x = \frac{-b \pm \sqrt{b^2 - 4ac}}{2a}$$

$$a = 2 \qquad\qquad\qquad = \frac{-(-5) \pm \sqrt{(-5)^2 - 4(2)(1)}}{2(\ )}$$

$$b = -5$$

$$c = 1 \qquad\qquad\qquad = \frac{5 \pm \sqrt{17}}{4}$$

Thus

$$x = \frac{5 + \sqrt{17}}{4} \quad \text{or} \quad x = \frac{5 - \sqrt{17}}{4}$$

When the radicand $b^2 - 4ac$ in the quadratic formula is negative, then the roots will be imaginary numbers, since the square root of a negative number is imaginary. Recall that $\sqrt{-7} = i\sqrt{7}$ where $i = \sqrt{-1}$. Example 3 is an illustration of a quadratic equation whose roots are imaginary numbers.

EXAMPLE 3 Solve for x: $2x^2 = x - 1$.

Solution First rewrite the equation in the general form $ax^2 + bx + c = 0$.

$$2x^2 - x + 1 = 0$$

CAUTION

If the equation is not first written in the form $ax^2 + bx + c = 0$ it could lead to incorrect assignments of the constants a, b, and c.

Use the quadratic formula with $a = 2$, $b = -1$, and $c = 1$.

$$x = \frac{-(-1) \pm \sqrt{(-1)^2 - 4(2)(1)}}{2(2)}$$

$$= \frac{1 \pm \sqrt{-7}}{4} \qquad\qquad (b^2 - 4ac = -7)$$

$$= \frac{1 \pm i\sqrt{7}}{4} \qquad \left(\text{or } \frac{1}{4} \pm \frac{\sqrt{7}}{4}i\right) \qquad\blacksquare$$

As in Example 3, if $b^2 - 4ac < 0$, then no real square roots are possible. Geometrically, this means that the parabola $y = ax^2 + bx + c$ does not meet the x-axis; there are no real solutions for $ax^2 + bx + c = 0$.

When $b^2 - 4ac = 0$, only the solution $x = -\dfrac{b}{2a}$ is possible; the x-axis is tangent to the parabola. Finally, when $b^2 - 4ac > 0$, we have two solutions that are the x-intercepts of the parabola. Since $b^2 - 4ac$ tells us how many (if any) x-intercepts the graph of $y = ax^2 + bx + c$ has, it is called the **discriminant.**

The following equations and graphs illustrate the three possible cases for the value of the discriminant.

Note that for parabolas opening downward, we have the same three possibilities for the x-intercepts.

$y = x^2 + 4x + 3$

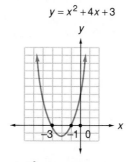

$y = x^2 - 4x + 4$

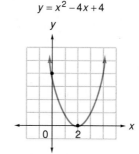

$y = 2x^2 - 4x + 5$

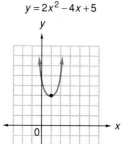

$b^2 - 4ac = 4 > 0$
Two x-intercepts.
The curve crosses
the x-axis at two
points.

$b^2 - 4ac = 0$
One x-intercept.
The curve touches
the x-axis at
one point.

$b^2 - 4ac = -24 < 0$
No x-intercept.
The curve does
not cross or
touch the x-axis.

In summary:

USING THE DISCRIMINANT WHEN a, b, c, ARE REAL NUMBERS

Note that if $b^2 - 4ac$ is a perfect square, then the given quadratic will be factorable.

1. **If $b^2 - 4ac > 0$,** then $ax^2 + bx + c = 0$ has two real solutions and the graph of $y = ax^2 + bx + c$ crosses the x-axis at two points. If a, b, c are rational numbers and the discriminant is a perfect square, these roots will be rational numbers; if not, they will be irrational.

2. **If $b^2 - 4ac = 0$,** the solution for $ax^2 + bx + c = 0$ is only one real number (a double root), and the graph of $y = ax^2 + bx + c$ touches the x-axis at one point.

3. **If $b^2 - 4ac < 0$,** $ax^2 + bx + c = 0$ has no real solutions, and the graph of $y = ax^2 + bx + c$ does not cross the x-axis. (The solutions are two imaginary numbers.)

The x-intercepts of a parabola can also be used to solve a **quadratic inequality in one variable,** such as $x^2 - x - 6 < 0$. To solve this inequality, first examine the graph of $y = f(x) = x^2 - x - 6$. This is a parabola with x-intercepts at -2 and 3, as can be seen by writing the equation in the factored form $y = (x + 2)(x - 3)$. Note that $y = x^2 - x - 6$ is below the x-axis between the x-intercepts, and above it outside the x-intercepts. That is,

In practice this method can be used without constructing the graph. All you really need to know are the x-intercepts and whether the parabola opens up or down.

$$x^2 - x - 6 < 0 \qquad \text{for } -2 < x < 3$$

$$x^2 - x - 6 > 0 \qquad \text{for } x < -2 \quad \text{or} \quad x > 3$$

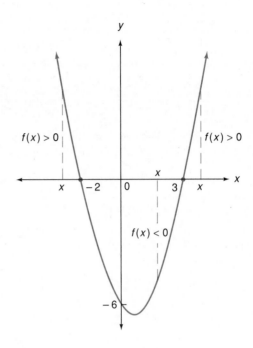

EXERCISES 3.3

Solve for x by factoring.

1. $x^2 - 5x + 6 = 0$
2. $x^2 - 4x + 4 = 0$
3. $x^2 - 10x = 0$
4. $2x^2 = x$
5. $10x^2 - 13x - 3 = 0$
6. $4x^2 - 15x = -9$
7. $4x^2 = 32x - 64$
8. $(x + 1)(x + 2) = 30$
9. $2x(x + 6) = 22x$

Solve for x by completing the square.

10. $x^2 - 2x - 15 = 0$
11. $x^2 - 4x + 1 = 0$
12. $2x^2 - 2x = 15$

 Use the quadratic formula to solve for x. When the roots are irrational numbers, give their radical forms and also use a calculator to find approximations to two decimal places.

13. $x^2 - 3x - 10 = 0$
14. $2x^2 + 3x - 2 = 0$
15. $x^2 - 2x - 4 = 0$
16. $2x^2 - 3x - 1 = 0$
17. $3x^2 + 7x + 2 = 0$
18. $x^2 + 4x + 1 = 0$
19. $x^2 - 6x + 6 = 0$
20. $x^2 - 2x + 2 = 0$
21. $-x^2 + 6x - 14 = 0$
22. $2 - 4x - x^2 = 0$
23. $3x + 1 = 2x^2$
24. $6x^2 + 2x = -3$
25. $4 + 3x + x^2 = 0$
26. $2 - 5x + 2x^2 = 0$
27. $5x^2 = 8x - 8$

Find the x-intercepts.

28. $y = x^2 - 3x - 4$

29. $y = 2x^2 - 5x - 3$

30. $y = x^2 - 10x + 25$

31. $y = x^2 - x + 3$

32. $y = 9x^2 - 4$

33. $y = 2x^2 - 7x + 6$

34. $y = x^2 + 4x + 1$

35. $y = -x^2 + 4x - 7$

36. $y = 3x^2 + x - 1$

Use the discriminant to describe the solutions as (a) a single real number, (b) two rational numbers, (c) two irrational numbers, or (d) two imaginary numbers.

37. $x^2 - 8x + 16 = 0$

38. $x^2 + 3x + 5 = 0$

39. $x^2 + 2x - 8 = 0$

40. $9x^2 - 6x + 1 = 0$

41. $2x^2 - x + 1 = 0$

42. $-x^2 + 2x + 15 = 0$

43. $x^2 + 3x - 1 = 0$

44. $2x^2 + x - 5 = 0$

45. $6x^2 + 7x - 4 = 0$

46. $4x^2 + x = 3$

47. $2x^2 + 5 = x$

48. $2x + 3 = -4x^2$

Use the discriminant to predict how many times, if any, the parabola will cross the x-axis. Then find (a) the vertex, (b) the y-intercept, and (c) the x-intercepts.

49. $y = x^2 - 4x + 4$

50. $y = x^2 - 6x + 13$

51. $y = 9x^2 - 6x + 1$

52. $y = x^2 - 9x$

53. $y = 2x^2 - 4x + 3$

54. $y = -x^2 - 4x + 3$

Find the values of b so that the x-axis will be tangent to the parabola. (Hint: Let $b^2 - 4ac = 0$.)

55. $f(x) = x^2 + bx + 9$

56. $f(x) = 4x^2 + bx + 25$

57. $f(x) = x^2 - bx + 7$

58. $f(x) = 9x^2 + bx + 14$

Find the values of k so that the parabola will intersect the x-axis at two points. (Hint: Let $b^2 - 4ac > 0$.)

59. $f(x) = -x^2 + 4x + k$

60. $f(x) = 2x^2 - 3x + k$

61. $f(x) = kx^2 - x - 1$

62. $f(x) = kx^2 + 3x - 2$

Find the values of t so that the parabola will not cross the x-axis. (Hint: Let $b^2 - 4ac < 0$.)

63. $f(x) = x^2 - 6x + t$

64. $f(x) = 2x^2 + tx + 8$

65. $f(x) = tx^2 - x - 1$

66. $f(x) = tx^2 + 3x + 7$

67. Show that the sum of the two roots of $ax^2 + bx + c = 0$ is $-\dfrac{b}{a}$ and that their product *is* $\dfrac{c}{a}$.

Use the results of Exercise 67 to find the sum and product of the roots of the given equation. Then verify your answers by solving for the roots and forming their sum and product.

68. $x^2 - 3x - 10 = 0$

69. $6x^2 + 5x - 4 = 0$

70. $x^2 = 25$

71. $3x^2 + 35 = 26x$

72. $x^2 + 2x - 5 = 0$

73. $2x^2 + 6x - 9 = 0$

Solve each quadratic inequality.

74. $8 + 2x - x^2 < 0$

75. $x^2 - 2x - 3 < 0$

76. $2x^2 + 3x - 2 \le 0$

77. $x^2 + 3x - 10 \ge 0$

78. $3 - x^2 > 0$

79. $x^2 + 6x + 9 < 0$

Graph. Show all intercepts.

80. $f(x) = (x - \frac{1}{2})(5 - x)$

81. $f(x) = (x - 3)(x + 1)$

82. $f(x) = x^2 + 1$

83. $f(x) = 4 - 4x - x^2$

84. $f(x) = -x^2 + 1$

85. $f(x) = 3x^2 - 4x + 1$

86. $f(x) = x^2 + 2x$

87. $f(x) = -x^2 + 2x - 4$

Solve for x.

88. $x^4 - 5x^2 + 4 = 0$ *(Hint: Use $u = x^2$ and solve $u^2 - 5u + 4 = 0$.)*

89. $2x^4 - 13x^2 - 7 = 0$ (*Hint: As in Exercise 88, use* $u = x^2$.)

90. $\left(x - \dfrac{2}{x}\right)^2 - 3\left(x - \dfrac{2}{x}\right) + 2 = 0$ $\left(\textit{Hint: Let } u = x - \dfrac{2}{x}.\right)$

91. $x^3 + 3x^2 - 4x - 12 = 0$ (*Hint: Factor by grouping.*)

92. $a^3x^2 - 2ax - 1 = 0$ $(a > 0)$

93. $\dfrac{4x}{2x - 1} + \dfrac{3}{2x + 1} = 0$ **94.** $x^2 + 2\sqrt{3}x + 2 = 0$

✏️ **Written Assignment:** Use examples of your own and describe how the discriminant can be used to tell the nature of the roots of a quadratic equation that is of the form $ax^2 + bx + c = 0$.

CHALLENGE
Think Creatively

Assume that a, b, and c in the quadratic equation $ax^2 + bx + c = 0$ are integers. Explain why the discriminant $b^2 - 4ac$ cannot have the value 23. (*Hint:* Either b is even or b is odd. Let $b = 2n$ or $b = 2n + 1$.)

EXPLORATIONS
Think Critically

1. How can you determine the maximum or minimum value of a function given by $f(x) = ax^2 + bx + c$?
2. Suppose that a function is increasing on one side of a point P with coordinates (a, b) and decreasing on the other side of this point. What can be said about the point P? Explain.
3. Consider the graph of $y = (x + 2)^2 - 1$. How can you use this to determine the graph of $x = (y + 2)^2 - 1$?
4. Describe the graph of $y = a(x - h)^2 + k$ for various values of a, h, and k. For example, consider $a > 0$, $h < 0$, and $k > 0$ as one case.
5. Sketch the graph of a function such that any horizontal line drawn through a range value will intersect the curve exactly once. Such functions are called *one-to-one functions*. If f is a one-to-one function, and if $f(x_1) = f(x_2)$, what can you say about x_1 and x_2?

3.4
APPLICATIONS OF QUADRATIC FUNCTIONS

The algebraic and graphical techniques developed thus far in this chapter can be used to solve a variety of problems that involve quadratic equations. The first two examples illustrate situations that call for the solution of a quadratic equation of the form $ax^2 + bx + c = 0$.

EXAMPLE 1 The length of a rectangular piece of cardboard is 2 inches more than its width. As in the following figure, an open box is formed by cutting out 4-inch squares from each corner and folding up the sides. If the volume of the box is 672 cubic inches, find the dimensions of the original cardboard.

Solution Let x be the width of the cardboard. Then $x + 2$ is the length.

Consider the steps used to solve this problem:
(a) Select the necessary information from the statement of the problem.
(b) Represent the information algebraically.
(c) Draw a figure.
(d) Write and solve an equation.
(e) Interpret the solution to meet the given conditions of the problem.

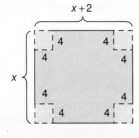

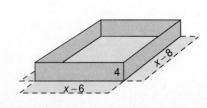

After the squares are cut off and the sides are folded up, the dimensions of the box are:

$$\text{length} = \ell = x + 2 - 8 = x - 6$$
$$\text{width} = w = x - 8$$
$$\text{height} = h = 4$$

Since the volume is to be 672 cubic inches, and $v = \ell w h$, we have

$$(x - 6)(x - 8)4 = 672$$
$$(x - 6)(x - 8) = 168$$
$$x^2 - 14x + 48 = 168$$
$$x^2 - 14x - 120 = 0$$
$$(x - 20)(x + 6) = 0$$
$$x = 20 \quad \text{or} \quad x = -6$$

Note: We reject the solution $x = -6$ since the dimensions must be positive numbers.

The dimensions of the original cardboard are: $w = 20$ inches, $\ell = 20 + 2 = 22$ inches. ■

EXAMPLE 2   Find two consecutive positive integers such that the sum of their squares is 113.

Solution   Let $x =$ the first integer. Then $x + 1 =$ the next consecutive integer. The sum of their squares can be represented as $x^2 + (x + 1)^2$. Thus

$$x^2 + (x + 1)^2 = 113$$
$$x^2 + x^2 + 2x + 1 = 113$$
$$2x^2 + 2x + 1 = 113$$
$$2x^2 + 2x - 112 = 0 \qquad \text{(Divide each side by 2.)}$$
$$x^2 + x - 56 = 0 \qquad \text{(The left side can be factored.)}$$
$$(x - 7)(x + 8) = 0$$
$$x = 7 \quad \text{or} \quad x = -8 \qquad \text{(If } A \cdot B = 0, \text{ then } A = 0 \text{ or } B = 0.)$$

If it were not required that the integers be positive, then another solution consists of the consecutive integers -8, -7.

Since the integers were said to be positive, we must reject the solution $x = -8$. Therefore, $x = 7$ and $x + 1 = 8$. The two consecutive integers are 7 and 8. ■

Some problems require the minimum or maximum value of a quadratic function. Such problems can be solved by using the standard form $y = f(x) = a(x - h)^2 + k$, which shows the vertex of the parabola to be (h, k). As observed in Section 3.1, this point is the lowest point or the highest point on the parabola, depending on the sign of a. When $a > 0$, the vertex is the lowest point on the parabola; when $a < 0$, it is the highest point. These special points will be useful in solving certain applied problems.

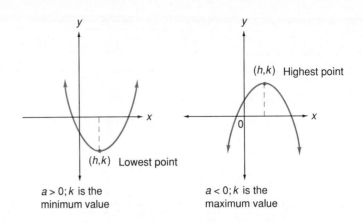

$a > 0; k$ is the
minimum value

$a < 0; k$ is the
maximum value

In summary, we have the following:

MINIMUM AND MAXIMUM VALUES OF QUADRATIC FUNCTIONS

$$y = f(x) = a(x - h)^2 + k \qquad (a \neq 0)$$

If $a > 0$, the graph is a parabola that opens upward and has a **minimum value** $f(h) = k$.

If $a < 0$, the graph is a parabola that opens downward and has a **maximum value** $f(h) = k$.

The examples that follow involve the concepts of minimum and maximum values of quadratic functions.

EXAMPLE 3 Find the maximum value or minimum value of the function $f(x) = 2(x + 3)^2 + 5$.

Solution Since $2(x + 3)^2 + 5 = 2[x - (-3)]^2 + 5$, we note that $(-3, 5)$ is the turning point. Also since $a = 2 > 0$, the parabola opens upward and $f(-3) = 5$ is the minimum value. ∎

EXAMPLE 4 Find the maximum value of the quadratic function $f(x) = -\frac{1}{3}x^2 + x + 2$. At which value x does f achieve this maximum?

The purpose of converting to the form $a(x - h)^2 + k$ is to find the vertex (h, k): $h = \frac{3}{2}$ and $k = \frac{11}{4}$.

Solution Convert to the form $a(x - h)^2 + k$:

$$y = f(x) = -\frac{1}{3}(x^2 - 3x) + 2$$
$$= -\frac{1}{3}(x^2 - 3x + \tfrac{9}{4}) + 2 + \tfrac{3}{4}$$
$$= -\frac{1}{3}(x - \tfrac{3}{2})^2 + \tfrac{11}{4}$$

From this form we have $a = -\frac{1}{3}$. Since $a < 0$, $(\frac{3}{2}, \frac{11}{4})$ is the highest point of the parabola. Thus f has a maximum value of $\frac{11}{4}$ when $x = \frac{3}{2}$. ∎

Find the maximum or minimum value of each quadratic function and state the x-value at which this occurs.

1. $f(x) = x^2 - 10x + 21$ **2.** $f(x) = x^2 + \frac{4}{3}x - \frac{7}{18}$

3. $f(x) = 10x^2 - 20x + \frac{21}{2}$ **4.** $f(x) = -8x^2 - 64x + 3$

5. $f(x) = -2x^2 - 1$ **6.** $f(x) = x^2 - 6x + 9$

7. $f(x) = 25x^2 + 70x + 49$ **8.** $f(x) = (x - 3)(x + 4)$

EXAMPLE 5 Suppose that 60 meters of fencing is available to enclose a rectangular garden, one side of which will be against the side of a house. What dimensions of the garden will guarantee a maximum area?

Solution Use the given information to write a quadratic function. Then write the function in standard form so as to determine the vertex of the corresponding parabola. The maximum or minimum value always occurs at the vertex.

For each x between 0 and 30, such a rectangle is possible. Here are a few.

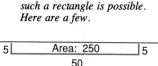

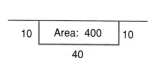

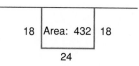

Example 5 shows how to select the rectangle of maximum area from such a vast collection of possibilities. Can you explain why the domain of A(x) is 0 < x < 30?

From the sketch you can see that the 60 meters need only be used for three sides, two of which are of the same length x.

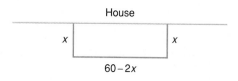

Let x represent the width of the garden. Then $60 - 2x$ represents the length, and the area A is given by

$$A(x) = x(60 - 2x)$$
$$= 60x - 2x^2$$

To "maximize" A, convert to the form $a(x - h)^2 + k$. Thus

$$A(x) = -2(x^2 - 30x)$$
$$= -2(x^2 - 30x + 225) + 450$$
$$= -2(x - 15)^2 + 450$$

Therefore, the maximum area of 450 square meters is obtained when the dimensions are $x = 15$ meters by $60 - 2x = 30$ meters. ∎

EXAMPLE 6 The sum of two numbers is 24. Find the two numbers if their product is to be a maximum.

Solution Let x represent one of the numbers. Since the sum is 24, the other number is $24 - x$. Now let p represent the product of these numbers.

$$p = x(24 - x)$$
$$= -x^2 + 24x$$
$$= -(x^2 - 24x)$$
$$= -(x^2 - 24x + 144) + 144$$
$$= -(x - 12)^2 + 144$$

Try to solve Example 6 if the product is to be a minimum.

The product has a maximum value of 144 when $x = 12$. Hence the numbers are 12 and 12. ∎

EXAMPLE 7   A ball is thrown straight upward from ground level with an initial velocity of 32 feet per second. The formula $s = 32t - 16t^2$ gives its height in feet, s, after t seconds.
(a)   What is the maximum height reached by the ball?
(b)   When does the ball return to the ground?

Solution
(a) First complete the square in t.

$$s = 32t - 16t^2$$
$$= -16t^2 + 32t$$
$$= -16(t^2 - 2t)$$
$$= -16(t^2 - 2t + 1) + 16$$
$$= -16(t - 1)^2 + 16$$

You should now recognize this as describing a parabola with vertex at $(1, 16)$. Because the coefficient of t^2 is negative, the curve opens downward. The maximum height, 16 feet, is reached after 1 second.

The motion of the ball is straight up and down.

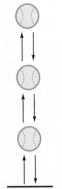

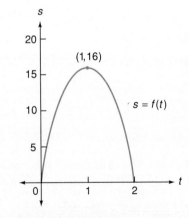

This parabolic arc is the graph of the relation between time t and distance s. It is *not* the path of the ball.

(b) The ball hits the ground when the distance $s = 0$ feet. Thus

$$s = 32t - 16t^2 = 0$$

$$16t(2 - t) = 0$$

$$t = 0 \quad \text{or} \quad t = 2$$

Since time $t = 0$ is the starting time, the ball returns to the ground 2 seconds later, when $t = 2$. ∎

For example, if $x = 2$ the reduction is $10, the sales would be 700, and the income would be ($90)(700) = $63,000.

EXAMPLE 8   The marketing department of the TENRAQ Tennis Company found that, on the average, 600 tennis rackets will be sold monthly at the unit price of $100. The department also observed that for each $5 reduction in price, an extra 50 rackets will be sold monthly. What price will bring the largest monthly income?

Solution   Let x be the number of $5 reductions in price for the racket. Then $5x$ is the total reduction and

$$100 - 5x = \text{reduced unit price}$$

Also, $50x$ is the increase in sales per month and

$$600 + 50x = \text{number of rackets sold monthly}$$

The monthly income, R, will be the unit price times the number of units sold. Then

$$R = (\text{unit price})(\text{units sold})$$
$$R(x) = (100 - 5x)(600 + 50x)$$
$$= 60{,}000 + 2000x - 250x^2$$
$$= -250(x^2 - 8x) + 60{,}000$$
$$= -250(x^2 - 8x + 16) + 60{,}000 + 4000$$
$$= -250(x - 4)^2 + 64{,}000 \qquad \text{(Change to the form } a(x - h)^2 + k.)$$

Since $a = -250 < 0$, the maximum monthly income of $64,000 is obtained when $x = 4$. The unit price should be set at $100 - 5(4)$ or $80. ∎

EXERCISES 3.4

Write and solve a quadratic equation for each exercise.

1. Find two consecutive positive integers whose product is 210.
2. The sum of a number and its square is 56. Find the number. (There are two possible answers.)
3. One positive integer is 3 greater than another. The sum of the squares of the two integers is 89. Find the integers.

4. Find two integers whose sum is 26 and whose product is 165. (*Hint:* Let the two integers be represented by x and $26 - x$.)

5. How wide a border of uniform width should be added to a rectangle that is 8 feet by 12 feet in order to double the area?

6. The length of a rectangle is 3 centimeters greater than its width. The area is 70 square centimeters. Find the dimensions of the rectangle.

7. The area of a rectangle is 15 square centimeters and the perimeter is 16 centimeters. What are the dimensions of the rectangle? (*Hint:* If x represents the width, then $\dfrac{16 - 2x}{2} = 8 - x$ represents the length.)

8. The altitude of a triangle is 5 centimeters less than the base to which it is drawn. The area of the triangle is 21 square centimeters. Find the length of the base.

9. A backyard swimming pool is rectangular in shape, 10 meters wide and 18 meters long. It is surrounded by a walk of uniform width whose area is 52 square meters. How wide is the walk?

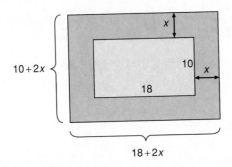

10. A square piece of tin is to be used to form a box without a top by cutting off a 2-inch square from each corner and then folding up the sides. The volume of the box will be 128 cubic inches. Find the length of a side of the original square.

11. The sum S of the first n consecutive positive integers, $1, 2, 3, \ldots, n$, is given by the formula $S = \frac{1}{2}n(n + 1)$. Find n when $S = 120$ and check your answer by addition.

12. The sum S of the first n consecutive even positive integers, $2, 4, 6, \ldots, 2n$, is given by the formula $S = n(n + 1)$. How many such consecutive even positive integers must be added to get a sum of 342?

13. A boat travels downstream (with the current) for 36 miles and then makes the return trip upstream (against the current). The trip downstream took $\frac{3}{4}$ of an hour less than the trip upstream. If the rate of the current is 4 mph, find the rate of the boat in still water and the time for each part of the trip.

14. Two motorcycles each make the same 220-mile trip, but one of them travels 5 mph faster than the other. Find the rate of each motorcycle if the slower one takes 24 minutes longer to complete the trip than the faster one.

Find the maximum or minimum value of the quadratic function and state the x-value at which this occurs.

15. $f(x) = -x^2 + 10x - 18$

16. $f(x) = x^2 + 18x + 49$

17. $f(x) = 16x^2 - 64x + 100$

18. $f(x) = -\frac{1}{2}x^2 + 3x - 6$

19. $f(x) = 49 - 28x + 4x^2$

20. $f(x) = x(x - 10)$

21. $f(x) = -x(\frac{2}{3} + x)$

22. $f(x) = (x - 4)(2x - 7)$

23. A manufacturer is in the business of producing small statues called Heros. He finds that the daily cost in dollars, C, of producing n Heros is given by the quadratic formula $C = n^2 - 120n + 4200$. How many Heros should be produced per day so that the cost will be minimum? What is the minimal daily cost?

24. A company's daily profit, P, in dollars, is given by $P = -2x^2 + 120x - 800$, where x is the number of articles produced per day. Find x so that the daily profit is a maximum.

25. The sum of two numbers is 12. Find the two numbers if their product is to be a maximum. (*Hint:* Find the maximum value for $y = x(12 - x)$.)

26. The sum of two numbers is n. Find the two numbers such that their product is a maximum. Are there two numbers that will give a minimum product? Explain.

27. The difference of two numbers is 22. Find the numbers if their product is to be a minimum and also find this product.

28. A homeowner has 40 feet of wire and wishes to use it to enclose a rectangular garden. What should be the dimensions of the garden so as to enclose the largest possible area?

29. Repeat Exercise 28, but this time assume that the side of the house is to be used as one boundary for the garden. Thus the wire is needed only for the other three sides.

The formula $h = 128t - 16t^2$ gives the distance in feet above the ground, h, reached by an object in t seconds. Use this formula for Exercises 30 through 33.

30. What is the maximum height reached by the object?

31. How long does it take for the object to reach its maximum height?

32. How long does it take for the object to return to the ground?

33. In how many seconds will the object be at a height of 192 feet? (There are two possible answers.)

34. Suppose it is known that if 65 apple trees are planted in a certain orchard, the average yield per tree will be 1500 apples per year. For each additional tree planted in the same orchard, the annual yield per tree drops by 20 apples. How many trees should be planted in order to produce the maximum crop of apples per year? (*Hint:* If n trees are added to the 65 trees, then the yield per tree is $1500 - 20n$.)

35. It is estimated that 14,000 people will attend a basketball game when the admission price is $7.00. For each 25¢ added to the price, the attendance will decrease by 280. What admission price will produce the largest gate receipts? (*Hint:* If x quarters are added, the attendance will be $14,000 - 280x$.)

36. The sum of the lengths of the two perpendicular sides of a right triangle is 30 centimeters. What are the lengths if the square of the hypotenuse is a minimum?

37. Each point P on the line segment between endpoints $(0, 4)$ and $(2, 0)$ determines a rectangle with dimensions x by y as shown in the figure. Find the coordinates of P that give the rectangle of maximum area.

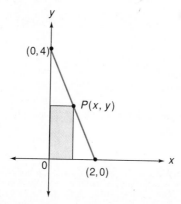

38. Verify that the coordinates of the vertex of the parabola $y = ax^2 + bx + c$ are

$$\left(-\frac{b}{2a}, \quad \frac{4ac - b^2}{4a}\right)$$

Use the result in Exercise 38 to find the coordinates of the vertex of each parabola, and decide whether $\dfrac{4ac - b^2}{4a}$ is a maximum or minimum value.

39. $y = 2x^2 - 6x + 9$

40. $y = -3x^2 + 24x - 41$

41. $y = -\frac{1}{2}x^2 - \frac{1}{3}x + 1$

41. $y - x^2 + 5x = 0$

43. $y + \frac{2}{3}x^2 = 9$

44. $y = 10x^2 + 100x + 1000.$

 45. The following quadratic formula can be used to approximate the distance in feet, d, that it takes to stop a car after the brakes are applied for a car traveling on a dry road at the rate of r miles per hour: $d = 0.045r^2 + 1.1r$. To the nearest foot, how far will a car travel once the brakes are applied at (a) 40 mph, (b) 55 mph, and (c) 65 mph?

 46. The police measured the skid marks made by a car that crashed into a tree. If the measurement gave a braking distance of 250 feet, was the driver exceeding the legal speed limit of 55 miles per hour? To the nearest mph, what was the speed of the car before the brakes were applied? (Use the formula in Exercise 45.)

47. When a department store sold a certain style of shirt for \$20, the average number of shirts sold per week was 100. The store observed that with each \$1 decrease in price, 10 more shirts were sold weekly. What unit price should be set for the shirts in order to realize the maximum weekly revenue?

**3.5
CONIC
SECTIONS:
THE CIRCLE**

A **conic section** is a curve formed by the intersection of a plane with a double right circular cone. These curves, also called **conics,** are known as the **circle, ellipse, parabola,** and **hyperbola.**

The figures indicate that the inclination of the plane in relation to the vertical axis of the cone determines the nature of the curve. These four curves have played a vital role in mathematics and its applications from the time of the ancient Greeks until the present day.

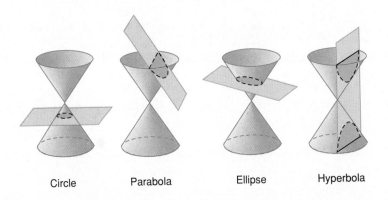

Circle   Parabola   Ellipse   Hyperbola

In order to study the conic sections, we first need to know how to find the distance AB between two points $A(x_1, y_1)$ and $B(x_2, y_2)$ in a rectangular system, given by the **distance formula:**

DISTANCE FORMULA

The distance AB between points $A = (x_1, y_1)$ and $B = (x_2, y_2)$ is given by

$$AB = \sqrt{(x_1 - x_2)^2 + (y_1 - y_2)^2}$$

The proof that follows is based on this typical figure:

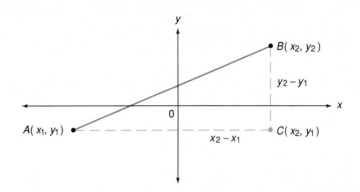

Note that the diagram was set up so that $x_2 - x_1 > 0$ and $y_2 - y_1 > 0$. Other situations may have negative values, but it makes no difference because $(x_2 - x_1)^2 = (x_1 - x_2)^2$ and $(y_2 - y_1)^2 = (y_1 - y_2)^2$.

Since AB is the hypotenuse of the right triangle ABC, the Pythagorean theorem gives

$$AB^2 = AC^2 + CB^2$$

But $AC = x_2 - x_1$ and $CB = y_2 - y_1$. Thus

$$AB^2 = (x_2 - x_1)^2 + (y_2 - y_1)^2$$

If A and B are on the same horizontal line, then $y_1 = y_2$ and $AB = \sqrt{(x_1 - x_2)^2} = |x_1 - x_2|$.

Taking the positive square root gives the stated result.

EXAMPLE 1 Find the length of the line segment determined by points $A(-2, 2)$ and $B(6, -4)$.

Solution

$$\begin{aligned} AB &= \sqrt{(-2 - 6)^2 + [2 - (-4)]^2} \\ &= \sqrt{(-8)^2 + 6^2} \\ &= \sqrt{64 + 36} \\ &= \sqrt{100} \\ &= 10 \end{aligned}$$

∎

Now we are ready to study the circle and its properties and will be able to use the distance formula to derive the equation of a circle.

DEFINITION OF A CIRCLE

A **circle** is the set of all points in the plane, each of which is at a fixed distance r from a given point called the **center** of the circle; r is the **radius** of the circle.

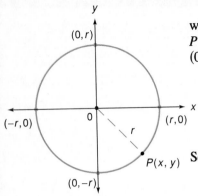

The figure at the left is a circle with center at the origin and radius r. A point will be on this circle *if and only if* its distance from the origin is equal to r. That is, $P(x, y)$ is on this circle if and only if $OP = r$. Since the origin has coordinates $(0, 0)$, the distance formula gives

$$r = \sqrt{(x - 0)^2 + (y - 0)^2}$$
$$= \sqrt{x^2 + y^2}$$

Squaring produces this result:

$$r^2 = x^2 + y^2$$

The words "if and only if" here mean that if P is on the circle, then OP = r and if OP = r then P is on the circle.

We conclude that $P(x, y)$ is on the circle with center at $(0, 0)$ and radius r if and only if the coordinates of P satisfy the preceding equation.

> ### STANDARD FORM FOR THE EQUATION OF A CIRCLE WITH CENTER AT THE ORIGIN AND RADIUS r
>
> $$x^2 + y^2 = r^2$$

EXAMPLE 2   What is the equation of the circle with center $(0, 0)$ and radius 3?

Solution   Using the equation $x^2 + y^2 = r^2$, we get

$$x^2 + y^2 = 3^2 = 9$$ ■

Now consider any circle of radius r, not necessarily one with the origin as center. Let the center C have coordinates (h, k). Then, using the distance formula, a point $P(x, y)$ is on this circle if and only if

$$CP = r = \sqrt{(x - h)^2 + (y - k)^2}$$

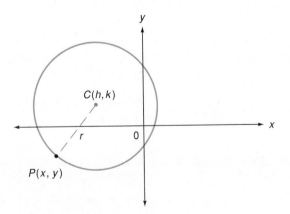

By squaring CP, we obtain the following:

Observe that this form includes the earlier case where $(h, k) = (0, 0)$.

STANDARD FORM FOR THE EQUATION OF A CIRCLE
WITH CENTER AT (h, k) AND RADIUS r

$$(x - h)^2 + (y - k)^2 = r^2$$

Circles are not graphs of functions. When $x = 2$ is substituted into the equation of the circle in Example 3, we get two y-values, -1 and -5. This is contrary to the definition of a function that calls for just one range value y for each domain value x.

EXAMPLE 3 Find the center and radius of the circle with this equation: $(x - 2)^2 + (y + 3)^2 = 4$.

Solution Using $y + 3 = y - (-3)$, rewrite the equation in this form:

$$(x - 2)^2 + [(y - (-3)]^2 = 2^2$$

Thus the radius $r = 2$ and the center is at $(2, -3)$. ■

EXAMPLE 4 Write the equation of the circle with center $(-3, 5)$ and radius $\sqrt{2}$.

Solution Use $h = -3$, $k = 5$, and $r = \sqrt{2}$ in the standard form to obtain the following:

$$[x - (-3)]^2 + (y - 5)^2 = (\sqrt{2})^2$$

or

$$(x + 3)^2 + (y - 5)^2 = 2$$ ■

TEST YOUR UNDERSTANDING
Think Carefully

Find the length of the line segment determined by the two points.

1. $(4, 0)$; $(-8, -5)$ **2.** $(9, -1)$; $(2, 3)$ **3.** $(-7, -5)$; $(3, -13)$

Find the center and radius of each circle.

4. $x^2 + y^2 = 100$ **5.** $x^2 + y^2 = 10$
6. $(x - 1)^2 + (y + 1)^2 = 25$ **7.** $(x + \frac{1}{2})^2 + y^2 = 256$
8. $(x + 4)^2 + (y + 4)^2 = 50$

Write the equation of the circle with the given center and radius in standard form.

(Answers: Page 202)

9. Center at $(0, 4)$; $r = 5$ **10.** Center at $(1, -2)$; $r = \sqrt{3}$

The equation in Example 3 can be written in another form.

$$(x - 2)^2 + (y + 3)^2 = 4$$
$$x^2 - 4x + 4 + y^2 + 6y + 9 = 4$$
$$x^2 - 4x + y^2 + 6y = -9$$

Note: The major reason for writing the equation of a circle in standard form is that this form enables us to identify the center and the radius of the circle. This information is sufficient to allow us to draw the circle.

This last equation no longer looks like the equation of a circle. Starting with such an equation we can convert it back into the standard form of a circle by completing the square in both variables, if necessary. For example, let us begin with

$$x^2 - 4x + y^2 + 6y = -9$$

To complete the squares
$(\frac{1}{2} \cdot 4)^2 = 4$ and $(\frac{1}{2} \cdot 6)^2 = 9$
are added to each side.

Then complete the squares in x and y:

$$(x^2 - 4x + 4) + (y^2 + 6y + 9) = -9 + 4 + 9$$

$$(x - 2)^2 + (y + 3)^2 = 4$$

EXAMPLE 5 Find the center and radius of the circle with equation $9x^2 + 12x + 9y^2 = 77$.

Solution First divide by 9 so that the x^2 and y^2 terms each have a coefficient of 1.

$$x^2 + \tfrac{4}{3}x + y^2 = \tfrac{77}{9}$$

Complete the square in x; add $\tfrac{4}{9}$ to both sides of the equation.

$$(x^2 + \tfrac{4}{3}x + \tfrac{4}{9}) + y^2 = \tfrac{77}{9} + \tfrac{4}{9}$$

$$(x + \tfrac{2}{3})^2 + y^2 = 9$$

In standard form:

$$[x - (-\tfrac{2}{3})]^2 + (y - 0)^2 = 3^2$$

Observe that in these expanded forms of circles the coefficients of the x^2 and y^2 terms are equal.

The center is at $(-\tfrac{2}{3}, 0)$ and $r = 3$. ∎

**TEST YOUR UNDERSTANDING
Think Carefully**

Find the center and radius of each circle.

1. $x^2 - 6x + y^2 - 10y = 2$

2. $x^2 + y^2 + y = \tfrac{19}{4}$

3. $x^2 - x + y^2 + 2y = \tfrac{23}{4}$

4. $16x^2 + 16y^2 - 8x + 32y = 127$

(Answers: Page 202)

A line that is tangent to a circle touches the circle at only one point and is perpendicular to the radius drawn to the point of tangency.

When the equation of a circle is given and the coordinates of a point P on the circle are known, then the equation of the tangent line to the circle at P can be found. For example, the circle $(x + 3)^2 + (y + 1)^2 = 25$ has center $(-3, -1)$ and $r = 5$. The point $P(1, 2)$ is on this circle because its coordinates satisfy the equation of the circle.

$$(1 + 3)^2 + (2 + 1)^2 = 4^2 + 3^2 = 25$$

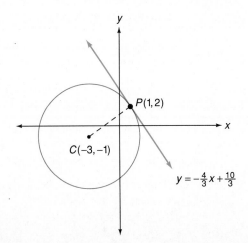

$$y = -\tfrac{4}{3}x + \tfrac{10}{3}$$

Recall the perpendicular lines have slopes that are negative reciprocals of one another.

The slope of radius CP is $\dfrac{2 - (-1)}{1 - (-3)} = \dfrac{3}{4}$. Then since the tangent at P is perpendicular to the radius CP, its slope is the negative reciprocal of $\frac{3}{4}$, namely $-\frac{4}{3}$. Now using the point-slope form we get this equation of the tangent at P:

$$y - 2 = -\tfrac{4}{3}(x - 1)$$

In slope-intercept form this becomes

$$y = -\tfrac{4}{3}x + \tfrac{10}{3}$$

The midpoint of a segment on a number line is discussed on page 24.

The center of a circle is the midpoint of each of its diameters. So if the endpoints of a diameter are given, the coordinates of the center can be found using the *midpoint formula,* as shown below. In the figure, PQ is a line segment with midpoint M having coordinates (x', y'). Since x_1, x', and x_2 are on the x-axis, $x' = \dfrac{x_1 + x_2}{2}$. Similarly, $y' = \dfrac{y_1 + y_2}{2}$.

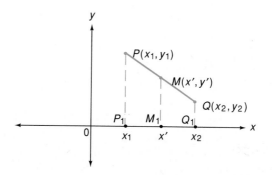

MIDPOINT FORMULA

If (x_1, x_2) and (y_1, y_2) are the endpoints of a line segment, the midpoint of the segment has coordinates

$$\left(\frac{x_1 + x_2}{2}, \frac{y_1 + y_2}{2} \right)$$

EXAMPLE 6   Points $P(2, 5)$ and $Q(-4, -3)$ are the endpoints of a diameter of a circle. Find the center, radius, and equation of the circle.

Solution   The center is the midpoint of PQ whose coordinates (x', y') are given by

$$x' = \frac{2 + (-4)}{2} = -1, \qquad y' = \frac{5 + (-3)}{2} = 1$$

The center is located at $C(-1, 1)$. To find the radius, use the distance formula between $C(-1, 1)$ and $P(2, 5)$.

$$r = \sqrt{(-1 - 2)^2 + (1 - 5)^2} = \sqrt{25} = 5$$

The equation of the circle is

$$(x + 1)^2 + (y - 1)^2 = 25$$ ■

CAUTION: Learn to Avoid These Mistakes	
WRONG	RIGHT
The circle $(x + 3)^2 + (y - 2)^2 = 7$ has center $(3, -2)$ and radius 7.	The circle has center $(-3, 2)$ and radius $\sqrt{7}$.
The equation of the circle with center $(-1, 0)$ and the radius 5 has equation $x^2 + (y + 1)^2 = 5$.	The circle has equation $(x + 1)^2 + y^2 = 25$.

EXERCISES 3.5

1. Graph these circles in the same coordinate system.
   (a) $x^2 + y^2 = 25$ (b) $x^2 + y^2 = 16$ (c) $x^2 + y^2 = 4$ (d) $x^2 + y^2 = 1$

2. Graph these circles in the same coordinate system.
   (a) $(x - 3)^2 + (y - 3)^2 = 9$ (b) $(x + 3)^2 + (y - 3)^2 = 9$
   (c) $(x + 3)^2 + (y + 3)^2 = 9$ (d) $(x - 3)^2 + (y + 3)^2 = 9$

Write the equation of each circle in the standard form $(x - h)^2 + (y - k)^2 = r^2$. Find the center and radius for each.

3. $x^2 - 4x + y^2 = 21$
4. $x^2 + y^2 + 8y = -12$
5. $x^2 - 2x + y^2 - 6y = -9$
6. $x^2 + 4x + y^2 + 10y + 20 = 0$
7. $x^2 - 4x + y^2 - 10y = -28$
8. $x^2 - 10x + y^2 - 14y = -25$
9. $x^2 - 8x + y^2 = -14$
10. $x^2 + y^2 + 2y = 7$
11. $x^2 - 20x + y^2 + 20y = -100$
12. $4x^2 - 4x + 4y^2 = 15$
13. $16x^2 + 24x + 16y^2 - 32y = 119$
14. $36x^2 - 48x + 36y^2 + 180y = -160$

Write the equation of each circle in standard form.

15. Center at $(2, 0)$; $r = 2$
16. Center at $(\frac{1}{2}, 1)$; $r = 10$
17. Center at $(-3, 3)$; $r = \sqrt{7}$
18. Center at $(-1, -4)$; $r = 2\sqrt{2}$

19. Draw the circle $x^2 + y^2 = 25$ and the tangent lines at the points $(3, 4)$, $(-3, 4)$, $(3, -4)$, and $(-3, -4)$. Write the equations of these tangent lines.

20. Where are the tangents to the circle $x^2 + y^2 = 4$ whose slopes equal 0? Write their equations.

21. Write the equations of the tangents to the circle $x^2 + y^2 = 4$ whose slopes are undefined.

Draw the given circle and the tangent line at the indicated point for each of the following. Write the equation of the tangent line.

22. $x^2 + y^2 = 80$; $(-8, 4)$
23. $x^2 - 2x + y^2 - 2y = 8$; $(4, 2)$

24. $(x - 4)^2 + (y + 5)^2 = 45; (1, 1)$
26. $x^2 + 14x + y^2 + 18y = 39; (5, -4)$

25. $x^2 + 4x + y^2 - 6y = 60; (6, 0)$
27. $x^2 + y^2 = 9; (-2, \sqrt{5})$

Find the coordinates of the midpoint of a line segment with endpoints as given.

28. $P(3, 2)$ and $Q(-2, 1)$
30. $P(-1, 0)$ and $Q(0, 5)$

29. $P(-2, 4)$ and $Q(3, -8)$
31. $P(-8, 7)$ and $Q(3, -6)$

32. Points $P(3, -5)$ and $Q(-1, 3)$ are the endpoints of a diameter of a circle. Find the center, radius, and equation of the circle.

33. Write the equation of the tangent line to the circle $x^2 + y^2 = 80$ at the point in the first quadrant where $x = 4$.

34. Write the equation of the tangent line to the circle $x^2 + y^2 = 9$ at the point in the third quadrant where $y = -1$.

35. Write the equation of the tangent line to the circle $x^2 + 14x + y^2 + 18y = 39$ at the point in the second quadrant where $x = -2$.

⚠ 36. Suppose a kite is flying at a height of 300 feet above a point P on the ground. The kite string is anchored in the ground x feet from P.

    (a) If s is the length of the string (assume the string forms a straight line) write s as a function of x.

    (b) Find s when $x = 400$ feet.

⚠ 37. A 13-foot-long board is leaning against the wall of a house so that its base is 5 feet from the wall. When the base of the board is pulled y feet further from the wall, the top of the board drops x feet.

    (a) Express y as a function of x.

    (b) Find y when $x = 7$.

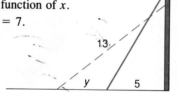
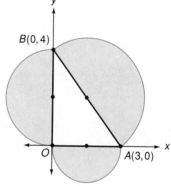

38. In the figure, each side of triangle OAB is the diameter of a semicircle as shown. Show that the area of the semicircle on AB is equal to the sum of the areas of the semicircles on the other two sides.

39. Generalize the result in Exercise 38 for any right triangle. (*Hint:* Use c as the length of the hypotenuse and a, b as the lengths of the other sides, and place the triangle in a coordinate system as in Exercise 38.)

⚠ 40. Let s be the square of the distance from the origin to point $P(x, y)$ on the line through points $(0, 4)$ and $(2, 0)$. Find the coordinates of P such that s is a minimum value.

⚠ 41. Let s be the square of the distance from point $(6, 6)$ to point $P(x, y)$ on the line $y = -2x + 4$. Find the coordinates of P such that s is a minimum value.

CHALLENGE
Think Creatively
The points $A(-3, -3)$, $B(-1, 11)$, and $C(5, 13)$ are all on the same circle. What is the radius of this circle? (*Hint:* Find the center of the circle by making use of any two of the chords AB, BC, and AC.)

In Section 3.5, the circle was defined in terms of sets of points in a plane equidistant from a fixed point. The other conic sections can be defined in a similar manner, as in the following definition of an *ellipse*.

This definition suggests that an ellipse can be drawn as follows. Take a loop of string and place it around two thumbtacks. Use a pencil to pull the string taut to form a triangle, and move the pencil around the loop.

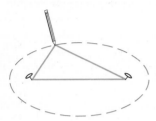

DEFINITION OF AN ELLIPSE

An **ellipse** is the set of all points in a plane such that the sum of the distances from two fixed points (called the **foci**) is a constant.

The reason that the construction shown in the margin gives an ellipse can be seen from the figure. From any position of the pencil the sum of the distances to the thumbtacks equals the length of the loop minus the distance between the thumbtacks. Since both the length of the loop and the distance between the thumbtacks are constants, their difference must also be a constant.

We first consider an ellipse whose foci, F_1 and F_2, are symmetric about the origin along the x-axis. Thus we let F_1 have coordinates $(-c, 0)$ and F_2 have coordinates $(c, 0)$, where c is some positive number.

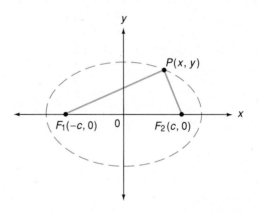

Since the sum of the distances PF_1 and PF_2 must be constant, we choose some positive number a and let this constant equal $2a$. (The form $2a$ will prove to be useful to simplify the algebraic computation.) Thus a point $P(x, y)$ is on the ellipse if and only if $PF_1 + PF_2 = 2a$. (Note $a > c$. Why?)

Using the distance formula gives

$$PF_1 = \sqrt{(x + c)^2 + y^2} \quad \text{and} \quad PF_2 = \sqrt{(x - c)^2 + y^2}$$

Thus

$$PF_1 + PF_2 = \sqrt{(x + c)^2 + y^2} + \sqrt{(x - c)^2 + y^2} = 2a$$

which implies the following:

$$\sqrt{(x + c)^2 + y^2} = 2a - \sqrt{(x - c)^2 + y^2}$$

Squaring both sides and collecting terms, we have

$$a\sqrt{(x - c)^2 + y^2} = a^2 - cx$$

Square both sides again and simplify:

$$(a^2 - c^2)x^2 + a^2y^2 = a^2(a^2 - c^2)$$

Since $a^2 - c^2 > 0$ we may let $b = \sqrt{a^2 - c^2}$ so that $b^2 = a^2 - c^2$. Therefore,

$$b^2x^2 + a^2y^2 = a^2b^2$$

Now divide through by a^2b^2 to obtain the following standard form:

STANDARD FORM FOR THE EQUATION OF AN ELLIPSE WITH FOCI AT $(-c, 0)$ AND $(c, 0)$

$$\frac{x^2}{a^2} + \frac{y^2}{b^2} = 1, \qquad \text{where } b^2 = a^2 - c^2$$

The geometric interpretations of a and b can be found from this last equation. Letting $y = 0$ produces the x-intercepts, $x = \pm a$. The points $V_1(-a, 0)$ and $V_2(a, 0)$ are called the **vertices** of the ellipse. The **major axis** of the ellipse is the chord V_1V_2, which has length $2a$. Letting $x = 0$ produces the y-intercepts, $y = \pm b$. The points $(0, -b)$ and $(0, b)$ are the endpoints of the **minor axis.** The intersection of the major and minor axes is the **center** of the ellipse; in this case the center is the origin.

Note: The minor axis has length $2b$, and $2b < 2a$ since $b = \sqrt{a^2 - c^2} < a$.

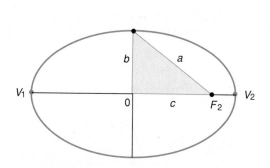

This figure shows the Pythagorean relationship of a, b, and c, namely $a^2 = b^2 + c^2$.

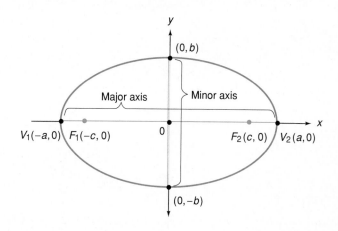

EXAMPLE 1 Sketch the graph of the ellipse $\dfrac{x^2}{25} + \dfrac{y^2}{9} = 1$. Find the coordinates of the foci.

Solution For $y = 0$, $x = \pm 5$. Therefore, the vertices are $(-5, 0)$ and $(5, 0)$, which are also the endpoints of the major axis. The endpoints of the minor axis are $(0, -3)$ and $(0, 3)$ since $y = \pm 3$ when $x = 0$. Locate these points, and several others, and draw the ellipse as shown on the following page.

If $x = \pm 4$, *then*

$$\frac{(\pm 4)^2}{25} + \frac{y^2}{9} = 1$$

$$\frac{y^2}{9} = 1 - \frac{16}{25} = \frac{9}{25}$$

$$y^2 = \frac{81}{25}$$

$$y = \pm \frac{9}{5}$$

This gives the four points shown.

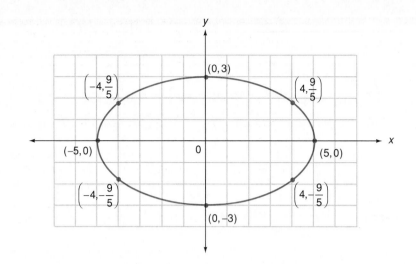

To find the foci, use $b^2 = a^2 - c^2$ with $a^2 = 25$ and $b^2 = 9$.

$$9 = 25 - c^2$$

$$c^2 = 16$$

$$c = +4 \qquad (c \text{ is a positive number.})$$

The foci are located at the points $(\pm 4, 0)$. ∎

EXAMPLE 2 Write the equation of the ellipse in standard form having vertices $(\pm 10, 0)$ and foci $(\pm 6, 0)$.

Solution Since $a = 10$ and $c = 6$ we get

$$b^2 = a^2 - c^2 = 100 - 36 = 64$$

Now substitute into the standard form.

$$\frac{x^2}{a^2} + \frac{y^2}{b^2} = 1 \Rightarrow \frac{x^2}{100} + \frac{y^2}{64} = 1$$ ∎

When the foci of the ellipse are on the y-axis, a similar development produces the following standard form:

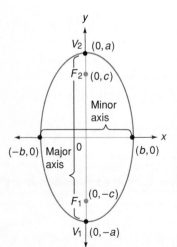

> **STANDARD FORM FOR THE EQUATION OF AN ELLIPSE**
> **WITH FOCI AT $(0, -c)$ AND $(0, c)$**
>
> $$\frac{x^2}{b^2} + \frac{y^2}{a^2} = 1, \qquad \text{where } b^2 = a^2 - c^2$$

The major axis is on the y-axis, and its endpoints are the vertices $(0, \pm a)$. The minor axis is on the x-axis and has endpoints $(\pm b, 0)$. Center: $(0, 0)$.

EXAMPLE 3   Change $25x^2 + 16y^2 = 400$ into standard form and graph.

Solution   Divide both sides of $25x^2 + 16y^2 = 400$ by 400 to obtain the form $\dfrac{x^2}{16} + \dfrac{y^2}{25} = 1$. Since $a^2 = 25$, $a = 5$ and the major axis is on the y-axis with length $2a = 10$. Similarly, $b^2 = 16$ gives $b = 4$, and the minor axis has length $2b = 8$. Also, $c^2 = a^2 - b^2 = 25 - 16 = 9$, so that $c = 3$, which locates the foci at $(0, \pm 3)$.

When the equation of an ellipse is in standard form, the major axis is horizontal if the x^2-term has the largest denominator. It will be vertical if the y^2-term has the largest denominator. Another way to determine this is to locate the endpoints of the two axes. The longer of the two is the major axis.

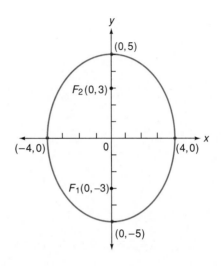

EXAMPLE 4   A satellite follows an elliptical orbit around the earth such that the center of the earth, E, is one of the foci. The figure indicates that the highest point that the satellite will be from the earth's surface is 2500 miles and the lowest will be 1000 miles. Observe that these distances are measured along the major axis, which is assumed to be on the y-axis. Use 4000 miles as the radius of the earth and find the equation of the orbit.

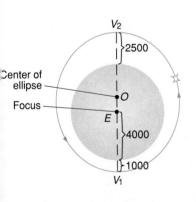

Solution   Since $2a = V_1 V_2$ is the length of the major axis, we have

$$2a = 2500 + 8000 + 1000 = 11{,}500$$

and

$$a = 5750$$

Also,

$$c = OE = OV_1 - EV_1 = a - 5000 = 5750 - 5000 = 750$$

Now find b^2.

$$b^2 = a^2 - c^2$$
$$= 5750^2 - 750^2$$
$$\approx 5701 \qquad \text{(by calculator)}$$

The symbol $\approx$ stands for "is approximately equal to."

Since the major axis is vertical, the equation of the orbit becomes

$$\frac{x^2}{5701^2} + \frac{y^2}{5750^2} = 1$$ ∎

Let us now consider the equation of an ellipse where the center is at a point (h, k), not necessarily the origin. If the major axis is horizontal, then the foci have coordinates $(h - c, k)$ and $(h + c, k)$, and it can be shown that the equation has this standard form:

Note: When $a = b$, the standard form produces $(x - h)^2 + (y - k)^2 = a^2$, which is the equation of a circle with center (h, k) and radius a. Thus a circle may be regarded as a special kind of ellipse, one for which the foci and center coincide.

> ### STANDARD FORM FOR THE EQUATION OF AN ELLIPSE WITH CENTER (h, k) AND FOCI $(h \pm c, k)$
>
> $$\frac{(x - h)^2}{a^2} + \frac{(y - k)^2}{b^2} = 1, \qquad \text{where } b^2 = a^2 - c^2$$

Similarly, an ellipse with center (h, k) whose major axis is vertical has this standard form:

> ### STANDARD FORM FOR THE EQUATION OF AN ELLIPSE WITH CENTER (h, k) AND FOCI $(h, k \pm c)$
>
> $$\frac{(x - h)^2}{b^2} + \frac{(y - k)^2}{a^2} = 1, \qquad \text{where } b^2 = a^2 - c^2$$

Observe that in this expanded form of an ellipse, the coefficients of x^2 and y^2 are not equal, but they have the same sign.

EXAMPLE 5 Write in standard form and graph:

$$4x^2 - 16x + 9y^2 + 18y = 11$$

Solution We follow a procedure much like that used in Section 3.5 for circles; that is, complete the square in both variables.

$$4x^2 - 16x + 9y^2 + 18y = 11$$

$$4(x^2 - 4x) + 9(y^2 + 2y) = 11$$

$$4(x^2 - 4x + 4) + 9(y^2 + 2y + 1) = 11 + 16 + 9$$

$$4(x - 2)^2 + 9(y + 1)^2 = 36$$

Divide both sides by 36:

$$\frac{(x - 2)^2}{9} + \frac{(y + 1)^2}{4} = 1$$

This is the equation of an ellipse having center at $(2, -1)$, with major axis $2a = 6$ and minor axis $2b = 4$. Since $c^2 = a^2 - b^2 = 5$, $c = \sqrt{5}$, which gives the foci $(2 \pm \sqrt{5}, -1)$.

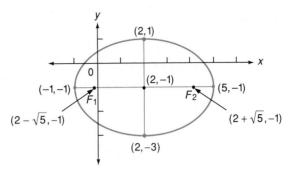

Find the coordinates of the center, vertices, and foci for each ellipse.

1. $\dfrac{x^2}{100} + \dfrac{y^2}{64} = 1$ 2. $16x^2 + y^2 = 16$

Write in standard form the equation of the ellipse having the given properties.

3. Vertices: $(\pm 8, 0)$
   Foci: $(\pm 7, 0)$

4. Vertices: $(0, \pm 9)$
   Length of minor axis: 12

5. Write the standard form of the ellipse $x^2 - 10x + 4y^2 + 16y = -37$, and give the center, vertices, and foci.

The ellipse has an interesting reflecting property. If a source of sound (or light) is positioned at one focus, the sound (or light) waves will reflect off the ellipse and pass through the other focus. Now suppose the ceiling of a large room is shaped like part of an ellipsoid (an ellipsoid is obtained by revolving an ellipse around its major axis), and you and a friend are each standing at a focus. If you whisper so that others in the room are unable to hear you, your friend will hear you because the sound waves will bounce off the ellipsoid and pass through the other focus. The Capitol building in Washington, D.C. has such a "whispering gallery."

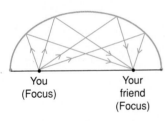

You
(Focus)

Your
friend
(Focus)

EXERCISES 3.6

Graph each ellipse. State the coordinates of the center, vertices, and foci.

1. $\dfrac{x^2}{25} + \dfrac{y^2}{16} = 1$ 2. $\dfrac{x^2}{16} + \dfrac{y^2}{25} = 1$ 3. $\dfrac{x^2}{16} + \dfrac{y^2}{9} = 1$ 4. $\dfrac{x^2}{9} + \dfrac{y^2}{16} = 1$

5. $\dfrac{x^2}{4} + \dfrac{y^2}{36} = 1$ 6. $\dfrac{x^2}{36} + \dfrac{y^2}{4} = 1$ 7. $9x^2 + y^2 = 9$ 8. $x^2 + 9y^2 = 9$

9. $4x^2 + 9y^2 = 36$ 10. $x^2 + 4y^2 = 4$ 11. $25x^2 + 9y^2 = 225$ 12. $16x^2 + 9y^2 = 144$

13. $\dfrac{(x-1)^2}{9} + \dfrac{(y+2)^2}{4} = 1$ 14. $\dfrac{(x+2)^2}{16} + \dfrac{(y-1)^2}{9} = 1$

15. $\dfrac{(x+2)^2}{4} + \dfrac{(y-3)^2}{9} = 1$ 16. $\dfrac{(x-3)^2}{9} + \dfrac{(y+2)^2}{16} = 1$

Write, in standard form, the equation of the ellipse having the given properties.

17. Center (0, 0); horizontal major axis of length 10; minor axis of length 6.
18. Center (0, 0); foci (±2, 0); vertices (±5, 0).
19. Center (2, 3); foci (−2, 3) and (6, 3); minor axis of length 8.
20. Center (2, −3); vertical major axis of length 12; minor axis of length 8.
21. Vertices (0, ±5); foci (0, ±3)
22. Center (−5, 0); foci (−5, ±2); $b = 3$.
23. Endpoints of the major and minor axes are (−8, 0), (8, 0), (0, −4), and (0, 4).
24. Endpoints of the major and minor axes are (−1, 0), (1, 0), (0, −3), and (0, 3).
25. Endpoints of the major and minor axes are (3, 1), (9, 1), (6, −1), and (6, 3).
26. Endpoints of the major and minor axes are (−4, 3), (2, 3), (−1, −2), and (−1, 8).

Write each ellipse in standard form. Find the coordinates of the center, vertices, and foci.

27. $25x^2 + y^2 − 12y = −11$
28. $x^2 + 4x + 9y^2 = 5$
29. $4x^2 + 24x + 13y^2 − 26y = 3$
30. $16x^2 − 32x + 9y^2 − 72y = −16$

31. In 1957, the Russians launched the first man-made satellite, Sputnik. Its orbit around the earth was elliptical with the center of the earth as one focus. The maximum height above the earth was about 580 miles, and the minimum height was approximately 130 miles.

   (a) Assuming that the earth's radius is 4000 miles, find the equation of Sputnick's orbit. (Leave the value b^2 in unsimplified form.)

    (b) Find the value of b to the nearest mile, and rewrite the equation of the ellipse using this result.

32. The orbit of the earth around the sun is elliptical with the sun being one of the foci. The earth's maximum distance from the sun is approximately 94.6 million miles, and the minimum distance is about 91.5 million miles.

   (a) Find a and b in millions of miles. (*Hint:* Use $a + c$ and $a − c$.)

   (b) Compare a and b and comment on the comparison in view of Exercise 34(a).

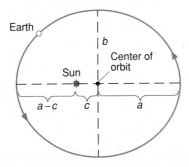

33. The underside of a bridge over a two-lane roadway is in the shape of a semiellipse. The elliptical arch spans 60 feet (the length of the major axis) and the height at the center is 20 feet.

   (a) The outsides of the driving lanes are marked by lines that are 10 feet from the base of the bridge. What is the bridge clearance y above these lines? (Write the exact numerical expressions.)

    (b) Find the clearance rounded to the nearest tenth of a foot.

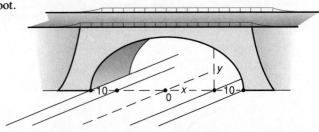

34. (a) The area of an ellipse with semimajor axis a and semiminor axis b is πab. Begin with an ellipse with such semiaxes, hold a fixed, and allow the length b to get closer and closer to the length a. As b gets close to a, what curve do the ellipses seem to approach? What is the area of this figure?

(b) Find the areas of the ellipses given in Exercises 1 and 13.

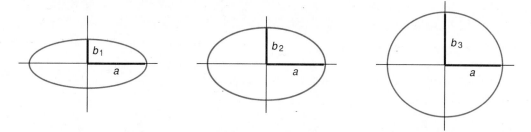

**3.7
CONIC
SECTIONS:
THE HYPERBOLA**

The hyperbola is another one of the conic sections. (See page 164.) The definition of the hyperbola is similar to that of the ellipse, except that we now make use of the *difference* of the distances from the foci.

DEFINITION OF A HYPERBOLA

A **hyperbola** is the set of all points in the plane such that the difference of the distances from two fixed points (called the **foci**) is a constant.

The two fixed points, F_1 and F_2, are called the **foci** of the hyperbola, and its **center** is the midpoint of the segment F_1F_2. It turns out that a hyperbola consists of two congruent branches which open in opposite directions.

We begin with a hyperbola with foci on the x-axis at $F_1(-c, 0)$ and $F_2(c, 0)$, where $c > 0$. Select a number $a > 0$ so that for any point P on the right branch of the hyperbola we have $PF_1 - PF_2 = 2a$. Also, for any point P on the left branch $PF_2 - PF_1 = 2a$, as in the figure below.

The form $2a$ is used to simplify computations.

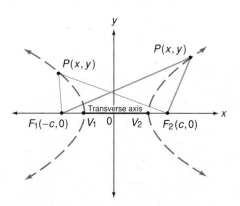

Note that $a < c$, since from triangle F_1PF_2 (P on the right branch) we have $PF_1 < PF_2 + F_1F_2$, which gives $PF_1 - PF_2 < F_1F_2$, or $2a < 2c$.

If we now follow the same type of analysis used to derive the equation of an ellipse, it can be shown that the equation of a hyperbola may be written in this standard form:

STANDARD FORM FOR THE EQUATION OF A HYPERBOLA WITH FOCI AT $(-c, 0)$ AND $(c, 0)$

$$\frac{x^2}{a^2} - \frac{y^2}{b^2} = 1, \qquad \text{where } b^2 = c^2 - a^2$$

Letting $y = 0$ gives $x = \pm a$; the points $V_1(-a, 0)$ and $V_2(a, 0)$ are the **vertices** of the hyperbola. The segment $V_1 V_2$ is called **transverse axis** of the hyperbola.

EXAMPLE 1   Write in standard form and identify the foci and vertices: $16x^2 - 25y^2 = 400$.

Solution   Divide through by 400 to place in standard form.

$$\frac{16x^2}{400} - \frac{25y^2}{400} = \frac{400}{400}$$

$$\frac{x^2}{25} - \frac{y^2}{16} = 1$$

Note that $a^2 = 25$ and $b^2 = 16$, so that $a = 5$ and $b = 4$. Then $c^2 = a^2 + b^2 = 25 + 16 = 41$ and $c = \pm\sqrt{41}$. The vertices of the hyperbola are located at $(-5, 0)$ and $(5, 0)$; the foci are at $(-\sqrt{41}, 0)$ and $(\sqrt{41}, 0)$.   ∎

Now return to the standard form for the equation of a hyperbola and solve for y:

$$\frac{x^2}{a^2} - \frac{y^2}{b^2} = 1$$

$$y^2 = \frac{b^2}{a^2}(x^2 - a^2)$$

$$y = \pm\frac{b}{a}\sqrt{x^2 - a^2}$$

Since $x^2 - a^2 \geq 0$, $x^2 \geq a^2$, which gives $|x| \geq |a|$.

Consequently, $|x| \geq a$, which means that there are no points of the hyperbola for $-a < x < a$.

An efficient way to sketch a hyperbola is first to draw the rectangle that is $2a$ units wide and $2b$ units high, as shown in the following figure. Note that the center of the hyperbola is also the center of this rectangle. Draw the diagonals of the rectangle and extend them in both directions; these are the **asymptotes.** Now sketch the hyperbola by beginning at the vertices $(\pm a, 0)$ so that the lines are asymptotes to the curve and the branches are between the asymptotes whose equations are

The distance between the curve and an asymptote becomes smaller and smaller and approaches 0; the curve approaches the line.

$$y = \pm\frac{b}{a}x.$$

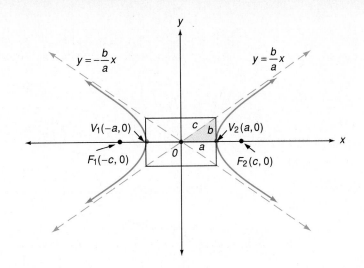

Since $b^2 = c^2 - a^2$, or $a^2 + b^2 = c^2$, it follows that b can be used as a side of a right triangle having hypotenuse c. Thus if we construct a perpendicular at V_2 of the length b, the resulting right triangle has hypotenuse c. We also see that this hypotenuse lies on the line $y = \dfrac{b}{a}x$.

EXAMPLE 2   Sketch the graph of $\dfrac{x^2}{9} - \dfrac{y^2}{4} = 1$.

Solution   Since $a^2 = 9$, the vertices are $(\pm 3, 0)$. Also,

$$b^2 = c^2 - a^2 \quad \text{or} \quad c^2 = a^2 + b^2 = 9 + 4 = 13$$

Therefore $(\pm\sqrt{13}, 0)$ are the foci. To sketch the hyperbola, first note that $2a = 6$, $2b = 4$ and draw the 6 by 4 rectangle with center $(0, 0)$. Draw the asymptotes by extending the diagonals and sketch the two branches.

The branches approach the asymptotes in the sense that as x gets larger (toward the right) the hyperbola gets closer to the asymptotes, but never touches them; similarly, for x, toward the left.

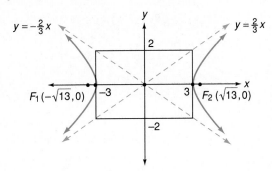

When the foci of a hyperbola are on the y-axis the equation has this standard form:

When the equation of a hyperbola is in standard form the branches open vertically if the minus sign precedes the x^2-term. They will open horizontally if the minus sign precedes the y^2-term.

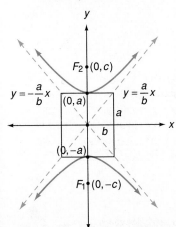

STANDARD FORM FOR THE EQUATION OF A HYPERBOLA WITH FOCI AT $(0, -c)$ AND $(0, c)$

$$\frac{y^2}{a^2} - \frac{x^2}{b^2} = 1, \qquad \text{where } b^2 = c^2 - a^2$$

The vertices are $(0, \pm a)$, the transverse axis has length $2a$, and the asymptotes are the lines $y = \pm\dfrac{a}{b}x$. The branches of this hyperbola open upward and downward.

Find the coordinates of the center, vertices, and foci, and also give the equations of the asymptotes.

1. $\dfrac{x^2}{9} - \dfrac{y^2}{16} = 1$

2. $\dfrac{y^2}{64} - \dfrac{x^2}{36} = 1$

Write in standard form the equation of the hyperbola having the given properties.

3. Vertices: $(0, \pm 5)$

   Foci: $(0, \pm 7)$

4. Vertices: $(\pm 2, 0)$

   Asymptotes: $y = \pm \dfrac{3}{2}x$

When the center of the hyperbola is at some point (h, k) and the transverse axis is horizontal, we have the following standard form:

STANDARD FORM FOR THE EQUATION OF A HYPERBOLA WITH CENTER AT (h, k) AND FOCI $(h \pm c, k)$

$$\frac{(x - h)^2}{a^2} - \frac{(y - k)^2}{b^2} = 1, \qquad \text{where } b^2 = c^2 - a^2$$

For a hyperbola with center at (h, k) and with a vertical transverse axis, the standard form is as follows:

STANDARD FORM FOR THE EQUATION OF A HYPERBOLA WITH CENTER AT (h, k) AND FOCI $(h, k \pm c)$

$$\frac{(y - k)^2}{a^2} - \frac{(x - h)^2}{b^2} = 1, \qquad \text{where } b^2 = c^2 - a^2$$

Observe that in this expanded form of a hyperbola the coefficients of x^2 and y^2 have opposite signs.

EXAMPLE 3   Write in standard form and graph:

$$4x^2 + 16x - 9y^2 + 18y = 29$$

Solution   Complete the square in x and y.

$$4(x^2 + 4x) - 9(y^2 - 2y) = 29$$

$$4(x^2 + 4x + 4) - 9(y^2 - 2y + 1) = 29 + 16 - 9$$

$$4(x + 2)^2 - 9(y - 1)^2 = 36$$

Divide both sides by 36:

$$\frac{(x + 2)^2}{9} - \frac{(y - 1)^2}{4} = 1$$

This is the standard form for a hyperbola with center at $(-2, 1)$. Since $a^2 = 9$, we have $a = 3$ and the vertices are located 3 units from the center at $(-5, 1)$ and

$(1, 1)$. Since $c^2 = a^2 + b^2 = 13$, the foci are located at $\sqrt{13}$ units from the center, namely at $(-2 \pm \sqrt{13}, 1)$.

To sketch the hyperbola, first draw the 6 by 4 rectangle with center at $(-2, 1)$ as shown. Draw the asymptotes by extending the diagonals and sketch the branches.

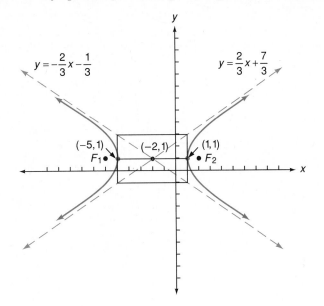

Observe that the asymptotes pass through the center $(-2, 1)$, their slopes are $\pm\dfrac{b}{a} = \pm\dfrac{2}{3}$, and their equations are

$$y - 1 = \pm\frac{2}{3}(x + 2)$$

In general, for this type of hyperbola, the equations of the asymptotes are

$$y - k = \pm\frac{b}{a}(x - h).$$

EXERCISES 3.7

Graph each hyperbola. State the coordinates of the center, vertices, and foci. Give the equations of the asymptotes.

1. $\dfrac{x^2}{25} - \dfrac{y^2}{9} = 1$

2. $\dfrac{x^2}{4} - \dfrac{y^2}{9} = 1$

3. $\dfrac{x^2}{16} - \dfrac{y^2}{25} = 1$

4. $\dfrac{x^2}{16} - \dfrac{x^2}{4} = 1$

5. $\dfrac{y^2}{36} - \dfrac{x^2}{9} = 1$

6. $\dfrac{y^2}{4} - \dfrac{x^2}{9} = 1$

7. $9x^2 - y^2 = 9$

8. $x^2 - 9y^2 = 9$

9. $4x^2 - 9y^2 = 36$

10. $9y^2 - 4x^2 = 36$

11. $25y^2 - 9x^2 = 225$

12. $9y^2 - 16x^2 = 144$

13. $\dfrac{(x - 2)^2}{9} - \dfrac{(y + 1)^2}{4} = 1$

14. $\dfrac{(x + 3)^2}{16} - \dfrac{(y - 2)^2}{9} = 1$

15. $\dfrac{(y + 2)^2}{25} - \dfrac{(x - 3)^2}{16} = 1$

16. $\dfrac{(y + 2)^2}{16} - \dfrac{(x - 1)^2}{4} = 1$

Write, in standard form, the equation of the hyperbola having the given properties.

17. Center $(0, 0)$; foci $(\pm6, 0)$; vertices $(\pm4, 0)$.

18. Center $(0, 0)$; foci $(0, \pm4)$; vertices $(0, \pm1)$.

19. Center $(-2, 3)$; vertical transverse axis of length 6; $c = 4$.

20. Center $(4, 4)$; vertex $(4, 7)$; $b = 2$.

21. Center $(0, 0)$; asymptotes $y = \pm\dfrac{1}{2}x$; vertices $(\pm4, 0)$.

22. Asymptotes $y = \pm\dfrac{5}{12}x$; foci $(\pm13, 0)$.

23. Asymptotes $y = \pm\dfrac{8}{15}x$; foci $(0, \pm17)$.

Identify the center, vertices, and foci of each hyperbola.

24. $\dfrac{(x-2)^2}{25} - \dfrac{(y+1)^2}{16} = 1$

25. $\dfrac{(x+1)^2}{16} - \dfrac{(y-3)^2}{9} = 1$

26. $\dfrac{(y-3)^2}{16} - \dfrac{(x+1)^2}{4} = 1$

27. $\dfrac{(y+1)^2}{64} - \dfrac{(x-3)^2}{36} = 1$

Write in standard form and identify the center, vertices, and foci of each hyperbola.

28. $9x^2 + 36x - 16y^2 - 96y = 252$

29. $4x^2 - 8x - 9y^2 - 36y = 68$

30. $16y^2 + 32y - 9x^2 - 90x = 353$

31. $y^2 + 4y - 4x^2 + 8x = 4$

32. Let P be on the right branch of the hyperbola with foci $F_1(-c, 0)$ and $F_2(c, 0)$, and let

$PF_1 - PF_2 = 2a$ for $a < c$. Derive the equation $\dfrac{x^2}{a^2} - \dfrac{y^2}{b^2} = 1$, where $b^2 = c^2 - a^2$.

3.8 CONIC SECTIONS: THE PARABOLA

The parabola, like the other conic sections, can be defined as a specific set of points in the plane and has numerous applications that are directly related to this definition.

Note: A considerable amount of work has already been done with parabolas in Sections 3.1 and 3.2.

DEFINITION OF A PARABOLA

A **parabola** is the set of all points in a plane equidistant from a given fixed line called the **directrix** and a given fixed point called the **focus.**

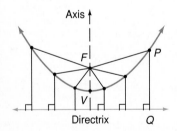

For each point P on the parabola at the left $PF = PQ$, where F is the focus and Q is the point on the directrix. The line through F and perpendicular to the directrix is called the **axis** of the parabola, and the point V, which is the intersection of the parabola with its axis, is called the **vertex.**

The parabolas we will consider will have either vertical or horizontal axes. We begin with parabolas whose axes are the y-axis.

In the figures that follow, let focus F have coordinates $(0, p)$, and let the directrix have equation $y = -p$ as indicated.

Observe that in each case the focus is within the parabola and the directrix is outside.

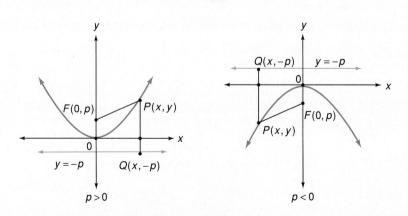

The origin, which is also the vertex, is on this parabola because it is the same distance from the focus and the directrix. In general, for any point $P(x, y)$ on the parabola the distance formula can be used to find PF and PQ. Then set these two distances equal to each other as indicated in the definition.

$$PF = \sqrt{(x - 0)^2 + (y - p)^2} = \sqrt{x^2 + (y - p)^2}$$
$$PQ = \sqrt{(x - x)^2 + (y - (-p))^2} = \sqrt{(y + p)^2}$$
$PF = PQ$: $\quad \sqrt{x^2 + (y - p)^2} = \sqrt{(y + p)^2}$ $\qquad$ (Square each side.)
$$x^2 + y^2 - 2py + p^2 = y^2 + 2py + p^2$$
$$x^2 = 4py$$

In summary, we have the following:

This equation can also be written as $y = \dfrac{1}{4p}x^2$. Letting $a = \dfrac{1}{4p}$ gives $y = ax^2$, the form we used in our earlier work with parabolas.

> ### STANDARD FORM FOR THE EQUATION OF A PARABOLA WITH FOCUS $(0, p)$ AND DIRECTRIX $y = -p$
>
> $$x^2 = 4py$$
>
> The axis is the y-axis and the vertex is the origin.

This form for the equation of a parabola can be used to determine the coordinates of the focus and the equation of the directrix, as in Example 1.

EXAMPLE 1 Find the coordinates of the focus and the equation of the directrix for the parabola $x^2 = 4y$ and sketch the graph.

Solution Consider the general form $x^2 = 4py$, and let $4p = 4$. Thus $p = 1$ and we can locate the focus and directrix. The parabola has its focus at $(0, 1)$ and the equation of the directrix is $y = -p = -1$.

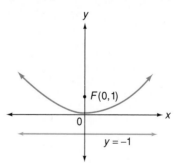

EXAMPLE 2 The focus of a parabola has coordinates $(0, -\frac{5}{2})$ and the vertex is at the origin. Find the equation of the parabola.

Solution Both the focus and vertex are on the y-axis. Therefore, the y-axis is the axis of the parabola. Since such parabolas have focus at $(0, p)$, we get $p = -\frac{5}{2}$. Then, using $x^2 = 4py$, the equation is

$$x^2 = 4(-\tfrac{5}{2})y = -10y$$

A parabola with vertex at the origin whose axis is the x-axis has an equation of the form $y^2 = 4px$. The derivation for this standard form is very similar to the derivation used to obtain the form $x^2 = 4py$.

STANDARD FORM FOR THE EQUATION OF A PARABOLA WITH FOCUS $(p, 0)$ AND DIRECTRIX $x = -p$

$$y^2 = 4px$$

The axis is the x-axis and the vertex is the origin.

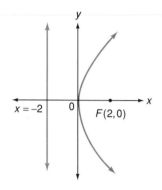

EXAMPLE 3   The directrix of a parabola is $x = -2$ and the focus is $(2, 0)$. Find the equation and graph.

Solution   Since the directrix is a vertical line, the axis of the parabola is horizontal. Also, since the focus $(2, 0)$ is on the x-axis, the parabola's axis is the x-axis. Such parabolas have focus $(p, 0)$ and therefore $p = 2$. Thus, using $y^2 = 4px$, the equation is

$$y^2 = 4(2)x = 8x$$

■

Observe that in the algebraic form of parabolas only one of the variables has a squared term and the other variable is to the first power.

The parabola $x^2 = 4py$ has vertex $(0, 0)$ and vertical axis $x = 0$. When this parabola is translated h units horizontally and k units vertically, a congruent parabola is obtained having equation $(x - h)^2 = 4p(y - k)$. The vertex $(0, 0)$ has been translated to (h, k), and the focus, directrix, and axis have been translated as follows:

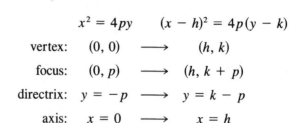

Translate the parabola $y^2 = 4px$ to obtain the parabola with vertex (h, k), and write a similar chart. (Also see Exercises 27–31.)

$$x^2 = 4py \qquad (x - h)^2 = 4p(y - k)$$

vertex: $(0, 0)$ $\longrightarrow$ (h, k)

focus: $(0, p)$ $\longrightarrow$ $(h, k + p)$

directrix: $y = -p$ $\longrightarrow$ $y = k - p$

axis: $x = 0$ $\longrightarrow$ $x = h$

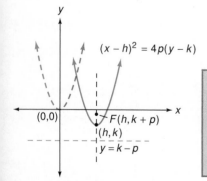

STANDARD FORM FOR THE EQUATION OF A PARABOLA WITH FOCUS $(h, k + p)$ AND DIRECTRIX $y = k - p$

$$(x - h)^2 = 4p(y - k)$$

The axis is $x = h$ and the vertex is (h, k).

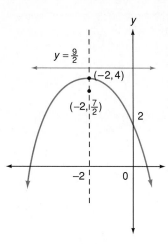

EXAMPLE 4 A parabola has vertex $(-2, 4)$ and focus $(-2, \frac{7}{2})$. Write the equations of the parabola, the directrix, and the axis.

Solution Since the vertex is $(-2, 4)$, $h = -2$ and $k = 4$. Also, since the first coordinate of the vertex and focus is -2, the axis is the vertical line $x = h = -2$, or $x = -2$. The focus for such a parabola is $(h, k + p)$. Therefore

$$k + p = 4 + p = \tfrac{7}{2}$$
$$p = -\tfrac{1}{2}$$

and the directrix is $y = k - p = 4 - (-\frac{1}{2}) = \frac{9}{2}$, or $y = \frac{9}{2}$. The equation of the parabola is

$$(x - (-2))^2 = 4(-\tfrac{1}{2})(y - 4)$$
$$(x + 2)^2 = -2(y - 4)$$ ∎

The last equation in Example 4 can be converted to $x^2 + 4x + 2y - 4 = 0$. Beginning with this form we can get back to the standard form by completing the square in x as follows:

$$x^2 + 4x + 2y - 4 = 0$$
$$x^2 + 4x = -2y + 4$$
$$x^2 + 4x + 4 = -2y + 8$$
$$(x + 2)^2 = -2(y - 4)$$
$$(x + 2)^2 = 4(-\tfrac{1}{2})(y - 4)$$

Parabolic reflectors are used in a wide variety of instruments, including reflecting telescopes, searchlights, microwave antennae, and solar energy devices. The surface of a parabolic reflector is obtained by revolving a parabola around its axis of symmetry.

The following figure at the left illustrates that when light rays, parallel to the axis, strike the surface they are reflected to the focus. Conversely, the figure at the right shows that when there is a light source at the focus of a parabolic reflector the light rays will reflect off the surface, forming a beam of light parallel to the axis.

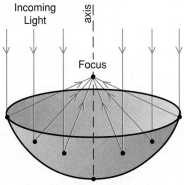

Parabolic Reflector

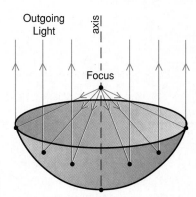

Parabolic Reflector

EXAMPLE 5 Suppose that the reflecting surface of a television antenna was formed by rotating the parabola $y = \frac{1}{15}x^2$ about its axis of symmetry for the interval $-5 \le x \le 5$. Assuming that the measurements are made in feet, how far from the bottom or vertex of this "dish" antenna should the receiver be placed? How deep is this antenna?

Solution The receiver must be at the focus and the distance the focus is from the vertex is the value p in the standard form $x^2 = py$. Since $y = \frac{1}{15}x^2$, $x^2 = 15y$. Then

$$4p = 15$$
$$p = \tfrac{15}{4} = 3.75$$

The receiver is 3.75 feet above the vertex along the axis.

The depth of the antenna is the y-value when $x = 5$. Thus, using 5 in $y = \frac{1}{15}x^2$, we get $y = \frac{25}{15} = \frac{5}{3}$. The antenna is $1\frac{2}{3}$ feet deep.

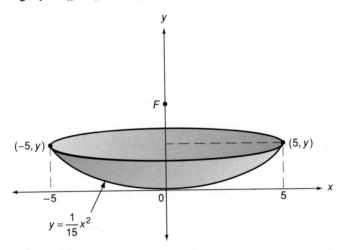

EXERCISES 3.8

Find the coordinates of the focus and the equation of the directrix for each of the following parabolas.

1. $x^2 = \frac{1}{4}y$
2. $y = \frac{1}{2}x^2$
3. $x^2 = \frac{1}{2}y$
4. $x^2 = -4y$
5. $y^2 = 2x$
6. $y^2 = -2x$
7. $y = -4x^2$
8. $y = -\frac{1}{8}x^2$
9. $-9x = 6y^2$
10. $\frac{1}{2}x = \frac{1}{8}y^2$
11. $y^2 = -\frac{1}{3}x$
12. $x = -\frac{3}{2}y^2$

Write the equation of each parabola having the given properties and sketch the graph.

13. Focus $(0, -3)$; directrix $y = 3$.
14. Focus $(0, 3)$; directrix $y = -3$.
15. Directrix $y = -\frac{2}{3}$; vertex $(0, 0)$.
16. Vertex $(0, 0)$; vertical axis; $(2, -2)$ is on the parabola.
17. Focus $(-\frac{3}{4}, 0)$; directrix $x = \frac{3}{4}$.
18. Focus $(\frac{5}{8}, 0)$; vertex $(0, 0)$.
19. Vertex $(0, 0)$; horizontal axis; $(12, -6)$ is on the parabola.
20. Directrix $x = \frac{3}{2}$; vertex $(0, 0)$.

21. A parabola has vertex $(2, -5)$, focus at $(2, -3)$, and directrix $y = -7$. Write its equation in the form $(x - h)^2 = 4p(y - k)$. Then write its equation in the forms $y = a(x - h)^2 + k$ and $y = ax^2 + bx + c$.

22. A parabola has directrix $y = -1$ and focus $(3, 3)$. Write the equations of the parabola, and the axis of symmetry, and the coordinates of the vertex.

Identify the vertex, the axis of symmetry, the focus, and the directrix.

23. $(x - 2)^2 = 4(y + 5)$

24. $(x + \frac{3}{2})^2 = -\frac{1}{2}(y - 1)$

Write the equation of each parabola in the standard form $(x - h)^2 = 4p(y - k)$ and identify the vertex, focus, and directrix.

25. $y = \frac{1}{4}x^2 - x + 4$

26. $y = 2x^2 - 12x + 16$

27. The origin is the vertex of each parabola having equation $y^2 = 4px$. What is the equation of the parabola with vertex (h, k) and focus $(h + p, k)$?

28. A parabola has vertex $(2, -5)$ and focus $(-1, -5)$. Write its equation in the form of Exercise 27. Then write its equation in the two forms $x = a(y - k)^2 + h$ and $x = ay^2 + by + c$.

29. A parabola has vertex $(-3, -2)$ and directrix $x = -\frac{9}{2}$. Find the focus and write the equations of the parabola and the axis of symmetry.

Identify the vertex, axis of symmetry, focus, and directrix.

30. $(y + 5)^2 = \frac{4}{3}(x - 6)$

31. $y^2 = -9(x - 4)$

32. Convert $y^2 + 6y - 2x + 9 = 0$ into the standard form $(y - k)^2 = 4p(x - h)$ and identify the vertex, axis of symmetry, focus, and directrix.

33. The reflecting surface of a radar antenna is generated by revolving the parabola $y = \frac{2}{9}x^2$ about its axis of symmetry for $-4 \le x \le 4$. Assuming the measurements are done in feet, how far from the bottom of the dish antenna is the receiver? What is the circumference of the antenna?

34. The reflecting surface of an antenna, as in Exercise 33, is generated by revolving a parabola of the form $x^2 = 4py$ about its axis of symmetry. If the antenna is 8 feet across the top (this is the length of a diameter) and $1\frac{1}{2}$ feet deep, where must the receiver be located?

35. The center cable of a suspension bridge forms a parabolic arc. The cable is suspended from the tops of the two support towers, which are 800 feet apart. If the tops of the towers are 160 feet above the road and the cable just touches the road midway between the towers, find the height of the cable at a distance of 100 feet from a tower.

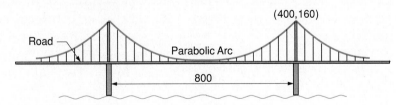

Name the conic section and, where applicable, give the coordinates of the center, vertices, and foci, the radius, and the equations of the asymptotes and of the directrix.

36. $y = (x + 2)^2$

37. $\dfrac{x^2}{25} + \dfrac{y^2}{16} = 1$

38. $\dfrac{x^2}{16} + \dfrac{y^2}{25} = 1$

39. $\dfrac{x^2}{36} - \dfrac{y^2}{25} = 1$

40. $\dfrac{y^2}{36} - \dfrac{x^2}{25} = 1$

41. $x^2 + y^2 = 16$

Name the conic and graph.

42. $\dfrac{x^2}{16} + \dfrac{y^2}{9} = 1$ **43.** $16y^2 - 9x^2 = 144$ **44.** $y = 2(x + 1)^2 - 1$

45. $\dfrac{(x - 1)^2}{64} + \dfrac{(y - 2)^2}{36} = 1$ **46.** $\dfrac{(y - 1)^2}{64} - \dfrac{(x - 3)^2}{36} = 1$ **47.** $16(y - 3)^2 - 9(x + 2)^2 = -144$

48. $(x - 1)^2 + (y + 1)^2 = 25$

Identify each conic and write in standard form.

49. $x^2 + y^2 - 2x + 4y + 1 = 0$ **50.** $x^2 + y^2 + 6x - 4y + 4 = 0$

51. $x^2 + 4y^2 + 2x - 3 = 0$ **52.** $x^2 - 9y^2 - 2x - 8 = 0$

53. $9x^2 + 18x - 16y^2 + 96y = 279$ **54.** $4x^2 - 16x + y^2 + 8y = -28$

55. $y^2 + 10y = 6x - 1$ **56.** $y = 2x^2 - 4x + 5$

EXPLORATIONS
Think Critically

1. Begin with a rectangular piece of paper or cardboard, 9 inches by 12 inches. Cut off squares of 1 inch from each corner and fold to form a box without a top, such as that shown on page 156. Repeat this with other pieces of paper for corner squares of 2 inches and 3 inches. Look at the three boxes formed and guess which has the maximum volume. Then do the necessary computations to verify the answer.

2. Sketch the graph of the ellipse $\dfrac{x^2}{25} + \dfrac{y^2}{16} = 1$. Join the endpoints of the major axis to those of the minor axis. Find the equation of this new figure, said to be *analogous* to the ellipse.

3. Sketch the graph for $|x| - |y| = 1$. Find the equation of the hyperbola that is analogous to this figure.

4. The graphs of the following are said to be *conjugate hyperbolas*:

$$\frac{x^2}{25} - \frac{y^2}{16} = 1 \qquad \frac{x^2}{16} - \frac{y^2}{25} = 1$$

Graph both on the same set of axes and describe the relationship between the two graphs.

5. All of the conic sections that we have studied may be written in the form $Ax^2 + Cy^2 + Dx + Ey + F = 0$, where A and C are not both equal to 0. What type of conic is represented if (a) $A = C \neq 0$; (b) $A = 0$ or $C = 0$; (c) $A \neq C$ and both are either positive or negative; (d) $A \neq C$ and are of opposite signs?

**3.9
SOLVING
NONLINEAR
SYSTEMS**

When you study calculus you will learn how to find the areas of the regions between curves. For example, the areas of the shaded regions in the first two diagrams can be found once the coordinates of the points of intersection are known. Here we will address ourselves only to this part of the problem: finding the points of intersection.

A straight line will intersect a parabola or a circle twice, or once, or not at all. Two parabolas of the form $y = ax^2 + bx + c$ can intersect at most two times; the same is true for two circles. A circle and a parabola can intersect at most four times. These diagrams illustrate some of these possibilities.

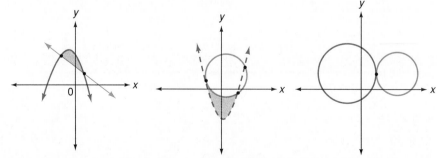

In each of the examples that follow, at least one of the two equations will not be linear. Thus we will be learning how to solve certain types of *nonlinear systems*. The underlying strategy in solving such systems will be the same as it was for linear systems, namely to first eliminate one of the two variables so as to obtain an equation in one unknown.

A Parabola and a Line

EXAMPLE 1   Solve the system of two equations and graph:

$$y = x^2$$
$$y = -2x + 8$$

Solution   Let (x, y) represent the points of intersection. Since these x- and y-values are the same in both equations, we may set the two values for y equal to each other and solve for x.

$$x^2 = -2x + 8$$
$$x^2 + 2x - 8 = 0$$
$$(x + 4)(x - 2) = 0$$
$$x = -4 \quad \text{or} \quad x = 2$$

To find the corresponding y-values, either of the original equations may be used. Using $y = -2x + 8$, we have:

For $x = -4$: $y = -2(-4) + 8 = 16$
For $x = 2$: $y = -2(2) + 8 = 4$

The solution of the system consists of the two ordered pairs $(-4, 16)$ and $(2, 4)$. The other equation can be used as a check of these results.

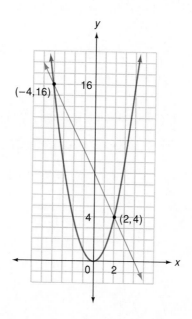

Two Parabolas

EXAMPLE 2 Solve the system and graph:

$$y = x^2 - 2$$
$$y = -2x^2 + 6x + 7$$

Solution   Set the two values for y equal to each other and solve for x.

$$x^2 - 2 = -2x^2 + 6x + 7$$
$$3x^2 - 6x - 9 = 0$$
$$x^2 - 2x - 3 = 0$$
$$(x + 1)(x - 3) = 0$$
$$x = -1 \quad \text{or} \quad x = 3$$

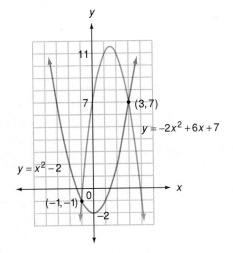

Use $y = x^2 - 2$ to solve for y.

$$y = (-1)^2 - 2 = -1 \qquad y = 3^2 - 2 = 7$$

Check these points in the second equation of the given system.

The points of intersection are $(-1, -1)$ and $(3, 7)$.

A Circle and a Parabola

EXAMPLE 3   Solve the system and graph.

$$x^2 + y^2 - 8y = -7$$
$$y - x^2 = 1$$

Solution   Solve the second equation for x^2.

$$x^2 = y - 1$$

Substitute into the first equation and solve for y.

Note the use of the substitution method here. Alternative methods can also be used to solve this example. Another easy way begins by adding the given equations. Try it. We may also solve $y - x^2 = 1$ for y and substitute into the first equation. You will find that the latter method is more difficult. With practice you will learn how to find the easier methods.

$$(y - 1) + y^2 - 8y = -7$$
$$y^2 - 7y + 6 = 0$$
$$(y - 1)(y - 6) = 0$$
$$y = 1 \quad \text{or} \quad y = 6$$

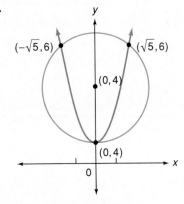

Use $x^2 = y - 1$ to solve for x.

For $y = 1$: $x^2 = 1 - 1 = 0$ $x = 0$

For $y = 6$: $x^2 = 6 - 1 = 5$ $x = \pm\sqrt{5}$

Check these points in the given system.

The points of intersection are $(0, 1)$, $(\sqrt{5}, 6)$, and $(-\sqrt{5}, 6)$. ■

A Circle and an Ellipse

EXAMPLE 4   Solve the system and graph.

$$x^2 + y^2 = 9$$

$$\frac{x^2}{25} + \frac{y^2}{16} = 1$$

Solution   You should recognize the equations as those of a circle and an ellipse. First rewrite the second equation in this form:

$$16x^2 + 25y^2 = 400$$

Then solve the first equation for either x^2 or y^2, say y^2, and substitute into the second equation:

$$y^2 = 9 - x^2$$
$$16x^2 + 25(9 - x^2) = 400$$
$$16x^2 + 225 - 25x^2 = 400$$
$$-9x^2 = 175$$
$$x^2 = -\frac{175}{9}$$
$$x = \pm\sqrt{-\frac{175}{9}}$$

The attempted solution produces the square root of a negative number, which is imaginary. Thus there are no real solutions and the two curves do not intersect.

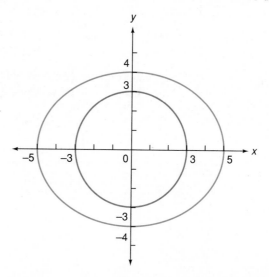

An Ellipse and a Hyperbola

EXAMPLE 5 Solve the system and graph.

$$x^2 - y^2 = 1$$
$$9x^2 + y^2 = 9$$

Solution Add the two equations:

$$
\begin{aligned}
x^2 - y^2 &= 1 \\
9x^2 + y^2 &= 9 \\
\hline
10x^2 \quad\;\; &= 10 \\
x^2 &= 1 \\
x = \pm 1 \quad &\text{and} \quad y = 0
\end{aligned}
$$

The points of intersection are located at $(\pm 1, 0)$.

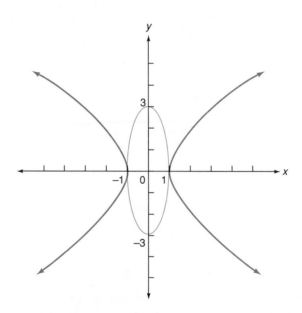

The graphing of quadratic inequalities in two variables was introduced in Section 3.1. We can also graph *systems* of such inequalities as demonstrated in the next example.

EXAMPLE 6 Graph the system of inequalities.

$$y \geq x^2$$
$$y \leq -2x + 8$$

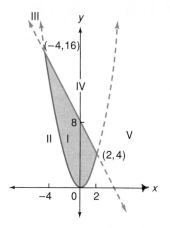

Solution First draw the graph of the equations $y = x^2$ and $y = -2x + 8$ (see Example 1). Now use test points from the numbered regions determined by the two graphs. If a test point satisfies both inequalities, then the region it comes from is part of the graph of the system. Otherwise, the region is not part of the graph. Using $(0, 4)$ from region I gives true statements:

$$4 \geq 0^2 = 0$$
$$4 \leq -2(0) + 8 = 8$$

Therefore, region I is included. Using $(3, 4)$ from region V gives

$$4 \geq 3^2 = 9$$
$$4 \leq -2(3) + 8 = 2$$

Since $(3, 4)$ does not satisfy *both* inequalities, region V is excluded. You should verify that regions II, III, IV are also excluded. The graph of the system is the shaded part as shown. Note that the boundaries of region I are included since the given system uses the inequality symbols $\leq$ and $\geq$. ∎

Another way to obtain this graph is to observe that $y \geq x^2$ calls for all points on and above the parabola $y = x^2$, and $y \leq -2x + 8$ calls for all points on or below the line $y = -2x + 8$.

EXERCISES 3.9

Solve each system and graph.

1. $y = -x^2 - 4x + 1$
   $y = 2x + 10$

2. $3x - 4y = -5$
   $(x + 3)^2 + (y + 1)^2 = 25$

3. $(x + 4)^2 + (y - 1)^2 = 16$
   $(x + 4)^2 + (y - 3)^2 = 4$

4. $y = x^2 - 6x + 9$
   $(x - 3)^2 + (y - 9)^2 = 9$
   (*Hint:* Factor $x^2 - 6x + 9$ and substitute.)

5. $y = x^2 + 6x + 6$
   $y = -x^2 - 6x + 6$

6. $y = \frac{1}{3}(x - 3)^2 - 3$
   $(x - 3)^2 + (y + 2)^2 = 1$
   (*Hint:* Solve the first equation for $(x - 3)^2$ and substitute into the second.)

Solve each system.

7. $y = (x + 1)^2$
   $y = (x - 1)^2$

8. $x^2 + y^2 = 9$
   $y = x^2 - 3$

9. $y = x^2$
   $y = -x^2 + 8x - 16$

10. $y = x^2$
    $y = x^2 - 8x + 24$

11. $7x + 3y = 42$
    $y = -3x^2 - 12x - 15$

12. $y + 2x = 1$
    $x^2 + 4x = 6 - 2y$

13. $y - x = 0$
    $(x - 2)^2 + (y + 5)^2 = 25$

14. $y - 2x = 0$
    $(x - 2)^2 + (y + 5)^2 = 25$

15. $x - 2y^2 = 0$
    $x^2 - y^2 = 3$

16. $x^2 + y^2 = 25$
    $2x^2 + y^2 = 34$

17. $4x^2 + y^2 = 4$
    $2x - y = 2$

18. $4x^2 + y^2 = 4$
    $x + y = 3$

19. $4x^2 - 9y^2 = 36$
    $9x^2 + 4y^2 = 36$

20. $2x^2 - y^2 = 1$
    $y^2 - x^2 = 3$

21. $2x^2 + y^2 = 11$
    $x^2 - 2y^2 = -2$

22. $x^2 - 2y^2 = 8$
    $3x + 4y = 4$

23. $y = -x^2 + 2x$
    $x^2 - 2x + y^2 - 2y = 0$

24. $x^2 + 4x + y^2 - 4y = -4$
    $(x - 2)^2 + (y - 2)^2 = 4$

25. $(x - 1)^2 + y^2 = 1$
    $x^2 + (y - 1)^2 = 1$

26. $4x + 3y = 25$
    $x^2 + y^2 = 25$

27. $y = \frac{1}{3}(x - 3)^2 - 3$
    $x^2 - 6x + y^2 + 2y = -6$

28. $y = x^2 - 6x + 9$
    $(x - 3)^2 + (y - 2)^2 = 58$

Graph each system.

29. $y \geq -x^2 - 4x + 1$
$y \leq 2x + 10$
(See Exercise 1.)

30. $3x - 4y \geq -5$
$(x + 3)^2 + (y + 1)^2 \leq 25$
(See Exercise 2.)

31. $(x + 4)^2 + (y - 1)^2 \leq 16$
$(x + 4)^2 + (y - 3)^2 \geq 4$
(See Exercise 3.)

32. $y \geq x^2 - 6x + 9$
$(x - 3)^2 + (y - 9)^2 \geq 9$
(See Exercise 4.)

33. $y \geq x^2 + 6x + 6$
$y \leq -x^2 - 6x + 6$
(See Exercise 5.)

34. $y \geq \frac{1}{3}(x - 3)^2 - 3$
$(x - 3)^2 + (y + 2)^2 \geq 1$
(See Exercise 6.)

35. $9x^2 + 25y^2 \leq 225$
$x^2 + 4y^2 \geq 16$

36. $\dfrac{x^2}{16} + \dfrac{y^2}{9} \leq 1$
$\dfrac{x^2}{9} + \dfrac{y^2}{16} \leq 1$

37. $\dfrac{x^2}{16} + \dfrac{y^2}{9} \leq 1$
$\dfrac{x^2}{9} + \dfrac{y^2}{16} \geq 1$

KEY TERMS

*Review these **key terms** so that you are able to define or describe them. A clear understanding of these terms will be very helpful when reviewing the developments of this chapter.*

Quadratic function • Parabola • Axis of symmetry and vertex • Increasing and decreasing • Concave up and concave down • Vertical shift of parabolas • Horizontal shift of parabolas • Multiplying ordinates • Relation • Vertical line test for functions • Quadratic inequality in two variables • Completing the square • Quadratic formula • Roots of a quadratic equation • Discriminant • Quadratic inequality in one variable • Maximum or minimum of a quadratic function • Distance formula • Midpoint formula • Conic section • Circle: center, radius • Tangent to a circle • Ellipse: major and minor axes, foci, vertices • Hyperbola: transverse axis, foci, vertices, asymptotes • Focus • Directrix • Vertex • Axis of symmetry (of a parabola) • Standard form for a circle, ellipse, hyperbola, parabola • Nonlinear system of equations or inequalities

REVIEW EXERCISES

The solutions to the following exercises can be found within the text of Chapter 3.
Try to answer each question before referring to the text.

Section 3.1

Graph each of the following.

1. $y = x^2$

2. $y = x^2 + 2$

3. $y = x^2 - 2$

4. $y = (x + 2)^2$

5. $y = (x - 2)^2$

6. $y = -x^2$

7. $y = \frac{1}{2}x^2$

8. $y = 2x^2$

9. $y = -x^2 + 2$

10. $y = f(x) = -(x - 2)^2 + 1$

11. $y = f(x) = -2(x - 3)^2 + 4$

12. $x = y^2$

13. $x = (y + 2)^2 - 4$

14. $y \geq x^2$

15. $y + (x - 2)^2 \leq 5$

Section 3.2

Write in standard form.

16. $y = x^2 + 4x + 3$

17. $y = 2x^2 - 12x + 11$

18. $y = -\frac{1}{3}x^2 - 2x + 1$

Write in standard form, graph, and give the vertex and axis of symmetry.

19. $y = x^2 - 4x + 4$

20. $y = -2 - 4x + 3x^2$

21. Graph the function $y = f(x) = |x^2 - 4|$ and find the range.

22. State the conditions on the values a and k so that the parabola $y = a(x - h)^2 + k$ opens downward and intersects the x-axis in two points. What are the domain and range of this function?

Section 3.3

23. Find the x-intercepts of $f(x) = 25x^2 + 30x + 9$.
24. State the quadratic formula.
25. Solve for x: $2x^2 - 5x + 1 = 0$.
26. Solve for x: $2x^2 = x - 1$.
27. Find the value of the discriminant and decide how many x-intercepts there are.
    (a) $y = x^2 + 4x + 3$ (b) $y = x^2 - 4x + 4$ (c) $y = 2x^2 - 4x + 5$
28. Solve: $x^2 - x - 6 < 0$.

Section 3.4

29. The length of a rectangular piece of cardboard is 2 inches more than its width. An open box is formed by cutting out 4-inch squares from each corner and folding up the sides. If the volume of the box is 672 cubic inches, find the dimensions of the original cardboard.
30. Find two consecutive positive integers such that the sum of their squares is 113.
31. Find the maximum value or minimum value of the function $f(x) = 2(x + 3)^2 + 5$.
32. Find the maximum value of the quadratic function $f(x) = -\frac{1}{3}x^2 + x + 2$. At which value x does f achieve this maximum?
33. Suppose that 60 meters of fencing is available to enclose a rectangular garden, one side of which will be against the side of a house. What dimensions of the garden will guarantee a maximum area?
34. The sum of two numbers is 24. Find the two numbers such that their product is to be a maximum.
35. A ball is thrown straight upward from ground level with an initial velocity of 32 feet per second. The formula $s = 32t - 16t^2$ gives its height in feet, s, after t seconds.
    (a) What is the maximum height reached by the ball?
    (b) When does the ball return to the ground?

Section 3.5

36. Find the length of the line segment determined by the points $(-2, 2)$ and $(6, -4)$.
37. What is the equation of the circle with center at $(0, 0)$ and radius 3?
38. Write the equation of the circle with center $(-3, 5)$ and radius $\sqrt{2}$.

Find the center and radius of the circle for each equation

39. $(x - 2)^2 + (y + 3)^2 = 4$ 40. $x^2 - 4x + y^2 + 6y = -9$ 41. $9x^2 + 12x + 9y^2 = 77$

42. Find the equation of the tangent line to the circle $(x + 3)^2 + (y + 1)^2 = 25$ at the point $(1, 2)$. Sketch.
43. Points $P(2, 5)$ and $Q(-4, -3)$ are the endpoints of a diameter of a circle. Find the center, radius, and equation of the circle.

Section 3.6

44. Sketch the graph of the ellipse $\dfrac{x^2}{25} + \dfrac{y^2}{9} = 1$.
45. Write the equation of the ellipse in standard form having vertices $(\pm 10, 0)$ and foci $(\pm 6, 0)$.

Write in standard form and graph.

46. $25x^2 + 16y^2 = 400$ 47. $4x^2 - 16x + 9y^2 + 18y = 11$

48. A satellite follows an elliptical orbit around the earth such that the center of the earth is one of the foci. The highest point that the satellite will be from the earth's surface is 2500 miles, and the lowest is 1000 miles. Use 4000 miles as the radius of the earth and

find the equation of the orbit. Let the major axis be on the y-axis and let the origin be the center.

Section 3.7

49. Write in standard form and identify the foci and vertices of the following hyperbola: $16x^2 - 25y^2 = 400$.

50. Sketch the graph of $\dfrac{x^2}{9} - \dfrac{y^4}{4} = 1$. Give the equations of the asympotes.

51. Write in standard form and graph: $4x^2 + 16x - 9y^2 + 18y = 29$. Give the equations of the asymptotes.

Section 3.8

52. Find the coordinates of the focus and the equation of the directrix for the parabola $x^2 = 4y$.

53. The focus of a parabola has coordinates $(0, -\frac{5}{2})$ and the vertex is at the origin. Find the equation of the parabola.

54. The directrix of a parabola is $x = -2$ and the focus is $(2, 0)$. Find the equation of the parabola and graph.

55. A parabola has vertex $(-2, 4)$ and focus $(-2, \frac{7}{2})$. Write the equations of the parabola, the directrix, and the axis.

56. Suppose that the reflecting surface of a television antenna was formed by rotating the parabola $y = \frac{1}{15}x^2$ about its axis of symmetry for $-5 \le x \le 5$. Assuming that the measurements are made in feet, how far from the bottom of this "dish" antenna should the receiver be placed? How deep is this antenna?

Section 3.9

Solve each system and graph.

57. $y = x^2$
    $y = -2x + 8$

58. $y = x^2 - 2$
    $y = -2x^2 + 6x + 7$

59. $x^2 + y^2 - 8y = -7$
    $y - x^2 = 1$

60. $x^2 + y^2 = 9$
    $\dfrac{x^2}{25} + \dfrac{y^2}{16} = 1$

61. $x^2 - y^2 = 1$
    $9x^2 + y^2 = 9$

62. Graph the system of inequalities:
    $y \ge x^2$
    $y \le -2x + 8$

CHAPTER 3 TEST: STANDARD ANSWER

Use these questions to test your knowledge of the basic skills and concepts of Chapter 3. Then check your answers with those given at the back of the book.

1. Graph $y = (x - 2)^2 + 3$.
2. Graph the function $y = f(x) = x^2 - 9$.
3. Let $y = -5x^2 + 20x - 1$.
   (a) Write the quadratic in the standard form $y = a(x - h)^2 + k$.
   (b) Give the coordinates of the vertex.
   (c) Write the equation of the axis of symmetry.
   (d) State the domain and range of the quadratic function.
4. Solve for x: $3x^2 - 8x - 3 = 0$.
5. Find the x-intercepts of $y = -x^2 + 4x + 7$.

Give the value of the discriminant and use this result to describe the x-intercepts, if any.

6. $y = x^2 + 3x + 1$ **7.** $y = 6x^2 + 5x - 6$

8. Find the maximum or minimum value of the quadratic function and state the x-value at which this occurs: $f(x) = -\frac{1}{2}x^2 - 6x + 2$.

9. In the figure the altitude BC of triangle ABC is 4 feet. The part of the perimeter $PQRCB$ is to be a total of 28 feet. How long should x be so that the area of rectangle $PCRQ$ is a maximum?

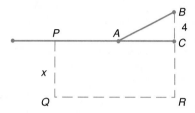

10. The formula $h = 64t - 16t^2$ gives the distance in feet above ground, h, reached by an object in t seconds. What is the maximum height reached by the object?

11. **(a)** Draw the circle $(x - 3)^2 + (y + 4)^2 = 25$.

   (b) Write the equation of the tangent line to this circle at the point $(6, 0)$.

12. Find the center and radius of the circle $4x^2 + 4x + 4y^2 - 56y = -97$.

13. Find the length of the line segment determined by points $A(-2, 5)$ and $B(3, -4)$.

14. Points $P(1, -5)$ and $Q(-3, 3)$ are the endpoints of a diameter of a circle. Find the center, radius, and equation of the circle.

Identify each conic and write in standard form. Find the coordinates of the center, the vertices, and the foci. If applicable, give the equations of the asymptotes.

15. $4x^2 + 16x - y^2 + 6y = -3$ **16.** $9x^2 - 18x + 4y^2 + 16y = 11$

17. Write the equation of the ellipse in standard form with center at $(0, 0)$, horizontal major axis of length 8, and minor axis of length 6.

18. Write the equation of the hyperbola in standard form with center at $(0, 0)$, foci at $(\pm 8, 0)$, and vertices at $(\pm 6, 0)$.

19. Find the coordinates of the focus and the equation of the directrix for the parabola $y = -\frac{1}{8}x^2$.

20. A parabola has directrix $x = \frac{2}{3}$ and vertex at the origin. Write the equation of the parabola and give the coordinates of the focus.

21. A parabola has vertex $(-3, 1)$ and focus $(-3, 3)$. Write the equations of the parabola, the directrix, and the axis.

22. Solve the system and graph: $y = (x - 2)^2 - 2$
$$(x - 2)^2 + (y - 2)^2 = 4$$

Solve each system.

23. $2x^2 - 3y^2 = 15$ **24.** $y^2 - 2x^2 = 16$
   $3x^2 + 2y^2 = 29$ $y - x = 2$

25. Graph the system of inequalities: $y \geq x^2 - 4$
$$y < x + 2$$

1. The coordinates of the vertex and the equation of the axis of symmetry for the graph of $y = (x - 2)^2 - 5$ are

   (a) $(2, 5)$; $x = -2$ (b) $(-2, -5)$; $x = 2$ (c) $(2, -5)$; $x = 2$ (d) $(-2, 5)$; $x = -2$

   (e) None of the preceding

2. The range of the function given by $y = -2(x + 1)^2 - 3$ is

   (a) all real numbers   (b) all $x \geq 2$ (c) all $y \leq 3$ (d) all $y \geq -3$ (e) None of the preceding

3. Which of the following is the standard form for the equation $y = x^2 - 4x + 1$?

   (a) $y = (x - 2)^2 - 3$ (b) $y = (x + 2)^2 - 3$ (c) $y = x(x - 4) + 1$ (d) $y = (x - 2)^2 + 1$

   (e) None of the preceding

4. The graph of $y = 3(x + 1)^2 - 2$ is congruent to that for $y = 3x^2$ but is

   (a) shifted 1 unit to the right and 2 units down

   (b) shifted 1 unit to the left and 2 units down

   (c) shifted 1 unit to the right and 2 units up

   (d) shifted 1 unit to the left and 2 units up

   (e) None of the preceding

5. The function $f(x) = x^2 - 8x + 10$ has

   (a) the minimum vaue -6 when $x = 4$

   (b) the minimum value 6 when $x = -4$

   (c) the minimum value 26 when $x = 4$

   (d) the minimum value 10 when $x = 8$

   (e) None of the preceding

6. When a border of uniform width is put around a 6 foot by 10 foot rectangle, the area is enlarged by 80 square feet. Which equation can be used to find the uniform width x?

   (a) $x(10 + 2x) + x(6 + 2x) = 80$

   (b) $x^2 + 16x - 20 = 0$

   (c) $x^2 + 16x - 80 = 0$

   (d) $x^2 + 8x - 20 = 0$

   (e) None of the preceding

7. Which of the following is true for the parabola $y = -4x^2 + 20x - 25$?

   (a) It opens downward and has two x-intercepts.

   (b) It opens downward and has no x-intercepts.

   (c) It opens downward and has one x-intercept.

   (d) It opens to the left and has two y-intercepts.

   (e) None of the preceding

8. The solutions of $3x^2 + 6x + 2 = 0$ are:

   (a) $\dfrac{-3 \pm \sqrt{3}}{3}$ (b) $\dfrac{-18 \pm \sqrt{3}}{3}$ (c) $-1 \pm \sqrt{2}$ (d) $-6 \pm 2\sqrt{3}\,i$ (e) None of the preceding

9. The circle $x^2 + y^2 + 3y = \frac{1}{4}$ has

   (a) center $(0, \frac{3}{2})$ and radius $r = \frac{1}{2}$

   (b) center $(0, -3)$ and radius $r = 2$

   (c) center $(3, 0)$ and radius $r = \frac{1}{2}$

   (d) center $(0, -\frac{3}{2})$ and radius $r = \sqrt{2}$

   (e) None of the preceding

10. Which of the following is the equation of the ellipse with foci $(0, \pm 3)$ and major axis of length 10?

    (a) $\dfrac{x^2}{9} + \dfrac{y^2}{100} = 1$ (b) $\dfrac{x^2}{100} + \dfrac{y^2}{9} = 1$ (c) $\dfrac{x^2}{16} + \dfrac{y^2}{25} = 1$ (d) $\dfrac{x^2}{25} + \dfrac{y^2}{9} = 1$ (e) None of the preceding

11. The graph of $\dfrac{x^2}{10} - \dfrac{y^2}{15} = 1$ is

   (a) a hyperbola with asymptotes $y = \pm\frac{3}{2}x$

   (b) a hyperbola with foci $(\pm5, 0)$

   (c) an ellipse with a vertical major axis

   (d) an ellipse with vertices $(0, \pm5)$

   (e) None of the preceding

12. The focus and directrix of the parabola $-2y^2 = x$ are

   (a) $(-\frac{1}{8}, 0)$; $x = \frac{1}{8}$ (b) $(0, \frac{1}{8})$; $y = -\frac{1}{8}$ (c) $(-2, 0)$; $x = 2$ (d) $(0, 0)$; $x = -2$ (e) None of the preceding

13. Which of the following is the equation of an ellipse with center $(-3, 2)$, minor axis of length 4, and foci at $(-4, 2)$ and $(-2, 2)$?

   (a) $\dfrac{(x-3)^2}{5} + \dfrac{(y+2)^2}{4} = 1$

   (b) $\dfrac{(x+3)^2}{4} + \dfrac{(y-2)^2}{5} = 1$

   (c) $\dfrac{(x+3)^2}{5} + \dfrac{(y-2)^2}{4} = 1$

   (d) $\dfrac{(x-3)^2}{4} + \dfrac{(y+2)^2}{5} = 1$

   (e) None of the preceding

14. Which of the following is the equation of a hyperbola with center $(-2, 1)$, vertices at $(1, 1)$ and $(-5, 1)$, and foci located $\sqrt{13}$ units from the center?

   (a) $\dfrac{(x+2)^2}{4} - \dfrac{(y-1)^2}{9} = 1$

   (b) $\dfrac{(x-2)^2}{9} - \dfrac{(y+1)^2}{4} = 1$

   (c) $\dfrac{(x-2)^2}{4} - \dfrac{(y+1)^2}{9} = 1$

   (d) $\dfrac{(x+2)^2}{9} - \dfrac{(y-1)^2}{4} = 1$

   (e) None of the preceding

15. The number of solutions for the system given below is which of the following?

$$x^2 + 8y = 80$$

$$x^2 + (y - 5)^2 = 25$$

   (a) None   (b) One   (c) Two   (d) Three   (e) None of the preceding

ANSWERS TO THE TEST YOUR UNDERSTANDING EXERCISES

Page 137

  1. (b)   2. (c)   3. (e)   4. (f)   5. (d)   6. (a)

Page 146

  1. 16   2. 25   3. 9   4. 36   5. $\frac{9}{4}$ 6. $\frac{25}{4}$ 7. $y = (x + 2)^2 - 7$

  8. $y = (x - 3)^2 - 2$ 9. $y = (x - 1)^2 + 8$ 10. $y = 2(x + 2)^2 - 9$

  11. $y = -(x - \frac{1}{2})^2 - \frac{7}{4}$ 12. $y = \frac{1}{2}(x - 3)^2 - \frac{5}{2}$

Page 153

  1. $x = -5$ or $x = 2$ 2. $x = -3$ or $x = 4$ 3. $x = -3$ or $x = 3$ 4. $x = 1 \pm \sqrt{3}$

  5. $x = -3 \pm \sqrt{3}$ 6. $-3 \pm \sqrt{3}i$ 7. $\frac{1}{2} \pm \frac{3}{2}i$ 8. $x = \dfrac{-1 \pm \sqrt{33}}{4}$

1. Minimum value $= -4$ at $x = 5$.   2. Minimum value $= -\frac{5}{6}$ at $x = -\frac{2}{3}$.
3. Minimum value $= \frac{1}{2}$ at $x = 1$.   4. Maximum value $= 131$ at $x = -4$.
5. Maximum value $= -1$ at $x = 0$.   6. Minimum value $= 0$ at $x = 3$.
7. Minimum value $= 0$ at $x = -\frac{7}{5}$.   8. Minimum value $= -\frac{49}{4}$ at $x = -\frac{1}{2}$.

1. 13   2. $\sqrt{65}$ 3. $2\sqrt{41}$ 4. $(0, 0)$; 10   5. $(0, 0)$; $\sqrt{10}$ 6. $(1, -1)$; 5
7. $(-\frac{1}{2}, 0)$; 16   8. $(-4, -4)$; $5\sqrt{2}$ 9. $x^2 + (y - 4)^2 = 25$
10. $(x - 1)^2 + (y + 2)^2 = 3$

1. $(3, 5)$; 6   2. $(0, -\frac{1}{2})$; $\sqrt{5}$ 3. $(\frac{1}{2}, -1)$; $\sqrt{7}$ 4. $(\frac{1}{4}, -1)$; 3

1. Center: $(0, 0)$ 2. Center: $(0, 0)$ 3. $\dfrac{x^2}{64} + \dfrac{y^2}{15} = 1$ 4. $\dfrac{x^2}{36} + \dfrac{y^2}{81} = 1$
   Vertices: $(\pm 10, 0)$ Vertices: $(0, \pm 4)$
   Foci: $(\pm 6, 0)$ Foci: $(0, \pm\sqrt{15})$

5. $\dfrac{(x - 5)^2}{4} + (y + 2)^2 = 1$;   Center: $(5, -2)$
   Vertices: $(3, -2), (7, -2)$
   Foci: $(5 - \sqrt{3}, -2), (5 + \sqrt{3}, -2)$

1. Center: $(0, 0)$ 2. Center: $(0, 0)$ 3. $\dfrac{y^2}{25} - \dfrac{x^2}{24} = 1$
   Vertices: $(\pm 3, 0)$ Vertices: $(0, \pm 8)$
   Foci: $(\pm 5, 0)$ Foci: $(0, \pm 10)$
   Asymptotes: $y = \pm\frac{4}{3}x$ Asymptotes: $y = \pm\frac{4}{3}x$

4. $\dfrac{x^2}{4} - \dfrac{y^2}{9} = 1$

*4.1
HINTS
FOR GRAPHING*

The concept of symmetry was used in Chapter 3 when graphing parabolas. Recall that the graph of the function given by $y = f(x) = x^2$ is said to be *symmetric about the y-axis*. (The *y*-axis is the axis of symmetry.) Observe that points such as $(-a, a^2)$ and (a, a^2) are symmetric points about the axis of symmetry.

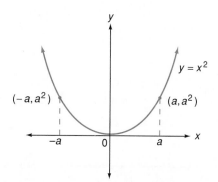

Now we turn our attention to a curve that is *symmetric with respect to a point*. As an illustration of a curve that has symmetry through a point we consider the function given by $y = f(x) = x^3$. This function may be referred to as the *cubing function* because for each domain value x, the corresponding range value is the cube of x.

A table of values is a very helpful aid for drawing the graph of a function. Several specific points are located and a smooth curve is drawn through them to show the graph of $y = x^3$.

$y = f(x) = x^3$

x	y
−2	−8
−1	−1
$-\frac{1}{2}$	$-\frac{1}{8}$
0	0
$\frac{1}{2}$	$\frac{1}{8}$
1	1
2	8

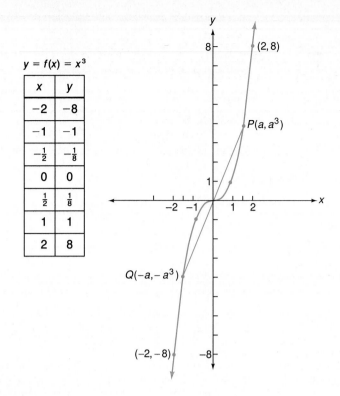

$f(x) = x^3$ is increasing for all x. The curve is concave down on $(-\infty, 0)$ and concave up on $(0, \infty)$.

Domain: all real numbers x
Range: all real numbers y

For the function $f(x) = x^3$, we have $f(-x) = (-x)^3 = -x^3 = -f(x)$. Thus $f(-x) = -f(x)$ and we have symmetry with respect to the origin.

*A function that has this symmetric property is also called an **odd function**.*

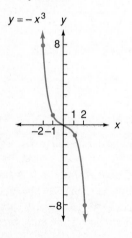

The table and the graph reveal that the curve is symmetric through the origin. Geometrically, this means that whenever a line through the origin intersects the curve at a point P, this line will also intersect the curve in another point Q (on the opposite side of the origin) so that the lengths of OP and OQ are equal. This means that both points (a, a^3) and $(-a, -a^3)$ are on the curve for each value $x = a$. These are said to be **symmetric points** through the origin. In particular, since $(2, 8)$ is on the curve, then $(-2, -8)$ is also on the curve.

In general, the graph of a function $y = f(x)$ is said to be *symmetric through the origin* if for all x in the domain of f, we have

$$f(-x) = -f(x)$$

The techniques used for graphing quadratic functions in Chapter 3 can be used for other functions as well. For example, the graph of $y = 2x^3$ can be obtained from the graph of $y = x^3$ by multiplying each of its ordinates by 2. Also, the graph of $y = -x^3$ can be obtained by reflecting the graph of $y = x^3$ through the x-axis (or by multiplying the ordinates of $y = x^3$ by -1), as shown in the figure in the margin.

Translations (shifting), as done in Chapter 3, can be applied to the graph of $y = x^3$ as well as to the graph of other functions.

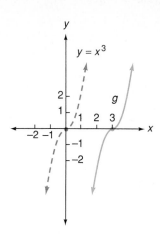

y = x³

g

EXAMPLE 1 Graph $y = g(x) = (x - 3)^3$.

Solution The graph of g is obtained by translating $y = x^3$ by 3 units to the right, as shown in the margin at the left.

EXAMPLE 2 Graph $y = f(x) = |(x - 3)^3|$.

Solution First graph $y = (x - 3)^3$ as in Example 1. Then take the part of this curve that is below the x-axis $[(x - 3)^3 < 0]$ and reflect it through the x-axis.

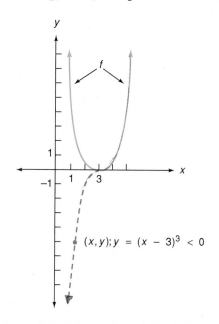

$(x, y); y = (x - 3)^3 < 0$

The domain of $f(x)$ is all real numbers, and the range is all $y \geq 0$.

Explain how you can use the graph of f to draw the graph of $y = |(x - 3)^3| + 2$.

Try to supply a specific example that illustrates each of these items.

GUIDELINES FOR GRAPHING

1. If $f(x) = f(-x)$, the curve is symmetric about the y-axis. *parabola*
2. If $f(-x) = -f(x)$, the curve is symmetric through the origin. $y = x^3$
3. The graph of $y = af(x)$ can be obtained by multiplying the ordinates of the curve $y = f(x)$ by the value a. The case $a = -1$ gives $y = -f(x)$, which is the reflection of $y = f(x)$ through the x-axis.
4. The graph of $y = |f(x)|$ can be obtained from the graph of $y = f(x)$ by taking the part of $y = f(x)$ that is below the x-axis and reflecting it through the x-axis.

In each of the following, h and k are positive.

5. The graph of $y = f(x - h)$ can be obtained by shifting $y = f(x)$ h units to the right.
6. The graph of $y = f(x + h)$ can be obtained by shifting $y = f(x)$ h units to the left.
7. The graph of $y = f(x) + k$ can be obtained by shifting $y = f(x)$ k units upward.
8. The graph of $y = f(x) - k$ can be obtained by shifting $y = f(x)$ k units downward.

Note: *To shift a curve means the same as to translate a curve.*

EXAMPLE 3 In the figure the curve C_1 is obtained by shifting the curve C with equation $y = x^3$ horizontally and C_2 is obtained by shifting C_1 vertically. What are the equations of C_1 and C_2?

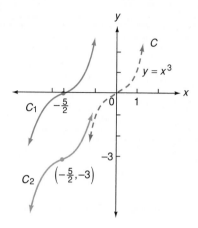

Solution

$$C_1: \quad y = (x + \tfrac{5}{2})^3 \qquad \longleftarrow C \text{ is shifted } \tfrac{5}{2} \text{ units left.}$$

$$C_2: \quad y = (x + \tfrac{5}{2})^3 - 3 \longleftarrow C_1 \text{ is shifted 3 units down.} \qquad \blacksquare$$

EXAMPLE 4 Graph: **(a)** $y = f(x) = x^4$; **(b)** $y = 2x^4$.

Solution

(a) Since $f(-x) = (-x)^4 = x^4 = f(x)$, the graph is symmetric about the y-axis. Use the table of values to locate the right half of the curve; the symmetry gives the rest, as shown next.

(b) Multiply each of the ordinates of $y = x^4$ by 2 to obtain the graph of $y = 2x^4$.

x	$y = x^4$	$y = 2x^4$
0	0	0
$\frac{1}{2}$	$\frac{1}{16}$	$\frac{1}{8}$
1	1	2
$\frac{3}{2}$	$\frac{81}{16}$	$\frac{81}{8}$
2	16	32

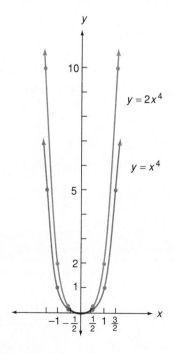

Observe that both functions are decreasing on $(-\infty, 0]$ and increasing on $[0, \infty)$. Also, both curves are concave up for all x.

EXERCISES 4.1

For Exercises 1–8, graph each set of curves in the same coordinate system. For each exercise use a dashed curve for the first equation and a solid curve for each of the others. For the last function given, state the domain and range, find where it is increasing or decreasing, and describe the concavity.

1. $y = x^2$, $y = (x - 3)^2$
2. $y = x^3$, $y = (x + 2)^3$
3. $y = x^3$, $y = -x^3$
4. $y = x^4$, $y = (x - 4)^4$
5. $f(x) = x^3$, $g(x) = \frac{1}{2}x^3$, $h(x) = \frac{1}{4}x^3$
6. $f(x) = x^3$, $g(x) = (x - 3)^3 - 3$, $h(x) = (x + 3)^3 + 3$
7. $f(x) = x^4$, $g(x) = (x - 1)^4 - 1$, $h(x) = (x - 2)^4 - 2$
8. $f(x) = x^4$, $g(x) = -\frac{1}{8}x^4$, $h(x) = -x^4 - 4$
9. Graph $y = |(x + 1)^3|$.
10. Graph $y = |[x]|$ for $-3 \le x < 4$.
11. Graph $y = f(x) = |x|$, $y = g(x) = |x - 3|$, and $y = h(x) = |x - 3| + 2$, on the same axes.

Graph each of the following by using translations and reflections.

12. $y = f(x) = (x + 1)^3 - 2$
13. $y = f(x) = (x - 1)^3 + 2$
14. $y = f(x) = 2(x - 3)^3 + 3$
15. $y = f(x) = 2(x + 3)^3 - 3$
16. $y = f(x) = -(x - 1)^3 + 1$
17. $y = f(x) = -(x + 1)^3 - 1$

Find the equation of the curve C which is obtained from the dashed curve by a horizontal or vertical shift, or by a combination of the two.

18.

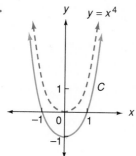

19.

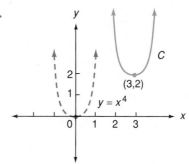

20.

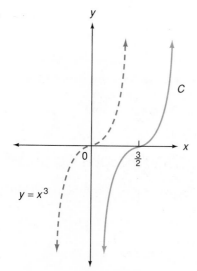

21.

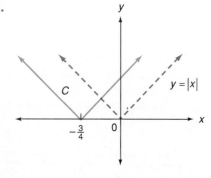

Find the equation of the curve C which is obtained from the dashed curve by using the indicated graphing procedures.

22. Translation *followed* by absolute value.

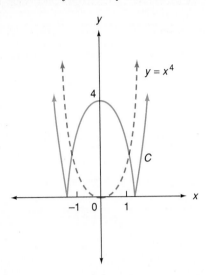

23. Vertical translation *followed* by absolute value.

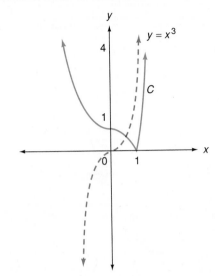

Graph each of the following.

24. $y = |x^4 - 16|$ **25.** $y = |x^3 - 1|$ **26.** $y = |-1 - x^3|$

Graph each of the following. (Hint: Consider the expansion of $(a \pm b)^n$ for appropriate values of n.)

27. $y = x^3 + 3x^2 + 3x + 1$
29. $y = -x^3 + 3x^2 - 3x + 1$

28. $y = x^3 - 6x^2 + 12x - 8$
30. $y = x^4 - 4x^3 + 6x^2 - 4x + 1$

 Evaluate and simplify the difference quotients.

31. $\dfrac{f(x) - f(3)}{x - 3}$ where $f(x) = x^3$

32. $\dfrac{f(-2 + h) - f(-2)}{h}$ where $f(x) = x^3$

33. $\dfrac{f(1 + h) - f(1)}{h}$ where $f(x) = x^4$

34. $\dfrac{f(x) - f(-1)}{x + 1}$ where $f(x) = x^4 + 1$

 Written Assignment: Consider the list "Guidelines for Graphing" on page 205. Supply a specific example of your own to illustrate each of the items listed.

**4.2
GRAPHING
SOME SPECIAL
RATIONAL
FUNCTIONS**

A *rational expression* is a ratio of polynomials. Such expressions may be used to define rational functions and are considered in more detail in later sections. At this point we introduce the topic by exploring several special rational functions.

Consider the function $y = f(x) = \dfrac{1}{x}$, a rational function whose domain consists of all numbers except zero. The denominator x is a polynomial of degree 1, and the numerator $1 = 1x^0$ is a (constant) polynomial of degree zero.

To draw the graph of this function, first observe that it is symmetric through the origin because of the following:

$$f(-x) = \frac{1}{-x} = -\frac{1}{x} = -f(x)$$

Moreover, both variables must have the same sign, since $xy = 1$, a positive number. That is, x and y must both be positive or both negative. Thus the graph will appear only in quadrants I and III. Next, we use a table of values to obtain points for the curve in the first quadrant. Finally, use the symmetry with respect to the origin to obtain the remaining portion of the graph in the third quadrant.

x	y
10	1
100	.01
1000	.001
10,000	.0001
⋮	⋮
↓	↓

Getting very large | Getting close to 0

Horizontal asymptote
$y = 0$

x	y
$\frac{1}{2}$	2
$\frac{1}{10}$	10
$\frac{1}{100}$	100
$\frac{1}{1000}$	1000
⋮	⋮
↓	↓

Getting close to 0 | Getting very large

Vertical asymptote
$x = 0$

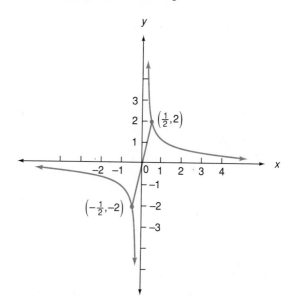

$y = f(x) = \dfrac{1}{x}$

x	y
$\frac{1}{5}$	5
$\frac{1}{2}$	2
1	1
2	$\frac{1}{2}$
3	$\frac{1}{3}$
4	$\frac{1}{4}$

Domain: all $x \neq 0$
Range: all $y \neq 0$

Observe that the curve approaches the x-axis in quadrant I. That is, as the values for x become large, the values for y approach zero. Also, as the values for x approach zero in the first quadrant, the y-values become very large. A similar observation can be made about the curve in the third quadrant. We say that the axes are **asymptotes** for the curve; in particular, the x-axis is a horizontal asymptote and the y-axis is a vertical asymptote.

EXAMPLE 1   Sketch the graph of $g(x) = \dfrac{1}{x - 3}$. Find the asymptotes.

Solution

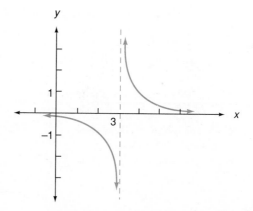

Using $f(x) = \dfrac{1}{x}$, we have $f(x - 3) = \dfrac{1}{x - 3} = g(x)$. Therefore, the graph of g can be drawn by shifting the graph of $f(x) = \dfrac{1}{x}$ by 3 units to the right. The x-axis is the horizontal asymptote and the line $x = 3$ is the vertical asymptote. The domain is all $x \neq 3$, and the range is all $y \neq 0$. ■

EXAMPLE 2 Graph $y = \dfrac{1}{x^2}$ and find the asymptotes.

Solution First note that $x \neq 0$. For all other values of x we have $x^2 > 0$, so that the curve will appear in quadrants I and II only. In quadrant I, as the values for x become large, the values for $y = \dfrac{1}{x^2}$ get close to 0. Moreover, as x approaches zero, y becomes very large. Thus the axes are asymptotes to the curve. Note that the curve is symmetric about the y-axis. That is,

$$f(x) = \frac{1}{x^2} = \frac{1}{(-x)^2} = f(-x)$$

In symbols, we write "As $x \to \infty$, $y \to 0$." That is, as x becomes larger and larger in value, the values of y approach 0. Similarly, as $x \to 0$, $y \to \infty$.

x	y
$\frac{1}{2}$	4
$\frac{1}{10}$	100
$\frac{1}{100}$	10,000

As x is getting close to 0, y is getting very large.
Vertical asymptote $x = 0$

x	y
2	$\frac{1}{4}$
10	$\frac{1}{100}$
100	$\frac{1}{10,000}$

As x is getting very large, y is getting close to 0.
Horizontal asymptote $y = 0$.

The domain is all $x \neq 0$, and the range is all $y > 0$. ■

EXAMPLE 3 Graph $y = \dfrac{1}{(x + 2)^2} - 3$. What are the asymptotes? Find the domain and the range, describe the concavity of the curve, and state where it is increasing or decreasing.

Solution Shift the graph of $y = \dfrac{1}{x^2}$ in Example 2 by 2 units left and then 3 units down.

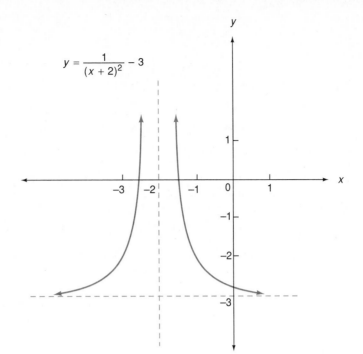

$$y = \frac{1}{(x + 2)^2} - 3$$

Find the x-intercepts by solving the equation
$$\frac{1}{(x + 2)^2} - 3 = 0$$
Use a calculator to show that, to one decimal place, the solutions are $x = -1.4$ and $x = -2.6$.

The curve is concave up and increasing on $(-\infty, -2)$ and concave up and decreasing on $(-2, \infty)$.

As $x \to -2$, $y \to \infty$; as $x \to \infty$, $y \to -3$; as $x \to -\infty$, $y \to -3$.

Vertical asymptote:　　$x = -2$　　*Domain*: all $x \ne -2$

Horizontal asymptote: $y = -3$　　*Range*:　all $y > -3$　　∎

The preceding examples suggest that *a vertical asymptote occurs when the denominator of the rational expression is 0*. Thus, for $g(x) = \dfrac{1}{x - 3}$, the line $x = 3$ is a vertical asymptote. However, this observation does not apply to $f(x) = \dfrac{x + 3}{x^2 - 9}$. The denominator of $f(x)$ is 0 when $x = \pm 3$, but we can reduce as follows:

$$f(x) = \frac{x + 3}{x^2 - 9} = \frac{\cancel{x + 3}}{(x - 3)\cancel{(x + 3)}} = \frac{1}{x - 3} \qquad (x \ne 3)$$

The point $(-3, -\frac{1}{6})$ is on the graph of $y = \dfrac{1}{x - 3}$, but not on the graph of $f(x)$.

This shows that there is no vertical asymptote when x is -3. In fact, the graph of f is the same as the graph in Example 1, except that there would be an open circle at point $(-3, -\frac{1}{6})$ to indicate that this point is *not* on the graph of f. In general

A graph of a rational function has a vertical asymptote at each value a where the denominator is 0, and the numerator is not 0.

Note: This is an example of a rational function that has no asymptotes.

EXAMPLE 4 Graph $y = f(x) = \dfrac{x^2 + x - 6}{x - 2}$

Solution Factor the numerator and simplify the fraction.

$$f(x) = \frac{x^2 + x - 6}{x - 2} = \frac{(x + 3)(x - 2)}{x - 2}$$

$$= x + 3 \quad (x \neq 2)$$

This simplified form indicates that the original rational function is the same as $y = x + 3$, except that $x \neq 2$ since division by 0 is not possible.

Thus the graph of the function is the line $y = x + 3$, with an open circle at $(2, 5)$ to show that this point is *not* part of the graph.

Observe that the reduced form $y = x + 3$, $x \neq 2$, makes it easy to see that the graph of this function has no asymptotes. This was not obvious from the given form.

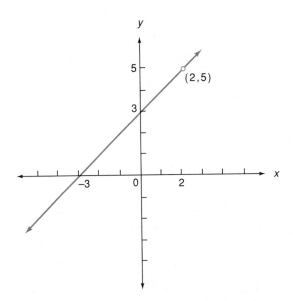

See the inside front cover for these and other useful curves.

The curves studied in this chapter, as well as the parabolas and straight lines discussed in earlier chapters, are very useful in the study of calculus. Having an almost instant recall of the graphs of the following functions will be helpful in future work. Not only should you know what these curves look like, but just as important, you should be able to obtain other curves from them by appropriate translations and reflections.

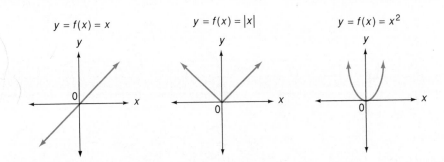

$y = f(x) = x$ $y = f(x) = |x|$ $y = f(x) = x^2$

$$y = f(x) = x^3 \qquad\qquad y = f(x) = \frac{1}{x} \qquad\qquad y = f(x) = \frac{1}{x^2}$$

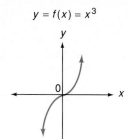 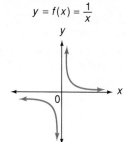 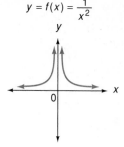

EXERCISES 4.2

Graph each set of curves in the same coordinate system. For each exercise use a dashed curve for the first equation and a solid curve for the other.

1. $y = \dfrac{1}{x}, y = \dfrac{2}{x}$

2. $y = \dfrac{1}{x}, y = -\dfrac{1}{x}$

3. $y = \dfrac{1}{x}, y = \dfrac{1}{x + 2}$

4. $y = \dfrac{1}{x}, y = \dfrac{1}{x} + 5$

5. $y = \dfrac{1}{x}, y = \dfrac{1}{2x}$

6. $y = \dfrac{1}{x}, y = \dfrac{1}{x} - 5$

7. $y = \dfrac{1}{x^2}, y = \dfrac{1}{(x - 3)^2}$

8. $y = \dfrac{1}{x^2}, y = -\dfrac{1}{x^2}$

For Exercises 9–17, graph each of the following. Find all asymptotes, if any. State the domain and range of the function, describe the concavity, and say where it is increasing or decreasing.

9. $y = -\dfrac{1}{x} + 2$

10. $y = -\dfrac{1}{x} - 2$

11. $y = \dfrac{1}{x + 4} - 2$

12. $y = \dfrac{1}{x - 4} + 2$

13. $y = -\dfrac{1}{x - 2} + 1$

14. $y = -\dfrac{1}{x + 1} - 2$

15. $y = \dfrac{1}{(x + 1)^2} - 2$

16. $y = \dfrac{1}{(x - 2)^2} + 3$

17. $y = \dfrac{1}{|x - 2|}$

For Exercises 18–30, graph and find the asymptotes, if any.

18. $y = \dfrac{1}{x^3}$

19. $xy = 3$

20. $xy = -2$

21. $xy - y = 1$

22. $xy - 2x = 1$

23. $y = f(x) = \dfrac{x^2 - 9}{x - 3}$

24. $y = f(x) = \dfrac{x^2 - 9}{x + 3}$

25. $y = f(x) = \dfrac{x^2 - x - 6}{x - 3}$

26. $y = f(x) = \dfrac{1 + x - 2x^2}{x - 1}$

27. $y = f(x) = \dfrac{x + 1}{x^2 - 1}$

28. $y = f(x) = \dfrac{x - 1}{x^2 + x - 2}$

29. $y = f(x) = \dfrac{x^3 - 8}{x - 2}$

30. $y = f(x) = \dfrac{x^3 - 2x^2 - 3x + 6}{2 - x}$

31. Graph $y = \left| \dfrac{1}{x} - 1 \right|$.

32. Graph $x = \dfrac{1}{y^2}$. Why is y not a function of x?

△ *Find the difference quotients and simplify.*

33. $\dfrac{f(x) - f(3)}{x - 3}$ where $f(x) = \dfrac{1}{x}$.

34. $\dfrac{f(1 + h) - f(1)}{h}$ where $f(x) = \dfrac{1}{x^2}$.

Factoring methods were studied in Chapter 1. We will now use factored forms to de-termine the signs of polynomial and rational functions, which, in turn, will aid in sketching their graphs. First, let us consider the inequality $x^2 - 7x + 10 < 0$ to see how the factored form is used to find the signs when the polynomial is negative.

$$(x - 2)(x - 5) < 0$$

Since $(x - 2)(x - 5) = 0$ if and only if $x = 2$ or $x = 5$, we need to consider the other possible values of x. First observe that the numbers 2 and 5 determine three in-tervals.

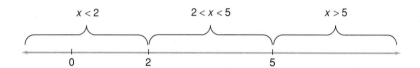

These three intervals are listed, left to right, at the top of the following **table of signs.** Now select a value in the first interval $(-\infty, 2)$, such as $x = 0$, and use it as a *test value* to determine the signs of the factors $x - 2$ and $x - 5$. The minus signs next to the factors $x - 2$ and $x - 5$ under the heading $x < 2$ indicate that these fac-tors are negative for this interval. Furthermore, since the product of two negative factors is positive, a plus sign is written next to $(x - 2)(x - 5)$ for this interval.

Note that 2 and 5 are not within any of the intervals listed at the top of the table. When $x = 2$ or $x = 5$, then $(x - 2)(x - 5) = 0$.

Interval	$x < 2$	$2 < x < 5$	$x > 5$
Sign of $x - 2$	−	+	+
Sign of $x - 5$	−	−	+
Sign of $(x - 2)(x - 5)$	+	−	+

Using a test value such as $x = 3$ in the interval $(2, 5)$ shows that $x - 2$ is posi-tive, $x - 5$ is negative, and consequently $(x - 2)(x - 5)$ is negative. This explains the signs in the column under $2 < x < 5$.

The entries for the last column can be checked using $x = 6$ as a specific exam-ple. The results in the last row of the table show the product $(x - 2)(x - 5) < 0$ only for the interval $(2, 5)$, which is the solution to the given inequality. The solu-tion set for the given inequality can be written as $\{x \mid 2 < x < 5\}$.

Now let us use a table of signs to assist in graphing the polynomial function given by $f(x) = x^3 + 4x^2 - x - 4$. To obtain the factored form, use factoring by grouping.

$$\begin{aligned} f(x) &= x^3 + 4x^2 - x - 4 \\ &= (x + 4)x^2 - (x + 4) \\ &= (x + 4)(x^2 - 1) \\ &= (x + 4)(x + 1)(x - 1) \end{aligned}$$

STEP 1 for graphing polyno-mial functions. Factor the polynomial and find the x-in-tercepts.

Set each factor equal to 0 to get the roots -4, -1, 1 of $f(x) = 0$, which are also the x-intercepts of $y = f(x)$. Since there are no other x-intercepts, the curve must stay above or below the x-axis for each of the intervals determined by -4, -1, 1.

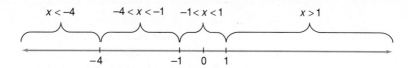

When $f(x) > 0$ the curve is above the x-axis, and it is below for $f(x) < 0$. The following table of signs contains this information and is used to sketch the graph.

STEP 2. *Form a table of signs for* $f(x)$.

Choose **convenient** *test values from each interval. For example, use* -5 *in* $(-\infty, -4)$, -2 *in* $(-4, -1)$, 0 *in* $(-1, 1)$, *and* 2 *in* $(1, \infty)$.

Interval	$(-\infty, -4)$	$(-4, -1)$	$(-1, 1)$	$(1, \infty)$
Sign of $x + 4$	−	+	+	+
Sign of $x + 1$	−	−	+	+
Sign of $x - 1$	−	−	−	+
Sign of $f(x)$	−	+	−	+
Position of curve relative to x-axis	below	above	below	above

The negative sign next to $x + 4$ and under $(-\infty, -4)$ means that $x + 4$ is negative for all x in this interval. This is true because if $x < -4$, then $x + 4 < -4 + 4$ or $x + 4 < 0$. It may be more efficient to decide these signs by using a specific test value from each interval. Thus, for the interval $(-\infty, -4)$ select a convenient value such as $x = -5$. Then $x + 4 = -5 + 4 = -1$, a negative number, and we conclude that $x + 4 < 0$ for all x in this interval. Similarly, $x + 1$ and $x - 1$ are negative for this interval. Therefore, since the product of three negative factors is negative, $f(x) < 0$. This is shown by a minus sign $(-)$ in the row for $f(x)$ under $(-\infty, -4)$.

$$f(x) = x^3 + 4x^2 - x - 4$$

STEP 3. *Sketch the graph using the x-intercepts, the signs of* $f(x)$, *and some additional points including the y-intercept.*

x	y
−4	0
−3	8
−2	6
−1	0
0	−4
1	0

Turning point

y - intercept; $f(0) = -4$

Turning point

Note: *We can only approximate the coordinates of the turning points. In the study of calculus you will learn how to determine the coordinates of such points accurately.*

In general, a polynomial of degree n has at most n − 1 turning points and at most n x-intercepts.

Notice that there are two turning points and three x intercepts. It turns out that for a third-degree polynomial there are at most two turning points and at most three x-intercepts. There may be less of either; for example, $y = x^3$ has no turning points and one x-intercept (see page 204).

To graph rational functions, first locate the asymptotes if there are any. For example, note that $y = \dfrac{1}{x - 2}$ has a vertical asymptote at $x = 2$, since as x is taken close to 2, y gets very large in absolute value. It has the horizontal asymptote $y = 0$ because as $|x|$ is taken very large, the y-values get close to 0. The graph of this function is shown in the margin.

The method for finding the horizontal asymptote (if any) for a rational function f can be summarized by referring to the *general form of a rational function*.

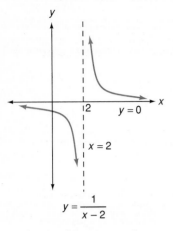

$$f(x) = \frac{a_n x^n + a_{n-1} x^{n-1} + \cdots + a_0}{b_m x^m + b_{m-1} x^{m-1} + \cdots + b_0}, \qquad a_n \neq 0 \neq b_m$$

$$y = \frac{1}{x - 2}$$

A rational function may have numerous vertical asymptotes, but at most one horizontal asymptote.

The numerator and denominator are polynomials in x of degree n and m, respectively (see page 43).

1. If $n < m$, then $y = 0$ is the horizontal asymptote.

2. If $n = m$, then $y = \dfrac{a_n}{b_m}$ is the horizontal asymptote.

3. If $n > m$, there is no horizontal asymptote.

EXAMPLE 1   Find the asymptotes for each rational function.

(a) $f(x) = \dfrac{x}{x^2 - 2x}$ **(b)** $g(x) = \dfrac{3x^2}{2x^2 + 1}$ **(c)** $h(x) = \dfrac{x^2}{x + 1}$

Solution

(a) $f(x) = \dfrac{x}{x^2 - 2x} = \dfrac{x}{x(x - 2)} = \dfrac{1}{x - 2}$ $(x \neq 0)$

Then $x = 2$ is a vertical asymptote since the denominator is 0 when x is 2 but the numerator is not 0.

$y = 0$ (the x-axis) is the horizontal asymptote since

degree of numerator $= 1 < 2 =$ degree of denominator

Recall that the graph of a rational function has a vertical asymptote at each value a where the denominator is 0 and the numerator is not 0.

(b) Since the denominator $2x^2 + 1 \neq 0$ for all x, there is no vertical asymptote. Also, $y = \dfrac{3}{2}$ is the horizontal asymptote since

degree of numerator $= 2 =$ degree of denominator

(c) $x = -1$ is a vertical asymptote. There is no horizontal asymptote since

degree of numerator $= 2 > 1 =$ degree of denominator   ∎

When graphing a rational function of the form $f(x) = \dfrac{p(x)}{q(x)}$, where $p(x)$ and $q(x)$ are polynomials, it will be useful to find the x- and y-intercepts, if there are any. To find the x-intercepts, let $y = 0$ and solve for x. This is done by setting the numerator $p(x) = 0$, since the only way that a fraction can be 0 is when the numerator is 0. To find the y-intercept, let $x = 0$ and solve for y.

Guidelines for Graphing Rational Functions:

EXAMPLE 2 Graph $y = f(x) = \dfrac{x-2}{x^2-3x-4}$.

STEP 1. Factor the numerator and denominator.

Solution

$$f(x) = \frac{x-2}{(x+1)(x-4)}$$

STEP 2. Find the x- and y-intercepts, if any.

$f(x) = 0$ when the numerator $x - 2 = 0$, so the x-intercept is 2. When $x = 0$, $y = \dfrac{0-2}{(0+1)(0-4)} = \dfrac{1}{2}$, so the y-intercept is $\dfrac{1}{2}$.

STEP 3. Find the vertical asymptotes.

Setting the denominator $(x+1)(x-4)$ equal to 0 gives the vertical asymptotes $x = -1$ and $x = 4$.

STEP 4. Find the horizontal asymptote.

The horizontal asymptote is $y = 0$ since the degree of the numerator is 1, which is less than 2, the degree of the denominator.

The numbers for which the numerator or denominator equals 0 are -1, 2, and 4. They determine the four intervals in the table of signs for f.

STEP 5. Form a table of signs for f in the intervals determined by the numbers for which the numerator or denominator equals 0.

Interval	$(-\infty, -1)$	$(-1, 2)$	$(2, 4)$	$(4, \infty)$
Sign of $x + 1$	−	+	+	+
Sign of $x - 2$	−	−	+	+
Sign of $x - 4$	−	−	−	+
Sign of $f(x)$	−	+	−	+
Positive of curve relative to x-axis	below	above	below	above

STEP 6. Graph the curve. First draw the asymptotes and locate the intercepts. Use the table of signs and some selected points to complete the graph.

When graphing, remember that the closer the curve is to a vertical asymptote, the steeper it gets. And the larger $|x|$ gets, the closer the curve is to the horizontal asymptote; the curve gets "flatter." Selected points on the curve will help to draw the correct shape as shown on the following page.

The decimals in the table are rounded off to tenths. A calculator can be used. Note that it is easier to substitute into the form

$$\frac{x - 2}{(x + 1)(x - 4)} \text{ than into}$$

$$\frac{x - 2}{x^2 - 3x - 4}.$$

x	y
-2	$-\frac{2}{3}$
$-\frac{5}{4}$	-2.5
$-.9$	5.9
0	$\frac{1}{2}$
2	0
3	$-\frac{1}{4}$
3.9	-3.9
4.2	2.1
5	$\frac{1}{2}$

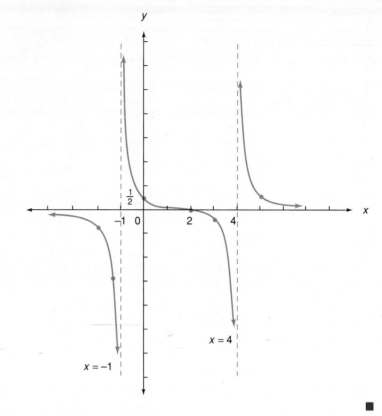

EXAMPLE 3 Graph: $y = f(x) = \dfrac{2}{x^2 + 1}$.

Solution No vertical asymptotes since the denominator $x^2 + 1 \neq 0$.

$y = 0$ is the horizontal asymptote, since the degree of the numerator is less than the degree of the denominator.

Symmetry through the y-axis since

$$f(-x) = \frac{2}{(-x)^2 + 1} = \frac{2}{x^2 + 1} = f(x)$$

The graph is above the x-axis since $y = \dfrac{2}{x^2 + 1} > 0$ for all x.

2 is the y-intercept since $\dfrac{2}{0^2 + 1} = 2$.

$(0, 2)$ is the highest point on the graph because the denominator is the smallest when $x = 0$; $2 = \dfrac{2}{0^2 + 1} \geq \dfrac{2}{x^2 + 1}$.

The preceding observations are used to obtain the following graph.

$$y = f(x) = \frac{2}{x^2 + 1}$$

x	y
0	2
$\frac{1}{2}$	1.6
1	1
2	0.4
3	0.4

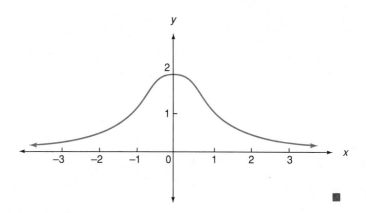

Part of our curve sketching is being done on "faith." After all, what is the shape of a curve connecting two points? Below are some possibilities. Locating more points between P and Q will not really answer the question, no matter how close the chosen points are, since the question as to how to connect them still remains. More advanced methods studied in calculus will help answer such questions, and curve sketching can then be done with less ambiguity.

EXERCISES 4.3

Use the table method to determine the signs of f.

1. $f(x) = (x - 1)(x - 2)(x - 3)$ **2.** $f(x) = x(x + 2)(x - 2)$

3. $f(x) = \dfrac{(3x - 1)(x + 4)}{x^2(x - 2)}$ **4.** $f(x) = \dfrac{x(2x + 3)}{(x + 2)(x - 5)}$

Use the table method to solve each inequality.

5. $(x^2 + 2)(x - 4)(x + 1) > 0$ **6.** $x(x + 1)(x + 2) < 0$

7. $\dfrac{x - 10}{3(x + 1)(5x - 1)} < 0$ **8.** $\dfrac{3}{(x + 2)(4x + 3)} > 0$

Sketch the graph of each polynomial function. Indicate all intercepts.

9. $f(x) = (x + 3)(x + 1)(x - 2)$ **10.** $f(x) = (x - 1)(x - 3)(x - 5)$
11. $f(x) = (x + 3)(x + 1)(x - 1)(x - 3)$ **12.** $f(x) = -x(x + 4)(x^2 - 4)$
13. $f(x) = x^3 - 4x$ **14.** $f(x) = 9x - x^3$
15. $f(x) = x^3 + 3x$ **16.** $f(x) = x^3 + x^2 - 6x$
17. $f(x) = -x^3 - x^2 + 6x$ **18.** $f(x) = x^3 + 2x^2 - 9x - 18$
19. $f(x) = x^4 - 4x^2$ **20.** $f(x) = (x - 1)^2(3 + 2x - x^2)$
21. $f(x) = x^4 - 6x^3 + 8x^2$ **22.** $f(x) = x(x^2 - 1)(x^2 - 4)$

Sketch the graph of each rational function. Indicate the asymptotes and the intercepts, if any.

23. $y = f(x) = \dfrac{1}{(x - 1)(x + 2)}$

24. $y = f(x) = \dfrac{1}{x^2 - 4}$

25. $y = f(x) = \dfrac{1}{4 - x^2}$

26. $y = f(x) = \dfrac{2}{x^2 - x - 6}$

27. $y = f(x) = \dfrac{x}{x^2 - 1}$

28. $y = f(x) = \dfrac{x + 1}{x^2 + x - 2}$

29. $y = f(x) = \dfrac{3}{x^2 + 1}$

30. $y = f(x) = \dfrac{2}{x + x^2}$

31. $y = f(x) = \dfrac{x + 2}{x - 2}$

32. $y = f(x) = \dfrac{x}{1 - x}$

33. $y = f(x) = \dfrac{x - 1}{x + 3}$

34. $y = f(x) = \dfrac{x^2 + x}{x^2 - 4}$

35. $y = f(x) = \dfrac{x^2 + x - 2}{x^2 + x - 12}$

36. $y = f(x) = \dfrac{1 - x^2}{x^2 - 9}$

37. $y = f(x) = \dfrac{9}{3x^2 + 6}$

38. $y = f(x) = -\dfrac{3}{x^2 + 2}$

39. $y = f(x) = \dfrac{2x^2}{x^2 + 1}$

40. $y = f(x) = \dfrac{2x^2}{x^2 + 1} - 1$

Simplify each function; obtain factored forms, and determine the signs of g. (The combination of such skills will be needed in the study of calculus.)

41. $g(x) = 2(2x - 1)(x - 2) + 2(x - 2)^2$

42. $g(x) = 2(x - 3)(x + 2)^2 + 2(x + 2)(x - 3)^2$

43. $g(x) = \dfrac{x^2 - 2x(x + 2)}{x^4}$

44. $g(x) = \dfrac{(x - 3)(2x) - (x^2 - 5)}{(x - 3)^2}$

45. $g(x) = 2\left(\dfrac{x - 5}{x + 2}\right)\dfrac{(x + 2) - (x - 5)}{(x + 2)^2}$

46. $g(x) = \dfrac{5(x + 10)^2 - 10x(x + 10)}{(x + 10)^4}$

 Written Assignment: Use specific examples of your own to describe the conditions for the graph of a rational function to have vertical and horizontal asymptotes.

**EXPLORATIONS
Think Critically**

1. Sketch the graph of a function having the following properties: $x = 1$ is a vertical asymptote, $y = -1$ is a horizontal asymptote, the domain is all $x \neq 1$, the range is all $y \neq -1$, the origin is the only x-intercept, f is increasing on both $(-\infty, 1)$ and $(1, \infty)$, and the graph is concave up on $(-\infty, 1)$ and concave down on $(1, \infty)$.

2. Explain, with a specific example, why the following statement is *not* true:

   A graph of a rational function has a vertical asymptote at each value a where the denominator is 0.

3. The graph of the function $f(x) = 2x + \dfrac{1}{x}$ has $x = 0$ as a vertical asymptote. In addition, it has another asymptote. Plot a few points and try to find the equation of this asymptote. Then sketch the graph.

4. Suppose $f(x) = mx + b + \dfrac{a}{x}$. Aside from a vertical asymptote, there is another kind of asymptote. Describe this asymptote by considering what happens to $f(x)$ as $|x|$ gets very large.

5. Explain the difference between the graph of $f(x) = \dfrac{(x - a)(x - b)}{x - a}$ and that of $g(x) = x - b$.

4.4
EQUATIONS AND INEQUALITIES WITH FRACTIONS

To find the x-intercepts of the graph of a rational function such as

$$y = f(x) = \frac{3x}{x + 7} - \frac{8}{5}$$

we need to let $y = 0$ and solve for x, as shown in Example 1.

EXAMPLE 1 Find the x-intercepts of $y = f(x) = \frac{3x}{x + 7} - \frac{8}{5}$ and state the domain.

Solution First set $f(x) = 0$ and then multiply both sides of the equation by $5(x + 7)$, the least common denominator (LCD); then solve for x.

$$5(x + 7)\left(\frac{3x}{x + 7} - \frac{8}{5}\right) = 5(x + 7) \cdot 0$$

$$5(x + 7) \cdot \frac{3x}{x + 7} - 5(x + 7) \cdot \frac{8}{5} = 5(x + 7) \cdot 0 \qquad \text{(distributive property)}$$

$$5 \cdot 3x - (x + 7) \cdot 8 = 0$$

$$15x - 8x - 56 = 0$$

$$7x = 56$$

$$x = 8 \qquad \text{(The } x\text{-intercept is 8.)}$$

Notice that multiplication by the LCD transforms the original equation into one that does not involve any fractions. Thereafter we are able to finish the solution using methods we have previously studied.

Since division by 0 is undefined, the domain consists of all $x \neq 7$.

Check: $\dfrac{3(8)}{8 + 7} - \dfrac{8}{5} = \dfrac{24}{15} - \dfrac{8}{5} = \dfrac{8}{5} - \dfrac{8}{5} = 0$ ∎

In this section, we focus primarily on the solutions of equations involving rational expressions. Keep in mind, however, that the equation in this example has the form $f(x) = 0$, whose solutions are the x-intercepts of the graph of $y = f(x)$.

EXAMPLE 2 Solve for x: $\dfrac{2x - 5}{x + 1} - \dfrac{3}{x^2 + x} = 0$

Solution Factor the denominator in the second fraction.

$$\frac{2x - 5}{x + 1} - \frac{3}{x(x + 1)} = 0$$

Multiply by the LCD, which is $x(x + 1)$.

At times, multiplication by the LCD can give rise to a quadratic equation, as in Example 2. Try to explain each step in the solution.

$$x(x + 1) \cdot \frac{2x - 5}{x + 1} - x(x + 1) \cdot \frac{3}{x(x + 1)} = x(x + 1) \cdot 0$$

$$x(2x - 5) - 3 = 0$$

$$2x^2 - 5x - 3 = 0$$

$$(2x + 1)(x - 3) = 0$$

$$2x + 1 = 0 \quad \text{or} \quad x - 3 = 0$$

You should check these answers in the original equation.

$$x = -\tfrac{1}{2} \quad \text{or} \quad x = 3$$ ∎

It is especially important to check each solution of a fractional equation. The reason for this can be explained through the use of another example.

Recall that $x^2 - 16$ can be factored as the difference of two squares: $x^2 - 16 = (x + 4)(x - 4)$

EXAMPLE 3   Solve for x: $\dfrac{24}{x^2 - 16} - \dfrac{5}{x + 4} = \dfrac{3}{x - 4}$

Solution   Begin by multiplying each side by $(x + 4)(x - 4)$.

$$(x + 4)(x - 4) \cdot \frac{24}{x^2 - 16} - (x + 4)(x - 4) \cdot \frac{5}{x + 4} = (x + 4)(x - 4) \cdot \frac{3}{x - 4}$$

$$24 - (x - 4)5 = (x + 4)3$$

$$24 - 5x + 20 = 3x + 12$$

$$-8x = -32$$

$$x = 4$$

In this example we could have noticed at the outset that $x - 4 \neq 0$, and thus $x \neq 4$. In other words, it is wise to notice such restrictions on the variable before starting the solution. These are the values of the variable that would cause division by zero.

Note that we began with the *assumption* that there was a value x for which the equation was true. This led to the value $x = 4$; that is, we argued that *if* there is a solution, then it must be 4. But if x is replaced by 4 in the given equation, we obtain

$$\frac{24}{0} - \frac{5}{8} = \frac{3}{0}$$

Since division by 0 is meaningless, we have an impossible equation and therefore 4 cannot be a solution. We conclude that there is no replacement of x for which the equation is true. Therefore, the solution set is the **empty set**, the set that contains no elements.   ∎

The equation in Example 1 can also be written as the **proportion** $\dfrac{3x}{x + 7} = \dfrac{8}{5}$.

A proportion is a statement that two ratios are equal, such as the following:

$$\frac{a}{b} = \frac{c}{d}$$

The proportion is in the form of a fractional equation and can be simplified by multiplying both sides by bd.

This is often read "a is to b as c is to d" and may be written in this form:

$$a : b = c : d$$

$$\frac{a}{b} = \frac{c}{d}$$

$$(bd)\frac{a}{b} = (bd)\frac{c}{d}$$

$$ad = bc$$

PROPORTION PROPERTY

If $\frac{a}{b} = \frac{c}{d}$, then $ad = bc$.

Observe how the proportion property can be used after the equation in Example 1 is written as a proportion.

$$\frac{3x}{x+7} \diagdown \frac{8}{5}$$

$$3x \cdot 5 = 8 \cdot (x + 7) \qquad \text{(proportion property)}$$

$$15x = 8x + 56$$

$$7x = 56$$

$$x = 8$$

TEST YOUR UNDERSTANDING
Think Carefully

Solve for x and check your results. Use the proportion property where appropriate.

1. $\frac{x}{3} + \frac{x}{2} = 10$ 2. $\frac{3}{x} + \frac{2}{x} = 10$ 3. $\frac{x-3}{8} = 4$

4. $\frac{8}{x-3} = 4$ 5. $\frac{x+3}{x} = \frac{2}{3}$ 6. $\frac{2}{x+3} = \frac{3}{x+3}$

7. $\frac{3x}{2} - \frac{2x}{3} = \frac{1}{4}$ 8. $\frac{2}{3x} + \frac{3}{2x} = 4$

9. $\frac{3x-1}{4} - \frac{x-1}{2} = 1$ 10. $\frac{x}{x-6} + \frac{2}{x} = \frac{1}{x-6}$

11. $\frac{x}{x-1} - \frac{3}{4x} = \frac{3}{4} - \frac{1}{x-1}$ 12. $\frac{2x+10}{x+18} = \frac{x+3}{4-x}$

(Answers: Page 260)

The general procedures for solving problems outlined in Section 1.2 apply to problems that involve fractions as well. The reader is advised to review that material at this time. The first illustrative problem makes use of the proportion property.

EXAMPLE 4   How high is a tree that casts an 18-foot shadow at the same time that a 3-foot stick casts a 2-foot shadow?

Similar triangles have the same shape, but not necessarily the same size. If two triangles are similar, their corresponding angles are equal and their corresponding sides are proportional.

Solution   From the diagram we use similar triangles to write a proportion, as shown on the following page.

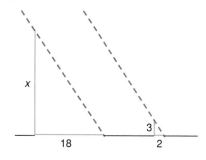

Let x represent the height of the tree. Then, since the triangles are similar,

$$\frac{x}{18} = \frac{3}{2}$$

$$2x = 54 \qquad \text{(proportion property)}$$

$$x = 27$$

The tree is 27 feet high. ∎

EXAMPLE 5 A wildlife conservation team wants to determine the deer population in a state park. They capture a sample of 80 deer, tag each one, and then release them. After a sufficient amount of time has elapsed for the tagged deer to mix thoroughly with the others in the park, the team captures another sample of 100 deer, five of which had previously been been tagged. Estimate the number of deer in the park.

Solution Let x be the total deer population. The 80 tagged deer were thoroughly mixed with the others. Thus it is reasonable to expect the ratio of 80 tagged deer to the total deer population to be the same as the ratio of the tagged deer, captured a second time, to the total number in the second sample.

$$\frac{\text{total tagged deer}}{\text{total deer population}} = \frac{\text{tagged deer in second sample}}{\text{total deer in second sample}}$$

$$\frac{80}{x} = \frac{5}{100}$$

$$5x = 8000$$

$$x = 1600$$

There are approximately 1600 deer in the park. ∎

Suppose that you are able to work at a steady rate and can paint a room in 6 hours. This means that in 1 hour you have painted $\frac{1}{6}$ of the room. In general, for a job that takes you n hours, the part of the job done in 1 hour is $\frac{1}{n}$. This idea is used to solve certain types of work problems that call for the solution of fractional equations, as in Example 6.

*Working together should enable the boys to do the job in less time than it would take either of them to do the job alone. If Elliot can do the job alone in 2 hours, together the two boys should take less than this time. This kind of reasoning can be **critical**, since an answer of 2 or more indicates that an error has been made in the solution.*

EXAMPLE 6 Working alone, Harry can mow a lawn in 3 hours. Elliot can complete the same job in 2 hours. How long will it take them working together, assuming they both start at the same time?

Solution To solve work problems of this type, we first consider the part of the job that can be done in 1 hour.

Let x = time (in hours) to do the job together

Then $\frac{1}{x}$ = portion of job done in 1 hour working together

Also: $\dfrac{1}{2}$ = portion of job done by Elliot in 1 hour

$\dfrac{1}{3}$ = portion of job done by Harry in 1 hour

Now we have that *each* of the expressions $\dfrac{1}{2} + \dfrac{1}{3}$ and $\dfrac{1}{x}$ represents the part of the job done in 1 hour. Therefore, we may set these expressions equal to each other to form the equation needed to solve this problem.

$$\frac{1}{2} + \frac{1}{3} = \frac{1}{x}$$

$$3x + 2x = 6$$

$$5x = 6$$

$$x = 1\tfrac{1}{5}$$

Together they need $1\tfrac{1}{5}$ hours, or 1 hour and 12 minutes, to complete the job. ■

Numerous formulas involving fractions are found in mathematics as well as in areas such as science, business, and industry. When working with such formulas it is sometimes useful to solve for one of the variables in terms of the others. This is illustrated in the next example.

EXAMPLE 7 $r = \dfrac{ab}{a + b + c}$ is the formula for the radius of the inscribed circle of a right triangle in terms of the sides of the triangle. Solve for side b of the adjacent figure.

Solution Begin by clearing fractions:

$$r = \frac{ab}{a + b + c}$$

$$r(a + b + c) = ab$$

$$ra + rb + rc = ab$$

$$ra + rc = ab - rb \qquad \text{(Bring all terms involving b to one side.)}$$

$$r(a + c) = (a - r)b \qquad \text{(Factor.)}$$

$$\frac{r(a + c)}{a - r} = b \qquad \text{(Divide by } a - r.\text{)}$$ ■

Check this answer by substituting for b in the given equation and simplify.

For solving equations with fractions the LCD was used to eliminate the denominators, resulting in a simpler equation. This idea is also used to solve inequalities with fractions.

EXAMPLE 8   Solve: $\dfrac{x}{3} - \dfrac{2x + 1}{4} > 1$

Solution

$$12\left(\frac{x}{3}\right) - 12\left(\frac{2x + 1}{4}\right) > 12(1)$$

$$4x - 3(2x + 1) > 12$$

$$4x - 6x - 3 > 12$$

$$-2x > 15$$

$$x < -\frac{15}{2} \qquad \text{(Why?)}$$

EXAMPLE 9   Graph on a number line: $\dfrac{1 - 2x}{x - 3} \le 0$

Note here that $x \ne 3$. For $x = 3$, $x - 3 = 0$ and division by 0 is not possible.

Solution   First find the values of x for which the numerator or the denominator is zero.

$$1 - 2x = 0 \qquad\qquad x - 3 = 0$$

$$-2x = -1 \qquad\qquad x = 3$$

$$x = \tfrac{1}{2}$$

Now locate the three intervals determined by these two points.

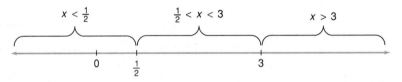

Next, select a specific value to test in each interval and record the results in a *table of signs* as follows. As an example, for $\frac{1}{2} < x < 3$, let $x = 1$, then $1 - 2x = -1$; negative. Also, $x - 3 = -2$; negative. Thus the quotient is positive.

Since $\frac{1}{2}$ precedes 3 across the top, write $1 - 2x$ above $x - 3$ at the left. It is advantageous to use such patterns when forming these tables.

Interval	$x < \frac{1}{2}$	$\frac{1}{2} < x < 3$	$x > 3$
Sign of $1 - 2x$	$+$	$-$	$-$
Sign of $x - 3$	$-$	$-$	$+$
Sign of $\dfrac{1 - 2x}{x - 3}$	$-$	$+$	$-$

*This solution **cannot** be described using a single inequality. For example, it makes no sense to write $3 < x \le \frac{1}{2}$ since this implies the false result $3 \le \frac{1}{2}$.*

The table shows that the fraction will be negative when $x < \frac{1}{2}$ or when $x > 3$. Also note that $x = \frac{1}{2}$ satisfies the given inequality but that $x = 3$ is excluded from the solution. Can you explain why this is so? The solution for the given inequality consists of all x such that $x \le \frac{1}{2}$ or $x > 3$, that is, $\{x \mid x \le \frac{1}{2} \text{ or } x > 3\}$.

EXAMPLE 10   Solve for x: $\dfrac{x+1}{x} < 3$

Solution   First rewrite as follows:

$$\frac{x+1}{x} - 3 < 0$$

An alternative method begins by multiplying through by x. However, this is more difficult because it calls for two cases, $x > 0$ and $x < 0$.

$$\frac{1-2x}{x} < 0$$

Now use a table of signs to obtain the solution $x < 0$ or $x > \frac{1}{2}$.   ∎

The next example illustrates the solution of a nonlinear system that involves a fractional equation.

EXAMPLE 11   Solve the system

$$y = \frac{2}{x}$$

$$y = -2x + 5$$

Solution   Set both values for y equal to each other, and multiply by x.

$$\frac{2}{x} = -2x + 5$$

$$2 = -2x^2 + 5x$$

$$2x^2 - 5x + 2 = 0$$

$$(2x - 1)(x - 2) = 0$$

Sketch the graph of this system.

$$x = \tfrac{1}{2} \quad \text{or} \quad x = 2$$

For $x = \frac{1}{2}$, $y = -2(\frac{1}{2}) + 5 = 4$. For $x = 2$, $y = -2(2) + 5 = 1$. The points of intersection are $(\frac{1}{2}, 4)$ and $(2, 1)$. Check both of these points by substituting in the original system.   ∎

EXERCISES 4.4

Solve for x and check your results.

1. $\dfrac{x}{2} - \dfrac{x}{5} = 6$

2. $\dfrac{2}{x} - \dfrac{5}{x} = 6$

3. $\dfrac{x-1}{2} = \dfrac{x+2}{4}$

4. $\dfrac{2}{x-1} = \dfrac{4}{x+2}$

5. $\dfrac{5}{x} - \dfrac{3}{4x} = 1$

6. $\dfrac{3x}{4} - \dfrac{3}{2} = \dfrac{x}{6} + \dfrac{4x}{3}$

7. $\dfrac{2x+1}{5} - \dfrac{x-2}{3} = 1$

8. $\dfrac{3x-2}{4} - \dfrac{x-1}{3} = \dfrac{1}{2}$

9. $\dfrac{x+4}{2x-10} = \dfrac{8}{7}$

10. $\dfrac{10}{x} - \dfrac{1}{2} = \dfrac{15}{2x}$

11. $\dfrac{2}{x+6} - \dfrac{2}{x-6} = 0$

12. $\dfrac{1}{3x-1} + \dfrac{1}{3x+1} = 0$

13. $\dfrac{x+1}{x+10} = \dfrac{1}{2x}$

14. $\dfrac{3}{x+1} = \dfrac{9}{x^2-3x-4}$

15. $\dfrac{x^2}{2} - \dfrac{3x}{2} + 1 = 0$

16. $\dfrac{x+1}{x-1} - \dfrac{2}{x(x-1)} = \dfrac{4}{x}$

17. $\dfrac{5}{x^2-9} = \dfrac{3}{x+3} - \dfrac{2}{x-3}$

18. $\dfrac{3}{x^2-25} = \dfrac{5}{x-5} - \dfrac{5}{x+5}$

19. $\dfrac{1}{x^2+4} + \dfrac{1}{x^2-4} = \dfrac{18}{x^4-16}$

20. $\dfrac{3}{x^3-8} = \dfrac{1}{x-2}$

Find the x-intercepts of $y = f(x)$ and state the domain.

21. $f(x) = \dfrac{2x-5}{x+1} - \dfrac{3}{x^2+x}$

22. $f(x) = \dfrac{2}{x} + 2x - 5$

23. $f(x) = \dfrac{10-5x}{3x} - \dfrac{2}{x+5} - \dfrac{8-4x}{x+5}$

24. $f(x) = \dfrac{2-x}{x+1} + \dfrac{x+8}{x-2} - \dfrac{4-x}{x^2-x-2}$

25. $f(x) = \dfrac{3}{2x^2-3x-2} - \dfrac{x+2}{2x+1} - \dfrac{2x}{10-5x}$

26. $f(x) = \dfrac{2x-2}{x^2+2x} - \dfrac{5x-6}{6x} - \dfrac{1-x}{x+5}$

Solve for the indicated variable.

27. $\dfrac{v^2}{K} = \dfrac{2g}{m}$, for m

28. $A = \dfrac{h}{2}(b + B)$, for B

29. $S = \pi(r_1 + r_2)s$, for r_1

30. $V = \dfrac{1}{3}\pi r^2 h$, for h

31. $d = \dfrac{s-a}{n-1}$, for s

32. $S = \dfrac{n}{2}[2a + (n-1)d]$, for d

33. $\dfrac{1}{f} = \dfrac{1}{m} + \dfrac{1}{p}$, for m

34. $c = \dfrac{2ab}{a+b}$, for b

Solve each inequality:

35. $\dfrac{x}{2} - \dfrac{x}{3} \le 5$

36. $\dfrac{x}{3} - \dfrac{x}{2} \le 5$

37. $\dfrac{x+3}{4} - \dfrac{x}{2} > 1$

38. $\dfrac{x}{3} - \dfrac{x-1}{2} < 2$

39. $\dfrac{1}{2}(x+1) - \dfrac{2}{3}(x-2) < \dfrac{1}{6}$

40. $\dfrac{1}{2}(x-2) - \dfrac{1}{5}(x-1) > 1$

41. $\dfrac{2x+3}{x} < 1$

42. $\dfrac{x}{x-1} > 2$

43. $\dfrac{x+4}{x+2} > \dfrac{1}{3}$

44. $\dfrac{x-2}{x+3} \ge 0$

45. $\dfrac{x+5}{2-x} < 0$

46. $\dfrac{(6-x)(3+x)}{x+1} \le 0$

47. What number must be subtracted from both the numerator and denominator of the fraction $\frac{11}{23}$ to give a fraction whose value is $\frac{2}{5}$?

48. The denominator of a fraction is 3 more than the numerator. If 5 is added to the numerator and 4 is subtracted from the denominator, the value of the new fraction is 2. Find the original fraction.

49. One pipe can empty a tank in 3 hours. A second pipe takes 4 hours to complete the same job. How long will it take to empty the tank if both pipes are used?

50. Working together, Amy and Julie can paint their room in 3 hours. If it takes Amy 5 hours to do the job alone, how long would it take Julie to paint the room working by herself?

51. Find two fractions whose sum is $\frac{2}{3}$ if the smaller fraction is $\frac{1}{2}$ the larger one. (*Hint:* Let x represent the larger fraction.)

52. A rope that is 20 feet long is cut into two pieces. The ratio of the smaller piece to the larger piece is $\frac{3}{5}$. Find the length of the shorter piece.

53. The shadow of a tree is 20 feet long at the same time that a 1-foot-high flower casts a 4-inch shadow. How high is the tree?

54. The area A of a triangle is given by $A = \dfrac{bh}{2}$, where b is the length of the base and h is the length of the altitude.

   (a) Solve for h in terms of A and b.   (b) Find h when $A = 100$ and $b = 25$.

55. If 561 is divided by a certain number, the quotient is 29 and the remainder is 10. Find the number.
   (*Hint:* For $N \div D$ we have $\dfrac{N}{D} = Q + \dfrac{R}{D}$, where Q is the quotient and where R is the remainder.)

56. A student received grades of 72, 75, and 78 on three tests. What must her score on the next test be for her to have an average grade of 80 for all four tests?

57. The denominator of a certain fraction is 1 more than the numerator. If the numerator is increased by $2\frac{1}{2}$, the value of the new fraction will be equal to the reciprocal of the original fraction. Find the original fraction.

58. Dan takes twice as long as George to complete a certain job. Working together, they can complete the job in 6 hours. How long will it take Dan to complete the job by himself?

59. Prove:

   If $\dfrac{a}{b} = \dfrac{c}{d}$,   then $\dfrac{a + b}{b} = \dfrac{c + d}{d}$

60. Prove:

   If $\dfrac{a}{b} = \dfrac{c}{d}$,   then $\dfrac{a}{a + b} = \dfrac{c}{c + d}$

61. A bookstore has a stock of 30 paperback copies of *Algebra*, as well as 50 hardcover copies of the same book. They wish to increase their stock for the new semester. Based on past experience, they want their final numbers of paperback and hardcover copies to be in the ratio 4 to 3. However, the publisher stipulates that they will sell the store only 2 copies of the paperback edition for each copy of the hardcover edition ordered. Under these conditions, how many of each edition should the store order to achieve the $4:3$ ratio?

62. To find out how wide a certain river is, a pole 20 feet high is set straight up on one of the banks. Another pole 4 feet long is also set straight up, on the same side, some distance away from the embankment. The observer waits until the shadow of the 20-foot pole just reaches the other side of the river. At that time he measures the shadow of the 4-foot pole and finds it to be 34 feet. Use this information to determine the width of the river.

63. Two hundred fish are caught, tagged, and returned to the lake from which they were caught. Some time later, when the 200 tagged fish had thoroughly mixed with the others, another 160 fish were caught and contained 4 that were tagged previously. Estimate the number of fish in the lake.

64. Estimate the total number of fish in the lake in Exercise 63 if the first catch contained 150 fish and the second catch had 120 fish, 3 of which were tagged previously.

65. A certain college gives 4 points per credit for a grade of A, 3 points per credit for a B, 2 points for a C, 1 point per credit for a D, and 0 for an F. A student is taking 15 credits for the semester. She expects A's in a 4 credit course and in a 3 credit course. She also expects a B in another 3 credit course and a D in a 2 credit course. What grade must she get in the fifth course in order to earn a 3.4 grade point average for the term?
   $\left(\textit{Hint: } \text{Grade point average} = \dfrac{\text{total points}}{\text{total credits}}. \right)$

66. For electric circuits, when two resistances R_1 and R_2 are wired in parallel, the single equivalent resistance R can be found by using this relationship: $\dfrac{1}{R} = \dfrac{1}{R_1} + \dfrac{1}{R_2}$. Find the single equivalent resistance for two resistances wired in parallel if these measure 2000 and 3000 ohms, respectively.

67. What resistance, wired in parallel with a 20,000-ohm resistance, will give a single equivalent resistance of 4000 ohms? (See Exercises 66.)

68. A formula used in optics relates the focal length of a lens, f, the distance of an object

from the lens, p, and the distance from the lens to the image, q, in this way: $\frac{1}{f} = \frac{1}{p} + \frac{1}{q}$.

If the focal length of a lens is 15 centimeters and the distance of an object from the lens is 20 centimeters, find the distance of the object from the lens.

69. Find the focal length of a lens if the distance of an object from the lens is 30 centimeters and the distance of the image from the lens is 15 centimeters. (See Exercise 68.)

70. Solve $f(x) < 0$, where $f(x) = \dfrac{1}{x^3 + 6x^2 + 12x + 8}$, and graph $y = f(x)$. (*Hint:* Consider the expansion of $(a + b)^3$. See Example 6, page 46.)

Solve each system and graph.

71. $2x + 3y = 7$

$\quad y = \dfrac{1}{x}$

72. $y = x^2 - 2x - 4$

$\quad y = -\dfrac{8}{x}$

4.5 VARIATION

If a car is traveling at a constant rate of 40 miles per hour, then the distance d traveled in t hours is given by $d = 40t$. The change in the distance is "directly" affected by the change in the time; as t increases, so does d. We say that d *is directly proportional to* t. This is because $d = 40t$ converts to the proportion $\dfrac{d}{t} = 40$. We also say that d *varies directly as* t and that 40 is the *constant of variation, or the constant of proportionality.*

*Direct variation is a functional relationship in the sense that $y = kx$ **defines** y to be a function of x.*

> ### DIRECT VARIATION
> y varies directly as x if $\quad y = kx \quad$ for some constant of variation k.

EXAMPLE 1

(a) Write the equation that expresses this direct variation; y varies directly as x, and y is 8 when x is 12.

(b) Find y for $x = 30$.

Solution

(a) Since y varies directly as x, we have $y = kx$ for some constant k. To find k, substitute the given values for the variables and solve.

$$8 = k(12)$$

$$\frac{2}{3} = k$$

Thus $y = \dfrac{2}{3}x$.

(b) For $x = 30$, $y = \dfrac{2}{3}(30) = 20$. ∎

When y varies directly as x, the equation $y = kx$ is equivalent to $\dfrac{y}{x} = k\,(x \neq 0)$.

Therefore, if each of the pairs x_1, y_1 and x_2, y_2 satisfies $y = kx$, then each of the ratios y_1 to x_1 and y_2 to x_2 is equal to k. Thus

$$\frac{y_1}{x_1} = \frac{y_2}{x_2}$$

This proportion can be used in some direct variation problems if the constant of proportionality is not required. Thus part (b) of Example 1 can be answered without first writing the equation of variation. In particular, using $x_1 = 12$ with $y_1 = 8$, and $x_2 = 30$, we can solve for y_2 as follows:

$$\frac{y_2}{x_2} = \frac{y_1}{x_1}$$

$$\frac{y_2}{30} = \frac{8}{12}$$

$$y_2 = 30\left(\frac{8}{12}\right) = 20$$

Numerous examples of direct variation can be found in geometry. Here are some illustrations.

Circumference of a circle of radius r:

$C = 2\pi r$;　　C varies directly as the radius r;
　　　　　　　2π is the constant of variation.

Area of a circle of radius r:

$A = \pi r^2$;　　A varies directly as the square of r;
　　　　　　π is the constant of variation.

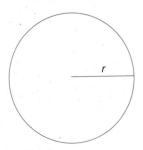

Area of an equilateral triangle of side s:

$A = \dfrac{\sqrt{3}}{4} s^2$;　　A varies directly as s^2;
　　　　　　$\dfrac{\sqrt{3}}{4}$ is the constant of proportionality.

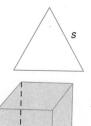

Volume of a cube of side e:

$V = e^3$;　　V varies directly as the cube of e (as e^3);
　　　　　　1 is the constant of variation.

EXAMPLE 2 According to Hooke's law, the force F required to hold a spring stretched x units beyond its natural length is directly proportional to x. If a force of 20 pounds is needed to hold a certain spring stretched 3 inches beyond its natural length, how far will 60 pounds of force hold the spring stretched beyond its natural length?

The same principle applies to the force required to hold a spring when compressed x units within its natural length.

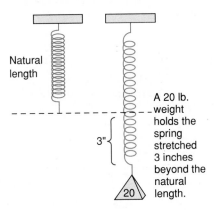

Natural length

A 20 lb. weight holds the spring stretched 3 inches beyond the natural length.

3"

20

Solution Since F varies directly as x, we have $F = kx$. Solve for k by substituting the known values for F and x.

$$20 = k(3)$$

Note: The force F increases as x increases. This will always be the case in a direct variation situation provided that the constant of variation is positive.

$$\tfrac{20}{3} = k$$

Thus $F = \tfrac{20}{3}x$. Now let $F = 60$ and solve for x.

$$60 = \tfrac{20}{3}x$$

$$60(\tfrac{3}{20}) = x$$

$$9 = x$$

Thus 60 pounds is needed to hold the spring stretched 9 inches beyond its natural length. ∎

When y varies directly as x and the constant of variation is positive, the variables x and y increase or decrease together. There are other situations where one variable increases as the other decreases. We refer to this as *inverse variation*.

INVERSE VARIATION

y varies inversely as x if $y = \dfrac{k}{x}$ for some constant of variation k.

EXAMPLE 3 According to Boyle's law, the pressure P of a compressed gas is inversely proportional to the volume V. Suppose that there is a pressure of 25 pounds per square inch when the volume of gas is 400 cubic inches. Find the pressure when the gas is compressed to 200 cubic inches.

Solution   Since P varies inversely as V, we have

$$P = \frac{k}{V}$$

Substitute the known values for P and V and solve for k.

$$25 = \frac{k}{400}$$

$$10{,}000 = k$$

Thus $P = \dfrac{10{,}000}{V}$, and when $V = 200$ we have

Note that the pressure increases as the volume decreases.

$$P = \frac{10{,}000}{200} = 50$$

The pressure is 50 pounds per square inch.   ∎

When y varies inversely as x the equation $y = \dfrac{k}{x}$ is equivalent to $xy = k$. Therefore, when each of the pairs x_1, y_1 and x_2, y_2 satisfies $y = \dfrac{k}{x}$, we get each of the products $x_1 y_1$ and $x_2 y_2$ equal to k. Thus

$$x_1 y_1 = x_2 y_2$$

This equation provides an alternative way to solve some inverse variation problems, if the equation of the variation is not required. Thus in Example 3, using $V_1 = 400$ with $P_1 = 25$ and $V_2 = 200$, we can solve for P_2 as follows:

$$V_2 P_2 = V_1 P_1$$

$$200 P_2 = (400)(25)$$

$$P_2 = \frac{(400)(25)}{200} = 50$$

TEST YOUR UNDERSTANDING
Think Carefully

y varies directly as x.

1. $y = 5$ when $x = 4$.
   Find y for $x = 20$.

2. $y = 2$ when $x = 7$.
   Find y for $x = 21$.

y varies inversely as x.

3. $y = 3$ when $x = 5$.
   Find y when $x = 3$.

4. $y = 4$ when $x = 8$.
   Find y when $x = 2$.

(Answers: Page 260)

The variation of a variable may depend on more than one other variable. Here are some illustrations:

Observe that the word
"jointly" is used to imply that
z varies directly as the product
of the two factors.

$$z = kxy \qquad \text{z varies } jointly \text{ as } x \text{ and } y.$$

$$z = kx^2y \qquad \text{z varies } jointly \text{ as } x^2 \text{ and } y.$$

$$z = \frac{k}{xy} \qquad \text{z varies } inversely \text{ as } x \text{ and } y.$$

$$z = \frac{kx}{y} \qquad \text{z varies } directly \text{ as } x \text{ and } inversely \text{ as } z.$$

$$w = \frac{kxy^3}{z} \qquad \text{w varies } jointly \text{ as } x \text{ and } y^3 \text{ and } inversely \text{ as } z.$$

EXAMPLE 4 Describe the variation given by these equations:

(a) $z = kx^2y^3$ **(b)** $z = \dfrac{kx^2}{y}$ **(c)** $V = \pi r^2 h$

Solution
(a) z varies jointly as x^2 and y^3.
(b) z varies directly as x^2 and inversely as y.
(c) V varies jointly as r^2 and h. ∎

EXAMPLE 5 Suppose that z varies directly as x and inversely as the square of y.
If $z = \frac{1}{3}$ when $x = 4$ and $y = 6$, find z when $x = 12$ and $y = 4$.

Solution

$$z = \frac{kx}{y^2}$$

$$\frac{1}{3} = \frac{k(4)}{6^2}$$

$$\frac{1}{3} = \frac{k}{9}$$

$$3 = k$$

Thus $z = \dfrac{3x}{y^2}$. When $x = 12$ and $y = 4$ we have

$$z = \frac{3(12)}{4^2} = \frac{9}{4} \qquad ∎$$

EXERCISES 4.5

Write the equation for the given variation and identify the constant of variation.

1. The perimeter P of a square varies directly as the side s.
2. The area of a circle varies directly as the square of the radius.
3. The area of a rectangle 5 centimeters wide varies directly as its length.
4. The volume of a rectangular-shaped box 10 centimeters high varies jointly as the length and width.

Write the equation for the given variation using k as the constant of variation.

5. z varies jointly as x and y^3.
6. z varies inversely as x and y^3.
7. z varies directly as x and inversely as y^3.
8. w varies jointly as x and y and z.
9. w varies directly as x^2 and inversely as y and z.

Find the constant of variation for Exercises 10–17.

10. y varies directly as x; $y = 4$ when $x = \frac{2}{3}$.
11. s varies directly as t^2; $s = 50$ when $t = 10$.
12. y varies inversely as x; $y = 15$ when $x = \frac{1}{3}$.
13. u varies jointly as v and w; $u = 2$ when $v = 15$ and $w = \frac{2}{3}$.
14. z varies directly as x and inversely as the square of y; $z = \frac{7}{2}$ when $x = 14$ and $y = 6$.
15. a varies inversely as the square of b, and $a = 10$ when $b = 5$. Find a when $b = 25$.
16. z varies jointly as x and y; $z = \frac{3}{2}$ when $x = \frac{5}{6}$ and $y = \frac{9}{20}$. Find z when $x = 2$ and $y = 7$.
17. s varies jointly as ℓ and the square of w; $s = \frac{10}{3}$ when $\ell = 12$ and $w = \frac{5}{6}$. Find s when $\ell = 15$ and $w = \frac{9}{4}$.

18. The cost C of producing x number of articles varies directly as x. If it costs \$560 to produce 70 articles, what is C when $x = 400$?

19. If a ball rolls down an inclined plane, the distance traveled varies directly as the square of the time. If the ball rolls 12 feet in 2 seconds, how far will it roll in 3 seconds?

20. The volume V of a right circular cone varies jointly as the square of the radius r of the base, and the altitude h. If $V = 8\pi$ cubic centimeters when $r = 2$ centimeters and $h = 6$ centimeters, find the formula for the volume V.

△ 21. A force of 2.4 pounds is needed to hold a spring stretched 1.8 inches beyond its natural length. Use Hooke's law to determine the force required to hold the spring stretched 3 inches beyond its natural length.

△ 22. The force required to hold a metal spring compressed from its natural length is directly proportional to the change in the length of the spring. If 235 pounds is required to hold the spring compressed within its natural length of 18 inches to a length of 15 inches, how much force is required to hold it compressed to a length of 12 inches?

23. Fifty pounds per square inch is the pressure exerted by 150 cubic inches of a gas. Use Boyle's law to find the pressure if the gas is compressed to 100 cubic inches.

24. The gas in Exercise 23 expands to 500 cubic inches. What is the pressure?

25. If we neglect air resistance, the distance that an object will fall from a height near the surface of the earth is directly proportional to the square of the time it falls. If the object falls 256 feet in 4 seconds, how far will it fall in 7 seconds?

26. The volume of a right circular cylinder varies jointly as its height and the square of the radius of the base. The volume is 360π cubic centimeters when the height is 10 centimeters and the radius is 6 centimeters. Find V when $h = 18$ centimeters and $r = 5$ centimeters.

27. If the volume of a sphere varies directly as the cube of its radius and $V = 288\pi$ cubic inches when $r = 6$ inches, find V when $r = 2$ inches.

28. The resistance to the flow of electricity through a wire depends on the length and thickness of the wire. The resistance R is measured in *ohms* and varies directly as the length l and inversely as the square of the diameter d. If a wire 200 feet long with diameter 0.16 inch has a resistance of 64 ohms, how much resistance will there be if only 50 feet of wire is used?

 29. A wire made of the same material as the wire in Exercise 28 is 100 feet long. Find R if $d = 0.4$ inch. Find R if $d = 0.04$ inch.

30. If z varies jointly as x and y, how does x vary with respect to y and z?

31. A rectangular-shaped beam is to be cut from a round log with a 2.5-foot diameter. The strength s of the beam varies jointly as its height y and the square of its width x.

(a) Write s as a function of y.
(b) Write s as a function of x.
(c) Find s to two decimal places when $x = y$. Express the answer in terms of the constant of proportionality.

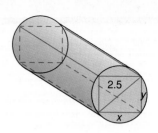

**4.6
DIVISION OF
POLYNOMIALS
AND SYNTHETIC
DIVISION**

A special case of the division of polynomials, called **synthetic division,** will be studied in this section. It will then be used in the following section to help us find the roots of polynomial equations. Before considering this procedure, let us first review the long division process by means of the following example.

EXAMPLE 1   Divide: $(2x^3 + 3x^2 - x + 16) \div (x^2 + 2x - 3)$

Note the steps in the process: Divide $2x^3$ by x^2 to get $2x$. Multiply $2x$ by the divisor and subtract. Then divide the remainder by x^2 again, and so forth. Stop when the remainder has degree less than that of the divisor.

Solution

$$
\begin{array}{r}
2x - 1 \\
x^2 + 2x - 3 \overline{\smash{\big)}\, 2x^3 + 3x^2 - x + 16} \\
\underline{2x^3 + 4x^2 - 6x} \\
-\ x^2 + 5x + 16 \\
\underline{-\ x^2 - 2x + 3} \\
7x + 13
\end{array}
$$

(Remainder has degree *less* than degree of divisor.)

The result of this division example can be stated in this form:

$$\frac{2x^3 + 3x^2 - x + 16}{x^2 + 2x - 3} = (2x - 1) + \frac{7x + 13}{x^2 + 2x - 3}$$

To check the solution, show that the following is correct:

$$\underbrace{2x^3 + 3x^2 - x + 16}_{f(x)} = \underbrace{(2x - 1)}_{q(x)} \underbrace{(x^2 + 2x - 3)}_{\times \quad g(x)} + \underbrace{(7x + 13)}_{+ \quad r(x)}$$

Dividend $=$ quotient $\times$ divisor $+$ remainder ∎

To develop the synthetic division process, we will make use of the following illustration in which the divisor has the form $x - c$.

Check this result by showing that $f(x) = q(x) \cdot g(x) + r$.

$$
\begin{array}{r}
x^2 + 3x - 5 \\
x - 2 \overline{\smash{\big)}\, x^3 + x^2 - 11x + 12} \\
\underline{x^3 - 2x^2} \\
+ 3x^2 - 11x \\
\underline{+ 3x^2 - 6x} \\
-\ 5x + 12 \\
\underline{-\ 5x + 10} \\
+2
\end{array}
$$

Now it should be clear that all of the work done involved the coefficients of the variables and the constants. Thus we could just as easily complete the division by omitting the variables, as long as we write the coefficients in the proper places. The division problem would then look like this:

$$
\begin{array}{r}
1 + 3 - 5 \\
1 - 2\overline{\smash{\big)}\,1 + 1 - 11 + 12} \\
①\ - 2 \\
\hline
+ 3 - 11 \\
+ ③ - 6 \\
\hline
- 5 + 12 \\
- ⑤ + 10 \\
\hline
+ 2
\end{array}
$$

Since the circled numerals are repetitions of those immediately above them, this process can be further shorted by deleting them. Moreover, since these circled numbers are the products of the numbers in the quotient by the 1 in the divisor, we may also eliminate this 1. Thus we have the following:

$$
\begin{array}{r}
1 + 3 - 5 \\
-2\overline{\smash{\big)}\,1 + 1 - 11 + 12} \\
- 2 \\
\hline
+ 3 - 11 \\
- 6 \\
\hline
- 5 + 12 \\
+ 10 \\
\hline
+ 2
\end{array}
$$
→ It is not necessary to bring down the −11 and 12.

$$
\begin{array}{r}
1 + 3 - 5 \\
-2\overline{\smash{\big)}\,1 + 1 - 11 + 12} \\
- 2 \\
\hline
+ 3 \\
- 6 \\
\hline
- 5 \\
+ 10 \\
\hline
+ 2
\end{array}
$$
→ Move the numerals upward.

$$
\begin{array}{r}
1 + 3 - 5 \\
-2\overline{\smash{\big)}\,1 + 1 - 11 + 12} \\
- 2 - 6 + 10 \\
\hline
+ 3 - 5 + 2
\end{array}
$$

When the top numeral 1 is brought down, the last line contains the coefficients of the quotient and the remainder. So eliminate the line above the dividend.

$$
\begin{array}{r}
-2\,\underline{\big|\,1 + 1 - 11 + 12} \\
\underline{- 2 - 6 + 10} \\
1 + 3 - 5\,\big|\,+ 2 \quad \text{Remainder}
\end{array}
$$

Quotient: $x^2 + 3x - 5$

We can further simplify this process by changing the sign of the divisor, making it +2 instead of −2. This change allows us to add throughout rather than subtract, as follows.

$$
\begin{array}{r}
+2\,\underline{\big|\,1 + 1 - 11 + 12\}} \quad \text{Coefficients of dividend} \\
\underline{+ 2 + 6 - 10}
\end{array}
$$

Coefficients of quotient: $1 + 3 - 5\,,\,+ 2$ Remainder

Quotient: $x^2 + 3x - 5$

The long-division process has now been condensed to this short form. Doing a division problem by this short form is called **synthetic division,** as illustrated in the examples that follow.

EXAMPLE 2 Use synthetic division to find the quotient and the remainder.

$$(2x^3 - 9x^2 + 10x - 7) \div (x - 3)$$

Solution Write the coefficients of the dividend in descending order. Change the sign of the divisor (change -3 to $+3$).

$$+3\underline{\big|\,2 - 9 + 10 - 7}$$

Now bring down the first term, 2, and multiply by $+3$.

$$+3\begin{array}{|r}2 - 9 + 10 - 7 \\ \underline{+ 6} \\ 2\end{array}$$

Add -9 and $+6$ to obtain the sum -3. Multiply this sum by $+3$ and repeat the process to the end. The completed example should look like this:

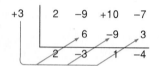

$$+3\begin{array}{|r}2 - 9 + 10 - 7 \\ \underline{+ 6 - 9 + 3}\end{array}$$

coefficients of quotient: $2 - 3 + 1\,\underline{\big|-4}$ remainder

Since the original dividend began with x^3 (third degree), the quotient will begin with x^2 (second degree). Thus we read the last line as implying a quotient of $2x^2 - 3x + 1$ and a remainder of -4. Check this result by using the long-division process. ∎

The synthetic division process has been developed for divisors of the form $x - c$. (Thus, in Example 1, $c = 3$.) A minor adjustment also permits divisors by polynomials of the form $x + c$. For example, a divisor of $x + 2$ may be written as $x - (-2)$; $c = -2$.

Note: The quotient in a synthetic division problem is always a polynomial of degree one less than that of the dividend. This is so because the divisor has degree 1. The bottom line in the synthetic division process, except for the last entry on the right, gives the coefficients of the quotient: a polynomial in standard form.

EXAMPLE 3 Use synthetic division to find the quotient and the remainder.

$$\left(-\tfrac{1}{3}x^4 + \tfrac{1}{6}x^2 - 7x - 4\right) \div (x + 3)$$

Solution Write $x + 3$ as $x - (-3)$. Since there is no x^3 term in the dividend, use $0x^3$.

$$-3\begin{array}{|r} -\tfrac{1}{3} + 0 + \tfrac{1}{6} - 7 - 4 \\ \underline{+ 1 - 3 + \tfrac{17}{2} - \tfrac{9}{2}} \\ -\tfrac{1}{3} + 1 - \tfrac{17}{6} + \tfrac{3}{2}\,\underline{\big|-\tfrac{17}{2}} = \text{remainder}\end{array}$$

Quotient: $-\tfrac{1}{3}x^3 + x^2 - \tfrac{17}{6}x + \tfrac{3}{2}$ Remainder: $-\tfrac{17}{2}$
Check this result. ∎

TEST YOUR UNDERSTANDING
Think Carefully

Use synthetic division, as indicated, to find each quotient and remainder.

1. $(x^3 - x^2 - 5x + 6) \div (x - 2)$
$$+2\underline{\big|\,1 - 1 - 5 + 6}$$

2. $(2x^3 + 7x^2 + 5x - 2) \div (x + 2)$
$$-2\underline{\big|\,2 + 7 + 5 - 2}$$

3. $(2x^3 - 3x^2 + 7) \div (x - 3)$
$$+3\underline{\big|\,2 - 3 + 0 + 7}$$

4. $(3x^3 - 7x + 12) \div (x + 1)$
$$-1\underline{\big|\,3 + 0 - 7 + 12}$$

(Answers: Page 260)

EXERCISES 4.6

As in Example 1, page 236, use the division algorithm to find the quotient and remainder. Check each result.

1. $(x^3 - 2x^2 - 13x + 6) \div (x + 3)$
2. $(x^3 + 4x^2 + 3x - 2) \div (x + 2)$
3. $(x^3 - x^2 + 7) \div (x - 1)$
4. $(5x + 2x^3 - 3) \div (x + 2)$
5. $(5x^2 - 7x + x^3 + 8) \div (x - 2)$
6. $(2x^3 + 9x^2 - 3x - 1) \div (2x - 1)$
7. $(4x^3 - 5x^2 + x - 7) \div (x^2 - 2x)$
8. $(8x^4 - 8x^2 + 6x + 6) \div (2x^2 - x)$
9. $(x^3 - x^2 - x + 10) \div (x^2 - 3x + 5)$
10. $(3x^3 + 4x^2 - 13x + 6) \div (x^2 + 2x - 3)$

Use synthetic division to find the quotient and remainder. Check each result.

11. $(x^3 - 2x^2 - 5x + 6) \div (x - 3)$
12. $(x^3 - x^2 - 5x + 2) \div (x + 2)$
13. $(2x^3 + x^2 - 3x + 7) \div (x + 1)$
14. $(3x^3 - 2x^2 + x - 1) \div (x - 1)$
15. $(x^3 + 5x^2 - 7x + 8) \div (x - 2)$
16. $(x^3 - 3x^2 + x - 5) \div (x + 3)$
17. $(x^4 - 3x^3 + 7x^2 - 2x + 1) \div (x + 2)$
18. $(x^4 + x^3 - 2x^2 + 3x - 1) \div (x - 2)$
19. $(2x^4 - 3x^2 + 4x - 2) \div (x - 1)$
20. $(3x^4 + x^3 - 2x + 3) \div (x + 1)$

Divide.

21. $(x^3 - 27) \div (x - 3)$
22. $(x^3 - 27) \div (x + 3)$
23. $(x^3 + 27) \div (x + 3)$
24. $(x^3 + 27) \div (x - 3)$
25. $(x^4 - 16) \div (x - 2)$
26. $(x^4 - 16) \div (x + 2)$
27. $(x^4 + 16) \div (x + 2)$
28. $(x^4 + 16) \div (x - 2)$
29. $(x^4 - \frac{1}{2}x^3 + \frac{1}{3}x^2 - \frac{1}{4}x + \frac{1}{5}) \div (x - 1)$
30. $(4x^5 - x^3 + 5x^2 + \frac{3}{2}x - \frac{1}{2}) \div (x + \frac{1}{2})$

Use synthetic divison to find each quotient by following the procedure shown in this example.

    Example:

$$\frac{2x^3 - 7x^2 + 8x + 6}{2x - 3} = \frac{2x^3 - 7x^2 + 8x + 6}{2(x - \frac{3}{2})}$$

$$= \frac{1}{2}\left(\frac{2x^3 - 7x^2 + 8x + 6}{x - \frac{3}{2}}\right) \quad \text{(Divisor within the parentheses has the form } x - c.)$$

$$= \frac{1}{2}\left(2x^2 - 4x + 2 + \frac{9}{x - \frac{2}{3}}\right) \quad \text{(By synthetic division)}$$

$$= x^2 - 2x + 1 + \frac{9}{2x - 3} \quad \text{(Multiplying by } \frac{1}{2})$$

31. $(6x^3 - 5x^2 - 3x + 4) \div (2x - 1)$
32. $(10x^3 - 3x^2 + 4x + 7) \div (2x + 1)$
33. $(6x^3 + 7x^2 + x + 8) \div (2x + 3)$
34. $(9x^4 + 6x^3 - 4x + 5) \div (3x - 1)$
35. $(15x^7 - x^6 + 8x^5 + 21x^4 - 9x^2 - 8x + 4) \div (5x - 2)$

4.7
THE REMAINDER FACTOR, AND RATIONAL ROOT THEOREMS

When the polynomial $p(x) = 2x^3 - 9x^2 + 10x - 7$ is divided by $x - 3$, the quotient is $q(x) = 2x^2 - 3x + 1$ and the remainder $r = -4$. (See Example 2, page 238.) As a check we see that

$$\underbrace{\frac{2x^3 - 9x^2 + 10x - 7}{p(x)}} = \underbrace{(2x^2 - 3x + 1)}_{q(x)} \underbrace{(x - 3)}_{(x - 3)} + \underbrace{(-4)}_{r}$$

In general, whenever a polynomial $p(x)$ is divided by $x - c$ we have

Another form of this is

$$\frac{p(x)}{x - c} = q(x) + \frac{r}{x - c}.$$

$$p(x) = q(x)(x - c) + r$$

where $q(x)$ is the quotient and r is the (constant) remainder. Since this equation holds for all x, we may let $x = c$ and obtain

$$p(c) = q(c)(c - c) + r$$
$$= q(c) \cdot 0 + r$$
$$= r$$

This result may be summarized as follows:

REMAINDER THEOREM

If a polynomial $p(x)$ is divided by $x - c$, the remainder is equal to $p(c)$.

EXAMPLE 1 Find the remainder when $p(x) = 3x^3 - 5x^2 + 7x + 5$ is divided by $x - 2$.

Solution By the remainder theorem, the answer is $p(2)$.

$$p(x) = 3x^3 - 5x^2 + 7x + 5$$
$$p(2) = 3(2)^3 - 5(2)^2 + 7(2) + 5$$

Check this result by dividing $p(x)$ by $x - 2$.

$$= 23 \qquad \blacksquare$$

EXAMPLE 2 Let $p(x) = x^3 - 2x^2 + 3x - 1$. Use synthetic division and the remainder theorem to find $p(3)$.

Solution According to the remainder theorem, $p(3)$ is equal to the remainder when $p(x)$ is divided by $x - 3$.

Check this result by substituting $x = 3$ in $p(x)$.

$$\begin{array}{r|rrrr} 3 & 1 & -2 & +3 & -1 \\ & & +3 & +3 & +18 \\ \hline & 1 & +1 & +6 & \boxed{+17} = \text{remainder} = p(3) \end{array} \qquad \blacksquare$$

TEST YOUR UNDERSTANDING
Think Carefully

(Answers: Page 260)

Use synthetic division to find the remainder r when $p(x)$ is divided by $x - c$. Verify that $r = p(c)$ by substituting $x = c$ into $p(x)$.

1. $p(x) = x^5 - 7x^4 + 4x^3 + 10x^2 - x - 5$; $x - 1$
2. $p(x) = x^4 + 11x^3 + 11x^2 + 11x + 10$; $x + 10$
3. $p(x) = x^4 + 11x^3 + 11x^2 + 11x + 10$; $x - 10$
4. $p(x) = 6x^3 - 40x^2 + 25$; $x - 6$

Once again, we are going to consider the division of a polynomial $p(x)$ by a divisor of the form $x - c$. First recall that

$$p(x) = q(x)(x - c) + r$$

where $q(x)$ is the quotient and r is the (constant) remainder. Now suppose that $r = 0$. Then the remainder theorem gives $p(c) = r = 0$, and the preceding equation becomes

$$p(x) = q(x)(x - c)$$

*If $u = vw$, then v and w are said to be **factors** of u.*

It follows that $x - c$ is a *factor* of $p(x)$. Conversely, suppose that $x - c$ is a factor of $p(x)$. This means there is another polynomial, say $q(x)$, so that

$$p(x) = q(x)(x - c)$$

or

$$p(x) = q(x)(x - c) + 0$$

which tells us that when $p(x)$ is divided by $x - c$ the remainder is zero. These observations comprise the following result:

*If $p(c) = 0$, then c is said to be a **zero of the polynomial** **$p(x)$**, and it is also a root of $p(x) = 0$.*

FACTOR THEOREM

A polynomial $p(x)$ has a factor $x - c$ if and only if $p(c) = 0$.

EXAMPLE 3   Show that $x - 2$ is a factor of $p(x) = x^3 - 3x^2 + 7x - 10$.

Solution   By the factor theorem we can state that $x - 2$ is a factor of $p(x)$ if $p(2) = 0$.

$$p(2) = 2^3 - 3(2)^2 + 7(2) - 10$$
$$= 0 \qquad \blacksquare$$

EXAMPLE 4   **(a)** Use the factor theorem to show that $x + 3$ is a factor of $p(x) = x^3 - x^2 - 8x + 12$.   **(b)** Factor $p(x)$ completely.

Solution
(a) First write $x + 3 = x - (-3)$, so that $c = -3$. Then use synthetic division.

$$
\begin{array}{r|rrrr}
-3 & 1 & -1 & -8 & +12 \\
   &   & -3 & +12 & -12 \\
\hline
   & 1 & -4 & +4 & +0 \\
\end{array}
$$

Since $p(-3) = 0$, the factor theorem tells us that $x + 3$ is a factor of $p(x)$.

(b) Synthetic division has produced the quotient $x^2 - 4x + 4$. Therefore, since $x + 3$ is a factor of $p(x)$, we may write

$$x^3 - x^2 - 8x + 12 = (x^2 - 4x + 4)(x + 3)$$

To find the complete factored form, observe that $x^2 - 4x + 4 = (x - 2)^2$. Thus

Note: $p(-3) = 0$. *We use synthetic division here in order to be able to factor $p(x)$ as shown in part (b).*

$$x^3 - x^2 - 8x + 12 = (x - 2)^2(x + 3) \qquad \blacksquare$$

In order to see how the factor and remainder theorems may be applied to the solution of equations, let us first consider this polynomial equation:

$$(3x + 2)(5x - 4)(2x - 3) = 0$$

To find the roots, set each factor equal to zero.

$$3x + 2 = 0 \qquad 5x - 4 = 0 \qquad 2x - 3 = 0$$

$$x = -\frac{2}{3} \qquad\qquad x = \frac{4}{5} \qquad\qquad x = \frac{3}{2}$$

Now multiply the original three factors and keep careful note of the details of this multiplication. Your result should be

$$(3x + 2)(5x - 4)(2x - 3) = 30x^3 - 49x^2 - 10x + 24 = 0$$

which must have the same three rational roots.

As you analyze this multiplication it becomes clear that the constant 24 is the product of the three constants in the binomials, 2, −4, and −3. Also, the leading coefficient, 30, is the product of the three original coefficients of x in the binomials, namely 3, 5, and 2.

Furthermore, 3, 5, and 2 are also the denominators of the roots $-\frac{2}{3}, \frac{4}{5},$ and $\frac{3}{2}$. Therefore, the denominators of the rational roots are all factors of 30, and their numerators are all factors of 24.

These results are not accidental. It turns out that we have been discussing the following general result:

RATIONAL ROOT THEOREM

Let $f(x) = a_n x^n + a_{n-1} x^{n-1} + \cdots + a_1 x + a_0 \ (a_0 \neq 0)$ be an nth-degree polynomial with integer coefficients. If $\dfrac{p}{q}$ is a rational root of $f(x) = 0$, where $\dfrac{p}{q}$ is in lowest terms, then p is a factor of a_0 and q is a factor of a_n.

Let us see how this theorem can be applied to find the rational roots of

$$f(x) = 4x^3 - 16x^2 + 11x + 10 = 0$$

Begin by listing all factors of the constant 10 and of the leading coefficient 4.

Note: The theorem allows only these 16 numbers as possible rational roots: no other rational numbers can be roots.

Possible numerators (factors of 10): $\pm 1, \pm 2, \pm 5, \pm 10$
Possible denominators (factors of 4): $\pm 1, \pm 2, \pm 4$
Possible rational roots (take each number in the first row and divide by each number in the second row):

$$\pm 1, \quad \pm\tfrac{1}{2}, \quad \pm\tfrac{1}{4}, \quad \pm 2, \quad \pm 5, \quad \pm\tfrac{5}{2}, \quad \pm\tfrac{5}{4}, \quad \pm 10$$

To decide which (if any) of these are roots of $f(x) = 0$, we could substitute the values directly into $f(x)$. However, it is easier to use synthetic division because in most cases it leads to easier computations and also makes quotients available. Therefore, we proceed by using synthetic division with divisors c, where c is a possible rational root.

If $f(c) = 0$, then c is a root; if $f(c) \neq 0$, then c is not a root.

Try $c = 1$:

$$\begin{array}{r|rrrr} 1 & 4 & -16 & +11 & +10 \\ & & +4 & -12 & -1 \\ \hline & 4 & -12 & -1 & \boxed{+9} \end{array}$$

Since $f(1) = 9 \neq 0$, 1 is *not* a root.

Try $c = -1$:

$$\begin{array}{r|rrrr} -1 & 4 & -16 & +11 & +10 \\ & & -4 & +20 & -31 \\ \hline & 4 & -20 & +31 & \boxed{-21} \end{array}$$

Since $f(-1) = -21 \neq 0$, -1 is *not* a root.

Try $c = \frac{1}{2}$:

$$\begin{array}{r|rrrr} \frac{1}{2} & 4 & -16 & +11 & +10 \\ & & +2 & -7 & +2 \\ \hline & 4 & -14 & +4 & \boxed{+12} \end{array}$$

Since $f(\frac{1}{2}) = 12 \neq 0$, $\frac{1}{2}$ is *not* a root.

Try $c = -\frac{1}{2}$:

$$\begin{array}{r|rrrr} -\frac{1}{2} & 4 & -16 & +11 & +10 \\ & & -2 & +9 & -10 \\ \hline & 4 & -18 & +20 & \boxed{+0} \end{array}$$

Since $f(-\frac{1}{2}) = 0$, $-\frac{1}{2}$ *is* a root.

By the factor theorem it follows that $x - (-\frac{1}{2}) = x + \frac{1}{2}$ is a factor of $f(x)$, and synthetic division gives the other factor, $4x^2 - 18x + 20$.

$$f(x) = (x + \tfrac{1}{2})(4x^2 - 18x + 20)$$

To find other roots of $f(x) = 0$ we could proceed by using the rational root theorem for $4x^2 - 18x + 20 = 0$. But this is unnecessary because the quadratic expression is factorable.

$$\begin{aligned} f(x) &= (x + \tfrac{1}{2})(4x^2 - 18x + 20) \\ &= (x + \tfrac{1}{2})(2)(2x^2 - 9x + 10) \\ &= 2(x + \tfrac{1}{2})(x - 2)(2x - 5) \\ &= (2x + 1)(x - 2)(2x - 5) \end{aligned}$$

The solution of $f(x) = 0$ can now be found by setting each factor equal to zero. The roots are $x = -\frac{1}{2}$, $x = 2$, and $x = \frac{5}{2}$.

EXAMPLE 5 Factor $f(x) = x^3 + 6x^2 + 11x + 6$.

Whenever a polynomial has ± 1 as the leading coefficient, then any rational root will be an integer that is a factor of the constant term of the polynomial.

Solution Since the leading coefficient is 1, whose only factors are ± 1, the possible denominators of a rational root of $f(x) = 0$ can only be ± 1. Hence the possible rational roots must all be factors of ± 6, namely ± 1, ± 2, ± 3, and ± 6. Use synthetic division to test these cases.

$$\begin{array}{r|rrrr} 1 & 1 & +6 & +11 & +6 \\ & & +1 & +7 & +18 \\ \hline & 1 & +7 & +18 & \boxed{+24} = r \end{array}$$

Since $r = f(1) \neq 0$, $x - 1$ is *not* a factor of $f(x)$.

$$\begin{array}{r|rrrr} -1 & 1 & +6 & +11 & +6 \\ & & -1 & -5 & -6 \\ \hline & 1 & +5 & +6 & \boxed{+0} = r \end{array}$$

Since $r = f(-1) = 0$, $x - (-1) = x + 1$ is a factor of $f(x)$.

$$x^3 + 6x^2 + 11x + 6 = (x + 1)(x^2 + 5x + 6)$$

Now factor the trinomial:

$$x^3 + 6x^2 + 11x + 6 = (x + 1)(x + 2)(x + 3)$$ ■

TEST YOUR UNDERSTANDING
Think Carefully

(Answers: Page 260)

For each $p(x)$, find (a) the possible rational roots of $p(x) = 0$, (b) the factored form of $p(x)$, and (c) the roots of $p(x) = 0$.

1. $p(x) = x^3 - 3x^2 - 10x + 24$
2. $p(x) = x^4 + 6x^3 + x^2 - 24x + 16$
3. $p(x) = 4x^3 + 20x^2 - 23x + 6$
4. $p(x) = 3x^4 - 13x^3 + 7x^2 - 13x + 4$

EXAMPLE 6 Find all real roots of $p(x) = 2x^5 + 7x^4 - 18x^2 - 8x + 8 = 0$.

Solution Begin by searching for rational roots. The possible rational roots are ± 1, $\pm\frac{1}{2}$, ± 2, ± 4, and ± 8. Testing these possibilities (left to right), the first root we find is $\frac{1}{2}$, as shown.

$$\begin{array}{r} \tfrac{1}{2}\,|\,2 + 7 + 0 - 18 - 8 + 8 \\ \underline{+ 1 + 4 + 2 - 8 - 8} \\ 2 + 8 + 4 - 16 - 16\,|\,+ 0 \end{array}$$

Therefore, $x - \frac{1}{2}$ is a factor of $p(x)$.

$$p(x) = (x - \tfrac{1}{2})(2x^4 + 8x^3 + 4x^2 - 16x - 16)$$
$$= 2(x - \tfrac{1}{2})(x^4 + 4x^3 + 2x^2 - 8x - 8)$$

To find other roots of $p(x) = 0$ it now becomes necessary to solve the **depressed equation**

$$x^4 + 4x^3 + 2x^2 - 8x - 8 = 0$$

The possible rational roots for this equation are ± 1, ± 2, ± 4, and ± 8. However, values like ± 1 that were tried before, and produced nonzero remainders, need not be tried again. Why? We find that $x = -2$ is a root:

$$\begin{array}{r} -2\,|\,1 + 4 + 2 - 8 - 8 \\ \underline{- 2 - 4 + 4 + 8} \\ 1 + 2 - 2 - 4\,|\,+ 0 \end{array}$$

$$\begin{aligned} x^4 + 4x^3 + 2x^2 - 8x - 8 &= (x + 2)(x^3 + 2x^2 - 2x - 4) \\ &= (x + 2)[x^2(x + 2) - 2(x + 2)] \\ &= (x + 2)(x^2 - 2)(x + 2) \\ &= (x + 2)^2(x^2 - 2) \end{aligned}$$

This gives

This result can also be written as
$(2x - 1)(x + 2)^2(x^2 - 2).$

$$p(x) = 2(x - \tfrac{1}{2})(x^4 + 4x^3 + 2x^2 - 8x - 8)$$
$$= 2(x - \tfrac{1}{2})(x + 2)^2(x^2 - 2)$$
$$= 2(x - \tfrac{1}{2})(x + 2)^2(x + \sqrt{2})(x - \sqrt{2})$$

These four roots of $p(x) = 0$ are also the x-intercepts of the graph of the polynomial function $p(x)$.

Setting each factor equal to zero produces the real roots of $p(x) = 0$, namely two rational and two irrational roots.

$$x = \tfrac{1}{2}, \qquad x = -2, \qquad x = -\sqrt{2}, \qquad x = \sqrt{2} \qquad \blacksquare$$

The roots of the equation $p(x) = 0$ in Example 6 are $\tfrac{1}{2}$, -2, $-\sqrt{2}$, and $\sqrt{2}$. Since the factored form of $p(x)$ contains the factor $(x + 2)^2$, we say that -2 is a *double root*. Therefore, counting the double root -2 as two roots, there are five roots for $p(x) = 0$, which is the same as the degree of $p(x)$. This is no coincidence. Such a result is true for any polynomial equation and is based upon the following results.

FUNDAMENTAL THEOREM OF ALGEBRA

If $p(x)$ is a polynomial of degree $n \geq 1$, then $p(x) = 0$ has at least one real or complex root.

This was proved by Carl Freidrich Gauss in 1797 when he was 20 years old. He is recognized as one of the greatest mathematicians of all time. (See page 563 for another marginal note about Gauss.)

The proof of this theorem is beyond the level of this course; however, it leads to the next result that tells us that a polynomial equation of degree n cannot have more than n distinct roots.

THE N-ROOTS THEOREM

If $p(x)$ is a polynomial of degree $n \geq 1$, then $p(x) = 0$ has exactly n roots, provided that a root of multiplicity k is counted k times.

For example, the fifth degree equation in Example 6 has five roots, counting -2 twice, because it is a double root.

Even though it can be very difficult to find all the roots of a polynomial equation, the preceding theorem does help us to the extent that we know that there can be no more than n roots when $p(x)$ has degree n. However, for illustrative purposes, we will consider polynomial equations whose roots can be found using the methods that have been studied.

EXAMPLE 7 Find all real and imaginary roots: $x^4 - 6x^2 - 8x + 24 = 0$.

Solution The possible rational roots are

$$\pm 1, \ \pm 2, \ \pm 3, \ \pm 4, \ \pm 6, \ \pm 8, \ \pm 12, \ \pm 24$$

Trying these possibilities, left to right, shows that 2 is a root:

$$\begin{array}{r} 2\,|\,1 + 0 - 6 - 8 + 24 \\ \underline{+\ 2 + 4 - 4 - 24} \\ 1 + 2 - 2 - 12 \,\underline{|\,+\ 0} \end{array}$$

Now, by the factor theorem, we have

$$p(x) = (x - 2)(x^3 + 2x^2 - 2x - 12)$$

and any remaining rational root of the given equation must be a root of the *depressed equation*

$$x^3 + 2x^2 - 2x - 12 = 0$$

This work can be condensed as follows:

$$\begin{array}{l} 2\,|\,1 + 0 - 6 - 8 + 24 \\ \quad\underline{+\ 2 + 4 - 4 - 24} \\ 2\,|\,1 + 2 - 2 - 12\,\underline{|\,+\ 0} \\ \quad\underline{+\ 2 + 8 + 12} \\ \quad\ 1 + 4 + 6\,\underline{|\,+\ 0} \end{array}$$

Since there is the possibility of a double root, try 2 again.

$$\begin{array}{l} 2\,|\,1 + 2 - 2 - 12 \\ \quad\underline{+\ 2 + 8 + 12} \\ \quad\ 1 + 4 + 6\,\underline{|\,+\ 0} \end{array}$$

Then

$$x^3 + 2x^2 - 2x - 12 = (x - 2)(x^2 + 4x + 6)$$

and

$$\begin{aligned} p(x) &= (x - 2)(x - 2)(x^2 + 4x + 6) \\ &= (x - 2)^2(x^2 + 4x + 6) \end{aligned}$$

Using the quadratic formula, the roots of $x^2 + 4x + 6 = 0$ are found to be $-2 \pm \sqrt{2}\,i$. Thus $p(x) = 0$ has two imaginary roots and the double root 2, a total of four roots. Since the degree of $p(x)$ is 4, we know from the *n*-roots theorem that there can be no other roots; we have found them all. ■

In Example 7, such numbers as $-2 + \sqrt{2}\,i$ and $-2 - \sqrt{2}\,i$ are said to be **conjugates** of one another. In general, the conjugate of a complex number $z = a + bi$ is sometimes denoted as $\bar{z}$, so that $\bar{z} = \overline{a + bi} = a - bi$. In the exercises you will be asked to prove that the conjugate of a sum is equal to the sum of the conjugates, $\overline{z + w} = \bar{z} + \bar{w}$, and similarly for differences, products, and quotients. Example 8 demonstrates such results for specific complex numbers.

Observe that the pair of conjugates $-2 + \sqrt{2}\,i$ and $-2 - \sqrt{2}\,i$ are both roots of $p(x) = 0$ in Example 7. This is a special case of the result that states that if a polynomial equation with real coefficients has nonreal complex roots, then they occur in conjugate pairs.

EXAMPLE 8   For $z = -2 + 5i$ and $w = 4 - 7i$, verify the following:
(a) $\overline{z + w} = \bar{z} + \bar{w}$ and **(b)** $\overline{zw} = \bar{z} \cdot \bar{w}$.

Solution

(a) $\overline{z + w} = \overline{(-2 + 5i) + (4 - 7i)}$ $\qquad \bar{z} + \bar{w} = \overline{-2 + 5i} + \overline{4 - 7i}$

$\phantom{\textbf{(a)}\ \overline{z + w}} = \overline{2 - 2i}$ $\qquad\qquad\qquad\qquad\ = (-2 - 5i) + (4 + 7i)$

$\phantom{\textbf{(a)}\ \overline{z + w}} = 2 + 2i$ $\qquad\qquad\qquad\qquad\ = 2 + 2i$

(b) $\overline{zw} = \overline{(-2 + 5i)(4 - 7i)}$ $\bar{z} \cdot \bar{w} = \overline{(-2 + 5i)}\,\overline{(4 - 7i)}$

$= \overline{27 + 34i}$ $= (-2 - 5i)(4 + 7i)$

$= 27 - 34i$ $= 27 - 34i$ ■

EXERCISES 4.7

Use synthetic division and the remainder theorem.

1. $f(x) = x^3 - x^2 + 3x - 2$; find $f(2)$.
2. $f(x) = 2x^3 + 3x^2 - x - 5$; find $f(-1)$.
3. $f(x) = x^4 - 3x^2 + x + 2$; find $f(3)$.
4. $f(x) = x^4 + 2x^3 - 3x - 1$; find $f(-2)$.
5. $f(x) = x^5 - x^3 + 2x^2 + x - 3$; find $f(1)$.
6. $f(x) = 3x^4 + 2x^3 - 3x^2 - x + 7$; find $f(-3)$.

Find the remainder for each division by substitution, using the remainder theorem. That is, in
Exercise 7 (for example) let $f(x) = x^3 - 2x^2 + 3x - 5$ and find $f(2) = r$.

7. $(x^3 - 2x^2 + 3x - 5) \div (x - 2)$
8. $(x^3 - 2x^2 + 3x - 5) \div (x + 2)$
9. $(2x^3 + 3x^2 - 5x + 1) \div (x - 3)$
10. $(3x^4 - x^3 + 2x^2 - x + 1) \div (x + 3)$
11. $(4x^5 - x^3 - 3x^2 + 2) \div (x + 1)$
12. $(3x^5 - 2x^4 + x^3 - 7x + 1) \div (x - 1)$

Show that the given binomial $x - c$ is a factor of $p(x)$, and then factor $p(x)$ completely.

13. $p(x) = x^3 + 6x^2 + 11x + 6$; $x + 1$
14. $p(x) = x^3 - 6x^2 + 11x - 6$; $x - 1$
15. $p(x) = x^3 + 5x^2 - 2x - 24$; $x - 2$
16. $p(x) = -x^3 + 11x^2 - 23x - 35$; $x - 7$
17. $p(x) = -x^3 + 7x + 6$; $x + 2$
18. $p(x) = x^3 + 2x^2 - 13x + 10$; $x + 5$
19. $p(x) = 6x^3 - 25x^2 - 29x + 20$; $x - 5$
20. $p(x) = 12x^3 - 22x^2 - 100x - 16$; $x + 2$
21. $p(x) = x^4 + 4x^3 + 3x^2 - 4x - 4$; $x + 2$
22. $p(x) = x^4 - 8x^3 + 7x^2 + 72x - 144$; $x - 4$
23. $p(x) = x^6 + 6x^5 + 8x^4 - 6x^3 - 9x^2$; $x + 3$

Find all real roots.

24. $x^3 + 2x^2 - 29x + 42 = 0$
25. $x^3 + x^2 - 21x - 45 = 0$
26. $2x^3 - 15x^2 + 24x + 16 = 0$
27. $3x^3 + 2x^2 - 75x - 50 = 0$
28. $x^3 + 3x^2 + 3x + 1 = 0$
29. $x^4 + 3x^3 + 3x^2 + x = 0$
30. $x^4 + 6x^3 + 7x^2 - 12x - 18 = 0$
31. $x^4 + 6x^3 + 2x^2 - 18x - 15 = 0$
32. $x^4 - x^3 - 5x^2 - x - 6 = 0$
33. $x^4 + 2x^3 - 7x^2 - 18x - 18 = 0$
34. $x^4 - 5x^3 + 3x^2 + 15x - 18 = 0$
35. $-x^5 + 5x^4 - 3x^3 - 15x^2 + 18x = 0$
36. $x^4 + 4x^3 - 7x^2 - 36x - 18 = 0$
37. $2x^3 - 5x - 3 = 0$
38. $6x^3 - 25x^2 + 21x + 10 = 0$
39. $3x^4 - 11x^3 - 3x^2 - 6x + 8 = 0$

Factor.

40. $x^3 - 3x^2 - 10x + 24$
41. $-x^3 - 3x^2 + 24x + 80$
42. $x^3 - 28x - 48$
43. $6x^4 + 9x^3 + 9x - 6$
44. Show that $2x^3 - 5x^2 - x + 8 = 0$ has no rational roots.

Find all real and imaginary roots.

45. $p(x) = x^4 - 3x^3 + 5x^2 - x - 10 = 0$
46. $p(x) = x^5 - 3x^4 - 3x^3 + 9x^2 - 10x + 30 = 0$
47. $p(x) = 3x^3 - 5x^2 + 2x - 8 = 0$
48. $p(x) = 2x^4 - 5x^3 + x^2 + 4x - 4 = 0$
49. $p(x) = x^5 - 9x^4 + 31x^3 - 49x^2 + 36x - 10 = 0$
50. $p(x) = 2x^6 + 5x^5 + x^4 + 10x^3 - 4x^2 + 5x - 3 = 0$

Use synthetic division to answer the following.

51. When $x^2 + 5x - 2$ is divided by $x + n$, the remainder is -8. Find all possible values of n and check by division.
52. Find d so that $x + 6$ is a factor of $x^4 + 4x^3 - 21x^2 + dx + 108$.
53. Find b so that $x - 2$ is a factor of $x^3 + bx^2 - 13x + 10$.
54. Find a so that $x - 10$ is a factor of $ax^3 - 25x^2 + 47x + 30$.
55. Find the complete factored form of $p(x) = x^5 + x^4 + 5x^2 - x - 6$ if $p(-2) = p(-1) = p(1) = 0$.
56. $p(x)$ is a fifth-degree polynomial in which 1 is the coefficient of x^5. Write the polynomial in standard form, given that $p(0) = p(1) = p(2) = p(3) = p(4) = 0$.

Solve the system.

57. $y = x^3 - 3x^2 + 3x - 1$
$y = 7x - 13$

58. $y = -x^3$
$y = -3x^2 + 4$

59. $y = 4x^3 - 7x^2 + 10$
$y = x^3 + 43x - 5$

60. Solve the system and graph:

$$y = 2x - 5$$

$$y = \frac{-1}{x^2 - 2x + 1}$$

Use $z = -6 + 8i$ and $w = \frac{1}{2} + \frac{1}{2}i$ to verify the following.

61. $\overline{z + w} = \overline{z} + \overline{w}$
62. $\overline{z - w} = \overline{z} - \overline{w}$
63. $\overline{zw} = \overline{z} \cdot \overline{w}$
64. $\overline{\left(\dfrac{z}{w}\right)} = \dfrac{\overline{z}}{\overline{w}}$

Use $z = a + bi$ and $w = c + di$ to prove the following.

65. $\overline{z + w} = \overline{z} + \overline{w}$
66. $\overline{z - w} = \overline{z} - \overline{w}$
67. $\overline{\left(\dfrac{z}{w}\right)} = \dfrac{\overline{z}}{\overline{w}}$
68. $\overline{zw} = \overline{z} \cdot \overline{w}$

69. Let $p(x) = a_n x^n + a_{n-1} x^{n-1} + \cdots + a_1 x + a_0$, where the a_i are real numbers. Prove that if the complex (imaginary) number $z = a + bi$ is a root of the polynomial equation $p(x) = 0$, then so is $\overline{z}$; $p(\overline{z}) = 0$.
70. Use the result in Exercise 69 and the n-roots theorem to list the possible combinations of real and imaginary roots of each equation. For example, the combinations for $x^4 - 2x^3 + x + 7 = 0$ are four complex (imaginary), or two complex (imaginary) and two real, or four real
(a) $x^3 - 7x^2 + 4x - 9 = 0$ (b) $2x^5 - 6x^4 + x^2 - 3 = 0$
(c) $5x^6 - 2x^5 + x^4 - x^3 - 4x + 1 = 0$

▸ **Written Assignment:** Use your own words, and your own specific examples, to explain both the remainder theorem and the factor theorem.

CHALLENGE
Think Creatively

Use the rational root theorem to prove that the real number $\sqrt{5}$ is an irrational number.

1. To solve an equation with fractions, we multiply through by the LCD. What happens if, instead, we just multiply by a common denominator? Thus, in Example 3 on page 222, is it possible to solve this equation by multiplying by the common denominator $(x^2 - 16)(x + 4)(x - 4)$? Explain your answer.

2. Suppose that y varies directly as x and that x varies inversely as z. How does y vary in relation to z?

3. Show that $x + y$ is a factor of $x^n - y^n$, where n is a positive even integer.

4. A sample of 100 deer is captured, tagged, and released. Later, a second sample is captured and the number previously tagged is counted. This count is used as the basis for estimating the total number of deer in the park. Under what conditions will this be a good estimate? When might it not be a valid basis for estimating the total number in the park?

5. In *base 5*, the numeral 23423_{five} means $2(5^4) + 3(5^3) + 4(5^2) + 2(5) + 3$ and can be thought of as the polynomial $2x^4 + 3x^3 + 4x^2 + 2x + 3$ where $x = 5$. Use synthetic division to find the value of this number in base ten.

6. Assume that $a_n x^n + a_{n-1} x^{n-1} + \cdots + a_1 x + a_0 = 0$ is an nth-degree polynomial equation with integer coefficients, having the root $\frac{6}{8}$. What can you say about the factors of a_0 and a_n? Explain.

4.8 DECOMPOSING RATIONAL FUNCTIONS

In Chapter 1 we learned how to combine rational expressions. For example, combining the fractions in

$$(1) \quad \frac{6}{x - 4} + \frac{3}{x - 2} \qquad \text{produces} \qquad (2) \quad \frac{9x - 24}{(x - 4)(x - 2)}$$

It is now our goal to start with a rational expression such as (2) and *decompose* it into the form (1). When this is accomplished we say that $\dfrac{9x - 24}{(x - 4)(x - 2)}$ has been decomposed into (simpler) **partial fractions.**

$$\frac{9x - 24}{(x - 4)(x - 2)} = \frac{6}{x - 4} + \frac{3}{x - 2}$$

First observe that each factor in the denominator on the left serves as a denominator of a partial fraction on the right. Let us assume, for the moment, that the numerators 6 and 3 are not known. Then it is reasonable to begin by writing

$$\frac{9x - 24}{(x - 4)(x - 2)} = \frac{A}{x - 4} + \frac{B}{x - 2}$$

We will only consider examples that involve linear factors in the denominator. Examples involving nonfactorable quadratic factors are considered in the Exercises.

where A and B are the constants to be found. To find these values, first clear fractions by multiplying both sides by $(x - 4)(x - 2)$.

$$(x - 4)(x - 2) \cdot \frac{9x - 24}{(x - 4)(x - 2)} = (x - 4)(x - 2)\left[\frac{A}{x - 4} + \frac{B}{x - 2}\right]$$

$$9x - 24 = A(x - 2) + B(x - 4)$$

Since we want this equation to hold for all values of x, we may select specific values for x that will produce the constants A and B. Observe that when $x = 4$ the term $B(x - 4)$ will become zero.

$$9(4) - 24 = A(4 - 2) + B(4 - 4)$$
$$12 = 2A + 0$$
$$6 = A$$

Similarly, B can be found by letting $x = 2$.

$$9(2) - 24 = A(2 - 2) + B(2 - 4)$$
$$-6 = 0 - 2B$$
$$3 = B$$

EXAMPLE 1   Decompose $\dfrac{6x^2 + x - 37}{(x - 3)(x + 2)(x - 1)}$ into partial fractions.

Solution   Since there are three linear factors in the denominator, we begin with the form

$$\frac{6x^2 + x - 37}{(x - 3)(x + 2)(x - 1)} = \frac{A}{x - 3} + \frac{B}{x + 2} + \frac{C}{x - 1}$$

Multiply by $(x - 3)(x + 2)(x - 1)$ to clear fractions.

$$6x^2 + x - 37 = A(x + 2)(x - 1) + B(x - 3)(x - 1) + C(x - 3)(x + 2)$$

Since the second and third terms on the right have the factor $x - 3$, the value $x = 3$ will make these two terms zero.

$$6(3)^2 + 3 - 37 = A(3 + 2)(3 - 1) + B(3 - 3)(3 - 1) + C(3 - 3)(3 + 2)$$
$$54 + 3 - 37 = A(5)(2) + 0 + 0$$
$$20 = 10A$$
$$2 = A$$

To find B, use $x = -2$.

$$6(-2)^2 + (-2) - 37 = A(-2 + 2)(-2 - 1) + B(-2 - 3)(-2 - 1)$$
$$+ C(-2 - 3)(-2 + 2)$$
$$24 - 2 - 37 = 0 + B(-5)(-3) + 0$$
$$-15 = 15B$$
$$-1 = B$$

To find C, let $x = 1$.

$$6(1)^2 + 1 - 37 = 0 + 0 + C(1 - 3)(1 + 2)$$
$$-30 = -6C$$
$$5 = C$$

Substituting the values for A, B, and C into the original form produces the desired decomposition.

Check this result by combining the fractions on the right side.

$$\frac{6x^2 + x - 37}{(x - 3)(x + 2)(x - 1)} = \frac{2}{x - 3} + \frac{-1}{x + 2} + \frac{5}{x - 1}$$
$$= \frac{2}{x - 3} - \frac{1}{x + 2} + \frac{5}{x - 1}$$ ∎

TEST YOUR UNDERSTANDING
Think Carefully

Decompose into partial fractions.

1. $\dfrac{8x - 19}{(x - 2)(x - 3)}$

2. $\dfrac{1}{(x + 2)(x - 4)}$

3. $\dfrac{6x^2 - 22x + 18}{(x - 1)(x - 2)(x - 3)}$

Factor the denominator and decompose into partial fractions.

4. $\dfrac{4x + 6}{x^2 + 5x + 6}$

5. $\dfrac{23x - 1}{6x^2 + x - 1}$

(Answers: Page 260)

Note that the least common denominator is the highest power of the linear factor in either denominator.

Let us look at a somewhat different example.

$$\frac{7}{x + 3} - \frac{4}{(x + 3)^2} = \frac{7x + 17}{(x + 3)^2}$$

Now assume that the specific numerators are not known, and begin the decomposition process in this way:

$$\frac{7x + 17}{(x + 3)^2} = \frac{A}{x + 3} + \frac{B}{(x + 3)^2}$$

Clear fractions.

$$(1) \quad 7x + 17 = A(x + 3) + B$$

To find B, let $x = -3$.

$$7(-3) + 17 = A(0) + B$$
$$-4 = B$$

Substitute this value for B into Equation (1).

$$(2) \quad 7x + 17 = A(x + 3) - 4$$

Now find A by substituting some easy value for x, say $x = 0$, into (2).

$$7(0) + 17 = A(0 + 3) - 4$$
$$17 = 3A - 4$$
$$7 = A$$

Note: If the original denominator had been $(x + 3)^3$, then we would have used the additional fraction $\dfrac{C}{(x + 3)^3}$ to start with.

Substituting these values for A and B into our original form produces the decomposition.

$$\frac{7x + 17}{(x + 3)^2} = \frac{7}{x + 3} + \frac{-4}{(x + 3)^2} = \frac{7}{x + 3} - \frac{4}{(x + 3)^2}$$

EXAMPLE 2 Decompose $\dfrac{6 + 26x - x^2}{(2x - 1)(x + 2)^2}$ into partial fractions.

Solution Begin with this form:

$$\frac{6 + 26x - x^2}{(2x - 1)(x + 2)^2} = \frac{A}{2x - 1} + \frac{B}{x + 2} + \frac{C}{(x + 2)^2}$$

Clear fractions.

(1) $6 + 26x - x^2 = A(x + 2)^2 + B(2x - 1)(x + 2) + C(2x - 1)$

Find A by substituting $x = \frac{1}{2}$.

$$6 + 13 - \tfrac{1}{4} = A\left(\tfrac{5}{2}\right)^2 + 0 + 0$$
$$\tfrac{75}{4} = \tfrac{25}{4}A$$
$$3 = A$$

Find C by letting $x = -2$.

$$6 - 52 - 4 = 0 + 0 + C(-5)$$
$$-50 = -5C$$
$$10 = C$$

Substitute $A = 3$ and $C = 10$ into Equation (1).

(2) $6 + 26x - x^2 = 3(x + 2)^2 + B(2x - 1)(x + 2) + 10(2x - 1)$

To find B, substitute a simple value like $x = 1$ into (2).

$$6 + 26 - 1 = 3(9) + B(1)(3) + 10(1)$$
$$-6 = 3B$$
$$-2 = B$$

Then the decomposition is

$$\frac{6 + 26x - x^2}{(2x - 1)(x + 2)^2} = \frac{3}{2x - 1} - \frac{2}{x + 2} + \frac{10}{(x + 2)^2}$$

You can check this by combining the right side. ∎

Thus far, in every decomposition problem the degree of the polynomial in the numerator has been less than the degree in the denominator. Here is an example where this is not the case.

$$\frac{2x^3 + 3x^2 - x + 16}{x^2 + 2x - 3}$$

In such cases the *first* step is to divide.

$$\frac{2x^3 + 3x^2 - x + 16}{x^2 + 2x - 3} = \text{quotient} + \frac{\text{remainder}}{\text{divisor}}$$

$$= 2x - 1 + \frac{7x + 13}{x^2 + 2x - 3}$$

The problem will be completed by decomposing $\dfrac{7x + 13}{x^2 + 2x - 3}$. You can verify that

$$\frac{7x + 13}{x^2 + 2x - 3} = \frac{7x + 13}{(x - 1)(x + 3)} = \frac{5}{x - 1} + \frac{2}{x + 3}$$

Therefore, the final decomposition is

$$\frac{2x^3 + 3x^2 - x + 16}{x^2 + 2x - 3} = 2x - 1 + \frac{5}{x - 1} + \frac{2}{x + 3}$$

CAUTION

When the degree of the numerator is greater than or equal to the degree in the denominator, you *must* divide first. If this step is ignored, the resulting decomposition will be wrong. For example, suppose that you started *incorrectly* in this way:

$$\frac{2x^3 + 3x^2 - x + 16}{(x - 1)(x + 3)} = \frac{A}{x - 1} + \frac{B}{x + 3}$$

This approach will produce the following *incorrect* answer:

$$\frac{2x^3 + 3x^2 - x + 16}{(x - 1)(x + 3)} = \frac{5}{x - 1} + \frac{2}{x + 3}$$

EXERCISES 4.8

Decompose into partial fractions.

1. $\dfrac{2x}{(x + 1)(x - 1)}$

2. $\dfrac{x}{x^2 - 4}$

3. $\dfrac{x + 7}{x^2 - x - 6}$

4. $\dfrac{4x^2 + 16x + 4}{(x + 3)(x^2 - 1)}$

5. $\dfrac{5x^2 + 9x - 56}{(x - 4)(x - 2)(x + 1)}$

6. $\dfrac{x}{(x - 3)^2}$

7. $\dfrac{3x - 3}{(x - 2)^2}$

8. $\dfrac{2 - 3x}{x^2 + x}$

9. $\dfrac{3x - 30}{15x^2 - 14x - 8}$

10. $\dfrac{2x + 1}{(2x + 3)^2}$

11. $\dfrac{x^2 - x - 4}{x(x + 2)^2}$

12. $\dfrac{x^2 + 5x + 8}{(x - 3)(x + 1)^2}$

First divide and then complete the decomposition into partial fractions.

13. $\dfrac{x^3 - x + 2}{x^2 - 1}$

14. $\dfrac{4x^2 - 14x + 2}{4x^2 - 1}$

15. $\dfrac{12x^4 - 12x^3 + 7x^2 - 2x - 3}{4x^2 - 4x + 1}$

Decompose into partial fractions.

16. $\dfrac{10x^2 - 16}{x^4 - 5x^2 + 4}$

17. $\dfrac{10x^3 - 15x^2 - 35x}{x^2 - x - 6}$

18. $\dfrac{25x^3 + 10x^2 + 31x + 5}{25x^2 + 10x + 1}$

19. $\dfrac{5x^2 - 24x - 173}{x^3 + 4x^2 - 31x - 70}$

The method of this section can be adjusted to fractions involving nonfactorable quadratics in the denominator. Study this example and apply it in Exercises 20–23 to decompose the fractions.

$$\frac{2x}{(x - 1)(x^2 + 1)} = \frac{A}{x - 1} + \frac{Bx + C}{x^2 + 1} \qquad \longleftarrow \text{Allow for a linear numerator over the quadratic factor}$$

$$2x = A(x^2 + 1) + (x - 1)(Bx + C)$$

$$2 = 2A \qquad \longleftarrow \text{Use } x = 1.$$

$$1 = A$$

$$0 = 1 + (-1)C \qquad \longleftarrow \text{Let } A = 1 \text{ and use } x = 0.$$

$$1 = C$$

$$-2 = 2 + (-2)(-B + 1) \qquad \longleftarrow \text{Let } A = 1, C = 1, \text{ and use } x = -1.$$

$$-1 = B$$

$$\frac{2x}{(x - 1)(x^2 + 1)} = \frac{1}{x - 1} + \frac{-x + 1}{x^2 + 1}$$

20. $\dfrac{x - 3}{(x + 2)(x^2 + 1)}$

21. $\dfrac{-4x}{(x^2 + 3)(x - 1)}$

22. $\dfrac{x^2 - 5x + 19}{(x - 3)(x^2 + 4)}$

23. $\dfrac{4x^2 + 5}{(x - 1)(x^2 + x + 1)}$

KEY TERMS

*Review these **key terms** so that you are able to define or describe them. A clear understanding of these terms will be very helpful when reviewing the developments in this chapter.*

Symmetry about the y-axis • Symmetry through the origin • Multiplying ordinates • Reflection through the x-axis • Horizontal and vertical translations • Vertical and horizontal asymptotes • Rational function • Polynomial function • Equations with fractions • Proportion property • Inequalities with fractions • Direct variation • Inverse variation • Constant of variation • Division of polynomials • Synthetic division • Remainder theorem • Factor theorem • Root of multiplicity k • Fundamental theorem of algebra • The n-roots theorem • Depressed equation • Complex conjugates • Partial fractions • Decomposition into partial fractions

REVIEW EXERCISES

The solutions to the following exercises can be found within the text of Chapter 4. Try to answer each question before referring to the text.

Section 4.1

Graph each of the following.

1. $y = f(x) = x^3$

2. $y = f(x) = -x^3$

3. $y = f(x) = (x - 3)^3$

4. $y = f(x) = |(x - 3)^3|$

5. $y = 2x^4$

6. Under what conditions is the graph of a function $y = f(x)$ said to be symmetric through the origin?

7. Under what conditions is the graph of $y = f(x)$ symmetric about the y-axis?

8. Explain how the graph of $y = f(x + h)$ can be obtained from the graph of $y = f(x)$ when $h > 0$, and also when $h < 0$.

9. Explain how the graph of $y = f(x) + k$ can be obtained from the graph of $y = f(x)$ when $k > 0$, and also when $k < 0$.

Section 4.2

10. Draw the graph of the function $y = f(x) = \dfrac{1}{x}$.

11. Draw the graph of $y = \dfrac{1}{x^2}$ and describe the symmetry of the curve.

Sketch each graph. Find the asymptotes, if any.

12. $g(x) = \dfrac{1}{x - 3}$

13. $y = \dfrac{1}{(x + 2)^2} - 3$

14. Graph: $y = f(x) = \dfrac{x^2 + x - 6}{x - 2}$

15. What are the conditions needed for the graph of a rational function to have a vertical asymptote?

Section 4.3

16. Construct a table of signs for the inequality $x^2 - 7x + 10 < 0$.

17. Construct a table of signs and graph the function $f(x) = x^3 + 4x^2 - x - 4$.

Find the asymptotes.

18. $f(x) = \dfrac{x}{x^2 - 2x}$

19. $g(x) = \dfrac{3x^2}{2x^2 + 1}$

20. $h(x) = \dfrac{x^2}{x + 1}$

21. Graph: $y = f(x) = \dfrac{x - 2}{x^2 - 3x - 4}$

22. Graph: $y = f(x) = \dfrac{2}{x^2 + 1}$

Section 4.4

23. Find the x-intercepts of $y = f(x) = \dfrac{3x}{x + 7} - \dfrac{8}{5}$ and state the domain.

Solve for x.

24. $\dfrac{2x - 5}{x + 1} - \dfrac{3}{x^2 + x} = 0$

25. $\dfrac{24}{x^2 - 16} - \dfrac{5}{x + 4} = \dfrac{3}{x - 4}$

26. State the proportion property.

27. How high is a tree that casts an 18-foot shadow at the same time that a 3-foot stick casts a 2-foot shadow?

28. Working alone, Harry can mow a lawn in 3 hours. Elliot can complete the job in 2 hours. How long will it take them working together, assuming they both start at the same time?

29. A sample of 80 deer is tagged and released. Later, another sample of 100 is captured, of which 5 had previously been tagged. Estimate the number of deer in the park.

30. Solve for b: $r = \dfrac{ab}{a + b + c}$.

Solve for x.

31. $\dfrac{x}{3} - \dfrac{2x + 1}{4} > 1$

32. $\dfrac{x + 1}{x} < 3$

33. Graph on a number line: $\dfrac{1 - 2x}{x - 3} \le 0$

34. Solve the system: $y = \dfrac{2}{x}$

$y = -2x + 5$

Section 4.5

35. Write the equation that expresses this direct variation: y varies directly as x, and y is 8 when $x = 12$. Find y for $x = 30$.

36. According to Hooke's law, the force F required to hold a spring stretched x units beyond its natural length is directly proportional to x. If a force of 20 pounds is needed to hold a certain spring stretched 3 inches beyond its natural length, how far will 60 pounds of force hold the spring stretched beyond its natural length?

37. According to Boyle's law, the pressure of a compressed gas is inversely proportional to the volume V. Suppose that there is a pressure of 25 pounds per square inch when the volume of gas is 400 cubic inches. Find the pressure when the gas is compressed to 200 cubic inches.

Describe the variation given by each equation.

38. $z = kx^2y^3$

39. $z = \dfrac{kx^2}{y}$

40. $V = \pi r^2 h$

41. Suppose that z varies directly as x and inversely as the square of y. If $z = \frac{1}{3}$ when $x = 4$ and $y = 6$, find z when $x = 12$ and $y = 4$.

Section 4.6

42. Divide $2x^3 + 3x^2 - x + 16$ by $x^2 + 2x - 3$.

Use synthetic division to find each quotient.

43. $(x^3 + x^2 - 11x + 12) \div (x - 2)$

44. $(2x^3 - 9x^2 + 10x - 7) \div (x - 3)$

45. $(-\frac{1}{3}x^4 + \frac{1}{6}x^2 - 7x - 4) \div (x + 3)$

46. State the rule for checking a division problem.

Section 4.7

47. State the remainder theorem.   **48.** State the factor theorem.

49. Find the remainder if $p(x) = 3x^3 - 5x^2 + 7x + 5$ is divided by $x - 2$.

50. Let $p(x) = x^3 - 2x^2 + 3x - 1$. Use synthetic division and the remainder theorem to find $p(3)$.

51. Show that $x - 2$ is a factor of $p(x) = x^3 - 3x^2 + 7x - 10$.

52. Use the factor theorem to show that $x + 3$ is a factor of $p(x) = x^3 - x^2 + 8x + 12$. Then factor $p(x)$ completely.

53. State the rational root theorem.

54. What are the possible rational roots of $f(x) = 4x^3 - 16x^2 + 11x + 10 = 0$?

55. Factor $f(x) = x^3 + 6x^2 + 11x + 6$.

56. Find all real roots of $p(x) = 2x^5 + 7x^4 - 18x^2 - 8x + 8 = 0$.

57. Find all real and imaginary roots: $p(x) = x^4 - 6x^2 - 8x + 24 = 0$.

58. For $z = -2 + 5i$ and $w = 4 - 7i$, verify that **(a)** $\overline{z + w} = \overline{z} + \overline{w}$ and **(b)** $\overline{zw} = \overline{z} \cdot \overline{w}$.

Section 4.8

Decompose into partial fractions.

59. $\dfrac{9x - 24}{(x - 4)(x - 2)}$

60. $\dfrac{6x^2 + x - 37}{(x - 3)(x + 2)(x - 1)}$

61. $\dfrac{7x + 17}{(x + 3)^2}$

62. $\dfrac{6 + 26x - x^2}{(2x - 1)(x + 2)^2}$

63. $\dfrac{2x^3 + 3x^2 - x + 16}{x^2 + 2x - 3}$

CHAPTER 4 TEST: STANDARD ANSWER

Use these questions to test your knowledge of the basic skills and concepts of Chapter 4. Then check your answers with those given at the back of the book.

Graph each function and write the equation of the asymptotes if there are any.

1. $f(x) = (x + 2)^3 - \frac{3}{2}$ **2.** $y = f(x) = x^3$ **3.** $y = f(x) = -\dfrac{1}{x - 2}$ **4.** $y = f(x) = \dfrac{x - 1}{x^2 - x - 2}$

5. Graph: $y = x^3 - x^2 - 4x + 4$

6. Construct a table of signs for $f(x) = \dfrac{x^2 - 2x}{x + 3}$.

7. Graph: $y = f(x) = \dfrac{2}{x^2 + 2}$

8. Graph on a number line: $\dfrac{x + 2}{x - 1} \le 0$

9. Solve for a: $S = \dfrac{n}{2}[2a + (n - 1)d]$

Solve for x.

10. $\dfrac{6}{x} = 2 + \dfrac{3}{x + 1}$ **11.** $\dfrac{x}{2} - \dfrac{3x + 1}{3} > 2$ **12.** $\dfrac{6}{x^2 - 9} - \dfrac{2}{x - 3} = \dfrac{1}{x + 3}$

13. A piece of wire that is 10 feet long is cut into two pieces. The ratio of the smaller piece to the larger piece is $\frac{3}{4}$. Find the length of the larger piece.

14. Working alone, Dave can wash his car in 45 minutes. If Ellen helps him, they can do the job together in 30 minutes. How long would it take Ellen to wash the car by herself?

15. z varies directly as x and inversely as y. If $z = \frac{2}{3}$ when $x = 2$ and $y = 15$, find z when $x = 4$ and $y = 10$.

16. If the volume of a sphere varies directly as the cube of its radius and $V = 36\pi$ cubic inches when $r = 3$ inches, find V when $r = 6$ inches.

17. Use synthetic division to divide $2x^5 + 5x^4 - x^2 - 21x + 7$ by $x + 3$.

18. Let $p(x) = 27x^4 - 36x^3 + 18x^2 - 4x + 1$. Use the remainder theorem to evaluate $p(\frac{1}{3})$.

19. Use the result of Exercise 18 and the factor theorem to determine whether or not $x - \frac{1}{3}$ is a factor of $p(x)$.

20. Show that $x - 2$ is a factor of $p(x) = x^4 - 4x^3 + 7x^2 - 12x + 12$, and factor $p(x)$ completely.

21. Make use of the rational root theorem to factor $f(x) = x^4 + 5x^3 + 4x^2 - 3x + 9$.

22. Find the roots of $p(x) = 0$ for $p(x) = x^4 + 3x^3 - 3x^2 - 11x - 6$.

23. Find all real and imaginary roots for $p(x) = x^4 - 2x^3 - 2x^2 - 2x - 3$.

ompose into partial fractions.

24. $\dfrac{x - 15}{x^2 - 25}$ 25. $\dfrac{6x^2 - 2x + 2}{x^3 - 2x^2 - 5x + 6}$

CHAPTER 4 TEST: MULTIPLE CHOICE

1. Which of the following are true?

   I. The graph of $f(x) = x^2$ is symmetric about the y-axis.

   II. The graph of $f(x) = x^3$ is symmetric through the origin.

   III. The graph of $f(x) = \dfrac{1}{x}$ is symmetric about the x-axis.

   (a) Only I   **(b)** Only II   **(c)** Only III   **(d)** Only I and II   **(e)** None of the preceding

2. The equation of the horizontal asymptote for $f(x) = \dfrac{1}{x - 3}$ is

   (a) $x = 0$ **(b)** $y = 0$ **(c)** $x = 3$ **(d)** $x = -3$ **(e)** None of the preceding

3. For $h > 0$, the graph of $y = f(x - h)$ can be obtained by shifting the graph of $y = f(x)$

   (a) h units to the right   **(b)** h units to the left   **(c)** h units upward   **(d)** h units downward

   (e) None of the preceding

4. Which of the following are true for the graph of $y = -\dfrac{1}{x^2}$?

   I. The horizontal asymptote is $x = 0$.

   II. The vertical asymptote is $y = 0$.

   III. The graph is symmetric about the y-axis.

   (a) Only I   **(b)** Only II   **(c)** Only III   **(d)** I, II, and III   **(e)** None of the preceding

5. The table below can be used to determine the position of the curve $f(x) = (x + 3)(x + 1)(x - 2)$ relative to the x-axis. Reading left to right, these positions for the given intervals, are

   (a) above, below, above, below

   (b) below, above, below, above

   (c) below, below, above, above

   (d) above, below, below, above

   (e) None of the preceding

Interval	$(-\infty, -3)$	$(-3, -1)$	$(-1, 2)$	$(2, \infty)$
Sign of $f(x)$				

6. What is the equation of the vertical asymptote for $f(x) = \dfrac{3x^2}{x^3 + x}$?

   (a) $x = 0$ (b) $x = -1$ (c) $x = 3$ (d) $y = 3$ (e) None of the preceding

7. Working alone, Amy can complete a job in 5 hours. Julie can complete the same job in 4 hours. Which of the following equations can be used to find out how long it will take to complete the job together? (Use x for the time it takes them to complete the job working together.)

   (a) $\dfrac{1}{4} + \dfrac{1}{x} = \dfrac{1}{5}$ (b) $\dfrac{1}{4} - \dfrac{1}{5} = \dfrac{1}{x}$ (c) $\dfrac{x}{4} - \dfrac{x}{5} = 1$ (d) $\dfrac{1}{4} + \dfrac{1}{5} = \dfrac{1}{x}$

   (e) None of the preceding

8. If $c = \dfrac{a + 2b}{ab}$, then $b = $

   (a) $\dfrac{a}{ac - 2}$ (b) $\dfrac{1}{c} - \dfrac{a}{2}$ (c) $a(c - a)$ (d) $c - 1$ (e) None of the preceding

9. The equation $\dfrac{9x + 14}{2x^2 - x - 6} - \dfrac{1}{x - 2} = \dfrac{2}{2x + 3}$ has

   (a) no solution   (b) just one solution   (c) exactly two solutions

   (d) more than two solutions   (e) None of the preceding

10. Neglecting air resistance, the distance an object falls from rest is directly proportional to the square of the time it has fallen. If an object falls 64 feet in the first 2 seconds, how far will it fall in the first 6 seconds?

    (a) 576 ft.   (b) 512 ft.   (c) 384 ft.   (d) 192 ft.   (e) None of the preceding

11. z varies directly as x and inversely as y^2. If $z = 60$ when $x = 3$ and $y = \frac{1}{2}$, then the value of z when $x = 6$ and $y = 4$ is:

    (a) 1920   (b) 30   (c) $\frac{45}{2}$ (d) $\frac{15}{8}$ (e) None of the preceding

12. When the synthetic division at the right is completed, the remainder is

    $-2 \lfloor 1 + 3 - 5 + 7$

    (a) 0   (b) $+7$ (c) $+21$

    (d) -7 (e) None of the preceding

13. Which of the following is a set of possible rational roots of $f(x) = 2x^3 - 8x^2 + 7x - 10$?

    (a) $\pm 1, \pm 2, \pm 5, \pm 10$ (b) $\pm 1, \pm \frac{1}{2}, \pm 2, \pm \frac{5}{2}, \pm 5, \pm 10$

    (c) $\pm 2, \pm 10,$ (d) $\pm 1, \pm 2, \pm 10$ (e) None of the preceding

14. Consider the decomposition of $\dfrac{6x^2 + x - 37}{(x - 3)(x + 2)(x - 1)}$ into the partial fractions

    $\dfrac{A}{x - 3} + \dfrac{B}{x + 2} + \dfrac{C}{x - 1}$. Then $A = $

    (a) -2 (b) 1   (c) 2   (d) 3   (e) None of the preceding

15. The equation $x^4 - 6x^3 + 14x^2 - 16x + 8 = 0$ has 2 as a double root. The remaining roots are

    (a) $1 \pm i$ (b) $1 \pm \sqrt{3}$ (c) $5 \pm 5i$ (d) ± 8 (e) None of the preceding

ANSWERS TO THE TEST YOUR UNDERSTANDING EXERCISES

Page 217

1. Vertical asymptotes: $x = 2$, $x = 3$; horizontal asymptote: $y = 0$
2. Vertical asymptote: $x = -3$; horizontal asymptote: $y = 1$
3. Vertical asymptote: $x = 2$; no horizontal asymptote
4. Vertical asymptote: $x = 4$; horizontal asymptote: $y = 0$

5. Vertical asymptotes: $x = -3$, $x = 4$; horizontal asymptote: $y = 0$
6. Vertical asymptotes: $x = 1$; no horizontal asymptote

Page 223
1. 12   2. $\frac{1}{2}$ 3. 35   4. 5   5. -9 6. No solution   7. $\frac{3}{10}$ 8. $\frac{13}{24}$ 9. 3
10. $-4, 3$ 11. $-3, -1$ 12. $-7, -\frac{2}{3}$

Page 233
1. 25   2. 6   3. 5   4. 16

Page 238
1. $x^2 + x - 3 = \; ; r = 0$ 2. $2x^2 + 3x - 1; r = 0$ 3. $2x^2 + 3x + 9; r = 34$
4. $3x^2 - 3x - 4; r = 16$

Page 240

1.
$$1 \,\underline{\big|\, 1 - 7 + 4 + 10 - 1 - 5}$$
$$\; + 1 - 6 - 2 + 8 + 7$$
$$\; 1 - 6 - 2 + 8 + 7 \,\underline{\big|\, + 2} = r$$
$$p(1) = 1 - 7 + 4 + 10 - 1 - 5 = 2$$

2.
$$-10 \,\underline{\big|\, 1 + 11 + 11 + 11 + 10}$$
$$\; - 10 - 10 - 10 - 10$$
$$\; 1 + 1 + 1 + 1 \,\underline{\big|\; 0} = r$$
$$p(-10) = 10{,}000 - 11{,}000 + 1100 - 110 + 10 = 0$$

3.
$$10 \,\underline{\big|\, 1 + 11 + 11 + 11 + 10}$$
$$\; + 10 + 210 + 2210 + 22210$$
$$\; 1 + 21 + 221 + 2221 \,\underline{\big|\, + 22220} = r$$
$$p(10) = 10{,}000 + 11{,}000 + 1100 + 110 + 10$$
$$= 22{,}220$$

4.
$$6 \,\underline{\big|\, 6 - 40 + 0 + 25}$$
$$\; + 36 - 24 - 144$$
$$\; 6 - 4 - 24 \,\underline{\big|\, - 119} = r$$
$$p(6) = 6(6)^3 - 40(6)^2 + 25 = -119$$

Page 244
1. **(a)** $\pm 1, \pm 2, \pm 3, \pm 4, \pm 6, \pm 8, \pm 12, \pm 24$ **(b)** $(x + 3)(x - 2)(x - 4)$ **(c)** $-3, 2, 4$
2. **(a)** $\pm 1, \pm 2, \pm 4, \pm 8, \pm 16$ **(b)** $(x + 4)^2(x - 1)^2$ **(c)** $-4, 1$
3. **(a)** $\pm 1, \pm\frac{1}{2}, \pm\frac{1}{4}, \pm 2, \pm 3, \pm\frac{3}{2}, \pm\frac{3}{4}, \pm 6$ **(b)** $(x + 6)(2x - 1)^2$ **(c)** $-6, \frac{1}{2}$
4. **(a)** $\pm 1, \pm\frac{1}{3}, \pm 2, \pm\frac{2}{3}, \pm 4, \pm\frac{4}{3}$ **(b)** $(3x - 1)(x - 4)(x^2 + 1)$ **(c)** $\frac{1}{3}, 4$

Page 251
1. $\dfrac{3}{x - 2} + \dfrac{5}{x - 3}$ 2. $-\dfrac{\frac{1}{6}}{x + 2} + \dfrac{\frac{1}{6}}{x - 4}$ 3. $\dfrac{1}{x - 1} + \dfrac{2}{x - 2} + \dfrac{3}{x - 3}$
4. $\dfrac{6}{x + 3} - \dfrac{2}{x + 2}$ 5. $\dfrac{4}{3x - 1} + \dfrac{5}{2x + 1}$

CHAPTER 5

THE ALGEBRA OF FUNCTIONS

5.1 GRAPHING SOME SPECIAL RADICAL FUNCTIONS

A radical expression in x, such as $\sqrt{x}$, may be used to define a function f, where $f(x) = \sqrt{x}$, the *square root function*. The domain of f consists of all real numbers $x \geq 0$, since the square root of a negative number is not a real number. To graph $y = \sqrt{x}$, it is helpful first to square both sides to obtain $y^2 = x$; that is, $x = y^2$. Recall the graph of $y = x^2$ and obtain the graph of $x = y^2$ by reversing the role of the variables.

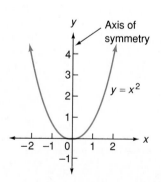

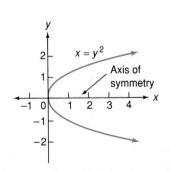

y	$x = y^2$
-2	4
-1	1
0	0
1	1
2	4

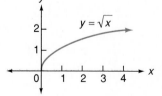

The graph of the parabola $x = y^2$ is *not* the graph of a function having x as the independent variable because for $x > 0$ there are two corresponding y-values. But if the bottom branch is removed, we have the correct graph of the **radical function** $y = \sqrt{x}$, for $x \geq 0$. (What is the equation for the bottom branch?)

Except for the origin, this graph is in the first quadrant, as could have been predicted by observing that for all $x > 0$, $\sqrt{x} > 0$. You can also verify that the specific points given in the following table are on the graph.

x	0	$\dfrac{1}{4}$	$\dfrac{9}{16}$	1	$\dfrac{9}{4}$	2	3	4
$y = \sqrt{x}$	0	$\dfrac{1}{2}$	$\dfrac{3}{4}$	1	$\dfrac{3}{2}$	$\sqrt{2}$	$\sqrt{3}$	2

Such a table of values presents us with an alternative method for sketching the graph of $y = \sqrt{x}$; plot the points in the table and connect with a smooth curve.

In summary, the graph of $y = f(x) = \sqrt{x}$ is the upper half of the horizontal parabola $x = y^2$. The curve is increasing and concave down on $(0, \infty)$. The domain of f is all $x \geq 0$, and, as indicated by the graph, the range is all $y \geq 0$.

Note: $x - 2 \geq 0$; thus $x \geq 2$.

EXAMPLE 1 Find the domain and range of $y = g(x) = \sqrt{x - 2}$ and graph.

Solution Since the square root of a negative number is not a real number, the expression $x - 2$ must be nonnegative; therefore, the domain of g consists of all $x \geq 2$. The graph of g may be found by shifting the graph of $y = \sqrt{x}$ by 2 units to the right.

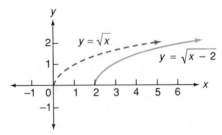

The range of $y = \sqrt{x}$ is all $y \geq 0$, and since the graph of g was obtained from $y = \sqrt{x}$ by a horizontal translation, the range of g is also all $y \geq 0$. ∎

EXAMPLE 2 Find the domain of $y = f(x) = \sqrt{|x|}$ and graph.

Solution Since $|x| \geq 0$ for all x, the domain of f consists of all real numbers. To graph f, first note that the graph is symmetric about the y-axis:

$$f(-x) = \sqrt{|-x|} = \sqrt{|x|} = f(x)$$

It is also helpful to locate a few specific points on the curve as an aid to graphing.

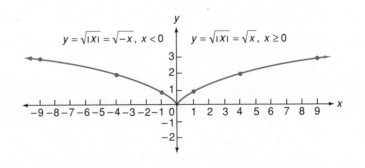

x	y
0	0
1	1
4	2
9	3
-1	1
-4	2
-9	3

f is decreasing and concave down for $x < 0$, and it is increasing and concave down for $x > 0$.

In drawing the graph of $f(x)$, we first found the graph for $x \geq 0$ and used symmetry to obtain the rest. For $x \geq 0$, we get $|x| = x$ and $f(x) = \sqrt{|x|} = \sqrt{x}$. ■

EXAMPLE 3 Find the domain of $y = h(x) = x^{-1/2}$ and graph by using a table of values.

Solution Note that $h(x) = x^{-1/2} = \dfrac{1}{x^{1/2}} = \dfrac{1}{\sqrt{x}}$. Thus the domain consists of all $x > 0$. Furthermore, $\dfrac{1}{\sqrt{x}} > 0$ for all x, so we know that the graph must be in the first quadrant only. Plot the points in the table and connect them with a smooth curve.

Explain why $x \neq 0$ in Example 3.

x	$\frac{1}{9}$	$\frac{1}{4}$	1	4	9
y	3	2	1	$\frac{1}{2}$	$\frac{1}{3}$

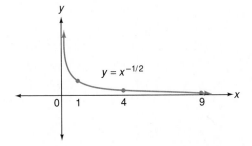

The equations of the two asymptotes are $x = 0$ and $y = 0$.

Observe that the closer x is to zero, the larger are the corresponding y-values. Also, as the values of x get larger, the corresponding y-values get closer to 0. These observations suggest that the coordinate axes are asymptotes to the curve $y = x^{-1/2}$. ■

EXAMPLE 4 Use the graph of $y = \dfrac{1}{\sqrt{x}}$ to graph $y = \dfrac{1}{\sqrt{x}} + 1$ and $y = \dfrac{1}{\sqrt{x+1}}$ and find the asymptotes.

Solution The graphing procedures developed earlier can be used here. Thus the graph of $y = \dfrac{1}{\sqrt{x}} + 1$ can be obtained from the graph for Example 3 by shifting up one unit. Then the asymptotes will be the y-axis and the line $y = 1$. The graph of $y = \dfrac{1}{\sqrt{x+1}}$ is found by shifting the graph for $y = \dfrac{1}{\sqrt{x}}$ to the left one unit. The equations of the asymptotes are $x = -1$ and $y = 0$.

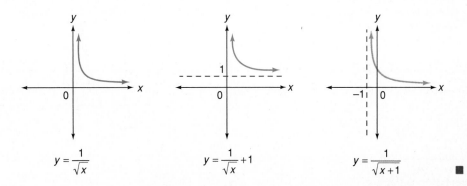

Graph each pair of functions on the same set of axes and state the domain of g.

1. $f(x) = \sqrt{x}, g(x) = \sqrt{x} - 2$ **2.** $f(x) = -\sqrt{x}, g(x) = -\sqrt{x - 1}$

3. $f(x) = \sqrt{x}, g(x) = 2\sqrt{x}$ **4.** $f(x) = \dfrac{1}{\sqrt{x}}, g(x) = \dfrac{1}{\sqrt{x + 2}}$

The graph of the *cube root function* $y = \sqrt[3]{x}$ can be found by a process similar to that used for graphing $y = \sqrt{x}$. First take $y = \sqrt[3]{x}$ and cube both sides to get $y^3 = x$. Now recall the graph of $y = x^3$ and obtain the graph of $x = y^3$ by reversing the role of the variables.

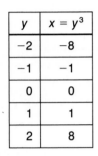

y	$x = y^3$
-2	-8
-1	-1
0	0
1	1
2	8

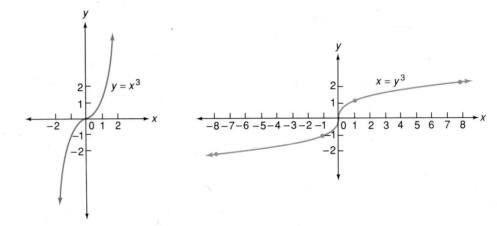

The next example illustrates a function that is defined by using both a polynomial and a radical expression.

EXAMPLE 5   Graph f defined on the domain $-2 \le x < 5$ as follows:

$$f(x) = \begin{cases} x^2 - 1 & \text{for } -2 \le x < 2 \\ \sqrt{x - 2} & \text{for } 2 \le x < 5 \end{cases}$$

Solution   The first part of f is given by $f(x) = x^2 - 1$ for $-2 \le x < 2$. This is an arc of a parabola obtained by shifting the graph of $y = x^2$ downward one unit.

The second part of f is given by $f(x) = \sqrt{x - 2}$ for $2 \le x < 5$. This is an arc of the square root curve obtained by shifting the graph of $y = \sqrt{x}$ two units to the right.

When $x = 2$, the radical part of f is used; $f(2) = \sqrt{2 - 2} = 0$. So there is a solid dot at $(2, 0)$ and an open dot at $(2, 3)$. Also, there is an open dot for $x = 5$ since 5 is not a domain value of f.

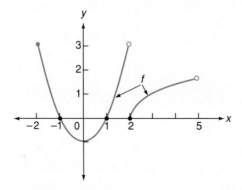

EXERCISES 5.1

Graph each set of curves on the same coordinate system. Use a dashed curve for the first equation and a solid curve for the second.

1. $f(x) = \sqrt{x}$, $g(x) = \sqrt{x} - 1$ 2. $f(x) = -\sqrt{x}$, $g(x) = -2\sqrt{x}$
3. $f(x) = \sqrt{x}$, $g(x) = \sqrt{x} + 1$ 4. $f(x) = -\sqrt{x}$, $g(x) = -\sqrt{x} - 2$
5. $f(x) = \sqrt[3]{x}$, $g(x) = \sqrt[3]{x} + 2$ 6. $f(x) = -\sqrt[3]{x}$, $g(x) = -\sqrt[3]{x} - 3$
7. $f(x) = \sqrt[3]{x}$, $g(x) = \sqrt[3]{x} - 1$ 8. $f(x) = \sqrt[3]{x}$, $g(x) = -\sqrt[3]{x} + 1$

Find the domain of f, sketch the graph, and give the equations of the asymptotic lines if there are any. Also state where f is increasing or decreasing and describe the concavity.

9. $f(x) = \sqrt{x + 2}$ 10. $f(x) = x^{1/2} + 2$ 11. $f(x) = \sqrt{x - 3} - 1$ 12. $f(x) = -\sqrt{x} + 3$
13. $f(x) = \sqrt{-x}$ 14. $f(x) = \sqrt{(x - 2)^2}$ 15. $f(x) = 2\sqrt[3]{x}$ 16. $f(x) = |\sqrt[3]{x}|$

17. $f(x) = -x^{1/3}$ 18. $f(x) = \sqrt[3]{-x}$ 19. $f(x) = \dfrac{1}{\sqrt{x}} - 1$ 20. $f(x) = \dfrac{1}{\sqrt{x - 2}}$

21. **(a)** Explain why the graph of $f(x) = \dfrac{1}{\sqrt[3]{x}}$ is symmetric through the origin.

    (b) What is the domain of f?

    (c) Use a table of values to graph f.

    (d) What are the equations of the asymptotes?

22. Find the domain of $f(x) = \dfrac{1}{\sqrt[3]{x + 1}}$, sketch the graph, and give the equations of the asymptotes.

23. Find the graph of the function $y = \sqrt[4]{x}$ by raising both sides of the equation to the fourth power and comparing to the graph of $y = x^4$.

24. Reflect the graph of $y = x^2$, for $x \geq 0$, through the line $y = x$. Obtain the equation of this new curve by interchanging variables in $y = x^2$ and solving for y.

25. Follow the instruction of Exercise 24 with $y = x^3$ for all values x.

In Exercises 26 and 27, the function f is defined by using more than one expression. Graph f on its given domain.

26. $f(x) = \begin{cases} \sqrt{x} & \text{for } 0 \leq x \leq 4 \\ 10 - \frac{1}{2}x^2 & \text{for } 4 < x < 6 \end{cases}$

27. $f(x) = \begin{cases} -2x - 1 & \text{for } -3 \leq x < 0 \\ \sqrt[3]{x - 1} & \text{for } 0 \leq x \leq 2 \end{cases}$

 Verify the equation involving the difference quotient for the given radical function (see Section 2.1).

28. $f(x) = \sqrt{x}$; $\dfrac{f(x) - f(25)}{x - 25} = \dfrac{1}{\sqrt{x} + 5}$ (Factor $x - 25$ as the difference of squares.)

29. $f(x) = \sqrt{x}$; $\dfrac{f(4 + h) - f(4)}{h} = \dfrac{1}{\sqrt{4 + h} + 2}$ (Rationalize the numerator.)

30. $f(x) = -\sqrt{x}$; $\dfrac{f(x) - f(9)}{x - 9} = -\dfrac{1}{\sqrt{x} + 3}$

 Exercises 31–34 call for the extraction of radical functions from geometric situations (see Section 2.7).

31. A runner starts at point A, goes to point P that is x miles from B, and then runs to D. (Angles at B and C are right angles.)

(a) Write the total distance d traveled as a function of x.

(b) The runner averages 12 miles per hour from A to P and 10 miles per hour from P to D. Write the time t for the trip as a function of x. (*Hint:* Use time = distance/rate.)

 (c) Approximate, to the nearest tenth of an hour, the time for the trip when $x = 5$ miles.

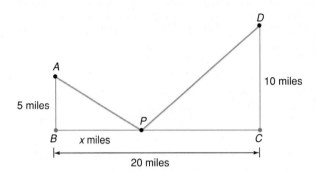

A — 5 miles — B — x miles — P — C — 20 miles; D — 10 miles

32. In the figure, AC is along the shoreline of a lake, and the distance from A to C is 12 miles. P represents the starting point of a swimmer who swims at 3 miles per hour along the hypotenuse PB. P is 5 miles from point A on the shoreline. After reaching B, he walks at 6 miles per hour to C.

(a) Express the total time t of the trip as a function of x, where x is the distance from A to B. (*Hint:* Use time = distance/rate and assume angle A is a right angle.)

 (b) Approximate, to the nearest tenth of an hour, the time for the trip when $x = 4$ miles.

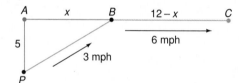

A — x — B — $12 - x$ — C; 6 mph; 5; 3 mph; P

33. Express the distance d from the origin to a point (x, y) on the curve $y = \dfrac{1}{\sqrt{x}}$ as a function of x.

34. Express the distance d from the point $(2, 5)$ to a point (x, y) on the line $3x + y = 6$ as a function of x.

 Written Assignment: Discuss how the graph of $y = f(x) = \sqrt{2 - x}$ can be obtained from the graph of $y = g(x) = \sqrt{x - 2}$ in Example 1.

**5.2
RADICAL
EQUATIONS**

An equation in which a variable occurs in a radicand, as illustrated by $\sqrt{x + 2} + 2 = 5$, is called a **radical equation.** To solve this equation we first isolate the radical on one side and then square both sides to eliminate the radical.

$$\sqrt{x + 2} + 2 = 5$$
$$\sqrt{x + 2} = 3$$
$$(\sqrt{x + 2})^2 = 3^2 \qquad \text{(Square both sides.)}$$

You can check this result by substituting 7 for x in the original equation.

$$x + 2 = 9$$
$$x = 7$$

In general, when solving radical equations, we will be making use of the following principle:

This statement says that every solution of a = b will also be a solution of $a^n = b^n$.

$$\text{If } a = b, \text{ then } a^n = b^n$$

Sometimes the method of raising both sides of an equation to the same power can produce a new equation that has more solutions than the original equation, as in the following example.

EXAMPLE 1   Solve for x: $\sqrt{x + 4} + 2 = x$

Solution

Note: If the radical in $\sqrt{x + 4} + 2 = x$ is not first isolated, it is still possible to solve the equation, but the work will be more involved. Try it. Also observe that $x = 5$ is the x-intercept of the curve $y = \sqrt{x + 4} + 2 - x$.

$$\sqrt{x + 4} + 2 = x$$
$$\sqrt{x + 4} = x - 2$$
$$(\sqrt{x + 4})^2 = (x - 2)^2 \quad \text{(Square both sides.)}$$
$$x + 4 = x^2 - 4x + 4$$
$$0 = x^2 - 5x$$
$$0 = x(x - 5)$$

From the last step we conclude that $x = 0$ or $x = 5$. (This is based on the *zero-product property;* see page 3). Now check each result.

Check:

$$\text{Let } x = 0: \qquad\qquad \text{Let } x = 5:$$
$$\sqrt{0 + 4} + 2 = 0 \qquad\qquad \sqrt{5 + 4} + 2 = 5$$
$$2 + 2 = 0 \quad \text{No} \qquad\qquad 3 + 2 = 5 \quad \text{Yes}$$

We conclude that the only solution for the given equation is 5. The number 0 is called an extraneous solution.   ∎

How did the *extraneous solution* in Example 1 arise? In going from

$$\sqrt{x + 4} = x - 2 \quad \text{to} \quad (\sqrt{x + 4})^2 = (x - 2)^2$$

we used the basic principle: If $a = b$, then $a^n = b^n$. Therefore, every solution of the first equation is also a solution for the second. But *this principle is not always reversible*. In particular, both 0 and 5 are solutions of the second equation, but only 5 is a solution of the first. In summary:

When raising both sides of an equation to the same power, **extraneous solutions** might be introduced. Therefore, whenever this method is used, all solutions must be checked in the original equation.

We have been looking at the solutions of radical equations in one variable. Keep in mind, however, that the solutions of $f(x) = 0$ can be regarded as the x-intercepts of the graph of the related equation in two variables, $y = f(x)$. This is illustrated in the following example.

EXAMPLE 2   Find the x-intercepts and the domain of the function

$$y = f(x) = \sqrt{x^2 + x - 6}$$

Solution   The x-intercepts occur when $f(x) = 0$. Let $\sqrt{x^2 + x - 6} = 0$ and solve for x.

Check: *for* $x = -3$,
$\sqrt{(-3)^2 + (-3) - 6} =$
$\sqrt{9 - 3 - 6} = \sqrt{0} = 0;$
for $x = 2$, $\sqrt{2^2 + 2 - 6} =$
$\sqrt{0} = 0.$

$$(\sqrt{x^2 + x - 6})^2 = (0)^2 \qquad \text{(Square both sides.)}$$

$$x^2 + x - 6 = 0$$

$$(x + 3)(x - 2) = 0$$

$$x = -3 \quad \text{or} \quad x = 2$$

You can solve this inequality by forming a table of signs or by noting where the parabola $y = x^2 + x - 6$ is above the x-axis. See pages 154 and 214.

Therefore, the x-intercepts are -3 and 2.

The domain of f is the solution of the inequality $(x + 3)(x - 2) \geq 0$, which consists of the intervals $(-\infty, -3]$ and $[2, \infty)$. ■

Radical equations may contain more than one radical. For such cases it is usually best to transform the equation first into one with as few radicals on each side as possible. Consider, for example, the following equation:

$$\sqrt{x - 7} - \sqrt{x} = 1$$

$$\sqrt{x - 7} = \sqrt{x} + 1$$

CAUTION
$(\sqrt{x} + 1)^2 \neq x + 1$. *Use*
$(a + b)^2 = a^2 + 2ab + b^2$
with $a = \sqrt{x}$ *and* $b = 1$.

$$(\sqrt{x - 7})^2 = (\sqrt{x} + 1)^2 \qquad \text{(Square both sides.)}$$

$$x - 7 = x + 2\sqrt{x} + 1$$

$$-8 = 2\sqrt{x}$$

$$-4 = \sqrt{x}$$

$$16 = x \qquad \text{(Square again.)}$$

Check: $\sqrt{16 - 7} - \sqrt{16} = \sqrt{9} - \sqrt{16} = 3 - 4 = -1 \neq 1$.   Therefore,   this equation has no solution, which could have been observed at an earlier stage as well. For example, $-4 = \sqrt{x}$ has no solution. (Why not?)

TEST YOUR UNDERSTANDING
Think Carefully

(Answers: Page 294)

Find the x-intercepts and the domain of f.

1. $f(x) = \sqrt{x} - 5$ **2.** $f(x) = \sqrt{x^2 - 3x}$

Solve for x.

3. $\sqrt{x + 1} = 3$

4. $\sqrt{x} - 2 = 3$

5. $\sqrt{x + 2} = \sqrt{2x - 5}$

6. $\sqrt{x^2 + 9} = -5$

7. $\dfrac{1}{\sqrt{x}} = 3$

8. $\dfrac{4}{\sqrt{x - 1}} = 2$

9. $\sqrt{x^2 - 5} = 2$

10. $\sqrt{x} + \sqrt{x - 5} = 5$

11. $\sqrt{x + 16} - x = 4$

12. $2x = 1 + \sqrt{1 - 2x}$

All the examples thus far have involved square roots. However, the method of raising both sides of an equation to the same power can also be applied to radical equations involving cube roots or fourth roots, and so on.

EXAMPLE 3   Solve: $\sqrt[3]{x + 3} = 2$

Solution   In this case, cube each side of the equation.

$$(\sqrt[3]{x + 3})^3 = (2)^3$$
$$x + 3 = 8 \qquad [(\sqrt[3]{a})^3 = a]$$
$$x = 5$$

Check: $\sqrt[3]{5 + 3} = \sqrt[3]{8} = 2$ ■

The equation in Example 3, as well as any other radical equation, can also be written using fractional exponents. Thus, for Example 3 we have the equation $(x + 3)^{1/3} = 2$, and to solve it we use the same steps as before.

$$(x + 3)^{1/3} = 2$$
$$[(x + 3)^{1/3}]^3 = 2^3 \qquad \text{(Cube both sides.)}$$
$$x + 3 = 8$$
$$x = 5$$

EXAMPLE 4   Solve for x:

$$\frac{3(2x - 1)^{1/2}}{x - 3} - \frac{(2x - 1)^{3/2}}{(x - 3)^2} = 0$$

Solution   Multiply through by $(x - 3)^2$ to clear fractions, as shown next.

$$(x - 3)^2\left[\frac{3(2x - 1)^{1/2}}{x - 3}\right] - (x - 3)^2\left[\frac{(2x - 1)^{3/2}}{(x - 3)^2}\right] = (x - 3)^2(0)$$

$$3(x - 3)(2x - 1)^{1/2} - (2x - 1)^{3/2} = 0$$

$$(2x - 1)^{1/2}[3(x - 3) - (2x - 1)] = 0$$

$$(2x - 1)^{1/2}(x - 8) = 0$$

Factor out $(2x - 1)^{1/2}$. This is easier to see if we let $a = (2x - 1)^{1/2}$ and factor a out of $3(x - 3)a - a^3$.

Set each factor equal to zero.

$$(2x - 1)^{1/2} = 0 \quad \text{or} \quad x - 8 = 0$$

$$2x - 1 = 0 \quad \text{or} \qquad x = 8$$

$$x = \tfrac{1}{2} \quad \text{or} \qquad x = 8$$

Check in the original equation to show that both $x = \tfrac{1}{2}$ and $x = 8$ are solutions for Example 4. ■

As you may have noticed, the algebraic techniques developed earlier involving other kinds of expressions often carry over into this work. Following is an example that uses our knowledge of quadratics in conjunction with radicals.

EXAMPLE 5 Solve for x: $\sqrt[3]{x^2} + \sqrt[3]{x} - 20 = 0$

Solution First rewrite the equation by using rational exponents.

$$x^{2/3} + x^{1/3} - 20 = 0$$

Then think of $x^{2/3}$ as the square of $x^{1/3}$, $x^{2/3} = (x^{1/3})^2$, and use the substitution $u = x^{1/3}$ as follows:

$$x^{2/3} + x^{1/3} - 20 = 0$$

$$(x^{1/3})^2 + x^{1/3} - 20 = 0$$

$$u^2 + u - 20 = 0$$

Alternatively, we may keep the radical sign and proceed with $u = \sqrt[3]{x}$.

$$(u + 5)(u - 4) = 0 \qquad \text{(factoring the quadratic)}$$

$$u + 5 = 0 \quad \text{or} \quad u - 4 = 0$$

$$u = -5 \quad \text{or} \qquad u = 4$$

Now replace u by $x^{1/3}$.

$$x^{1/3} = -5 \quad \text{or} \quad x^{1/3} = 4$$

Thus:

$$x = -125 \quad \text{or} \qquad x = 64$$

Check to show that both values are solutions of the given equation. ■

EXAMPLE 6 Solve the system and graph:

$$y = \sqrt[3]{x}$$

$$y = \tfrac{1}{4}x$$

Solution For the points of intersection the x- and y-values are the same in both equations. Thus, for such points, we set the y-values equal to one another.

$$\tfrac{1}{4}x = \sqrt[3]{x} \quad\longleftarrow\quad x^{1/3} \text{ can be used in place of } \sqrt[3]{x}$$

Cube both sides and solve for x.

CAUTION
A common error is to take
$x^3 - 64x = 0$ *and divide*
through by x to get
$x^2 - 64 = 0$. *This step pro-*
duces the roots ± 8. The root
0 has been lost because we di-
vided by x, and 0 is the num-
ber for which the factor x in
$x(x^2 - 64)$ *is zero. You may*
always divide by a nonzero ex-
pression and get an equivalent
form of the equation. But when
you divide by a variable quan-
tity there is the danger of los-
ing some roots, those for
which the divisor is 0.

$$\tfrac{1}{64}x^3 = x$$

$$x^3 = 64x$$

$$x^3 - 64x = 0$$

$$x(x^2 - 64) = 0 \qquad \text{(factoring out } x)$$

$$x(x + 8)(x - 8) = 0$$

$$x = 0 \quad \text{or} \quad x = -8 \quad \text{or} \quad x = 8$$

Substitute these values into either of the given equations to obtain the corresponding y-values. The remaining equation can be used for checking. The solutions are $(-8, -2)$, $(0, 0)$, and $(8, 2)$.

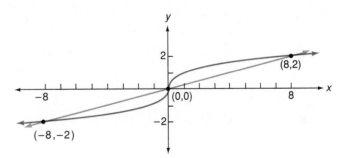

$\blacksquare$

EXERCISES 5.2

Find the x-intercepts of each curve and find the domain of f. (Hint: In Exercise 5, for example,
note that $x^2 - 5x - 6$ cannot be negative. Thus, solve $x^2 - 5x - 6 \geq 0$.

1. $f(x) = \sqrt{x - 1} - 4$
4. $f(x) = \sqrt{x - x^2}$

2. $f(x) = \sqrt{3x - 2} - 5$
5. $f(x) = \sqrt{x^2 - 5x - 6}$

3. $f(x) = \sqrt{x^2 + 2x}$
6. $f(x) = \sqrt{2x^2 - x - 2} - 1$

Solve each equation.

7. $\sqrt{4x + 9} - 7 = 0$

10. $\sqrt{x - 1} - \sqrt{2x - 11} = 0$

8. $2 + \sqrt{7x - 3} = 7$

11. $\sqrt{x^2 - 36} = 8$

9. $(3x + 1)^{1/2} = (2x + 6)^{1/2}$

12. $3x = \sqrt{3 - 5x - 3x^2}$

13. $\sqrt{x^2 + \dfrac{1}{2}} = \dfrac{1}{\sqrt{3}}$

14. $3\sqrt{x} = 2\sqrt{3}$

15. $\dfrac{8}{\sqrt{x + 2}} = 4$

16. $\left(\dfrac{5x + 4}{2}\right)^{1/3} = 3$

17. $\dfrac{1}{\sqrt{2x - 1}} = \dfrac{3}{\sqrt{5 - 3x}}$

18. $\sqrt[3]{2x + 7} = 3$

19. $\sqrt[4]{1 - 3x} = \frac{1}{2}$

20. $2 + \sqrt{7x - 4} = 0$

21. $\sqrt{x} + \sqrt{x - 5} = 5$

22. $\sqrt{x - 7} = 7 - \sqrt{x}$

23. $\sqrt{x - 1} + \sqrt{3x - 2} = 3$

24. $\sqrt{10 - x} - \sqrt{x + 3} = 1$

25. $\sqrt{4x + 1} + \sqrt{x + 7} = 6$

26. $\sqrt{x + 4} + \sqrt{3x + 1} = 7$

27. $x\sqrt{4 - x} - \sqrt{9x - 36} = 0$

28. $2x - 5\sqrt{x} - 3 = 0$

29. $x = 8 - 2\sqrt{x}$

30. $\dfrac{(2x + 2)^{3/2}}{(x + 9)^2} - \dfrac{(2x + 2)^{1/2}}{x + 9} = 0$

31. $\sqrt{x^2 - 6x} = x - \sqrt{2x}$

32. $x^{1/3} - 3x^{1/6} + 2 = 0$

33. $4x^{2/3} - 12x^{1/3} + 9 = 0$

34. $\sqrt[4]{3x^2 + 4} = x$

35. $(5x - 6)^{1/5} + \dfrac{x}{(5x - 6)^{4/5}} = 0$

36. $\frac{3}{2}x^{1/2} - \frac{3}{2}x^{-1/2} = 0$

37. $x^{-3/2} - \frac{1}{9}x^{-1/2} = 0$

38. $x^{1/2} + \frac{1}{2}x^{-1/2}(x - 9) = 0$

39. $x^{2/3} + \frac{2}{3}x^{-1/3}(x - 10) = 0$

40. $\frac{1}{2}x^2(x + 5)^{-1/2} + 2x(x + 5)^{1/2} = 0$

41. The radius of a sphere with surface area A is given by the formula $r = \dfrac{1}{2}\sqrt{\dfrac{A}{\pi}}$. Use this formula to solve for A in terms of π and r. Then find A when $r = 2$ centimeters.

42. The radius of a right circular cylinder with height h and volume V is given by this formula: $r = \sqrt{\dfrac{V}{\pi h}}$. Use this formula to solve for V in terms of r and h. Then find V when $r = 2$ centimeters and $h = 3$ centimeters.

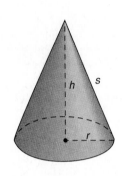

 43. The slant height s of a right circular cone is given by the formula $s = \sqrt{r^2 + h^2}$, where r is the radius of the base and h is the altitude. Solve for h in terms of r and s, and then find h when $s = 17.23$ cm and $r = 8.96$ cm, to two decimal places.

44. In the figure, rectangle $ABCD$ is inscribed in a circle of radius 10. Express the area of the rectangle as a function of x. (*Hint:* solve for y in terms of x.)

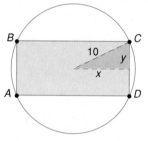

45. The volume of a right circular cylinder with altitude h and radius r is 5 cm³. Solve for r in terms of h and express the surface area of the cylinder as a function of h.

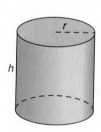

46. A right circular cone with height h and radius r is inscribed in a sphere of radius 1 as shown. Solve for r in terms of h and express the volume of the cone as a function of h.

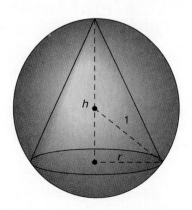

47. The distance from the point $(3, 0)$ to a point $P(x, y)$ on the curve $y = \sqrt{x}$ is $3\sqrt{5}$. Find the coordinates of P. (*Hint:* Use the formula for the distance between two points.)

Solve the system and then sketch each graph.

48. $y = \dfrac{2}{\sqrt{x}}$ **49.** $y = \sqrt{x}$

     $x + 3y = 7$ $y = \frac{1}{2}x$

Solve for the indicated variable in terms of the others.

50. $y = \sqrt{16 - x^2}$, for x, $0 < x < 4$ **51.** $x = \frac{3}{5}\sqrt{25 - y^2}$, for y, $0 < y < 5$

52. $\dfrac{1}{2\sqrt{x}} + \dfrac{f}{2\sqrt{y}} = 0$, for f **53.** $\dfrac{1}{\sqrt{xy}}(xf + y) = f$, for f

 Written Assignment: Explain what is meant by an extraneous solution to an equation and how such a solution can be generated.

 5.3
DETERMINING
THE SIGNS
OF RADICAL
FUNCTIONS

The methods used for determining the signs of rational functions are easily modified to determine the signs of functions defined in terms of radical expressions. As a first example, consider this function:

$$f(x) = (x - 5)\sqrt{x - 3}$$

Observe that the factor $\sqrt{x - 3}$ indicates $x \geq 3$. (Why?) Therefore, we do not need to consider any values for $x < 3$. The following *table of signs* shows that $f(x)$ is negative on the interval $(3, 5)$ and positive on the interval $(5, \infty)$.

Review the construction of a table of signs, as shown in Section 4.3.

Interval	(3, 5)	(5, ∞)
Sign of $\sqrt{x - 3}$	+	+
Sign of $x - 5$	−	+
Sign of $f(x)$	−	+

EXAMPLE 1   Find the domain of $f(x) = \dfrac{\sqrt{x^2 + 9}}{x + 2}$ and determine the signs of f.

Solution   Since the denominator of a fraction cannot equal 0, $x + 2 \neq 0$ or $x \neq -2$. Also, since $\sqrt{x^2 + 9}$ is a real number for all x, the domain of f consists of all $x \neq -2$.

     The signs of $f(x)$ depend entirely on the denominator because the numerator is positive for all x. Therefore, since the denominator $x + 2 > 0$ when $x > -2$, and $x + 2 < 0$ when $x < -2$, we have

$$f(x) > 0 \qquad \text{on } (-2, \infty)$$
$$f(x) < 0 \qquad \text{on } (-\infty, -2)$$

■

You should note immediately
that $x \neq 2$ and $x \neq -2$. Also
note that $4 - x^2 > 0$ is equivalent to $x^2 - 4 < 0$, or
$(x + 2)(x - 2) < 0$, which
can be solved by using a table
of signs. (Also see page 154.)

EXAMPLE 2   Find the domain of $f(x) = \dfrac{x}{\sqrt{4 - x^2}}$ and determine its signs.

Solution   The domain consists of those x for which $4 - x^2 > 0$. One way to solve this inequality is to note that $y = 4 - x^2$ is a parabola that opens downward and crosses the x-axis at ± 2. Hence $4 - x^2 > 0$ on the interval $(-2, 2)$, which is the domain of f. The signs of $f(x)$ depends only on the numerator since the denominator is always positive on the interval $(-2, 2)$. Thus $f(x) > 0$ on $(0, 2)$, and $f(x) < 0$ on $(-2, 0)$. ∎

**TEST YOUR
UNDERSTANDING
Think Carefully**

(Answers: Page 294)

Find the domain of each function and determine its signs.

1. $f(x) = \dfrac{x - 4}{\sqrt{x}}$ **2.** $f(x) = \dfrac{\sqrt{x + 2}}{x}$ **3.** $f(x) = \dfrac{2}{3\sqrt[3]{x}}$

Algebraic procedures developed earlier can be helpful with this type of work. For example, the signs of

$$f(x) = x^{1/2} + \tfrac{1}{2}x^{-1/2}(x - 3)$$

can be determined after combining and simplifying as follows:

The domain of f consists of all
$x > 0$; in other words, the domain is the interval $(0, \infty)$.

$$f(x) = x^{1/2} + \tfrac{1}{2}x^{-1/2}(x - 3)$$

$$= x^{1/2} + \frac{x - 3}{2x^{1/2}}$$

$$= \frac{x^{1/2}(2x^{1/2})}{2x^{1/2}} + \frac{x - 3}{2x^{1/2}} \qquad \text{(The common denominator is } 2x^{1/2}.)$$

$$= \frac{2x + x - 3}{2x^{1/2}}$$

$$= \frac{3x - 3}{2x^{1/2}} = \frac{3(x - 1)}{2\sqrt{x}}$$

Since $\sqrt{x} > 0$ for all $x > 0$,
the signs of $f(x)$ depend entirely on $x - 1$. Therefore you
should be able to determine
the signs mentally without constructing a table.

$$f(x) = \frac{3(x - 1)}{2\sqrt{x}}$$

Interval	$(0, 1)$	$(1, \infty)$
Sign of $\sqrt{x}$	$+$	$+$
Sign of $x - 1$	$-$	$+$
Sign of $f(x)$	$-$	$+$

Hence $f(x) < 0$ on $(0, 1)$; $f(x) > 0$ on $(1, \infty)$.

EXAMPLE 3   Find the domain of f, the roots of $f(x) = 0$, and determine the signs of f where

$$f(x) = \frac{x^2 - 1}{2\sqrt{x - 1}} + 2x\sqrt{x - 1}$$

Solution The radical in the denominator calls for $x - 1 > 0$. Therefore, the domain consists of all $x > 1$. Now simplify as follows:

$$f(x) = \frac{x^2 - 1}{2\sqrt{x - 1}} + 2x\sqrt{x - 1}$$

$$= \frac{x^2 - 1}{2\sqrt{x - 1}} + \frac{(2x\sqrt{x - 1})(2\sqrt{x - 1})}{2\sqrt{x - 1}}$$

$$= \frac{x^2 - 1 + 4x(x - 1)}{2\sqrt{x - 1}}$$

$$= \frac{5x^2 - 4x - 1}{2\sqrt{x - 1}}$$

$$= \frac{(5x + 1)(x - 1)}{2\sqrt{x - 1}}$$

Any of the last three forms is an acceptable simplification, but the last form is more advantageous for determining the signs of $f(x)$.

$$= \frac{5(x + \frac{1}{5})(x - 1)}{2\sqrt{x - 1}} \qquad \blacksquare$$

The numerator is zero when $x = -\frac{1}{5}$ or $x = 1$. But these values are *not* in the domain of f. Since a fraction can be zero only when the numerator is zero, it follows that $f(x) = 0$ has no solutions. The signs of $f(x)$ are given in this brief table:

$$f(x) = \frac{5(x + \frac{1}{5})(x - 1)}{2\sqrt{x - 1}}$$

Interval	$(1, \infty)$
Sign of $x + \frac{1}{5}$	+
Sign of $x - 1$	+
Sign of $\sqrt{x - 1}$	+
Sign of $f(x)$	+

Thus $f(x) > 0$ on its domain $(1, \infty)$.

EXAMPLE 4 Graph the function $y = f(x) = \sqrt{x^2 - 4}$.

Solution The domain of f is the set of all $x \le -2$ or $x \ge 2$. This is found by solving the inequality $x^2 - 4 \ge 0$.

The x-intercepts are -2 and 2, found by solving $\sqrt{x^2 - 4} = 0$.

$f(x) > 0$ for all domain values x except the x-intercepts. Therefore, the graph is above the x-axis on the intervals $(-\infty, -2)$ and $(2, \infty)$.

The graph is symmetric about the y-axis since $f(-x) = \sqrt{(-x)^2 - 4} = \sqrt{x^2 - 4} = f(x)$.

Use the preceding information and some selected points to draw the graph on $[2, \infty)$. Complete the graph by using the symmetry through the y-axis.

x	2	4	6	8
y	0	$\sqrt{12} = 3.5$	$\sqrt{32} = 5.7$	$\sqrt{60} = 7.7$

← Use Table I or a calculator to get the square roots to one decimal place.

This graph can also be obtained by squaring

$y = \sqrt{x^2 - 4}$ *to get*

$y^2 = x^2 - 4$ *or* $\dfrac{x^2}{4} - \dfrac{y^2}{4} = 1.$

Since this is a hyperbola in which $y \geq 0$, *the graph is the upper half of the two branches of the hyperbola.*

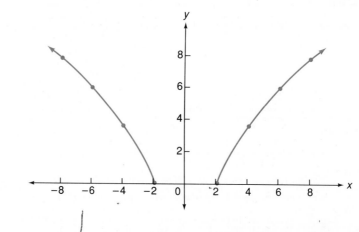

CAUTION: Learn to Avoid These Mistakes

WRONG	RIGHT
$(9 - x^2)^{1/2}(9 - x^2)^{3/2}$ $= (9 - x^2)^{3/4}$	$(9 - x^2)^{1/2}(9 - x^2)^{3/2}$ $= (9 - x^2)^{(1/2)+(3/2)}$ $= (9 - x^2)^2$
$\dfrac{2}{\sqrt{9 - x^2}} + 3\sqrt{9 - x^2}$ $= \dfrac{2 + 3\sqrt{9 - x^2}}{\sqrt{9 - x^2}}$	$\dfrac{2}{\sqrt{9 - x^2}} + 3\sqrt{9 - x^2}$ $= \dfrac{2}{\sqrt{9 - x^2}} + \dfrac{3(9 - x^2)}{\sqrt{9 - x^2}}$ $= \dfrac{2 + 3(9 - x^2)}{\sqrt{9 - x^2}}$
$x^{-1/3} + x^{2/3} = x^{-1/3}(1 + x^{1/3})$	$x^{-1/3} + x^{2/3} = x^{-1/3}(1 + x)$
$x^2\sqrt{1 + x} = \sqrt{x^2 + x^3}$	$x^2\sqrt{1 + x} = \sqrt{x^4}\sqrt{1 + x}$ $= \sqrt{x^4 + x^5}$

EXERCISES 5.3

In the following exercises **(a)** *find the domain of* f, **(b)** *determine the signs of* f, *and* **(c)** *find the roots of* $f(x) = 0$.

1. $f(x) = (x + 4)\sqrt{x - 2}$ **2.** $f(x) = (x + 4)\sqrt[3]{x - 2}$ **3.** $f(x) = (x - 2)^{2/3}$

4. $f(x) = \frac{2}{3}(x - 2)^{-1/3}$ **5.** $f(x) = \dfrac{9 + x}{9\sqrt{x}}$ **6.** $f(x) = \dfrac{9 - x}{9x^{3/2}}$

7. $f(x) = \dfrac{(x + 4)\sqrt[3]{x - 2}}{3\sqrt[3]{x}}$ 8. $f(x) = \dfrac{-5x(x - 4)}{2\sqrt{5 - x}}$ 9. $f(x) = \dfrac{\sqrt{9 - x^2}}{x}$

10. $f(x) = \dfrac{x}{\sqrt{x^2 - 4}}$

Take the expression at the left and change it into the equivalent form given at the right.

11. $x^{1/2} + (x - 4)\frac{1}{2}x^{-1/2}; \ \dfrac{3x - 4}{2\sqrt{x}}$ 12. $x^{1/3} + (x - 1)\frac{1}{3}x^{-2/3}; \ \dfrac{4x - 1}{3\sqrt[3]{x^2}}$

13. $\frac{1}{2}(4 - x^2)^{-1/2}(-2x); \ -\dfrac{x}{\sqrt{4 - x^2}}$ 14. $\dfrac{-\frac{1}{2}(25 - x^2)^{-1/2}(-2x)}{25 - x^2}; \ \dfrac{x}{\sqrt{(25 - x^2)^3}}$

15. $\dfrac{x}{3(x - 1)^{2/3}} + (x - 1)^{1/3}; \ \dfrac{4x - 3}{3(\sqrt[3]{x - 1})^2}$ 16. $(x + 1)^{1/2}(2) + (2x + 1)\frac{1}{2}(x + 1)^{-1/2}; \ \dfrac{6x + 5}{2\sqrt{x + 1}}$

17. $\dfrac{x}{(5x - 6)^{4/5}} + (5x - 6)^{1/5}; \ \dfrac{6(x - 1)}{(5x - 6)^{4/5}}$ 18. $x^{-3/2} - \dfrac{1}{9}x^{-1/2}; \ \dfrac{9 - x}{9x^{3/2}}$

Simplify and determine the signs of f.

19. $f(x) = \sqrt{x} - \dfrac{1}{\sqrt{x}}$ 20. $f(x) = \frac{3}{2}x^{1/2} - \frac{3}{2}x^{-1/2}$

21. $f(x) = x^{1/2} + \frac{1}{2}x^{-1/2}(x - 9)$ 22. $f(x) = x^{2/3} + \frac{2}{3}x^{-1/3}(x - 10)$

23. $f(x) = \dfrac{x^2}{2}(x - 2)^{-1/2} + 2x(x - 2)^{1/2}$ 24. $f(x) = \dfrac{\sqrt{x - 1} - \dfrac{x}{2\sqrt{x - 1}}}{x - 1}$

Find the domain and graph the function.

25. $y = f(x) = \sqrt{x^2 - 9}$ 26. $y = f(x) = \dfrac{1}{\sqrt{x^2 - 9}}$ 27. $y = f(x) = \sqrt{9 - x^2}$

28. $y = f(x) = \dfrac{1}{\sqrt{9 - x^2}}$ 29. $y = f(x) = (x - 3)\sqrt{x}$ 30. $y = (x + 3)\sqrt{x}$

5.4
COMBINING
AND
DECOMPOSING
FUNCTIONS

What does it mean to add two functions f and g? The answer turns out to be relatively simple. To illustrate, let us use f and g, where

$$f(x) = \dfrac{1}{x^3 - 1} \quad \text{and} \quad g(x) = \sqrt{x}$$

The domain of f consists of all $x \neq 1$, and g has domain all $x \geq 0$. The sum of f and g will be symbolized by $f + g$. We will take as the domain of $f + g$ all values x that are in *each* of the domains of f and g simultaneously. Therefore $f + g$ has domain $x \geq 0$ and $x \neq 1$.

We then define $(f + g)(x)$ as follows:

Note that the plus sign in $f(x) + g(x)$ is addition of real numbers.

$$(f + g)(x) = f(x) + g(x) = \dfrac{1}{x^3 - 1} + \sqrt{x}$$

Specifically, for $x = 4$ we have

$$(f + g)(4) = f(4) + g(4) = \frac{1}{64 - 1} + \sqrt{4} = \frac{1}{63} + 2 = \frac{127}{63}$$

The difference, product, and quotient are found in a similar manner. Using f and g as defined, we get

$$(f - g)(x) = f(x) - g(x) = \frac{1}{x^3 - 1} - \sqrt{x}$$

$$(f \cdot g)(x) = f(x)g(x) = \frac{1}{x^3 - 1} \cdot \sqrt{x} = \frac{\sqrt{x}}{x^3 - 1}$$

$$\frac{f}{g}(x) = \frac{f(x)}{g(x)} = \frac{\frac{1}{x^3 - 1}}{\sqrt{x}} = \frac{1}{(x^3 - 1)\sqrt{x}}$$

The domains of $f - g$ and $f \cdot g$ are the same as for $f + g$, namely $x \geq 0$ and $x \neq 1$. The domain of $\frac{f}{g}$ also has those x common to the domains of f and g except those for which $g(x) = 0$; $x > 0$ and $x \neq 1$.

EXAMPLE 1 Let $f(x) = \dfrac{1}{x^3 - 1}$ and $g(x) = \sqrt{x}$.

(a) Evaluate $f(4)$, and $g(4)$, and compute $f(4) \cdot g(4)$.
(b) Use the expression for $(f \cdot g)(x)$, as in the preceding discussion, to evaluate $(f \cdot g)(4)$.

Solution

(a) $f(4) = \dfrac{1}{4^3 - 1} = \dfrac{1}{63}$, $g(4) = \sqrt{4} = 2$ **(b)** $(f \cdot g)(x) = \dfrac{\sqrt{x}}{x^3 - 1}$

$f(4) \cdot g(4) = \dfrac{1}{63} \cdot 2 = \dfrac{2}{63}$ $(f \cdot g)(4) = \dfrac{\sqrt{4}}{4^3 - 1} = \dfrac{2}{63}$ ∎

We are ready to state the general definition for forming the sum, difference, product, and quotient of two functions.

COMBINING FUNCTIONS

For functions f and g:

$$(f + g)(x) = f(x) + g(x)$$
$$(f - g)(x) = f(x) - g(x)$$
$$(f \cdot g)(x) = f(x)g(x)$$
$$\frac{f}{g}(x) = \frac{f(x)}{g(x)}$$

The domains of $f + g$, $f - g$, and $f \cdot g$ are all the same and consist of all x common to the domains of f and g. The domain of $\dfrac{f}{g}$ has all x common to the domains of f and g except for those x where $g(x) = 0$.

Another way of forming new functions from given functions is referred to as the **composition of functions.** Let f and g be given by

$$f(x) = \frac{1}{x-2} \quad \text{and} \quad g(x) = \sqrt{x}$$

Take a specific value in the domain of f, say $x = 6$. Then the corresponding range value is $f(6) = \frac{1}{4}$. Take this range value and use it as a domain value for g to produce $g(\frac{1}{4}) = \sqrt{\frac{1}{4}} = \frac{1}{2}$. This work may be condensed in this way.

The output of f becomes the input of g: $f(6) = \frac{1}{4}$ comes out of f and goes into g.

$$g(f(6)) = g\left(\frac{1}{4}\right) = \sqrt{\frac{1}{4}} = \frac{1}{2} \qquad \text{Read this as ``g of f of 6.''}$$

The roles of f and g may be interchanged. Thus

$$f(g(6)) = f(\sqrt{6}) = \frac{1}{\sqrt{6}-2} \qquad \text{Read this as ``f of g of 6.''}$$

The composition of functions does not always produce a real number. For instance, if $x = -3$, then $f(-3) = -\frac{1}{5}$; $g(-\frac{1}{5}) = \sqrt{-\frac{1}{5}}$ is not a real number. We therefore say that $g(f(-3))$ is undefined or that it does not exist.

TEST YOUR UNDERSTANDING
Think Carefully

(Answers: Page 294)

For each pair of functions f and g, evaluate (if possible) each of the following:

(a) $g(f(1))$ (b) $f(g(1))$ (c) $f(g(0))$ (d) $g(f(-2))$

1. $f(x) = 3x - 1$; $g(x) = x^2 + 4$ 2. $f(x) = \sqrt{x}$; $g(x) = x^2$

3. $f(x) = \sqrt[3]{3x - 1}$; $g(x) = 5x$ 4. $f(x) = \dfrac{x+2}{x-1}$; $g(x) = x^3$

The preceding computations for specific values of x can be stated in terms of any allowable x. For instance, using $f(x) = \dfrac{1}{x-2}$ and $g(x) = \sqrt{x}$, we have

$$g(f(x)) = g\left(\frac{1}{x-2}\right) = \sqrt{\frac{1}{x-2}} = \frac{1}{\sqrt{x-2}}$$

This new correspondence between a domain value x and the range value $\dfrac{1}{\sqrt{x-2}}$ is referred to as the **composite function of g by f** (or the *composition of g by f*). This composite function is denoted by $g \circ f$. That is:

$g(f(x))$ is read as "g of f of x."

$$(g \circ f)(x) = g(f(x)) = \frac{1}{\sqrt{x-2}}$$

The domain of $g \circ f$ will consist of all values x in the domain of f such that $f(x)$ is in the domain of g. Since $f(x) = \dfrac{1}{x-2}$ has domain all $x \neq 2$ and the domain of

$g(x) = \sqrt{x}$ is all $x \geq 0$, the domain of $g \circ f$ is all $x \neq 2$ for which $\dfrac{1}{x-2}$ is positive, that is, all $x > 2$.

Reversing the roles of the two functions gives the composite of f by g, where

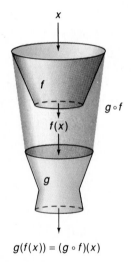

$f(g(x))$ is read as "f of g of x."

$$(f \circ g)(x) = f(g(x)) = f(\sqrt{x}) = \frac{1}{\sqrt{x} - 2}$$

The domain of $f \circ g$ consists of all $x \geq 0$ and $x \neq 4$.

Following is the definition of composite functions:

DEFINITION OF COMPOSITE FUNCTION

For functions f and g the composite function of g by f, denoted $g \circ f$, has range values defined by

$$(g \circ f)(x) = g(f(x))$$

and domain consisting of all x in the domain of f for which $f(x)$ is in the domain of g.

It may help you to remember the construction of composites by looking at the following schematic diagram, as well as the figure in the margin.

$$x \xrightarrow{\;f\;} f(x) \xrightarrow{\;g\;} g(f(x)) = (g \circ f)(x)$$
$$g \circ f$$

It is also helpful to view the composition $(g \circ f)(x) = g(f(x))$ as consisting of an "inner" function f and an "outer" function g.

$g(f(x)) = (g \circ f)(x)$

EXAMPLE 2   Form the composite functions $f \circ g$ and $g \circ f$ and give their domains, where $f(x) = \dfrac{1}{x^2 - 1}$ and $g(x) = \sqrt{x}$.

Solution   We find that $f \circ g$ is given by

$$(f \circ g)(x) = f(g(x)) = f(\sqrt{x}) = \frac{1}{(\sqrt{x})^2 - 1} = \frac{1}{x - 1}$$

The domain of $f \circ g$ excludes $x < 0$ since such x are not in the domain of g. And it excludes $x = 1$ since $g(1) = 1$, which is not in the domain of f. Therefore, the domain of $f \circ g$ is $x \geq 0$ and $x \neq 1$. Also, $g \circ f$ is given by

$$(g \circ f)(x) = g(f(x)) = g\!\left(\frac{1}{x^2 - 1}\right) = \sqrt{\frac{1}{x^2 - 1}} = \frac{1}{\sqrt{x^2 - 1}}$$

From Example 2 it follows that $(f \circ g)(x) \neq (g \circ f)(x)$. We conclude that, in general, the composite of f by g is not equal to the composite of g by f, that is, $f(g(x)) \neq g(f(x))$. Composition is not commutative.

The domain of $g \circ f$ excludes $x = \pm 1$ since these are not in the domain of f. And it excludes $-1 < x < 1$, since negative values of $f(x)$ are not in the domain of g. Therefore, the domain of $g \circ f$ is seen to be $x < -1$ or $x > 1$. ∎

The composition of functions may be extended to include more than two functions.

EXAMPLE 3   Let $f(x) = \sqrt{x}$, $g(x) = x^2 + 1$, and $h(x) = \dfrac{1}{x}$. Find the composition of f by g by h, denoted $f \circ g \circ h$,

Solution

$$
\begin{aligned}
(f \circ g \circ h)(x) &= f(g(h(x))) &&\text{h is the ``inner'' function.} \\
&= f\left(g\left(\frac{1}{x}\right)\right) &&\text{g is the ``middle'' function.} \\
&= f\left(\frac{1}{x^2} + 1\right) &&\text{f is the ``outer'' function.} \\
&= \sqrt{\frac{1}{x^2} + 1}
\end{aligned}
$$

∎

One of the most useful skills needed in the study of calculus is the ability to recognize that a given function may be viewed as the **composition** of two or more functions. For instance, let h be given by

$$h(x) = \sqrt{x^2 + 2x + 2}$$

If we let $f(x) = x^2 + 2x + 2$ and $g(x) = \sqrt{x}$, then the composite g by f is

$$
\begin{aligned}
(g \circ f)(x) &= g(f(x)) \\
&= g(x^2 + 2x + 2) \\
&= \sqrt{x^2 + 2x + 2} = h(x)
\end{aligned}
$$

Thus the given function h has been *decomposed* into the composition of the two functions f and g. Such decompositions are not unique. More than one decomposition is possible. For h we may also use

$$t(x) = x^2 + 2x \qquad s(x) = \sqrt{x + 2}$$

Then

You would most likely agree that the first of these decompositions is more "natural." Just which decomposition one is to choose will, in later work, depend on the situation.

$$
\begin{aligned}
(s \circ t)(x) &= s(t(x)) \\
&= s(x^2 + 2x) \\
&= \sqrt{x^2 + 2x + 2} = h(x)
\end{aligned}
$$

EXAMPLE 4 Find f and g such that $h = f \circ g$, where $h(x) = \left(\dfrac{1}{3x - 1}\right)^5$ and the inner function g is rational.

Solution Let $g(x) = \dfrac{1}{3x - 1}$ and $f(x) = x^5$.

$$(f \circ g)(x) = f(g(x))$$

$$= f\left(\dfrac{1}{3x - 1}\right)$$

$$= \left(\dfrac{1}{3x - 1}\right)^5 = h(x) \qquad \blacksquare$$

As another example of a decomposition where the inner function is a binomial, let $g(x) = 3x - 1$ and $f(x) = \dfrac{1}{x^5}$.

EXAMPLE 5 Decompose $h(x) = \sqrt{(x^2 - 3x)^5}$ into two functions so that the outer function is a monomial.

Solution Write $h(x) = (\sqrt{x^2 - 3x})^5$. Let $f(x) = \sqrt{x^2 - 3x}$ and $g(x) = x^5$.

$$(g \circ f)(x) = g(f(x))$$

$$= g(\sqrt{x^2 - 3x})$$

$$= (\sqrt{x^2 - 3x})^5$$

$$= h(x) \qquad \blacksquare$$

TEST YOUR UNDERSTANDING
Think Carefully

For each function h, find the functions f and g so that $h = g \circ f$.

1. $h(x) = (8x - 3)^5$ **2.** $h(x) = \sqrt[5]{8x - 3}$

3. $h(x) = \sqrt{\dfrac{1}{8x - 3}}$ **4.** $h(x) = \left(\dfrac{5}{7 + 4x^2}\right)^3$

5. $h(x) = \dfrac{(2x + 1)^4}{(2x - 1)^4}$ **6.** $h(x) = \sqrt{(x^4 - 2x^2 + 1)^3}$

(Answers: Page 294)

Example 5 showed a way of decomposing $h(x) = \sqrt{(x^2 - 3x)^5}$ into the composition of two functions. It is also possible to express h as the composition of three functions. For example, we may use these functions.

$$f(x) = x^2 - 3x \qquad g(x) = x^5 \qquad t(x) = \sqrt{x}$$

Then

$$(t \circ g \circ f)(x) = t(g(f(x))) \qquad \text{\textit{f} is the "inner" function.}$$

$$= t(g(x^2 - 3x)) \qquad \text{\textit{g} is the "middle" function.}$$

$$= t((x^2 - 3x)^5) \qquad \text{\textit{t} is the "outer" function.}$$

$$= \sqrt{(x^2 - 3x)^5}$$

$$= h(x)$$

EXERCISES 5.4

In Exercises 1–6, let $f(x) = 2x - 3$ and $g(x) = 3x + 2$.

1. **(a)** Find $f(1)$, $g(1)$, and $f(1) + g(1)$.
   (b) Find $(f + g)(x)$ and state the domain of $f + g$.
   (c) Use the result in part (b) to evaluate $(f + g)(1)$.
2. **(a)** Find $g(2)$, $f(2)$, and $g(2) - f(2)$.
   (b) Find $(g - f)(x)$ and state the domain of $g - f$.
   (c) Use the result in part (b) to evaluate $(g - f)(2)$.
3. **(a)** Find $f(\frac{1}{2})$, $g(\frac{1}{2})$, and $f(\frac{1}{2}) \cdot g(\frac{1}{2})$.
   (b) Find $(f \cdot g)(x)$ and state the domain of $f \cdot g$.
   (c) Use the result in part (b) to evaluate $(f \cdot g)(\frac{1}{2})$.
4. **(a)** Find $g(-2)$, $f(-2)$, and $\dfrac{g(-2)}{f(-2)}$.

   (b) Find $\dfrac{g}{f}(x)$ and state the domain of $\dfrac{g}{f}$.

   (c) Use the result in part (b) to evaluate $\dfrac{g}{f}(-2)$.

5. **(a)** Find $g(0)$ and $f(g(0))$.
   (b) Find $(f \circ g)(x)$ and state the domain of $f \circ g$.
   (c) Use the result in part (b) to evaluate $(f \circ g)(0)$.
6. **(a)** Find $f(0)$ and $g(f(0))$.
   (b) Find $(g \circ f)(x)$ and state the domain of $g \circ f$.
   (c) Use the result in part (b) to evaluate $(g \circ f)(0)$.

For each pair of functions, find the following:

   (a) $(f + g)(x)$; domain of $f + g$.
   (b) $\left(\dfrac{f}{g}\right)(x)$; domain $\dfrac{f}{g}$.
   (c) $(f \circ g)(x)$; domain $f \circ g$.

7. $f(x) = x^2$, $g(x) = \sqrt{x}$

8. $f(x) = 5x - 1$, $g(x) = \dfrac{5}{1 + 3x}$

9. $f(x) = x^3 - 1$, $g(x) = \dfrac{1}{x}$

10. $f(x) = 3x - 1$, $g(x) = \frac{1}{3}x + \frac{1}{3}$

11. $f(x) = x^2 + 6x + 8$, $g(x) = \sqrt{x - 2}$

12. $f(x) = \sqrt[3]{x}$, $g(x) = x^2$

For each pair of functions, find the following:

   (a) $(g - f)(x)$; domain of $g - f$.
   (b) $(g \cdot f)(x)$; domain of $g \cdot f$.
   (c) $(g \circ f)(x)$; domain of $g \circ f$.

13. $f(x) = -2x + 5$, $g(x) = 4x - 1$

14. $f(x) = |x|$, $g(x) = 3|x|$

15. $f(x) = 2x^2 - 1$, $g(x) = \dfrac{1}{2x}$

16. $f(x) = \sqrt{2x + 3}$, $g(x) = x^2 - 1$

Find $(f \circ g)(x)$ and $(g \circ f)(x)$.

17. $f(x) = x^2$, $g(x) = x - 1$

18. $f(x) = |x - 3|$, $g(x) = 2x + 3$

19. $f(x) = \dfrac{x}{x-2}$, $g(x) = \dfrac{x+3}{x}$

20. $f(x) = x^3 - 1$, $g(x) = \dfrac{1}{x^3 + 1}$

21. $f(x) = \sqrt{x+1}$, $g(x) = x^4 - 1$

22. $f(x) = 2x^3 - 1$, $g(x) = \sqrt[3]{\dfrac{x+1}{2}}$

23. $f(x) = \sqrt{x}$, $g(x) = 4$

24. $f(x) = \sqrt[3]{1-x}$, $g(x) = 1 - x^3$

25. Let $f(x) = \dfrac{1}{x}$, $g(x) = 2x - 1$, and $h(x) = x^{1/3}$. Find the following:

   (a) $(f \circ g \circ h)(x)$ **(b)** $(g \circ f \circ h)(x)$ **(c)** $(h \circ f \circ g)(x)$

26. Let $f(x) = x + 2$, $g(x) = \sqrt{x}$, and $h(x) = x^3$. Find the following:

   (a) $(f \circ g \circ h)(x)$ **(b)** $(f \circ h \circ g)(x)$ **(c)** $(g \circ f \circ h)(x)$

   (d) $(g \circ h \circ f)(x)$ **(e)** $(h \circ f \circ g)(x)$ **(f)** $(h \circ g \circ f)(x)$

27. Let $f(x) = \dfrac{1}{x}$. Find $(f \circ f)(x)$ and $(f \circ f \circ f)(x)$.

28. Let $f(x) = x^2$, $g(x) = \dfrac{1}{x-1}$, and $h(x) = 1 + \dfrac{1}{x}$. Find the following:

   (a) $(f \circ h \circ g)(x)$ **(b)** $(g \circ h \circ f)(x)$ **(c)** $(h \circ g \circ f)(x)$

 Find functions f and g so that $h(x) = (f \circ g)(x)$. In each case let the inner function g be a polynomial or a rational function.

29. $h(x) = (3x + 1)^2$

30. $h(x) = (x^2 - 2x)^3$

31. $h(x) = \sqrt{1 - 4x}$

32. $h(x) = \sqrt[3]{x^2 - 1}$

33. $h(x) = \left(\dfrac{x+1}{x-1}\right)^2$

34. $h(x) = \left(\dfrac{1 - 2x}{1 + 2x}\right)^3$

35. $h(x) = (3x^2 - 1)^{-3}$

36. $h(x) = \left(1 + \dfrac{1}{x}\right)^{-2}$

37. $h(x) = \sqrt{\dfrac{x}{x-1}}$

38. $h(x) = \sqrt[3]{\dfrac{x-1}{x}}$

39. $h(x) = \sqrt{(x^2 - x - 1)^3}$

40. $h(x) = \sqrt[3]{(1 - x^4)^2}$

41. $h(x) = \dfrac{2}{\sqrt{4 - x^2}}$

42. $h(x) = -\left(\dfrac{3}{x-1}\right)^5$

 Find three functions, f, g, and h such that $k(x) = (h \circ g \circ f)(x)$.

43. $k(x) = (\sqrt{2x + 1})^3$

44. $k(x) = \sqrt[3]{(2x - 1)^2}$

45. $k(x) = \sqrt{\left(\dfrac{x}{x+1}\right)^5}$

46. $k(x) = \left(\sqrt[7]{\dfrac{x^2 - 1}{x^2 + 1}}\right)^4$

47. $k(x) = (x^2 - 9)^{2/3}$

48. $k(x) = (5 - 3x)^{5/2}$

49. $k(x) = -\sqrt{(x^2 - 4x + 7)^3}$

50. $k(x) = -\left(\dfrac{2x}{1-x}\right)^{2/5}$

51. $k(x) = (1 + \sqrt{2x - 11})^2$

52. $k(x) = \sqrt[3]{(x^2 - 4)^5} - 1$

53. Find f so that $((x + 1)^2 + 1)^2 = (f \circ f)(x)$.

54. Find f so that $\sqrt{1 + \sqrt{1 + \sqrt{1 + x}}} = (f \circ f \circ f)(x)$.

55. Let $f(x) = x^3$. Find a function g so that $(f \circ g)(x) = x$ and $(g \circ f)(x) = x$.

56. Let $f(x) = x$. Find $(f \circ g)(x)$ and $(g \circ f)(x)$ for any function g.

57. If $f(x) = 2x + 1$, find $g(x)$ so that $(f \circ g)(x) = 2x^2 - 4x + 1$.

 Written Assignment: Explain, with specific examples, the distinction between the composition and the decomposition of functions.

1. Certain operations on an equation can produce extraneous roots. Others can have the effect of losing a root. Explain, with examples, how each of these situations can occur.

2. If $f(x) = x^2$, what is $f(f(x))$? Find a formula for $f(f(\cdots f(x))\cdots)$ where f appears n times.

3. In general, for two functions $f(x)$ and $g(x)$, $f(g(x)) \neq g(f(x))$. Show that for the functions $f(x) = 3x - 2$ and $g(x) = \dfrac{x + 2}{3}$ the equality does hold. Explain what that means in terms of the graphs of the two functions.

4. Sketch two graphs of functions $f(x)$ and $g(x)$ such that the domain of $f(x)$ is equal to the range of $g(x)$ and the domain of $g(x)$ is equal to the range of $f(x)$. Furthermore, for each function, the domain and range are not to be the same set of numbers.

5. Is it true that $\sqrt[3]{|x|} = |\sqrt[3]{x}|$ for all real numbers x? Justify your answer.

6. The graph of a function f is symmetric about the vertical line $x = c$ provided that $f(c + h) = f(c - h)$. Make a sketch that demonstrates this condition; then use it to prove that $g(x) = \sqrt{|x - 2|}$ is symmetric about $x = 2$.

5.5
INVERSE
FUNCTIONS

By definition, a function has each domain value x corresponding to exactly one range value y. Some (but not all) functions have the additional property that to every range value y there corresponds exactly one domain value x. Such functions are said to be **one-to-one functions.** To understand this concept, consider the functions $y = x^2$ and $y = x^3$.

For a one-to-one function you can start at a range value y and trace it back to exactly one domain value x.

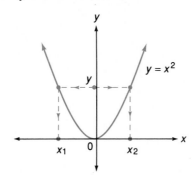

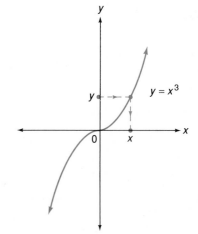

$y = x^2$ *is not* a one-to-one function. There are two domain values for a range value $y > 0$.

$y = x^3$ *is* a one-to-one function. There is exactly one domain value x for each range value y.

DEFINITION OF A ONE-TO-ONE FUNCTION

A function f is a one-to-one function if and only if for each range value there corresponds exactly one domain value.

Once a graph of a function is known, there is a simple *horizontal line test* for determining the one-to-one property. Consider a horizontal line through each range value y, as in the preceding figures. If the line meets the curve exactly once, then we have a one-to-one function; otherwise, it is not one-to-one.

If the variables in $y = x^3$ are interchanged, we obtain $x = y^3$. Here are the graphs for these equations, together with the two graphs on the same axes. Because of this interchange of coordinates, the two curves are *reflections* of each other through the line $y = x$, as in part (c) of the figure. Another way to describe this relationship is to say that they are *mirror images* of one another through the "mirror line" $y = x$.

If the paper were folded along the line $y = x$, the two curves would coincide.

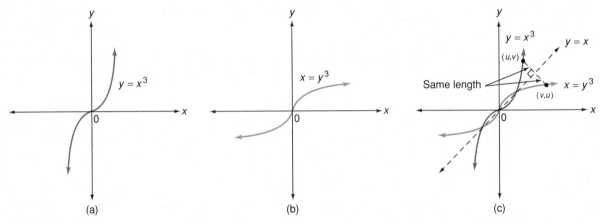

(a)   (b)   (c)

The equation $x = y^3$ may be solved for y by taking the cube root of both sides: $y = \sqrt[3]{x}$. The equations $x = y^3$ and $y = \sqrt[3]{x}$ are equivalent and therefore define the same function of x. However, since $y = \sqrt[3]{x}$ shows *explicitly* how y depends on x, it is the preferred form.

We began with the one-to-one function $y = f(x) = x^3$ and, by interchanging variables, arrived at the new function $y = g(x) = \sqrt[3]{x}$. If the composites $f \circ g$ and $g \circ f$ are formed, something surprising happens.

Function f cubes and function g "uncubes"; f and g are inverse functions.

$$(f \circ g)(x) = f(g(x)) = f(\sqrt[3]{x}) = (\sqrt[3]{x})^3 = x$$
$$(g \circ f)(x) = g(f(x)) = g(x^3) = \sqrt[3]{x^3} = x$$

In each case we obtained the same value x that we started with; whatever one of the functions does to a value x, the other function undoes. Whenever two functions act on each other in such a manner, we say that they are **inverse functions** or that either function is the inverse of the other.

f^{-1} is read as "the inverse of f" or as "f inverse."

The notation f^{-1} is also used to represent the inverse of f. Thus if $f(x) = x^3$, then $f^{-1}(x) = \sqrt[3]{x}$. Note that if $f(a) = b$, then $f^{-1}(b) = a$. For example, let $a = 2$:

$$f(x) = x^3: \qquad f(2) = 2^3 = 8$$
$$f^{-1}(x) = \sqrt[3]{x}: \qquad f^{-1}(8) = \sqrt[3]{8} = 2$$

CAUTION
*In the expression $f^{-1}(x)$, -1 is **not** a negative exponent. Thus $f^{-1}(x) \neq \dfrac{1}{f(x)}$. The reciprocal of $f(x)$ can be correctly written in this way:*

$$\frac{1}{f(x)} = [f(x)]^{-1}$$

Also, using this notation and the definition of inverse functions, we have

$$f(f^{-1}(x)) = x \quad \text{and} \quad f^{-1}(f(x)) = x$$

Thus

$$f(f^{-1}(8)) = 8 \quad \text{and} \quad f^{-1}(f(2)) = 2$$

$y = x^2$ is not one-to-one. Its reflection in the line $y = x$ has equation $x = y^2$, or $y = \pm\sqrt{x}$, which is not a function of x.

It turns out (as suggested by our work with $y = x^3$) that *every one-to-one function f has an inverse g and that their graphs are reflections of each other through the line $y = x$.* The technique of interchanging variables, used to obtain $y = \sqrt[3]{x}$ from $y = x^3$, can be applied to many situations, as illustrated in the following examples.

To find the inverse g of f, begin with
$$y = f(x)$$
Then interchange variables
$$x = f(y)$$
and solve for y in terms of x, producing
$$y = g(x) = f^{-1}(x).$$

EXAMPLE 1 Find the inverse g of $y = f(x) = 2x + 3$. Then show that $(f \circ g)(x) = x$ and $(g \circ f)(x) = x$ and graph both functions on the same axes.

Solution Interchange variables in $y = 2x + 3$ and solve for y.

$$x = 2y + 3$$
$$2y = x - 3$$
$$y = f^{-1}(x) = \tfrac{1}{2}x - \tfrac{3}{2}$$

Using $y = g(x) = \tfrac{1}{2}x - \tfrac{3}{2}$, we have

$$(f \circ g)(x) = f(g(x)) = f(\tfrac{1}{2}x - \tfrac{3}{2})$$
$$= 2(\tfrac{1}{2}x - \tfrac{3}{2}) + 3$$
$$= x - 3 + 3$$
$$= x$$

The procedure used in Examples 1 and 2 to find the inverse has its limitations. Thus, if $y = f(x)$ is complicated, it may be algebraically difficult, or even impossible, to interchange variables and solve for the inverse. We will avoid such situations by limiting our work in this section to functions for which the procedure can be applied.

$$(g \circ f)(x) = g(f(x)) = g(2x + 3)$$
$$= \tfrac{1}{2}(2x + 3) - \tfrac{3}{2}$$
$$= x + \tfrac{3}{2} - \tfrac{3}{2}$$
$$= x$$

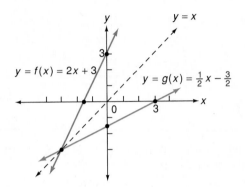

EXAMPLE 2 Follow the instructions in Example 1 for $y = f(x) = \sqrt{x}$.

Solution Interchange variables in $y = \sqrt{x}$ to get $x = \sqrt{y}$ and note that x cannot be negative: $x \geq 0$. Solving for y by squaring produces $y = x^2$. Therefore the inverse function is $y = g(x) = x^2$ with domain $x \geq 0$.

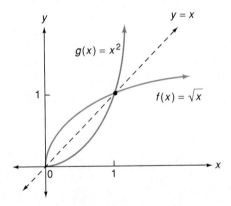

Using $f(x) = \sqrt{x}$ and $g(x) = x^2$, we have:

$$(f \circ g)(x) = f(g(x)) = f(x^2) = \sqrt{x^2} = |x| = x \qquad \text{(since } x \geq 0\text{)}$$
$$(g \circ f)(x) = g(f(x)) = g(\sqrt{x}) = (\sqrt{x})^2 = x$$

EXERCISES 5.5

Use the horizontal line test to decide if the function is one-to-one.

1. $f(x) = (x - 1)^2$ **2.** $f(x) = x^3 - 3x^2$ **3.** $f(x) = \dfrac{1 - x}{x}$

4. $f(x) = \dfrac{1}{x - 2}$ **5.** $f(x) = \sqrt{x} + 3$ **6.** $f(x) = \dfrac{1}{x^2 - x - 2}$

Show that f and g are inverse functions according to the criteria $(f \circ g)(x) = x$ and $(g \circ f)(x) = x$. Then graph both functions and the line $y = x$ on the same axes.

7. $f(x) = \frac{1}{3}x - 3;\quad g(x) = 3x + 9$ 8. $f(x) = 2x - 6;\quad g(x) = \frac{1}{2}x + 3$

9. $f(x) = (x + 1)^3;\quad g(x) = \sqrt[3]{x} - 1$ 10. $f(x) = -(x + 2)^3;\quad g(x) = -\sqrt[3]{x} - 2$

11. $f(x) = \dfrac{1}{x - 1};\quad g(x) = \dfrac{1}{x} + 1$ 12. $f(x) = x^2 + 2$ for $x \geq 0;\quad g(x) = \sqrt{x - 2}$

Find the inverse function g of the given function f.

13. $y = f(x) = (x - 5)^3$ 14. $y = f(x) = x^{1/3} - 3$ 15. $y = f(x) = \frac{2}{3}x - 1$

16. $y = f(x) = -4x + \frac{2}{5}$ 17. $y = f(x) = (x - 1)^5$ 18. $y = f(x) = -x^5$

19. $y = f(x) = x^{3/5}$ 20. $y = f(x) = x^{5/3} + 1$

Find $f^{-1}(x)$ and show that $f(f^{-1}(x)) = x$ and $f^{-1}(f(x)) = x$.

21. $f(x) = \dfrac{2}{x - 2}$ 22. $f(x) = -\dfrac{1}{x} - 1$ 23. $f(x) = \dfrac{3}{x + 2}$

24. $f(x) = \dfrac{x}{x + 1}$ 25. $f(x) = x^{-5}$ 26. $f(x) = \dfrac{1}{\sqrt[3]{x - 2}}$

Verify that the function is its own inverse by showing that $(f \circ f)(x) = x$.

27. $f(x) = \dfrac{1}{x}$ 28. $f(x) = \sqrt{4 - x^2}; 0 \leq x \leq 2$ 29. $f(x) = \dfrac{x}{x - 1}$

30. $f(x) = \dfrac{3x - 8}{x - 3}$

Find the inverse g of the given function f, and graph both in the same coordinate system.

31. $y = f(x) = (x + 1)^2; x \geq -1$ 32. $y = f(x) = x^2 - 4x + 4; x \geq 2$

33. $y = f(x) = \dfrac{1}{\sqrt{x}}$ 34. $y = f(x) = -\sqrt{x}$

35. $y = f(x) = x^2 - 4; x \geq 0$ 36. $y = f(x) = 4 - x^2; 0 \leq x \leq 2$

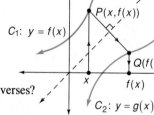

37. Aside from the linear function $y = x$, what other linear functions are their own inverses? (*Hint:* Inverse functions are reflections of one another through the line $y = x$.)

38. In the figure, curve C_1 is the graph of a one-to-one function $y = f(x)$ and curve C_2 is the graph of $y = g(x)$. The points on C_2 have been obtained by interchanging the first and second coordinates of the points on curve C_1. For a specific value x in the domain of f, the point $P(x, f(x))$ on C_1 produces $Q(f(x), x)$ on C_2. How does this figure demonstrate that $(g \circ f)(x) = x$?

CHALLENGE
Think Creatively

(a) The one-to-one property for a function f can be stated as follows:

$$\text{If } f(x_1) = f(x_2), \text{ then } x_1 = x_2.$$

Use this condition to prove that $f(x) = \dfrac{3}{4x + 5}$ is a one-to-one function.

(b) Use the letter g for the inverse of f and make direct use of the property $f(g(x)) = x$ to find g, without the use of the variable y.

REVIEW EXERCISES

The solutions to the following exercises can be found within the text of Chapter 5.
Try to answer each question before referring to the text.

Section 5.1

Find the domain and graph.

1. $y = \sqrt{x}$ 2. $y = \sqrt{x - 2}$ 3. $y = \sqrt{|x|}$

4. $y = x^{-1/2}$ 5. $y = \dfrac{1}{\sqrt{x}} + 1$ 6. $y = \dfrac{1}{\sqrt{x + 1}}$

7. Find the equations of the asymptotes for the curves in Exercises 4, 5, and 6.
8. Graph $y = x^3$ and $x = y^3$.
9. Graph $f(x) = \begin{cases} x^2 - 1 & \text{if } -2 \le x < 2 \\ \sqrt{x - 2} & \text{if } 2 \le x < 5 \end{cases}$

Section 5.2

Solve for x.

10. $\sqrt{x + 2} + 2 = 5$ 11. $\sqrt{x + 4} + 2 = x$
12. $\sqrt{x - 7} - \sqrt{x} = 1$ 13. $\sqrt[3]{x + 3} = 2$
14. $\dfrac{3(2x - 1)^{1/2}}{x - 3} - \dfrac{(2x - 1)^{3/2}}{(x - 3)^2} = 0$ 15. $\sqrt[3]{x^2} + \sqrt[3]{x} - 20 = 0$
16. Find the *x*-intercepts and the domain of $y = f(x) = \sqrt{x^2 + x - 6}$.
17. Solve the system and graph: $y = \sqrt[3]{x}$
$$y = \tfrac{1}{4}x$$

Section 5.3

Find the domain of f(x) and determine its signs.

18. $f(x) = \dfrac{\sqrt{x^2 + 9}}{x + 2}$ 19. $f(x) = \dfrac{x}{\sqrt{4 - x^2}}$

20. $f(x) = x^{1/2} + \tfrac{1}{2}x^{-1/2}(x - 3)$ 21. $f(x) = \dfrac{x^2 - 1}{2\sqrt{x - 1}} + 2x\sqrt{x - 1}$

22. Graph the function $f(x) = \sqrt{x^2 - 4}$.

Section 5.4

23. Determine $(f + g)(x)$, $(f - g)(x)$, $(f \cdot g)(x)$, and $\frac{g}{f}(x)$, where $f(x) = \dfrac{1}{x^3 - 1}$ and $g(x) = \sqrt{x}$.

24. What are the domains of $f + g$, $f - g$, and $f \cdot g$ given in Exercise 23?

25. What is the domain of $\dfrac{f}{g}$ given in Exercise 23?

26. Let $f(x) = \dfrac{1}{x^3 - 1}$ and $g(x) = \sqrt{x}$. Evaluate $f(4)$, $g(4)$, and $(f \cdot g)(4)$.

27. Let $f(x) = \dfrac{1}{x - 2}$ and $g(x) = \sqrt{x}$. Evaluate $g(f(6))$ and $f(g(6))$.

28. Form the composites $g \circ f$ and $f \circ g$ and give their domains, where $f(x) = \dfrac{1}{x^2 - 1}$ and $g(x) = \sqrt{x}$.

29. Find $(f \circ g \circ h)(x)$, where $f(x) = \sqrt{x}$, $g(x) = x^2 + 1$, and $h(x) = \dfrac{1}{x}$.

30. Find functions f and g so that $h(x) = \sqrt{x^2 + 2x + 2} = (g \circ f)(x)$.

31. Find functions f and g so that $h = f \circ g$, where $h(x) = \left(\dfrac{1}{3x - 1}\right)^5$ and the inner function g is rational.

32. Decompose $h(x) = \sqrt{(x^2 - 3x)^5}$ so that the outer function is a monomial.

33. Show that $h(x) = \sqrt{(x^2 - 3x)^5}$ can be written as the composition of three functions.

Section 5.5

34. State the definition of a one-to-one function.

35. Describe the horizontal line test for one-to-one functions.

36. State the definition of inverse functions.

37. Find the inverse g of $y = f(x) = 2x + 3$. Then show that $(f \circ g)(x) = x$ and $(g \circ f)(x) = x$, and graph both functions on the same axes.

38. Find the inverse of $y = f(x) = \sqrt{x}$. Graph both functions on the same axes.

CHAPTER 5 TEST: STANDARD ANSWER

Use these questions to test your knowledge of the basic skills and concepts of Chapter 5. Then check your answers with those given at the back of the book.

Graph each function, state its domain, and then give the equations of the asymptotes if there are any.

1. $f(x) = \sqrt[3]{x + 2}$ 2. $g(x) = \dfrac{1}{\sqrt{x}} + 2$ 3. $f(x) = -\sqrt[3]{x - 2}$ 4. $g(x) = \dfrac{1}{\sqrt{x - 3}}$

5. Graph: $f(x) = \begin{cases} 1 - x^2 & \text{for } -3 \le x < 0 \\ \sqrt{x} + 1 & \text{for } 0 \le x < 2 \end{cases}$

Solve each equation.

6. $\sqrt{18x + 5} - 9x = 1$ 7. $6x^{2/3} + 5x^{1/3} - 4 = 0$ 8. $\sqrt{x - 7} + \sqrt{x + 9} = 8$

9. Convert $(x + 1)^{1/2}(2x) + (x + 1)^{-1/2}\left(\dfrac{x^2}{2}\right)$ into the equivalent form $\dfrac{x(5x + 4)}{2\sqrt{x + 1}}$. Show your work.

Determine the signs of f.

10. $f(x) = \dfrac{\sqrt[3]{x-4}}{x-2}$ **11.** $f(x) = x^{1/2} - \frac{1}{2}x^{-1/2}(x+4)$

12. Find the x-intercepts and domain of the function $f(x) = \sqrt{x^2 + x - 2}$.

Let $f(x) = \dfrac{1}{x^2 - 1}$ and $g(x) = \sqrt{x+2}$. Find and state the domain for each of the following.

13. **(a)** $(f+g)(x)$ **(b)** $(f-g)(x)$ **14.** **(a)** $\dfrac{f}{g}(x)$ **(b)** $(f \cdot g)(x)$

15. Describe how the graph of $g(x) = -\sqrt[3]{x} - 4$ can be obtained from the graph of $f(x) = \sqrt[3]{x}$.

Let $f(x) = \dfrac{1}{x^2 + 1}$ and $g(x) = 2\sqrt{x}$.

16. Find $f(4)$, $g(4)$, and $(f \cdot g)(4)$.

17. Evaluate $f(g(9))$ and $g(f(9))$.

18. For $f(x) = \dfrac{1}{1 - x^2}$ and $g(x) = \sqrt{x}$ find the composites $(f \circ g)(x)$ and $(g \circ f)(x)$ and state their domains.

19. Find the functions f and g so that $h(x) = \sqrt[3]{(x-2)^2} = (f \circ g)(x)$, where g is a binomial.

20. Let $F(x) = \dfrac{1}{(2x-1)^{3/2}}$. Find functions f, g, and h so that $(f \circ g \circ h)(x) = F(x)$.

Complete the following for the function $f(x) = 3x - 2$.

21. Find the inverse function, $f^{-1}(x)$.   **22.** Graph $f(x)$ and $f^{-1}(x)$ on the same axes.

23. Find the inverse g of $y = f(x) = \sqrt[3]{x} - 1$ and show that $(f \circ g)(x) = x$.

24. State the domain of $f(x) = \sqrt[3]{3x+1} - x - 1$ and find the x-intercepts.

25. Solve the system: $y = 2\sqrt{x} - 1$
$$y = \tfrac{1}{2}x + 1$$

CHAPTER 5 TEST: MULTIPLE CHOICE

1. The equation $\sqrt{x^2 - 9x} + \sqrt{3x} = x$ has

    (a) no solution   **(b)** one solution   **(c)** two solutions

    (d) three solutions   **(e)** None of the preceding

2. The graph of $y = \dfrac{1}{\sqrt{x-1}}$ is found by shifting the graph of $y = \dfrac{1}{\sqrt{x}}$

    (a) One unit downward   **(b)** One unit upward   **(c)** One unit to the left

    (d) One unit to the right   **(e)** None of the preceding

3. For which of the following functions is both the domain and the range the set of all real numbers greater than or equal to 0?

(a) $y = \sqrt{x}$ (b) $y = x^3$ (c) $y = |x|$ (d) $y = \sqrt[3]{x}$ (e) None of the preceding

4. What are the x-intercepts for the graph of $y = \sqrt{2x^2 - 8x - 24}$?

(a) 2, −6   (b) −4, 3   (c) −6, 4   (d) 0   (e) None of the preceding

5. Which of the following are the solutions for the equation $\sqrt[3]{x^2} - \sqrt[3]{x} - 6 = 0$?

(a) 2, −3   (b) −2, 3   (c) −8, 27   (d) 8, −27   (e) None of the preceding

6. The radius of a right circular cylinder of height h and volume V is given by $r = \sqrt{\dfrac{V}{\pi h}}$.

Using this formula to solve for h in terms of r and V, $h =$

(a) $\dfrac{\pi r^2}{V}$ (b) $\dfrac{V}{\pi r^2}$ (c) $\dfrac{(\pi r)^2}{V}$ (d) $\dfrac{V}{\pi r}$ (e) None of the preceding

7. Which of the following are the signs of $f(x) = \dfrac{x - 2}{\sqrt{x^2 + 4}}$?

(a) For $x \neq \pm 2$, $f(x) > 0$ (b) For $x \neq 2$, $f(x) > 0$

(c) For $x < 2$, $f(x) < 0$; for $x > 2$, $f(x) > 0$

(d) For $x < -2$, $f(x) < 0$; for $x > -2$, $f(x) > 0$ (e) None of the preceding

8. Which of the following intervals represents the domain of the function

$$f(x) = \dfrac{x^2}{\sqrt{1 - x^2}}?$$

(a) $(1, \infty)$ (b) $(-\infty, -1)$ (c) $(-\infty, -1)$ and $(1, \infty)$ (d) $(-1, 1)$ (e) None of the preceding

9. Which of the following are correct?

I. $x\sqrt{x^2 + 1} = \sqrt{x^3 + x}$

II. $x^{-1/5} + x^{3/5} = x^{-1/5}(1 + x^{2/5})$

III. $(x^2 - 3)^{1/3}(x^2 - 3)^{2/3} = (x^2 - 3)^{2/9}$

(a) Only I   (b) Only II   (c) Only III   (d) Only I and II   (e) None of the preceding

10. Let $f(x) = \dfrac{1}{x^2 - 1}$ and $g(x) = \sqrt{x}$. Then $\dfrac{f}{g}(4) =$

(a) $\dfrac{2}{15}$ (b) $\dfrac{15}{2}$ (c) $\dfrac{1}{30}$ (d) 30   (e) None of the preceding

11. $x^{1/3} + (x - 8)\dfrac{1}{3}x^{-2/3} =$

(a) $\dfrac{x^{1/3} + x - 8}{3x^{2/3}}$ (b) $\dfrac{3x^{1/3} + x - 8}{x^{2/3}}$ (c) $\dfrac{x - 8}{3\sqrt[3]{x}}$ (d) $\dfrac{4(x - 2)}{3\sqrt[3]{x^2}}$ (e) None of the preceding

12. Let $f(x) = \sqrt[3]{x}$ and $g(x) = \dfrac{1}{x^3 - 1}$. Then $g(f(x)) =$

(a) $\dfrac{1}{\sqrt[3]{x^3} - 1}$ (b) $\dfrac{1}{\sqrt[3]{x} - 1}$ (c) $\sqrt[3]{x^3 - 1}$ (d) $\dfrac{1}{x - 1}$ (e) None of the preceding

13. Which of the following is a one-to-one function of x?

(a) $y = x^4$ (b) $y = \dfrac{1}{x}$ (c) $y = |x|$ (d) $x = y^2$ (e) None of the preceding

14. Which of the following is the inverse of $y = f(x) = 3x + 2$?

(a) $y = 2x + 3$ (b) $y = \frac{1}{3}x - \frac{2}{3}$ (c) $y = 2 + 3x$ (d) $y = \frac{1}{3}x + 2$ (e) None of the preceding

15. Which of the functions f and g are such that $h(x) = \sqrt{(4 - x^2)^3} = (f \circ g)(x)$?

(a) $f(x) = 4 - x^2$; $g(x) = x^{3/2}$ (b) $f(x) = x^{2/3}$; $g(x) = 4 - x^2$

(c) $f(x) = x^{3/2}$; $g(x) = 4 - x^2$ (d) $f(x) = x^{1/3}$; $g(x) = (4 - x^2)^2$ (e) None of the preceding

Page 264

1. Domain of g: $x \geq 0$

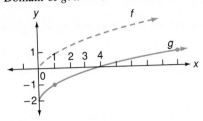

2. Domain of g: $x \geq 1$

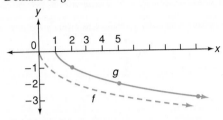

3. Domain of g: $x \geq 0$

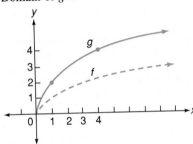

4. Domain of g: $x > -2$

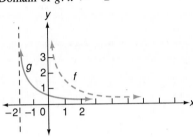

Page 269

1. 25; all $x \geq 0$ **2.** 0, 3; all $x \leq 0$ or all $x \geq 3$ **3.** $x = 8$ **4.** $x = 25$ **5.** $x = 7$

6. No solution **7.** $x = \frac{1}{9}$ **8.** $x = 5$ **9.** $x = -3$ or $x = 3$ **10.** $x = 9$ **11.** $x = 0$

12. $x = \frac{1}{2}$

Page 274

1. Domain: $x > 0$
$f(x) > 0$ on $(4, \infty)$
$f(x) < 0$ on $(0, 4)$

2. Domain: $x \geq -2$ and $x \neq 0$
$f(x) > 0$ on $(0, \infty)$
$f(x) < 0$ on $(-2, 0)$

3. Domain: all $x \neq 0$
$f(x) > 0$ on $(0, \infty)$
$f(x) < 0$ on $(-\infty, 0)$

Page 279

1. (a) $g(f(1)) = g(2) = 8$
(b) $f(g(1)) = f(5) = 14$
(c) $f(g(0)) = f(4) = 11$
(d) $g(f(-2)) = g(-7) = 53$

2. (a) $g(f(1)) = g(1) = 1$
(b) $f(g(1)) = f(1) = 1$
(c) $f(g(0)) = f(0) = 0$
(d) $g(f(-2))$ is undefined.

3. (a) $g(f(1)) = g(\sqrt[3]{2}) = 5\sqrt[3]{2}$
(b) $f(g(1)) = f(5) = \sqrt[3]{14}$
(c) $f(g(0)) = f(0) = -1$
(d) $g(f(-2)) = g(\sqrt[3]{-7}) = -5\sqrt[3]{7}$

4. (a) $g(f(1))$ is undefined.
(b) $f(g(1))$ is undefined.
(c) $f(g(0)) = f(0) = -2$
(d) $g(f(-2)) = g(0) = 0$

Page 282

(Other answers are possible.)

1. $f(x) = 8x - 3; g(x) = x^5$ **2.** $f(x) = 8x - 3; g(x) = \sqrt[5]{x}$ **3.** $f(x) = \dfrac{1}{8x - 3}; g(x) = \sqrt{x}$

4. $f(x) = \dfrac{5}{7 + 4x^2}; g(x) = x^3$ **5.** $f(x) = \dfrac{2x + 1}{2x - 1}; g(x) = x^4$ **6.** $f(x) = x^4 - 2x^2 + 1; g(x) = \sqrt{x^3}$

Page 286

The functions given in 2, 3, 5, and 6 are one-to-one. The others are not.

6

EXPONENTIAL AND LOGARITHMIC FUNCTIONS

6.1 EXPONENTIAL FUNCTIONS

$20,000 = (10,000)2^1$

$40,000 = (10,000)2^2$

$80,000 = (10,000)2^3$

We use b > 0 in order to avoid even roots of negative numbers, such as $(-4)^{1/2} = \sqrt{-4}$.

Imagine that a bacterial culture is growing at a rate such that after each hour the number of bacteria has doubled. Thus, if there were 10,000 bacteria when the culture started to grow, then after 1 hour the number would have grown to 20,000, after 2 hours there would be 40,000, and so on. It becomes reasonable to say that

$$y = f(x) = (10,000)2^x$$

gives the number y of bacteria present after x hours. This equation defines an *exponential function* with independent variable x and dependent variable y.

A function such as $f(x) = b^x$, with the variable as an exponent, is known as an **exponential function.** We shall study such functions, with the assumption that the *base number* $b > 0$. For example, let us consider the function $y = f(x) = 2^x$ and its graph. Note the following:

1. The function is defined for all real values of x. When x is negative, we may apply the definition for negative exponents. Thus, for $x = -2$,

$$2^x = 2^{-2} = \frac{1}{2^2} = \frac{1}{4}$$

   The domain of the function is the set of real numbers.

2. For all replacements of x, the function takes on a positive value. That is, 2^x can never represent a negative number, nor can 2^x be equal to 0. The range of the function is the set of positive real numbers.

3. Finally, a few specific ordered pairs of numbers can be located as an aid to graphing, as shown on the following page.

The function is increasing and the curve is concave up. The x-axis is a horizontal asymptote toward the left.

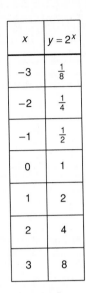

x	$y = 2^x$
-3	$\frac{1}{8}$
-2	$\frac{1}{4}$
-1	$\frac{1}{2}$
0	1
1	2
2	4
3	8

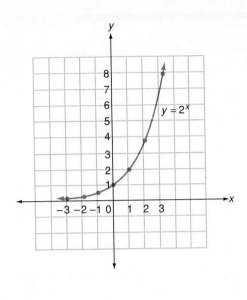

If desired, the accuracy of this graph can be improved by using more points. For example, consider such rational values of x as $\frac{1}{2}$ or $\frac{3}{2}$:

The value of $\sqrt{2}$ is given correct to two decimal places from Table I, page 636, or from a calculator.

$$2^{1/2} = \sqrt{2} = 1.41$$
$$2^{3/2} = (\sqrt{2})^3 = 2.80$$

Using irrational values for x, such as $\sqrt{2}$ or π, is another matter entirely. (Remember that our development of exponents stopped with the rationals.) To give a precise meaning of such numbers is beyond the scope of this course. However, using a scientific calculator you can obtain approximations such as $(\sqrt{2}, 2^{\sqrt{2}}) \approx (1.41, 2.67)$ and $(\sqrt{5}, 2^{\sqrt{5}}) \approx (2.24, 4.71)$ and then verify that these points "fit" the curve.

When graphing $y = 2^x$ we made use of the property $2^{-n} = \dfrac{1}{2^n}$. This and other properties listed below have been studied previously for rational exponents r and s.

$$b^r b^s = b^{r+s} \qquad \frac{b^r}{b^s} = b^{r-s} \qquad (b^r)^s = b^{rs}$$

$$a^r b^r = (ab)^r \qquad b^0 = 1 \qquad b^{-r} = \frac{1}{b^r}$$

In more advanced work it can be shown that these rules hold for positive bases a and b and all real numbers r and s.

EXAMPLE 1 Graph the curve $y = 8^x$ on the interval $[-1, 1]$ by using a table of values.

Solution

x	y
−1	$\frac{1}{8}$
$-\frac{2}{3}$	$\frac{1}{4}$
$-\frac{1}{3}$	$\frac{1}{2}$
0	1
$\frac{1}{3}$	2
$\frac{2}{3}$	4
1	8

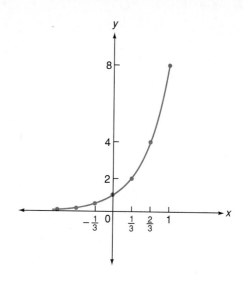

■

Thus far we have restricted our attention to exponential functions of the form $y = f(x) = b^x$, where $b > 1$. All of these have graphs that are of the same general shape as that for $y = 2^x$. For $b = 1$, $y = b^x = 1^x = 1$ for all x. Since this is a constant function, $f(x) = 1$, we do not use the base $b = 1$ in the classification of exponential functions.

Now let us explore exponential functions $y = f(x) = b^x$ for which $0 < b < 1$. In particular, if $b = \frac{1}{2}$, we get $y = \left(\frac{1}{2}\right)^x = \frac{1}{2^x}$ or $y = 2^{-x}$.

x	$y = 2^{-x}$
−3	8
−2	4
−1	2
0	1
1	$\frac{1}{2}$
2	$\frac{1}{4}$
3	$\frac{1}{8}$

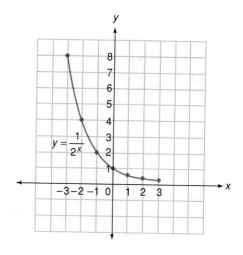

All curves $y = b^x$ for $0 < b < 1$ have this same basic shape. The curve is concave up, the function is decreasing, and the y-axis is a horizontal asymptote toward the right as x becomes large.

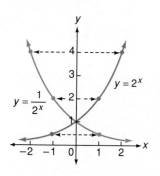

As shown in the margin, the graph of $y = g(x) = \frac{1}{2^x}$ can also be found by comparing it to the graph of $y = f(x) = 2^x$. Since $g(x) = \frac{1}{2^x} = 2^{-x} = f(-x)$, the y-values for g are the same as the y-values for f *on the opposite side of the y-axis.* In other words, the graph of g is the *reflection* of the graph of f through the y-axis. The next example demonstrates how the graphs of exponential functions can be translated.

EXAMPLE 2 Use the graph of $y = f(x) = 2^x$ to sketch the curves

$$y = g(x) = 2^{x-3} \quad \text{and} \quad y = h(x) = 2^x - 1$$

Even though functions g and h are not exactly of the form b^x, they are also classified as being exponential functions because the variable is within the exponent.

Solution Since $g(x) = f(x - 3)$, the graph of g can be obtained by shifting (translating) $y = 2^x$ by 3 units to the right. Moreover, since $h(x) = f(x) - 1$ the graph of h can be found by shifting $y = 2^x$ down 1 unit.

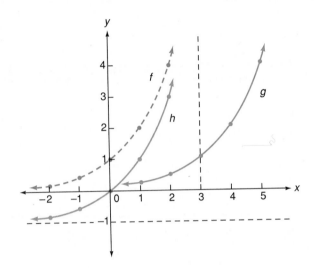

We have explored functions of the form $y = f(x) = b^x$ for specific values of b. In each case you should note that the graphs went through the point $(0, 1)$ since $y = b^0 = 1$. Also, each such graph has the x-axis as a one-sided asymptote and there are no x-intercepts. These and other properties of $y = f(x) = b^x$ for $b > 0$ and $b \neq 1$ are summarized below.

PROPERTIES OF $y = f(x) = b^x$

1. The domain consists of all real numbers x.
2. The range consists of all positive numbers y.
3. The function is increasing (the curve is rising) when $b > 1$, and it is decreasing (the curve is falling) when $0 < b < 1$.
4. The curve is concave up for $b > 1$ and for $0 < b < 1$.
5. It is a one-to-one function.
6. The point $(0, 1)$ is on the curve. There is no x-intercept.
7. The x-axis is a horizontal asymptote to the curve, toward the left for $b > 1$ and toward the right for $0 < b < 1$.
8. $b^{x_1} b^{x_2} = b^{x_1+x_2}$; $b^{x_1}/b^{x_2} = b^{x_1-x_2}$; $(b^{x_1})^{x_2} = b^{x_1 x_2}$.

This form of the one-to-one property can sometimes be applied to the solutions of equations.

The one-to-one property of a function f may be stated in this way:

$$\text{If } f(x_1) = f(x_2), \text{ then } x_1 = x_2.$$

That is, since $f(x_1)$ and $f(x_2)$ represent the same range value, there can only be one corresponding domain value; consequently, $x_1 = x_2$. Using $f(x) = b^x$, this statement means the following:

$$\text{If } b^{x_1} = b^{x_2}, \text{ then } x_1 = x_2.$$

This property can be used to solve certain **exponential equations,** such as $5^{x^2} = 625$. First note that 625 can be written as 5^4.

$$5^{x^2} = 625$$
$$5^{x^2} = 5^4$$

CAUTION
a^{b^c} *means* $a^{(b^c)}$, *and* $(a^b)^c = a^{bc}$. *Thus, in general* $a^{b^c} = a^{(b^c)} \neq (a^b)^c$.

By the one-to-one property applied to the function $f(t) = 5^t$, we may equate the exponents and solve for x.

$$x^2 = 4$$
$$x = \pm 2 \qquad (x = 2 \quad \text{or} \quad x = -2)$$

To check these solutions, note that $5^{2^2} = 5^4 = 625$ and $5^{(-2)^2} = 5^4 = 625$.

EXAMPLE 3 Solve for x: $\dfrac{1}{3^{x-1}} = 81$

Solution Write 81 as 3^4 and $\dfrac{1}{3^{x-1}}$ as $3^{-(x-1)}$.

Note that the one-to-one property here is being applied to the function $f(t) = 3^t$.

$$3^{-(x-1)} = 3^4$$
$$-(x - 1) = 4 \qquad \text{(by the one-to-one property)}$$
$$-x + 1 = 4$$
$$-x = 3$$
$$x = -3$$

∎

TEST YOUR UNDERSTANDING
Think Carefully

Solve for x.

1. $2^{x-1} = 32$　　**2.** $2^{x^2} = 16$　　**3.** $8^{2x+1} = 64$

4. $\dfrac{1}{2^x} = 64$　　**5.** $\dfrac{1}{5^{x+1}} = 125$　　**6.** $\dfrac{1}{4^{x-2}} = 64$

7. $27^x = 3$　　**8.** $27^x = 9$　　**9.** $125^x = 25$

10. $\left(\dfrac{1}{4}\right)^x = 32$　　**11.** $\left(\dfrac{3}{5}\right)^x = \dfrac{27}{125}$　　**12.** $\left(\dfrac{9}{25}\right)^x = \dfrac{5}{3}$

(Answers: Page 340)

EXERCISES 6.1

Graph the exponential function f by making use of a brief table of values. Then use this curve to sketch the graph of g. Indicate the horizontal asymptotes.

1. $f(x) = 2^x$; $g(x) = 2^{x+3}$
2. $f(x) = 3^x$; $g(x) = 3^x - 2$
3. $f(x) = 4^x$; $g(x) = -(4^x)$
4. $f(x) = 5^x$; $g(x) = (\frac{1}{5})^x$
5. $f(x) = (\frac{3}{2})^x$; $g(x) = (\frac{3}{2})^{-x}$
6. $f(x) = 8^x$; $g(x) = 8^{x-2} + 3$
7. $f(x) = 3^x$; $g(x) = 2(3^x)$
8. $f(x) = 3^x$; $g(x) = \frac{1}{2}(3^x)$
9. $f(x) = 2^{x/2}$; $g(x) = 2^{x/2} - 3$
10. $f(x) = 4^x$; $g(x) = 4^{1-x}$

Sketch the curves on the same axes.

11. $y = (\frac{3}{2})^x$, $y = 2^x$, $y = (\frac{5}{2})^x$
12. $y = (\frac{1}{4})^x$, $y = (\frac{1}{3})^x$, $y = (\frac{1}{2})^x$
13. $y = 2^{|x|}$, $y = -(2^{|x|})$
14. $y = 2^x$, $y = 2^{-x}$, $y = 2^x - 2^{-x}$ (*Hint:* Subtract ordinates.)

Use the one-to-one property of an appropriate exponential function to solve the indicated equation.

15. $2^x = 64$
16. $3^x = 81$
17. $2^{x^2} = 512$
18. $3^{x-1} = 27$
19. $5^{2x+1} = 125$

20. $2^{x^3} = 256$
21. $7^{x^2+x} = 49$
22. $b^{x^2+x} = 1$
23. $\frac{1}{2^x} = 32$
24. $\frac{1}{10^x} = 10{,}000$

25. $9^x = 3$
26. $64^x = 8$
27. $9^x = 27$
28. $64^x = 16$
29. $(\frac{1}{49})^x = 7$

30. $5^x = \frac{1}{125}$
31. $(\frac{27}{8})^x = \frac{9}{4}$
32. $(0.01)^x = 1000$

33. Graph the functions $y = 2^x$ and $y = x^2$ in the same coordinate system for the interval $[0, 5]$. (Use a larger unit on the x-axis than on the y-axis.) What are the points of intersection?

34. Solve for x: $(6^{2x})(4^x) = 1728$.   35. Solve for x: $(5^{2x+1})(7^{2x}) = 175$.

36. Use a calculator to verify that $\sqrt{3} = 1.732050 \ldots$. Now fill in the powers of 2 in the table, rounding off each entry to four decimal places. (Note that the numbers across the top of the table are successive decimal approximations of $\sqrt{3}$.)

x	1.7	1.73	1.732	1.7320	1.73205
2^x					

On the basis of the results, what is your estimate of $2^{\sqrt{3}}$ to three decimal places? Now find $2^{\sqrt{3}}$ directly on the calculator and compare.

37. Follow the instructions in Exercise 36 for these numbers:
(a) $3^{\sqrt{2}}$ (b) $3^{\sqrt{3}}$ (c) $2^{\sqrt{5}}$ (d) 4^{π}

6.2 LOGARITHMIC FUNCTIONS

One of the properties of the exponential function $y = b^x$ is that it is a one-to-one function. This means that it has an inverse function whose graph can be obtained by reflecting the graph of $y = b^x$ through the line $y = x$. This has been done for the case when $b > 1$ and also when $0 < b < 1$, as shown next.

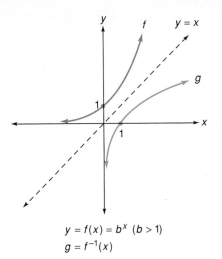

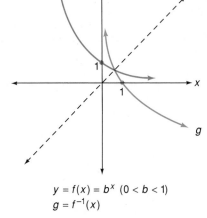

$y = f(x) = b^x \ (b > 1)$
$g = f^{-1}(x)$

$y = f(x) = b^x \ (0 < b < 1)$
$g = f^{-1}(x)$

Recall from Section 5.5 that $f^{-1}(x)$ is a notation used to represent the inverse of a function f.

An equation for the inverse function g can be obtained by interchanging the roles of the variables as follows:

$$\text{Function, } f: \quad y = f(x) = b^x$$
$$\text{Inverse function, } g: \quad x = g(y) = b^y$$

Thus $x = b^y$ is an equation for g. Unfortunately, we have no way of solving $x = b^y$ to get y explicitly in terms of x. To overcome this difficulty we create some new terminology.

The equation $x = b^y$ tells us that y *is the exponent on b that produces x*. In situations like this the word **logarithm** is used in place of *exponent*. A logarithm, then, is an exponent. Now we may say that y *is the logarithm on b that produces x*. This description can be abbreviated to $y = \text{logarithm}_b \ x$, and abbreviating further we reach the final form

$$y = \log_b x$$

which is read "y equals log x to the base b" or "y equals log x base b."

It is important to realize that we are only defining (not proving) the equation $y = \log_b x$ to have the same meaning as $x = b^y$. In other words, these two forms are equivalent:

$$\text{Exponential form:} \quad x = b^y$$
$$\text{Logarithmic form:} \quad y = \log_b x$$

And since they are equivalent they define the same function g:

$$y = g(x) = \log_b x$$

Now we know that the exponential function $y = f(x) = b^x$ and the **logarithmic function** $y = g(x) = \log_b x$ are inverse functions. Consequently, we have the following:

Recall from Section 5.5 that for inverse functions, $f(g(x)) = x$, and $g(f(x)) = x$.

$$f(g(x)) = f(\log_b x) = b^{\log_b x} = x \quad \text{and} \quad g(f(x)) = g(b^x) = \log_b(b^x) = x$$

EXAMPLE 1   Write the equation of the inverse function g of $y = f(x) = 2^x$ and graph both on the same axes.

Solution   The inverse g has equation $y = g(x) = \log_2 x$, and its graph can be obtained by reflecting $y = f(x) = 2^x$ through the line $y = x$.

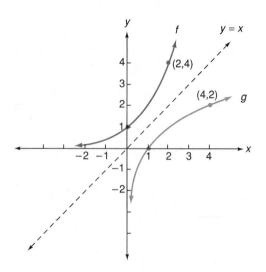

We found $y = \log_b x$ by interchanging the role of the variables in $y = b^x$. As a consequence of this switching, the domains and ranges of the two functions are also interchanged. Thus

Domain of $y = \log_b x$ is the same as the range of $y = b^x$.
Range of $y = \log_b x$ is the same as the domain of $y = b^x$.

These results are incorporated into the following list of important properties of the function $y = \log_b x$, where $b > 0$ and $b \neq 1$.

EXAMPLE 2   Find the domain for $y = \log_2(x - 3)$.

Solution   In $y = \log_2(x - 3)$ the quantity $x - 3$ plays the role that x does in $\log_2 x$. Thus $x - 3 > 0$ and the domain consists of all $x > 3$.   ∎

As noted earlier, the forms $\log_b x = y$ and $b^y = x$ are equivalent. The following table lists specific cases illustrating this equivalence.

CAUTION
*Do not confuse $x = b^y$ with its **inverse** $y = b^x$. These two forms are **not** equivalent.*

Logarithmic form $\log_b x = y$	Exponential form $b^y = x$
$\log_5 25 = 2$	$5^2 = 25$
$\log_{27} 9 = \frac{2}{3}$	$27^{2/3} = 9$
$\log_6 \frac{1}{36} = -2$	$6^{-2} = \frac{1}{36}$

Of the two forms $y = \log_b x$ and $x = b^y$, the exponential form is usually easier to work with. Consequently, when there is a question concerning $y = \log_b x$ it is often useful to convert to the exponential form. For instance, here are two logarithmic results that are easily verified using the exponential form:

$$\log_b 1 = 0 \qquad \text{since } b^0 = 1$$

$$\log_b b = 1 \qquad \text{since } b^1 = b$$

The next example shows how to solve for the variables x, y or b in the form $\log_b x = y$.

EXAMPLE 3   Solve for the indicated variable:
(a) $y = \log_9 27$ **(b)** $\log_b 8 = \frac{3}{4}$ **(c)** $\log_{49} x = -\frac{1}{2}$

Solution

(a) Convert $y = \log_9 27$ to exponential form as shown on the following page.

$$9^y = 27$$

$$(3^2)^y = 3^3 \qquad \text{(Express each side as a power of 3.)}$$

$$3^{2y} = 3^3$$

Check: $\log_9 27 = \frac{3}{2}$ is correct
since $9^{3/2} = 27$.

$$2y = 3 \qquad \text{(by the one-to-one property)}$$

$$y = \tfrac{3}{2}$$

(b) Convert $\log_b 8 = \frac{3}{4}$ to exponential form.

$$b^{3/4} = 8$$

Check: $\log_{16} 8 = \frac{3}{4}$ since
$16^{3/4} = (\sqrt[4]{16})^3 = 8$.

$$(b^{3/4})^{4/3} = 8^{4/3} \qquad \text{(Raise both sides to the } \tfrac{4}{3} \text{ power.)}$$

$$b = 16 \qquad 8^{4/3} = (\sqrt[3]{8})^4 = 2^4$$

(c) Convert $\log_{49} x = -\frac{1}{2}$ to exponential form.

Check this solution.

$$x = 49^{-1/2} = \frac{1}{\sqrt{49}} = \frac{1}{7}$$

■

EXERCISES 6.2

Sketch the graph of the function f. Reflect this curve through the line $y = x$ to obtain the graph of the inverse function g, and write the equation for g.

1. $y = f(x) = 4^x$ **2.** $y = f(x) = 5^x$ **3.** $y = f(x) = (\tfrac{1}{3})^x$ **4.** $y = f(x) = (0.2)^x$

Describe how the graph of h can be obtained from the graph of g. Find the domain of h, and write the equation of the vertical asymptote.

5. $g(x) = \log_3 x$; $h(x) = \log_3(x + 2)$ **6.** $g(x) = \log_5 x$; $h(x) = \log_5(x - 1)$
7. $g(x) = \log_8 x$; $h(x) = 2 + \log_8 x$ **8.** $g(x) = \log_{10} x$; $h(x) = 2 \log_{10} x$

Sketch the graph of f and state its domain.

9. $f(x) = \log_{10} x$ **10.** $f(x) = -\log_{10} x$ **11.** $f(x) = |\log_{10} x|$
12. $f(x) = \log_{10}(-x)$ **13.** $f(x) = \log_{10}|x|$ **14.** $f(x) = \log_{1/10}(x + 1)$

Convert from the exponential to the logarithmic form.

15. $2^8 = 256$ **16.** $5^{-3} = \frac{1}{125}$ **17.** $(\tfrac{1}{3})^{-1} = 3$
18. $81^{3/4} = 27$ **19.** $17^0 = 1$ **20.** $(\tfrac{1}{49})^{-1/2} = 7$

Convert from the logarithmic form to the exponential form.

21. $\log_{10} 0.0001 = -4$ **22.** $\log_{64} 4 = \frac{1}{3}$ **23.** $\log_{\sqrt{2}} 2 = 2$
24. $\log_{13} 13 = 1$ **25.** $\log_{12} \frac{1}{1728} = -3$ **26.** $\log_{27/8} \frac{9}{4} = \frac{2}{3}$

Solve for the indicated quantity: y, x, or b.

27. $\log_2 16 = y$ **28.** $\log_{1/2} 32 = y$ **29.** $\log_{1/3} 27 = y$ **30.** $\log_7 x = -2$
31. $\log_{1/6} x = 3$ **32.** $\log_8 x = -\frac{2}{3}$ **33.** $\log_b 125 = 3$ **34.** $\log_b 8 = \frac{3}{2}$
35. $\log_b \frac{1}{8} = -\frac{3}{2}$ **36.** $\log_{100} 10 = y$ **37.** $\log_{27} 3 = y$ **38.** $\log_{1/16} x = \frac{1}{4}$
39. $\log_b \frac{16}{81} = 4$ **40.** $\log_8 x = -3$ **41.** $\log_b \frac{1}{27} = -\frac{3}{2}$ **42.** $\log_{\sqrt{3}} x = 2$
43. $\log_{\sqrt{8}}(\tfrac{1}{8}) = y$ **44.** $\log_b \frac{1}{128} = -7$ **45.** $\log_{0.001} 10 = y$ **46.** $\log_{0.2} 5 = y$
47. $\log_9 x = 1$

> ## PROPERTIES OF $y = f(x) = \log_b x$ for $b > 0$ and $b \neq 1$
>
> 1. The domain consists of all positive numbers x.
> 2. The range consists of all real numbers y.
> 3. The function increases (the curve is rising) for $b > 1$, and it decreases (the curve is falling) for $0 < b < 1$.
> 4. The curve is concave down for $b > 1$, and it is concave up for $0 < b < 1$.
> 5. It is a one-to-one function; if $\log_b(x_1) = \log_b(x_2)$, then $x_1 = x_2$.
> 6. The point $(1, 0)$ is on the graph. There is no y-intercept.
> 7. The y-axis is a vertical asymptote to the curve in the downward direction for $b > 1$ and in the upward direction for $0 < b < 1$.
> 8. $\log_b(b^x) = x$ and $b^{\log_b x} = x$.

EXAMPLE 2 Find the domain for $y = \log_2(x - 3)$.

Solution In $y = \log_2(x - 3)$ the quantity $x - 3$ plays the role that x does in $\log_2 x$. Thus $x - 3 > 0$ and the domain consists of all $x > 3$. ∎

As noted earlier, the forms $\log_b x = y$ and $b^y = x$ are equivalent. The following table lists specific cases illustrating this equivalence.

Logarithmic form $\log_b x = y$	Exponential form $b^y = x$
$\log_5 25 = 2$	$5^2 = 25$
$\log_{27} 9 = \frac{2}{3}$	$27^{2/3} = 9$
$\log_6 \frac{1}{36} = -2$	$6^{-2} = \frac{1}{36}$

CAUTION
*Do not confuse $x = b^y$ with its **inverse** $y = b^x$. These two forms are **not** equivalent.*

Of the two forms $y = \log_b x$ and $x = b^y$, the exponential form is usually easier to work with. Consequently, when there is a question concerning $y = \log_b x$ it is often useful to convert to the exponential form. For instance, here are two logarithmic results that are easily verified using the exponential form:

$$\log_b 1 = 0 \quad \text{since } b^0 = 1$$
$$\log_b b = 1 \quad \text{since } b^1 = b$$

The next example shows how to solve for the variables x, y or b in the form $\log_b x = y$.

EXAMPLE 3 Solve for the indicated variable:
(a) $y = \log_9 27$ **(b)** $\log_b 8 = \frac{3}{4}$ **(c)** $\log_{49} x = -\frac{1}{2}$

Solution

(a) Convert $y = \log_9 27$ to exponential form as shown on the following page.

$$9^y = 27$$
$$(3^2)^y = 3^3 \qquad \text{(Express each side as a power of 3.)}$$
$$3^{2y} = 3^3$$

Check: $\log_9 27 = \frac{3}{2}$ is correct
since $9^{3/2} = 27$.

$$3^{2y} = 3^3$$
$$2y = 3 \qquad \text{(by the one-to-one property)}$$
$$y = \tfrac{3}{2}$$

(b) Convert $\log_b 8 = \frac{3}{4}$ to exponential form.

$$b^{3/4} = 8$$

Check: $\log_{16} 8 = \frac{3}{4}$ since
$16^{3/4} = (\sqrt[4]{16})^3 = 8.$

$$(b^{3/4})^{4/3} = 8^{4/3} \qquad \text{(Raise both sides to the } \tfrac{4}{3} \text{ power.)}$$
$$b = 16 \qquad 8^{4/3} = (\sqrt[3]{8})^4 = 2^4$$

(c) Convert $\log_{49} x = -\frac{1}{2}$ to exponential form.

Check this solution.

$$x = 49^{-1/2} = \frac{1}{\sqrt{49}} = \frac{1}{7}$$

EXERCISES 6.2

Sketch the graph of the function f. Reflect this curve through the line $y = x$ to obtain the graph of the inverse function g, and write the equation for g.

1. $y = f(x) = 4^x$ **2.** $y = f(x) = 5^x$ **3.** $y = f(x) = (\tfrac{1}{3})^x$ **4.** $y = f(x) = (0.2)^x$

Describe how the graph of h can be obtained from the graph of g. Find the domain of h, and write the equation of the vertical asymptote.

5. $g(x) = \log_3 x$; $h(x) = \log_3(x + 2)$ **6.** $g(x) = \log_5 x$; $h(x) = \log_5(x - 1)$
7. $g(x) = \log_8 x$; $h(x) = 2 + \log_8 x$ **8.** $g(x) = \log_{10} x$; $h(x) = 2 \log_{10} x$

Sketch the graph of f and state its domain.

9. $f(x) = \log_{10} x$ **10.** $f(x) = -\log_{10} x$ **11.** $f(x) = |\log_{10} x|$
12. $f(x) = \log_{10}(-x)$ **13.** $f(x) = \log_{10}|x|$ **14.** $f(x) = \log_{1/10}(x + 1)$

Convert from the exponential to the logarithmic form.

15. $2^8 = 256$ **16.** $5^{-3} = \frac{1}{125}$ **17.** $(\tfrac{1}{3})^{-1} = 3$
18. $81^{3/4} = 27$ **19.** $17^0 = 1$ **20.** $(\tfrac{1}{49})^{-1/2} = 7$

Convert from the logarithmic form to the exponential form.

21. $\log_{10} 0.0001 = -4$ **22.** $\log_{64} 4 = \frac{1}{3}$ **23.** $\log_{\sqrt{2}} 2 = 2$
24. $\log_{13} 13 = 1$ **25.** $\log_{12} \frac{1}{1728} = -3$ **26.** $\log_{27/8} \frac{9}{4} = \frac{2}{3}$

Solve for the indicated quantity: y, x, or b.

27. $\log_2 16 = y$ **28.** $\log_{1/2} 32 = y$ **29.** $\log_{1/3} 27 = y$ **30.** $\log_7 x = -2$
31. $\log_{1/6} x = 3$ **32.** $\log_8 x = -\frac{2}{3}$ **33.** $\log_b 125 = 3$ **34.** $\log_b 8 = \frac{3}{2}$
35. $\log_b \frac{1}{8} = -\frac{3}{2}$ **36.** $\log_{100} 10 = y$ **37.** $\log_{27} 3 = y$ **38.** $\log_{1/16} x = \frac{1}{4}$
39. $\log_b \frac{16}{81} = 4$ **40.** $\log_8 x = -3$ **41.** $\log_b \frac{1}{27} = -\frac{3}{2}$ **42.** $\log_{\sqrt{3}} x = 2$
43. $\log_{\sqrt{8}}(\tfrac{1}{8}) = y$ **44.** $\log_b \frac{1}{128} = -7$ **45.** $\log_{0.001} 10 = y$ **46.** $\log_{0.2} 5 = y$
47. $\log_9 x = 1$

Evaluate each expression.

48. $\log_2(\log_4 256)$ **49.** $\log_{3/4}\left(\log_{1/27}\dfrac{1}{81}\right)$

By interchanging the roles of the variables, find the inverse function g.
Show that $(f \circ g)(x) = x$ and $(g \circ f)(x) = x$.

50. $y = f(x) = 2^{x+1}$ **51.** $y = f(x) = \log_3(x + 3)$

6.3
THE LAWS OF LOGARITHMS

$\log_2 8 = 3; \ (2^3 = 8)$
$\log_2 16 = 4; \ (2^4 = 16)$
$\log_2 128 = 7; \ (2^7 = 128)$

From our knowledge of logarithms we have

$$\log_2 8 + \log_2 16 = 3 + 4$$
$$= 7 = \log_2 128 = \log_2 8 \cdot 16$$

Thus

$$\log_2 8 \cdot 16 = \log_2 8 + \log_2 16$$

It turns out that this equation is a special case of the first law of logarithms.

Law 1 says that the log of a product is the sum of the logs of the factors. Can you give similar interpretations for Laws 2 and 3?

LAWS OF LOGARITHMS

If M and N are positive, $b > 0$, and $b \neq 1$, then

LAW 1. $\log_b MN = \log_b M + \log_b N$

LAW 2. $\log_b \dfrac{M}{N} = \log_b M - \log_b N$

LAW 3. $\log_b(N^k) = k \log_b N$

Since logarithms are exponents, it is not surprising that these laws can be proved by using the appropriate rules of exponents. Following is a proof of Law 1; the proofs of Laws 2 and 3 are left as exercises.

Let

$$\log_b M = r \quad \text{and} \quad \log_b N = s$$

Recall: if $\log_b x = y$ then $b^y = x$.

Convert to exponential form:

$$M = b^r \quad \text{and} \quad N = b^s$$

Multiply the two equations:

$$MN = b^r b^s = b^{r+s}$$

Then convert to logarithmic form:

$$\log_b MN = r + s$$

Substitute for r and s to get the final result:

$$\log_b MN = \log_b M = \log_b N$$

EXAMPLE 1   For positive numbers A, B, and C, show that

$$\log_b \frac{AB^2}{C} = \log_b A + 2 \log_b B - \log_b C$$

Solution

$$\log_b \frac{Ab^2}{C} = \log_b(AB^2) - \log_b C \qquad \text{(Law 2)}$$

$$= \log_b A + \log_b B^2 - \log_b C \qquad \text{(Law 1)}$$

$$= \log_b A + 2 \log_b B - \log_b C \qquad \text{(Law 3)} \qquad \blacksquare$$

EXAMPLE 2   Express $\frac{1}{2} \log_b x - 3 \log_b(x - 1)$ as the logarithm of a single expression in x.

Identify the laws of logarithms being used in Examples 2 and 3.

Solution

$$\frac{1}{2} \log_b x - 3 \log_b(x - 1) = \log_b x^{1/2} - \log_b(x - 1)^3$$

$$= \log_b \frac{x^{1/2}}{(x - 1)^3}$$

$$= \log_b \frac{\sqrt{x}}{(x - 1)^3} \qquad \blacksquare$$

EXAMPLE 3   Given: $\log_b 2 = 0.6931$ and $\log_b 3 = 1.0986$; find $\log_b \sqrt{12}$.

Solution

This example indicates that for some base number b, $b^{0.6931} = 2$, $b^{1.0986} = 3$, and $b^{1.2424} = \sqrt{12}$.

$$\log_b \sqrt{12} = \log_b 12^{1/2} = \tfrac{1}{2} \log_b 12$$

$$= \tfrac{1}{2} \log_b(3 \cdot 4) = \tfrac{1}{2} [\log_b 3 + \log_b 4]$$

$$= \tfrac{1}{2} [\log_b 3 + \log_b 2^2]$$

$$= \tfrac{1}{2} [\log_b 3 + 2 \log_b 2]$$

$$= \tfrac{1}{2} \log_b 3 + \log_b 2$$

$$= \tfrac{1}{2}(1.0986) + 0.6931$$

$$= 1.2424 \qquad \blacksquare$$

Convert the given logarithms into expressions involving $\log_b A$, $\log_b B$, and $\log_b C$.

1. $\log_b ABC$ **2.** $\log_b \dfrac{A}{BC}$ **3.** $\log_b \dfrac{(AB)^2}{C}$

4. $\log_b AB^2C^3$ **5.** $\log_b \dfrac{A\sqrt{B}}{C}$ **6.** $\log_b \dfrac{\sqrt[3]{A}}{(BC)^3}$

Change each expression into the logarithm of a single expression in x.

7. $\log_b x + \log_b x + \log_b 3$ **8.** $2\log_b(x-1) + \frac{1}{2}\log_b x$
9. $\log_b(2x-1) - 3\log_b(x^2+1)$
10. $\log_b x - \log_b(x-1) - 2\log_b(x-2)$

Use the information given in Example 3 to find these logarithms.

11. $\log_b 18$ **12.** $\log_b \dfrac{16}{27}$

---

*Examples 4 to 6 illustrate how the laws of logarithms can be used to solve **logarithmic equations**.*

CAUTION
$\log_8(x^2 - 36) \neq$
$\log_8 x^2 - \log_8 36$

EXAMPLE 4   Solve for x: $\log_8(x-6) + \log_8(x+6) = 2$

Solution   First note that in $\log_8(x-6)$ we must have $x - 6 > 0$, or $x > 6$. Similarly, $\log_8(x+6)$ calls for $x > -6$. Therefore, the only solutions, if there are any, must satisfy $x > 6$.

$$\log_8(x-6) + \log_8(x+6) = 2$$
$$\log_8(x-6)(x+6) = 2 \qquad \text{(Law 1)}$$
$$\log_8(x^2 - 36) = 2$$
$$x^2 - 36 = 8^2 \qquad \text{(converting to exponential form)}$$
$$x^2 - 100 = 0$$
$$(x+10)(x-10) = 0$$
$$x = -10 \quad \text{or} \quad x = 10$$

The only possible solutions are -10 and 10. Our initial observation that $x > 6$ automatically eliminates -10. (If that initial observation had not been made, -10 could still have been eliminated by checking in the given equation.) The value $x = 10$ can be checked as follows:

$$\log_8(10-6) + \log_8(10+6) = \log_8 4 + \log_8 16$$
$$= \tfrac{2}{3} + \tfrac{4}{3} = 2 \qquad \blacksquare$$

EXAMPLE 5   Solve for x: $\log_{10}(x^3 - 1) - \log_{10}(x^2 + x + 1) = 1$

Solution $\log_{10}(x^3 - 1) - \log_{10}(x^2 + x + 1) = 1$

$$\log_{10} \frac{x^3 - 1}{x^2 + x + 1} = 1 \qquad \text{(Law 2)}$$

See page 49 for the factorization of the difference of two cubes; $a^3 - b^3$.

Then factor as follows:

$$\log_{10} \frac{(x - 1)(x^2 + x + 1)}{x^2 + x + 1} = 1 \qquad \text{(by factoring)}$$

$$\log_{10}(x - 1) = 1$$

$$x - 1 = 10^1 \qquad \text{(Why?)}$$

$$x = 11$$

$$\text{\textit{Check}: } \log_{10}(11^3 - 1) - \log_{10}(11^2 + 11 + 1) = \log_{10} 1330 - \log_{10} 133$$

$$= \log_{10} \tfrac{1330}{133}$$

$$= \log_{10} 10 = 1 \qquad \blacksquare$$

Sometimes it is convenient to solve a logarithmic equation using the one-to-one property of logarithmic functions. This property (stated on page 303) says:

$$\text{If } \log_b M = \log_b N, \text{ then } M = N.$$

Without the one-to-one property the solution begins with

$$\log_3 \frac{2x}{x + 5} = 0$$

$$\frac{2x}{x + 5} = 3^0 = 1$$

EXAMPLE 6   Solve for x: $\log_3 2x - \log_3 (x + 5) = 0$

Solution

$$\log_3 2x - \log_3(x + 5) = 0$$

$$\log_3 2x = \log_3(x + 5)$$

$$2x = x + 5 \qquad \text{(by the one-to-one property)}$$

$$x = 5$$

$$\text{\textit{Check}: } \log_3 2(5) - \log_3(5 + 5) = \log_3 10 - \log_3 10 = 0 \qquad \blacksquare$$

CAUTION: Learn to Avoid These Mistakes	
WRONG	RIGHT
$\log_b A + \log_b B = \log_b(A + B)$	$\log_b A + \log_b B = \log_b AB$
$\log_b(x^2 - 4) = \log_b x^2 - \log_b 4$	$\log_b(x^2 - 4)$ $= \log_b(x + 2)(x - 2)$ $= \log_b(x + 2) + \log_b(x - 2)$
$(\log_b x)^2 = 2 \log_b x$	$(\log_b x)^2 = (\log_b x)(\log_b x)$
$\log_b A - \log_b B = \dfrac{\log_b A}{\log_b B}$	$\log_b A - \log_b B = \log_b \dfrac{A}{B}$
If $2 \log_b x = \log_b(3x + 4)$, then $2x = 3x + 4$.	If $2 \log_b x = \log_b(3x + 4)$, then $\log_b x^2 = \log_b(3x + 4)$ and $x^2 = 3x + 4$.
$\log_b \dfrac{x}{2} = \dfrac{\log_b x}{2}$	$\log_b \dfrac{x}{2} = \log_b x - \log_b 2$
$\log_b(x^2 + 2) = 2 \log_b(x + 2)$	$\log_b(x^2 + 2)$ cannot be simplified further.

EXERCISES 6.3

Use the laws of logarithms (as much as possible) to convert the given logarithms into expressions involving sums, differences, and multiples of logarithms.

1. $\log_b \dfrac{3x}{x+1}$ **2.** $\log_b \dfrac{x^2}{x-1}$ **3.** $\log_b \dfrac{\sqrt{x^2-1}}{x}$ **4.** $\log_b \dfrac{1}{x}$ **5.** $\log_b \dfrac{1}{x^2}$ **6.** $\log_b \sqrt{\dfrac{x+1}{x-1}}$

Convert each expression into logarithm of a single expression in x.

7. $\log_b(x+1) - \log_b(x+2)$ **8.** $\log_b x + 2\log_b(x-1)$ **9.** $\frac{1}{2}\log_b(x^2-1) - \frac{1}{2}\log_b(x^2+1)$

10. $\log_b(x+2) - \log_b(x^2-4)$ **11.** $3\log_b x - \log_b 2 - \log_b(x+5)$ **12.** $\frac{1}{3}\log_b(x-1) + \log_b 3 - \frac{1}{3}\log_b(x+1)$

Use the appropriate laws of logarithms to explain why each statement is correct.

13. $\log_b 27 + \log_b 3 = \log_b 243 - \log_b 3$ **14.** $\log_b 16 + \log_b 4 = \log_b 64$

15. $-2\log_b \frac{4}{9} = \log_b \frac{81}{16}$ **16.** $\frac{1}{2}\log_b 0.0001 = -\log_b 100$

Find the logarithms by using the laws of logarithms and the given information that $\log_b 2 = 0.3010$, $\log_b 3 = 0.4771$, *and* $\log_b 5 = 0.6990$. *Assume that all logs have base b.*

17. (a) $\log 4$ (b) $\log 8$ (c) $\log \frac{1}{2}$ **18.** (a) $\log \sqrt{2}$ (b) $\log 9$ (c) $\log 12$

19. (a) $\log 48$ (b) $\log \frac{2}{3}$ (c) $\log 125$ **20.** (a) $\log 50$ (c) $\log 10$ (c) $\log \frac{25}{6}$

21. (a) $\log \sqrt[3]{5}$ (b) $\log \sqrt{20^3}$ (c) $\log \sqrt{900}$ **22.** (a) $\log 0.2$ (b) $\log 0.25$ (c) $\log 2.4$

Solve for x and check.

23. $\log_{10} x + \log_{10} 5 = 2$ **24.** $\log_{10} x + \log_{10} 5 = 1$

25. $\log_{10} 5 - \log_{10} x = 2$ **26.** $\log_{10}(x+21) + \log_{10} x = 2$

27. $\log_{12}(x-5) + \log_{12}(x-5) = 2$ **28.** $\log_3 x + \log_3(2x+51) = 4$

29. $\log_{16} x + \log_{16}(x-4) = \frac{5}{4}$ **30.** $\log_2(x^2) - \log_2(x-2) = 3$

31. $\log_{10}(3-x) - \log_{10}(12-x) = -1$ **32.** $\log_{10}(3x^2-5x-2) - \log_{10}(x-2) = 1$

33. $\log_{1/7} x + \log_{1/7}(5x-28) = -2$ **34.** $\log_{1/3} 12x^2 - \log_{1/3}(20x-9) = -1$

35. $\log_{10}(x^3-1) - \log_{10}(x^2+x+1) = -2$ **36.** $2\log_{10}(x-2) = 4$

37. $2\log_{25} x - \log_{25}(25-4x) = \frac{1}{2}$ **38.** $\log_3(8x^3+1) - \log_3(4x^2-2x+1) = 2$

39. Prove Law 2. $\left(Hint: \text{Follow the proof of Law 1 using } \dfrac{b^r}{b^s} = b^{r-s}.\right)$

40. Prove Law 3. (*Hint*: Use $(b^r)^k = b^{rk}$.)   **41.** Solve for x: $(x+2)\log_b b^x = x$.

42. Solve for x: $\log_{N^2} N = x$.   **43.** Solve for x: $\log_x(2x)^{3x} = 4x$.

44. (a) Explain why $\log_b b = 1$.
   (b) Show that $(\log_b a)(\log_a b) = 1$.
   (*Hint*: Use Law 3 and the result $b^{\log_b x} = x$.)

45. Use $B^{\log_B N} = N$ to derive $\log_B N = \dfrac{\log_b N}{\log_b B}$. (*Hint*: Begin by taking the log base b of both sides.)

 Written Assignment: State Laws 2 and 3 of logarithms in your own words, that is, without symbols.

CHALLENGE
Think Creatively

Prove that $\log_9 16 = \log_3 4$.
Do not use log tables or a calculator.

1. In order to make the graphing of $y = (\sqrt{5})^{2x}$ easier, what should first be done with the form $(\sqrt{5})^{2x}$?

2. Consider the inequalities $2^x < 3^x$, $2^x > 3^x$ and the equation $2^x = 3^x$. Decide where each of them is true and where they are false.

3. A simplified form of $\sqrt{3^{2x} + 2 + 3^{-2x}}$ that does not involve radicals or fractional exponents can be obtained after factoring the radicand. Find this simplification.

4. For the function $f(x) = \log_2 x$, form the composition $f \circ f \circ f$ and determine x for which $f(f(f(x))) = 2$. Also, generalize the preceding by replacing the base 2 by b and solving for x.

5. Make use of your knowledge of logarithmic functions to determine the values of c and d, in $y = f(x) = \log_b(cx + d)$, so that the graph of f has the vertical asymptote $x = -\frac{1}{2}$ and x-intercept 0.

6. The function $f(x) = \log_{10} x$ is increasing on the interval $(0, \infty)$. It is also true that the value of $\log_{10} x$ becomes arbitrarily large with increasing values of x. Find the values of x for which each of the following inequalities is true:

$$\log_{10} x > 3, \quad \log_{10} x > 30, \quad \log_{10} x > 300.$$

As x increases, would you describe the growth of $\log_{10} x$ as fast, moderate, or slow?

**6.4
THE BASE e**

The most important base number for exponential and logarithmic functions is the number denoted by the letter e. This is an irrational number approximately equal to 2.72. Here is the number e showing its first 15 decimal places.

$$e = 2.718281828459045 \ldots$$

Like the irrational number π, e plays an important role in mathematics and its applications. One way to approximate e is to evaluate the expression $\left(1 + \dfrac{1}{n}\right)^n$ for larger and larger values of n. Some of these calculations are shown in the margin. Since $2 < e < 3$, the graph of $y = e^x$ will be between the graphs of $y = 2^x$ and $y = 3^x$ as shown below.

The larger n is taken, the closer $\left(1 + \dfrac{1}{n}\right)^n$ gets to e. For example:

$(1 + \frac{1}{10})^{10} = 2.59374$

$(1 + \frac{1}{100})^{100} = 2.70481$

$(1 + \frac{1}{1000})^{1000} = 2.71692$

(See Exercise 57.)

In the study of calculus, the concept of a tangent to a circle is extended to other curves. There you will learn that the tangent to the curve $y = e^x$ at point $(0, 1)$ has slope equal to 1. If you carefully draw the line that touches this curve at $(0, 1)$, and at no other point, you should find its slope to be close 1.

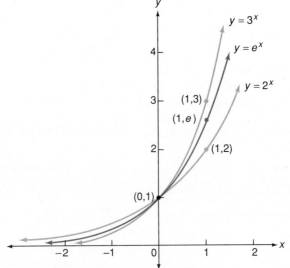

The inverse of $y = e^x$ is given by $y = \log_e x$. In place of $\log_e x$ we will now write **ln x**, which is called the **natural log of x**. Thus $x = e^y$ and $y = \ln x$ are equivalent. As before, the graph of $y = \ln x$ can be obtained by reflecting the graph of $y = e^x$ through the line $y = x$.

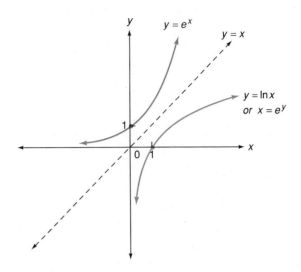

Since $e > 1$, the properties of $y = b^x$ and $y = \log_b x(b > 1)$ carry over to $y = e^x$ and $y = \ln x$. We collect these properties next for easy reference.

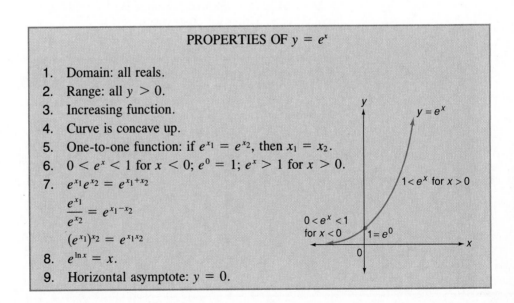

PROPERTIES OF $y = e^x$

1. Domain: all reals.
2. Range: all $y > 0$.
3. Increasing function.
4. Curve is concave up.
5. One-to-one function: if $e^{x_1} = e^{x_2}$, then $x_1 = x_2$.
6. $0 < e^x < 1$ for $x < 0$; $e^0 = 1$; $e^x > 1$ for $x > 0$.
7. $e^{x_1}e^{x_2} = e^{x_1+x_2}$

    $\dfrac{e^{x_1}}{e^{x_2}} = e^{x_1-x_2}$

    $(e^{x_1})^{x_2} = e^{x_1 x_2}$
8. $e^{\ln x} = x$.
9. Horizontal asymptote: $y = 0$.

In this list, as well as in the following one, property 8 is a direct consequence of the fact that $f(x) = e^x$ and $g(x) = \ln x$ are inverse functions. Thus,

$$f(g(x)) = f(\ln x) = e^{\ln x} = x$$

and

$$g(f(x)) = g(e^x) = \ln e^x = x$$

Also, see the general case on page 298.

PROPERTIES OF $y = \ln x$

1. Domain: all $x > 0$.
2. Range: all reals.
3. Increasing function.
4. Curve is concave down.
5. One-to-one function;
   if $\ln x_1 = \ln x_2$, then $x_1 = x_2$.
6. $\ln x < 0$ for $0 < x < 1$;
   $\ln 1 = 0$; $\ln x > 0$ for $x > 1$.
7. $\ln x_1 x_2 = \ln x_1 + \ln x_2$

   $\ln \dfrac{x_1}{x_2} = \ln x_1 - \ln x_2$

   $\ln x_1{}^{x_2} = x_2 \ln x_1$

8. $\ln e^x = x$.
9. Vertical asymptote; $x = 0$.

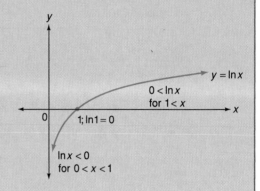

The examples that follow utilize the base e and are solved in a manner similar to those done earlier for other bases.

EXAMPLE 1 (a) Find the domain of $y = \ln(x + 2)$.
(b) Sketch $y = \ln x^2$ for $x > 0$.

Solution

(a) Since the domain of $y = \ln x$ is all $x > 0$, the domain of $y = \ln(x + 2)$ consists of all x for which $x + 2 > 0$; all $x > -2$.

(b) Since $y = \ln x^2 = 2 \ln x$, we obtain the graph by multiplying the ordinates of $y = \ln x$, by 2.

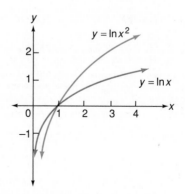

EXAMPLE 2 Let $f(x) = \dfrac{3x^2}{x^2 + 4}$. Use the laws of logarithms to write $\ln f(x)$ as an expression involving sums, differences, and multiples of natural logarithms.

Solution Since $f(x) = \dfrac{3x^2}{x^2 + 4}$, we can proceed as follows.

Whenever $M = N$ it follows from the definition of a function that $\ln M = \ln N$. That is, for equal domain values M and N, there can be only one range value.

$$\ln f(x) = \ln \frac{3x^2}{x^2 + 4}$$

$$= \ln 3x^2 - \ln(x^2 + 4) \qquad \text{(by Law 2 of logarithms)}$$

$$= \ln 3 + \ln x^2 - \ln(x^2 + 4) \qquad \text{(by Law 1 of logarithms)}$$

$$= \ln 3 + 2 \ln x - \ln(x^2 + 4) \qquad \text{(by Law 3 of logarithms)} \qquad \blacksquare$$

EXAMPLE 3 Solve for t: $e^{\ln(2t-1)} = 5$

Solution

$$e^{\ln(2t-1)} = 5$$

$$2t - 1 = 5 \qquad \text{(Using } e^{\ln x} = x \text{ with } x = 2t - 1)$$

$$2t = 6$$

$$t = 3 \qquad\qquad\qquad\qquad\qquad \blacksquare$$

EXAMPLE 4 Solve for t: $e^{2t-1} = 5$

Solution Rewrite the exponential expression in logarithmic form.

Using a calculator, to three decimal places
$t = \frac{1}{2}(1 + \ln 5) = 1.305.$

$$e^{2t-1} = 5$$

$$2t - 1 = \ln 5 \qquad (\log_e 5 = \ln 5 = 2t - 1)$$

$$2t = 1 + \ln 5$$

$$t = \tfrac{1}{2}(1 + \ln 5)$$

Check: $e^{2[1/2(1+\ln 5)]-1} = e^{1+\ln 5 - 1} = e^{\ln 5} = 5 \qquad \blacksquare$

EXAMPLE 5 Solve for x: $\ln(x + 1) = 1 + \ln x$

Solution

$$\ln(x + 1) - \ln x = 1$$

$$\ln \frac{x + 1}{x} = 1$$

Now convert to exponential form:

Recall: If $\log_b x = y$, then $b^y = x$.

$\ln \dfrac{x + 1}{x} = \log_e \dfrac{x + 1}{x} = 1$

So, $e^1 = \dfrac{x + 1}{x}$.

$$\frac{x + 1}{x} = e$$

$$ex = x + 1$$

$$(e - 1)x = 1$$

Try to check this result before turning to the next page.

$$x = \frac{1}{e - 1}$$

$$\text{Check: } \ln\left(\frac{1}{e-1} + 1\right) = \ln\frac{e}{e-1} = \ln e - \ln(e-1)$$

$$= 1 + \ln(e-1)^{-1} = 1 + \ln\frac{1}{e-1} \qquad \blacksquare$$

EXAMPLE 6 **(a)** Show $h(x) = \ln(x^2 + 5)$ as the composite of two functions.
(b) Show $F(x) = e^{\sqrt{x^2-3x}}$ as the composite of three functions.

Solution
(a) Let $f(x) = \ln x$ and $g(x) = x^2 + 5$. Then

$$(f \circ g)(x) = f(g(x)) = f(x^2 + 5) = \ln(x^2 + 5) = h(x)$$

(b) Let $f(x) = e^x$, $g(x) = \sqrt{x}$, and $h(x) = x^2 - 3x$. Then

$$
\begin{aligned}
(f \circ g \circ h)(x) &= f(g(h(x))) &&\text{h is the "inner" function.} \\
&= f(g(x^2 - 3x)) &&\text{g is the "middle" function.} \\
&= f(\sqrt{x^2 - 3x}) &&\text{f is the "outer" function.} \\
&= e^{\sqrt{x^2-3x}} = F(x)
\end{aligned}
$$

(Other solutions are possible.) $\qquad \blacksquare$

EXAMPLE 7 Determine the signs of $f(x) = x^2 e^x + 2xe^x$.

Solution We find that $f(x) = x^2 e^x + 2xe^x = xe^x(x + 2)$ in which $e^x > 0$ for all x, and the other factors are zero when $x = 0$ or $x = -2$.

Use specific test values in each interval to determine the sign of $f(x)$ for that interval. For example, let $x = -1$ in the interval $(-2, 0)$. Also, see page 214.

Interval	$(-\infty, -2)$	$(-2, 0)$	$(0, \infty)$
Sign of $x + 2$	−	+	+
Sign of x	−	−	+
Sign of e^x	+	+	+
Sign of $f(x)$	+	−	+

$f(x) > 0$ on the intervals $(-\infty, -2)$ and $(0, \infty)$

$f(x) < 0$ on the interval $(-2, 0)$ $\qquad \blacksquare$

EXERCISES 6.4

Sketch each pair of functions on the same axes.

1. $y = e^x$; $y = e^{x-2}$
2. $y = e^x$; $y = 2e^x$
3. $y = \ln x$; $y = \frac{1}{2} \ln x$
4. $y = \ln x$; $y = \ln(x + 2)$
5. $y = \ln x$; $y = \ln(-x)$
6. $y = e^x$; $y = e^{-x}$
7. $y = e^x$; $y = e^x + 2$
8. $y = \ln x$; $y = \ln|x|$
9. $f(x) = -e^{-x}$; $g(x) = 1 - e^{-x}$
10. $g(x) = 1 - e^{-x}$; $s(x) = 1 - e^{-2x}$
11. $g(x) = 1 - e^{-x}$; $t(x) = 1 - e^{(-1/2)x}$
12. $u(x) = 1 - e^{-3x}$; $v(x) = 1 - e^{(-1/3)x}$

Explain how the graph of f can be obtained from the curve $y = \ln x$. *(Hint: First apply the appropriate laws of logarithms.)*

13. $f(x) = \ln ex$ **14.** $f(x) = \ln \dfrac{x}{e}$ **15.** $f(x) = \ln \sqrt{x}$ **16.** $f(x) = \ln \dfrac{1}{x}$

17. $f(x) = \ln(x^2 - 1) - \ln(x + 1)$ **18.** $f(x) = \ln x^{-3}$

Find the domain.

19. $f(x) = \ln(x + 2)$ **20.** $f(x) = \ln|x|$ **21.** $f(x) = \ln(2x - 1)$

22. $f(x) = \dfrac{1}{\ln x}$ **23.** $f(x) = \dfrac{\ln(x - 1)}{x - 2}$ **24.** $f(x) = \ln(\ln x)$

Use the laws of logarithms (as much as possible) to write $\ln f(x)$ as an expression involving sums, differences, and multiples of natural logarithms.

25. $f(x) = \dfrac{5x}{x^2 - 4}$ **26.** $f(x) = x\sqrt{x^2 + 1}$ **27.** $f(x) = \dfrac{(x - 1)(x + 3)^2}{\sqrt{x^2 + 2}}$

28. $f(x) = \sqrt{\dfrac{x + 7}{x - 7}}$ **29.** $f(x) = \sqrt{x^3(x + 1)}$ **30.** $f(x) = \dfrac{x}{\sqrt[3]{x^2 - 1}}$

Convert each expression into the logarithm of a single expression.

31. $\frac{1}{2}\ln x + \ln(x^2 + 5)$ **32.** $\ln 2 + \ln x - \ln(x - 1)$

33. $3 \ln(x + 1) + 3 \ln(x - 1)$ **34.** $\ln(x^3 - 1) - \ln(x^2 + x + 1)$

35. $\frac{1}{2}\ln x - 2 \ln(x - 1) - \frac{1}{3}\ln(x^2 + 1)$

Simplify.

36. $\ln(e^{3x})$ **37.** $e^{\ln \sqrt{x}}$ **38.** $\ln(x^2 e^3)$ **39.** $e^{-2\ln x}$ **40.** $(e^{\ln x})^2$ **41.** $\ln\left(\dfrac{e^x}{e^{x-1}}\right)$

Solve for x.

42. $e^{3x+5} = 100$ **43.** $e^{-0.01x} = 27$ **44.** $e^{x^2} = e^x e^{3/4}$

45. $e^{\ln(1-x)} = 2x$ **46.** $\ln x + \ln 2 = 1$ **47.** $\ln(x + 1) = 0$

48. $\ln x = -2$ **49.** $\ln e^{\sqrt{x+1}} = 3$ **50.** $e^{\ln(6x^2-4)} = 5x$

51. $\ln(x^2 - 4) - \ln(x + 2) = 0$ **52.** $(e^{x+2} - 1)\ln(1 - 2x) = 0$ **53.** $\ln x = \frac{1}{2}\ln 4 + \frac{2}{3}\ln 8$

54. $\frac{1}{2}\ln(x + 4) = \ln(x + 2)$ **55.** $\ln x = 2 + \ln(1 - x)$ **56.** $\ln(x^2 + x - 2) = \ln x + \ln(x - 1)$

 57. Use a calculator to complete the table. Round off the entries of five decimal places.

N	2	10	100	500	1000	5000	10,000	100,000	1,000,000
$\left(1 + \dfrac{1}{n}\right)^n$									

 Show that each function is the composite of two functions.

58. $h(x) = e^{2x+3}$ **59.** $h(x) = e^{-x^2+x}$ **60.** $h(x) = \ln(1 - 2x)$

61. $h(x) = \ln \dfrac{x}{x + 1}$ **62.** $h(x) = (e^x + e^{-x})^2$ **63.** $h(x) = \sqrt[3]{\ln x}$

 Show that each function is the composite of three functions.

64. $F(x) = e^{\sqrt{x+1}}$ **65.** $F(x) = e^{(3x-1)^2}$ **66.** $F(x) = [\ln(x^2 + 1)]^3$ **67.** $F(x) = \ln \sqrt{e^x + 1}$

△ *Determine the signs of each function.*

68. $f(x) = xe^x + e^x$ **69.** $f(x) = e^{2x} - 2xe^{2x}$ **70.** $f(x) = -3x^2e^{-3x} + 2xe^{-3x}$

71. $f(x) = 1 + \ln x$ **72.** Show that $(e^x + e^{-x})^2 - (e^x - e^{-x})^2 = 4$.

△ **73.** Show that $\ln\left(\dfrac{x}{4} - \dfrac{\sqrt{x^2 - 4}}{4}\right) = -\ln(x + \sqrt{x^2 - 4})$.   **74.** Solve for x: $\dfrac{e^x + e^{-x}}{2} = 1$.

△ **75.** Solve for x in terms of y if $y = \dfrac{e^x}{2} - \dfrac{1}{2e^x}$. (*Hint:* Let $u = e^x$ and solve the resulting quadratic in u.)

CHALLENGE	Solve for x:	$\log_b(\log_b nx) = 1 \ (n > 0)$
Think Creatively		

6.5 EXPONENTIAL GROWTH AND DECAY

The number e to three decimal places is 2.718. This can be written as $e^1 = 2.718$. Other powers of e are given in Table II, page 637, rounded to various decimal places. Thus, from Table II,

$$e^{2.1} = 8.17 \quad \text{and} \quad e^{-2.1} = 0.122$$

In Table III, page 638, natural logarithms of x are given to three decimal places. For example, $\ln 2.1 = 0.742$.

Such values can also be obtained using a calculator. If the calculator has an $\boxed{e^x}$ key, then $e^{2.1}$ can be found using this calculator sequence

$$\underset{\underset{\text{enter}}{\uparrow}}{2.1} \ \underset{\underset{\text{press}}{\uparrow}}{\boxed{e^x}} = \underset{\underset{\text{answer}}{\uparrow}}{8.166} \qquad \text{(to 3 decimal places)}$$

A large variety of calculators are available. For this work, one of the keys $\boxed{e^x}$ or $\boxed{y^x}$ and the key $\boxed{\ln x}$ are needed. (Sometimes the letter x is not included in the last key.)

For $e^{-2.1}$ use the sequence

$$\underset{\underset{\text{enter}}{\uparrow}}{2.1} \ \underset{\substack{\uparrow \\ \text{press} \\ \text{(to} \\ \text{change} \\ \text{the} \\ \text{sign)}}}{\boxed{\pm}} \ \underset{\underset{\text{press}}{\uparrow}}{\boxed{e^x}} = \underset{\underset{\text{answer}}{\uparrow}}{0.122} \qquad \text{(to 3 decimal places)}$$

If the calculator does not have an $\boxed{e^x}$ key, but it has the exponential key $\boxed{y^x}$, use the following sequence.

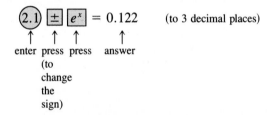

$$\underset{\substack{\uparrow \\ \text{enter} \\ \text{(e to 4} \\ \text{decimal} \\ \text{places)}}}{2.7183} \ \underset{\underset{\text{press}}{\uparrow}}{\boxed{y^x}} \ \underset{\underset{\text{enter}}{\uparrow}}{2.1} \ \underset{\underset{\text{press}}{\uparrow}}{\boxed{=}} \ \underset{\underset{\text{answer}}{\uparrow}}{8.166}$$

To find $\ln 2.1$ on a calculator, use the sequence

$$\underset{\underset{\text{enter}}{\uparrow}}{2.1} \ \underset{\underset{\text{press}}{\uparrow}}{\boxed{\ln x}} = \underset{\underset{\text{answer}}{\uparrow}}{0.742} \qquad \text{(to 3 decimal places)}$$

In the preceding calculator operations, we have circled a number to be entered and placed a rectangle around a key to be pressed. From now on, we will follow this convention without using the words "enter" and "press."

EXAMPLE 1   Solve the exponential equation $2^x = 35$ for x.

Solution

(a) (Using Table III)

$$2^x = 35$$

$$\ln 2^x = \ln 35 \qquad \text{(If } A = B, \text{ then } \ln A = \ln B.\text{)}$$

$$x \ln 2 = \ln 35 \qquad \text{(Why?)}$$

$$x = \frac{\ln 35}{\ln 2}$$

Even though ln 35 is not given in the table, the first law of logarithms in conjunction with Table III can be used to find ln 35 as follows:

$$\ln 35 = \ln(3.5)(10) = \ln 3.5 + \ln 10$$

$$= 1.253 + 2.303 \qquad \text{(Table III)}$$

$$= 3.556$$

Note that the values found in the tables are approximations. For simplicity, however, we will use the equals sign (=).

Now we have

$$x = \frac{\ln 35}{\ln 2} = \frac{3.556}{0.693} = 5.13$$

Solution

(b) (Using a calculator)

From solution (a) we have

$$x = \frac{\ln 35}{\ln 2}$$

Now use this calculator sequence:

As a rough check note that 5.13 is a reasonable answer since $2^5 = 32$.

$$x = \boxed{35}\;\boxed{\ln x}\;\boxed{\div}\;\boxed{2}\;\boxed{\ln x}\;\boxed{=}\;5.13$$

Therefore $x = 5.13$, rounded to two decimal places.   ∎

As a variation of Example 1, consider the calculator solution of $2^{3x} = 35$. First isolate x as follows:

$$3x \ln 2 = \ln 35 \qquad \text{(Apply ln and use Law 3 for logarithms.)}$$

$$x = \frac{\ln 35}{3 \ln 2}$$

Now use either of the following sequences. Note that the second sequence calls for the parentheses keys $\boxed{(}$ and $\boxed{)}$.

$$x = \boxed{35}\ \boxed{\ln x}\ \boxed{\div}\ \boxed{3}\ \boxed{\div}\ \boxed{2}\ \boxed{\ln x}\ \boxed{=}\ 1.7098$$

These two calculator solutions indicate that more than one correct procedure may be possible.

or

$$x = \boxed{35}\ \boxed{\ln x}\ \boxed{\div}\ \boxed{(}\ \boxed{3}\ \boxed{\times}\ \boxed{2}\ \boxed{\ln x}\ \boxed{)}\ \boxed{=}\ 1.7098$$

Thus $x = 1.7098$ (to 4 decimal places.)

TEST YOUR UNDERSTANDING
Think Carefully

(Answers: Page 340)

Solve each equation for x in terms of natural logarithms. Approximate the answer to three decimal places by using Table III or a calculator.

1. $4^x = 5$ 2. $4^{-x} = 5$ 3. $(\frac{1}{2})^x = 12$
4. $2^{3x} = 10$ 5. $4^x = 15$ 6. $67^{5x} = 4$

At the beginning of Section 6.1 we developed the formula $y = (10,000)2^x$, which gives the number of bacteria present after x hours of growth; 10,000 is the initial number of bacteria. How long will it take for this bacterial culture to grow to 100,000? To answer this question we let $y = 100,000$ and solve for x.

$$(10,000)2^x = 100,000$$

$$2^x = 10 \qquad \text{(Divide by 10,000.)}$$

$$x \ln 2 = \ln 10$$

$$x = \frac{\ln 10}{\ln 2}$$

$$= 3.32 \qquad \text{(by Table III or calculator)}$$

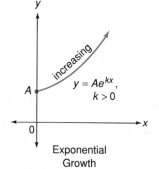

$y = Ae^{kx}$, $k > 0$

Exponential Growth

It will take about 3.3 hours.

In the preceding illustration the exponential and logarithmic functions were used to solve a problem of *exponential growth*. Many problems involving **exponential growth** or **exponential decay** can be solved by using the general formula.

$$y = f(x) = Ae^{kx}$$

which shows how the amount of a substance y depends on the time x. Since $f(0) = A$, A represents the initial amount of the substance and k is a constant. In a given situation $k > 0$ signifies that y is growing (increasing) with time. For $k < 0$ the substance is decreasing. (Compare to the graphs of $y = e^x$ and $y = e^{-x}$.)

$y = Ae^{kx}$, $k < 0$

Exponential Decay

The preceding bacterial problem also fits this general form. This can be seen by substituting $2 = e^{\ln 2}$ into $y = (10,000)2^x$:

$$y = (10,000)2^x = (10,000)(e^{\ln 2})^x = 10,000e^{(\ln 2)x}$$

EXAMPLE 2 A radioactive substance is decaying (it is changing into another element) according to the formula $y = Ae^{-0.2x}$, where y is the amount of material remaining after x years.

(a) If the initial amount $A = 80$ grams, how much is left after 3 years?

(b) The **half-life** of a radioactive substance is the time it takes for half of it to decompose. Find the half-life of this substance in which $A = 80$ grams.

Solution

(a) Since $A = 80$, $y = 80e^{-0.2x}$. We need to solve for the amount y when $x = 3$.

$$y = 80e^{-0.2x}$$
$$= 80e^{-0.2(3)}$$
$$= 80e^{-0.6} = 80(0.549) = 43.920 \qquad \text{(using Table II)}$$

The results using a calculator are usually more accurate than those obtained by using the tables. To one decimal place, however, the answers are the same in part (a).

By calculator we have the following solution:

$$y = \boxed{80} \boxed{\times} \boxed{2.7183} \boxed{y^x} \boxed{0.6} \boxed{\pm} \boxed{=} 43.90$$

There will be about 43.9 grams after 3 years.

(b) This question calls for the time x at which only half of the initial amount is left. Consequently, the half-life x is the solution to $40 = 80e^{-0.2x}$. Divide each side by 80:

$$\frac{1}{2} = e^{-0.2x}$$

Take the natural log of both sides, or change to logarithmic form, to obtain $-0.2x = \ln \frac{1}{2}$. Since $\ln \frac{1}{2} = \ln 1 - \ln 2 = -\ln 2$, we solve for x as follows.

$$-0.2x = -\ln 2$$

$$x = \frac{\ln 2}{0.2} = \frac{0.693}{0.2} = 3.465 \qquad \text{(using Table III)}$$

By calculator:

$$x = \boxed{2} \boxed{\ln x} \boxed{\div} \boxed{0.2} \boxed{=} 3.466$$

The half-life is approximately 3.47 years. ∎

This process of finding the age of the remains is referred to as radioactive carbon dating.

Carbon-14, also written as ^{14}C, is a radioactive isotope of carbon with a half-life of about 5750 years. By finding how much ^{14}C is contained in the remains of a formerly living organism, it becomes possible to determine what percentage this is of the original amount of ^{14}C at the time of death. Once this information is given, the formula $y = Ae^{kx}$ will enable us to date the age of the remains. The dating will be done after we solve for the constant k. Since the amount of ^{14}C after 5750 years will be $\frac{A}{2}$, we have the following:

A calculator gives
$k = -0.0001205$. This
verifies that the amount, y,
of carbon-14 is decreasing
(decaying).

$$\frac{A}{2} = Ae^{5750k}$$

$$\frac{1}{2} = e^{5750k}$$

$$5750k = \ln \frac{1}{2}$$

$$k = \frac{\ln 0.5}{5750}$$

Substitute this value for k into $y = Ae^{kx}$ to obtain the following formula for the amount of carbon-14 remaining after x years:

$$y = Ae^{(\ln 0.5/5750)x}$$

EXAMPLE 3   An animal skeleton is found to contain one-third of its original amount of ^{14}C. How old is the skeleton?

Solution   Let x be the age of the skeleton. Then

$$\frac{1}{3}A = Ae^{(\ln 0.5/5750)x}$$

$$\frac{1}{3} = e^{(\ln 0.5/5750)x}$$

$$\left(\frac{\ln 0.5}{5750}\right)x = \ln \frac{1}{3} = -\ln 3$$

$$x = \frac{(5750)(-\ln 3)}{\ln 0.5}$$

$$= \frac{-5750 \ln 3}{\ln 0.5} = -\frac{5750(1.099)}{-0.693} = 9118.7 \qquad \text{(using Table III)}$$

By calculator we have the following solution:

$$x = \boxed{5750} \; \boxed{\pm} \; \boxed{\times} \; \boxed{3} \; \boxed{\ln x} \; \boxed{\div} \; \boxed{0.5} \; \boxed{\ln x} \; \boxed{=} \; 9113.5$$

The skeleton is about 9,000 years old, to the nearest 500 years.   ∎

The formulas used in the evaluation of **compound interest** are also applications of exponential growth. The statement that an investment earns compound interest means that the interest earned over a fixed period of time is added to the initial investment, and then the new total earns interest for the next investment period, and so on. For example, suppose an investment of P dollars earns interest at a yearly rate of r percent, and the interest is compounded annually. Then, after the first year the total value is the sum of the initial investment P plus the interest Pr (r is being used as a decimal fraction). Thus, the total after one year is

$$P + Pr = P(1 + r)$$

After the second year the total amount is $P(1 + r)$ plus the interest this amount earns, which is $P(1 + r)r$. Then the total after two years is

This expression is simplified by factoring out $P(1 + r)$.

$$P(1 + r) + P(1 + r)r = P(1 + r)(1 + r) = P(1 + r)^2$$

Similarly, after three years the total is

$$P(1 + r)^2 + P(1 + r)^2 r = P(1 + r)^2(1 + r) = P(1 + r)^3$$

and after t years the final amount A is given by

$$A = P(1 + r)^t$$

The interest periods for compound interest are usually less than one year. They could be quarterly (4 times per year), monthly, or daily, and so forth. For such cases the interest rate per period is the annual rate r divided by the number of interest periods per year. Thus, if the interest is compounded quarterly, the rate per period is $r/4$. Now, following the reasoning used to obtain $A = P(1 + r)^t$, the final amount A after one year (4 interest periods) is

$$A_1 = P\left(1 + \frac{r}{4}\right)^4$$

If there are n interest periods per year, the rate per period is r/n, and after one year we have

$$A_1 = P\left(1 + \frac{r}{n}\right)^n$$

This result can be derived from the preceding result. See Exercise 46.

Likewise, after t years the final amount A_t is given by

$$A_t = P\left(1 + \frac{r}{n}\right)^{nt}$$

EXAMPLE 4   A $5000 investment earns interest at the annual rate of 8.4% compounded monthly. Use a calculator to answer the following:
(a) What is the investment worth after one year?
(b) What is it worth after 10 years?
(c) How much interest was earned in 10 years?

Solution

Use the formula $I = pr$ to compute the simple interest that $5000 would earn in 1 year. Compare this to the compound interest earned in part (a).

(a) Since the annual rate $r = 8.4\% = 0.084$ and the compounding is done monthly, the interest rate per month is $r/n = 0.084/12 = 0.007$. Substitute this, together with $P = 5000$ and $n = 12$, into $A = P\left(1 + \frac{r}{n}\right)^n$.

$$A = 5000(1 + 0.007)^{12} = 5000(1.007)^{12}$$
$$= \boxed{5000} \times \boxed{1.007} \boxed{y^x} \boxed{12} \boxed{=} 5436.55$$

To the nearest dollar, the amount on deposit after one year is $5437.

(b) Use $A_t = P\left(1 + \dfrac{r}{n}\right)^{nt}$ where $P = 5000$, $\dfrac{r}{n} = 0.007$, $n = 12$, and $t = 10$.

$$A = 5000(1.007)^{12(10)} = 5000(1.007)^{120}$$

$$= \boxed{5000} \; \boxed{\times} \; \boxed{1.007} \; \boxed{y^x} \; \boxed{120} \; \boxed{=} \; 11{,}547.99$$

The amount after 10 years is approximately $11,548.

As an illustration, let $r = 0.2$ and use a calculator to verify the following computations, rounded to 5 decimal places. They demonstrate that

$\left(1 + \dfrac{0.2}{n}\right)^n$ *approaches*

$e^{0.2}$ *as n gets larger.*

$\left(1 + \dfrac{0.2}{10}\right)^{10} = 1.21899$

$\left(1 + \dfrac{0.2}{100}\right)^{100} = 1.22116$

$\left(1 + \dfrac{0.2}{1000}\right)^{1000} = 1.22138$

Also, $e^{0.2} = 1.22140$

(c) After 10 years the interest earned is

$$11{,}548 - 5000 = 6548 \text{ dollars} \qquad \blacksquare$$

The marginal note on page 310 points out that the values of $\left(1 + \dfrac{1}{n}\right)^n$ approach the number e as n gets larger and larger. It is also true that $\left(1 + \dfrac{r}{n}\right)^n$ approaches e^r as n gets larger and larger. These observations, when made mathematically precise, lead to the following formula for **continuous compound interest.**

$$A = Pe^{rt}$$

where P is the initial investment, r is the annual rate of interest, and t is the number of years. Thus if $1000 is invested at 10%, compounded continuously for 10 years, the amount on deposit will be

$$A = 1000e^{(0.10)(10)} = 1000e^1$$

$$= 1000(2.718)$$

$$= 2718$$

After 10 years, despite continuous compounding, the amount on deposit (to the nearest dollar) will not grow beyond $2718.

EXAMPLE 5 Suppose that $1000 is invested at 10% interest, compounded continuously. How long will it take for this investment to double?

Solution We wish the final amount on deposit to be $2000. Thus we have the following equation and need to solve for t:

$$2000 = 1000e^{(0.10)t}$$

$$2 = e^{(0.1)t} \qquad \text{(Divide by 1000.)}$$

$$\ln 2 = (0.1)t \qquad \text{(Write in logarithmic form.)}$$

$$\frac{\ln 2}{0.1} = t \qquad \text{(Divide by 0.1.)}$$

$$6.93 = t$$

It will take approximately 7 years for the investment to double in value. As a check, note from Table II that $e^{(0.1)(7)} = e^{0.7} = 2.01$, which is approximately equal to 2. $\qquad \blacksquare$

EXERCISES 6.5

Use Table III and evaluate each of the following to three decimal places.

1. $\dfrac{\ln 6}{\ln 2}$ 2. $\dfrac{\ln 10}{\ln 5}$ 3. $\dfrac{\ln 8}{\ln 0.2}$ 4. $\dfrac{\ln 0.8}{\ln 4}$ 5. $\dfrac{\ln 15}{2 \ln 3}$ 6. $\dfrac{\ln 25}{3 \ln 5}$

7. $\dfrac{\ln 100}{-4 \ln 10}$ 8. $\dfrac{\ln 80}{-5 \ln 8}$

Estimate the value of y in $y = Ae^{kx}$ for the given values of A, k, and x.

9. $A = 100,\ k = 0.75,\ x = 4$ 10. $A = 25,\ k = 0.5,\ x = 10$
11. $A = 1000,\ k = -1.8,\ x = 2$ 12. $A = 12.5,\ k = -0.04,\ x = 50$

Solve for k. Leave the answer in terms of natural logarithms.

13. $5000 = 50e^{2k}$ 14. $75 = 150e^{10k}$ 15. $\dfrac{A}{3} = Ae^{4k}$ 16. $\dfrac{A}{2} = Ae^{100k}$

17. A bacterial culture is growing according to the formula $y = 10,000e^{0.6x}$, where x is the time in days. Estimate the number of bacteria after 1 week. (*Note*: $A = 10,000$.)

18. Estimate the number of bacteria after the culture in Exercise 17 has grown for 12 hours.

19. How long will it take for the bacterial culture in Exercise 17 to triple in size?

20. How long will it take until the number of bacteria in Exercise 17 reaches 1,000,000?

21. A certain radioactive substance decays according to the exponential formula

$$S = S_0 e^{-0.04t}$$

where S_0 is the initial amount of the substance and S is the amount of the substance left after t years. If there were 50 grams of the radioactive substance to begin with, how long will it take for half of it to decay?

22. Show that when the formula in Exercise 21 is solved for t, the result is $t = -25 \ln \dfrac{S}{S_0}$.

23. A radioactive substance is decaying according to the formula $y = Ae^{kx}$, where x is the time in years. The initial amount $A = 10$ grams, and 8 grams remain after 5 years.
    (a) Find k. Leave the answer in terms of natural logs.
    (b) Estimate the amount remaining after 10 years.
    (c) Find the half-life to the nearest tenth of a year.

24. The half-life of radium is approximately 1690 years. A laboratory has 50 milligrams of radium.
    (a) Use the half-life to solve for k in $y = Ae^{kx}$. Answer in terms of natural logs.
    (b) To the nearest 10 years, how long does it take until there are 40 milligrams left?

25. Suppose that 5 grams of a radioactive substance decrease to 4 grams in 30 seconds. What is its half-life to the nearest tenth of a second?

26. How long does it take for two-thirds of the radioactive material in Exercise 25 to decay? Give your answer to the nearest tenth of a second.

27. When the population growth of a certain city was first studied, the population was 22,000. It was found that the population P grows with respect to time t (in years) by the exponential formula $P = (22,000)(10^{0.0163t})$. How long will it take for the city to double its population?

28. How long will it take for the population of the city described in Exercise 27 to triple?

29. An Egyptian mummy is found to contain 60% of its ^{14}C. To the nearest 100 years, how old is the mummy? (*Hint:* If A is the original amount of ^{14}C, then $\frac{3}{5}A$ is the amount left. Also, see Example 3.)

30. A skeleton contains one-hundredth of its original amount of ^{14}C. To the nearest 1000 years, how old is the skeleton?

31. Answer the question in Exercise 30 if one-millionth of its ^{14}C is left.

 Use a calculator having an exponential key and a natural logarithm key to answer the following questions.

32. Suppose that a $10,000 investment earns compound interest at the annual interest rate of 9%. If the time of deposit of the investment is one year ($t = 1$), find the value of the investment for each of the following types of compounding.

   (a) $n = 4$ (quarterly)   (b) $n = 12$ (monthly)   (c) $n = 52$ (weekly)
   (d) $n = 365$ (daily)   (e) continuously.

33. Follow the instructions in Exercise 32 but change the time of deposit to 5 years.

34. Compute the interest earned for each case in Exercise 32.

35. Follow the instructions in Exercise 32 but change the time of deposit to 3.5 years.

36. Suppose that $1500 is invested at the annual interest rate of 8%, compounded continuously. How much will be on deposit after 5 years? After 10 years?

37. Mrs. Kassner deposits $5000 at the annual interest rate of 9%. How long will it take her investment to double in value? How long would it take if the interest rate were 12%? Assume continuous compounding in both cases.

38. How long would it take for a $1000 investment to double if it earns 12% interest annually, compounded continuously? How long would it take to triple?

39. A $1000 investment earns interest at the annual rate of $r\%$, compounded continuously. If the investment doubles in 5 years, what is r?

40. How long does it take a $4000 investment to double if it earns interest at the annual rate of 8%, compounded quarterly?

41. In Exercise 40, how long would it take if the compounding were done monthly?

42. An investment P earns 9% interest per year, compounded continuously. After 3 years, the value of the investment is $5000. Find the initial amount P. (*Hint:* Solve $A = Pe^{rt}$ for P.)

43. Answer the question in Exercise 42 using 6 years as the time of deposit.

44. An investment P earns 8% interest per year, compounded quarterly. After one year the value of the investment is $5000. Find the initial amount P. (*Hint:* Solve $A_1 = P\left(1 + \frac{r}{n}\right)^n$ for P.)

45. How many dollars must be invested at the annual rate of 12%, compounded monthly, in order for the value of the investment to be $20,000 after 5 years? (*Hint:* Solve $A_t = P\left(1 + \frac{r}{n}\right)^{nt}$ for P.)

46. Explain how the result $A_t = P\left(1 + \frac{r}{n}\right)^{nt}$ can be obtained from $A_1 = P\left(1 + \frac{r}{n}\right)^n$. [*Hint:* A_2, the value of the investment after 2 years, is obtained when A_1 has earned interest for one year compounded n times. Thus $A_2 = A_1\left(1 + \frac{r}{n}\right)^n$.]

47. Solve for y: $x = -\dfrac{1}{k}(\ln A - \ln y)$. Use your result to describe the relationship between the variables x and y, assuming that x represents time, A is positive constant, and k is a nonzero constant.

 Written Assignment: Contact at least two local banks and determine how much interest you can earn on a certificate of deposit of $5000 for five years. Find out what the annual interest rate is and how it is compounded. Use appropriate results developed in this section to compute the interest earned over the five-year period and compare your results with the information obtained from the bank.

EXPLORATIONS
Think Critically

1. The solution of Example 5, page 313, was accomplished by converting from the logarithmic to the exponential form. Instead of making this conversion, what should be done so that the equation can be solved using the one-to-one property?

2. It is not true, in general, that $(\log_b x)^3$ is equal to $3 \log_b x$ for all $x > 0$. However, there are specific values of x for which they are equal. Find these values.

3. On each of three sheets of graph paper draw a coordinate system using a unit length of at least 1 inch. Use a sharpened pencil to make careful drawings of $y = 2^x$, $y = 3^x$, and $y = e^x$, for $-2 \leq x \leq 2$, one on each sheet. Do this using the x-values -2, -1, -0.5, -0.25, 0, 0.25, 0.5, 1, 2, and use a calculator, when necessary, for the y-values. Now use a straightedge to draw the line tangent to each curve at point $P(0, 1)$. This line should touch the curve *only* at P. Use the grid lines to estimate the slope of each tangent and decide which slope is closest to the value 1.

4. What is the half-life of a substance that decays exponentially according to $y = Ae^{kt}$, $k < 0$? Use the result to verify the answer in Exercise 21, page 323. What is the doubling time for a substance that grows exponentially according to $y = Ae^{kt}$, $k > 0$?

6.6
SCIENTIFIC NOTATION

To write very large or very small numbers, the scientist often makes use of a form called **scientific notation.** As you will see, scientific notation is helpful in simplifying certain kinds of computations. Here are some illustrations of scientific notation:

$$623,000 = 6.23 \times 10^5 \qquad 0.00623 = 6.23 \times 10^{-3}$$
$$6230 = 6.23 \times 10^3 \qquad 0.0000623 = 6.23 \times 10^{-5}$$

It is easy to verify that these are correct. For example:

$$6.23 \times 10^5 = 6.23 \times 100,000 = 623,000$$

$$6.23 \times 10^{-3} = 6.23 \times \frac{1}{10^3} = \frac{6.23}{1000} = 0.00623$$

The preceding illustrations indicate that *a number N has been put into scientific notation when it has been expressed as the product of a number between 1 and 10 and an integral power of 10.* Thus:

$$N = x(10^c)$$

where $1 \leq x < 10$ and c is an integer.

If the number is greater than 1, then the power of 10 is positive; if the number is less than 1, the power of 10 is negative.

WRITING A NUMBER IN SCIENTIFIC NOTATION

Place the decimal point behind the first nonzero digit. (This produces the number between 1 and 10.) Then determine the power of 10 by counting the number of places you moved the decimal point. If you moved the decimal point to the left, then the power is positive; and if you moved it to the right, it is negative.

Illustrations:

$$2{,}070{,}000. = 2.07 \times 10^6$$

six places left

$$0.00000084 = 8.4 \times 10^{-7}$$

seven places right

To convert a number given in scientific notation back into standard notation, all you need do is move the decimal point as many places as indicated by the exponent of 10. Move the decimal point to the right if the exponent is positive and to the left if it is negative.

EXAMPLE 1 Write 1.21×10^4 in standard notation.

Solution Move the decimal point in 1.21 four places to the right.

$$1.21 \times 10^4 = 12{,}100$$ ∎

EXAMPLE 2 Write 1.21×10^{-2} in standard notation.

Solution Move the decimal point in 1.21 two places to the left.

$$1.21 \times 10^{-2} = 0.0121$$ ∎

TEST YOUR UNDERSTANDING
Think Carefully

Convert into scientific notation.

1. 739   **2.** 73,900   **3.** 0.00739
4. 0.739   **5.** 73.9   **6.** 7.39

Convert into standard notation.

7. 4.01×10^3 **8.** 4.01×10^{-3} **9.** 1.11×10^{-2}
10. 1.11×10^5 **11.** 9.2×10^{-4} **12.** 4.27×10^0

(Answers: Page 340)

Scientific notation can help simplify arithmetic computations, as shown in the following examples.

EXAMPLE 3 Use scientific notation to compute $\dfrac{1}{800{,}000}$.

Solution

In scientific notation the solution to Example 3 is written 1.25×10^{-6}.

$$\frac{1}{800{,}000} = \frac{1}{8 \times 10^5} = \frac{1}{8} \times \frac{1}{10^5} = 0.125 \times 10^{-5}$$

$$= 0.00000125$$ ∎

EXAMPLE 4 Use scientific notation to evaluate

$$\frac{(2,310,000)^2}{(11,200,000)(0.000825)}$$

Solution

$$\frac{(2,310,000)^2}{(11,200,000)(0.000825)} = \frac{(2.31 \times 10^6)^2}{(1.12 \times 10^7)(8.25 \times 10^{-4})}$$

$$= \frac{(2.31)^2 \times (10^6)^2}{(1.12 \times 10^7)(8.25 \times 10^{-4})} \qquad [(ab)^n = a^n b^n]$$

$$= \frac{(2.31)^2}{(1.12)(8.25)} \times \frac{10^{12}}{(10^7)(10^{-4})} \qquad [(a^m)^n = a^{mn}]$$

$$= 0.5775 \times 10^9$$

$$= 577,500,000$$

■

EXERCISES 6.6

Write each number in scientific notation.

1. 4680 **2** 0.0092 **3.** 0.92 **4.** 0.9 **5.** 7,583,000 **6.** 93,000,000

7. 25 **8.** 36.09 **9.** 0.000000555 **10.** 0.57721 **11.** 202.4 **12.** 7.93

Write each number in standard form.

13. 7.89×10^4 **14.** 7.89×10^{-4} **15.** 3.0×10^3 **16.** 3.0×10^{-3} **17.** 1.74×10^{-1}

18. 1.74×10^0 **19.** 1.74×10^1 **20.** 2.25×10^5 **21.** 9.06×10^{-2}

Express each of the following as a single power of 10.

22. $\dfrac{10^{-3} \times 10^5}{10}$ **23.** $\dfrac{10^8 \times 10^4 \times 10^{-5}}{10^2 \times 10^3}$ **24.** $\dfrac{10^{-3}}{10^{-5}}$

25. $\dfrac{10^1 \times 10^2 \times 10^3 \times 10^4}{10^{10}}$ **26.** $\dfrac{10^9 \times 10^{-2}}{10^6 \times 10^{-9}}$ **27.** $\dfrac{(10^2)^3 \times 10^{-1}}{(10^{-3})^4}$

Compute, using scientific notation.

28. $\dfrac{2}{80,000}$ **29.** $\dfrac{1}{0.0005}$ **30.** $\dfrac{(6000)(720)}{12,000}$ **31.** $\dfrac{0.0064}{0.000016}$ **32.** $\dfrac{(240)(0.000032)}{(0.008)(12,000)}$ **33.** $\dfrac{4,860,000}{(0.081)(19,200)}$

Perform the indicated operation using scientific notation.

34. $\dfrac{(40)^4(0.015)^2}{24,000}$ **35.** $\dfrac{\sqrt{0.000625}}{3125}$ **36.** $(0.0006)^3$

37. Light travels at a rate of about 186,000 miles per second. The average distance from the sun to the earth is 93,000,000 miles. Use scientific notation to find how long it takes light to reach the earth from the sun.

38. Based on information given in Exercise 37, use scientific notation to show that 1 light-year (the distance light travels in 1 year) is approximately $5.87 \times 10^{12} = 5,870,000,000,000$ miles.

6.7
COMMON
LOGARITHMS
AND
APPLICATIONS
(OPTIONAL)

Logarithms were developed about 350 years ago. Since then, they have been widely used to simplify involved numerical computations. Much of this work can now be done more efficiently with the aid of computers and calculators. However, logarithmic computations will help us to better understand the theory of logarithms, which plays an important role in many parts of mathematics (including calculus) and in its applications.

For scientific and technical work, numbers are often written in scientific notation and we will therefore be using logarithms to the base 10, called **common logarithms.**

Below is an excerpt of Appendix Table IV. It contains the common logarithms of three-digit numbers from 1.00 to 9.99. To find a logarithm, say $\log_{10} 3.47$, first find the entry 3.4 in the left-hand column under the heading x. Now in the row for 3.4 and in the column headed by the digit 7 you will find the entry .5403. This is the common logarithm of 3.47. We write

$$\log_{10} 3.47 = 0.5403 \qquad \text{(Recall that this means } 3.47 = 10^{0.5403}.\text{)}$$

By reversing this process we can begin with $\log_{10} x = 0.5403$ and find x.

x	0	1	2	3	4	5	6	7	8	9
·	·	·	·	·	·	·	·	·	·	·
·	·	·	·	·	·	·	·	·	·	·
3.3	.5185	.5198	.5211	.5224	.5237	.5250	.5263	.5276	.5289	.5302
3.4	.5315	.5328	.5340	.5353	.5366	.5378	.5391	.5403	.5416	.5428
3.5	.5441	.5453	.5465	.5478	.5490	.5502	.5514	.5527	.5539	.5551
·	·	·	·	·	·	·	·	·	·	·
·	·	·	·	·	·	·	·	·	·	·
·	·	·	·	·	·	·	·	·	·	·

The common logarithms in Table IV are four-place decimals between 0 and 1. Except for the case $\log_{10} 1 = 0$, they are all approximations. The fact that they are between 0 and 1 will be taken up in the exercises.

Note: *The values found in the tables of logarithms are approximations. For simplicity, however, we will use the equals sign (=).*

Since common logarithms are always considered to be to the base 10, we can simplify the notation and drop the subscript 10 from the logarithmic statements. Thus we will write $\log N$ instead of $\log_{10} N$. Verify the following entries taken from Table IV:

$$\log 3.07 = 0.4871 \qquad \log 8.88 = 0.9484$$

$$\text{If } \log x = 0.7945, \text{ then } x = 6.23$$

To find $\log N$, where N may not be between 1 and 10, we first write N in scientific notation, $N = x(10^c)$. This form of N, in conjunction with Table IV, will allow us to find $\log N$. In general,

$$\log N = \log(x\,10^c)$$
$$= \log x + \log 10^c \qquad \text{(Law 1 for logs)}$$
$$= \log x + c \qquad \text{(Why?)}$$

The integer c is the **characteristic** of log N, and the four-place decimal fraction log x is its **mantissa.** Using $N = 62,300$, we have

Note the distinction:

log N ⟵ common
logarithm, base 10

ln N ⟵ natural
logarithm, base e

$$\log 62{,}300 = \log 6.23(10^4) = \log 6.23 + \log 10^4$$
$$= \log 6.23 + 4$$
$$= 0.7945 + 4 \qquad \text{(Table IV)}$$
$$= 4.7945$$

EXAMPLE 1   Find log 0.0419.

Solution

$$\log 0.0419 = \log 4.19(10^{-2})$$
$$= \log 4.19 + \log 10^{-2}$$
$$= 0.6222 + (-2) \qquad\blacksquare$$

Suppose that in Example 1 the mantissa 0.6222 and the negative characteristic are combined:

$$0.6222 + (-2) = -1.3778 = -(1 + 0.3778)$$
$$= -1 + (-0.3778)$$

Since Table IV does not have negative mantissas, like -0.3778, we avoid such combining and preserve the form of log 0.0419 so that its mantissa is positive. For computational purposes there are other useful forms of $0.6222 + (-2)$ in which the mantissa 0.6222 is preserved. Note that $-2 = 8 - 10$, $18 - 20$, and so forth. Thus

$$0.6222 + (-2) = 0.6222 + 8 - 10 = 8.6222 - 10 = 18.6222 - 20$$

Similarly,

$$\log 0.00569 = 7.7551 - 10 = 17.7551 - 20$$
$$\log 0.427 = 9.6304 - 10 = 29.6304 - 30$$

An efficient way to find N, if log $N = 6.1239$, is to find the three-digit number x from Table IV corresponding to the mantissa 0.1239. Then multiply x by 10^6. Thus, since log $1.33 = 0.1239$, we have

$$N = 1.33(10^6) = 1{,}330{,}000$$

In the following explanation you can discover why this technique works.

Note: Unless otherwise stated,
log N will always mean
$\log_{10} N$.

$$\log N = 6.1239$$
$$= 6 + 0.1239$$
$$= 6 + \log 1.33$$
$$= \log 10^6 + \log 1.33$$
$$= \log 10^6(1.33)$$
$$= \log 1{,}330{,}000$$

Therefore, log $N = \log 1{,}330{,}000$, and we conclude that $N = 1{,}330{,}000$.

EXAMPLE 2   Estimate $P = (963)(0.00847)$ by using (common) logarithms.

Solution

$$\log P = \log(963)(0.00847)$$
$$= \log 963 + \log 0.00847 \qquad \text{(Law 1)}$$

For easy reference:

Law 1. $\log MN = $

$$\log M + \log N$$

Law 2. $\log \dfrac{M}{N} = $

$$\log M - \log N$$

Law 3. $\log N^k = k \log N$

Now use Table IV.

$$\left. \begin{aligned} \log 963 &= 2.9836 \\ \log 0.00847 &= 7.9279 - 10 \end{aligned} \right\} \quad \text{(Add.)}$$

$$\log P = 10.9115 - 10 = 0.9115$$

$$P = 8.16(10^0) = 8.16$$

For a more accurate procedure, see Exercise 39. Exercise 38 shows how to find $\log x$ when $0 \le x < 1$ and x has more than three digits.

Note: The mantissa 0.9115 is not in Table IV. In this case we use the closest entry, namely 0.9117, corresponding to $x = 8.16$. Such approximations are good enough for our purposes. ■

EXAMPLE 3   Use logarithms to estimate $Q = \dfrac{0.00439}{0.705}$.

Solution   We find $\log Q = \log 0.00439 - \log 0.705$ (by Law 2). Now use the table.

(This form is used to avoid a negative mantissa when subtracting in the next step.)

$$\left. \begin{aligned} \log 0.00439 &= 7.6425 - 10 = 17.6425 - 20 \\ \log 0.705 &= 9.8482 - 10 = 9.8482 - 10 \end{aligned} \right\} \quad \text{(Subtract.)}$$

$$\log Q = 7.7943 - 10$$

$$Q = 6.23(10^{-3})$$

$$= 0.00623$$

■

EXAMPLE 4   Use logarithms to estimate $R = \sqrt[3]{0.0918}$.

Solution

$$\log R = \log(0.0918)^{1/3}$$
$$= \tfrac{1}{3} \log 0.0918 \qquad \text{(Law 3)}$$
$$= \tfrac{1}{3}(8.9628 - 10)$$
$$= \tfrac{1}{3}(28.9628 - 30) \quad \longleftarrow \quad \text{(We avoid a fractional characteristic by changing to } 28.9628 - 30.)$$
$$= 9.6543 - 10$$
$$R = 4.51(10^{-1}) = 0.451$$

■

EXAMPLE 5   To determine how much a paint dealer should charge for a gallon of paint, he needs to find out how much the paint cost him per gallon in the first place. The paint is stored in a cylindrical drum $2\frac{1}{2}$ feet in diameter and $3\frac{3}{4}$ feet high. If he paid \$400 for this quantity of paint, what did it cost him per gallon? (Use 1 cubic foot = 7.48 gallons.)

Solution   The volume of the drum is the area of the base times the height. Thus there are

Volume of a cylinder:
$V = \pi r^2 h$

$$\pi (1.25)^2(3.75)$$

cubic feet of paint in the drum. Then the number of gallons is

$$\pi (1.25)^2(3.75)(7.48)$$

Since the total cost was \$400, the cost per gallon is given by

$$C = \frac{400}{\pi (1.25)^2(3.75)(7.48)}$$

We use $\pi = 3.14$ to do the computation, using logarithms:

$$\log C = \log 400 - (\log 3.14 + 2 \log 1.25 + \log 3.75 + \log 7.48)$$

$$
\left.
\begin{array}{r}
\log 400 = 2.6021 \\[4pt]
\left.
\begin{array}{l}
\log 3.14 = 0.4969 \\
\log 1.25 = 0.0969 \rightarrow 2 \log 1.25 = 0.1938 \\
\log 3.75 = 0.5740 \\
\log 7.48 = 0.8739 \\
\hline
 2.1386
\end{array}
\right\} \text{(Add.)} \\
\rightarrow \quad 2.1386
\end{array}
\right\} \text{(Subtract.)}
$$

$$\log C = 0.4635$$
$$C = 2.91 \times 10^0$$
$$= 2.91$$

The paint cost the dealer approximately \$2.91 per gallon.   ■

EXERCISES 6.7

Find the common logarithm.

1. log 457   2. log 45.7   3. log 0.457   4. log 0.783   5. log 72.9   6. log 8.56

Find N.

7. $\log N = 0.5705$ 8. $\log N = 0.8904$ 9. $\log N = 1.8331$
10. $\log N = 2.9523$ 11. $\log N = 9.1461 - 10$ 12. $\log N = 8.6972 - 10$

In Exercises 13–24, estimate by using common logarithms.

13. (512)(84,000)   14. (906)(2330)(780)   15. $\dfrac{(927)(818)}{274}$ 16. $\dfrac{274}{(927)(818)}$

17. $\dfrac{(0.421)(81.7)}{(368)(750)}$

18. $\dfrac{(579)(28.3)}{\sqrt{621}}$

19. $\dfrac{(28.3)\sqrt{621}}{579}$

20. $\left[\dfrac{28.3}{(579)(621)}\right]^2$

21. $\sqrt{\dfrac{28.3}{(579)(621)}}$

22. $\dfrac{(0.0941)^3(0.83)}{(7.73)^2}$

23. $\dfrac{\sqrt[3]{(186)^2}}{(600)^{1/4}}$

24. $\dfrac{\sqrt[4]{600}}{(186)^{2/3}}$

Use common logarithms to solve the remaining problems.

25. After running out of gasoline, a motorist had her gas tank filled at a cost of $16.93. What was the cost per gallon if the gas tank's capacity is 14 gallons?

26. Suppose that a spaceship takes 3 days, 8 hours, and 20 minutes to travel from the earth to the moon. If the distance traveled was one-quarter of a million miles, what was the average speed of the spaceship in miles per hour?

27. A spaceship, launched from the earth, will travel 432,000,000 miles on its trip to the planet Jupiter. If its average velocity is 21,700 miles per hour, how long will the trip take? Give the answer in years.

Exercises 28–30 make use of the compound interest formulas studied in Section 6.5.

28. When P dollars is invested in a bank that pays compound interest at the rate of r percent (expressed as a decimal) per year, the amount A after t years is given by the formula

$$A = P(1 + r)^t$$

(a) Find A for $P = 2500$, $r = 0.09$ (9%), and $t = 3$.

(b) An investment of $3750 earns compound interest at the rate of 11.2% per year. Find the amount A after 5 years.

29. The formula $P = \dfrac{A}{(1 + r)^t}$ gives the initial investment P in terms of the current amount of money A, the annual compound interest rate r, and the number of years t. How much money was invested at 12.8% if after 6 years there is now $8440 in the bank?

30. If P dollars is invested at an interest rate r and the interest is compounded n times per year, the amount A after t years is given by

$$A = P\left(1 + \frac{r}{n}\right)^{nt}$$

(a) Use this formula to compute A for $P = \$5000$ and $r = 0.08$ if the interest is compounded semiannually for 3 years.

(b) Find A in part (a) with interest compounded quarterly.

(c) Find A in part (a) with $n = 8$.

31. An oil tanker carries 253,000 barrels of crude oil. This oil will produce 1,830,000 gallons of a certain kind of fuel. How many gallons of fuel are produced by 1 gallon of crude oil? (1 barrel = 31.5 gallons.)

32. The dimensions of a rectangular-shaped container are 2.75 by 5.35 by 4.4 feet. How many gallons can this container hold? (Use 1 cubic foot = 7.48 gallons.) If this container is filled with water, how many pounds of water does it hold? (Use 1 cubic foot water = 62.4 pounds.)

33. The volume V of a sphere with radius r is given by $V = \frac{4}{3}\pi r^3$. Use $\pi = 3.14$ to find the volume of a sphere with a radius of 12 centimeters.

34. The surface area S of a sphere is given by $S = 4\pi r^2$. What is the surface area of the sphere in Exercise 33?

35. The period P of a simple pendulum is the time (in seconds) it takes to make one full swing. The period is given by $P = 2\pi\sqrt{\dfrac{l}{32}}$ where l is the length of the pendulum. Find the period of a pendulum with a length of $3\frac{3}{4}$ feet.

36. The area of a triangle A can be given in terms of its three sides. The formula is

$$A = \sqrt{s(s - a)(s - b)(s - c)}$$

in which a, b, and c are the lengths of the three sides and s is half the perimeter; $s = \frac{1}{2}(a + b + c)$.

 (a) Use this formula to find the area of a triangle whose sides are 346, 330, and 104 cm.
 (b) Use the Pythagorean theorem to show that this is a right triangle. (*Do not* use logarithms to do this.)
 (c) Use the formula $A = \frac{1}{2}$(base)(height) and compare to the result in part (a).

37. Explain why the mantissas in Table IV are between 0 and 1. (*Hint:* Take $1 \le x < 10$ and now consider the common logarithms of 1, x, and 10.)

38. Here is a computation for finding log 6.477. Study this procedure carefully and then find the logarithms below in the same manner.

$$0.010\begin{cases}0.007\begin{cases}6.470 \longrightarrow 0.8109 \\ 6.477 \longrightarrow \ \ ? \end{cases}d \\ 6.480 \longrightarrow \overline{0.8116}\end{cases}0.0007$$

$$\frac{0.007}{0.010} = \frac{d}{0.0007}$$

$$0.7 = \frac{d}{0.0007}$$

$$d = (0.7)(0.0007) = 0.00049$$

$$\log N = 0.8109 + 0.00049$$

$$= 0.8114 \qquad \text{(rounded off to four decimal places)}$$

 (a) log 3.042 (b) log 7.849 (c) log 1.345 (d) log 5.444
 (e) log 6.803 (f) log 2.711 (g) log 4.986 (h) log 9.008

The method used here is called **linear interpolation.** The rationale behind this method is suggested by the accompanying figure.

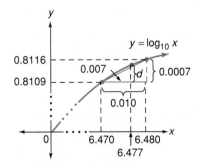

Note that the line segment approximates the curve of $y = \log_{10} x$.

39. The method in Exercise 38 can be adapted for finding the number when the given logarithm is not an exact table entry. Study the following procedure for finding N in $\log N = 0.7534$, and then find the numbers N below in the same manner.

$$0.0008 \begin{cases} 0.0006 \begin{cases} \begin{array}{ccc} & \log N & N \\ 0.7528 & \longrightarrow & 5.660 \\ 0.7534 & \longrightarrow & ? \\ \end{array} \end{cases} d \\ \begin{array}{ccc} 0.7536 & \longrightarrow & 5.670 \end{array} \end{cases} 0.010$$

$$\frac{0.0006}{0.0008} = \frac{d}{0.01}$$

$$0.75 = \frac{d}{0.01}$$

$$d = (0.01)(0.75) = 0.0075$$

$$N = 5.660 + 0.0075$$

$$= 5.668 \qquad \text{(rounded off to three decimal places)}$$

(a) $\log N = 0.4510$ **(b)** $\log N = 0.9672$ **(c)** $\log N = 0.1391$

(d) $\log N = 0.7395$ **(e)** $\log N = 0.6527$ **(f)** $\log N = 0.8749$

(g) $\log N = 0.0092$ **(h)** $\log N = 0.9781$ **(i)** $\log N = 0.3547$

KEY TERMS

Review these **key terms** *so that you are able to define or describe them. A clear understanding of these terms will be very helpful when reviewing the developments of this chapter.*

Exponential function • Properties of $y = b^x$ • Exponential equation • Logarithmic function • Properties of $y = \log_b x$ • Laws of logarithms • Logarithmic equation • The base e • The function $y = e^x$ and its properties • The natural logarithm function $y = \ln x$ and its properties • Exponential growth • Exponential decay • Radioactive carbon dating • Compound interest • Continuous compound interest • Scientific notation • Common logarithm • Characteristic • Mantissa

REVIEW EXERCISES

The solutions to the following exercises can be found within the text of Chapter 6. Try to answer each question before referring to the text.

Section 6.1

1. List the important properties of the exponential function $f(x) = b^x$ for $b > 1$ and for $0 < b < 1$.
2. Use a table of values to sketch $y = 8^x$ on the interval $[-1, 1]$.
3. Sketch $y = 2^x$ and $y = (\frac{1}{2})^x$ on the same axes.
4. Explain how to obtain the graphs of $y = 2^{x-3}$ and $y = 2^x - 1$ from $y = 2^x$.
5. If f is a one-to-one function and $f(x_1) = f(x_2)$, what can you say about x_1 and x_2?
6. Solve for x: $5^{x^2} = 625$
7. Solve for x: $\dfrac{1}{3^{x-1}} = 81$.

Section 6.2

8. Which of the following statements are true?
   (a) If $0 < b < 1$, the function $f(x) = b^x$ decreases.
   (b) The point $(0, 1)$ is on the curve $y = \log_b x$.
   (c) $y = \log_b x$, for $b > 1$, increases and the curve is concave down.
   (d) The domain of $y = b^x$ is the same as the range of $y = \log_b x$.
   (e) The x-axis is an asymptote to $y = \log_b x$ and the y-axis is an asymptote to $y = b^x$.
9. Write the equation of the inverse of $y = 2^x$ and graph both on the same axes.
10. Find the domain for $y = \log_2(x - 3)$.
11. Change to logarithmic form: $27^{2/3} = 9$.
12. Change to exponential form: $\log_6 \frac{1}{36} = -2$.
13. Solve for the indicated variable:
    (a) $y = \log_9 27$ (b) $\log_b 8 = \frac{3}{4}$ (c) $\log_{49} x = -\frac{1}{2}$

Section 6.3

14. Write the three laws of logarithms.
15. Express $\log_b \dfrac{AB^2}{C}$ in terms of $\log_b A$, $\log_b B$, and $\log_b C$.
16. Express $\frac{1}{2} \log_b x - 3 \log_b(x - 1)$ as the logarithm of a single expression in x.
17. Given that $\log_b 2 = 0.6931$ and $\log_b 3 = 1.0986$, find $\log_b \sqrt{12}$.
18. Solve for x: $\log_8(x - 6) + \log_8(x + 6) = 2$.
19. Solve for x: $\log_{10}(x^3 - 1) - \log_{10}(x^2 + x + 1) = 1$.
20. Solve for x: $\log_3 2x - \log_3(x + 5) = 0$.

Section 6.4

21. Sketch the curves $y = 2^x$, $y = e^x$, and $y = 3^x$ in the same coordinate system.
22. Graph $y = e^x$ and $y = \ln x$ on the same axes.
23. Find the domain of $y = \ln(x + 2)$.
24. Sketch $y = \ln x^2$ for $x > 0$.
25. Let $f(x) = \dfrac{3x^2}{x^2 + 4}$. Use the laws of logarithms to write $\ln f(x)$ as an expression involving sums, differences, and multiples of natural logarithms.
26. Solve for t: $e^{\ln(2t-1)} = 5$.
27. Solve for t: $e^{2t-1} = 5$.
28. Solve for x: $\ln(x + 1) = 1 + \ln x$.
29. Determine the signs of $f(x) = x^2 e^x + 2xe^x$.
30. (a) Show $h(x) = \ln(x^2 + 5)$ as the composite of two functions.
    (b) Show $F(x) = e^{\sqrt{x^2-3x}}$ as the composite of three functions.
31. Match the columns.
    (i) $\ln x < 0$ (a) $x < 0$
    (ii) $\ln x = 0$ (b) $x = 0$
    (iii) $\ln x > 0$ (c) $x > 0$
    (iv) $0 < e^x < 1$ (d) $0 < x < 1$
    (v) $e^x = 1$ (e) $x = 1$
    (vi) $e^x > 1$ (f) $x > 1$

Section 6.5

32. Approximate the value of x in (a) $2^x = 35$, and (b) $2^{3x} = 35$.

33. Solve for x: $(10,000)2^x = 100,000$

34. A radioactive material is decreasing according to the formula $y = Ae^{-0.2x}$, where y is the amount of material remaining after x years. If the initial amount $A = 80$ grams, how much is left after 3 years?

35. Find the half-life of the radioactive substance in Exercise 34.

36. Solve for k in $\dfrac{A}{2} = Ae^{5750k}$. Leave your answer in terms of natural logs.

37. Use the formula $y = Ae^{(\ln 0.5/5750)x}$ to estimate the age of a skeleton that is found to contain one-third of its original amount of carbon-14.

38. A \$5000 investment earns interest at the annual rate of 8.4% compounded monthly. What is the investment worth after one year? After 10 years? How much interest was earned in 10 years?

39. Suppose that \$1000 is invested at 10% interest, compounded continuously. How much will be on deposit in 10 years?

40. How long will it take for the investment in Exercise 39 to double?

Section 6.6

41. Convert to scientific notation:
    (a) 2,070,000   (b) 0.00000084

42. Convert to standard notation:
    (a) 1.21×10^4 (b) 1.21×10^{-2}

scientific notation to evaluate

43. $\dfrac{1}{800,000}$

44. $\dfrac{(2,310,000)^2}{(11,200,000)(0.000825)}$

Section 6.7

45. Find $\log 0.0419$.

46. Find N if $\log N = 6.1239$.

47. Use common logarithms to estimate $P = (963)(0.00847)$.

48. Use common logarithms to estimate $Q = \dfrac{0.00439}{0.705}$.

49. Use common logarithms to estimate $R = \sqrt[3]{0.0918}$.

50. To determine how much a paint dealer should charge for a gallon of paint, he needs to find out how much the paint cost him per gallon in the first place. The paint is stored in a cylindrical drum $2\frac{1}{2}$ feet in diameter and $3\frac{3}{4}$ feet high. If he paid \$400 for this quantity of paint, use common logarithms to find what it cost him per gallon. (Use 1 cubic foot = 7.48 gallons.)

Use these questions to test your knowledge of the basic skills and concepts of Chapter 6. Then test your answers with those given at the back of the book.

1. Match each curve with one of the given equations listed below.

(i)
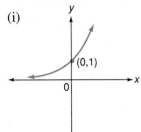

(ii)
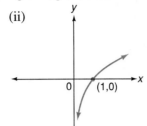

(iii)
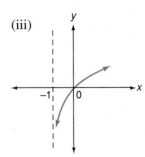

(iv)
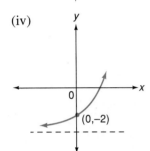

(v)
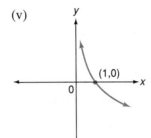

(vi)
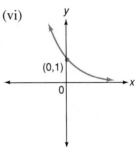

   (a) $y = b^x$; $b > 1$ **(b)** $y = b^x$; $0 < b < 1$ **(c)** $y = \log_b x$; $b > 1$

   (d) $y = \log_b x$; $0 < b < 1$ **(e)** $y = \log_b(x + 1)$; $b > 1$ **(f)** $y = \log_b(x - 1)$; $b > 1$

   (g) $y = b^{x+2}$; $b > 1$ **(h)** $y = b^x - 3$; $b > 1$

2. **(a)** Convert $\log_5 125 = 3$ into exponential form.

   (b) Convert $16^{3/4} = 8$ into logarithmic form.

3. Solve for x.

   (a) $81^x = 9$ **(b)** $e^{\ln(x^2 - x)} = 6$

4. Solve for x: $\dfrac{1}{2^{x+1}} = 64$

5. Simplify: $80e^{3(\ln 1/2)}$

6. **(a)** Solve for b: $\log_b \dfrac{27}{8} = -3$ **(b)** Evaluate: $\log_{10} 0.01$

7. Describe the graph of $y = 3^x$ by answering the following:

   (a) What are the domain and range?

   (b) Where is the function increasing or decreasing?

   (c) Describe the concavity.

   (d) Find the x- and y-intercepts, if any.

   (e) Write the equations of the asymptotes, if any.

8. Sketch the graphs of $y = e^{-x}$ and its inverse on the same axes. Write an equation of the inverse in the form $y = g(x)$.

Find the domain of each function and give the equation of the vertical or horizontal asymptote.

9. $y = f(x) = 2^x - 4$ 10. $y = f(x) = \log_3(x + 4)$

11. Use the laws of logarithms (as much as possible) to write the following as an expression involving sums, differences, and multiples of logarithms:

   (a) $\log_b x(x^2 + 1)^{10}$ **(b)** $\ln \dfrac{x^3}{(x + 1)\sqrt{x^2 + 2}}$

12. Express as the logarithm of a single expression in x:

   (a) $\log_7 x + \log_7 2x + \log_7 5$ **(b)** $\frac{1}{3} \log_b x - 2 \log_b(x + 2)$

13. Solve for x: $\log_{25} x^2 - \log_{25}(2x - 5) = \frac{1}{2}$.

14. Solve for x: $\log_{10} x + \log_{10}(3x + 20) = 2$.

15. Explain how the graph of $y = \ln e^2(x - 1)$ can be obtained from the graph of $y = \ln x$ by using translations.

16. Solve for x in $4^{2x} = 5$ and express the answer in terms of natural logs.

17. A radioactive substance decays according to the formula $y = Ae^{-0.04t}$, where t is the time in years. If the initial amount $A = 50$ grams, find the half-life. Leave the answer in terms of natural logs.

18. A \$2000 investment earns interest at the annual rate of 8%, compounded quarterly. Write the expression that equals the value of the investment after 6 years. (Leave the answer in exponential form.)

19. How long will it take the investment in Exercise 18 to double? Leave the answer in terms of natural logarithms.

20. If \$6000 is invested at the annual interest rate of 7%, compounded continuously, what is the investment worth in 10 years?

21. **(a)** Write 42,900 in scientific notation.

   (b) Write 3.25×10^{-3} in standard notation.

22. Use scientific notation to evaluate

$$\frac{\sqrt{144{,}000{,}000}}{(2000)^2(0.0005)}$$

23. If $\log_{10} 24 = 1.38$, $\log_{10} 84 = 1.92$, and $\log_{10} 29 = 1.46$, evaluate $\log_{10} \dfrac{(24)^2}{84\sqrt{29}}$.

24. Solve for k in $y = 120e^{kt}$ if $y = 180$ when $t = 3$.

25. Solve for x: $\ln 2 - \ln(1 - x) = 1 - \ln(x + 1)$.

CHAPTER 6 TEST: MULTIPLE CHOICE

1. Which of the following are true for the function $y = f(x) = b^x$; $b > 0$ and $b \neq 1$?

   I. The domain consists of all real numbers x.

   II. The range consists of all positive numbers y.

   III. It is a one-to-one function.

   (a) Only I   **(b)** Only II   **(c)** Only III   **(d)** I, II, and III   **(e)** None of the preceding

2. Solve for x: $b^{x^2 - 2x} = 1$

   (a) $x = 0$ or $x = 2$ **(b)** $x = 1$ or $x = 3$ **(c)** $x = 1$ **(d)** $x = 0$ **(e)** None of the preceding

3. For which of the following is the point $(-1, \frac{1}{4})$ on the graph of the function?

   (a) $y = f(x) = -\log_2(x - 1)$ **(b)** $y = f(x) = \log_2(x - 1)$
   (c) $y = f(x) = 2^{x-1}$ **(d)** $y = f(x) = -2^{x-1}$ **(e)** None of the preceding

4. Which of the following is the domain for the function $y = f(x) = \log_2(x + 3)$?

   (a) $x < 3$ **(b)** $x \geq 3$ **(c)** $x < -3$ **(d)** $x \geq -3$ **(e)** None of the preceding

5. Solve for x: $\log_3 x + \log_3(2x + 51) = 4$
   (a) $\frac{3}{2}$ (b) -27 (c) $\frac{47}{3}$ (d) 10 (e) None of the preceding

6. Solve for b: $\log_b \frac{1}{64} = -\frac{3}{2}$.
   (a) $\frac{1}{16}$ (b) 8 (c) 16 (d) 512 (e) None of the preceding

7. Which of the following are true?
   I. $(\log_b x)^2 = 2 \log_b x$
   II. $\log_b A + \log_b B = \log_b(A + B)$

   III. $\log_b A - \log_b B = \dfrac{\log_b A}{\log_b B}$

   (a) Only I   (b) Only II   (c) Only III   (d) I, II, and III   (e) None of the preceding

8. Which of the following are true for the function $y = f(x) = e^x$?
   I. The domain consists of all real numbers x.
   II. The curve is concave up.
   III. It is a one-to-one function.
   (a) Only I   (b) Only II   (c) Only III   (d) I, II, and III   (e) None of the preceding

9. Which of the following are true for the function $y = f(x) = \ln x$?
   I. The domain consists of all real numbers x.
   II. The range consists of all $y > 0$.
   III. The curve is concave down.
   (a) Only I   (b) Only II   (c) Only III   (d) I, II, and III   (e) None of the preceding

10. Solve for x: $\ln(x + 1) - 1 = \ln x$.

    (a) $e - 1$ (b) $\dfrac{1}{e - 1}$ (c) 1 (d) e (e) None of the preceding

11. Solve for x: $9^{x-1} = 4$

    (a) $\dfrac{\ln 4}{\ln 9} + 1$ (b) $\dfrac{\ln 5}{\ln 9}$ (c) $\ln 4 - \ln 9 + 1$ (d) $\dfrac{\ln 9}{\ln 4} + 1$ (e) None of the preceding

12. A radioactive substance decays according to the formula $y = Ae^{kx}$, where the initial amount $A = 40$ grams. If after 8 years 30 grams remain, then the amount y remaining after x years is given by
    (a) $y = 30e^{(x \ln 0.75)/8}$ (b) $y = 40e^{(x \ln 0.75)/8}$ (c) $y = 40e^{(x \ln 1.3)/8}$
    (d) $y = 40e^{(x \ln 0.75)/8}$ (e) None of the preceding

13. A \$3000 investment earns interest at the annual rate of 8.4% compounded monthly. After five years the total amount of the investment, A_5, is given by
    (a) $A_5 = 3000(1.084)^{60}$ (b) $A_5 = 3000(1.084)^5$ (c) $A_5 = 3000(1.007)^{60}$
    (d) $A_5 = 3000(1.007)^{12}$ (e) None of the preceding

14. Using scientific notation, the expression $\dfrac{(0.00054)(9200)}{370}$ can be rewritten as

    (a) $\dfrac{(5.4)(9.2)}{3.7} \times 10^3$ (b) $\dfrac{(5.4)(9.2)}{3.7} \times 10^{-3}$ (c) $\dfrac{(5.4)(9.2)}{3.7} \times 10^2$

    (d) $\dfrac{(5.4)(9.2)}{3.7} \times 10^{-2}$ (e) None of the preceding

15. If $Q = \dfrac{(\sqrt[5]{409})(0.0058)}{7.29}$, then $\log_b Q$ is which of the following?
    (a) $\frac{1}{5}(\log_b 409 + \log_b 0.0058) - \log 7.29$ (b) $5 \log_b 409 + \log_b 0.0058 - \log_b 7.29$
    (c) $\dfrac{\frac{1}{5} \log_b 409 + \log_b 0.0058}{\log_b 7.29}$ (d) $\frac{1}{5} \log_b 409 + \log_b 0.0058 - \log_b 7.29$
    (e) None of the preceding

ANSWERS TO THE TEST YOUR UNDERSTANDING EXERCISES

Page 299

1. 6   **2.** 2 or -2 **3.** $\frac{1}{2}$ **4.** -6 **5.** -4 **6.** -1 **7.** $\frac{1}{3}$ **8.** $\frac{2}{3}$ **9.** $\frac{2}{3}$
10. $-\frac{5}{2}$ **11.** 3   **12.** $-\frac{1}{2}$

Page 302

1.

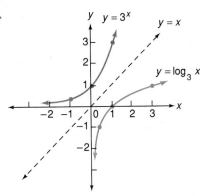

2.

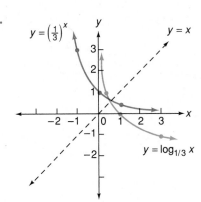

3. Shift 2 units left.   **4.** Shift 2 units up.   **5.** Reflect through x-axis.   **6.** Double the size of each ordinate.

Page 307

1. $\log_b A + \log_b B + \log_b C$ **2.** $\log_b A - \log_b B - \log_b C$ **3.** $2 \log_b A + 2 \log_b B - \log_b C$
4. $\log_b A + 2 \log_b B + 3 \log_b C$ **5.** $\log_b A + \frac{1}{2} \log_b B - \log_b C$ **6.** $\frac{1}{3} \log_b A + 3 \log_b B + 3 \log_b C$
7. $\log_b 3x^2$ **8.** $\log_b[\sqrt{x}(x-1)^2]$ **9.** $\log_b \dfrac{2x-1}{(x^2+1)^3}$ **10.** $\log_b \dfrac{x}{(x-1)(x-2)^2}$ **11.** 2.8903
12. -0.5234

Page 318

1. 1.161   **2.** -1.161 **3.** -3.585 **4.** 1.107   **5.** 1.953   **6.** 0.066

Page 326

1. 7.39×10^2 **2.** 7.39×10^4 **3.** 7.39×10^{-3} **4.** 7.39×10^{-1} **5.** 7.39×10^1
6. 7.39×10^0 **7.** 4010   **8.** 0.00401   **9.** 0.0111   **10.** 111,000   **11.** 0.00092   **12.** 4.27

**7.1
TRIGONOMETRIC
RATIOS**

The word *trigonometry* is based on the Greek words for triangle (*trigōnon*) and measure (*metron*). This study dates back more than 3000 years and over the centuries has been instrumental in developing knowledge in areas such as architecture, astronomy, navigation, and surveying. Our study of trigonometry will include work with triangles as well as more recent developments that have become important in the modern world. We begin with right triangles.

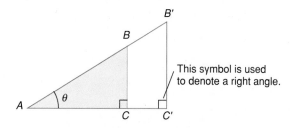

This symbol is used to denote a right angle.

Our first task is to establish the **trigonometric ratios** for an acute angle. We do this by considering an acute angle within a right triangle. In the figure angle A (also written as $\angle A$ or $\angle BAC$) is an acute angle of both right triangles ABC and $AB'C'$, in which $\angle C$ and $\angle C'$ are right angles. The Greek letter θ (theta) is used to denote the measure of $\angle A$. It is also used to name the angle. Since the two triangles have the measures of their angles respectively equal to one another, the triangles are similar and we may write the following:

An acute angle has measure between 0° and 90°.

341

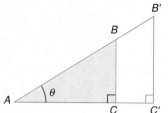

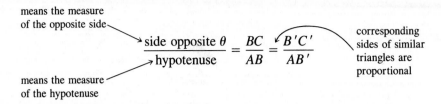

means the measure of the opposite side

$$\text{side opposite } \theta = \frac{BC}{AB} = \frac{B'C'}{AB'}$$
hypotenuse

means the measure of the hypotenuse

corresponding sides of similar triangles are proportional

The resulting proportion shows that the ratio of the side opposite θ to the hypotenuse *depends only on the size of the angle and not on the size of the triangle.* This constant ratio is called the sine of the measure of θ. It may be written as *sine of θ*, but we normally use the standard notation *sin θ*.

$$\sin \theta = \frac{\text{side opposite } \theta}{\text{hypotenuse}} = \frac{BC}{AB}$$

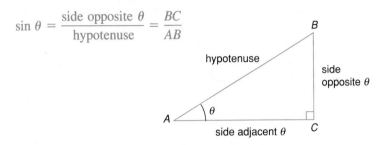

Another important ratio is formed by taking the side adjacent to θ and the hypotenuse. We call this ratio the *cosine of θ* and use the notation *cos θ*.

$$\cos \theta = \frac{\text{side adjacent } \theta}{\text{hypotenuse}} = \frac{AC}{AB}$$

When the side opposite θ is compared to the adjacent side, we call the resulting ratio the *tangent of θ* and use the notation *tan θ*.

You can use similar triangles to verify that cos θ and tan θ depend only on the size of θ, not on the size of the triangle.

$$\tan \theta = \frac{\text{side opposite } \theta}{\text{side adjacent } \theta} = \frac{BC}{AC}$$

EXAMPLE 1   Find sin θ, cos θ, and tan θ.

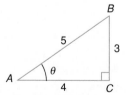

Solution   Relative to $\angle A$, BC is the side opposite and AC is the side adjacent.

$$\sin \theta = \frac{\text{side opposite}}{\text{hypotenuse}} = \frac{BC}{AB} = \frac{3}{5}$$

$$\cos \theta = \frac{\text{side adjacent}}{\text{hypotenuse}} = \frac{AC}{AB} = \frac{4}{5}$$

$$\tan \theta = \frac{\text{side opposite}}{\text{side adjacent}} = \frac{BC}{AC} = \frac{3}{4}$$

Three more ratios can be formed from the sides of a triangle by taking the reciprocals of the sine, cosine, and tangent ratios. All six of these ratios, called the **trigonometric ratios of an angle,** need to be studied carefully and remembered as defini-tions. Although they are given in the following table in terms of a specific reference triangle, they should be remembered as ratios of sides that can be applied to any acute angle of a right triangle.

Trigonometric ratio	Abbreviation	Definition	
			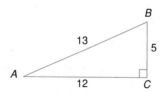
sine of θ	$\sin \theta$	$\dfrac{\text{side opposite}}{\text{hypotenuse}}$	$\dfrac{a}{c}$
cosine of θ	$\cos \theta$	$\dfrac{\text{side adjacent}}{\text{hypotenuse}}$	$\dfrac{b}{c}$
tangent of θ	$\tan \theta$	$\dfrac{\text{side opposite}}{\text{side adjacent}}$	$\dfrac{a}{b}$
cotangent of θ	$\cot \theta$	$\dfrac{\text{side adjacent}}{\text{side opposite}}$	$\dfrac{b}{a}$
secant of θ	$\sec \theta$	$\dfrac{\text{hypotenuse}}{\text{side adjacent}}$	$\dfrac{c}{b}$
cosecant of θ	$\csc \theta$	$\dfrac{\text{hypotenuse}}{\text{side opposite}}$	$\dfrac{c}{a}$

When it is convenient, the letter used for the vertex of an angle will also stand for its measure. Thus, in the preceding chart, θ could be replaced by A throughout.

EXAMPLE 2   Use the figure and write the six trigonometric ratios for angle B.

Solution   Note that we need to consider sides opposite and adjacent to $\angle B$. The following diagram will help orient you correctly.

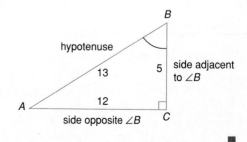

Observe that the six values are listed so that in each row the pair of values are reciprocals; sin and csc, cos and sec, tan and cot.

$$\sin B = \tfrac{12}{13} \qquad \csc B = \tfrac{13}{12}$$

$$\cos B = \tfrac{5}{13} \qquad \sec B = \tfrac{13}{5}$$

$$\tan B = \tfrac{12}{5} \qquad \cot B = \tfrac{5}{12}$$

Use the given figure to answer each question.

1. Write the six trigonometric ratios for angle A.
2. Write the six trigonometric ratios for angle B.
3. Find the value of $(\sin A)^2 + (\cos A)^2$.
4. Find the value of $(\sin A)(\csc A)$.
5. Find the value of $(\sec B)^2 - (\tan B)^2$.
6. Find the value of $(\tan B)(\cot B)$.

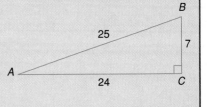

Note: In later work an expression such as $(\sin A)^2$ is usually written as $\sin^2 A$. Thus $\cos^2 A$ means $(\cos A)^2$, and so on.

When only two sides of a right triangle are given, all the trigonometric ratios can still be found by using the Pythagorean theorem to find the third side. This is illustrated in Example 3.

EXAMPLE 3 Triangle ABC is a right triangle with right angle at C. Then if $\tan A = \frac{3}{2}$, find $\sin B$.

Solution Since $\tan A = \dfrac{\text{side opposite}}{\text{side adjacent}} = \dfrac{3}{2}$, label the sides opposite and adjacent to $\angle A$ as shown, and use the Pythagorean theorem to find c.

$$c^2 = a^2 + b^2$$
$$= 9 + 4$$
$$= 13$$

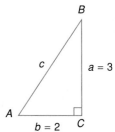

It is common notation to use the same letter, capital and lowercase, for an angle and its opposite side such as $\angle A$ and side a.

Then

$$c = \sqrt{13} \quad \text{and} \quad \sin B = \frac{AC}{AB} = \frac{2}{\sqrt{13}} \quad \text{or} \quad \frac{2\sqrt{13}}{13} \qquad \blacksquare$$

EXAMPLE 4 Use the Pythagorean theorem to solve for AC in terms of x. Then write each of the six trigonometric ratios of $\angle A$ in terms of x.

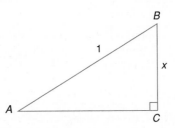

The positive square root of $1 - x^2$ is used since the side of the triangle has a positive measure.

Solution By the Pythagorean theorem, $(AC)^2 + x^2 = 1^2 = 1$. Thus $(AC)^2 = 1 - x^2$ and $AC = \sqrt{1 - x^2}$.

$$\sin A = x \qquad\qquad \csc A = \frac{1}{x}$$

$$\cos A = \sqrt{1 - x^2} \qquad \sec A = \frac{1}{\sqrt{1 - x^2}}$$

$$\tan A = \frac{x}{\sqrt{1 - x^2}} \qquad \cot A = \frac{\sqrt{1 - x^2}}{x}$$
∎

EXAMPLE 5 Show that $(\sin \theta)(\cos \theta) = \dfrac{x\sqrt{25 - x^2}}{25}$ where θ is an acute angle in a right triangle having one side of measure x.

Solution The difference $25 - x^2$, in conjunction with the Pythagorean theorem, suggests a right triangle with hypotenuse 5 as shown below.

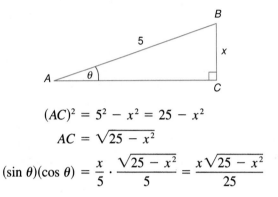

If a given expression in x involved the sum $25 + x^2$, rather than the difference $25 - x^2$, then the two perpendicular sides of the triangle would be labeled with 5 and x, and the hypotenuse would be $\sqrt{25 + x^2}$.

$$(AC)^2 = 5^2 - x^2 = 25 - x^2$$

$$AC = \sqrt{25 - x^2}$$

$$(\sin \theta)(\cos \theta) = \frac{x}{5} \cdot \frac{\sqrt{25 - x^2}}{5} = \frac{x\sqrt{25 - x^2}}{25}$$

(Observe that if $AC = x$, then $CB = \sqrt{25 - x^2}$, and the result would be the same.)
∎

Angles of $30°$, $45°$, and $60°$ will be of special importance to our work because their trigonometric ratios can be found using basic geometry. The trigonometric ratios of angles of other sizes will be studied in Sections 7.3 and 7.4.

Remember that the ratios are independent of the lengths of the sides of the triangle, so the choice of 1 for the equal sides is a matter of convenience.

Since an isosceles right triangle has acute angles of $45°$, it may be referred to as a $45°$–$45°$–$90°$ triangle. Now consider such a triangle whose equal sides are of unit length, solve for the hypotenuse c, and write the ratios.

$$c^2 = 1^2 + 1^2 = 2$$

$$c = \sqrt{2}$$

TRIGONOMETRIC RATIOS OF 45°

When necessary, $\dfrac{1}{\sqrt{2}}$ can be expressed with a rational denominator;

$$\frac{1}{\sqrt{2}} = \frac{1}{\sqrt{2}} \cdot \frac{\sqrt{2}}{\sqrt{2}} = \frac{\sqrt{2}}{2}$$

$$\sin 45° = \frac{1}{\sqrt{2}} \qquad \csc 45° = \frac{\sqrt{2}}{1} = \sqrt{2}$$

$$\cos 45° = \frac{1}{\sqrt{2}} \qquad \sec 45° = \frac{\sqrt{2}}{1} = \sqrt{2}$$

$$\tan 45° = \frac{1}{1} = 1 \qquad \cot 45° = \frac{1}{1} = 1$$

Angles of 30° and 60° are complementary since $30° + 60° = 90°$. Consequently, only one triangle is needed to find the ratios for each angle: a 30°–60°–90° triangle. If the side opposite 30° in such a triangle is assigned the unit length 1, then the lengths of the sides can be found as follows: Extend BC to D so that $DC = CB$. Then triangles ABC and ACD are congruent (SAS), so that $AB = 2$. Also, $(AC)^2 = 2^2 - 1^2 = 3$, $AC = \sqrt{3}$.

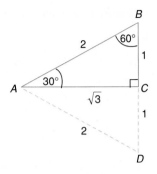

We may use this right triangle to write the trigonometric ratios of 30° and 60°.

Observe that $\dfrac{1}{\sqrt{3}}$ and $\dfrac{2}{\sqrt{3}}$ may be expressed with rational denominators. Thus

$$\frac{1}{\sqrt{3}} = \frac{1}{\sqrt{3}} \cdot \frac{\sqrt{3}}{\sqrt{3}} = \frac{\sqrt{3}}{3}$$

$$\frac{2}{\sqrt{3}} = \frac{2}{\sqrt{3}} \cdot \frac{\sqrt{3}}{\sqrt{3}} = \frac{2\sqrt{3}}{3}$$

TRIGONOMETRIC RATIOS OF 30° AND 60°			
$\sin 30° = \dfrac{1}{2}$	$\csc 30° = 2$	$\sin 60° = \dfrac{\sqrt{3}}{2}$	$\csc 60° = \dfrac{2}{\sqrt{3}}$
$\cos 30° = \dfrac{\sqrt{3}}{2}$	$\sec 30° = \dfrac{2}{\sqrt{3}}$	$\cos 60° = \dfrac{1}{2}$	$\sec 60° = 2$
$\tan 30° = \dfrac{1}{\sqrt{3}}$	$\cot 30° = \sqrt{3}$	$\tan 60° = \sqrt{3}$	$\cot 60° = \dfrac{1}{\sqrt{3}}$

EXERCISES 7.1

Find the six trigonometric ratios of θ.

1.

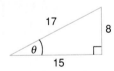

2.

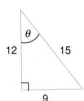

3.

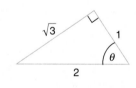

4.

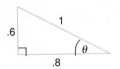

5.

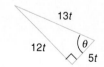

6.

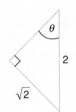

Draw an appropriate right triangle and find θ.

7. $\sin\theta = \dfrac{1}{\sqrt{2}}$ **8.** $\tan\theta = \dfrac{1}{\sqrt{3}}$ **9.** $\tan\theta = \sqrt{3}$

10. $\cos\theta = \frac{1}{2}$ **11.** $\csc\theta = 2$ **12.** $\cot\theta = 1$

Two sides of right △ABC are given in which ∠C is the right angle. Use the Pythagorean theorem to find the third side and then write the six trigonometric ratios of (a) angle A and (b) angle B.

13. $a = 6; b = 8$ **14.** $a = 13; b = 5$ **15.** $a = 4; b = 5$

16. $a = 1; b = 1$ **17.** $a = 1; b = 2$ **18.** $a = 4; b = 10$

19. $a = 3; c = 4$ **20.** $a = 20; c = 29$ **21.** $b = 2; c = 5$

22. $b = 4; c = 7$ **23.** $a = 21; c = 29$ **24.** $a = \frac{1}{3}; b = \frac{1}{4}$

Use triangle XYZ to find the value for each expression.

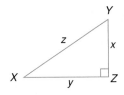

25. $(\tan X)(\cot X)$ **26.** $(\cos Y)(\sec Y)$ **27.** $(\sin X)\left(\dfrac{1}{\csc X}\right)$

28. $(\tan Y)\left(\dfrac{1}{\cot Y}\right)$ **29.** $\sin^2 X + \cos^2 X$ **30.** $\sin^2 Y + \cos^2 Y$

31. $\sec^2 X - \tan^2 X$ **32.** $\csc^2 Y - \cot^2 Y$

Angle C is the right angle of right △ABC. Use the given information to find the other five ratios of the indicated angle.

33. $\sin A = \frac{3}{4}$ **34.** $\cos B = \frac{2}{5}$ **35.** $\tan A = \frac{9}{40}$

36. $\cot A = 1$ **37.** $\sec B = \dfrac{\sqrt{11}}{\sqrt{2}}$ **38.** $\csc B = 10$

Use the information given for △ABC and verify the indicated equality.

39. $(\sin A)(\cos A) = x\sqrt{1 - x^2}$ **40.** $\tan^2 A = \dfrac{x^2}{9 - x^2}$

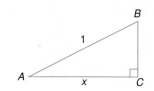   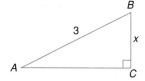

41. $\dfrac{4\sin^2 A}{\cos A} = \dfrac{x^2}{\sqrt{16 - x^2}}$ **42.** $(\sin A)(\cos A) = \dfrac{2x}{4 + x^2}$

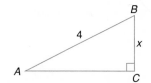   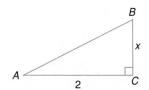

Show that the equation is true for an appropriate acute angle θ in a right triangle having one side whose measure is x.

43. $(\sin \theta)(\tan \theta) = \dfrac{x^2}{2\sqrt{4 - x^2}}$

44. $\sec^3 \theta = \dfrac{8}{(4 - x^2)^{3/2}}$

45. $(\tan^2 \theta)(\cos^2 \theta) = \dfrac{x^2}{x^2 + 16}$

46. $(\sin^3 \theta)(\cot^3 \theta) = \dfrac{(x^2 - 5)^{3/2}}{x^3}$

47. The side of an equilateral triangle is s centimeters long. Express the area as a function of s.

△ 48. The figure shows an equilateral triangle that is perpendicular to the plane of the circle $x^2 + y^2 = 9$ with one of its sides coinciding with a chord of the circle that is perpendicular to the x-axis. Write the area of the triangle as a function of x and give its domain. (*Hint:* $s = 2y$.)

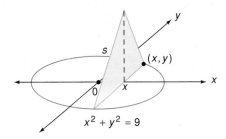

△ 49. Replace the equilateral triangle in Exercise 48 by a right isosceles triangle having one of its equal sides coinciding with the chord that is perpendicular to the x-axis, and find the area of the triangle.

△ 50. Replace the equilateral triangle in Exercise 48 by a right isosceles triangle with its hypotenuse coinciding with the chord that is perpendicular to the x-axis, and find the area of the triangle.

△ 51. The wire AB is 100 inches long and is cut into two pieces at point P as shown. If AP is used to form an equilateral triangle, and PB is used for a circle, express the total area of the two resulting figures as a function of x and state its domain.

| 7.2
ANGLE
MEASURE | Angles are often measured in degrees, the traditional method inherited from the Babylonians. They used a numeration system based upon groups of 60 and did much to influence the manner in which we measure. In this system a circle is divided into 360 equal parts, each of which is called a *degree*. |

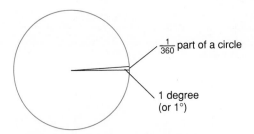

Each degree can be divided into 60 equal parts called min-utes, and each minute can be divided into 60 equal parts called seconds. In our work, however, degrees will be used to the nearest tenth of a degree.

The trigonometric ratios of nonacute angles will be stud-ied in Section 7.4.

Here are illustrations of some angles using degree measure.

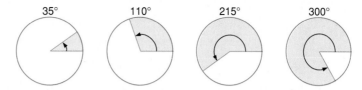

Just as straight-line distances can be measured using different kinds of units (inches, centimeters), there are also different ways to measure angles. The most important of these angular measurements in mathematics is the **radian.** Furthermore, just as there are conversion factors to change inches to centimeters, and centimeters to inches, you will see that there are also conversion factors to change degrees to radians and radians to degrees. We begin with the following definition of a radian.

DEFINITION OF A RADIAN

One **radian** is the measure of a central angle of a circle that is subtended by an arc whose length is equal to the radius of the circle.

In the following figure $\angle AOB$ has been doubled so that $\angle AOC$ has measure 2 radians and arc length $AC = 2r$. This demonstrates that

$$\text{Arc length} = \left(\begin{array}{c} \text{angle measure} \\ \text{in radians} \end{array} \right) \times (\text{radius})$$

$$2r = (2) \times (r)$$

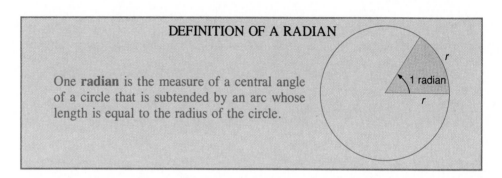

Greek letters such as θ are of-ten used to name an angle and to indicate the measure. Thus we write $\theta = 2$ to mean that the angle θ has measure 2 radians.

Using s for the arc length, θ for the angle measure, and r for the radius, the preceding can be generalized as follows.

$$s = r\theta \quad \text{or} \quad \theta = \frac{s}{r}$$

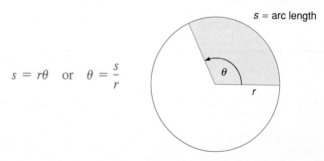

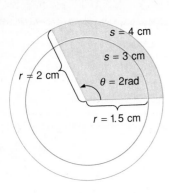

We then say that the measure of angle θ is given by the quotient $\frac{s}{r}$; that is, angle θ has a measure of $\frac{s}{r}$ radians. Note that the size of the circle does not affect the radian measure of the angle; the measure is the *ratio* of intercepted arc length to the radius. This is demonstrated in the figure in the margin, where $\theta = \frac{s}{r} = \frac{4}{2} = \frac{3}{1.5} = 2$ radians. In particular, when $s = r$ then $\theta = \frac{s}{r} = 1$, and the angle has a measure of 1 radian.

From geometry we know that the circumference C, of a circle is given by the formula $C = 2\pi r$. Thus $\frac{C}{r} = 2\pi$, so there are 2π radians in a complete rotation of 360° about a point. Therefore

Since $2r = d$, the diameter of the circle, $C = 2\pi r$, gives $\pi = C/d$, showing that π is the ratio of the circumference to the diameter of the circle. This ratio is the same for all circles.

$$2\pi \text{ radians} = 360°$$

Dividing by 2 produces this fundamental relationship between radians and degrees:

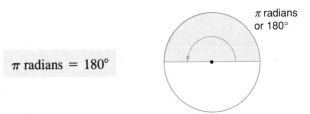
π radians or 180°

$$\pi \text{ radians} = 180°$$

When both sides of the equation $180° = \pi$ radians are divided by 180, we obtain this conversion formula:

$1°$ contains $\frac{\pi}{180}$ radians,

which is approximately 0.0175 radian. You can verify this using a calculator.

$$1° = \frac{\pi}{180} \text{ radian}$$

Since each degree equals $\frac{\pi}{180}$ radian, degrees can be converted into radians by multiplying the number of degrees by $\frac{\pi}{180}$.

EXAMPLE 1

We can also convert 30° to radians by using a proportion:

$$\frac{30}{180} = \frac{x}{\pi}$$

(a) Convert 30° and 135° into radians. Express the results in terms of π.
(b) Use the conversion factor 0.0175 to change the degree measures in part (a) into approximate radian measure to three decimal places.

Solution

(a) $30° = \left(30 \cdot \dfrac{\pi}{180}\right) \text{ radian} = \dfrac{\pi}{6} \text{ radian}$

For simplicity, we use the equals sign and write 30° = 0.525 radian, even though it is not an exact equality.

$135° = \left(135 \cdot \dfrac{\pi}{180}\right) \text{ radians} = \dfrac{3\pi}{4} \text{ radian}$

(b) $30° = (30)(0.0175) \text{ radian} = 0.525 \text{ radian}$

$135° = (135)(0.0175) \text{ radians} = 2.363 \text{ radians}$ ∎

From now on, radian measures will (in most instances) be stated without using the word "radian." Thus the angle measure 2 automatically means 2 radians unless degree measure is explicitly stated.

EXAMPLE 2   A central angle in a circle of radius 4 centimeters is 75°. Find the length of the intercepted arc to the nearest tenth of a centimeter.

Solution   First change to radians:

$$75° = \left(75 \cdot \frac{\pi}{180}\right)$$

$$= \frac{5\pi}{12}$$

Then

$$s = r\theta = 4\left(\frac{5\pi}{12}\right)$$

$$= \frac{5\pi}{3} = 5.2 \text{ centimeters} \qquad \text{(to one decimal place)}$$

The conversion formula to change from radians to degrees is obtained by dividing both sides of the equation π radians $= 180°$ by π.

A radian contains $\left(\frac{180}{\pi}\right)°$, which is approximately 57.296°.

$$\boxed{1 \text{ radian} = \frac{180}{\pi} \text{ degrees}}$$

Since each radian equals $\dfrac{180}{\pi}$ degrees, radians can be converted to degrees by multiplying the number of radians by $\dfrac{180}{\pi}$.

EXAMPLE 3

We can also express $\frac{\pi}{4}$ in degree measure by using a proportion:

$$\frac{\frac{\pi}{4}}{\pi} = \frac{x}{180}$$

(a) Express $\dfrac{\pi}{4}$ and $\dfrac{5\pi}{6}$ in degree measure.

(b) Convert $\frac{3}{5}$ radian to degrees. State the result in terms of π and also to the nearest tenth of a degree.

Solution

(a) $\dfrac{\pi}{4} = \left(\dfrac{\pi}{4} \cdot \dfrac{180}{\pi}\right)° = 45°$; $\dfrac{5\pi}{6} = \left(\dfrac{5\pi}{6} \cdot \dfrac{180}{\pi}\right)° = 150°$

(b) $\dfrac{3}{5} = \left(\dfrac{3}{5} \cdot \dfrac{180}{\pi}\right)° = \left(\dfrac{108}{\pi}\right)°$; $\dfrac{3}{5} = \left(\dfrac{3}{5}\right)(57.296) = 34.4°$ (to the nearest tenth of a degree)

The following discussion shows how angular measure in radians is used to determine the area of a **sector of a circle.**

In the figure notice that the area of the shaded sector depends on the central angle θ. That is, $A = k \cdot \theta$, where k is some constant to be determined. To find k, we take the special case where $\theta = 2\pi$. This means that we have the entire circle whose area $A = \pi r^2$. Using this information, we may then proceed as follows:

$$A = k \cdot \theta$$

$$\pi r^2 = k \cdot 2\pi$$

$$k = \tfrac{1}{2} r^2$$

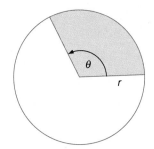

Thus the area of a sector of a circle with radius r and central angle θ in radians is given by

$$A = \tfrac{1}{2} r^2 \theta$$

EXAMPLE 4 Find the area of a sector of a circle of radius 6 centimeters if the central angle is $60°$.

Solution First convert $60°$ to radian measure:

$$60° = 60 \times \frac{\pi}{180} = \frac{\pi}{3}$$

Then $A = \dfrac{1}{2}(6)^2\left(\dfrac{\pi}{3}\right) = 6\pi$ square centimeters. ∎

Angular measure is also used to study motion on a circle. First, recall how speed is determined when the motion is linear. When a car travels 120 miles in 3 hours, its speed is $120 \div 3 = 40$ miles per hour (mph). Or, if an object travels 120 feet in 3 seconds, its speed is 40 feet per second (ft/sec). In general,

The speed is assumed to be constant.

Speed is the distance traveled per unit of time.

This concept also applies to circular motion. Suppose that one end of a level board is tangent to a circle at point P and that the board is pushed horizontally at a constant rate, causing the circle to rotate. Assuming that the board remains tangent to the circle and that there is no slipping, you can see that the arc length s that P has moved is the same as the straight line or linear distance the board has moved. Thus, for a point moving on a circle of radius r we define its **linear speed** v to be the arc length s it has traveled per unit of time.

$$v = \frac{\text{distance}}{\text{time}} = \frac{s}{t}$$

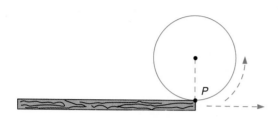

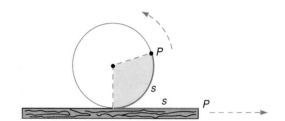

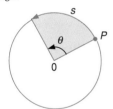

ω is the lowercase Greek letter omega.

Similarly, the **angular speed** ω is defined as the number of radians turned (angular distance) per unit of time. Therefore, when OP rotates through θ radians, the angular speed is given by

$$\omega = \frac{\text{angular rotation}}{\text{time}} = \frac{\theta}{t}$$

Substituting $s = r\theta$ into $v = s/t$ gives $v = (r\theta)/t = r(\theta/t) = r\omega$, and we have the following relationship between the linear and angular speeds;

$$v = r\omega$$

When applying these results, θ is in radians, the same linear units are used for r and s, and the same units of time are used for v and ω.

EXAMPLE 5   Assuming the earth's radius to be 4000 miles, find the linear speed of a point on the equator in mph and also in ft/sec. What is the angular speed in radians per hour (rad/hr)?

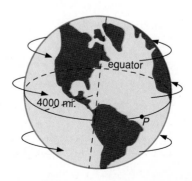

SECTION 7.2 Angle Measure   353

Solution   As the earth makes one complete rotation in 24 hours, the angular rotation is 2π radians. Thus

$$v = \frac{s}{t} = \frac{r\theta}{t} = \frac{(4000)(2\pi)}{24} = 1047$$

The linear speed is 1047 mph rounded to the nearest mph. Since 1 mile = 5280 feet and 1 hour = (60)(60) = 3600 seconds,

$$v = 1047 \text{ mph} = 1047(5280) \text{ ft/hr}$$

$$= \frac{1047(5280)}{3600} \text{ ft/sec}$$

$$= 1536 \text{ ft/sec} \qquad \text{(rounded to the nearest foot per second)}$$

Note:

$$\frac{\pi}{12} \, rad/hr = \frac{\pi}{12}\left(\frac{180}{\pi}\right)$$

$$= 15° \, per \, hour.$$

Also, the angular speed ω is given by:

$$\omega = \frac{\theta}{t} = \frac{2\pi}{24} = \frac{\pi}{12} \, \text{rad/hr} \qquad ∎$$

EXERCISES 7.2

Convert the degrees to radians. Leave the answers in terms of π.

1. 45°   2. 60°   3. 90°   4. 180°   5. 270°
6. 360°   7. 150°   8. 135°   9. 225°   10. 240°
11. 210°   12. 300°   13. 330°   14. 345°   15. 75°

Convert the degrees to radians and also use the conversion factor 0.0175 to give the results to two decimal places.

16. 10°   17. 100°   18. 220°   19. 340°

Convert the radians to degrees.

20. $\dfrac{\pi}{2}$ 21. π 22. $\dfrac{3\pi}{2}$ 23. 2π 24. $\dfrac{\pi}{3}$

25. $\dfrac{5\pi}{9}$ 26. $\dfrac{\pi}{6}$ 27. $\dfrac{2\pi}{3}$ 28. $\dfrac{3\pi}{4}$ 29. $\dfrac{5\pi}{4}$

30. $\dfrac{7\pi}{6}$ 31. $\dfrac{5\pi}{3}$ 32. $\dfrac{\pi}{12}$ 33. $\dfrac{5\pi}{18}$ 34. $\dfrac{11\pi}{36}$

Convert the radians to degrees and also use the conversion factor 57.296 to give the results to the nearest tenth of a degree.

35. $\dfrac{\pi}{15}$ 36. $\dfrac{1}{2}$ 37. 3   38. 6

Find the trigonometric ratios.

39. $\sin \dfrac{\pi}{4}$ 40. $\cos \dfrac{\pi}{3}$ 41. $\tan \dfrac{\pi}{6}$ 42. $\sec \dfrac{\pi}{6}$

Solve for the indicated part s, r, θ (in radians).

43.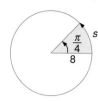

44.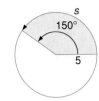

45.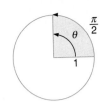

46.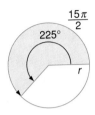

47.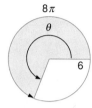

48.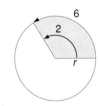

A circle has a radius of 12 centimeters. Find the area of a sector of this circle for each given central angle.

49. 30°   **50.** 90°   **51.** 135°   **52.** 225°   **53.** 315°

54. The area of a sector of a circle with radius 6 centimeters is 15 square centimeters. Find the measure of the central angle of the sector in degrees.

55. Find the area of a sector of a circle whose radius is 2 inches if the length of the intercepted arc is 8 inches.

56. The area of a sector of a circle with radius 4 centimeters is $\dfrac{16\pi}{3}$ square centimeters. Find the measure of the central angle of the sector in degrees.

57. Find the area of a circular sector with central angle 45° if the length of the intercepted arc is $\dfrac{\pi}{2}$ centimeters. Give the exact answer in terms of π, and approximate the area to the nearest tenth of a square centimeter.

58. A flower garden is a 270° sector with a 6-foot radius. Find the exact area of the garden in terms of π, and approximate the area to the nearest tenth of a square foot.

59. A curve along a highway is an arc of a circle with a 250-meter radius. If the curve is 50 meters long, by how many degrees does the highway change its direction? Give the answer to the nearest degree.

60. A 40-inch pendulum swings through an angle of 15°. Find the length of the arc through which the tip of the pendulum swings to the nearest tenth of an inch.

61. A cup is in the shape of a right circular cone made from a circular sector with an 8-inch radius and a central angle of 270°. Find the surface area of the cup to the nearest tenth of a square inch. (Use $\pi = 3.14$.)

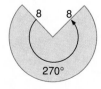

62. Find the area of the sector inscribed in the square $ABCD$.

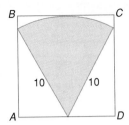

63. The equilateral triangle ABC in the following figure represents a wooden platform standing in a lawn. A goat is tied to a corner with a 15-foot rope. Assuming that the goat can not climb onto the platform, what is the maximum amount of grazing area available to the goat? (Assume that the rope does not go over the platform.)

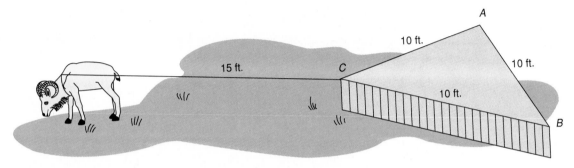

64. The shaded region is called a segment of the circle. Show that its area is given by $\frac{1}{2}r^2(\theta - \sin\theta)$.

65. (a) Suppose that a satellite has a circular orbit that is 300 miles above the earth's surface and that one revolution around the earth takes 1 hour and 45 minutes. Using $r = 4000$ miles as the radius of the earth, find the satellite's linear speed in mph and also in ft/sec. (Use a calculator and round the answers to the nearest mph and nearest ft/sec.)

(b) Find the angular speed of the satellite in rad/hr.

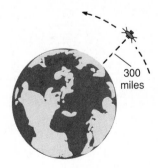

66. Answer the questions in Exercise 65 if the satellite has a circular orbit that is 1000 miles above the earth and if it remains above the same point on the earth throughout its orbit.

67. Suppose that a fly is sitting on the end of the minute hand of a clock. Find the angular speed of the fly in rad/min and in rad/hr.

68. A point on the rim of a wheel is turning with an angular speed of 220 rad/sec. If the wheel's diameter is 5 feet, find the linear speed in ft/sec.

69. (a) The outer diameter of each wheel on a bicycle is 22 inches. If the wheels are turning at a rate of 240 revolutions per minute, find the linear speed of the bicycle.

(b) Use a calculator to estimate how many minutes it takes the bicycle to travel 1 mile.

22 in.

70. A car is traveling at 50 mph. Find the revolutions per minute of the wheels if the car's tires have a 26-inch diameter.

71. A motorcycle is traveling on a curve along a highway. The curve is an arc of a circle with a radius of $\frac{1}{4}$ mile. If the motorcycle's speed is 42 mph, what is the angle through which the motorcycle will turn in $\frac{1}{2}$ minute?

 **Written Assignment:** Which is bigger, an angle of 1° or an angle of 1 radian? Discuss and justify your observations.

EXPLORATIONS
Think Critically

1. Consider all acute angle measures θ in a right triangle, and decide what the possible values for $\sin \theta$ can be.

2. Suppose in a right $\triangle ABC$, with right angle at C, $AB = 1$ unit. What single measurement would be an approximation of $\sin A$? of $\cos A$? For each of the remaining four ratios, decide which side should have length 1 so that one measurement will give that ratio.

3. Let $\triangle ABC$ be a right triangle with right angle of C and base $AC = 1$. Assume that point B is moving vertically upward from C, and let θ represent the increasing angle measures at point A. By comparing appropriate pairs of sides of the resulting triangles ABC, discuss the behavior of $\tan \theta$, $\sin \theta$, and $\cos \theta$ as θ increases toward 90°.

4. The measure of the central angle in a circle subtended by an arc that is $\frac{1}{100}$ part of a circle is called a gradient. (This is another way of measuring angles, and you may find that your calculator can be set in the gradient or GRAD mode.) Find the conversion factors for changing degrees into gradients and vice versa, and do the same for radians and gradients.

5. Explain why the linear speed on a circle increases as the radius increases for a fixed angular speed $\omega > 0$.

Can you find the distance across the lake between the two cottages at B and C? A surveyor can do this without getting wet. This, and many more problems involving distances that cannot be measured directly, can be solved with the aid of trigonometric ratios.

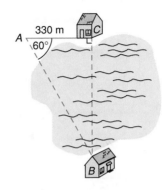

Tan A will be used because this ratio involves the unknown side and the given side. Cot A can also be used.

Here is one way the surveyor can find BC. First locate a point A so that the resulting triangle ABC has a right angle at C and a $60°$ angle at A. From A to C is along flat ground and can be measured directly. Assume that $AC = 330$ meters and solve for BC as follows:

A transit is an instrument that surveyors can use to determine such angles. You may have seen a transit being used in highway or building construction.

$$\tan A = \frac{BC}{AC}$$

$$\sqrt{3} = \frac{BC}{330} \qquad (\tan 60° = \sqrt{3}, AC = 330)$$

$$BC = 330\sqrt{3} = 571.6 \qquad \text{(to one decimal place)}$$

Thus the distance across the lake between the cottages is about 572 meters.

EXAMPLE 1   Solve for x. Give both the exact value and the approximation to the nearest tenth of a unit.

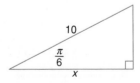

Solution   Both the cosine and secant ratios involve the sides labeled x and 10. The cosine is somewhat easier to use.

$$\cos \frac{\pi}{6} = \frac{x}{10}$$

$$\frac{\sqrt{3}}{2} = \frac{x}{10}$$

Note that $5\sqrt{3}$ is an exact answer, whereas 8.7 is an approximation to one decimal place.

$$x = 5\sqrt{3} = 8.7$$

■

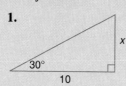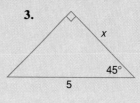
The trigonometric ratios given in Table VI are determined using more advanced mathematics. They are really decimal approximations. Rough approximations of some of these ratios can be made by constructing accurate figures and taking careful measurements.

The preceding problems called for trigonometric ratios of the special angles 30°, 45°, and 60°, angles whose ratios had already been determined. However, many problems do not involve such special acute angles. In such cases it will become necessary to refer to Table VI at the back of this book or to use a calculator. A portion of this table is reproduced here.

Deg	sin	cos	tan	cot	sec	csc	
30.0	.5000	.8660	.5774	1.7321	1.1547	2.0000	60.0
30.1	.5015	.8652	.5797	1.7251	1.1559	1.9940	59.9
30.2	.5030	.8643	.5820	1.7182	1.1570	1.9880	59.8
30.3	.5045	.8634	.5844	1.7113	1.1582	1.9821	59.7
30.4	.5060	.8625	.5867	1.7045	1.1594	1.9762	59.6
33.5	.5519	.8339	.6619	1.5108	1.1992	1.8118	56.5
33.6	.5534	.8329	.6644	1.5051	1.2006	1.8070	56.4
33.7	.5548	.8320	.6669	1.4994	1.2020	1.8023	56.3
33.8	.5563	.8310	.6694	1.4938	1.2034	1.7976	56.2
33.9	.5577	.8300	.6720	1.4882	1.2048	1.7929	56.1
34.0	.5592	.8290	.6745	1.4826	1.2062	1.7883	56.0
34.1	.5606	.8281	.6771	1.4770	1.2076	1.7837	55.9
34.2	.5621	.8271	.6796	1.4715	1.2091	1.7791	55.8
34.3	.5635	.8261	.6822	1.4659	1.2105	1.7745	55.7
34.4	.5650	.8251	.6847	1.4605	1.2120	1.7700	55.6
34.5	.5664	.8241	.6873	1.4550	1.2134	1.7655	55.5
34.6	.5678	.8231	.6899	1.4496	1.2149	1.7610	55.4
34.7	.5693	.8221	.6924	1.4442	1.2163	1.7566	55.3
34.8	.5707	.8211	.6950	1.4388	1.2178	1.7522	55.2
34.9	.5721	.8202	.6976	1.4335	1.2193	1.7478	55.1
35.0	.5736	.8192	.7002	1.4281	1.2208	1.7434	55.0
	cos	sin	cot	tan	csc	sec	Deg

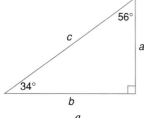

$$\tan 34° = \frac{a}{b} = \cot 56°$$

Similarly,

$$\sin 34° = \frac{a}{c} = \cos 56°$$

$$\sec 34° = \frac{c}{b} = \csc 56°$$

First notice that in the left-hand columns the angle measures are given in degrees to the nearest tenth of a degree, ranging from 0° through 45°. The right-hand columns, which are read from the bottom of each page to the top, then proceed from 45° through 90°. Thus to find the ratio of an angle measure from 0° through 45°, you read the angle measure at the left and go down the appropriate column according to the headings on top. To find the ratio of an angle measure between 45° and 90°, read the angle measure in the column on the right and use the headings that appear at the bottom of the table.

To find tan 34°, for example, we locate 34.0° at the left. Then locate the column headed "tan" at the top. This row and column intersect at the entry that gives the tangent of 34°, namely .6745. Notice that this same entry applies for the cotangent of 56°. Thus locate 56.0° in the column at the right and the column headed "cot" at the bottom. The intersection here is also .6745, which is what we should expect; that is, tan 34° = cot 56°.

A calculator, set in the degree mode, will give the same values when rounded to four decimal places. For example,

$$\cot 26° = \frac{1}{\tan 26°}$$

$$= 26 \boxed{\text{TAN}} \boxed{1/x}$$

$$= 2.0503$$

EXAMPLE 2 Use Table VI to find (approximate) each trigonometric ratio.
(a) cos 13°   **(b)** cot 26°   **(c)** sin 58.2°   **(d)** csc 83.7°

Solution   **(a)** .9744   **(b)** 2.0503   **(c)** .8496   **(d)** 1.0061   ■

EXAMPLE 3 Find θ to the nearest tenth of a degree using Table VI and by calculator.
(a) $\sin \theta = .4741$ **(b)** $\cot \theta = .4625$

Solution   Since the trigonometric ratio is given, we now need to reverse the process (used in Example 2) of finding the ratio for a given θ.

Just as the $\boxed{\text{SIN}}$ *key gives the sine ratio for a given* θ, *the* $\boxed{\text{INV}}$ $\boxed{\text{SIN}}$ *keys give* θ *for a given sine ratio. Some calculators would use* $\boxed{\text{2nd}}$ $\boxed{\text{SIN}^{-1}}$

(a) Examine the two columns headed "sin" at the top and bottom of the table. The entry .4741 is found in the column with the "sin" heading at the top, opposite 28.3°. Thus $\theta = 28.3°$. When doing this reverse process on a calculator, the inverse key $\boxed{\text{INV}}$ is used as follows to find θ.

$$\theta = \boxed{.4741} \boxed{\text{INV}} \boxed{\text{SIN}} = 28.300767° = 28.3°$$ (rounded to one decimal place)

When using this table to find an angle measure of a given ratio that is not in the table, we use the angle measure of the closest ratio.

Also, if sec θ *is given,* θ *is found using the keys* $\boxed{1/x}$ $\boxed{\text{INV}}$ $\boxed{\text{COS}}$. *Similarly for* θ *if* csc θ *is given.*

(b) Examine the columns headed by "cot" at the top and bottom. The closest entry to .4625 is .4621 in the column with the "cot" heading at the bottom. Since this entry is opposite 65.2° at the far right, $\theta = 65.2°$. If the calculator does not have a $\boxed{\text{COT}}$ key, then θ in cot $\theta = .4625$ can be found by first observing that $\tan \theta = \dfrac{1}{.4625}$, which makes it possible to use this sequence:

$$\theta = \boxed{.4625} \boxed{1/x} \boxed{\text{INV}} \boxed{\text{TAN}} = 65.179459° = 65.2°$$ (rounded to one decimal place)   ■

TEST YOUR UNDERSTANDING
Think Carefully

(Answers: Page 387)

Use Table VI for these exercises. You may use a calculator to confirm your results. Find each trigonometric ratio.

1. sin 27°   **2.** cos 38°   **3.** tan 72°   **4.** cot 58°

Find θ in degree measure.

5. $\cos \theta = .3907$ **6.** $\tan \theta = 1.4281$ **7.** $\sin \theta = .8960$ **8.** $\sec \theta = 1.0794$

Here are some general guidelines for solving trigonometric problems that involve geometric figures:

1. After carefully reading the problem, draw a figure that matches the given information. Try to draw it as close to scale as possible.
2. Record any given values directly on the corresponding parts, and label the required unknown parts with appropriate letters.
3. Determine the trigonometric ratios and geometric formulas that can be used to solve for an unknown part, and solve.

The examples that follow make use of these guidelines.

EXAMPLE 4 The **angle of elevation** of an 80-foot ramp leading to a bridge above a highway is 10.5°. Find the height of the bridge above the highway.

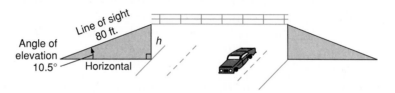

Applying the general guidelines will help you to develop your critical thinking skills.

1. After rereading the problem, draw a figure.
2. Label the known angle with its measure, the given side, and the unknown side.
3. Select an appropriate trigonometric ratio to solve for h. The sine or cosecant can be used.

Solution First note that the angle of elevation is the angle between the horizontal and the line of sight to the top of an object. Since the hypotenuse is given, and the side opposite the 10.5° angle is the unknown, use the sine ratio.

$$\sin 10.5° = \frac{h}{80}$$

By Table VI: $h = 80 \sin 10.5° = 80(.1822) = 14.576$

By calculator: $h = \boxed{80} \boxed{\times} \boxed{10.5} \boxed{\text{SIN}} \boxed{=} 14.578842$

Then $h = 14.6$ feet to the nearest tenth of a foot.   ∎

EXAMPLE 5 From the top of a house the **angle of depression** of a point on the ground is 25°. The point is 35 meters from the base of the building. How high is the building?

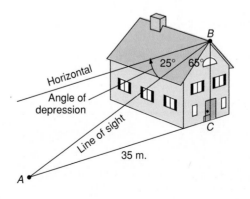

Solution By the **angle of depression** we mean the angle between the horizontal and the line of sight viewed from the top of an object to a point below. As you will notice in the figure, this angle of 25° is not within the triangle. However, the complement of the angle, 65°, is the measure of $\angle B$ inside the triangle. Thus we may write this ratio:

$$\cot 65° = \frac{x}{35} \qquad (x = BC)$$

$$x = 35 \cot 65° = 35(.4663) = 16.3205 \qquad \text{(by Table VI)}$$

$$= \boxed{35} \boxed{\times} \boxed{65} \boxed{\text{TAN}} \boxed{1/x} \boxed{=} \ 16.320768 \qquad \text{(by calculator)}$$

Then the height = 16 meters to the nearest meter. ∎

EXAMPLE 6 The top of a hill is 40 meters higher than a nearby airport, and the horizontal distance from the end of a runway is 325 meters from a point directly below the hilltop. An airplane takes off at the end of the runway in the direction of the hill, at an angle that is to be kept constant until the hill is passed. If the pilot wants to clear the hill by 30 meters, what should be the angle at takeoff?

Solution Sketch a figure and let the takeoff angle be θ. Since the side opposite θ is 40 + 30, the tangent ratio gives

$$\tan \theta = \frac{70}{325}$$
$$= .2154$$

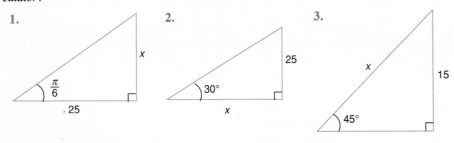

The closest entry in the tangent column is .2162, which corresponds to 12.2°. Therefore, $\theta = 12.2°$ to the nearest tenth of a degree. Using a calculator in the degree mode,

$$\theta = \boxed{70} \boxed{\div} \boxed{325} \boxed{=} \boxed{\text{INV}} \boxed{\text{TAN}} \ 12.154942 = 12.2° \qquad \text{(rounded to one decimal place)} \ ∎$$

EXERCISES 7.3

Solve for the indicated part of each right triangle without using trigonometric tables or a calculator.

1.

x

$\dfrac{\pi}{6}$

. 25

2.

25

30°

x

3.

x

15

45°

4.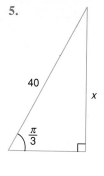

5.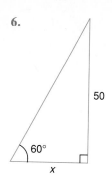

6.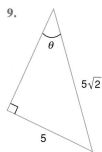

7.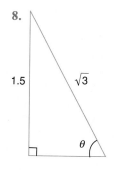

8.

9.

Find the six trigonometric ratios for each θ.

10. $\theta = 78.8°$ **11.** $\theta = 35.3°$ **12.** $\theta = 24.7°$

Find the trigonometric ratios.

13. $\sin 42°$	**14.** $\cos 48°$	**15.** $\tan 12°$
16. $\cot 78°$	**17.** $\sec 89°$	**18.** $\csc 1°$
19. $\sin 2.6°$	**20.** $\tan 45.1°$	**21.** $\cot 44.9°$
22. $\cos 75.4°$	**23.** $\sec 19.2°$	**24.** $\csc 5.5°$

Find the measure of θ in degrees to the nearest tenth of a degree.

25. $\tan \theta = 6.314$ **26.** $\sin \theta = .7214$ **27.** $\cot \theta = .4592$
28. $\sec \theta = 14.31$ **29.** $\cos \theta = .9940$ **30.** $\csc \theta = 2.763$

Triangle ABC is a right triangle with $C = 90°$. Solve for the indicated part. Give the angle measures and sides to the nearest tenth.

31. $A = 70°$, $a = 35$; find b.   **32.** $A = 70°$, $a = 35$; find c.
33. $B = 42.3°$, $a = 20$; find b.   **34.** $B = 42.3°$, $a = 20$; find c.
35. $a = 1$, $b = 3$; find B.   **36.** $a = 12$, $b = 9.5$; find A.
37. $b = 9$, $c = 25$; find B.   **38.** $A = 15.5°$, $c = 48$; find a.

39. The angle of elevation to the top of a flagpole is 35° from a point 50 meters from the base of the pole. What is the height of the pole to the nearest meter?

40. How high is a building whose horizontal shadow is 50 meters when the angle of elevation of the sun is 60.4°? Give the answer to the nearest tenth of a meter.

41. At a point 100 feet away from the base of a giant redwood tree, a surveyor measures the angle of elevation to the top of the tree to be 70°. How tall is the tree to the nearest tenth of a foot?

42. From the top of a 172-foot-high water tank, the angle of depression to a house is 13.3°. How far away is the house from the water tank to the nearest tenth of a foot?

43. An observation post along a shoreline is 225 feet above sea level. If the angle of depression from this post to a ship at sea is 6.7°, how far is the ship from the shore to the nearest foot?

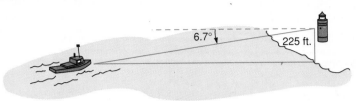

44. Suppose that a kite string forms a 42.5° angle with the ground when 740 feet of string are out. What is the altitude of the kite to the nearest foot? (Assume that the string forms a straight line.)

45. One of the cables that helps to stabilize a telephone pole is 82 feet long and is anchored into the ground 14.5 feet from the base of the pole. Find the angle that the cable makes with the ground.

46. One of the equal sides of an isosceles triangles is 18.7 units long and the vertex angle is 33°. Find the length of the base to the nearest tenth of a unit.

47. One side of an inscribed angle of a circle is a diameter of the circle, and the other side is a chord of length 10. If the inscribed angle is 66°, what is the length of the radius to the nearest tenth?

48. From the top of a 250-foot cliff, the angle of depression to the far side of a river is 12°, and the angle of depression to a point directly on the opposite side of the river is 62°. How wide is the river between the two points to the nearest foot?

49. A surveyor finds that from point A on the ground the angle of elevation to the top of a mountain is 23°. When he is at a point B that is $\frac{1}{4}$ mile closer to the base of the mountain, the angle of elevation is 43°. How high is the mountain? One mile = 5280 feet. (Assume that the base of the mountain and the two observation points are on the same line.) Give the answer to the nearest foot.

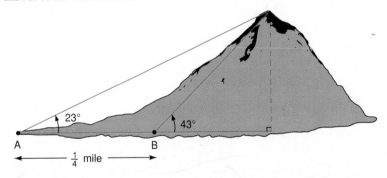

50. A 12-foot flagpole is standing vertically at the edge of the roof of a building. The angle of elevation to the top of the flagpole from a point on the ground that is 64 feet from the building is 78.6°. Find the height of the building to the nearest foot.

51. A circle with a 20-centimeter radius is inscribed in a regular pentagon. Find the perimeter of the pentagon to the nearest centimeter.

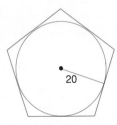

52. In the figure a wedge is formed by a plane cutting into a right circular cylinder of radius 2 at an angle of θ. The plane intersects the base of the cylinder in a diameter of the circular base. Express the area of the triangular cross section as a function of x if

    (a) $\theta = 30°$ **(b)** $\theta = 45°$ **(c)** $\theta = 60°$

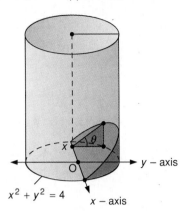

The six trigonometric ratios were defined only for acute angles: $0 < \theta < \dfrac{\pi}{2}$. Now we are going to extend these definitions to include all angles. First observe that an acute angle may be placed in a rectangular coordinate system so that its *initial side* lies on the positive part of the x-axis and its *terminal side* is in the first quadrant. Since the trigonometric ratios do not depend on the size of the sides of the right triangle, we may choose the sides so that the length of the terminal side (the hypotenuse) is 1. Consequently, the intersection of the terminal side of θ with the **unit circle** centered at the origin is a point $P(x, y)$ whose coordinates satisfy $x^2 + y^2 = 1$.

*An angle with its vertex at the origin and initial side on the positive part of the x-axis is said to be in **standard position**.*

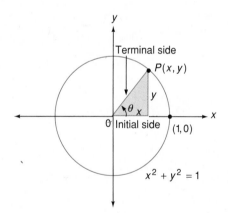

From the right triangle we have $\cos \theta = \dfrac{x}{1} = x$ and $\sin \theta = \dfrac{y}{1} = y$. Therefore, we see that the earlier definitions

$$\sin \theta = \frac{\text{opposite}}{\text{hypotenuse}} \quad \text{and} \quad \cos \theta = \frac{\text{adjacent}}{\text{hypotenuse}}$$

are equivalent to the following:

Note that sin θ is the second
coordinate of point P and
cos θ is the first coordinate of
P. That is,

$(x, y) = (\cos \theta, \sin \theta)$

$$\sin \theta = y \quad \text{and} \quad \cos \theta = x$$

where (x, y) are the coordinates of the point of intersection of the terminal side of θ with the unit circle whose center is at the origin.

Now look at angles of other sizes. Regardless of the size of θ, its terminal side will always intersect the unit circle at some point $P(x, y)$. For positive angles the terminal side is found by rotating counterclockwise; for negative angles we rotate clockwise. Some typical cases are shown below.

Observe that in each case angle θ is in standard position since its initial side is on the positive part of the x-axis, and its vertex is at the origin.

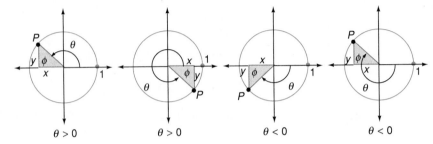

For each case the right triangle formed by drawing the perpendicular from P to the x-axis is called the **reference triangle,** and the acute angle between the terminal side and the x-axis is called the **reference angle.** The reference angle ϕ (phi) is a positive acute angle: $0 < \phi < \dfrac{\pi}{2}$. For example, the reference angle of $\dfrac{2\pi}{3}$ is $\dfrac{\pi}{3}$, and for $-\dfrac{5\pi}{4}$ it is $\dfrac{\pi}{4}$. We use the reference triangle to define the trigonometric ratios of θ as follows:

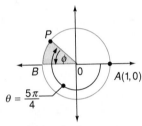

Reference angle *BOP*
of measure $\phi = \dfrac{\pi}{4}$

$$\text{vertical side} = y = \sin \theta$$

$$\text{horizontal side} = x = \cos \theta$$

The remaining trigonometric ratios are formed as before; thus

$$\tan \theta = \frac{y}{x} = \frac{\sin \theta}{\cos \theta} \qquad \sec \theta = \frac{1}{x} = \frac{1}{\cos \theta}$$

$$\cot \theta = \frac{x}{y} = \frac{\cos \theta}{\sin \theta} \qquad \csc \theta = \frac{1}{y} = \frac{1}{\sin \theta}$$

The signs of the ratios depend on the quadrant that $P(x, y)$ is in.

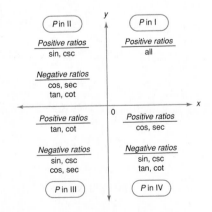

EXAMPLE 1 Find the six trigonometric ratios for the obtuse angle

$$\theta = 150° \left(\text{or } \frac{5\pi}{6} \right)$$

Solution Construct the 30°-60°-90° reference triangle in quadrant II and find the values of x and y for point P. $y = \frac{1}{2}$ because the side opposite the 30° reference angle is one-half the hypotenuse 1. Also, using the Pythagorean theorem,

$$x^2 = 1^2 - \left(\frac{1}{2} \right)^2 = \frac{3}{4}$$

$$x = -\frac{\sqrt{3}}{2} \qquad \longleftarrow \quad \text{Use the negative root since } P \text{ is in quadrant II.}$$

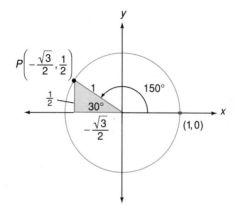

Then

$$\sin 150° = y = \frac{1}{2} \qquad\qquad \csc 150° = \frac{1}{y} = 2$$

$$\cos 150° = x = -\frac{\sqrt{3}}{2} \qquad\qquad \sec 150° = \frac{1}{x} = -\frac{2}{\sqrt{3}}$$

$$\tan 150° = \frac{y}{x} = -\frac{1}{\sqrt{3}} \qquad\qquad \cot 150° = \frac{x}{y} = -\sqrt{3} \qquad\qquad ■$$

The trigonometric ratios of an angle θ can also be found by making direct use of the reference angle ϕ. To do this, first find the ratios of the acute angle ϕ. Then determine the correct signs by noting the quadrant containing the terminal side of θ. In Example 1, $\phi = 30°$ and the terminal side of $\theta = 150°$ is in quadrant II. Thus,

$$\sin 150° = \sin 30° = \frac{1}{2}$$

$$\cos 150° = -\cos 30° = -\frac{\sqrt{3}}{2}$$

$$\tan 150° = -\tan 30° = -\frac{1}{\sqrt{3}}$$

Provided that there is a reference angle, this demonstrates how to find the ratios of an angle of any size by using the ratios of the acute reference angle.

EXAMPLE 2 Find $\tan\left(-\dfrac{5\pi}{4}\right)$.

Solution

(a) For $-\dfrac{5\pi}{4}$ (or $-225°$) we have a $45°$–$45°$–$90°$ reference triangle in quadrant II. Then the coordinates of P are $\left(-\dfrac{1}{\sqrt{2}}, \dfrac{1}{\sqrt{2}}\right)$ and

$$\tan\left(-\frac{5\pi}{4}\right) = \frac{\dfrac{1}{\sqrt{2}}}{-\dfrac{1}{\sqrt{2}}} = -1$$

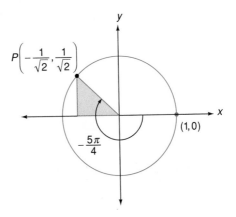

(b) Since $-\dfrac{5\pi}{4}$ is an quadrant II and the reference angle is $\dfrac{\pi}{4}$,

$$\tan\left(-\frac{5\pi}{4}\right) = -\tan\frac{\pi}{4} \quad\longleftarrow\quad \text{The tangent is negative in the second quadrant.}$$

$$= -1 \qquad\qquad\qquad\qquad \blacksquare$$

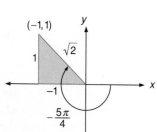

Since the trigonometric ratios do not depend on the size of the triangle, the ratios for an angle in standard position can also be found by using a reference triangle whose shortest side, rather than the hypotenuse, has measure other than 1. Thus, in Example 2, we could use the reference triangle that is shown in the figure at the left.

$$\tan\left(-\frac{5\pi}{4}\right) = \frac{1}{-1} = -1$$

You can see that there is an arithmetic advantage in using this size reference triangle in comparison to the one used in Example 2. In the next example, a similar style $30°$–$60°$–$90°$ reference triangle is used.

EXAMPLE 3 Find $\csc \theta$ if $\cos \theta = \dfrac{\sqrt{3}}{2}$ and $\cot \theta = -\sqrt{3}$.

Solution θ is in quadrant IV since $\cos \theta > 0$ and $\cot \theta < 0$. Also, since $\cos \theta = \dfrac{\sqrt{3}}{2} = \dfrac{\text{adjacent}}{\text{hypotenuse}}$, the reference angle must be $\dfrac{\pi}{6}$, and we may use the following reference triangle whose sides are twice those of a $30°$–$60°$–$90°$ reference triangle with hypotenuse 1.

In the figure $\theta = \dfrac{11\pi}{6}$. How-ever, any θ with the same ter-minal side could be used here.

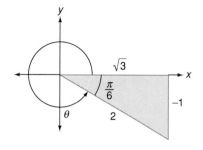

$$\csc \theta = \frac{2}{-1} = -2$$

When the reference angle is not one of the special angles, then Table VI can be used, as illustrated in Example 4. Also see the paragraph after Example 6 for the use of a calculator.

EXAMPLE 4 Find $\cos 200°$ using the reference angle.

Solution   The terminal side of $200°$ is in quadrant III. Consequently, $x < 0$. Since the reference angle is $20°$, we have

$$\cos 200° = -\cos 20° = -.9397 \qquad \text{(Table VI)}$$

When two angles have the same terminal side, they must have the same ratios. Such angles are said to be **coterminal.** Thus

$$\cos(-160°) = \cos 200° = -.9397$$

because the angles are coterminal, as shown in the figure at the left.

Coterminal angles

An angle that has its terminal side coinciding with one of the coordinate axes is called a **quadrantal angle.** As in the following figures, such angles do not have reference triangles. We use the coordinates of P to define the sine and cosine ratios.

Recall that for $P(x, y)$ on the unit circle we have $x = \cos\theta$ and $y = \sin\theta$.

Thus $\sin\dfrac{\pi}{2} = 1$, $\cos\dfrac{\pi}{2} = 0$, $\sin(-\pi) = 0$, and $\cos(-\pi) = -1$.

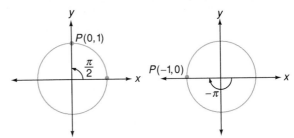

The remaining trigonometric ratios are defined by forming the appropriate fractions whenever possible. For example, $\tan\dfrac{\pi}{2}$ is undefined since $\dfrac{1}{0}$ is undefined, and

$$\cot\frac{\pi}{2} = \frac{x}{y} = \frac{0}{1} = 0.$$

It is unnecessary to memorize the ratios for quadrantal angles. Visualize the unit circle with the four points on the axes shown in the figure and use the coordinates of the appropriate point for the ratios.

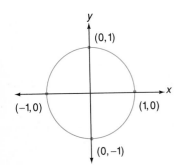

EXAMPLE 5 Determine the ratios of the quadrantal angles $\theta = \dfrac{3\pi}{2}$ and $\theta = -\dfrac{\pi}{2}$.

Solution Since the terminal side of $\dfrac{3\pi}{2}$ coincides with the negative part of the y-axis, use point $(0, -1)$ in the figure in the margin as follows:

$$\sin\frac{3\pi}{2} = -1 \qquad\qquad \csc\frac{3\pi}{2} = \frac{1}{-1} = -1$$

$$\cos\frac{3\pi}{2} = 0 \qquad\qquad \sec\frac{3\pi}{2} \text{ is undefined}$$

$$\tan\frac{3\pi}{2} \text{ is undefined} \qquad \cot\frac{3\pi}{2} = \frac{0}{-1} = 0$$

As demonstrated here, when determining the trigonometric ratios of a quadrantal angle, two of the ratios will be undefined.

The ratios for $-\dfrac{\pi}{2}$ are exactly the same as for $\dfrac{3\pi}{2}$ since the angles are coterminal. ∎

One way to find the trigonometric ratios of angles greater than 360° is to find the quadrant containing the terminal side and use the reference angle. For $\theta > 360°$, first *subtract* the largest multiple of 360° to get the coterminal angle between 0° and 360°. Now the quadrant of the terminal side has been found and the reference angle is used as before. For $\theta < -360°$, the process is similar except that we begin by *adding* an appropriate multiple of 360°.

EXAMPLE 6 Find **(a)** sec 932° and **(b)** csc(−383.6°) using the reference angles.

Solution

(a) Since 932° − 2(360°) = 212°, the terminal side is in quadrant III with reference angle 32°.

$$\sec 932° = -\sec 32° = -1.1792$$

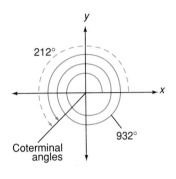

Coterminal angles

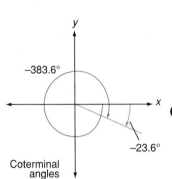

Coterminal angles

(b) Since −383.6° + 1(360°) = −23.6°, the terminal side is in quadrant IV with reference angle 23.6°.

$$\csc(-383.6°) = -\csc 23.6° = -2.4978$$ ■

The calculator is very efficient when the trigonometric ratios are not directly available in the tables or when the special reference angles are not involved. For instance, in Example 6(a), a calculator gives

The calculator has been constructed to give the ratios without having to enter the reference angles or quadrant locations.

$$\sec 932° = \frac{1}{\cos 932°} = \boxed{932}\ \boxed{\cos}\ \boxed{1/x} = -1.1792 \qquad \text{(to four decimal places)}$$

However, even though a calculator is recommended for such work, it is important for your understanding of trigonometry that you know how to use reference angles and triangles.

CAUTION: LEARN TO AVOID THESE MISTAKES	
WRONG	RIGHT
$\cos(-50°) = -\cos 50°$	$\cos(-50°) = \cos 50°$
$\cos 125° = \cos 55°$	$\cos 125° = -\cos 55°$
$\sin\left(-\dfrac{\pi}{3}\right) = \dfrac{\sqrt{3}}{2}$	$\sin\left(-\dfrac{\pi}{3}\right) = -\sin \dfrac{\pi}{3} = -\dfrac{\sqrt{3}}{2}$
$\sec \dfrac{\pi}{2} = \dfrac{1}{\cos \dfrac{\pi}{2}}$	$\cos \dfrac{\pi}{2} = 0$; $\sec \dfrac{\pi}{2}$ is undefined

EXERCISES 7.4

Assume that θ is in standard position.
(a) Find the quadrant containing the terminal side and find the reference angle φ.
(b) Find the coterminal angle that is between −360° and 0°.

1. $\theta = 73°$ **2.** $\theta = 300°$ **3.** $\theta = 500°$ **4.** $\theta = 1110.2°$ **5.** $\theta = \dfrac{5\pi}{4}$ **6.** $\theta = \dfrac{25\pi}{6}$

Assume that θ is in standard position.
(a) Find the quadrant containing the terminal side and find the reference angle φ.
(b) Find the coterminal angle that is between 0° and 360°.

7. $\theta = -73°$ **8.** $\theta = -201°$ **9.** $\theta = -850°$

10. $\theta = -735.9°$ **11.** $\theta = -\dfrac{11\pi}{3}$ **12.** $\theta = -\dfrac{25\pi}{6}$

Locate θ in a coordinate system, include the reference triangle, and find the reference angle.
Then find the coordinates of P (x, y) on the terminal side and on the unit circle, and write the
six trigonometric ratios.

13. $\theta = \dfrac{2\pi}{3}$ **14.** $\theta = \dfrac{29\pi}{6}$ **15.** $\theta = -\dfrac{7\pi}{4}$ **16.** $\theta = -30°$ **17.** $\theta = 405°$ **18.** $\theta = -930°$

Find the coordinates of P (x, y) on the unit circle and on the terminal side of θ. Write the
trigonometric ratios for θ.

19. $\theta = -\dfrac{\pi}{2}$ **20.** $\theta = 3\pi$ **21.** $\theta = -\dfrac{7\pi}{2}$

22. Complete this table of ratios that includes all quadrantal angles.

θ coterminal with	sin θ	cos θ	tan θ	cot θ	sec θ	csc θ
0						
$\dfrac{\pi}{2}$						
π						
$\dfrac{3\pi}{2}$						

Find each trigonometric ratio (if it exists). Use Table VI or a calculator only when necessary.

23. $\tan 220°$ **24.** $\sec(-72°)$ **25.** $\sin 261°$ **26.** $\sec(-\pi)$ **27.** $\csc(-\pi)$

28. $\tan \dfrac{7\pi}{2}$ **29.** $\cot\left(\dfrac{7\pi}{2}\right)$ **30.** $\tan 0$ **31.** $\cot 0$ **32.** $\cot\left(-\dfrac{5\pi}{2}\right)$

33. $\tan 8\pi$ **34.** $\sin \dfrac{9\pi}{2}$ **35.** $\cos(-275°)$ **36.** $\csc(-12°)$ **37.** $\cot 368°$

38. $\sin(242.4°)$ **39.** $\cos(-792.5°)$ **40.** $\tan 120°$ **41.** $\cot 1200°$ **42.** $\sec 420°$

43. $\csc(-80°)$ **44.** $\sin(94.3°)$ **45.** $\cos(-200.8°)$ **46.** $\tan 1.2°$ **47.** $\cos \dfrac{19\pi}{4}$

48. $\sec\left(-\dfrac{23\pi}{6}\right)$ **49.** $\csc(-480°)$ **50.** $\sin 585°$

Find the coordinates of P_1 through P_6 on the unit circle.

51.

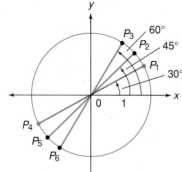

52.

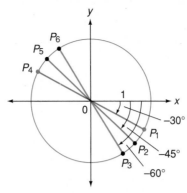

53. **(a)** Verify that $P\left(\dfrac{2}{3}, \dfrac{\sqrt{5}}{3}\right)$ is on the unit circle.

   (b) Locate θ, where $0 < \theta < 2\pi$, so that its terminal side intersects the unit circle at P.

   (c) Write the six trigonometric ratios for θ.

   (d) Use tables or a calculator to find an approximation for θ.

54. Follow the instructions in Exercise 53 for $P\left(-\dfrac{3}{4}, \dfrac{\sqrt{7}}{4}\right)$.

55. Find y so that $P\left(\dfrac{\sqrt{3}}{4}, y\right)$ is on the unit circle in the fourth quadrant. Use tables or a calculator to find θ, where $-2\pi < \theta < 0$, having OP as its terminal side.

56. The terminal side of θ coincides with the line $y = -x$ and lies in the second quadrant. Find $\cos \theta$.

Find all possible values for θ, $0 \le \theta < 2\pi$. Do not use tables or a calculator.

57. $\sin \theta = 1$ **58.** $\cos \theta = -1$ **59.** $\sin \theta = 0$ **60.** $\cos \theta = 0$

61. $\sin \theta = \frac{1}{2}$ **62.** $\cos \theta = -\frac{1}{2}$ **63.** $\cos \theta = -\dfrac{\sqrt{2}}{2}$ **64.** $\sin \theta = -\dfrac{\sqrt{3}}{2}$

Use the given information to determine the quadrant containing the terminal side of θ and find the remaining ratios.

65. $\tan \theta = 1$, $\sin \theta = -\dfrac{\sqrt{2}}{2}$

66. $\cos \theta = -\frac{1}{2}$, $\csc \theta = \dfrac{2}{\sqrt{3}}$

67. $\cot \theta = -\sqrt{3}$, $\sin \theta = -\frac{1}{2}$

68. $\sec \theta = \sqrt{2}$, $\cot \theta = 1$

69. Find $\cos \theta$ if $\dfrac{\pi}{2} < \theta < \pi$ and $\sin \theta = \frac{1}{3}$.

70. Find $\tan \theta$ if $-\dfrac{\pi}{2} < \theta < 0$ and $\sec \theta = \frac{5}{3}$.

71. Find $\sec \theta$ if the terminal side of θ is in quadrant I and $\csc \theta = 4$.

72. Find $\sin \theta$ if $\tan \theta = \sqrt{63}$ and $\cos \theta < 0$.

73. The terminal side of θ coincides with the line $y = \frac{1}{2}x$ and lies in the third quadrant. Find $\sin \theta$.

74. Explain why $\sin^2 \theta + \cos^2 \theta = 1$, where θ is any angle measure.

75. In the figure, AB is tangent to the circle at A and meets the terminal side of θ at B. Why is AB equal to $\tan \theta$? Which segment has measure equal to $\sec \theta$?

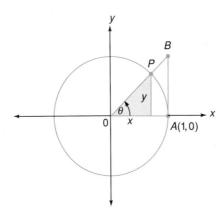

✏️ **Written Assignment:** Consider the definitions of $\sin \theta$ and $\cos \theta$ that were given using the unit circle on page 365. Use these definitions to find the possible range of values for $\sin \theta$ and $\cos \theta$, and justify your conclusions.

CHALLENGE
Think Creatively

Construct a figure, somewhat similar to the one in Exercise 75, that includes two line segments, one of which has length $\cot \theta$ and the other $\csc \theta$.

EXPLORATIONS
Think Critically

1. Suppose that the height of a building is known and that the angle of depression θ from a point Q at the top of the building to a point P on the ground has been measured. If point R is at ground level directly under Q, what trigonometric ratio of θ would be the simplest to use to find PR by using Table VI (not a calculator)? Explain.

2. The solution to Exercise 49, page 364, can be generalized after making some substitutions. Replace $23°$ by α, $43°$ by β, $\frac{1}{4}$ mile by d, and let $BC = x$. Now let h be the height of the mountain and show that $h = \dfrac{d \tan \alpha \tan \beta}{\tan \beta - \tan \alpha}$.

3. Study the figure in Exercise 75, and give a geometric explanation of why there is no tangent value when $\theta = \dfrac{\pi}{2}$.

4. For any value $x = \cos \theta$, $0 \leq \theta \leq \pi$, what is the value of $\sin \theta$ in terms of x?

5. Why are the tangent values for $\pi < \theta < \dfrac{3\pi}{2}$ the same as the tangent values for $0 < \theta < \dfrac{\pi}{2}$?

7.5
IDENTITIES

Identities will be useful later in solving equations, and eventually they will play an important role in a wide variety of mathematical situations.

Solving the equation $\dfrac{\sin^2 \theta}{\cos^2 \theta} = \dfrac{1}{3}$ for θ is easier to do if $\dfrac{\sin \theta}{\cos \theta}$ is first replaced by the equivalent expression $\tan \theta$ to get $\tan^2 \theta = \frac{1}{3}$. Such equations are discussed in detail in Section 8.5. Our objective here is to learn to recognize and work with **trigonometric identities**, such as $\dfrac{\sin \theta}{\cos \theta} = \tan \theta$. This equation is called an identity because it is a true statement for all values of θ except those when $\cos \theta = 0$.

> A trigonometric identity is an equation that is true for all values of the variable for which the expressions in the equation are defined.

The following list of fundamental identities, together with some alternative forms, will be used throughout our work. It is therefore important that you learn to recognize them.

Fundamental Identities

Fundamental Identity	Common Alternative Forms	Restrictions on θ
$\csc \theta = \dfrac{1}{\sin \theta}$	$\sin \theta = \dfrac{1}{\csc \theta}$ $\sin \theta \csc \theta = 1$	Not coterminal with 0, π.
$\sec \theta = \dfrac{1}{\cos \theta}$	$\cos \theta = \dfrac{1}{\sec \theta}$ $\cos \theta \sec \theta = 1$	Not coterminal with $\dfrac{\pi}{2}$, $\dfrac{3\pi}{2}$.
$\cot \theta = \dfrac{1}{\tan \theta}$	$\tan \theta = \dfrac{1}{\cot \theta}$ $\tan \theta \cot \theta = 1$	Not a quadrantal angle.
$\tan \theta = \dfrac{\sin \theta}{\cos \theta}$		Not coterminal with $\dfrac{\pi}{2}$, $\dfrac{3\pi}{2}$.
$\cot \theta = \dfrac{\cos \theta}{\sin \theta}$		Not coterminal with 0, π.
$\sin^2 \theta + \cos^2 \theta = 1$	$\sin^2 \theta = 1 - \cos^2 \theta$ $\cos^2 \theta = 1 - \sin^2 \theta$	None
$\tan^2 \theta + 1 = \sec^2 \theta$	$\tan^2 \theta = \sec^2 \theta - 1$	Not coterminal with $\dfrac{\pi}{2}$, $\dfrac{3\pi}{2}$.
$1 + \cot^2 \theta = \csc^2 \theta$	$\cot^2 \theta = \csc^2 \theta - 1$	Not coterminal with 0, π.

*The last three identities are called the **Pythagorean identities** because they are based on the Pythagorean theorem.*

Each of the preceding identities can be established by using the definitions of the trigonometric ratios on the unit circle $x^2 + y^2 = 1$, as shown next.

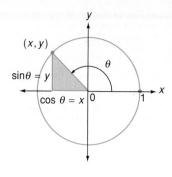

$$\csc \theta = \frac{1}{y} = \frac{1}{\sin \theta} \qquad \sec \theta = \frac{1}{x} = \frac{1}{\cos \theta}$$

$$\tan \theta = \frac{y}{x} = \frac{\sin \theta}{\cos \theta} \qquad \cot \theta = \frac{x}{y} = \frac{\cos \theta}{\sin \theta}$$

$$\tan \theta \cot \theta = \frac{y}{x} \cdot \frac{x}{y} = 1$$

$$\sin^2 \theta + \cos^2 \theta = x^2 + y^2 = 1 \qquad \text{(by the Pythagorean theorem)}$$

In practice, when working with identities, no special mention is made about the restrictions on the variable. However, you should be able to see what restrictions are necessary. In most cases such restrictions will occur when the denominator of an expression would be equal to 0.

Dividing the last equation by $\cos^2 \theta$ produces

$$\frac{\sin^2 \theta}{\cos^2 \theta} + \frac{\cos^2 \theta}{\cos^2 \theta} = \frac{1}{\cos^2 \theta}$$

$$\left(\frac{\sin \theta}{\cos \theta}\right)^2 + 1 = \left(\frac{1}{\cos \theta}\right)^2$$

$$\tan^2 \theta + 1 = \sec^2 \theta$$

Similarly, when $\sin^2 \theta + \cos^2 \theta = 1$ is divided by $\sin^2 \theta$, we get

$$1 + \cot^2 \theta = \csc^2 \theta$$

EXAMPLE 1 Use fundamental identities to convert $\dfrac{1 - \csc \theta}{\cot \theta}$ into an expression involving only sines and cosines, and simplify.

Solution

Here is another way to simplify the right side.

$$\frac{1 - \dfrac{1}{\sin \theta}}{\dfrac{\cos \theta}{\sin \theta}} = \frac{\sin \theta - 1}{\sin \theta} \cdot \frac{\sin \theta}{\cos \theta}$$

$$= \frac{\sin \theta - 1}{\cos \theta}$$

$$\frac{1 - \csc \theta}{\cot \theta} = \frac{1 - \dfrac{1}{\sin \theta}}{\dfrac{\cos \theta}{\sin \theta}} \longleftarrow \quad \begin{cases} \csc \theta = \dfrac{1}{\sin \theta} \\[2mm] \cot \theta = \dfrac{\cos \theta}{\sin \theta} \end{cases}$$

$$= \frac{1 - \dfrac{1}{\sin \theta}}{\dfrac{\cos \theta}{\sin \theta}} \cdot \frac{\sin \theta}{\sin \theta}$$

$$= \frac{\sin \theta - 1}{\cos \theta} \qquad \text{(multiplying fractions)} \quad \blacksquare$$

The result in Example 1 can be stated as

$$\frac{1 - \csc \theta}{\cot \theta} = \frac{\sin \theta - 1}{\cos \theta}$$

and the solution in Example 1 is the *proof* that this equation is an identity. The solution to Example 1 also suggests the following procedure.

"Verify the identity" is another way of saying "Prove that the given equation is an identity."

METHOD I FOR VERIFYING IDENTITIES

Use fundamental or previously proven identities to change one side of the equation into the form of the other side.

EXAMPLE 2   Verify this identity:

$$\frac{\tan^2 \theta + 1}{\tan \theta \csc^2 \theta} = \tan \theta$$

Solution   In most cases it is easier to work with the more complicated side. Supply a reason for each step of the proof.

$$\frac{\tan^2 \theta + 1}{\tan \theta \csc^2 \theta} = \frac{\sec^2 \theta}{\tan \theta \csc^2 \theta}$$

$$= \frac{\dfrac{1}{\cos^2 \theta}}{\dfrac{\sin \theta}{\cos \theta} \cdot \dfrac{1}{\sin^2 \theta}}$$

$$= \frac{1}{\cos^2 \theta} \cdot \frac{\cos \theta \sin^2 \theta}{\sin \theta}$$

$$= \frac{\sin \theta}{\cos \theta}$$

$$= \tan \theta \qquad \blacksquare$$

It is often helpful to convert the expression into one involving sines and cosines.

Here is another procedure that can be used to verify an identity.

METHOD II FOR VERIFYING IDENTITIES

Convert each side of the given equation until the same form is obtained on each side.

EXAMPLE 3   Verify the identity: $\dfrac{\csc\theta + 1}{\csc\theta - 1} = (\sec\theta + \tan\theta)^2$.

Solution   When working separately on each side, a vertical line helps to separate the work.

$$\dfrac{\csc\theta + 1}{\csc\theta - 1} \qquad\qquad (\sec\theta + \tan\theta)^2$$

$$\downarrow \qquad\qquad\qquad \downarrow$$

$$\dfrac{\dfrac{1}{\sin\theta} + 1}{\dfrac{1}{\sin\theta} - 1} \qquad\qquad \left(\dfrac{1}{\cos\theta} + \dfrac{\sin\theta}{\cos\theta}\right)^2 \qquad \text{(Convert to sines and cosines.)}$$

$$\downarrow \qquad\qquad\qquad \downarrow$$

$$\dfrac{\left(\dfrac{1}{\sin\theta} + 1\right)\sin\theta}{\left(\dfrac{1}{\sin\theta} - 1\right)\sin\theta} \qquad \left(\dfrac{1 + \sin\theta}{\cos\theta}\right)^2 \qquad \text{(Combine fractions.)}$$

$$\downarrow \qquad\qquad\qquad \downarrow$$

$$\dfrac{1 + \sin\theta}{1 - \sin\theta} \qquad\qquad \dfrac{(1 + \sin\theta)^2}{\cos^2\theta} \qquad \left(\dfrac{a}{b}\right)^2 = \dfrac{a^2}{b^2}$$

$$\downarrow$$

$$\dfrac{(1 + \sin\theta)^2}{1 - \sin^2\theta} \qquad \cos^2\theta = 1 - \sin^2\theta$$

$$\downarrow$$

$$\dfrac{(1 + \sin\theta)^2}{(1 - \sin\theta)(1 + \sin\theta)} \qquad a^2 - b^2 = (a - b)(a + b)$$

$$\downarrow$$

$$\dfrac{1 + \sin\theta}{1 - \sin\theta} \qquad \text{(Simplify.)}$$

Since the same form has been obtained for each side, the proof is complete.   ■

It takes practice to becomes good at verifying identities. You will have to call on a wide variety of algebraic skills to be successful. Example 4 calls on a process similar to the one used earlier in rationalizing denominators.

EXAMPLE 4   Verify the identity: $\dfrac{\sin\theta}{1 + \cos\theta} = \dfrac{1 - \cos\theta}{\sin\theta}$.

Solution   Begin by multiplying the numerator and denominator of the left side by $1 - \cos\theta$.

The motivation for this approach is that it gets $1 - \cos \theta$ into the numerator.

$$\frac{\sin \theta}{1 + \cos \theta} = \frac{(\sin \theta)(1 - \cos \theta)}{(1 + \cos \theta)(1 - \cos \theta)}$$

$$= \frac{(\sin \theta)(1 - \cos \theta)}{1 - \cos^2 \theta}$$

$$= \frac{(\sin \theta)(1 - \cos \theta)}{\sin^2 \theta}$$

$$= \frac{1 - \cos \theta}{\sin \theta} \qquad \blacksquare$$

*To disprove a statement that claims to be true in general, all that is needed is to find one exception, one case for which the statement is false. Such an exception is also called a **counterexample**.*

Notice that in all of the preceding solutions we avoided performing the same operation on each side of the given equality at the same time. Doing so could, at times, lead to false results. For example, $\sin \theta = -|\sin \theta|$ is certainly not an identity. Squaring both sides gives $\sin^2 \theta = \sin^2 \theta$, which might *incorrectly* suggest that the original statement is an identity.

Sometimes a trigonometric equation may have the appearance of an identity but really is not one. A quick way to discover this is to find one value of θ for which the expressions in the equation are defined but for which the equality is false.

EXAMPLE 5 Prove that $\sin^2 \theta - \cos^2 \theta = 1$ is *not* an identity.

Solution One counterexample is sufficient. Try $\theta = \dfrac{\pi}{4}$.

$$\sin^2 \frac{\pi}{4} - \cos^2 \frac{\pi}{4} = \left(\frac{1}{\sqrt{2}}\right)^2 - \left(\frac{1}{\sqrt{2}}\right)^2 = 0 \neq 1$$

Since the expressions in the equation are defined for $\dfrac{\pi}{4}$, but the equation is false for $\theta = \dfrac{\pi}{4}$, the equation cannot be an identity. $\qquad \blacksquare$

EXERCISES 7.5

Complete by using the fundamental identities.

1. $\dfrac{1}{\csc \theta} =$

2. $1 - \sin^2 \theta =$

3. $\csc^2 \theta - \cot^2 \theta =$

4. $\tan \theta \cos \theta =$

5. $\sin \theta \cot \theta =$

6. $-\dfrac{1}{\cos \theta} =$

Without using the trigonometric tables or a calculator, find the value of each of the following.

7. $\sin^2 39° + \cos^2 39°$

8. $\sin(-7°) \csc(-7°)$

9. $\tan^2(3.2°) - \sec^2(3.2°)$

10. $3 \cos^2 \dfrac{4\pi}{7} \sec^2 \dfrac{4\pi}{7}$

Show that each expression is equal to 1.

11. $(\sin \theta)(\cot \theta)(\sec \theta)$

12. $(\cos \theta)(\tan \theta)(\csc \theta)$

13. $\cos^2 \theta (\tan^2 \theta + 1)$

14. $\tan^2 \theta (\csc^2 \theta - 1)$

Find the values of θ (if any) for which the expressions are defined.

15. $\dfrac{1}{\cos \theta}$ **16.** $\tan \theta$ **17.** $\dfrac{1}{\tan \theta}$ **18.** $\cot \theta$ **19.** $\dfrac{\cos \theta}{\cot \theta}$ **20.** $\sec^2 \theta \csc \theta$

Use fundamental identities to convert the expressions into a form involving only sines or cosines of θ, and simplify. Also, find the restrictions on θ, if any.

21. $\dfrac{\tan \theta}{\cot \theta}$ **22.** $\cot \theta \sec^2 \theta$ **23.** $\dfrac{1 - \csc \theta}{\cot \theta}$

24. $\dfrac{1 - \cot^2 \theta}{\cot^2 \theta}$ **25.** $\dfrac{\sec \theta + \csc \theta}{\cos \theta + \sin \theta}$ **26.** $\dfrac{\sec \theta}{\tan \theta + \cot \theta}$

Express in terms of sin θ only.

27. $\cos^2 \theta - \sin^2 \theta$ **28.** $\tan^2 \theta \csc^2 \theta$

Express in terms of cos θ only.

29. $\dfrac{\cot \theta - \sin \theta}{\csc \theta}$ **30.** $\sec \theta - \sin \theta \tan \theta$

Verify each identity using Method I. Find the restrictions on θ, if any.

31. $\dfrac{\cos \theta}{\cot \theta} = \sin \theta$ **32.** $\cos^2 \theta - \sin^2 \theta = 1 - 2 \sin^2 \theta$

33. $(\tan \theta - 1)^2 = \sec^2 \theta - 2 \tan \theta$ **34.** $\dfrac{1}{\sec^2 \theta} = 1 - \dfrac{1}{\csc^2 \theta}$

Verify each identity using Method II. Find the restrictions on θ, if any.

35. $\sec \theta - \cos \theta = \sin \theta \tan \theta$ **36.** $(1 - \sin \theta)(1 + \sin \theta) = \dfrac{1}{1 + \tan^2 \theta}$

37. $\dfrac{\cot \theta - 1}{1 - \tan \theta} = \dfrac{\csc \theta}{\sec \theta}$ **38.** $\dfrac{1 + \sec \theta}{\csc \theta} = \sin \theta + \tan \theta$

Verify each identity using either method.

39. $\tan \theta + \cot \theta = \dfrac{1}{(\sin \theta)(\cos \theta)}$ **40.** $\cot^2 \theta - \cos^4 \theta \csc^2 \theta = \cos^2 \theta$

41. $(\sec \theta + \tan \theta)(1 - \sin \theta) = \cos \theta$ **42.** $\sec^4 \theta - \tan^4 \theta = 1 + 2 \tan^2 \theta$

43. $(\csc^2 \theta - 1) \sin^2 \theta = \cos^2 \theta$ **44.** $\sin^4 \theta - \cos^4 \theta + \cos^2 \theta = \sin^2 \theta$

45. $\tan^2 \theta - \sin^2 \theta = \sin^2 \theta \tan^2 \theta$ **46.** $\dfrac{\cot^2 \theta + 1}{\tan^2 \theta + 1} = \cot^2 \theta$

47. $\dfrac{1 + \sec \theta}{\csc \theta} = \sin \theta + \tan \theta$ **48.** $\dfrac{\sec^2 \theta}{\sec^2 \theta - 1} = \csc^2 \theta$

49. $\dfrac{1}{1 + \cos \theta} + \dfrac{1}{1 - \cos \theta} = 2 \csc^2 \theta$ **50.** $\dfrac{\tan \theta \sin \theta}{\sec^2 \theta - 1} = \cos \theta$

51. $\dfrac{1 + \tan^2 \theta}{1 + \cot^2 \theta} = \sec^2 \theta - 1$ **52.** $\dfrac{\sin \theta + \tan \theta}{1 + \sec \theta} = \sin \theta$

53. $\dfrac{\sin\theta + \cos\theta}{\sin\theta - \cos\theta} = \dfrac{\sec\theta + \csc\theta}{\sec\theta - \csc\theta}$

54. $\tan\theta\sec^2\theta = \dfrac{\sec\theta}{\csc\theta - \sin\theta}$

55. $\dfrac{1 - \cos\theta}{1 + \cos\theta} = (\csc\theta - \cot\theta)^2$

56. $\dfrac{1 + \tan\theta}{1 - \tan\theta} = \dfrac{\cot\theta + 1}{\cot\theta - 1}$

57. $(\sin^2\theta + \cos^2\theta)^5 = 1$

58. $\dfrac{2 + \cot^2\theta}{\csc^2\theta} - 1 = \sin^2\theta$

59. $\dfrac{1}{\cos^2\theta} + \dfrac{1}{\sin^2\theta} = \dfrac{1}{\sin^2\theta - \sin^4\theta}$

60. $\dfrac{1 + \sin\theta}{\cos\theta} = \dfrac{\cos\theta}{1 - \sin\theta}$

61. $\dfrac{\tan^2\theta + 1}{\tan^2\theta} = \csc^2\theta$

62. $\dfrac{\cot\theta}{\csc\theta + 1} = \sec\theta - \tan\theta$

63. $\dfrac{\tan\theta}{\sec\theta - 1} = \dfrac{\sec\theta + 1}{\tan\theta}$

64. $\dfrac{\cos\theta}{\csc\theta - 2\sin\theta} = \dfrac{\tan\theta}{1 - \tan^2\theta}$

65. $\dfrac{\cos^2\theta + 2\sin^2\theta}{\cos^3\theta} = \sec^3\theta + \dfrac{\sin^2\theta}{\cos^3\theta}$

66. $\dfrac{\cos\theta}{\csc\theta + 2\sin\theta} = \dfrac{\tan\theta}{1 + 3\tan^2\theta}$

67. $\dfrac{1}{(\csc\theta - \sec\theta)^2} = \dfrac{\sin^2\theta}{\sec^2\theta - 2\tan\theta}$

68. $\csc\theta - \cot\theta = \dfrac{1}{\csc\theta + \cot\theta}$

69. $\dfrac{\sec\theta}{\tan\theta - \sin\theta} = \dfrac{1 + \cos\theta}{\sin^3\theta}$

70. $\dfrac{1 - \cos\theta}{\sin\theta} + \dfrac{\sin\theta}{1 - \cos\theta} = 2\csc\theta$

△ **71.** $-\ln|\cos x| = \ln|\sec x|$

△ **72.** $\ln|\sin x| = -\ln|\csc x|$

△ **73.** $-\ln|\sec x - \tan x| = \ln|\sec x + \tan x|$

△ **74.** $\ln\left|\dfrac{\sin x}{1 - \cos x}\right| = -\ln\left|\dfrac{\sin x}{1 + \cos x}\right|$

Show that each equation is not an identity by finding one value for which the expressions in the equation are defined but for which the equation is false.

75. $\tan(-x) = \tan x$

76. $\tan\left(x + \dfrac{\pi}{2}\right) = \cot x$

77. $\sin x = -\sqrt{1 - \cos^2 x}$

78. $\tan x = \sqrt{\sec^2 x - 1}$

79. $\dfrac{\cot\theta - 1}{1 - \tan\theta} = \dfrac{\sec\theta}{\csc\theta}$

80. $\dfrac{1 + \sin\theta}{\cos\theta} = \dfrac{\sin\theta}{1 + \cos\theta}$

△ **81. (a)** Let θ be an acute angle of a triangle where $\tan\theta = \dfrac{x}{3}$. Show that

$$\ln(\sec\theta + \tan\theta) = \ln\left(\dfrac{\sqrt{x^2 + 9} + x}{3}\right)$$

(b) Let $F(x) = \ln\left(\dfrac{\sqrt{x^2 + 9} + x}{3}\right)$ and evaluate $F(4) - F(0)$.

CHALLENGE
Think Creatively

When it is unknown whether or not a trigonometric equation is an identity, it would be useless to try to prove that it is an identity when, in fact, it isn't. Likewise, it would be futile to try and prove that it isn't an identity when it really is. Thus, not knowing ahead of time what type of equation it is makes it more challenging to decide. You may have to try each possibility. Prove or disprove that the given equation is an identity.

(a) $\dfrac{\sin x}{\csc x} - \dfrac{\cos x}{\sec x} = 1$ **(b)** $\dfrac{1 + \csc x}{\sec x} = \cos x - \tan x$ **(c)** $\dfrac{\sec^2 x - 2}{(1 + \tan x)^2} = \dfrac{1 - \cot x}{1 + \cot x}$

REVIEW EXERCISES

The solutions to the following exercises can be found within the text of Chapter 7. Try to answer each question before referring to the text.

Section 7.1

1. Use the figure to explain why sin θ does not depend on the size of the triangle.

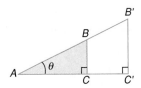

2. For a right triangle having sides of length 3, 4, and 5, find the sine, cosine, and tangent of the angle opposite the side of length 3.

3. For a right triangle having sides 5, 12, and 13, find the six trigonometric ratios of the angle opposite the side of length 12.

4. Triangle ABC has a right angle at C. If $\tan A = \frac{3}{2}$, find $\sin B$.

5. Using "side opposite," "side adjacent," and "hypotenuse," state the definitions of the trigonometric ratios of an acute angle θ in a right triangle.

6. Find the six trigonometric ratios for angle A in terms of x, where x is the side opposite acute angle A in a right triangle with hypotenuse 1.

7. Show that $(\sin A)(\cos A) = \dfrac{x\sqrt{25 - x^2}}{25}$ where $\angle A$ is an acute angle in a right triangle having one side of measure x.

8. Write the six trigonometric ratios of 45°.

9. Write the six trigonometric ratios of 30°.

10. Write the six trigonometric ratios of 60°.

Section 7.2

11. State the definition of a radian.   12. Convert 30° and 135° into radians.

13. Use the conversion factor 0.0175 to change the degree measures in Exercise 12 into approximate radian measure.

14. Convert $\dfrac{\pi}{4}$ and $\dfrac{5\pi}{6}$ into degree measure.

15. Convert $\frac{3}{5}$ radian into degrees and state the answer in terms of π. Also, use the conversion factor 57.296 to express $\frac{3}{5}$ radian to the nearest tenth of a degree.

16. If an arc length $s = 3$ cm subtends a central angle θ in a circle of radius 1.5 cm, find the measure of θ in terms of radians.

17. A central angle in a circle of radius 4 centimeters is 75°; find the length of the intercepted arc to the nearest tenth of a centimeter.

18. Find the area of a sector of a circle of radius 6 centimeters if the central angle is 60°.

19. State the relationship between the linear and angular speeds for a point moving on a circle.

20. Assuming the earth's radius to be 4000 miles, find the linear speed of a point on the equator in mph and also in ft/sec. What is the angular speed in rad/hr?

Section 7.3

21. Points B and C are on opposite sides of a lake and point A is situated so that triangle ABC has a right angle at C. If $\angle A = 60°$ and $AC = 330$ meters, find the distance across the lake from B to C.

22. Solve for the adjacent side of a $\dfrac{\pi}{6}$ angle in a right triangle with hypotenuse 10. Give the exact answer and the approximation to the nearest tenth.

23. Use Table VI to find these values:
    (a) cos 13°   (b) cot 26°   (c) sin 58.2°   (d) csc 83.7°

24. Use Table VI to solve for θ to the nearest tenth of a degree:
    (a) $\sin \theta = .4741$ (b) $\cot \theta = .4625$

25. The angle of elevation of an 80-foot ramp leading to a bridge above a highway is 10.5°. Find the height of the bridge above the highway.

26. From the top of a house the angle of depression of a point on the ground is 25°. The point is 35 meters from the base of the building. How high is the building?

27. The top of a hill is 40 meters higher than a nearby airport, and the horizontal distance from the end of a runway is 325 meters from a point directly below the hilltop. An airplane takes off at the end of the runway in the direction of the hill, at an angle that is to be kept constant until the hill is passed. If the pilot wants to clear the hill by 30 meters, what should be the angle at takeoff?

Section 7.4

28. If the terminal side of an angle in standard position is within one of the four quadrants, then some of its trigonometric ratios are positive. List the positive ratios for each quadrant.

29. Find the six trigonometric ratios for $\dfrac{5\pi}{6}$.

30. Express the sine, cosine, and tangent of 150° in terms of the sine, cosine, and tangent of its reference angle.

31. Find $\tan\left(-\dfrac{5\pi}{4}\right)$.

32. Find $\csc \theta$ if $\cos \theta = \dfrac{\sqrt{3}}{2}$ and $\cot \theta = -\sqrt{3}$.

33. Use Table VI and the reference angle to find cos 200°.

34. Determine the trigonometric ratios of $\dfrac{3\pi}{2}$ and $-\dfrac{\pi}{2}$.

35. Find sec 932° and csc(−383.6°).

36. Write two equivalent forms of $\sin \theta \csc \theta = 1$ and state the restrictions on θ.

37. Write two equivalent forms of $\cot \theta = \dfrac{1}{\tan \theta}$ and state the restrictions on θ.

38. Use the unit circle to explain why $\cos^2 \theta + \sin^2 \theta = 1$ is true for all θ.

39. Use the identity in Exercise 38 to derive $1 + \tan^2 \theta = \sec^2 \theta$ and state the restrictions on θ.

40. Convert $\dfrac{1 - \csc \theta}{\cot \theta}$ into an expression involving only $\sin \theta$ and $\cos \theta$.

Verify the identity.

41. $\dfrac{\tan^2 \theta + 1}{\tan \theta \csc^2 \theta} = \tan \theta$ 42. $\dfrac{\csc \theta + 1}{\csc \theta - 1} = (\sec \theta + \tan \theta)^2$ 43. $\dfrac{\sin \theta}{1 + \cos \theta} = \dfrac{1 - \cos \theta}{\sin \theta}$

44. Prove that $\sin^2 \theta - \cos^2 \theta = 1$ is *not* an identity.

CHAPTER 7 TEST: STANDARD ANSWER

Use these questions to test your knowledge of the basic skills and concepts of Chapter 7. Then check your answers with those given at the end of the book.

1. Use the given figure to write the following trigonometric ratios:
   (a) $\tan A$ **(b)** $\sin B$ **(c)** $\sec A$

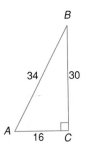

2. If $\angle A$ is an acute angle such that $\sec A = \frac{3}{2}$, find $\sin A$, $\cos A$, and $\tan A$.

3. Use the given triangle to show that

$$3(\sin^2 A)(\sec A) = \dfrac{x^2}{\sqrt{x^2 + 9}}$$

4. Right $\triangle ABC$, with right angle at C, has sides $b = 5$ and $c = 7$. Find the trigonometric ratios of the angle opposite side b.

5. Convert $\dfrac{27\pi}{20}$ radians into degrees and $160°$ into radians.

6. If the arc of a circle of radius 6 cm has length $s = 15$ cm, find the radian measure of the angle subtended by the arc. What is the degree measure of this angle?

7. A sector of a circle with radius 10 inches has a central angle of $120°$. Find the arc length of the sector.

8. Find each ratio (if it exists):
   (a) $\sin \dfrac{\pi}{3}$ **(b)** $\cot \dfrac{3\pi}{2}$ **(c)** $\sin(-135°)$

9. Find each ratio (if it exists): **(a)** $\tan 510°$ **(b)** $\cos 3\pi$ **(c)** $\csc\left(-\dfrac{7\pi}{6}\right)$

10. Find the trigonometric ratios of $\theta = -\dfrac{7\pi}{2}$.

11. Find the trigonometric ratios of $\theta = \dfrac{7\pi}{4}$.

12. From the top of a cliff, the angle of depression to a point 120 meters from the base of the cliff (and on the same level as the base) is 47.8°. Find the height of the cliff to the nearest tenth of a meter.

13. Solve for the hypotenuse of an isosceles right triangle, one of whose legs has length 20.

14. A building casts a shadow of 100 feet. The angle of elevation from the tip of the shadow to the top of the building is 75°. Find the height of the building correct to the nearest foot.

15. Right $\triangle ABC$, with right angle at C, has $a = 3.2$ and $b = 9.6$. Find the measure of $\angle B$ to the nearest tenth of a degree.

16. Express $\dfrac{\tan \theta + \cot \theta}{\sec \theta}$ in terms of $\sin \theta$ only.

17. Verify the identity $\dfrac{1}{1 - \sin \theta} + \dfrac{1}{1 + \sin \theta} = 2 \sec^2 \theta$ and find the restrictions on θ.

18. Find the exact value of $\cot \theta$ if the terminal side of θ is in quadrant III and $\cos \theta = -\frac{3}{8}$.

In Questions 19–21, verify the identities.

19. $\cot^2 \theta - \cos^2 \theta = \cos^2 \theta \cot^2 \theta$ 20. $\dfrac{1 - \sin \theta}{1 + \sin \theta} = (\sec \theta - \tan \theta)^2$

21. $\dfrac{\tan \theta - \sin \theta}{\csc \theta - \cot \theta} = \sec \theta - \cos \theta$

22. The terminal side of θ coincides with the line $y = 2x$ and lies in the first quadrant. Find $\cos \theta$.

23. A sector is inscribed in a 30°–60°–90° triangle as in the following figure. The side opposite the 30° angle is 2 units. Find the area of the shaded part.

24. Find $\sec \theta$ if $\sin \theta = -\dfrac{\sqrt{3}}{2}$ and $\tan \theta = \sqrt{3}$.

25. The wheels on a bicycle have an outer diameter of 26 inches. If the wheels are revolving 246 times per minute, find each of the following. (State the answers in terms of π.)
    (a) Find the angular speed in rad/min.   **(b)** Find the linear speed in ft/min.
    (c) How many miles will the bicycle travel in 30 minutes?

CHAPTER 7 TEST: MULTIPLE CHOICE

Do not use tables or calculators for this test.

1. Triangle ABC is a right triangle with right angle at C. If $\tan A = \frac{2}{3}$, find $\sec B$.
   (a) $\dfrac{\sqrt{13}}{3}$ **(b)** $\dfrac{\sqrt{13}}{2}$ **(c)** $\dfrac{2\sqrt{13}}{13}$ **(d)** $\dfrac{3\sqrt{13}}{13}$ **(e)** None of the preceding

2. Evaluate: $\sin 30° + \cos 60° - \tan 45°$

   (a) 1   (b) −1   (c) 0   (d) $1 + \dfrac{\sqrt{2}}{2}$ (e) None of the preceding

3. Express $\dfrac{11\pi}{6}$ radians in degree measure:

   (a) 330°   (b) 660°   (c) $\dfrac{180}{\pi}$ (d) $\left(\dfrac{11\pi}{6}\right)\left(\dfrac{\pi}{180}\right)$ (e) None of the preceding

4. What is the area in square centimeters of a sector of a circle of radius 6 cm if the central angle is 30°?

   (a) $\dfrac{\pi}{2}$ (b) 12π (c) 3π (d) 540   (e) None of the preceding

5. From the top of a building the angle of depression of a point on the ground is 35°. The point is 50 meters from the base of the building. Which of the following can be used to determine the height x of the building?

   (a) $\tan 35° = \dfrac{50}{x}$ (b) $\tan 55° = \dfrac{50}{x}$ (c) $\tan 55° = \dfrac{x}{50}$ (d) $\cot 35 = \dfrac{x}{50}$ (e) None of the preceding

6. Find $\tan\left(-\dfrac{3\pi}{4}\right)$

   (a) 1   (b) −1   (c) $\dfrac{\sqrt{2}}{2}$ (d) $-\dfrac{\sqrt{2}}{2}$ (e) None of the preceding

7. Find $\cos 330°$.

   (a) $\dfrac{1}{2}$ (b) $-\dfrac{1}{2}$ (c) $\dfrac{2}{\sqrt{3}}$ (d) $-\dfrac{\sqrt{3}}{2}$ (e) None of the preceding

8. Which of the following are correct?

   I. $\sin\left(-\dfrac{3\pi}{2}\right) = 1$

   II. $\cos\left(-\dfrac{3\pi}{2}\right) = 0$

   III. $\tan\left(-\dfrac{3\pi}{2}\right)$ is undefined.

   (a) Only I   (b) Only II   (c) Only III   (d) I, II, and III   (e) None of the preceding

9. Find $\sin(-420°)$.

   (a) $-\dfrac{1}{2}$ (b) $-\dfrac{\sqrt{3}}{2}$ (c) $\dfrac{\sqrt{3}}{2}$ (d) $-\dfrac{2}{\sqrt{3}}$ (e) None of the preceding

10. Which of the following are *correct*?

    I. $\cos(-30°) = \cos 30°$
    II. $\sin 125° = \sin 55°$
    III. $\tan(-45°) = -\tan 45°$
    (a) Only I   (b) Only II   (c) Only III   (d) I, II, and III   (e) None of the preceding

11. The point $P\left(\dfrac{1}{3}, y\right)$ where $y < 0$ is on the unit circle having center at the origin. Find $\tan \theta$ where θ is an angle in standard position whose terminal side intersects the unit circle at point P.

    (a) $-2\sqrt{2}$ (b) -2 (c) $\dfrac{-\sqrt{2}}{9}$ (d) $2\sqrt{2}$ (e) None of the preceding

12. Which of the following are true for all acute angles θ?
   I. $\sec^2 \theta - \tan^2 \theta = 1$
   II. $\csc^2 \theta - \cot^2 \theta = 1$
   III. $(\cos \theta)(\tan \theta) = \sin \theta$
   (a) Only I   **(b)** Only II   **(c)** Only III   **(d)** I, II, and III   **(e)** None of the preceding

13. Which of the following values of θ can be used to show that $\sin^2 \theta - \cos^2 \theta = 1$ is *not* an identity?
   (a) $\dfrac{\pi}{2}$ **(b)** $\dfrac{3\pi}{2}$ **(c)** $\dfrac{\pi}{4}$ **(d)** $-\dfrac{\pi}{2}$ **(e)** None of the preceding

14. A wheel having a 3-foot diameter is rotating on its axis at a speed of 280 rad/sec. How far has a point on the rim of the wheel traveled after 1 minute?
   (a) 420 ft   **(b)** 840 ft   **(c)** 25200 ft   **(d)** 50400 ft   **(e)** None of the preceding

15. When the expression $\sec^2 \theta + 2 \sec \theta \tan \theta + \tan^2 \theta$ is converted into an expression involving only $\sin \theta$, the result is
   (a) $\dfrac{1 + \sin^2 \theta}{1 - \sin^2 \theta}$ **(b)** $\dfrac{1 + \sin \theta}{1 - \sin \theta}$ **(c)** $\dfrac{1}{1 - \sin \theta}$ **(d)** $\dfrac{1 + 2 \sin \theta + \sin^2 \theta}{\sqrt{1 - \sin^2 \theta}}$
   (e) None of the preceding

ANSWERS TO THE TEST YOUR UNDERSTANDING EXERCISES

Page 344

1. and 2.
$\sin A = \frac{7}{25} = \cos B$ $\cot A = \frac{24}{7} = \tan B$
$\cos A = \frac{24}{25} = \sin B$ $\sec A = \frac{25}{24} = \csc B$
$\tan A = \frac{7}{24} = \cot B$ $\csc A = \frac{25}{7} = \sec B$

3. $\frac{49}{625} + \frac{576}{625} = \frac{625}{625} = 1$ 4. $\left(\frac{7}{25}\right)\left(\frac{25}{7}\right) = 1$ 5. $\frac{625}{49} - \frac{576}{49} = \frac{49}{49} = 1$ 6. $\left(\frac{24}{7}\right)\left(\frac{7}{24}\right) = 1$

Page 352

1. $\dfrac{\pi}{12}$ 2. $\dfrac{\pi}{5}$ 3. $\dfrac{\pi}{2}$ 4. $\dfrac{5\pi}{6}$ 5. $\dfrac{13\pi}{12}$ 6. $\dfrac{7\pi}{4}$ 7. $\dfrac{16\pi}{9}$ 8. 2π

9. $12°$ 10. $80°$ 11. $210°$ 12. $240°$ 13. $153°$ 14. $276°$ 15. $300°$ 16. $330°$

Page 359

1. $\dfrac{10\sqrt{3}}{3}$; 5.8   2. $10\sqrt{3}$; 17.3   3. $\dfrac{5\sqrt{2}}{2}$; 3.5

Page 360

1. .4540   2. .7880   3. 3.0777   4. .6249   5. $67°$ 6. $55°$ 7. $63.6°$ 8. $22.1°$

Page 369

1. $\dfrac{\sqrt{2}}{2}$ 2. $-\sqrt{2}$ 3. 1   4. $\dfrac{2\sqrt{3}}{3}$ 5. -1.5399 6. $-\sqrt{2}$ 7. $\dfrac{\sqrt{3}}{2}$ 8. 4.3315

8

THE TRIGONOMETRIC FUNCTIONS

8.1
THE SINE
AND COSINE
FUNCTIONS

On the unit circle, arc length is numerically the same as the radian measure of the central angle subtended by the arc. Therefore, the sine of θ radians, sin θ, may also be thought of as the sine of the arc length θ. But arc lengths may be regarded as real numbers, just as the lengths on a number line are given by real numbers. Consequently, sin θ can be interpreted as the sine of the *number* θ.

Remember that for a circle of radius r, $s = r\theta$ is the length of an arc with a central angle of θ radians (see page 349).

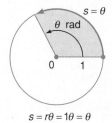

$s = r\theta = 1\theta = \theta$

Since sin θ is a unique value for any real number (radian), it is now correct to say that the equation $y = \sin \theta$ defines y to be a function of θ, with domain all real numbers. Also, since $y = \sin \theta$ is the second coordinate of the point on the terminal side of θ and on the unit circle, the range consists of all y such that $-1 \leq y \leq 1$.

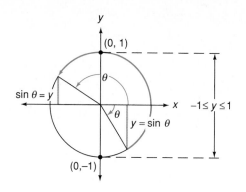

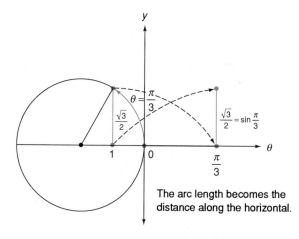

The arc length becomes the
distance along the horizontal.

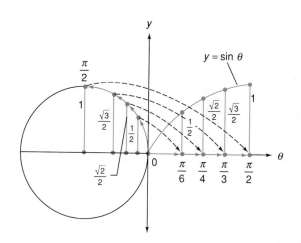

*You may think of the circle as
being "unrolled" to get the θ
values on the horizontal.*

One way to graph $y = \sin \theta$ on a rectangular system is to use the θ values on the horizontal axis. Then the corresponding values $y = \sin \theta$ become the ordinates.

Now that you see the connection between $y = \sin \theta$, defined on the unit circle, and its graph in a rectangular system, we will obtain a more accurate graph using a table of values. First recall that $\sin \theta$ is positive for θ in quadrants I and II; $0 < \theta < \pi$. Also, $\sin \theta$ is negative for $\pi < \theta < 2\pi$. Let us form a table of values by using intervals of $\dfrac{\pi}{6}$ for $0 \le \theta \le 2\pi$.

θ	0	$\dfrac{\pi}{6}$	$\dfrac{\pi}{3}$	$\dfrac{\pi}{2}$	$\dfrac{2\pi}{3}$	$\dfrac{5\pi}{6}$	π	$\dfrac{7\pi}{6}$	$\dfrac{4\pi}{3}$	$\dfrac{3\pi}{2}$	$\dfrac{5\pi}{3}$	$\dfrac{11\pi}{6}$	2π
$y = \sin \theta$	0	$\dfrac{1}{2}$	$\dfrac{\sqrt{3}}{2}$	1	$\dfrac{\sqrt{3}}{2}$	$\dfrac{1}{2}$	0	$-\dfrac{1}{2}$	$-\dfrac{\sqrt{3}}{2}$	-1	$-\dfrac{\sqrt{3}}{2}$	$-\dfrac{1}{2}$	0

We plot these points (using the approximation $\dfrac{\sqrt{3}}{2} = 0.87$) and connect the points with a smooth curve to find the graph of $y = \sin \theta$ for $0 \le \theta \le 2\pi$ as shown on the following page. The segment from zero to 2π on the horizontal axis may be viewed as the unit circle after it has been "unrolled."

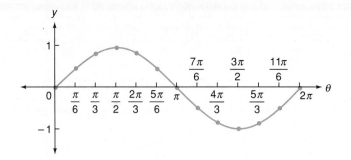

For values of θ, where $2\pi \leq \theta \leq 4\pi$, we know that the terminal sides of θ (in the unit circle) are the same as for those angles from 0 to 2π. Hence everything repeats. Similarly for $-2\pi \leq \theta \leq 0$, and so on. Thus we say that the sine function is *periodic,* with **period** 2π; $\sin(\theta + 2\pi) = \sin \theta$. That is, the sine curve repeats itself to the right and left as shown in the figure below.

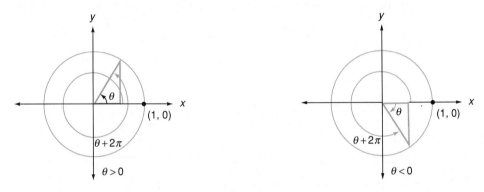

These two unit circle diagrams show coterminal angles θ and $\theta + 2\pi$. At the left θ is positive and at the right θ is negative. In either case, $\sin \theta = \sin(\theta + 2\pi)$.

A function f is said to be periodic with period p provided that p is the smallest positive constant, if any, such that $f(x + p) = f(x)$ for all x in the domain of f.

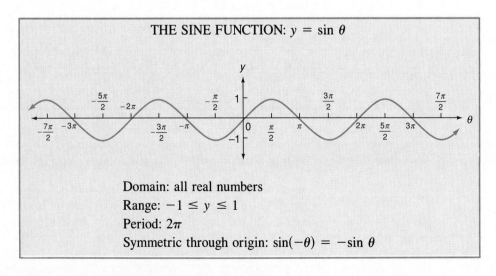

THE SINE FUNCTION: $y = \sin \theta$

Domain: all real numbers

Range: $-1 \leq y \leq 1$

Period: 2π

Symmetric through origin: $\sin(-\theta) = -\sin \theta$

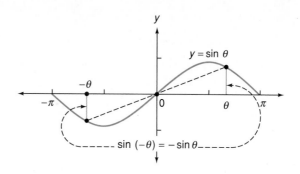

The graph of $y = \sin \theta$ indicates that the sine function is symmetric through the origin. This symmetry can be observed in the figure above and it can be verified by returning to the unit circle below.

Letting $f(\theta) = \sin \theta$, we have

$$f(-\theta) = \sin(-\theta) = -\sin(\theta) = -f(\theta)$$

$$f(-\theta) = -f(\theta)$$

A similar unit circle diagram can be drawn when the terminal sides of θ and $-\theta$ are in quadrants II and III.

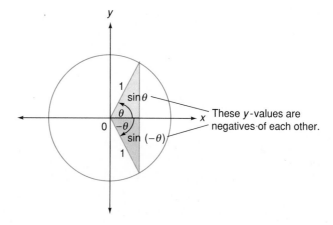

Observe that the table on page 389 uses θ-values in radians, with which we are already familiar. Most of our work in graphing trigonometric functions will use such special radian values. However, additional trigonometric function values, such as $\sin \theta$ for θ in radians, are given in Table V at the back of this book. For example, for $\theta = 0.15$ radians locate 0.15 under the radian heading at the left on page 641, then in the same row as the 0.15, and under the sin heading the entry 0.1494 gives $\sin 0.15 = 0.1494$. Similarly, from Table V,

$$\cos 1 = 0.5403 \quad \text{and} \quad \tan 1.54 = 32.461$$

The same information can be obtained using a calculator that has been set in the radian mode. On many calculators this is done by pressing the DRG key until RAD is shown on the display. For example, after the RAD has been set, $\tan 1.54$ can be found using this calculator sequence:

We use the equal sign here for simplicity, even though the table entries are (close) approximations.

$$\tan 1.54 = \boxed{1.54} \; \boxed{\text{TAN}} = 32.461139 = 32.461 \qquad \text{(rounded to three decimal places)}$$

The graph of the cosine function can be obtained from the graph of $y = \sin\theta$ by observing that $\cos\theta = \sin\left(\theta + \dfrac{\pi}{2}\right)$. To see why this is true, consider these two typical situations:

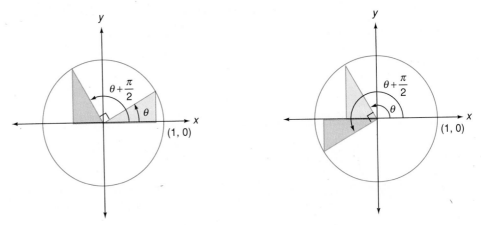

Similar diagrams can be drawn for the cases where the terminal side is in the 3rd or 4th quadrant, as well as for the cases when $\theta < 0$.

In each case the reference triangles for θ and $\theta + \dfrac{\pi}{2}$ are congruent. Consequently, the side adjacent to the reference angle for θ has the same measure as the side opposite the reference angle for $\theta + \dfrac{\pi}{2}$. It follows that $\cos\theta = \sin\left(\theta + \dfrac{\pi}{2}\right)$.

*Due to the relationship of the sine and cosine to the unit circle, as well as to right triangles, we refer to them as **circular** or **trigonometric** functions.*

The graph of $y = \sin\left(\theta + \dfrac{\pi}{2}\right)$ can be obtained by shifting the graph of $y = \sin\theta$ by $\dfrac{\pi}{2}$ units to the left. Thus the cosine curve can be obtained by shifting the sine curve $\dfrac{\pi}{2}$ units to the left.

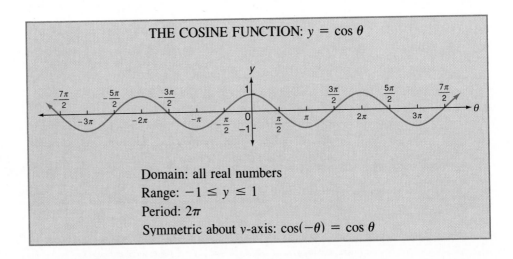

THE COSINE FUNCTION: $y = \cos\theta$

Domain: all real numbers
Range: $-1 \le y \le 1$
Period: 2π
Symmetric about y-axis: $\cos(-\theta) = \cos\theta$

In Section 7.4 we used the unit circle to arrive at $\cos\theta = x$. However, to be consistent with the usual labeling of the vertical axis in a rectangular system, we have written the equation in the form $y = \cos\theta$. Furthermore, from now on we will use x instead of θ for the horizontal axis. We use the letter θ when making direct reference to the unit circle.

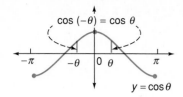

$\cos(-\theta) = \cos\theta$

$y = \cos\theta$

The cosine curve may be regarded as being $\dfrac{\pi}{2}$ units ahead of (or behind) the sine curve, and vice versa. Both functions have the same period, 2π. The symmetry about the y-axis can be observed in the figure in the margin and it can be verified by studying the following unit circle diagram. (A similar diagram can be drawn when the terminal sides of θ and $-\theta$ are in quadrants I and IV.)

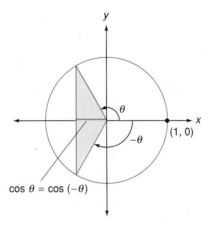

$\cos\theta = \cos(-\theta)$

The sine function has a maximum value of $M = 1$ and a minimum value of $m = -1$. One-half of the difference is called the **amplitude.**

$$\text{Amplitude} = \frac{M - m}{2}$$

For $y = \sin x$, the amplitude $= \dfrac{1 - (-1)}{2} = 1$. Thus, for both $y = 2 \sin x$, $M = 2$, $m = -2$, and the amplitude is $\dfrac{2 - (-2)}{2} = 2$. In general, the amplitude of both $y = a \sin x$ and $y = a \cos x$ is equal to $|a|$.

In the next figure the idea of amplitude is illustrated for a few cases. Notice that each of these functions has period 2π.

Note that $-1 \le \sin x \le 1$ is equivalent to $-2 \le 2 \sin x \le 2$.

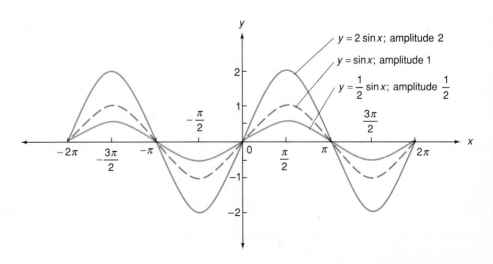

$y = 2 \sin x$; amplitude 2

$y = \sin x$; amplitude 1

$y = \dfrac{1}{2} \sin x$; amplitude $\dfrac{1}{2}$

Graph the curve for the values $-2\pi \le x \le 2\pi$.

1. $y = -\sin x$ **2** $y = 3 \cos x$ **3.** $y = \sin(-x)$

Find the amplitude.

4. $y = 10 \sin x$ **5.** $y = -\frac{2}{3} \cos x$ **6.** $y = \frac{1}{2} - \cos x$

For $y = \sin x$, the coefficient of x is 1. By changing the coefficient, we alter the period of the function. Consider, for example, the graph of $y = \sin 2x$. As x assumes values from 0 through π, $2x$ takes on values from 0 through 2π. That is, $0 \le 2x \le 2\pi$ is equivalent to $0 \le x \le \pi$. Thus the graph goes through a complete cycle for $0 \le x \le \pi$ and has period π. This information is shown in the following table of values and graph. Note that the graph completes two full cycles in the interval $0 \le x \le 2\pi$.

x	0	$\frac{\pi}{4}$	$\frac{\pi}{2}$	$\frac{3\pi}{4}$	π
$2x$	0	$\frac{\pi}{2}$	π	$\frac{3\pi}{2}$	2π
$y = \sin 2x$	0	1	0	-1	0

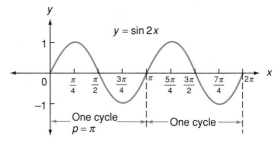

The equivalence of $0 \le 2x \le 2\pi$ and $\frac{0}{2} \le \frac{2x}{2} \le \frac{2\pi}{2}$ gave the period $p = \pi$ for $y = \sin 2x$. In a similar manner you can show that both functions $y = a \sin bx$ and $y = a \cos bx$ have period $p = \frac{2\pi}{|b|}$.

The preceding observations regarding the amplitude and period are included in the following five guidelines for graphing $y = a \sin bx$ and $y = a \cos bx$.

GUIDELINES FOR GRAPHING $y = a \sin bx$, $y = a \cos bx$

1. Find the period $p = \frac{2\pi}{|b|}$ and the amplitude $|a|$.

2. Divide the segment $[0, p]$ into four equal parts:

$$0 \quad \frac{p}{4} \quad \frac{p}{2} \quad \frac{3p}{4} \quad p$$

3. For $y = a \sin bx$, $y = 0$ at $x = 0$, $\frac{p}{2}$, and p; $y = a$ at $x = \frac{p}{4}$; and $y = -a$ at $x = \frac{3p}{4}$.

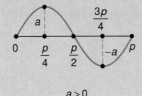

$a > 0$

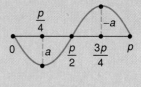

$a < 0$

---

4. For $y = a \cos bx$, $y = 0$ at $x = \dfrac{p}{4}$ and $\dfrac{3p}{4}$; $y = a$ at $x = 0$ and at $x = p$; $y = -a$ at $x = \dfrac{p}{2}$.

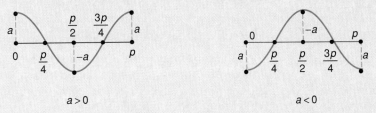

$a > 0$ $a < 0$

5. Connect the five points as shown and repeat the basic cycle as required in either direction.

EXAMPLE 1   Graph $y = 2 \cos \dfrac{x}{2}$ on the interval $[-p, p]$, where p is the period, and compare to the graph of $y = \cos x$.

Solution   The amplitude is $|a| = |2| = 2$, and $p = \dfrac{2\pi}{|b|} = \dfrac{2\pi}{\frac{1}{2}} = 4\pi$. Now divide $[0, 4\pi]$ into four equal parts and note that $a = 2$ is positive to get the following:

$$y = 0 \qquad \text{at } x = \frac{4\pi}{4} = \pi \text{ and } \frac{3(4\pi)}{4} = 3\pi$$

$$y = 2 \qquad \text{at } x = 0 \text{ and } 4\pi$$

$$y = -2 \qquad \text{at } x = \frac{4\pi}{2} = 2\pi$$

Connect the five points as shown. Complete the problem by repeating the cycle to -4π at the left and include the graph of $y = \cos x$.

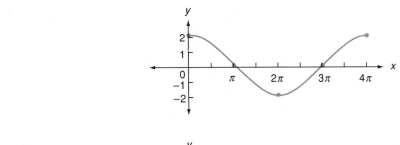

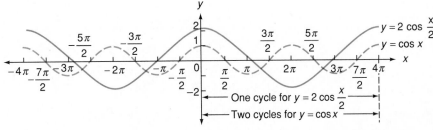

The methods of shifting or translating the graph of a given equation to obtain the graph of a more complicated equation can be applied to the circular functions. This was done earlier (page 392) where the graph of $y = \sin \theta$ was shifted $\dfrac{\pi}{2}$ units to the left to obtain the graph of $y = \sin\left(\theta + \dfrac{\pi}{2}\right) = \sin\left(\theta - \left(-\dfrac{\pi}{2}\right)\right) = \cos x$. The number $-\dfrac{\pi}{2}$ is called the **phase shift.**

In general, the graph of $y = a \sin b(x - h)$ is the same as the graph of $y = a \sin bx$ except that it has been shifted h units to the right if $h > 0$ and h units to the left if $h < 0$. In either case, h is called the phase shift. Similar observations hold for the graph of $y = a \cos b(x - h)$.

In general, for
$y = a \sin(bx + c)$ *or*
$y = a \cos(bx + c)$ *the phase shift is* $-\dfrac{c}{b}$, *since* $bx + c$ *can be written as* $b\left(x - \left(-\dfrac{c}{b}\right)\right)$.

EXAMPLE 2 For $y = 3 \sin\left(2x - \dfrac{\pi}{2}\right)$ determine the amplitude, period, and phase shift, and sketch the graph for one period.

Solution Since $y = 3 \sin\left(2x - \dfrac{\pi}{2}\right) = 3 \sin 2\left(x - \dfrac{\pi}{4}\right)$, the amplitude is 3, the period is $\dfrac{2\pi}{2} = \pi$, and the phase shift is $\dfrac{\pi}{4}$. Therefore, the graph is obtained by shifting the graph of $y = 3 \sin 2x$ to the right $\dfrac{\pi}{4}$ units, as shown in the graph.

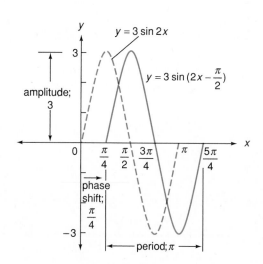

The next example demonstrates how to graph the sum of two circular functions by adding the ordinates.

EXAMPLE 3 Graph $y = \sin x + \cos x$ on $[-2\pi, 2\pi]$.

Solution First graph $y = \sin x$ and $y = \cos x$ on the same set of axes as shown by the dashed curves in the following graph.

x	$\sin x$	$\cos x$	$\sin x + \cos x$
0	0	1	1
$\dfrac{\pi}{4}$	$\dfrac{\sqrt{2}}{2}$	$\dfrac{\sqrt{2}}{2}$	$\sqrt{2}$
$\dfrac{\pi}{2}$	1	0	1
$\dfrac{3\pi}{4}$	$\dfrac{\sqrt{2}}{2}$	$-\dfrac{\sqrt{2}}{2}$	0
π	0	-1	-1
$\dfrac{5\pi}{4}$	$-\dfrac{\sqrt{2}}{2}$	$-\dfrac{\sqrt{2}}{2}$	$-\sqrt{2}$
$\dfrac{3\pi}{2}$	-1	0	-1
$\dfrac{7\pi}{4}$	$-\dfrac{\sqrt{2}}{2}$	$\dfrac{\sqrt{2}}{2}$	0
2π	0	1	1

With practice you can learn to locate points, like those marked with the dots, by visual inspection and avoid using a table of values.

Select specific values of x in $[0, 2\pi]$ for which $\sin x$ and $\cos x$ are easy to find and add these ordinates. The preceding table contains such values, and the resulting points have been indicated by the crosses. After connecting the crosses by a smooth curve, the graph is completed either by repeating the process described on $[-2\pi, 0]$ or by copying the cycle obtained on $[0, 2\pi]$ onto $[-2\pi, 0]$.

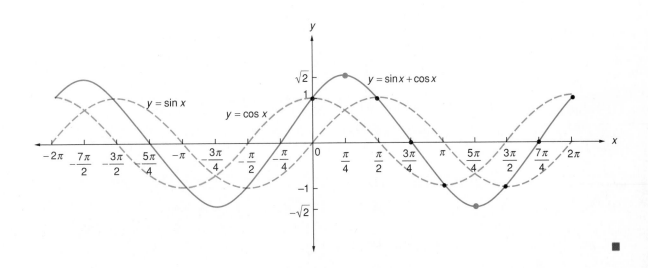

EXERCISES 8.1

1. Complete the table and use these points and the symmetry through the origin to graph $y = \sin x$ for $-\pi \le x \le \pi$.

x	$-\pi$	$-\dfrac{5\pi}{6}$	$-\dfrac{2\pi}{3}$	$-\dfrac{\pi}{2}$	$-\dfrac{\pi}{3}$	$-\dfrac{\pi}{6}$	0
$y = \sin x$							

2. **(a)** Complete the table and use these points and the symmetry about the y-axis to graph $y = \cos x$ for $-\pi \le x \le \pi$.

x	0	$\dfrac{\pi}{4}$	$\dfrac{\pi}{2}$	$\dfrac{3\pi}{4}$	π
$y = \sin x$					

(b) On which intervals is the cosine increasing? decreasing? On which intervals is the graph of the cosine concave up? down?

Explain how the graph of g can be obtained from the graph of f by using an appropriate phase shift. Graph both functions on the same axes for $0 \le x \le 2\pi$.

3. $f(x) = \sin x$, $g(x) = \sin\left(x - \dfrac{\pi}{2}\right)$ 4. $f(x) = \sin x$, $g(x) = \sin\left(x + \dfrac{\pi}{4}\right)$

5. $f(x) = \cos x$, $g(x) = \cos\left(x - \dfrac{\pi}{3}\right)$ 6. $f(x) = \cos x$, $g(x) = \cos\left(x + \dfrac{\pi}{3}\right)$

7. $f(x) = \sin x$, $g(x) = 2\sin(x + \pi)$ 8. $f(x) = \cos x$, $g(x) = -\cos(x - \pi)$

Graph for $-2\pi \le x \le 2\pi$.

9. $y = |\sin x|$ 10. $y = -|\cos x|$

11. Graph $y = 3\sin x$, $y = \frac{1}{3}\sin x$, and $y = -3\sin x$ on the same axes for $0 \le x \le 2\pi$. Find the amplitudes.

12. Graph $y = 2\cos x$, $y = \frac{1}{2}\cos x$, and $y = -2\cos x$ on the same axes for $-\pi \le x \le \pi$. Find the amplitudes.

Sketch the curve on the interval $0 \le x \le 2\pi$. Find the amplitude and period.

13. $y = \cos 2x$ 14. $y = -\sin 2x$ 15. $y = -\frac{3}{2}\sin 4x$

16. $y = \cos 4x$ 17. $y = -\cos\frac{1}{2}x$ 18. $y = -2\sin\frac{1}{2}x$

19. Find the period p of $y = \frac{1}{2}\cos\frac{1}{4}x$. Graph this curve and the curve $y = \cos x$ for $-p \le x \le p$ on the same axes.

20. Find the period p of $y = 3\sin\frac{1}{3}x$. Graph this curve and the curve $y = \sin x$ for $-p \le x \le p$ on the same axes.

21. Find the period p of $y = 2\sin \pi x$ and sketch the curve for $0 \le x \le p$.

22. Find the period p of $y = -\dfrac{3}{4}\cos\dfrac{\pi}{2}x$ and sketch the curve for $0 \le x \le p$.

In the following exercises, the curve has equation of the form $y = a \sin bx$ or $y = a \cos bx$.
Find a and b and write the equation.

23.

24.

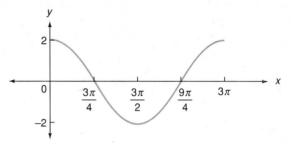

25.

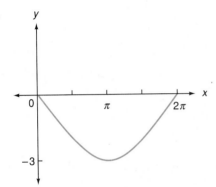

26.

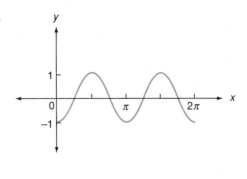

Determine the amplitude, period, and phase shift.

27. $y = 5 \cos 3\left(x - \dfrac{\pi}{6}\right)$ 28. $y = \dfrac{1}{2} \sin 2(x + \pi)$ 29. $y = -2 \sin(2x + \pi)$

30. $y = -\cos\left(\dfrac{x}{2} - \dfrac{\pi}{3}\right)$ 31. $y = \dfrac{3}{2} \cos\left(\dfrac{x}{4} - 1\right)$ 32. $y = 4 \sin\left(\pi x + \dfrac{\pi}{2}\right)$

Determine the amplitude, period, and phase shift, and sketch the graph for one period.

33. $y = \sin(4x - \pi)$ 34. $y = \sin\left(2x + \dfrac{\pi}{2}\right)$ 35. $y = 2 \cos\left(2x - \dfrac{\pi}{2}\right)$

36. $y = \frac{1}{2} \cos(3x + \pi)$ 37. $y = -\dfrac{5}{2} \sin\left(\dfrac{x}{2} + \dfrac{\pi}{4}\right)$ 38. $y = -2 \cos\left(\dfrac{\pi x}{2} - \dfrac{\pi}{2}\right)$

Explain how the graph of the first equation can be obtained from the graph of the second equation. Sketch the graph for one period.

39. $y = 3 + \sin\left(\dfrac{x}{2} + \dfrac{\pi}{2}\right);\ y = \sin\left(\dfrac{x}{2} + \dfrac{\pi}{2}\right)$ 40. $y = -3 + 2 \cos\left(x - \dfrac{\pi}{6}\right);\ y = 2 \cos\left(x - \dfrac{\pi}{6}\right)$

Sketch the graphs of the functions for $0 \le x \le 2\pi$.

41. $f(x) = \sin x - \cos x$ 42. $f(x) = 2 \sin x + \cos x$ 43. $f(x) = \cos x + \cos 2x$

44. $f(x) = \sin 2x + \frac{1}{2} \cos x$ 45. $f(x) = |\sin x| + |\cos x|$ 46. $f(x) = x + \sin x$

47. Let $f(x) = 2x + 5$ and $g(x) = \cos x$. Form the composites $f \circ g$ and $g \circ f$.

48. Let $f(x) = \sqrt{x^2 + 1}$ and $g(x) = \sin x$. Form the composites $f \circ g$ and $g \circ f$.

49. Let $h(x) = \cos(5x^2)$. Find f and g so that $h = f \circ g$, where the inner function g is quadratic.

50. Let $h(x) = \sin(\ln x)$. Find f and g so that $h = f \circ g$, where the outer function is trigonometric.

51. Let $F(x) = \cos \sqrt[3]{1 - 2x}$. Find f, g, and h so that $F = f \circ g \circ h$, where h is linear.

52. Let $F(x) = \ln(\sin(x^2 - 1))$. Find f, g, and h so that $F = f \circ g \circ h$, where h is a binomial.

53. Prove that if p is the period of the function f, then $f(x + 2p) = f(x)$ for x in the domain of f.

✏️ **Written Assignment:** Explain how the range of the cosine curve can be found using the unit circle.

8.2 GRAPHING OTHER TRIGONOMETRIC FUNCTIONS

Since $\tan x = \dfrac{\sin x}{\cos x}$, the properties of the tangent function depend on the sine and cosine functions. First observe that $\cos x = 0$ for $x = \pm\dfrac{\pi}{2}, \pm\dfrac{3\pi}{2}, \pm\dfrac{5\pi}{2}, \ldots$

Therefore, the domain of $\tan x = \dfrac{\sin x}{\cos x}$ consists of all real numbers x except those of the form $x = \dfrac{\pi}{2} + k\pi$, where k is any integer.

Recall that a function f is symmetric through the origin provided that $f(-t) = -f(t)$.

The graph of $y = \tan x$ is symmetric through the origin, since for any x in the domain we have

$$\tan(-x) = \frac{\sin(-x)}{\cos(-x)} = \frac{-\sin x}{\cos x} = -\frac{\sin x}{\cos x} = -\tan x$$

The period of $y = \tan x$ is π, as can be observed by returning to the unit circle. Consider, for example, the case where the terminal side of an angle θ is in quadrant I; in particular, we assume that $0 < \theta < \dfrac{\pi}{2}$. Adding π to θ gives the angle $\theta + \pi$, whose terminal side is in quadrant III. Since the two terminal sides are on the same line, the reference triangles are congruent, and the ratio $\dfrac{\text{opposite}}{\text{adjacent}}$ is positive in each case. Therefore, $\tan(\theta + \pi) = \tan \theta$.

It is also true that $\tan(\theta + 2\pi) = \tan \theta$. Why, then, is the period not 2π?

Similar diagrams can be used when the terminal side of θ is in any of the other quadrants, as well as for the cases where $\theta < 0$.

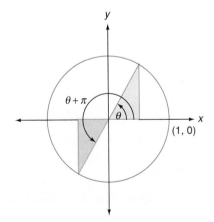

It is possible to construct the graph of $y = \tan x$ after considering the geometric interpretation of the tangent ratio as suggested in the following figure. (See also Exercise 75, page 374.) Note that AB is perpendicular to the circle at A and meets the terminal side of $\angle AOP$ at B.

Since triangles OAB and OQP are similar

$$\frac{AB}{OA} = \frac{QP}{OQ}$$

$$\frac{AB}{1} = \frac{y}{x}$$

$$AB = \tan \theta$$

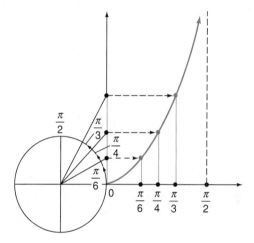

For $0 \le \theta < \dfrac{\pi}{2}$ put θ on the horizontal axis and place the corresponding tangent segment vertically at the end of θ. As the θ-values "unroll" along the horizontal axis, the tangent line to the circle at 0 is "stretched" and "curved" to become the tangent curve in a rectangular system.

Now reflect this branch of the curve through the origin to get one full cycle on $\left(-\dfrac{\pi}{2}, \dfrac{\pi}{2}\right)$. In doing so, the variable θ has been replaced by x.

The tangent function is increasing on $\left(-\dfrac{\pi}{2}, \dfrac{\pi}{2}\right)$. It is concave down on $\left(-\dfrac{\pi}{2}, 0\right)$ and concave up on $\left(0, \dfrac{\pi}{2}\right)$. There is no amplitude.

x	$\tan x$
0	0
$\dfrac{\pi}{6}$	$\dfrac{1}{\sqrt{3}} \approx 0.6$
$\dfrac{\pi}{4}$	1
$\dfrac{\pi}{3}$	$\sqrt{3} \approx 1.7$

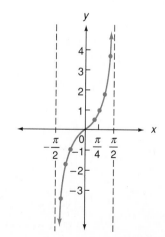

Notice that the vertical lines $x = \pm\dfrac{\pi}{2}$ are asymptotes to the curve. You can observe this "growth" of $y = \tan x$ by noting that as x gets close to $\dfrac{\pi}{2}$, the sine gets close to 1 and the cosine gets close to 0. Hence the fraction $\dfrac{\sin x}{\cos x}$ gets very large. This is also demonstrated by the following table values obtained by using a calculator set in the radian mode.

θ	1.4	1.5	1.57	1.5704	$\longrightarrow$ getting close to $\dfrac{\pi}{2} = 1.57079\cdots$
$\sin\theta$	.9854	.9974	.9999	.9999	$\longrightarrow$ getting close to 1
$\cos\theta$	.1699	.0707	.0007	.0003	$\longrightarrow$ getting close to 0
$\tan\theta$	5.797	14.10	1255	2523	$\longrightarrow$ getting very large

Since the period of $\tan x$ is π, the preceding figure shows one cycle of the tangent function, which repeats to the left and right. The range of $y = \tan x$ consists of all real numbers.

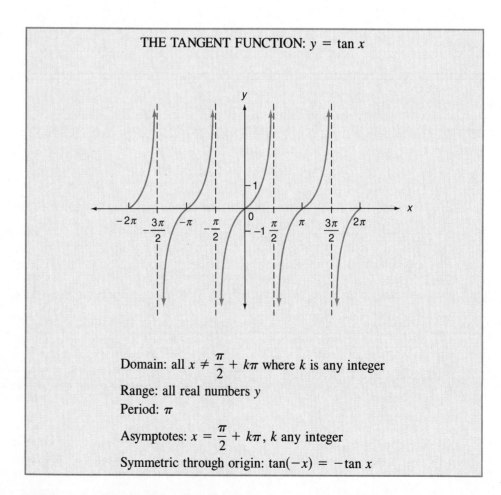

THE TANGENT FUNCTION: $y = \tan x$

Domain: all $x \neq \dfrac{\pi}{2} + k\pi$ where k is any integer

Range: all real numbers y

Period: π

Asymptotes: $x = \dfrac{\pi}{2} + k\pi$, k any integer

Symmetric through origin: $\tan(-x) = -\tan x$

EXAMPLE 1 Graph $y = -\tan 3x$.

Solution

Observe that the period of $y = \tan x$ is π, and the asymptotes occur every π units along the x-axis. Likewise, the period here is $\dfrac{\pi}{3}$, and the asymptotes occur every $\dfrac{\pi}{3}$ units along the x-axis.

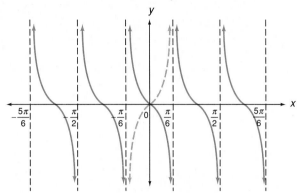

One complete cycle of the tangent curve takes place for x where $-\dfrac{\pi}{2} < x < \dfrac{\pi}{2}$, and the length of this interval, $\dfrac{\pi}{2} - \left(-\dfrac{\pi}{2}\right) = \pi$, is the period of $y = \tan x$. Then, since $-\dfrac{\pi}{2} < 3x < \dfrac{\pi}{2}$ is equivalent to $-\dfrac{\pi}{6} < x < \dfrac{\pi}{6}$ and $\dfrac{\pi}{6} - \left(-\dfrac{\pi}{6}\right) = \dfrac{\pi}{3}$, the period of $y = \tan 3x$ is $\dfrac{\pi}{3}$. Note that to graph $y = -\tan 3x$, we first graphed $y = \tan 3x$ using a dashed curve and then reflected this curve through the x-axis, as shown in the preceding graph. ∎

In general, the period of $y = \tan bx$ or of $y = \tan b(x - h)$ is $\dfrac{\pi}{|b|}$. This is also true for the cotangent graphed below.

By using an analysis similar to the one used to obtain the graph of $y = \tan x$, we can obtain the following graph of $y = \cot x = \dfrac{\cos x}{\sin x}$.

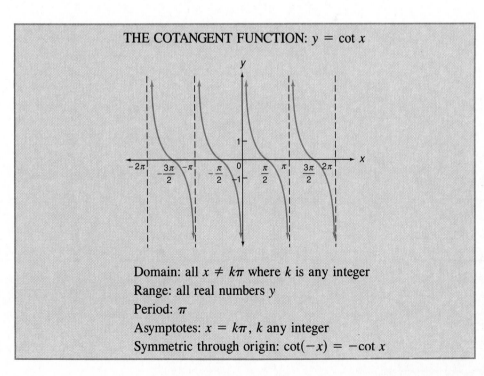

THE COTANGENT FUNCTION: $y = \cot x$

Domain: all $x \neq k\pi$ where k is any integer
Range: all real numbers y
Period: π
Asymptotes: $x = k\pi$, k any integer
Symmetric through origin: $\cot(-x) = -\cot x$

The secant function can be graphed by making use of the cosine because $\sec x = \dfrac{1}{\cos x}$ for $\cos x \neq 0$. We need only consider $x \geq 0$ since

$$\sec(-x) = \frac{1}{\cos(-x)} = \frac{1}{\cos x} = \sec x$$

shows that the graph of the secant function is symmetric with respect to the y-axis. Now consider $0 \leq x < \dfrac{\pi}{2}$. For such x, take the reciprocal of $\cos x$ to get $y = \dfrac{1}{\cos x} = \sec x$. Below are some specific cases to help graph the curve.

x	$\cos x$	$\sec x = \dfrac{1}{\cos x}$
0	1	1
$\dfrac{\pi}{6}$	$\dfrac{\sqrt{3}}{2}$	$\dfrac{2}{\sqrt{3}} \approx 1.2$
$\dfrac{\pi}{4}$	$\dfrac{1}{\sqrt{2}}$	$\sqrt{2} \approx 1.4$
$\dfrac{\pi}{3}$	$\dfrac{1}{2}$	2

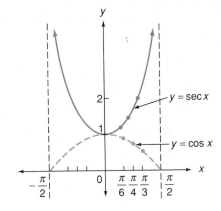

You can see that as x gets close to $\dfrac{\pi}{2}$, $\cos x$ gets close to 0, and therefore the reciprocals get very large. It follows that $x = \dfrac{\pi}{2}$ is a vertical asymptote and, by symmetry, so is $x = -\dfrac{\pi}{2}$. By similar analysis the graph of $y = \sec x$ can be found for $\dfrac{\pi}{2} < x < \dfrac{3\pi}{2}$. Then the periodicity gives the rest. Note that for all x in the domain of the secant, we have

$$\sec(x + 2\pi) = \frac{1}{\cos(x + 2\pi)} = \frac{1}{\cos x} = \sec x$$

If 2π were replaced by any positive number $a < 2\pi$, then the preceding would no longer be true. Therefore, 2π is the period of $y = \sec x$.

Also, just as the asymptotes above occurred where the cosine was 0, the asymptotes for $y = \sec x$ will occur whenever the $\cos x = 0$, where $x = \dfrac{\pi}{2} + k\pi$.

Using a calculator in the radian mode $\left(\dfrac{\pi}{2} = 1.570796 \ldots\right)$, we obtain

$$\sec 1.5 = 14$$
$$\sec 1.56 = 93$$
$$\sec 1.57 = 1256$$
$$\sec 1.5707 = 10381$$

As $x \to \dfrac{\pi}{2}$, $\sec x \to \infty$.

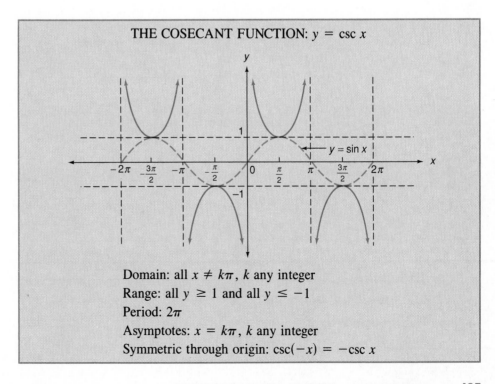

For what parts of $\left(-\dfrac{\pi}{2}, \dfrac{3\pi}{2}\right)$ is the secant increasing or decreasing? Where is the curve concave up or down? Is there an amplitude?

Domain: all $x \ne \dfrac{\pi}{2} + k\pi$, k any integer

Range: all $y \ge 1$ and all $y \le -1$

Period: 2π

Asymptotes: $x = \dfrac{\pi}{2} + k\pi$, k any integer

Symmetric about y-axis: $\sec(-x) = \sec x$

The properties of $y = \csc x = \dfrac{1}{\sin x}$ can be obtained from $y = \sin x$, just as the properties of the cosine were used for the secant.

THE COSECANT FUNCTION: $y = \csc x$

Domain: all $x \ne k\pi$, k any integer
Range: all $y \ge 1$ and all $y \le -1$
Period: 2π
Asymptotes: $x = k\pi$, k any integer
Symmetric through origin: $\csc(-x) = -\csc x$

Phase shifts can be used in any of the circular functions, not just for the sine and cosine. This is illustrated in the next example.

EXAMPLE 2   For $y = \csc\left(2x + \dfrac{\pi}{2}\right)$ determine the period and phase shift, and sketch the graph for one period.

Solution   Since $y = \csc\left(2x + \dfrac{\pi}{2}\right) = \csc 2\left(x + \dfrac{\pi}{4}\right)$, the period is $\dfrac{2\pi}{2} = \pi$ and the phase shift is $-\dfrac{\pi}{4}$. The graph is the same as the graph of $y = \csc 2x$ only

The period p for y = a csc b(x − h) is the same as for y = a csc bx, which, in turn, is the same as for y = a sin bx; p = $\dfrac{2\pi}{|b|}$. This is also true for y = a sec b(x − h).

shifted $\dfrac{\pi}{4}$ units to the left. Just as the asymptotes for $y = \csc x$ are at the endpoints and midpoint of the interval $[0, 2\pi]$, so are the asymptotes for $y = \csc 2\left(x + \dfrac{\pi}{4}\right)$ at the endpoints and midpoint of the interval $\left[-\dfrac{\pi}{4}, \dfrac{3\pi}{4}\right]$ as shown.

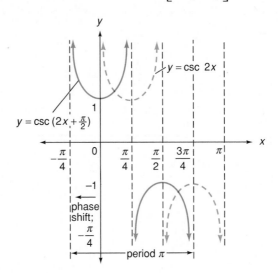

EXERCISES 8.2

1. Complete the following table (using Table V or a calculator only if necessary) and sketch the curve $y = \tan x$ for $-\dfrac{\pi}{2} < x \leq 0$. Then use the symmetry through the origin to obtain the graph for $-\dfrac{\pi}{2} < x < \dfrac{\pi}{2}$.

x	-1.4	-1.3	$-\dfrac{\pi}{3}$	$-\dfrac{\pi}{4}$	$-\dfrac{\pi}{6}$	0
$y = \tan x$						

2. **(a)** Verify that $\cot(\theta + \pi) = \cot\theta$ for a typical value θ in quadrant I or II, using a unit circle diagram.

   (b) Complete the table of values and sketch the graph of $y = \cot x$ for $0 < x < \pi$. Use the result in part (a) to extend the graph for $\pi < x < 2\pi$.

x	$\dfrac{\pi}{6}$	$\dfrac{\pi}{4}$	$\dfrac{\pi}{3}$	$\dfrac{\pi}{2}$	$\dfrac{2\pi}{3}$	$\dfrac{3\pi}{4}$	$\dfrac{5\pi}{6}$
$y = \cot x$							

   (c) Refer to part (b) and find the subintervals of $(0, \pi)$ for which the cotangent is increasing or decreasing and the subintervals where the curve is concave up or down.

3. **(a)** Show that the cotangent is symmetric through the origin by verifying that $\cot(-x) = -\cot x$.

   (b) Repeat part (a) for the cosecant.

   (c) Verify that $\csc(x + 2\pi) = \csc x$.

Sketch the graph of f by making an appropriate phase shift of the graph of g.

4. $f(x) = \tan\left(x - \dfrac{\pi}{2}\right)$; $g(x) = \tan x$

5. $f(x) = \cot\left(x + \dfrac{\pi}{2}\right)$; $g(x) = \cot x$

6. $f(x) = \sec\left(x + \dfrac{\pi}{4}\right)$; $g(x) = \sec x$

7. $f(x) = \csc\left(x - \dfrac{\pi}{3}\right)$; $g(x) = \csc x$

8. Compare the graph of the tangent and cotangent and decide which of the following equations are identities.

   (a) $\cot x = -\tan x$ **(b)** $\tan x = \cot\left(x + \dfrac{\pi}{2}\right)$ **(c)** $\cot x = -\tan\left(x - \dfrac{\pi}{2}\right)$

Sketch the graph of the equation, including at least two cycles. Indicate the period and vertical asymptotes.

9. $y = \cot 3x$

10. $y = \tan 2x$

11. $y = -2\cot\dfrac{x}{2}$

12. $y = \dfrac{1}{2}\tan\dfrac{x}{2}$

13. $y = \sec 4x$

14. $y = -\sec x$

15. $y = -\csc\frac{1}{3}x$

16. $y = \csc 2x$

17. $y = 2\sec\dfrac{3x}{2}$

Determine the period and phase shift, and sketch the graph for one period.

18. $y = \tan 2\left(x - \dfrac{\pi}{2}\right)$

19. $y = \dfrac{1}{2}\cot 2\left(x + \dfrac{\pi}{4}\right)$

20. $y = 2\sec\left(\dfrac{x}{2} + \dfrac{\pi}{8}\right)$

21. $y = -\csc\left(\dfrac{x}{2} - \dfrac{\pi}{4}\right)$

Sketch the graph of the equations.

22. $y = |\sec x|$ 23. $y = |\tan x|$ 24. $y = -|\cot x|$

25. Let $f(x) = x^2$ and $g(x) = \tan x$. Form the composites $f \circ g$ and $g \circ f$.

26. Let $f(x) = \dfrac{x}{x+1}$ and $g(x) = \sec x$. Form the composites $f \circ g$ and $g \circ f$.

27. Let $f(x) = e^x$, $g(x) = \sqrt{x}$, and $h(x) = \sec x$. Form the composites $f \circ g \circ h$, $g \circ f \circ h$, and $h \circ g \circ f$.

28. Let $h(x) = \cot^3 x$. Find f and g so that $h = f \circ g$, where the outer function f is a polynomial.

29. Let $F(x) = \sqrt{\tan(2x + 1)}$. Find f, g, and h so that $F = f \circ g \circ h$ and g is trigonometric.

30. Let $F(x) = \tan^2\left(\dfrac{x + 1}{x - 1}\right)$. Find f, g, h so that $F = f \circ g \circ h$, where h is rational and f is quadratic.

31. Refer to the figure and explain why $0 < \dfrac{\sin \theta}{\theta} < 1$ for $0 < \theta < \dfrac{\pi}{2}$. (*Hint:* In a unit circle the central angle in radians is the same as the length of the intercepted arc.)

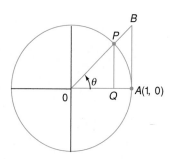

32. Refer to the figure in Exercise 31 and prove $\cos \theta < \dfrac{\sin \theta}{\theta} < \dfrac{1}{\cos \theta}$ for $0 < \theta < \dfrac{\pi}{2}$.

(*Hint:* Use the areas of the two right triangles and the sector of the circle.)

33. Use a calculator to complete this table where θ is given in radians.

θ	1	0.5	0.25	0.1	0.01
$\dfrac{\sin \theta}{\theta}$					

As θ approaches 0 in value, what appears to be happening to the ratio $\dfrac{\sin \theta}{\theta}$?

34. Find the subintervals of the given interval on which each function is increasing or decreasing. Also, find the subintervals where the graph is concave up or concave down.

(a) $y = \tan x$; $\left(-\dfrac{\pi}{2}, \dfrac{\pi}{2}\right)$ **(b)** $y = \cot x$; $(0, \pi)$

(c) $y = \sec x$; $\left(-\dfrac{\pi}{2}, \dfrac{3\pi}{2}\right)$ **(d)** $y = \csc x$; $(0, 2\pi)$

Written Assignment:

(a) The equation in Exercise 8(c) is true for all allowable values of x. Explain how this equation can be used to obtain the graph of $y = \cot x$ from the graph of $y = \tan x$.

(b) From the graphs of the sine and cosine functions explain why $\sin x = \cos\left(x - \dfrac{\pi}{2}\right)$ for all x. Use this to explain how to obtain the graph of $y = \csc x$ from the graph of $y = \sec x$.

Is it true that $\cos(30° + 60°) = \cos 30° + \cos 60°$? Some quick calculations show that it is not.

$$\cos(30° + 60°) = \cos 90° = 0 \quad \text{but} \quad \cos 30° + \cos 60° = \frac{\sqrt{3}}{2} + \frac{1}{2}$$

$\cos(30° + 60°)$
$= \cos 30° \cos 60°$
$\quad - \sin 30° \sin 60°$
$= \dfrac{\sqrt{3}}{2} \cdot \dfrac{1}{2} - \dfrac{1}{2} \cdot \dfrac{\sqrt{3}}{2}$
$= 0$
$= \cos 90°$

To write $\cos(\theta_1 + \theta_2) = \cos \theta_1 + \cos \theta_2$ would be to assume, incorrectly, that the cosine function obeys the distributive property. We emphasize that cos is the *name* of a function; it is not a number.

The cosine of the sum of two angles is correctly evaluated (see the computation at the left) by using formula (3) that is listed below; it is one of several important trigonometric identities in *two* variables that give the trigonometric values of sums and differences.

ADDITION AND SUBTRACTION FORMULAS	
(1)	$\sin(\alpha + \beta) = \sin \alpha \cos \beta + \cos \alpha \sin \beta$
(2)	$\sin(\alpha - \beta) = \sin \alpha \cos \beta - \cos \alpha \sin \beta$
(3)	$\cos(\alpha + \beta) = \cos \alpha \cos \beta - \sin \alpha \sin \beta$
(4)	$\cos(\alpha - \beta) = \cos \alpha \cos \beta + \sin \alpha \sin \beta$
(5)	$\tan(\alpha + \beta) = \dfrac{\tan \alpha + \tan \beta}{1 - \tan \alpha \tan \beta}$
(6)	$\tan(\alpha - \beta) = \dfrac{\tan \alpha - \tan \beta}{1 + \tan \alpha \tan \beta}$

These formulas are stated in terms of the variables α and β, which can represent angles measured in either degrees or radians (real numbers.) We now prove formula (4) by making use of the unit circle.

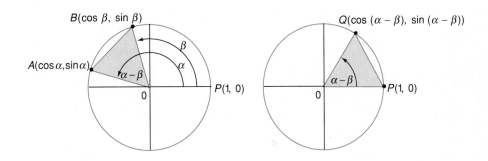

In the preceding figure the unit circle at the left contains a typical situation in which $\alpha > \beta$. Since point A is on the unit circle, we find that the coordinates of A are $(\cos \alpha, \sin \alpha)$. Similarly, B has coordinates $(\cos \beta, \sin \beta)$. Then, by the distance formula applied to points A and B, we have the following:

The distance formula is given on page 164.

$$(AB)^2 = (\cos \alpha - \cos \beta)^2 + (\sin \alpha - \sin \beta)^2$$
$$= (\cos^2 \alpha - 2 \cos \alpha \cos \beta + \cos^2 \beta) + (\sin^2 \alpha - 2 \sin \alpha \sin \beta + \sin^2 \beta)$$
$$= (\cos^2 \alpha + \sin^2 \alpha) + (\cos^2 \beta + \sin^2 \beta) - 2(\cos \alpha \cos \beta + \sin \alpha \sin \beta)$$
$$= 2 - 2(\cos \alpha \cos \beta + \sin \alpha \sin \beta)$$

In the unit circle at the right in the preceding figure, the central angle $\alpha - \beta$ is in standard position. Therefore, the coordinates of Q are $(\cos(\alpha - \beta), \sin(\alpha - \beta))$. Using the distance formula for points P and Q, we have

Recall that for any θ, $\cos^2 \theta + \sin^2 \theta = 1$. This is used here for $\theta = \alpha - \beta$:

$$\cos^2(\alpha - \beta) + \sin^2(\alpha - \beta) = 1$$

$$(PQ)^2 = [\cos(\alpha - \beta) - 1]^2 + [\sin(\alpha - \beta) - 0]^2$$
$$= \cos^2(\alpha - \beta) - 2 \cos(\alpha - \beta) + 1 + \sin^2(\alpha - \beta)$$
$$= 2 - 2 \cos(\alpha - \beta)$$

Triangles AOB and QOP are congruent (SAS). Then $AB = PQ$ and this implies $(AB)^2 = (PQ)^2$. We may therefore equate the preceding results and then simplify.

$$2 - 2 \cos(\alpha - \beta) = 2 - 2(\cos \alpha \cos \beta + \sin \alpha \sin \beta)$$
$$\cos(\alpha - \beta) = \cos \alpha \cos \beta + \sin \alpha \sin \beta$$

The proofs of formulas (1) and (2) will be called for in the exercises.

Formula (3) is now easy to prove by using formula (4). The trick is to write $\alpha + \beta = \alpha - (-\beta)$.

$$\cos(\alpha + \beta) = \cos(\alpha - (-\beta))$$
$$= \cos \alpha \cos(-\beta) + \sin \alpha \sin(-\beta) \qquad \text{(by (4))}$$
$$= \cos \alpha \cos \beta + (\sin \alpha)(-\sin \beta) \qquad (\cos(-\theta) = \cos \theta,$$
$$\sin(-\theta) = -\sin \theta)$$
$$= \cos \alpha \cos \beta - \sin \alpha \sin \beta$$

EXAMPLE 1 Evaluate $\sin \dfrac{7\pi}{12}$ and $\tan \dfrac{7\pi}{12}$ by using $\dfrac{7\pi}{12} = \dfrac{\pi}{4} + \dfrac{\pi}{3}$. State the answers in radical form, and check by calculator.

Solution Use formulas (1) and (5) with $\alpha = \dfrac{\pi}{4}$, $\beta = \dfrac{\pi}{3}$.

Check (use radian mode):

$\sin \dfrac{7}{12}\pi$

$= \boxed{7} \boxed{\div} \boxed{12} \boxed{\times} \boxed{\pi} \boxed{=} \boxed{\text{SIN}} = 0.9659$

$\frac{1}{4}(\sqrt{2} + \sqrt{6})$

$= \boxed{2} \boxed{\sqrt{}} \boxed{+} \boxed{6} \boxed{\sqrt{}} \boxed{=} \boxed{\div} \boxed{4} \boxed{=}$

$\qquad \qquad 0.9659$

$$\sin \frac{7}{12}\pi = \sin\left(\frac{\pi}{4} + \frac{\pi}{3}\right)$$
$$= \sin \frac{\pi}{4} \cos \frac{\pi}{3} + \cos \frac{\pi}{4} \sin \frac{\pi}{3}$$
$$= \frac{\sqrt{2}}{2} \cdot \frac{1}{2} + \frac{\sqrt{2}}{2} \cdot \frac{\sqrt{3}}{2}$$
$$= \frac{1}{4}(\sqrt{2} + \sqrt{6})$$

$$\tan \frac{7}{12}\pi = \tan\left(\frac{\pi}{4} + \frac{\pi}{3}\right)$$

$$= \frac{\tan \dfrac{\pi}{4} + \tan \dfrac{\pi}{3}}{1 - \tan \dfrac{\pi}{4} \tan \dfrac{\pi}{3}}$$

$$= \frac{1 + \sqrt{3}}{1 - \sqrt{3}} \qquad\blacksquare$$

EXAMPLE 2 Let $\cos \alpha = -\frac{4}{5}$ and $\cos \beta = -\frac{12}{13}$, where α and β have terminal sides in quadrants II and III, respectively. Find $\cos(\alpha - \beta)$ and $\sin(\alpha - \beta)$.

Solution Place α and β in standard position and draw reference triangles. Since $\cos \alpha = \dfrac{-4}{5} = \dfrac{\text{adjacent}}{\text{hypotenuse}}$, the triangle for α has hypotenuse 5 and adjacent side -4. The third side is 3 by the Pythagorean theorem. Similarly, the triangle for β has sides 13, -12, -5.

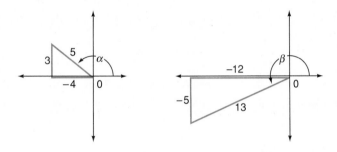

Although a unit circle is not used here, triangles such as these may be used since the ratios do not depend on the size of the triangle.

$$\cos(\alpha - \beta) = \cos \alpha \cos \beta + \sin \alpha \sin \beta$$

$$= \left(-\frac{4}{5}\right)\left(-\frac{12}{13}\right) + \left(\frac{3}{5}\right)\left(-\frac{5}{13}\right)$$

$$= \frac{48}{65} - \frac{15}{65}$$

$$= \frac{33}{65}$$

$$\sin(\alpha - \beta) = \sin \alpha \cos \beta - \cos \alpha \sin \beta$$

$$= \left(\frac{3}{5}\right)\left(-\frac{12}{13}\right) - \left(-\frac{4}{5}\right)\left(-\frac{5}{13}\right)$$

Can you determine the quadrant containing the terminal side of $\alpha - \beta$ using these two results?

$$= -\frac{36}{65} - \frac{20}{65}$$

$$= -\frac{56}{65} \qquad\blacksquare$$

Complete each equation to get a specific case of an addition or subtraction formula.

1. $\sin 5° \cos 12° + \cos 5° \sin 12° = ?$

2. $\cos \dfrac{3\pi}{10} \cos \dfrac{\pi}{5} + \sin \dfrac{3\pi}{10} \sin \dfrac{\pi}{5} = ?$

3. $\dfrac{\tan(-15°) - \tan(-20°)}{1 + \tan(-15°)\tan(-20°)} = ?$

4. $\sin(\theta + 5°)\cos(\theta - 5°) - \cos(\theta + 5°)\sin(\theta - 5°) = ?$

State the answers in radical form for Exercises 5–10.

5. Use $15° = 45° - 30°$ and appropriate subtraction formulas to evaluate $\sin 15°$, $\cos 15°$, and $\tan 15°$.

6. Repeat Exercise 5 with $15° = 60° - 45°$.

7. Use $\dfrac{11\pi}{12} = \dfrac{\pi}{6} + \dfrac{3\pi}{4}$ and appropriate addition formulas to evaluate $\sin \dfrac{11\pi}{12}$, $\cos \dfrac{11\pi}{12}$, and $\tan \dfrac{11\pi}{12}$.

8. Repeat Exercise 7 with $\dfrac{11\pi}{12} = \dfrac{7\pi}{6} - \dfrac{\pi}{4}$ using subtraction formulas.

9. Use an addition or subtraction formula to evaluate $\sin 195°$.

10. Use an addition or subtraction formula to evaluate $\cos \dfrac{5\pi}{12}$.

11. $\cos \alpha = \dfrac{3}{5}$ and the terminal side of α is in quadrant IV. $\sin \beta = \dfrac{15}{17}$ and the terminal side of β is in quadrant I. Find $\cos(\alpha - \beta)$.

(Answers: Page 449)

12. For α and β as in Exercise 11, find $\sin(\alpha - \beta)$.

Formulas (1) and (3) can be used to verify the addition formula for the tangent function.

$$\tan(\alpha + \beta) = \frac{\sin(\alpha + \beta)}{\cos(\alpha + \beta)} = \frac{\sin \alpha \cos \beta + \cos \alpha \sin \beta}{\cos \alpha \cos \beta - \sin \alpha \sin \beta}$$

Divide the numerator and denominator by $\cos \alpha \cos \beta$.

$$\tan(\alpha + \beta) = \frac{\dfrac{\sin \alpha \cos \beta}{\cos \alpha \cos \beta} + \dfrac{\cos \alpha \sin \beta}{\cos \alpha \cos \beta}}{\dfrac{\cos \alpha \cos \beta}{\cos \alpha \cos \beta} - \dfrac{\sin \alpha \sin \beta}{\cos \alpha \cos \beta}}$$

Thus

$$(5) \qquad \tan(\alpha + \beta) = \frac{\tan \alpha + \tan \beta}{1 - \tan \alpha \tan \beta}$$

A similar analysis will verify formula (6).

With the aid of the addition and subtraction formulas, we will be able to derive other useful formulas, some of which have already been encountered. For example, the result $\cos\theta = \sin\left(\theta + \dfrac{\pi}{2}\right)$ was derived in Section 8.1 by using the unit circle. This formula can be described as a *reduction formula* in the sense that a trigonometric function of $\theta + \dfrac{\pi}{2}$ is "reduced" to a trigonometric function of just θ.

Each of these results may be restated in degrees; for example, $\cos(180° - \theta) = -\cos\theta$. They are collected here for easy reference.

REDUCTION FORMULAS

$$\sin(\theta - 2\pi) = \sin\theta \qquad\qquad \sin(\theta + 2\pi) = \sin\theta$$

$$\cos(\theta - 2\pi) = \cos\theta \qquad\qquad \cos(\theta + 2\pi) = \cos\theta$$

$$\tan(\theta - \pi) = \tan\theta \qquad\qquad \tan(\theta + \pi) = \tan\theta$$

$$\sin(\pi - \theta) = \sin\theta \qquad\qquad \sin(\pi + \theta) = -\sin\theta$$

$$\cos(\pi - \theta) = -\cos\theta \qquad\qquad \cos(\pi + \theta) = -\cos\theta$$

$$\tan(\pi - \theta) = -\tan\theta \qquad\qquad \tan(\pi + \theta) = \tan\theta$$

$$\sin\left(\dfrac{\pi}{2} - \theta\right) = \cos\theta \qquad\qquad \sin\left(\dfrac{\pi}{2} + \theta\right) = \cos\theta$$

$$\cos\left(\dfrac{\pi}{2} - \theta\right) = \sin\theta \qquad\qquad \cos\left(\dfrac{\pi}{2} + \theta\right) = -\sin\theta$$

$$\tan\left(\dfrac{\pi}{2} - \theta\right) = \cot\theta \qquad\qquad \tan\left(\dfrac{\pi}{2} + \theta\right) = -\cot\theta$$

Formulas (3) and (4) can be used to prove two of the preceding reduction formulas. Others will be taken up in the exercises.

$$\cos\left(\dfrac{\pi}{2} - \theta\right) = \cos\dfrac{\pi}{2}\cos\theta + \sin\dfrac{\pi}{2}\sin\theta \qquad \text{formula (4)}$$

$$= 0\cdot\cos\theta + (1)\sin\theta$$

$$= \sin\theta$$

$$\cos(\pi + \theta) = \cos\pi\cos\theta - \sin\pi\sin\theta \qquad \text{formula (3)}$$

$$= (-1)\cos\theta - 0\cdot\sin\theta$$

$$= -\cos\theta$$

The reduction formulas involving the tangent can be obtained from the results for the sine and cosine. Thus

Why can't the formula for $\tan(\alpha - \beta)$ be used here?

$$\tan\left(\dfrac{\pi}{2} - \theta\right) = \dfrac{\sin\left(\dfrac{\pi}{2} - \theta\right)}{\cos\left(\dfrac{\pi}{2} - \theta\right)} = \dfrac{\cos\theta}{\sin\theta} = \cot\theta$$

EXAMPLE 3   Verify the identity $\sin\left(\theta + \dfrac{\pi}{3}\right) = \dfrac{1}{2}(\sin\theta + \sqrt{3}\cos\theta)$.

Solution   Use Formula (1).

$$\sin\left(\theta + \frac{\pi}{3}\right) = \sin\theta\cos\frac{\pi}{3} + \cos\theta\sin\frac{\pi}{3}$$

$$= \frac{1}{2}\sin\theta + \frac{\sqrt{3}}{2}\cos\theta$$

$$= \frac{1}{2}(\sin\theta + \sqrt{3}\cos\theta) \qquad\blacksquare$$

When θ is acute, then θ and $90° - \theta$ are complementary and the equation $\sin\theta = \cos(90° - \theta)$ shows that the sine of θ equals the cosine of its complement. The preceding equation is an example of a **cofunction identity,** and the sine and cosine are said to be **cofunctions** of one another. Likewise, the tangent and cotangent are cofunctions, as are the secant and cosecant. The following list of cofunction identities shows that *a trigonometric function of an acute angle θ is the same as the cofunction of its complement.*

If we let $\phi = 90° - \theta$, then $\theta = 90° - \phi$ and the first identity can be rewritten in this equivalent form:

$$\cos\phi = \sin(90° - \phi)$$

Similarly for the other identities.

COFUNCTION IDENTITIES
$\sin\theta = \cos(90° - \theta)$
$\tan\theta = \cot(90° - \theta)$
$\sec\theta = \csc(90° - \theta)$

EXERCISES 8.3

Use an addition or subtraction formula to evaluate the expression.
Do not use tables or a calculator.

1. $\cos 22° \cos 38° - \sin 22° \sin 38°$ 2. $\sin 52° \cos 7° - \cos 52° \sin 7°$

3. $\dfrac{\tan 25° - \tan 55°}{1 + \tan 25° \tan 55°}$ 4. $\cos\dfrac{5\pi}{12}\cos\dfrac{7\pi}{12} + \sin\dfrac{5\pi}{12}\sin\dfrac{7\pi}{12}$

Use the addition or subtraction formulas to evaluate $\sin\theta$, $\cos\theta$, and $\tan\theta$ for the specified value of θ, and check by calculator.

5. $\theta = 75°$ 6. $\theta = 105°$ 7. $\theta = \dfrac{\pi}{12}$ 8. $\theta = \dfrac{19\pi}{12}$

9. $\theta = 165°$ 10. $\theta = 345°$ 11. $\theta = \dfrac{17\pi}{12}$ 12. $\theta = \dfrac{7\pi}{3}$

Prove the reduction formula by using an appropriate addition or subtraction formula.

13. $\sin(\pi - \theta) = \sin\theta$ 14. $\cos\left(\dfrac{\pi}{2} + \theta\right) = -\sin\theta$

15. $\tan(\pi - \theta) = -\tan\theta$ 16. $\cos(\theta - 2\pi) = \cos\theta$

Prove the cofunction identity.

17. $\sin \theta = \cos(90° - \theta)$ 18. $\tan \theta = \cot(90° - \theta)$

19. If α and β are acute angles having $\sin \alpha = \frac{3}{5}$ and $\cos \beta = \frac{12}{13}$, find the exact values of $\sin(\alpha - \beta)$ and $\tan(\alpha + \beta)$.

20. Suppose that $\sin \alpha$ and $\cos \beta$ are the same as in Exercise 19, but α is in quadrant II and β is in quadrant IV. Find the exact values of $\sin(\alpha - \beta)$ and $\tan(\alpha + \beta)$.

21. Find the exact values of $\sin(\alpha + \beta)$ and $\cos(\alpha + \beta)$ for α in quadrant III and β in quadrant IV, where $\sin \alpha = -\frac{1}{3}$ and $\cos \beta = \frac{2}{5}$.

22. Let $\cos \alpha = \frac{24}{25}$ with α in quadrant I, and $\tan \beta = -\frac{15}{8}$ with β in quadrant II. Find the exact values of $\cos(\alpha - \beta)$ and $\tan(\alpha - \beta)$.

23. Use Table VI to evaluate the following ratios by using a reduction formula that calls for a ratio of an acute angle. Thus $\cos 155° = \cos(180° - 25°) = -\cos 25° = -.9063$.

    (a) $\cos 191°$ **(b)** $\sin 132°$ **(c)** $\tan 200°$ **(d)** $\cos 102°$

Use addition or subtraction formulas to verify the identity.

24. $\tan\left(\dfrac{\pi}{4} + \theta\right) = \dfrac{1 + \tan \theta}{1 - \tan \theta}$ 25. $\cos\left(\theta - \dfrac{\pi}{4}\right) = \dfrac{\sqrt{2}}{2}(\cos \theta + \sin \theta)$

26. $\sin\left(\dfrac{\pi}{6} - \theta\right) = \dfrac{1}{2}(\cos \theta - \sqrt{3} \sin \theta)$ 27. $\cos(\theta + 30°) + \cos(\theta - 30°) = \sqrt{3} \cos \theta$

28. $\cos\left(\theta + \dfrac{\pi}{6}\right) + \sin\left(\theta - \dfrac{\pi}{3}\right) = 0$ 29. $\dfrac{\sin\left(\theta + \dfrac{\pi}{2}\right)}{\cos\left(\theta + \dfrac{\pi}{2}\right)} = -\cot \theta$

30. $\sec\left(\dfrac{\pi}{2} - \theta\right) = \csc \theta$ 31. $\csc(\pi - \theta) = \csc \theta$

32. $\cos(\alpha + \beta) + \cos(\alpha - \beta) = 2 \cos \alpha \cos \beta$ 33. $\sin(\alpha + \beta) - \sin(\alpha - \beta) = 2 \cos \alpha \sin \beta$

34. $\dfrac{\sin(\alpha - \beta)}{\cos(\alpha + \beta)} = \dfrac{\cot \beta - \cot \alpha}{\cot \alpha \cot \beta - 1}$ 35. $\dfrac{\sin(\alpha + \beta)}{\sin(\alpha - \beta)} = \dfrac{\tan \alpha + \tan \beta}{\tan \alpha - \tan \beta}$

36. $\cos(\alpha + \beta) \cos(\alpha - \beta) = \cos^2 \alpha - \sin^2 \beta$ 37. $2 \sin\left(\dfrac{\pi}{4} - \theta\right) \sin\left(\dfrac{\pi}{4} + \theta\right) = \cos^2 \theta - \sin^2 \theta$

38. $\cos \alpha \cos(\alpha - \beta) + \sin \alpha \sin(\alpha - \beta) = \cos \beta$

39. Let $x = \dfrac{\pi}{2} - \theta$ in the equation $\cos\left(\dfrac{\pi}{2} - x\right) = \sin x$. Then prove that $\sin\left(\dfrac{\pi}{2} - \theta\right) = \cos \theta$.

40. Prove addition formula (1). $\left(\textit{Hint: } \text{Begin with } \sin(\alpha + \beta) = \cos\left[\dfrac{\pi}{2} - (\alpha + \beta)\right] \text{ and}\right.$

    $\left.\text{note that } \dfrac{\pi}{2} - (\alpha + \beta) = \left(\dfrac{\pi}{2} - \alpha\right) - \beta.\right)$

41. Use addition formula (1) to prove (2). (*Hint:* Use $\alpha - \beta = \alpha + (-\beta)$.)

42. Use addition formula (5) to prove (6). (*Hint:* Use $\alpha - \beta = \alpha + (-\beta)$.)

43. Derive $\cot(\alpha + \beta) = \dfrac{\cot \alpha \cot \beta - 1}{\cot \alpha + \cot \beta}$ by forming $\dfrac{\cos(\alpha + \beta)}{\sin(\alpha + \beta)}$ and using formulas (1) and (3).

44. Derive the formula in Exercise 43 by forming $\dfrac{1}{\tan(\alpha + \beta)}$ and using formula (5).

45. Explain how the graph of the sine can be obtained from the cosine curve by using the reduction formula $\cos\left(\dfrac{\pi}{2} + \theta\right) = -\sin \theta$.

46. Explain how the graph of the sine can be obtained from the cosine curve by using the reduction formula $\cos\left(\dfrac{\pi}{2} - \theta\right) = \sin\theta$.

47. Find $\tan\beta$ for β in the given figure. (*Hint:* Use the formula for $\tan(\alpha + \beta)$.)

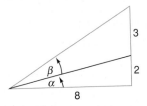

48. Let α and β be acute angles such that $\tan\alpha = \frac{3}{5}$ and $\tan\beta = \frac{1}{4}$. Find $\alpha + \beta$ without using a calculator or tables.

 49. Let $f(x) = \sin x$ and $g(x) = \cos x$. Prove the following:

  (a) $\dfrac{f(x + h) - f(x)}{h} = \left(\dfrac{\cos h - 1}{h}\right)\sin x + \left(\dfrac{\sin h}{h}\right)\cos x$

  (b) $\dfrac{g(x + h) - g(x)}{h} = \left(\dfrac{\cos h - 1}{h}\right)\cos x - \left(\dfrac{\sin h}{h}\right)\sin x$

 50. In the figure line ℓ_1 has equation $y = -\frac{1}{2}x + \frac{3}{2}$ and ℓ_2 has equation $y = -\frac{4}{5}x + \frac{6}{5}$. The lines intersect at P and the x-axis at A and B.

  (a) For angles α and β as shown, explain why $\tan\alpha = $ slope ℓ_1 and $\tan\beta = $ slope ℓ_2.

  (b) Explain why $\angle BPA = \alpha - \beta$.

  (c) Use the subtraction formula for $\tan(\alpha - \beta)$ to find the angle between the lines, $\alpha - \beta$, to the nearest tenth of a degree.

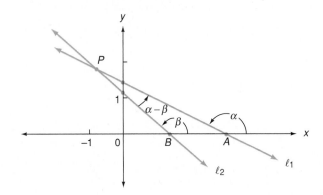

 For each pair of lines ℓ_1 and ℓ_2 let P be their point of intersection, let A be the intersection of ℓ_1 with the x-axis, and let B be the intersection of ℓ_2 with the x-axis. Follow the instructions in Exercise 50, using α, β, and $\alpha - \beta$, as in Exercise 50, to find the measure of $\angle BPA$ to the nearest tenth of a degree.

51. $\ell_1\colon 3x + 2y = 6$
    $\ell_2\colon 6x + y = -6$

52. $\ell_1\colon y = 5x - 10$
    $\ell_2\colon 5x - 2y = 5$

53. $\ell_1\colon x + 3y = 17$
    $\ell_2\colon 5x - 6y = -20$
    (*Hint:* $\tan(\alpha - \beta) < 0$ implies $\alpha - \beta$ is obtuse.)

54. $\ell_1\colon y - 1 = \sqrt{3}(x - 1)$
    $\ell_2\colon y = x$

1. Study the graph of $y = \sin \theta$ on page 390 and explain how the sine curve on the interval $\left[0, \dfrac{\pi}{2}\right]$ could be used to obtain the sine curve on $\left[\dfrac{\pi}{2}, \pi\right]$. Also, explain how the curve on $[0, \pi]$ can be used to obtain the curve on $[\pi, 2\pi]$.

2. Let a, b, c be any constants where a and b are not zero. Explain why the amplitude of $y = a \sin bx + c$ is $|a|$.

3. Find the equation of a sine curve that has two complete cycles on the interval $\left[0, \dfrac{3\pi}{4}\right]$ with amplitude 1.

4. For $-\dfrac{\pi}{2} < x \leq 0$, is it true or false that $|\tan x| = \tan|x|$? Explain.

5. The addition-subtraction formulas, on page 409, for the sine and cosine apply to all angle measures α and β. What are the restrictions on α and β for the tangent in formula (5)?

8.4
THE DOUBLE-
AND
HALF-ANGLE
FORMULAS

The formula $\sin(\alpha + \beta) = \sin \alpha \cos \beta + \cos \alpha \sin \beta$ is true for all values of α and β. If, in particular, we let $\beta = \alpha$, then

$$\sin 2\alpha = \sin(\alpha + \alpha) = \sin \alpha \cos \alpha + \cos \alpha \sin \alpha$$

Consequently, we have the following *double-angle formula:*

Formula 7 implies that, in general, $\sin 2\alpha \neq 2 \sin \alpha$. Verify this using $\alpha = 30°$.

(7) $$\sin 2\alpha = 2 \sin \alpha \cos \alpha$$

Next use $\alpha = \beta$ in the formula for $\cos(\alpha + \beta)$:

$$\cos(\alpha + \beta) = \cos \alpha \cos \beta - \sin \alpha \sin \beta$$

$$\cos 2\alpha = \cos(\alpha + \alpha) = \cos \alpha \cos \alpha - \sin \alpha \sin \alpha$$

Thus a double-angle formula for the cosine is

(8) $$\cos 2\alpha = \cos^2 \alpha - \sin^2 \alpha$$

Since $\cos^2 \alpha = 1 - \sin^2 \alpha$, formula (8) may be written as

(9) $$\cos 2\alpha = 1 - 2 \sin^2 \alpha$$

Similarly, using $\sin^2 \alpha = 1 - \cos^2 \alpha$, we have

(10) $$\cos 2\alpha = 2 \cos^2 \alpha - 1$$

Substituting $\alpha = \beta$ in $\tan(\alpha + \beta) = \dfrac{\tan \alpha + \tan \beta}{1 - \tan \alpha \tan \beta}$ gives

(11) $$\tan 2\alpha = \frac{2 \tan \alpha}{1 - \tan^2 \alpha}$$

Here is a summary of the double-angle formulas stated in terms of the variable θ.

	DOUBLE-ANGLE FORMULAS
(7)	$\sin 2\theta = 2 \sin \theta \cos \theta$
(8)	$\cos 2\theta = \cos^2 \theta - \sin^2 \theta$
(9)	$\cos 2\theta = 1 - 2 \sin^2 \theta$
(10)	$\cos 2\theta = 2 \cos^2 \theta - 1$
(11)	$\tan 2\theta = \dfrac{2 \tan \theta}{1 - \tan^2 \theta}$

Double-angle formulas can also be derived for other circular functions. (See Exercises 37, 38, and 40.)

EXAMPLE 1   Evaluate $\sin 15° \cos 15°$ using a double-angle formula.

Solution   Rewrite $2 \sin \theta \cos \theta = \sin 2\theta$ in the equivalent form

$$\sin \theta \cos \theta = \tfrac{1}{2} \sin 2\theta$$

Then

$$\sin 15° \cos 15° = \tfrac{1}{2} \sin 2(15°)$$
$$= \tfrac{1}{2} \sin 30°$$
$$= \tfrac{1}{2}(\tfrac{1}{2})$$
$$= \tfrac{1}{4}$$

TEST YOUR UNDERSTANDING
Think Carefully

Complete each equation to obtain a special case of a double-angle formula.

1. $2 \sin 5° \cos 5° = ?$

2. $\dfrac{2 \tan \dfrac{\theta}{2}}{1 - \tan^2 \dfrac{\theta}{2}} = ?$

3. $2 \cos^2 3\theta - 1 = ?$

4. $\cos^2 \dfrac{\alpha}{6} - \sin^2 \dfrac{\alpha}{6} = ?$

Evaluate the following using the double-angle formulas.

5. $\cos^2 15° - \sin^2 15°$

6. $2 \sin \dfrac{\pi}{8} \cos \dfrac{\pi}{8}$

7. $\dfrac{4 \tan(22.5°)}{1 - \tan^2(22.5°)}$

8. $2 \sin^2 \dfrac{\pi}{12} - 1$

(Answers: Page 449)

EXAMPLE 2   If θ is obtuse such that $\sin \theta = \frac{5}{13}$, find $\sin 2\theta$, $\cos 2\theta$, and $\tan 2\theta$.

Solution   Since θ is in quadrant II, and $\sin \theta = \dfrac{5}{13} = \dfrac{\text{opposite}}{\text{hypotenuse}}$, the third side of the reference triangle is found using the Pythagorean theorem; that is, $-\sqrt{13^2 - 5^2} = -12$.

Note: $\cos \theta$ and $\tan \theta$ can also be found without using a reference triangle. Thus

$$\cos \theta = -\sqrt{1 - \sin^2 \theta}$$

$$= -\sqrt{1 - \tfrac{25}{169}} = -\tfrac{12}{13}$$

$$\tan \theta = \frac{\sin \theta}{\cos \theta} = \frac{\frac{5}{13}}{-\frac{12}{13}}$$

$$= -\tfrac{5}{12}$$

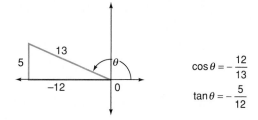

$$\cos \theta = -\frac{12}{13}$$

$$\tan \theta = -\frac{5}{12}$$

Then, from the reference triangle,

$$\sin 2\theta = 2(\tfrac{5}{13})(-\tfrac{12}{13}) = -\frac{120}{169} \qquad \text{formula (7)}$$

$$\cos 2\theta = 2(-\tfrac{12}{13})^2 - 1 = \frac{119}{169} \qquad \text{formula (10)}$$

$$\tan 2\theta = \frac{2(-\tfrac{5}{12})}{1 - (-\tfrac{5}{12})^2} = -\frac{120}{119} \qquad \text{formula (11)}$$

EXAMPLE 3   A photographer wants to take a picture of a 4-foot vase standing on a 3-foot pedestal. She wants to position the camera at a point C on the floor so that the angles subtended by the vase and the pedestal are the same size. How far away from the foot of the pedestal should the camera be placed?

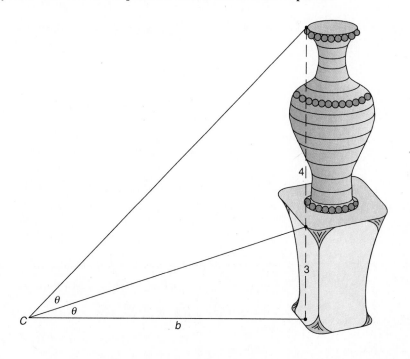

Solution From the figure we have $\tan \theta = \dfrac{3}{b}$ and $\tan 2\theta = \dfrac{4+3}{b} = \dfrac{7}{b}$. Now substitute for $\tan \theta$ in the formula for $\tan 2\theta$ as follows.

$$\frac{7}{b} = \tan 2\theta = \frac{2 \tan \theta}{1 - \tan^2 \theta} = \frac{2\left(\dfrac{3}{b}\right)}{1 - \left(\dfrac{3}{b}\right)^2}$$

Then

$$\frac{7}{b} = \frac{\dfrac{6}{b}}{1 - \dfrac{9}{b^2}}$$

$$\frac{7}{b} = \frac{6b}{b^2 - 9}$$

$$7b^2 - 63 = 6b^2$$

$$b^2 = 63$$

$$b = \sqrt{63} = 3\sqrt{7}$$

Since $3\sqrt{7} \approx 7.9$, the camera should be placed approximately 8 feet from the pedestal. ■

EXAMPLE 4 Let θ be an acute angle of a right triangle where $\sin \theta = \dfrac{x}{2}$. Show that $\sin 2\theta = \dfrac{x\sqrt{4 - x^2}}{2}$.

Solution Since $\sin \theta = \dfrac{x}{2}$, use x as the side opposite θ and 2 as the hypotenuse. The Pythagorean theorem gives the third side. Now use formula (7) to solve for $\sin 2\theta$.

$$\sin 2\theta = 2 \sin \theta \cos \theta$$

$$= 2\left(\frac{x}{2}\right)\left(\frac{\sqrt{4 - x^2}}{2}\right)$$

$$= \frac{x\sqrt{4 - x^2}}{2}$$

■

EXAMPLE 5 Show that $\cos 3\theta = 4 \cos^3 \theta - 3 \cos \theta$.

Solution

$$\begin{aligned}
\cos 3\theta &= \cos(2\theta + \theta) \\
&= \cos 2\theta \cos \theta - \sin 2\theta \sin \theta &&\text{formula (3)} \\
&= (2 \cos^2 \theta - 1) \cos \theta - (2 \sin \theta \cos \theta) \sin \theta &&\text{formulas (10) and (7)} \\
&= 2 \cos^3 \theta - \cos \theta - 2 \cos \theta \sin^2 \theta \\
&= 2 \cos^3 \theta - \cos \theta - 2 \cos \theta(1 - \cos^2 \theta) \\
&= 4 \cos^3 \theta - 3 \cos \theta
\end{aligned}$$

■

Formula (9) may be solved for $\sin^2 \alpha$ as follows:

$$\cos 2\alpha = 1 - 2 \sin^2 \alpha$$

$$2 \sin^2 \alpha = 1 - \cos 2\alpha$$

(12) $$\sin^2 \alpha = \frac{1 - \cos 2\alpha}{2}$$

Similarly, formula (10) produces a formula for $\cos^2 \alpha$ and the result for $\tan^2 \alpha$ is found using $\dfrac{\sin^2 \alpha}{\cos^2 \alpha}$.

(12) $$\sin^2 \theta = \frac{1 - \cos 2\theta}{2}$$

(13) $$\cos^2 \theta = \frac{1 + \cos 2\theta}{2}$$

(14) $$\tan^2 \theta = \frac{1 - \cos 2\theta}{1 + \cos 2\theta}$$

Notice that these identities convert the second power of a trigonometric function into an expression involving only the first power. This type of reduction makes these identities useful when it is necessary to reduce the powers within a given trigonometric expression.

EXAMPLE 6 Convert $\sin^4 \theta$ to an expression involving only the first power of the cosine.

Solution

$$\sin^4 \theta = (\sin^2 \theta)^2$$

$$= \left(\frac{1 - \cos 2\theta}{2}\right)^2 \qquad \text{formula (12)}$$

$$= \frac{1 - 2 \cos 2\theta + \cos^2 2\theta}{4}$$

$$= \tfrac{1}{4} - \tfrac{1}{2} \cos 2\theta + \tfrac{1}{4} \cos^2 2\theta$$

$$= \tfrac{1}{4} - \tfrac{1}{2} \cos 2\theta + \tfrac{1}{4}\left(\frac{1 + \cos 4\theta}{2}\right) \qquad \text{formula (13)}$$

$$= \tfrac{1}{4} - \tfrac{1}{2} \cos 2\theta + \tfrac{1}{8} + \tfrac{1}{8} \cos 4\theta$$

$$= \tfrac{3}{8} - \tfrac{1}{2} \cos 2\theta + \tfrac{1}{8} \cos 4\theta \qquad \blacksquare$$

Since (12) is an identity for all values of θ, we may change the form of θ by writing $\theta = \dfrac{\phi}{2}$ and substitute to obtain the **half-angle formula** for the sine function as follows:

$$\sin^2 \frac{\phi}{2} = \frac{1 - \cos \phi}{2}$$

If $x^2 = a$, then $x = \pm\sqrt{a}$.　　(15)

$$\sin \frac{\phi}{2} = \pm\sqrt{\frac{1 - \cos \phi}{2}}$$

When using this formula, the appropriate sign will depend on the location of the terminal side of $\dfrac{\phi}{2}$.

EXAMPLE 7　Evaluate $\sin 15°$ by using the half-angle formula.

Solution　Note that $15° = \dfrac{30°}{2}$ is in quadrant I. Therefore, we use the plus sign in the half-angle formula for $\sin \dfrac{\phi}{2}$:

$$\sin 15° = \sqrt{\frac{1 - \cos 30°}{2}} = \sqrt{\frac{1 - \dfrac{\sqrt{3}}{2}}{2}} = \frac{\sqrt{2 - \sqrt{3}}}{2} \qquad \blacksquare$$

Following the procedure used to get the formula for $\sin \dfrac{\phi}{2}$ from (12), the half-angle formula for the cosine is obtained from (13). These results are included in the summary below, written in terms of the variable θ. A formula for the tangent can be derived using (7) and (13).

$$\tan \frac{\theta}{2} = \frac{\sin \dfrac{\theta}{2}}{\cos \dfrac{\theta}{2}} = \frac{2 \sin \dfrac{\theta}{2} \cos \dfrac{\theta}{2}}{2 \cos^2 \dfrac{\theta}{2}} \qquad \text{multiplying numerator and denominator}$$
$$\text{by } 2 \cos \dfrac{\theta}{2}$$

Another variation is:
$$\tan \frac{\theta}{2} = \frac{1 - \cos \theta}{\sin \theta}$$
See Example 4, page 738.

$$= \frac{\sin \theta}{2\left(\dfrac{1 + \cos \theta}{2}\right)} \qquad \text{formulas (7) and (13)}$$

$$= \frac{\sin \theta}{1 + \cos \theta}$$

The plus or minus sign depends on the location of the terminal side of $\dfrac{\theta}{2}$.

HALF-ANGLE FORMULAS

(15)　　$$\sin \frac{\theta}{2} = \pm\sqrt{\frac{1 - \cos \theta}{2}}$$

(16)　　$$\cos \frac{\theta}{2} = \pm\sqrt{\frac{1 + \cos \theta}{2}}$$

(17)　　$$\tan \frac{\theta}{2} = \frac{\sin \theta}{1 + \cos \theta} = \frac{1 - \cos \theta}{\sin \theta}$$

EXAMPLE 8   Solve for *b* without using tables or a calculator.

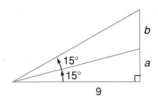

Solution   From the preceding figure $\tan 15° = \dfrac{a}{9}$. Also, using (17),

$$\tan 15° = \tan \frac{30°}{2} = \frac{1 - \cos 30°}{\sin 30°}$$

how that the same result is btained by using the form

$$\tan 15° = \frac{\sin 30°}{1 + \cos 30°}$$

$$= \frac{1 - \dfrac{\sqrt{3}}{2}}{\frac{1}{2}}$$

$$= 2 - \sqrt{3}$$

Then

$$\frac{a}{9} = 2 - \sqrt{3} \quad \text{or} \quad a = 18 - 9\sqrt{3}.$$

Also, since $\tan 30° = \dfrac{\sqrt{3}}{3}$, and since the given figure shows that $\tan 30° = \dfrac{a + b}{9}$, we have

$$\frac{a + b}{9} = \frac{\sqrt{3}}{3}$$

$$a + b = 3\sqrt{3}$$

$$b = 3\sqrt{3} - a$$

$$= 3\sqrt{3} - (18 - 9\sqrt{3}) \qquad (a = 18 - 9\sqrt{3})$$

$$= 12\sqrt{3} - 18$$

CAUTION: Learn to Avoid These Mistakes	
WRONG	**RIGHT**
$\sin 4x = 4 \sin x$	$\sin 4x = \sin 2(2x)$ $= 2 \sin 2x \cos 2x$
$\cos^4 x = \dfrac{1 + \cos^2 2x}{2}$	$\cos^4 x = \left(\dfrac{1 + \cos 2x}{2}\right)^2$

EXERCISES 8.4

Evaluate the expression using an appropriate double-angle formula. Do not use tables or a calculator.

1. $1 - 2\sin^2\dfrac{7\pi}{12}$

2. $\sin 105° \cos 105°$

3. $\dfrac{6\tan 75°}{1 - \tan^2 75°}$

4. $\dfrac{1}{4}\left(\cos^2\dfrac{3\pi}{8} - \sin^2\dfrac{3\pi}{8}\right)$

Use half-angle formulas to verify the equation.

5. $\cos 75° = \frac{1}{2}\sqrt{2 - \sqrt{3}}$

6. $\sin\left(-\dfrac{\pi}{8}\right) = -\dfrac{1}{2}\sqrt{2 - \sqrt{2}}$

7. $\tan\dfrac{3\pi}{8} = \sqrt{2} + 1$

8. $\sin\dfrac{\pi}{12} = \dfrac{1}{2}\sqrt{2 - \sqrt{3}}$

9. (a) Show that the half-angle formula in (17) gives $\tan 15° = 2 - \sqrt{3}$.

 (b) Use Tables I and VI, or a calculator, to verify that $2 - \sqrt{3}$ and $\tan 15°$ are the same.

10. (a) Use an appropriate formula to show that $\cos^2\dfrac{\pi}{8} = \dfrac{2 + \sqrt{2}}{4}$.

 (b) From part (a) show that $\cos\dfrac{\pi}{8} = \dfrac{1}{2}\sqrt{2 + \sqrt{2}}$.

 (c) Use a calculator to verify that $\dfrac{1}{2}\sqrt{2 + \sqrt{2}}$ and $\cos\dfrac{\pi}{8}$ are the same.

11. (a) Use an appropriate formula to show that $\tan(22.5°) = \sqrt{2} - 1$.

 (b) Use Tables I and VI, or a calculator, to show that $\sqrt{2} - 1$ and $\tan(22.5°)$ are the same.

12. If $\cos\theta = \frac{12}{13}$ and θ is in the first quadrant, use double-angle formulas to find:

 (a) $\sin 2\theta$ **(b)** $\cos 2\theta$ **(c)** $\tan 2\theta$

13. If θ is obtuse and $\tan\theta = -\frac{15}{8}$, use double-angle formulas to find:

 (a) $\sin 2\theta$ **(b)** $\cos 2\theta$ **(c)** $\tan 2\theta$

14. If $\sin\theta = -\frac{24}{25}$ and $\tan\theta > 0$, find $\cos 2\theta$.

15. If $\tan\theta = -\frac{2}{3}$ and $\cos\theta > 0$, find $\sin 2\theta$.

 16. Use the information from the triangle to derive $\cos 2\theta = 1 - \frac{2}{9}x^2$.

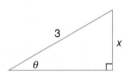

 17. Use the information in Exercise 16 to derive $9\sin 2\theta = 2x\sqrt{9 - x^2}$.

 18. Let θ be an acute angle of a right triangle so that $\tan\theta = x$. Find the expressions for $\sin 2\theta$, $\cos 2\theta$, and $\tan 2\theta$ in terms of x.

Solve for b without using tables or a calculator.

19.

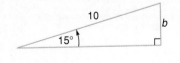

20.

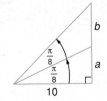

21. At a point A that is 50 meters from the base of a tower, the angle of elevation to the top of the tower is twice as large as is the angle of elevation from a point B that is 150 meters from the tower. Assuming that the base of the tower and the points A and B are in the same line on level ground, find the height of the tower h.

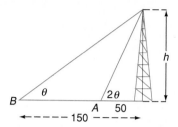

Use a double-angle formula to solve for b without using tables or a calculator.

22.

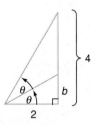

23.

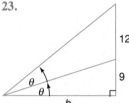

24.

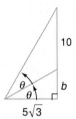

(*Hint:* Use the rational root theorem.)

Verify the identities.

25. $\cot x \sin 2x = 2 \cos^2 x$

26. $\dfrac{\sin 2x}{1 + \cos 2x} = \tan x$

27. $\sec^2 \dfrac{x}{2} = \dfrac{2}{1 + \cos x}$

28. $\sin^2 \dfrac{x}{2} = \dfrac{\sec x - 1}{2 \sec x}$

29. $\cos^2 \dfrac{x}{2} = \dfrac{\tan x + \sin x}{2 \tan x}$

30. $\tan^2 \dfrac{x}{2} = \dfrac{\sec x - 1}{\sec x + 1}$

31. $\cot \dfrac{x}{2} = \csc x + \cot x$

32. $\dfrac{1}{2} \tan 2x = \dfrac{1}{\cot x - \tan x}$

33. $(\sin x + \cos x)^2 = 1 + \sin 2x$

34. $\sin 3x = 3 \sin x - 4 \sin^3 x$

35. $\tan 3x = \dfrac{3 \tan x - \tan^3 x}{1 - 3 \tan^2 x}$

36. $\sin 2x = \dfrac{2 \tan x}{\sec^2 x}$

37. $\cot 2x = \dfrac{\cot^2 x - 1}{2 \cot x}$

38. $\csc 2x = \frac{1}{2} \csc x \sec x$

39. $\csc 2x - \cot 2x = \tan x$

40. $\sec 2\theta = \dfrac{\sec^2 \theta}{2 - \sec^2 \theta}$

41. $\cos 2x = \cos^4 x - \sin^4 x$

42. $\sin 4x = 8 \sin x \cos^3 x - 4 \sin x \cos x$

43. $\cos 4x = 8 \cos^4 x - 8 \cos^2 x + 1$

44. $\sin^4 \dfrac{x}{2} = \dfrac{1}{4} - \dfrac{1}{2} \cos x + \dfrac{1}{4} \cos^2 x$

45. $\cos^4 x = \frac{1}{8} \cos 4x + \frac{1}{2} \cos 2x + \frac{3}{8}$

46. $\tan \dfrac{x}{2} + \cot \dfrac{x}{2} = 2 \csc x$

47. **(a)** Add the formulas for $\cos(\alpha + \beta)$ and $\cos(\alpha - \beta)$ to derive this *product formula:*
$\cos \alpha \cos \beta = \frac{1}{2}[\cos(\alpha + \beta) + \cos(\alpha - \beta)]$
(b) Use a similar analysis to prove these product formulas:
$\sin \alpha \sin \beta = \frac{1}{2}[\cos(\alpha - \beta) - \cos(\alpha + \beta)]$
$\sin \alpha \cos \beta = \frac{1}{2}[\sin(\alpha + \beta) + \sin(\alpha - \beta)]$
$\cos \alpha \sin \beta = \frac{1}{2}[\sin(\alpha + \beta) - \sin(\alpha - \beta)]$

48. Use the results of Exercise 47 to express each of the following as a sum or difference:
(a) $\sin 6x \sin 2x$
(b) $2 \cos x \cos 4x$
(c) $3 \cos 5x \sin(-2x)$
(d) $4 \sin x \cos \dfrac{x}{2}$

49. Substitute $\alpha = \dfrac{u + v}{2}$ and $\beta = \dfrac{u - v}{2}$ into the formulas in Exercise 47 and derive these *sum formulas:*

$$\cos u + \cos v = 2 \cos \frac{u + v}{2} \cos \frac{u - v}{2}$$

$$\cos v - \cos u = 2 \sin \frac{u + v}{2} \sin \frac{u - v}{2}$$

$$\sin u + \sin v = 2 \sin \frac{u + v}{2} \cos \frac{u - v}{2}$$

$$\sin u - \sin v = 2 \cos \frac{u + v}{2} \sin \frac{u - v}{2}$$

50. Use the results of Exercise 49 to express each of the following as a product:

(a) $\sin 4x + \sin 2x$ **(b)** $\cos 6x - \cos 3x$

(c) $2 \sin 5x - 2 \sin x$ **(d)** $\dfrac{1}{2} \cos \dfrac{x}{2} + \dfrac{1}{2} \cos \dfrac{5x}{2}$

**CHALLENGE
Think Creatively**

(a) Explain how $\tan \dfrac{\theta}{2} = \dfrac{\sin \theta}{1 + \cos \theta}$ is demonstrated in the figure.

(b) Find a geometric interpretation of $\tan \dfrac{\theta}{2} = \dfrac{1 - \cos \theta}{\sin \theta}$.

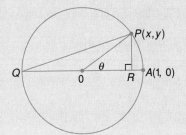

8.5
TRIGONOMETRIC
EQUATIONS

Can you find the points of intersection of the sine and cosine curves?

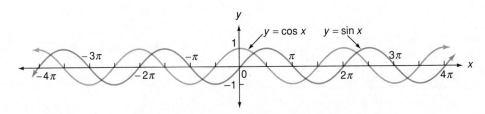

A trigonometric equation that is true for some values of the variables but is not an identity is a conditional equation.

Answering this question amounts to finding all values x for which $\sin x = \cos x$. To solve this **trigonometric equation** for x, first divide by $\cos x$.

$$\sin x = \cos x$$

$$\frac{\sin x}{\cos x} = 1$$

Now use $\dfrac{\sin x}{\cos x} = \tan x$ and solve $\tan x = 1$. For x in the interval $\left(-\dfrac{\pi}{2}, \dfrac{\pi}{2}\right)$, $\tan x = 1$ has only the one solution, $x = \dfrac{\pi}{4}$. Then, since the tangent has period π, all the solutions are obtained by adding on all the multiples of π:

$$x = \dfrac{\pi}{4} + k\pi, \qquad \text{where } k \text{ is any integer}$$

For $k = -2, -1, 0\ 1, 2$, the following specific solutions are produced:

$$-\dfrac{7\pi}{4}, \ -\dfrac{3\pi}{4}, \dfrac{\pi}{4}, \dfrac{5\pi}{4}, \dfrac{9\pi}{4}$$

or

$$-315°, \ -135°, \ 45°, \ 225°, \ 405°$$

$x = \dfrac{\pi}{4} + k\pi$, k any integer, is the **general solution** of $\sin x = \cos x$.

EXAMPLE 1 Find the general solution: $\sqrt{3} \csc x - 2 = 0$.

Solution Begin by isolating $\csc x$ on one side.

$$\sqrt{3} \csc x - 2 = 0$$
$$\sqrt{3} \csc x = 2$$
$$\csc x = \dfrac{2}{\sqrt{3}} \qquad \left(\text{or } \sin x = \dfrac{\sqrt{3}}{2}\right)$$

Unless otherwise indicated, the solutions for trigonometric equations are assumed to be the exact values, not approximations.

Thus, in the interval $(0, 2\pi)$, $x = \dfrac{\pi}{3}$ or $x = \dfrac{2\pi}{3}$. Since the cosecant has period 2π, we obtain the general solution by adding on the multiples of 2π. Then, for any integer k we have

$$x = \begin{cases} \dfrac{\pi}{3} + 2k\pi \\[2mm] \dfrac{2\pi}{3} + 2k\pi \end{cases} \quad \text{or} \quad x = \begin{cases} 60° + k\,(360°) \\[1mm] 120° + k\,(360°) \end{cases}$$

∎

EXAMPLE 2 Find the general solution: $\cos^2 x = \cos x$.

Solution Since the cosine has period 2π, first solve for x in the interval $0 \le x < 2\pi$. Begin by subtracting $\cos x$ from both sides. Thus

CAUTION
Do not begin by dividing $\cos^2 x = \cos x$ by $\cos x$. This step would produce $\cos x = 1$ and we would have lost all the roots of $\cos x = 0$. (This error is comparable to solving the equation $x^2 = x$ by first dividing each side by x.)

$$\cos^2 x - \cos x = 0$$

Factor out $\cos x$.

$$(\cos x)(\cos x - 1) = 0$$

Then

$$\cos x = 0 \quad \text{or} \quad \cos x - 1 = 0$$

Hence $\cos x = 0$ for $x = \dfrac{\pi}{2}$ or $\dfrac{3\pi}{2}$ and $\cos x = 1$ for $x = 0$. Since the period of the cosine is 2π, we obtain all the solutions by taking each solution in $[0, 2\pi)$ and adding all multiples of 2π. The general solution may be presented in these forms, where k is any integer.

$$x = \begin{cases} \dfrac{\pi}{2} + 2k\pi \\[2mm] \dfrac{3\pi}{2} + 2k\pi \\[2mm] 2k\pi \end{cases} \quad \text{or} \quad x = \begin{cases} 90° + k(360°) \\[1mm] 270° + k(360°) \\[1mm] k(360°) \end{cases} \qquad \blacksquare$$

EXAMPLE 3 Find the general solution: $\cos 2x = \frac{1}{2}$

Solution Reduce the given equation to a simpler one by substituting $\theta = 2x$ and solve $\cos \theta = \frac{1}{2}$. In $[0, 2\pi)$ the solutions for θ are $\dfrac{\pi}{3}$ and $\dfrac{5\pi}{3}$, so that for any integer k

$$\theta = \begin{cases} \dfrac{\pi}{3} + 2k\pi \\[3mm] \dfrac{5\pi}{3} + 2k\pi \end{cases}$$

Then, since $x = \dfrac{\theta}{2}$, divide the preceding equations by 2 to get the general solution.

$$x = \begin{cases} \dfrac{\pi}{6} + k\pi \\[3mm] \dfrac{5\pi}{6} + k\pi \end{cases} \qquad \blacksquare$$

Observe that in the general solution multiples of π, rather than of 2π, are added. This is due to the fact that $\cos 2x$ has period π.

Example 3 can also be solved using the identity $\cos 2x = 2\cos^2 x - 1$ and noting that the period is π.

$$2\cos^2 x - 1 = \frac{1}{2}$$

Check for $x = \dfrac{\pi}{6} + k\pi$:

$$\cos 2x = \cos 2\left(\frac{\pi}{6} + k\pi\right)$$

$$= \cos\left(\frac{\pi}{3} + 2k\pi\right)$$

$$= \cos \frac{\pi}{3} = \frac{1}{2}.$$

The check for $x = \dfrac{5\pi}{6} + k\pi$ is similar.

$$\cos^2 x = \frac{3}{4}$$

$$\cos x = \pm \frac{\sqrt{3}}{2}$$

Then, in $[0, 2\pi)$, $x = \dfrac{\pi}{6}, \dfrac{5\pi}{6}, \dfrac{7\pi}{6}, \dfrac{11\pi}{6}$.

Now, since $\dfrac{\pi}{6} + 1(\pi) = \dfrac{7\pi}{6}$ and $\dfrac{5\pi}{6} + 1(\pi) = \dfrac{11\pi}{6}$, the same general solution is obtained by adding multiples of π to $\dfrac{\pi}{6}$ and $\dfrac{5\pi}{6}$ as before.

Numerous situations in trigonometry only call for the solutions of an equation in the interval $[0, 2\pi)$. This is illustrated in the next example.

EXAMPLE 4   Solve $\sin 2x = \sin x$ for $0 \le x < 2\pi$.

Solution   Using the double-angle formula $\sin 2x = 2 \sin x \cos x$, we may write

$$2 \sin x \cos x = \sin x$$

or

$$2 \sin x \cos x - \sin x = 0$$

Factor the left side.

$$(\sin x)(2 \cos x - 1) = 0$$

Then

$$\sin x = 0 \quad \text{or} \quad 2 \cos x = 1$$
$$\sin x = 0 \quad \text{or} \quad \cos x = \tfrac{1}{2}$$

Check:

$x = 0$: $\sin (2 \cdot 0)$
$\quad = 0 = \sin 0$

$x = \dfrac{\pi}{3}$: $\sin \dfrac{2\pi}{3}$

$\quad = \dfrac{\sqrt{3}}{2} = \sin \dfrac{\pi}{3}$

$x = \pi$: $\sin 2\pi = 0$
$\quad = \sin \pi$

$x = \dfrac{5\pi}{3}$: $\sin \dfrac{10\pi}{3} = \sin \dfrac{4\pi}{3}$

$\quad = -\dfrac{\sqrt{3}}{2} = \sin \dfrac{5\pi}{3}$

Now $\sin x = 0$ for $x = 0$ and π; $\cos x = \dfrac{1}{2}$ for $x = \dfrac{\pi}{3}$ and $\dfrac{5\pi}{3}$. The solutions in $[0, 2\pi)$ are 0, $\dfrac{\pi}{3}$, π, and $\dfrac{5\pi}{3}$. ∎

EXAMPLE 5   Solve: $\dfrac{\sec x}{\cos x} - \dfrac{1}{2} \sec x = 0$.

Solution

$$\frac{\sec x}{\cos x} - \frac{1}{2} \sec x = 0$$

$$(\sec x)\left(\frac{1}{\cos x} - \frac{1}{2}\right) = 0$$

$$\sec x = 0 \quad \text{or} \quad \frac{1}{\cos x} - \frac{1}{2} = 0$$

$$\sec x = 0 \quad \text{or} \quad \frac{1}{\cos x} = \frac{1}{2}$$

$$\sec x = 0 \quad \text{or} \quad \cos x = 2$$

Some trigonometric equations have no solutions. This is comparable to an algebraic equation such as $x^2 + 1 = 0$, which has no real solutions.

Since $\sec x \ge 1$ or $\sec x \le -1$ for all x, $\sec x = 0$ has no roots. Also, since $|\cos x| \le 1$, $\cos x = 2$ has no roots. Thus the given equation has no solutions. ∎

EXAMPLE 6   Solve: $\cos^2 x + \tfrac{1}{2} \sin x - \tfrac{1}{2} = 0$ for $0 \le x \le 2\pi$.

Solution   Multiply by 2 to clear fractions.

$$2(\cos^2 x + \tfrac{1}{2} \sin x - \tfrac{1}{2}) = 2 \cdot 0$$
$$2 \cos^2 x + \sin x - 1 = 0$$

Convert to an equivalent form involving only $\sin x$ by using $\cos^2 x = 1 - \sin^2 x$.

$$2(1 - \sin^2 x) + \sin x - 1 = 0$$
$$2 \sin^2 x - \sin x - 1 = 0$$

Think of $u = \sin x$ and factor $2u^2 - u - 1$ as $(2u + 1)(u - 1)$.

The left side is quadratic in $\sin x$, which is factorable.

$$(2 \sin x + 1)(\sin x - 1) = 0$$
$$2 \sin x + 1 = 0 \quad \text{or} \quad \sin x - 1 = 0$$
$$\sin x = -\tfrac{1}{2} \quad \text{or} \quad \sin x = 1$$

Now $\sin x = -\dfrac{1}{2}$ for $x = \dfrac{7\pi}{6}$ and $\dfrac{11\pi}{6}$; $\sin x = 1$ for $x = \dfrac{\pi}{2}$. The solutions in $[0, 2\pi)$ are $\dfrac{\pi}{2}, \dfrac{7\pi}{6}$, and $\dfrac{11\pi}{6}$. ■

EXAMPLE 7 Use Table VI or a calculator to approximate the roots of $3 \tan x - 7 = 0$ in degree measure, where $0° \leq x < 360°$.

Solution Since $3 \tan x - 7 = 0$, we get $\tan x = \tfrac{7}{3} = 2.3333$ to four decimal places. We use the closest entry in Table VI to obtain $x = 66.8°$. Adding $180°$ produces the second answer, $246.8°$, which is in the third quadrant.

The calculator solution for an angle having a given trigonometric ratio was introduced on page 360.

Since $\tan x = \tfrac{7}{3}$, the following sequence can be used to find the first-quadrant solution, after the calculator has been set in the degree mode:

$$x = \boxed{7} \; \boxed{\div} \; \boxed{3} \; \boxed{=} \; \boxed{\text{INV}} \; \boxed{\text{TAN}} = 66.801409° = 66.8° \qquad \text{(to one decimal place)}$$

The third-quadrant solution is found as before; $66.8° + 180° = 246.8°$. ■

EXAMPLE 8 Solve $2 \sin^2 x + 5 \sin x - 2 = 0$ for $0° \leq x < 360°$, using tables or a calculator.

Solution Let $u = \sin x$ and substitute to get $2u^2 + 5u - 2 = 0$. Now the quadratic equation produces

$$u = \frac{-5 \pm \sqrt{5^2 - 4(2)(-2)}}{2(2)} = \frac{-5 \pm \sqrt{41}}{4}$$

$$\sin x = \frac{-5 \pm \sqrt{41}}{4}$$

$$= \frac{-5 \pm 6.403}{4} \qquad \text{(Table I)}$$

$$= \begin{cases} 0.3508 \\ -2.8508 \end{cases} \qquad \text{(to four decimal places)}$$

Since $|\sin x| \leq 1$, $\sin x = -2.8508$ has no solutions. For $\sin x = 0.3508$ Table VI gives $20.5°$ in the first quadrant. Then in the second quadrant we have $180° - 20.5° = 159.5°$. A calculator solution is shown on the next page.

By calculator (in degree mode) use

$$x = \boxed{5} \boxed{+/-} \boxed{+} \boxed{41} \boxed{\sqrt{}} \boxed{=} \boxed{\div} \boxed{4} \boxed{=} \boxed{\text{INV}} \boxed{\text{SIN}} = 20.535096° = 20.5°$$

(to one decimal place)

The solution in the second quadrant is found as before. ■

EXERCISES 8.5

Find the general solution and check.

1. $\cos x = 1$
2. $\sin x = 1$
3. $\sin x = \frac{1}{2}$
4. $\cos^2 x = \frac{1}{2}$
5. $\sec x = -1$
6. $\csc x = 2$
7. $\sin 2x = -\frac{1}{2}$
8. $2 \cos \frac{x}{3} = 1$
9. $\frac{1}{2} \sin^2 x = 1$
10. $\tan x = \sqrt{3}$
11. $\tan 2x = \sqrt{3}$
12. $2 \sin(x - 1) = \sqrt{2}$
13. $2 \cos(x + 1) = -2$
14. $\frac{1}{\sec x} = 2$
15. $\cos 3x = 1$

Solve the equation for θ in the interval $[0°, 360°)$ and check.

16. $2 \sin \theta = 1$
17. $\sqrt{3} \csc \theta = 2$
18. $2 \cos \theta + 3 = 0$
19. $2 \sec \theta - 2\sqrt{2} = 0$
20. $\tan^2 \theta - 3 = 0$
21. $\sin^2 \theta - \cos^2 \theta = 1$
22. $2 \csc^2 \theta - 1 = 0$
23. $2 \sin \frac{\theta}{2} - 1 = 0$
24. $1 + \sqrt{2} \sin 2\theta = 0$
25. $-1 + \tan \frac{3\theta}{2} = 0$
26. $\sin^2 \theta - \cos^2 \theta = 0$
27. $2 \tan \theta \cos^2 \theta = 1$

Solve the equation for x in the interval $0 \le x < 2\pi$.

28. $\sin x(\cos x + 1) = 0$
29. $(\cos x - 1)(2 \sin x + 1) = 0$
30. $\sin x + \cos x = 0$
31. $\sin x - \sqrt{3} \cos x = 0$
32. $\sec x + \tan x = 0$
33. $2 \cos^2 x - \sqrt{3} \cos x = 0$
34. $2 \tan x = \sin x$
35. $\sin^2 x + \sin x - 2 = 0$
36. $2(\cos^2 x - \sin^2 x) = \sqrt{2}$
37. $\sin 2x = \cos x$
38. $\sin^2 x + 2 \cos^2 x = 2$
39. $\sin^2 x + \cos^2 x = 1.5$
40. $\sec^2 x = 2 \tan x$
41. $2 \cos^2 x + 9 \cos x - 5 = 0$
42. $\cos 2x - \cos x = 0$
43. $\cos^2 2x = \cos 2x$
44. $3 \cos^4 x + 4 \cos^2 x = 0$
45. $2 \sin^4 x - 3 \sin^2 x + 1 = 0$
46. $2 \sin^2 x - 1 = \sin x$
47. $2 \tan x - 1 = \tan^2 x$
48. $\sin x \cot^2 x - 3 \sin x = 0$
49. $3 \tan^2 x = 7 \sec x - 5$
50. $\sin^3 2x - \sin 2x = 0$
51. $\cos^2 x - \sin^2 x + \sin x = 1$
52. $2 \cos x - 2 \sec x - 3 = 0$
53. $3 \cos 2x + 2 \sin^2 x = 2$
54. $\cos 4x = \sin 2x$
55. $\sin 2x + \sin 4x = 0$
56. $\sec x + \tan x = 1$ (*Hint:* Square both sides.)
57. $\sin x + \cos x = 1$ (*Hint:* Square both sides.)

Use Table VI to approximate the solutions of each equation to the nearest tenth of a degree for x in the interval $[0°, 360°)$.

58. $3 \sin x = 2$
59. $7 \sec x - 15 = 0$
60. $\cos 2x = .9033$
61. $\sin \frac{x}{2} = .8259$

62. $12 \cos^2 x + 5 \cos x - 3 = 0$ **63.** $4 \cot^2 x - 12 \cot x + 9 = 0$

64. $\tan^2 x + \tan x - 1 = 0$ (*Hint:* Use the quadratic formula.)

65. $\cos^2 x - \sin^2 x = -\frac{3}{4}$ **66.** $4 \sin x - 5 \cos x = 0$

67. In the figure α and β are acute angles.

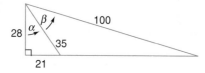

(a) Show that $\alpha < \frac{\pi}{4}$ by comparing $\cos \alpha$ and $\cos \frac{\pi}{4}$.

(b) Without using tables or a calculator, prove that $\alpha = \beta$. (*Hint:* Apply a double-angle formula to 2α.)

Find the coordinates of the points of intersection of the two curves for the interval $[0, 2\pi]$.

68. $y = \tan x$ **69.** $y = \sin x$ **70.** $y = \sin x$
$ y = \cot x$ $ y = -\cos x$ $ y = \cos 2x$

**CHALLENGE
Think Creatively**

Observe that in the general solution in Example 2 of this section, the first two forms differ by π for each k. Find a single form that can replace these two forms.

**EXPLORATIONS
Think Critically**

1. What are the restrictions on θ in formula (11) on page 418?

2. A purpose of solving Example 8 on page 423 without the use of tables or a calculator is to become familiar with the use of the half-angle formula for the tangent. Simpler, though approximate, calculator solutions are possible. Write a calculator sequence for finding b.

3. Verification of formulas in Section 8.4 can be made for specific cases of the variable by using known trigonometric ratios. In this regard, first find $\cos 15°$ using the half-angle formula; then use this result in the identity of Exercise 43, page 425, to verify that $\cos 60° = \frac{1}{2}$.

4. Two solutions for Example 3 are given in Section 8.5. Find a third way of solving this example.

5. When solving the equation $\sin x - \cos x = \sqrt{2}$ for $0 \le x < 2\pi$, we can begin by squaring both sides of the equation and simplify to obtain $\sin 2x = -1$. After solving this last equation for x, special care needs to be taken. Why?

**8.6
INVERSE
TRIGONOMETRIC
FUNCTIONS**

The sine function is not one-to-one. This fact is apparent from its graph, since a horizontal line through a range value y intersects the curve more than once; more than one x corresponds to y.

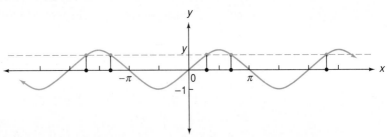

If we restrict the domain to $\left[-\dfrac{\pi}{2}, \dfrac{\pi}{2}\right]$, then $y = \sin x$ is one-to-one because for each range value there corresponds exactly one domain value.

This restricted function will have the same range as the original function: $-1 \le y \le 1$. The graph is shown on the next page.

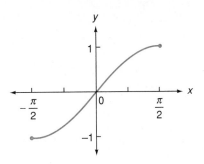

See Section 5.5 to review the inverse function concept.

Since $y = \sin x$ for $-\dfrac{\pi}{2} \le x \le \dfrac{\pi}{2}$ is one-to-one, we know that there is an inverse function. The graph of the inverse is obtained by reflecting the graph of $y = \sin x$ through the line $y = x$, and the equation of the inverse is obtained by interchanging the variables in $y = \sin x$. We also know that the domain and range of the restricted function $y = \sin x$ become the range and domain of the inverse, respectively.

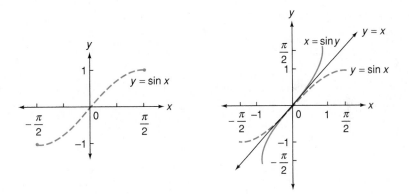

$$\left(\begin{array}{c}\text{Begin with } y = \sin x \text{ for} \\ -\dfrac{\pi}{2} \le x \le \dfrac{\pi}{2} \text{ and } -1 \le y \le 1\end{array}\right) \rightarrow \left(\begin{array}{c}\text{Reflect} \\ \text{through } y = x\end{array}\right) \rightarrow \left(\begin{array}{c}\text{To get the inverse } x = \sin y \\ \text{for } -1 \le x \le 1 \text{ and } -\dfrac{\pi}{2} \le y \le \dfrac{\pi}{2}\end{array}\right)$$

The equation of the inverse, $x = \sin y$, does not express y explicitly as a function of x. To do this we create some new terminology. First observe that $x = \sin y$ means that y is the radian whose sine is x, or

y is the *arc* length on the unit circle whose sine is x

To shorten this, we replace "arc length on the unit circle whose sine is x" by "arc sine of x," and this phrase is further abbreviated to "arcsin x." Thus $y =$ **arcsin x** is, by definition, equivalent to $x = \sin y$. To sum up, the basic information about this new function is shown on the next page.

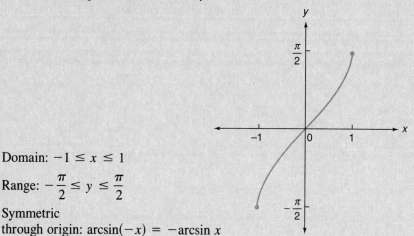

THE INVERSE SINE FUNCTION

$y = \arcsin x$ (equivalent to $x = \sin y$)

Domain: $-1 \leq x \leq 1$

Range: $-\dfrac{\pi}{2} \leq y \leq \dfrac{\pi}{2}$

Symmetric
through origin: $\arcsin(-x) = -\arcsin x$

The proof of the symmetric property is called for in Exercise 45.

Another common notation for the inverse sine function is $y = \sin^{-1} x$, in which the -1 is *not* an exponent; $\sin^{-1} x \neq \dfrac{1}{\sin x}$.

Since we are dealing with inverse functions, the following are true:

$$\arcsin(\sin x) = x \qquad \text{if } -\frac{\pi}{2} \leq x \leq \frac{\pi}{2}$$

$$\sin(\arcsin x) = x \qquad \text{if } -1 \leq x \leq 1$$

EXAMPLE 1   Evaluate $\arcsin \frac{1}{2}$.

Solution   Let $y = \arcsin \frac{1}{2}$. Then y is the radian whose sine is equal to $\frac{1}{2}$. Thus $y = \dfrac{\pi}{6}$. To check this, we use the fact that $y = \arcsin x$ is equivalent to $x = \sin y$. Hence

$$\sin y = \sin \frac{\pi}{6} = \frac{1}{2} = x \qquad \blacksquare$$

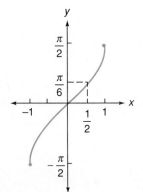

You will find it helpful in finding values like $\arcsin \frac{1}{2}$ to remember that $y = \arcsin x$ is negative for $-1 \leq x < 0$ and positive for $0 < x \leq 1$. You can see this from its graph.

CAUTION: Even though $\sin \dfrac{5\pi}{6} = \dfrac{1}{2}$, we do not have $\arcsin \dfrac{1}{2} = \dfrac{5\pi}{6}$ because the range of the arcsin function consists of the numbers in the interval $\left[-\dfrac{\pi}{2}, \dfrac{\pi}{2} \right]$.

EXAMPLE 2 Evaluate arcsin(−0.3981) by using the following:
(a) Table V and (b) a calculator

Solution

(a) The closest entry to 0.3981 in Table V in the sin column is 0.3986. We use this closest entry and observe that in the same row as 0.3986 and under the radian (real number) heading at the left we have 0.41, so that arcsin(0.3981) = 0.41. Then, since the arcsin is symmetric through the origin, we have

$$\arcsin(-0.3981) = -\arcsin(0.3981)$$

$$= -0.41$$

Let $x = \arcsin(-0.3981)$ so that $\sin x = -0.3981$. Thus we need to reverse the process of finding a ratio for a given angle. And since the table only has positive values, we begin with 0.3981.

(b) Writing $x = \arcsin(-0.3981)$ or $\sin x = -0.3981$ indicates that x can be found with a calculator using the inverse $\boxed{\text{INV}}$ key

$$\arcsin(-0.3981) = \boxed{0.3981}\,\boxed{+/-}\,\boxed{\text{INV}}\,\boxed{\text{SIN}} = -0.4094447$$

↑
answer
↓

$$= -0.41 \text{ (rounded to two places)} \quad \blacksquare$$

To define the inverse cosine function, known as *arccos*, we begin by restricting $y = \cos x$ to $0 \le x \le \pi$ to obtain a one-to-one function. Next, reflect the curve for $0 \le x \le \pi$ through the line $y = x$ to obtain the graph of the inverse whose equation is $x = \cos y$.

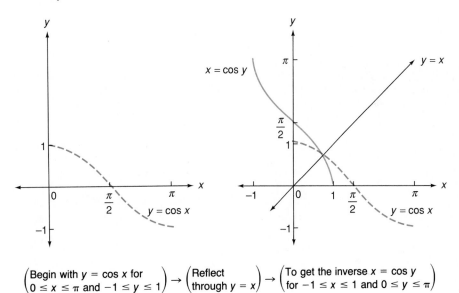

$$\begin{pmatrix} \text{Begin with } y = \cos x \text{ for} \\ 0 \le x \le \pi \text{ and } -1 \le y \le 1 \end{pmatrix} \rightarrow \begin{pmatrix} \text{Reflect} \\ \text{through } y = x \end{pmatrix} \rightarrow \begin{pmatrix} \text{To get the inverse } x = \cos y \\ \text{for } -1 \le x \le 1 \text{ and } 0 \le y \le \pi \end{pmatrix}$$

Now define $y = \mathbf{arccos}\ x$ to mean $x = \cos y$ and call it the *inverse cosine function*. (This function is also written in the form $y = \cos^{-1} x$.) Since arccos and cos are inverse functions, we have the following:

$$\arccos(\cos x) = x \qquad \text{if } 0 \le x \le \pi$$

$$\cos(\arccos x) = x \qquad \text{if } -1 \le x \le 1$$

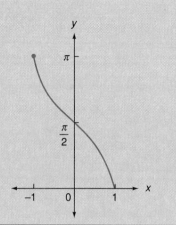
EXAMPLE 3 Find $\arccos(-\frac{1}{2})$.

Solution Since $\cos \dfrac{2\pi}{3} = -\frac{1}{2}$ and $\dfrac{2\pi}{3}$ is in the range of the arccos function, we have $\arccos(-\frac{1}{2}) = \dfrac{2\pi}{3}$. ∎

EXAMPLE 4 Find $\sin(\arccos \frac{1}{2})$.

Solution Since $\arccos \dfrac{1}{2} = \dfrac{\pi}{3}$, we have

$$\sin\left(\arccos \frac{1}{2}\right) = \sin \frac{\pi}{3} = \frac{\sqrt{3}}{2}$$

∎

EXAMPLE 5 Find the exact value of $\cos(\arcsin \frac{2}{3})$.

Solution We know that $y = \arcsin \frac{2}{3}$ is an acute angle. (Why?) Now construct a right triangle, as shown in the margin, with y an acute angle and $\sin y = \frac{2}{3}$. Then, from the triangle,

$$\cos\left(\arcsin \frac{2}{3}\right) = \cos y = \frac{\sqrt{5}}{3}$$

∎

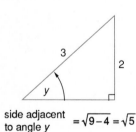

side adjacent
to angle y $= \sqrt{9-4} = \sqrt{5}$

The sine and cosine functions are "connected" through a variety of identities. The most fundamental of all these identities is $\sin^2 \theta + \cos^2 \theta = 1$; another is $\cos\left(\dfrac{\pi}{2} - x\right) = \sin x$. We almost expect the inverse sine and cosine functions to be connected by some identity. Such an identity does exist, as illustrated in the next example.

EXAMPLE 6 Show that $\arcsin x + \arccos x = \dfrac{\pi}{2}$ for all x in the common domain $-1 \le x \le 1$.

Solution Let $\arcsin x = y$; $-\dfrac{\pi}{2} \le y \le \dfrac{\pi}{2}$. Then $x = \sin y$. Now for any y value we have $\cos\left(\dfrac{\pi}{2} - y\right) = \sin y$. Therefore,

$$x = \sin y = \cos\left(\dfrac{\pi}{2} - y\right)$$

However, note that $x = \cos\left(\dfrac{\pi}{2} - y\right)$ is equivalent to $\dfrac{\pi}{2} - y = \arccos x$ because $0 \le \dfrac{\pi}{2} - y \le \pi$. Then, adding y yields

$$\dfrac{\pi}{2} = y + \arccos x$$

and substituting for y gives

$$\dfrac{\pi}{2} = \arcsin x + \arccos x$$

Note that the identity $\arcsin x + \arccos x = \dfrac{\pi}{2}$ says that the angle whose sine is x and the angle whose cosine is x must be complementary angles. ■

By using a similar argument to that in Example 6 (see Exercise 48), it can also be shown that

Solve Example 3 using this new result.

$$\arccos x + \arccos(-x) = \pi$$

or

$$\arccos(-x) = \pi - \arccos x \qquad \text{for} \qquad -1 \le x \le 1$$

EXAMPLE 7 Evaluate $\arccos(-0.7385)$.

Solution

(a) Solution using Table V:

$$\arccos(-0.7385) = \pi - \arccos(0.7385)$$
$$= \pi - 0.74 \qquad \text{(Table V)}$$
$$= 3.14 - 0.74 \qquad \text{(using } \pi = 3.14)$$
$$= 2.40$$

(b) Solution by calculator:
Set the calculator into radian mode. Then

$$\arccos(-0.7385) = \boxed{0.7385}\ \boxed{+/-}\ \boxed{\text{INV}}\ \boxed{\text{COS}} = 2.4016393 = 2.40$$

(rounded to two decimal places) ■

The tangent function has period π and completes a full cycle on the interval $\left(-\dfrac{\pi}{2}, \dfrac{\pi}{2}\right)$. Thus when $y = \tan x$ is restricted to $-\dfrac{\pi}{2} < x < \dfrac{\pi}{2}$ we have a one-to-one function whose range consists of all real numbers.

$y = \tan x$

Domain: $-\dfrac{\pi}{2} < x < \dfrac{\pi}{2}$

Range: all real numbers

Asymptotes: $x = \pm\dfrac{\pi}{2}$

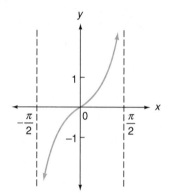

We define the inverse by $y = $ **arctan** x, which means that $x = \tan y$. Its graph is obtained by reflecting the curve in the figure through the line $y = x$. (The inverse tangent function is also written in the form $y = \tan^{-1} x$.) Also since arctan and tan are inverse functions, we have:

$$\arctan(\tan x) = x \qquad \text{if } -\dfrac{\pi}{2} < x < \dfrac{\pi}{2}$$

$$\tan(\arctan x) = x \qquad \text{for all } x$$

THE INVERSE TANGENT FUNCTION

$y = \arctan x$ (equivalent to $x = \tan y$)

Domain: all real numbers

Range: $-\dfrac{\pi}{2} < y < \dfrac{\pi}{2}$

Asymptotes: $y = \dfrac{\pi}{2}, y = -\dfrac{\pi}{2}$

The proof of the symmetric property is called for in Exercise 46.

Symmetric through origin: $\arctan(-x) = -\arctan x$.

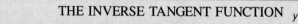

EXAMPLE 8 Evaluate arctan $\sqrt{3}$.

Solution Let $y = \text{arctan } \sqrt{3}$. Then y is the radian whose tangent is $\sqrt{3}$. Therefore, since $\tan \dfrac{\pi}{3} = \sqrt{3}$ and $-\dfrac{\pi}{2} < y < \dfrac{\pi}{2}$, we have $y = \dfrac{\pi}{3}$. ∎

EXAMPLE 9 Solve for x: $\text{arctan}\left(\dfrac{x-5}{\sqrt{3}}\right) = -\dfrac{\pi}{3}$.

Solution The given equation is equivalent to

$$\frac{x-5}{\sqrt{3}} = \tan\left(-\frac{\pi}{3}\right)$$

Since $\tan\left(-\dfrac{\pi}{3}\right) = -\sqrt{3}$, we get

Check:

$$\frac{x-5}{\sqrt{3}} = -\sqrt{3}$$

$\text{arctan}\left(\dfrac{2-5}{\sqrt{3}}\right)$

$$x - 5 = -(\sqrt{3})^2 = -3$$

$= \text{arctan}\left(-\dfrac{3}{\sqrt{3}}\right)$

$$x = 2 \qquad ∎$$

$= \text{arctan}(-\sqrt{3})$

$= -\dfrac{\pi}{3}$

EXAMPLE 10 In the figure, AB represents an 8-foot-high billboard that subtends the angle of θ radians. $BC = 22$ feet is the distance the billboard is above the ground, and point P is 40 feet from C at ground level. Use the arctan function to solve for θ.

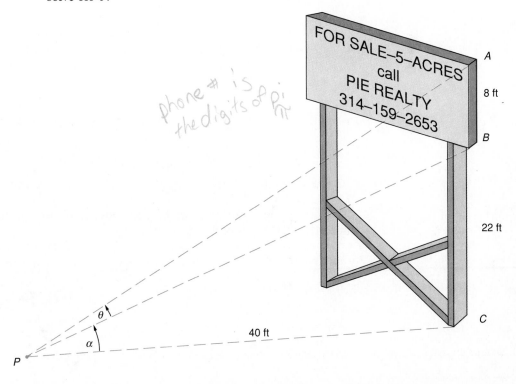

Solution Using α as shown, $\tan \alpha = \dfrac{22}{40} = \dfrac{11}{20}$ or $\alpha = \arctan \dfrac{11}{20}$. Also, $\tan(\alpha + \theta) = \dfrac{8 + 22}{40} = \dfrac{3}{4}$ so that

$$\alpha + \theta = \arctan \tfrac{3}{4}$$
$$\theta = \arctan \tfrac{3}{4} - \alpha$$
$$\theta = \arctan \tfrac{3}{4} - \arctan \tfrac{11}{20}$$

Using a calculator in the radian mode, we may use the following sequence:

$$\theta = \boxed{0.75}\ \fbox{INV}\ \fbox{TAN}\ \fbox{$-$}\ \fbox{(}\ 11\ \fbox{$\div$}\ \boxed{20}\ \fbox{)}\ \fbox{INV}\ \fbox{TAN}\ \fbox{$=$}\ 0.14$$

(rounded to two decimal places)

Thus, $\theta = 0.14$ radian. (This is approximately 8°.) ■

EXAMPLE 11 Find the exact value of $\tan(\arctan 3 - \arctan 2)$, and check by calculator.

Solution Let $\alpha = \arctan 3$ and $\beta = \arctan 2$. Now use the formula for $\tan(\alpha - \beta)$.

Note that $\dfrac{1}{7}$ is the exact answer and 0.142857 in the check is a close approximation.

$$\tan(\arctan 3 - \arctan 2) = \tan(\alpha - \beta)$$
$$= \frac{\tan \alpha - \tan \beta}{1 + \tan \alpha \tan \beta}$$
$$= \frac{3 - 2}{1 + 3 \cdot 2} \qquad \left(\begin{matrix} \tan \alpha = \tan(\arctan 3) = 3; \\ \tan \beta = \tan(\arctan 2) = 2 \end{matrix} \right)$$
$$= \frac{1}{7}$$

Check: (Use radian mode)

$$\tan(\arctan 3 - \arctan 2)$$
$$= \boxed{3}\ \fbox{INV}\ \fbox{TAN}\ \fbox{$-$}\ \boxed{2}\ \fbox{INV}\ \fbox{TAN}\ \fbox{$=$}\ \fbox{TAN}\ = 0.142857$$

Also,
$$\frac{1}{7} = \boxed{1}\ \fbox{$\div$}\ \boxed{7}\ \fbox{$=$}\ 0.142857$$ ■

The cotangent, secant, and cosecant functions also have inverses. The basic properties and graphs of the inverses are summarized here for reference.

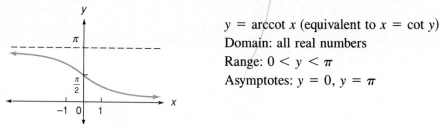

$y = \text{arccot}\ x$ (equivalent to $x = \cot y$)
Domain: all real numbers
Range: $0 < y < \pi$
Asymptotes: $y = 0, y = \pi$

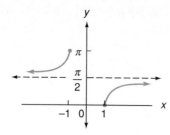

$y = \operatorname{arcsec} x$ (equivalent to $x = \sec y$)

Domain: $x \le -1$ or $x \ge 1$

Range: $0 \le y < \pi$, $y \ne \dfrac{\pi}{2}$

Asymptotes: $y = \dfrac{\pi}{2}$

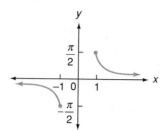

$y = \operatorname{arccsc} x$ (equivalent to $x = \csc y$)

Domain: $x \le -1$ or $x \ge 1$

Range: $-\dfrac{\pi}{2} \le y \le \dfrac{\pi}{2}$, $y \ne 0$

Asymptotes: $y = 0$

EXERCISES 8.6

Evaluate. Use Table V or a calculator only when necessary.

1. $\arcsin 0$
2. $\arccos 0$
3. $\arcsin(-1)$
4. $\arccos(-1)$

5. $\arctan(-1)$
6. $\arcsin \dfrac{1}{\sqrt{2}}$
7. $\arccos\left(-\dfrac{1}{\sqrt{2}}\right)$
8. $\arctan \dfrac{1}{\sqrt{3}}$

9. $\arctan 115$
10. $\arcsin(-\tfrac{3}{2})$
11. $\arcsin(.5562)$
12. $\arccos 2$

13. $\arccos(-0.6137)$
14. $\arctan(.6128)$
15. $\sin(\arctan \tfrac{1}{12})$
16. $\tan(\arcsin \tfrac{1}{2})$

17. $\cos[\arcsin(-1)]$
18. $\tan(\arccos \tfrac{4}{5})$
19. $\sin\left(\arcsin \dfrac{\sqrt{3}}{2}\right)$
20. $\cos(\arctan(-2.35))$

21. $\sin(\arccos \tfrac{13}{12})$
22. $\arcsin\left(\cos \dfrac{\pi}{3}\right)$
23. $\arccos\left(\tan \dfrac{\pi}{4}\right)$
24. $\arctan\left(\tan \dfrac{\pi}{6}\right)$

25. Explain why $\sin(\arcsin x) = x$ for $-1 \le x \le 1$.
26. Explain why $\tan(\arctan x) = x$ for all x.

Graph the curves. State the domain and range.

27. $y = 2 \arcsin x$
28. $y = \arcsin(x - 2)$
29. $y = 2 + \arctan x$
30. $y = -\arcsin x$

31. $y = \tfrac{1}{2} \arcsin 2x$
32. $y = 2 \arccos \dfrac{x}{2}$
33. $y = \sin(\arcsin x)$
34. $y = \arcsin x + \arccos x$

Use the formulas in Sections 8.3 and 8.4 to find the exact value of the expression.

35. $\tan(\arctan 3 + \arctan 4)$
36. $\sin(\arcsin \tfrac{2}{3} + \arctan \tfrac{4}{5})$
37. $\cos(\arcsin \tfrac{8}{17} - \arccos \tfrac{12}{13})$
38. $\tan(2 \arctan \tfrac{4}{5})$
39. $\cos(2 \arcsin \tfrac{1}{3})$
40. $\sin(\tfrac{1}{2} \arccos \tfrac{3}{5})$

Solve for x and check.

41. $\arcsin(x + 2) = \dfrac{\pi}{6}$
42. $\arctan \dfrac{x}{3} = \dfrac{\pi}{4}$

43. $\arctan(x^2 + 4x + 3) = -\dfrac{\pi}{4}$
44. $\arccos(2x - 1) = \dfrac{2\pi}{3}$

45. Prove that $\arcsin(-x) = -\arcsin x$. (*Hint:* begin with $\arcsin x = y$ and multiply by -1.)

46. Prove that $\arctan(-x) = -\arctan x$. (*Hint:* begin with $\arctan x = y$ and multiply by -1.)

47. In the figure, θ represents the angle subtended by a 5-foot picture when viewed from point P that is 7 feet below the picture and 14 feet away from the wall on which the picture hangs. Solve for θ using the arctan function. Use a calculator or Table V.

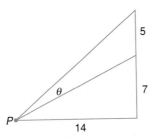

48. Verify the identity $\cos y = -\cos(y - \pi)$ using (4) on page 409 and use it to prove that $\arccos x + \arccos(-x) = \pi$ for $-1 \le x \le 1$. (*Hint:* Let $y = \arccos x$ for $0 \le y \le \pi$, and consider the proof in Example 6, page 437.)

49. (a) Let θ be the measure of an acute angle of a right triangle so that $\theta = \arcsin x$. Write the six trigonometric ratios of θ in terms of x.

    (b) Follow the instructions in part (a) using $\theta = \arctan x$.

50. (a) Prove that $\text{arccot}\, x = \arctan \dfrac{1}{x}$. $\left(\textit{Hint: Let } y = \arctan \dfrac{1}{x} \text{ so that } \tan y = \dfrac{1}{x}.\right)$

    (b) Draw an appropriate right triangle containing an acute angle of measure θ that demonstrates the equality $\text{arccot}\, x = \arctan \dfrac{1}{x}$, and describe the equation in words.

    (c) Write two similar results, one involving the inverse secant and inverse cosine functions and one involving the inverse cosecant and inverse sine functions.

51. Let $f(x) = \arcsin x$ and $g(x) = 3x + 2$ and form the composites $f \circ g$ and $g \circ f$.

52. Let $f(x) = \arctan x$ and $g(x) = x^2$ and form the composites $f \circ g$ and $g \circ f$.

△ 53. Let $h(x) = \ln(\arccos x)$. Find f and g so that $h = f \circ g$.

△ 54. Let $k(x) = \arctan \sqrt{x^2 - 1}$. Find f, g, and h so that $k = f \circ g \circ h$.

Written Assignment: The inverse sine function was defined after we restricted the domain of the sine function to the interval $\left[-\dfrac{\pi}{2}, \dfrac{\pi}{2}\right]$. Could an inverse sine function also have been defined if we had restricted the sine function to the interval $[0, \pi]$? Justify your answer.

KEY TERMS

*Review these **key terms** so that you are able to define or describe them. A clear understanding of these terms will be very helpful when reviewing the developments of this chapter.*

Trigonometric functions and their graphs: sine cosine tangent cotangent secant cosecant • Period of a function • Amplitude • Phase shift • Addition of ordinates • Addition formulas • Subtraction formulas • Reduction formulas • Cofunction identities • Double-angle formulas • Half-angle formulas • Trigonometric equation • General solution • Inverse trigonometric functions and their graphs: arcsin arccos arctan

REVIEW EXERCISES

The solutions to the following exercises can be found within the text of Chapter 8.
Try to answer each question before referring to the text.

Section 8.1

1. What does it mean to say that the period of a function f is the positive number p? What is the period of the sine and cosine functions?
2. Graph $y = \sin x$ for $-3\pi \le x \le 3\pi$.
3. Graph $y = \cos x$ for $-3\pi \le x \le 3\pi$.
4. Graph the curves $y = \sin x$, $y = \frac{1}{2} \sin x$, and $y = 2 \sin x$ on the same set of axes for $-2\pi \le x \le 2\pi$ and find the amplitudes.
5. For which subintervals of $[0, 2\pi]$ is the sine increasing? decreasing? concave down? concave up?
6. Find the period of $y = \sin 2x$ and graph for $0 \le x \le 2\pi$.
7. Find the period and amplitude of $y = 2 \cos \dfrac{x}{2}$ and graph for $0 \le x < 4\pi$.
8. What are the amplitude, period, and phase shift of $y = a \sin b(x - h)$?
9. For $y = 3 \sin\left(2x - \dfrac{\pi}{2}\right)$ determine the amplitude, period, and phase shift, and sketch the graph for one period.
10. Graph $y = \sin x + \cos x$ on $[-2\pi, 2\pi]$.

Section 8.2

11. Graph $y = \tan x$ for $\dfrac{-3\pi}{2} < x < \dfrac{3\pi}{2}$.
12. Prove that $y = \tan x$ is symmetric through the origin.
13. What is the domain of the tangent function? What is the range? What is the period? Write the equations of the asymptotes.
14. For which subintervals of $\left(-\dfrac{\pi}{2}, \dfrac{\pi}{2}\right)$ is the tangent increasing? decreasing? For which subintervals is the curve concave up? down?
15. Graph $y = \cot x$ for $-\pi < x < \pi$.
16. Graph $y = -\tan 3x$ for $\dfrac{-\pi}{2} < x < \dfrac{\pi}{2}$.
17. Graph $y = \sec x$ for one period.
18. Graph $y = \csc x$ for one period.
19. What are the domain and range of the cosecant function? Write the equations of the asymptotes.
20. For $y = \csc\left(2x + \dfrac{\pi}{2}\right)$ determine the period and phase shift, and graph for one period.

Section 8.3

21. Complete these addition and subtraction formulas:

$$\sin(\alpha - \beta) = \qquad \cos(\alpha + \beta) = \qquad \tan(\alpha - \beta) =$$

22. Use the formula $\cos(\alpha - \beta) = \cos \alpha \cos \beta + \sin \alpha \sin \beta$ to derive the formula for $\cos(\alpha + \beta)$.

23. Use an addition formula to evaluate $\sin\dfrac{7\pi}{12}$, and check using a calculator.

24. Use an addition formula to evaluate $\tan\dfrac{7\pi}{12}$.

25. Verify the identity $\cos\left(\dfrac{\pi}{2} - \theta\right) = \sin\theta$.

26. Verify the identity $\cos(\pi + \theta) = -\cos\theta$.

27. Verify the identity $\tan\left(\dfrac{\pi}{2} - \theta\right) = \cot\theta$.

28. Verify the identity $\sin\left(\theta + \dfrac{\pi}{3}\right) = \frac{1}{2}(\sin\theta + \sqrt{3}\cos\theta)$.

29. Derive the addition formula for $\tan(\alpha + \beta)$ using the addition formula for the sine and cosine.

30. Find $\cos(\alpha - \beta)$ for α in quadrant II with $\cos\alpha = -\frac{4}{5}$, and β in quadrant III and with $\cos\beta = -\frac{12}{13}$.

Section 8.4

31. Complete these double-angle formulas:

$$\sin 2\theta = \qquad \cos 2\theta = \qquad \tan 2\theta =$$

32. Use the formula for $\tan(\alpha + \beta)$ to derive the formula for $\tan 2\theta$.

33. Evaluate $\sin 15° \cos 15°$ using a double-angle formula.

34. If θ is obtuse such that $\sin\theta = \frac{5}{13}$, find $\sin 2\theta$, $\cos 2\theta$, and $\tan 2\theta$.

35. Solve for b using a double-angle formula.

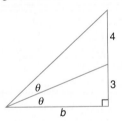

36. Let $\sin\theta = \dfrac{x}{2}$ for $0 < \theta < \dfrac{\pi}{2}$. Show that $\sin 2\theta = \dfrac{x\sqrt{4 - x^2}}{2}$.

37. Complete these formulas in terms of 2θ.

$$\sin^2\theta = \qquad \cos^2\theta = \qquad \tan^2\theta =$$

38. Complete these half-angle formulas: $\sin\dfrac{\theta}{2} = \qquad \cos\dfrac{\theta}{2} = \qquad \tan\dfrac{\theta}{2} =$

39. Use a half-angle formula to evaluate $\sin 15°$.

40. Verify the identity $\cos 3\theta = 4\cos^3\theta - 3\cos\theta$.

41. Verify the identity: $\sin^4\theta = \frac{3}{8} - \frac{1}{2}\cos 2\theta + \frac{1}{8}\cos 4\theta$

42. Solve for b without using tables or a calculator.

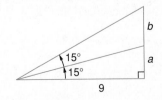

Section 8.5

Find the general solution.

43. $\sin x = \cos x$ 44. $\sqrt{3}\csc x - 2 = 0$ 45. $\cos^2 x = \cos x$

46. $\cos 2x = \frac{1}{2}$ 47. Solve $\sin 2x = \sin x$ for $0 \le x < 2\pi$.

48. Solve $\cos^2 x + \frac{1}{2}\sin x - \frac{1}{2} = 0$ for $0 \le x < 2\pi$.

49. Solve $\dfrac{\sec x}{\cos x} - \dfrac{1}{2}\sec x = 0$.

50. Approximate the solutions of $3\tan x - 7 = 0$ for $0° \le x < 360°$ using Table VI or a calculator.

51. Solve $2\sin^2 x + 5\sin x - 2 = 0$

Section 8.6

52. State the domain and range of $y = \arcsin x$ and graph.

53. State the domain and range of $y = \arccos x$ and graph.

54. State the domain and range of $y = \arctan x$ and graph. Write the equations of the asymptotes.

Evaluate without the use of tables or a calculator.

55. $\arcsin \frac{1}{2}$ 56. $\arccos(-\frac{1}{2})$ 57. $\arctan \sqrt{3}$ 58. $\sin(\arccos \frac{1}{2})$

59. Why is $\arcsin \dfrac{1}{2} = \dfrac{5\pi}{6}$ a false statement?

Evaluate using Table V and also by calculator.

60. $\arcsin(-0.3981)$ 61. $\arccos(-0.7385)$

62. Solve for x: $\arctan\left(\dfrac{x-5}{\sqrt{3}}\right) = -\dfrac{\pi}{3}$

63. Use the information in the figure to solve for θ using the arctan function.

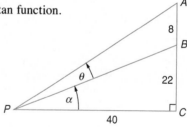

64. Prove that $\arcsin x + \arccos x = \dfrac{\pi}{2}$.

65. Find the exact value of $\tan(\arctan 3 - \arctan 2)$.

CHAPTER 8 TEST: STANDARD ANSWER

Use these questions to test your knowledge of the basic skills and concepts in Chapter 8. Then check your answers with those given at the end of the book.

1. Find the amplitude and period of the function $y = 2\sin 2x$ and graph for x in the interval $0 \le x \le 2\pi$.

2. (a) Find the subintervals of $\left(-\dfrac{\pi}{2}, \dfrac{\pi}{2}\right)$ where the tangent function is increasing or decreasing, and those for which the curve is concave up or down.

   (b) Explain how the graph of $y = \tan\left(x + \dfrac{\pi}{2}\right)$ can be obtained from the graph of $y = \tan x$.

3. State the domain and range of $y = \sec x$ and graph for x in the interval $-\dfrac{\pi}{2} < x < \dfrac{\pi}{2}$.

4. Use $165° = 135° + 30°$ and an appropriate addition formula to evaluate $\cos 165°$.

5. Use a half-angle formula to evaluate $\sin\left(-\dfrac{\pi}{12}\right)$.

6. If $\sin \theta = \frac{1}{3}$ and θ is acute, find $\sin 2\theta$.

7. Evaluate $\dfrac{\tan 115° - \tan 55°}{1 + \tan 115° \tan 55°}$ without using tables or a calculator.

Verify the identities.

8. $\sin\left(\dfrac{\pi}{4} - \theta\right) = \dfrac{\sqrt{2}}{2}(\cos \theta - \sin \theta)$.

9. $\csc 4x = \frac{1}{4}(\csc x)(\sec x)(\sec 2x)$.

10. $\sin^2 \dfrac{x}{2} = \dfrac{\tan x - \sin x}{2 \tan x}$

11. Solve $\sin^2 x - \cos^2 x + \sin x = 0$ for x in the interval $0 \le x < 2\pi$.

12. Find the general solution of $\cos 2x = 2 \sin^2 x - 2$.

13. The given curve has equation $y = a \cos bx$ or $y = a \sin bx$. Find a and b and write the equation of the curve.

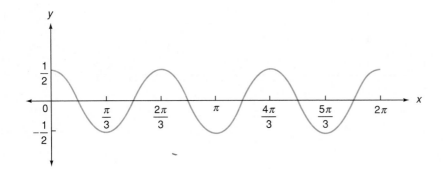

14. Find the period of $y = \tan 2x$ and write the equation of each asymptote that occurs on the interval $[0, \pi]$.

15. Find $\cos(\alpha + \beta)$ if α is in quadrant II with $\sin \alpha = \frac{3}{5}$, and β is in quadrant III with $\tan \beta = \frac{12}{5}$.

16. $\triangle ABC$ has $\angle C = 90°$, $\angle A = 15°$, and $a = 8$. Find the exact value of b using a half-angle formula.

17. **(a)** State the domain and range of $y = \arctan x$ and graph.

(b) Describe the symmetry and write the equations of the asymptotes.

18. Evaluate without the use of tables or a calculator:

(a) $\arcsin \dfrac{1}{\sqrt{2}}$ **(b)** $\arccos(-1)$ **(c)** $\arctan\left(-\dfrac{1}{\sqrt{3}}\right)$

19. Evaluate:

(a) $\arcsin(0.5398)$ **(b)** $\arccos(-0.1896)$ **(c)** $\arctan(-0.2341)$

20. Find the exact value of each expression.

(a) $\sin(\arccos \frac{3}{7})$ **(b)** $\cos(2 \arcsin \frac{1}{2})$ **(c)** $\arctan(\tan x^2)$

21. Solve for x: $\arcsin(4x + 1) = -\dfrac{\pi}{6}$

22. For $F(x) = \csc^3 2x$ find f, g, and h so that $F = f \circ g \circ h$.

23. For $y = -\dfrac{1}{2}\sin\left(3x - \dfrac{\pi}{2}\right)$ find the period, amplitude, and phase shift.

24. Graph $y = 2\cos(2x + \pi)$ for one period and indicate the period, amplitude, and phase shift.

25. Use a double-angle formula to solve for b.

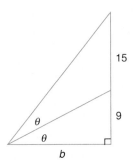

CHAPTER 8 TEST: MULTIPLE CHOICE

1. Which of the following are true for the graph of $y = \sin\theta$?

   I. Domain: all real numbers x

   II. Range: $-1 \le y \le 1$

   III. Period: 2π

   (a) Only I   (b) Only II   (c) Only III   (d) I, II, and III   (e) None of the preceding

2. The amplitude A and phase shift p for the graph of $y = -\frac{1}{2}\sin(3x - \pi)$ are which of the following?

   (a) $A = \dfrac{1}{2}, p = \dfrac{\pi}{3}$ (b) $A = -\dfrac{1}{2}, p = \pi$ (c) $A = \dfrac{3}{2}, p = \dfrac{\pi}{3}$ (d) $A = \dfrac{1}{2}, p = 3\pi$ (e) None of the preceding

3. What is the period of $y = 2\cos 4x$?

   (a) 8π (b) 4π (c) π (d) $\dfrac{\pi}{2}$ (e) None of the preceding

4. Which of the following are true for the graph of $y = \tan x$?

   I. The domain is the set of all real numbers x.

   II. The vertical lines $x = \pm\dfrac{\pi}{2}$ are asymptotes to the curve.

   III. The curve is symmetric with respect to the x-axis.

   (a) Only I   (b) Only II   (c) Only III   (d) I, II, and III   (e) None of the preceding

5. Which of the answers show the correct sequence of true (T) and false (F) for the following statements?

   I. The period of $y = \tan x$ is π.

   II. The period of $y = \cos\dfrac{x}{2}$ is π.

   III. The period of $y = \sec x$ is 2π.

   (a) F, F, F   (b) T, T, F   (c) F, T, T   (d) T, T, T   (e) None of the preceding

6. Complete: $\cos\dfrac{5\pi}{4}\cos\dfrac{\pi}{2} - \sin\dfrac{5\pi}{4}\sin\dfrac{\pi}{2} =$

   (a) $\cos\dfrac{7\pi}{4}$ (b) $\sin\dfrac{7\pi}{4}$ (c) $\cos\dfrac{3\pi}{4}$ (d) $\sin\dfrac{3\pi}{4}$ (e) None of the preceding

7. Let $\sin \alpha = \frac{3}{5}$ and $\sin \beta = -\frac{4}{5}$ where α and β have terminal sides in quadrants II and III respectively. Then $\sin (\alpha - \beta) =$
(a) -1 (b) 1 (c) $-\frac{7}{25}$ (d) $\frac{7}{25}$ (e) None of the preceding

8. Which of the following are true?

   I. The range of $y = \arcsin x$ is given by $-\dfrac{\pi}{2} \le y \le \dfrac{\pi}{2}$.

   II. The range of $y = \arctan x$ is given by $-\dfrac{\pi}{2} \le y \le \dfrac{\pi}{2}$.

   III. $\arccos\left(-\dfrac{\sqrt{2}}{2}\right) = -\dfrac{\pi}{4}$.

(a) I, II, and III   (b) Only I and II   (c) Only I   (d) Only II and III   (e) None of the preceding

9. Which of the following are true for $\cos 2\theta$?
   I. $\cos 2\theta = \sin^2 \theta - \cos^2 \theta$
   II. $\cos 2\theta = 2 \sin^2 \theta - 1$
   III. $\cos 2\theta = 2 \cos^2 \theta - 1$
(a) Only I   (b) Only II   (c) Only III   (d) I, II, and III   (e) None of the preceding

10. If $\cos \theta = -\frac{5}{13}$ and θ is in the second quadrant, then $\sin 2\theta =$
(a) $\frac{119}{169}$ (b) $\frac{24}{13}$ (c) $-\frac{60}{169}$ (d) $-\frac{120}{169}$ (e) None of the preceding

11. Use a half-angle formula to evaluate $\sin \dfrac{\pi}{8}$.

(a) $\dfrac{\sqrt{2}}{4}$ (b) $\sqrt{2 + \sqrt{2}}$ (c) $\frac{1}{2}\sqrt{2 - \sqrt{2}}$ (d) $\frac{1}{4}(2 - \sqrt{2})$ (e) None of the preceding

12. When the double-angle formula for the sine function is used to simplify $\dfrac{\sin 2x}{1 - \cos^2 x}$ the result is

(a) $-2 \cot x$ (b) $2 \cot x$ (c) $\dfrac{2}{\sin x}$ (d) $\tan 2x$ (e) None of the preceding

13. Which of the following is the general solution of $\sin 2x = 1$, where k represents any integer?

(a) $\dfrac{\pi}{4} + k\pi$ (b) $\dfrac{\pi}{2} + 2k\pi$ (c) $\dfrac{\pi}{4} + 2k\pi$ (d) $\dfrac{\pi}{2} + k\pi$ (e) None of the preceding

14. Solve for x in the interval $0 \le x < 2\pi$: $\sin^2 x = \sin x$

(a) $0, \pi$ (b) $\dfrac{\pi}{2}, \dfrac{3\pi}{2}$ (c) $0, \dfrac{\pi}{2}, \dfrac{3\pi}{2}$ (d) $\dfrac{\pi}{2}, \pi$ (e) None of the preceding

15. $\tan\left(\arc\sin \dfrac{5}{7}\right) =$

(a) $\dfrac{7}{12}$ (b) $\dfrac{5\sqrt{6}}{12}$ (c) $\dfrac{2\sqrt{6}}{5}$ (d) $\dfrac{7}{5}$ (e) None of the preceding

ANSWERS TO THE TEST YOUR UNDERSTANDING EXERCISES

Page 394

1.

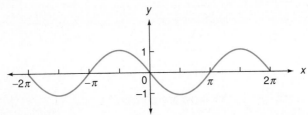

2.

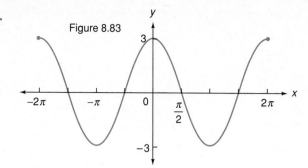

Figure 8.83

3. Same as in Exercise 1.   **4.** 10   **5.** $\frac{2}{3}$ **6.** 1

Page 412

1. $\sin 17°$ **2.** $\cos \dfrac{\pi}{10}$ **3.** $\tan 5°$ **4.** $\sin 10°$

5. $\frac{1}{4}(\sqrt{6} - \sqrt{2}); \frac{1}{4}(\sqrt{6} + \sqrt{2}); \dfrac{3 - \sqrt{3}}{3 + \sqrt{3}} = 2 - \sqrt{3}$

6. Same as Exercise 5. $\left(Note: \dfrac{\sqrt{3} - 1}{1 + \sqrt{3}} \cdot \dfrac{\sqrt{3}}{\sqrt{3}} = \dfrac{3 - \sqrt{3}}{3 + \sqrt{3}} = 2 - \sqrt{3}\right)$

7. $\frac{1}{4}(\sqrt{6} - \sqrt{2}); -\frac{1}{4}(\sqrt{6} + \sqrt{2}); \dfrac{1 - \sqrt{3}}{1 + \sqrt{3}} = -2 + \sqrt{3}$

8. Same as Exercise 7.   **9.** $\frac{1}{4}(\sqrt{2} - \sqrt{6})$ **10.** $\frac{1}{4}(\sqrt{6} - \sqrt{2})$ **11.** $-\frac{36}{85}$
12. $-\frac{77}{85}$

Page 418

1. $\sin 10°$ **2.** $\tan \theta$ **3.** $\cos 6\theta$ **4.** $\cos \dfrac{\alpha}{3}$

5. $\dfrac{\sqrt{3}}{2}$ **6.** $\dfrac{1}{\sqrt{2}}$ **7.** 2   **8.** $-\dfrac{\sqrt{3}}{2}$

Page 438

1. $-\dfrac{\pi}{2}$ rad   **2.** $\dfrac{\pi}{6}$ rad   **3.** undefined   **4.** $\dfrac{2\pi}{3}$ rad
5. 1   **6.** $\frac{1}{2}$ **7.** 1   **8.** x
9. $\frac{4}{5}$

The answers for 10–15, when rounded to the indicated decimal places, will be obtained either by calculator or by Table V.

10. 0.82 rad   **11.** −1.19 rad   **12.** 1.34 rad
13. 2.03 rad   **14.** 0.5396   **15.** 0.68 rad

CHAPTER 9

ADDITIONAL APPLICATIONS OF TRIGONOMETRY

9.1 *The Law of Cosines*
9.2 *The Law of Sines*
9.3 *Trigonometric Form of Complex Numbers*
9.4 *De Moivre's Theorem*
9.5 *Polar Coordinates*
9.6 *Graphing Polar Equations*

9.1 THE LAW OF COSINES

Suppose that the distance *AB* cannot be measured directly. It cannot be found using the methods studied previously, since that earlier work depended on acute angles within right triangles. However, since trigonometric ratios are available for other size angles, obtuse angles in particular, we will be able to solve this problem (see Exercise 17) after first establishing a new result known as the **Law of Cosines**.

As you will learn, the Law of Cosines will be helpful when solving for unknown parts of triangles that are not right triangles.

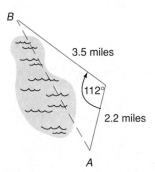

For the preceding figure, observe that the length of two sides and the measure of the included angle are given. In brief, we say that this is the case "side-angle-side" (SAS). The Law of Cosines, as you will learn, will be applicable in such cases, as well as in the case when the measures of all three sides (SSS) are given. The other cases will be considered in Section 9.2, where the *Law of Sines* is developed. A proof of the Law of Cosines is given on the following page.

In each of the following, vertex A is placed at the origin ($\angle A$ is in standard position) and side AC coincides with the x-axis. The measure of $\angle A$ is denoted by the Greek letter α (alpha). The coordinates of C are $(b, 0)$. To find the coordinates of B, first construct altitude BD. From right $\triangle ABD$ we have

$$\cos \alpha = \frac{AD}{c} \quad \text{or} \quad AD = c \cos \alpha \quad \longleftarrow \quad \text{the } x\text{-coordinate of } B$$

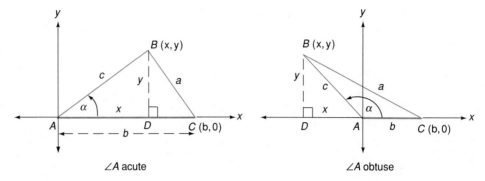

Observe that in each case $\triangle ABD$ would be similar to the reference triangle determined by the unit circle. Thus, since the trigonometric ratios are independent of the size of the reference triangle, we use the sides of $\triangle ABD$ to form the ratios.

$\angle A$ acute $\angle A$ obtuse

Similarly,

$$\sin \alpha = \frac{DB}{c} \quad \text{or} \quad DB = c \sin \alpha \quad \longleftarrow \quad \text{the } y\text{-coordinate of } B$$

Apply the distance formula (page 164) to the points $B\ (c \cos \alpha,\ c \sin \alpha)$ and $C\ (b, 0)$.

$$a^2 = (c \cos \alpha - b)^2 + (c \sin \alpha - 0)^2$$
$$= c^2 \cos^2 \alpha - 2bc \cos \alpha + b^2 + c^2 \sin^2 \alpha$$
$$= b^2 + c^2(\cos^2 \alpha + \sin^2 \alpha) - 2bc \cos \alpha$$

Since $\cos^2 \alpha + \sin^2 \alpha = 1$, we have

$$a^2 = b^2 + c^2 - 2bc \cos \alpha$$

If, in turn, $\triangle ABC$ is oriented so that B and C coincide with the origin, then similar formulas can be derived for b^2 and c^2. These results are summarized in the following. We use the Greek letters α (alpha), β (beta), and γ (gamma) for the measures of the angles whose vertices are A, B, and C, respectively, in $\triangle ABC$.

LAW OF COSINES

You may find it easier to remember this law by using this verbalized form: The square of any side of a triangle equals the sum of the squares of the remaining two sides minus twice their product times the cosine of their included angle.

For any $\triangle ABC$ with angle measures α, β, γ, and sides of length a, b, c,

$$a^2 = b^2 + c^2 - 2bc \cos \alpha$$
$$b^2 = a^2 + c^2 - 2ac \cos \beta$$
$$c^2 = a^2 + b^2 - 2ab \cos \gamma$$

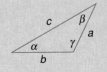

The Law of Cosines is used to solve for a side of a triangle when the other two sides and their included angle are given.

The Case Side-Angle-Side: SAS

EXAMPLE 1   Solve for c in $\triangle ABC$ if $a = 4$, $b = 7$, and $\gamma = 130°$.

Solution   **(a)** By calculator:

$$c^2 = 4^2 + 7^2 - 2(4)(7) \cos 130° = 100.99611$$

(b) By Table VI:

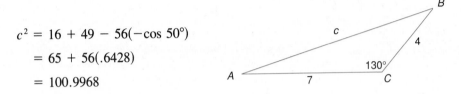

$$c^2 = 16 + 49 - 56(-\cos 50°)$$
$$= 65 + 56(.6428)$$
$$= 100.9968$$

Thus c is approximately 10.   ■

The Case Side-Side-Side: SSS

When the three sides of a triangle are known, the Law of Cosines can be used to find the angles. This is illustrated next.

EXAMPLE 2   Solve for α and β for $\triangle ABC$ in Example 1.

Solution   First solve $a^2 = b^2 + c^2 - 2bc \cos \alpha$ for $\cos \alpha$. Thus

$$\cos \alpha = \frac{b^2 + c^2 - a^2}{2bc}$$

Now substitute $a = 4$, $b = 7$, and $c = 10$,

$$\cos \alpha = \frac{7^2 + 10^2 - 4^2}{2(7)(10)} = .95$$

From Table VI, $\alpha = 18.2°$, or, using a calculator, $\alpha = \arccos(.95) = 18.2°$, rounded to one decimal place. The measure β can now be found using the fact that the sum of the angle measures of a triangle is 180°.

$$\beta = 180° - (\alpha + \gamma)$$
$$= 180° - (18.2° + 130°)$$
$$= 31.8°$$ ■

TEST YOUR UNDERSTANDING
Think Carefully

(Answers: Page 502)

Use the Law of Cosines, but not tables or a calculator, to solve for the indicated part of $\triangle ABC$. Leave the answer in radical form in Exercises 1, 2, and 3.

1. $a = 6, b = 4, \gamma = 60°$; find c.

2. $b = 15, c = 10\sqrt{2}, \alpha = \dfrac{\pi}{4}$; find a.

3. $a = 5\sqrt{3}, c = 12, \beta = 150°$; find b.
4. $a = \sqrt{67}, b = 7, c = 9$; find α.
5. $a = 8, b = \sqrt{34}, c = 3\sqrt{2}$; find β.
6. $a = 4, b = 6, c = \sqrt{76}$; find γ.

An obtuse angle has measure between 90° and 180°.

When the Law of Cosines is applied to a situation involving an *obtuse angle*, it is essential that you remember that the cosine of an obtuse angle is negative.

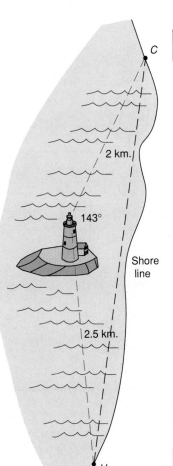

EXAMPLE 3 An offshore lighthouse is 2 kilometers from a Coast Guard station C and 2.5 kilometers from a hospital H near the shoreline. If the angle formed by light beams to C and H is 143°, what is the distance CH (by sea) between the Coast Guard station and the hospital?

Solution Use the Law of Cosines to solve for CH.
(a) By calculator:

$$(CH)^2 = 2^2 + (2.5)^2 - 2(2)(2.5)(\cos 143°) = 18.236355$$

(b) By Table VI:

$$(CH)^2 = 4 + 6.25 - 10(-\cos 37°) \qquad \text{(The reference angle is in quadrant II.)}$$
$$= 10.25 + 10(.7986)$$
$$= 18.236$$

Taking the square root and rounding off to the nearest tenth, we have

$$CH = 4.3 \text{ kilometers} \qquad \blacksquare$$

The Law of Cosines can be used to derive **Heron's formula** for the area of a triangle. This formula is based on the *semiperimeter* of the triangle $s = \frac{1}{2}(a + b + c)$. Before obtaining this result, observe that the area of a triangle can be found according to this rule:

The area K of a triangle is one-half the product of two sides and the sine of the included angle.

$$K = \tfrac{1}{2}bc \sin \alpha$$
$$K = \tfrac{1}{2}ac \sin \beta$$
$$K = \tfrac{1}{2}ab \sin \gamma$$

The proof for the first of these formulas is as follows:

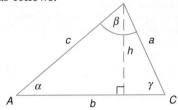

$$\sin \alpha = \frac{h}{c} \quad \text{or} \quad h = c \sin \alpha$$

$$K = \tfrac{1}{2}bh$$

$$= \tfrac{1}{2}bc \sin \alpha$$

Geometric formulas are given inside the back cover.

EXAMPLE 4 Find the area of the triangle shown in the margin:

Solution

$$K = \tfrac{1}{2}bc \sin \alpha$$

$$= \tfrac{1}{2}(12)(8)\sin 110° = 45.105246 \qquad \text{(by calculator)}$$

The area is approximately 45 cm². ■

To derive Heron's formula we begin by squaring $K = \tfrac{1}{2}bc \sin \alpha$ to obtain the following:

$$K^2 = \tfrac{1}{4}b^2c^2 \sin^2 \alpha$$

$$= \tfrac{1}{4}b^2c^2(1 - \cos^2 \alpha) \quad \longleftarrow \quad (\sin^2 \alpha = 1 - \cos^2 \alpha)$$

$$= \tfrac{1}{4}b^2c^2(1 + \cos \alpha)(1 - \cos \alpha)$$

By the Law of Cosines, $\cos \alpha = \dfrac{b^2 + c^2 - a^2}{2bc}$. Then, substituting for $\cos \alpha$ into the preceding, you can verify (see Exercise 38) that this result is equivalent to

$$K^2 = \tfrac{1}{16}(a + b + c)(b + c - a)(a + c - b)(a + b - c)$$

Then

Observe that the factor $b + c - a = a + b + c - 2a$; similarly, for the last two factors.

$$K^2 = \tfrac{1}{16}(a + b + c)(a + b + c - 2a)(a + b + c - 2b)(a + b + c - 2c)$$

Now since $s = \tfrac{1}{2}(a + b + c)$, substitute $2s$ for $a + b + c$. Thus

$$K^2 = \tfrac{1}{16}(2s)(2s - 2a)(2s - 2b)(2s - 2c)$$

$$K^2 = s(s - a)(s - b)(s - c)$$

Finally, take the positive square root to obtain Heron's formula.

HERON'S FORMULA

The area K of a triangle with sides of length a, b, and c is given by

$$K = \sqrt{s(s - a)(s - b)(s - c)}$$

where $s = \tfrac{1}{2}(a + b + c)$.

EXAMPLE 5   Find the area of the triangle using Heron's formula.

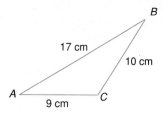

Solution   First find the semiperimeter, *s*.

$$s = \tfrac{1}{2}(a + b + c) = \tfrac{1}{2}(10 + 9 + 17) = 18$$

Then

$$K = \sqrt{s(s - a)(s - b)(s - c)}$$
$$= \sqrt{18(18 - 10)(18 - 9)(18 - 17)}$$
$$= \sqrt{1296}$$
$$= 36$$

The area is 36 cm².   ∎

EXERCISES 9.1

Use the Law of Cosines to solve for the indicated parts of △ABC.
Do not use tables or a calculator.

1. Solve for *a* if *b* = 4, *c* = 11, and *α* = 60°.
3. Solve for *α* if *a* = 5√3, *b* = 10√3, and *c* = 15.

2. Solve for *b* if *a* = 20, *c* = 8, and *β* = 45°.
4. Solve for *β* if *a* = 9√2, *b* = √337, and *c* = 7.

 Solve for the indicated parts of △ABC. Give the sides to the nearest tenth and the angle mea-
sure to the nearest tenth of a degree.

5. *α* = 20°, *b* = 8, *c* = 13; find *a* and *β*.
7. *a* = 12, *b* = 5, *c* = 13; find *γ* and *α*.
9. *a* = 18, *b* = 15, *c* = 4; find *β* and *γ*.
11. *a* = 18, *b* = 9, *γ* = 30.2°; find *c* and *α*.
13. *b* = 2.2, *c* = 6.4, *α* = 42°; find *a* and *β*.

6. *a* = 9, *c* = 14, *β* = 110°; find *b* and *γ*.
8. *a* = *b* = *c* = 10; find *α*, *β*, and *γ*.
10. *α* = 65.5°, *b* = 4, *c* = 11; find *a* and *γ*.
12. *a* = 15, *c* = 5, *β* = 157.5°, find *b* and *α*.
14. *a* = 60, *b* = 20, *c* = 75; find *α* and *β*.

15. Two points *A* and *B* are on the shoreline of a lake. A surveyor is located at a point *C* where *AC* = 180 meters and *BC* = 120 meters and finds that ∠*ACB* has measure 56.3°. What is the distance between *A* and *B* to the nearest meter?

16. Point *C* is 2.7 kilometers from a house *A* and 3 kilometers from house *B*, where *A* and *B* are on opposite sides of a valley. If ∠*ACB* has measure 130.1°, find the distance between the houses to the nearest kilometer.

17. Find the distance *AB*, discussed in the opening to this section, to the nearest tenth of a mile.

18. A diagonal of a parallelogram has length 80 and makes an angle of 20° with one of the sides. If this side has length 34, find the length of the other side of the parallelogram to the nearest unit.

19. The equal sides of an isosceles trapezoid are 10 cm, and the shorter base is 14 cm. If one pair of equal angles has a measure of 40° each, find the length of a diagonal to the nearest tenth of a centimeter.

20. The equal sides of an isosceles triangle are each 30 units long and the vertex angle is 27.7°. Find the base to the nearest tenth of a unit, and the measure of the base angles to the nearest tenth of a degree.

21. In the accompanying figure, solve for *AB* to the nearest tenth.

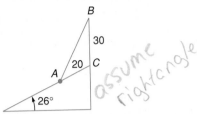

22. The four bases of a baseball diamond form a square 90 feet on a side. The shortstop *S* is in a position that is 50 feet from second base and forms a 15° angle with the base path as shown. Find the distance between the shortstop and first base to the nearest foot.

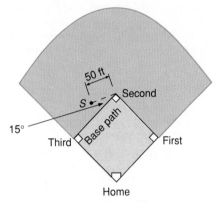

23. In Exercise 22, find the distance between the shortstop and home plate to the nearest foot.

24. Points *A* and *B* are the endpoints of a proposed tunnel through a mountain. From a point *P*, away from the mountain, a surveyor is able to see both points *A* and *B*. The surveyor finds that *PA* = 620 meters, *PB* = 450 meters, and ∠*APB* has measure 83.3°. Find the length *AB* of the tunnel to the nearest meter.

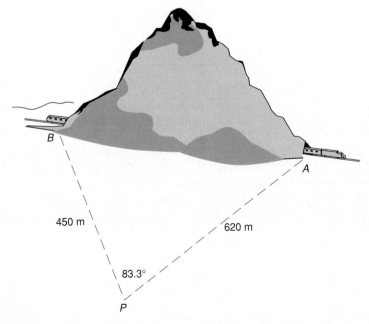

25. Two trains leave the same station at 11 A.M. and travel in straight lines at speeds of 54 miles per hour and 60 miles per hour, respectively. If the difference in their direction is 124°, how far apart are they at 11:20 A.M. to the nearest tenth of a mile?

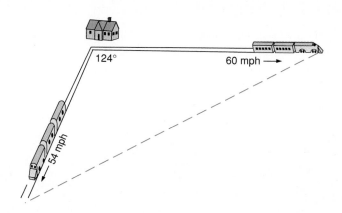

26. Prove that for any △ABC, $c = a \cos \beta + b \cos \alpha$. (*Hint:* consider the cases where $\angle A$ is acute, obtuse, or a right angle, and put $\angle A$ in standard position.)

Find the area of △ABC having the given parts.

27. $\alpha = 30°, b = 25, c = 18$

28. $\alpha = 82°, b = 14, c = 31$

29. $\beta = 40.5°, a = 8.4, c = 12.6$

30. Find the area of quadrilateral *ABCD*. (Give the answer to the nearest square unit.)

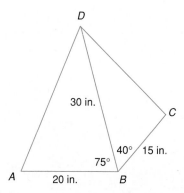

Find the area of the triangle using $K = \frac{1}{2}(base)(height)$; then verify that Heron's formula gives the same result.

31.

32.

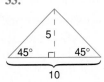

33.

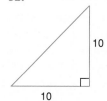

Use a calculator and Heron's formula to find the areas of the triangles with the given sides.

34. $a = 10$ ft, $b = 13$ ft, $c = 20.1$ ft

35. $a = 10$ ft, $b = 13$ ft, $c = 3.4$ ft

36. $a = 210$ cm, $b = 280$ cm, $c = 425$ cm

37. Use the Law of Cosines to find AD and CD in Exercise 30 and then use Heron's formula to find the area of quadrilateral $ABCD$ and compare to the result in Exercise 30.

38. Show that $K^2 = \frac{1}{4}b^2c^2(1 + \cos \alpha)(1 - \cos \alpha)$ is equivalent to

$$K^2 = \tfrac{1}{16}(a + b + c)(b + c - a)(a + c - b)(a + b - c)$$

by following these directions (see page 454):

(a) Substitute $\dfrac{b^2 + c^2 - a^2}{2bc}$ for $\cos \alpha$ and obtain the form

$$K^2 = \tfrac{1}{16}(2bc + b^2 + c^2 - a^2)(2bc - b^2 - c^2 + a^2)$$

(b) Rewrite the factor $2bc + b^2 + c^2 - a^2$ as the difference of two squares $u^2 - a^2$, and factor. Do the same for the last factor $2bc - b^2 - c^2 + a^2$ and obtain the stated result.

Written Assignment: The proof of the area formula $K = \frac{1}{2}bc \sin \alpha$ was done on page 454 using a triangle in which $\angle A$ was an acute angle. Explain why such a proof also applies to the case when $\angle A$ is obtuse.

CHALLENGE
Think Creatively

(a) Find the area of the equilateral triangle using $K = \frac{1}{2}$(base)(height); then verify that Heron's formula gives the same result.

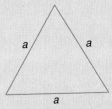

(b) Use Heron's formula to verify that the area of the isosceles triangle is $\frac{1}{2}a^2$.

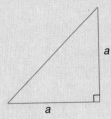

9.2
THE LAW OF SINES

Triangular problems in which the given information is either SAS or SSS can be solved using the Law of Cosines. However, if the given information is ASA, AAS, or SSA, then the Law of Cosines is not adequate and another property of triangles, called the **Law of Sines,** is needed. To establish this new result, first construct altitude DB for $\triangle ABC$ as shown. From right triangles ABD and CBD, we have

$$\sin \alpha = \frac{DB}{c} \quad \text{and} \quad \sin \gamma = \frac{DB}{a}$$

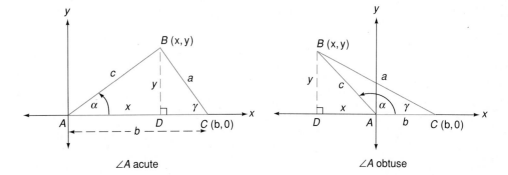

See Exercise 30 for the case
$\alpha = 90°$.

∠A acute   ∠A obtuse

Then $DB = c \sin \alpha$ and $DB = a \sin \gamma$, which gives

$$a \sin \gamma = c \sin \alpha$$

$$\frac{a}{\sin \alpha} = \frac{c}{\sin \gamma} \qquad \text{(Why?)}$$

Similar reasoning produces $\dfrac{a}{\sin \alpha} = \dfrac{b}{\sin \beta}$, which is combined with the first result into the following:

The "double" equality is an abbreviation for these three results:

$$\frac{a}{\sin \alpha} = \frac{b}{\sin \beta}$$

$$\frac{a}{\sin \alpha} = \frac{c}{\sin \gamma}$$

$$\frac{b}{\sin \beta} = \frac{c}{\sin \gamma}$$

THE LAW OF SINES

For any $\triangle ABC$ with angle measures α, β, γ and sides of length a, b, c,

$$\frac{a}{\sin \alpha} = \frac{b}{\sin \beta} = \frac{c}{\sin \gamma}$$

Our first two examples demonstrate the use of the Law of Sines when two angles and one side are given.

The Case Angle-Side-Angle: ASA

EXAMPLE 1   Solve $\triangle ABC$ if $\beta = 75°$, $a = 5$, and $\gamma = 41°$.

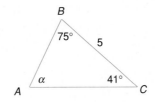

Solution   Since two angles are given, the third is easy to find because the sum of the angles of a triangle is 180°.

$$\alpha = 180° - (\beta + \gamma)$$
$$= 180° - (75° + 41°)$$
$$= 180° - 116°$$
$$= 64°$$

Now use the Law of Sines to solve for b, and substitute the appropriate values. From $\dfrac{a}{\sin \alpha} = \dfrac{b}{\sin \beta}$ we obtain $b = \dfrac{a \sin \beta}{\sin \alpha}$. Then, using Table VI and a calculator, we have

$$b = \frac{5 \sin 75°}{\sin 64°} = \frac{5(.9659)}{.8988} = 5.4$$

Or, just by calculator, we solve for b as follows:

$$b = \boxed{5} \boxed{\times} \boxed{75} \boxed{\text{SIN}} \boxed{\div} \boxed{64} \boxed{\text{SIN}} \boxed{=} \; 5.3734547$$

Thus, rounded to one decimal place, $b = 5.4$. We solve for c in a similar manner:

$$c = \frac{a \sin \gamma}{\sin \alpha} = \frac{5 \sin 41°}{\sin 64°} = 3.6 \qquad \blacksquare$$

EXAMPLE 2   To the nearest meter, find the distance $AB = c$ across the pond if $\beta = 108°$, $\gamma = 39°$, and $AC = 950$ meters.

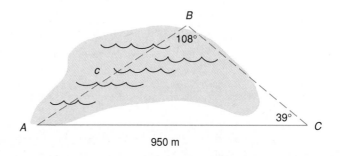

Observe that AAS also fits the case ASA, since two given angles of a triangle determine the third angle.

Solution $\dfrac{c}{\sin \gamma} = \dfrac{b}{\sin \beta}$ can be used to solve for c since only c is unknown. Thus

$$c = \frac{b \sin \gamma}{\sin \beta}$$

$$= \frac{950 \sin 39°}{\sin 108°} = 628.62129 \qquad \text{(by calculator)}$$

The distance AB is 629, to the nearest meter. $\qquad \blacksquare$

When two sides and an angle opposite one of the sides (SSA) are given, there are a number of possibilities. The following figures illustrate the four possible cases when $\angle A$ is acute. It is assumed that parts a, α, and b are given. Note that $h = b \sin \alpha$ is the altitude opposite $\angle A$.

CASE 1. No solution: $a < b \sin \alpha$

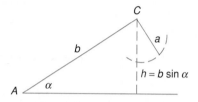

CASE 2. One solution: $a = b \sin \alpha$

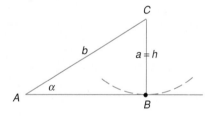

CASE 3. Two solutions: $b \sin \alpha < a < b$

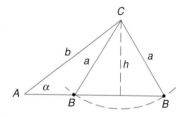

CASE 4. One solution: $a \geq b$

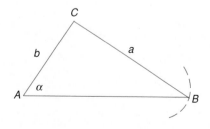

The preceding descriptions are independent of the symbols chosen. Any choice of symbols could be used as long as we have the case SSA.

When $\angle A$ is obtuse, there are three possibilities:

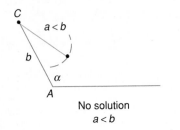

No solution
$a < b$

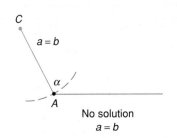

No solution
$a = b$

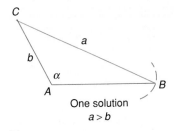

One solution
$a > b$

EXAMPLE 3 Solve $\triangle ABC$ if $a = 50$, $b = 65$, and $\alpha = 57°$.

Solution $b \sin \alpha = 65 \sin 57° = 54.5$. Since $a < 54.5$ we have $a < b \sin \alpha$ and there is no solution possible (case 1). ∎

The next example illustrates Case 3, in which $b \sin \alpha < a < b$. This is called the *ambiguous* case. The two angles obtained are supplementary, and therefore have the same sine value.

EXAMPLE 4 Approximate the remaining parts of $\triangle ABC$ if $a = 10$, $b = 13$, and $\alpha = 25°$.

Solution $b \sin \alpha = 13 \sin 25° = 5.5$. Since $5.5 < 10 < 13$, it follows that $b \sin \alpha < a < b$ (case 3) and there are two solutions for β. From the Law of Sines,

$$\sin \beta = \frac{b \sin \alpha}{a} = \frac{13(\sin 25°)}{10} = .5494$$

This solution has been done by the use of a calculator. Note that sin β has been rounded to four decimal places, and the measures of the angles and the sides are rounded to tenths.

Then, $\beta = \arcsin .5494 = 33.3°$ and the solution for the supplementary angle is $180° - 33.3° = 146.7°$. If $\beta = 33.3°$, then $\gamma = 180° - (\alpha + \beta) = 121.7°$ and

$$c = \frac{a \sin \gamma}{\sin \alpha} = \frac{10 \sin (121.7°)}{\sin 25°} = 20.1$$

If $\beta = 146.7°$, we get $\gamma = 8.3°$ and $c = 3.4$

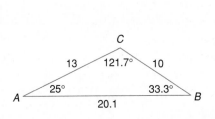

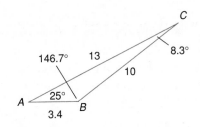

∎

The Laws of Sines and Cosines can be helpful in solving applied problems using vectors. A **vector** is a directed line segment such as the following:

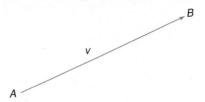

*Boldface is used to distinguish between the geometric vector **v** and the number or variable v. Since it is clumsy to reproduce boldface by hand, you might find it simpler to use the symbol $\vec{v}$ instead.*

A is the *initial point* and B is the *terminal point* of the vector denoted by $\overrightarrow{AB}$. A letter in boldface type, such as **v**, is also used to represent a vector.

Two vectors that have the same length and direction are said to be equal; the location of the vector in the plane is arbitrary. Thus the same vector **v** in three locations is shown in the margin.

A vector PQ is used in applications to represent a quantity that has both magnitude and direction. The length of $\overrightarrow{PQ}$, denoted by $|\overrightarrow{PQ}|$, is the magnitude, and the direction of $\overrightarrow{PQ}$ is the direction from the initial point P to the terminal point Q. When a vector **v** is used for a quantity with magnitude and direction, we locate **v** in a position that is convenient to the situation. For example:

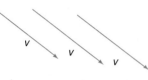

*In each case **v** has the same length and direction.*

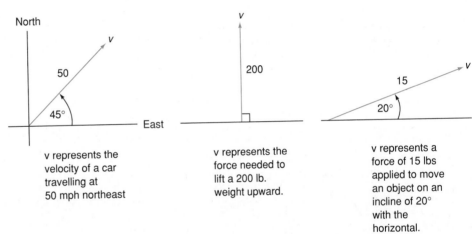

v represents the velocity of a car travelling at 50 mph northeast

v represents the force needed to lift a 200 lb. weight upward.

v represents a force of 15 lbs applied to move an object on an incline of 20° with the horizontal.

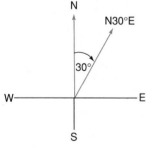

When two vectors are added, the sum is a vector called the **resultant.** It is found using the *parallelogram law* in which the sum is the diagonal of the parallelogram determined by two vectors such as $\overrightarrow{AB}$ and $\overrightarrow{AC}$, as shown.

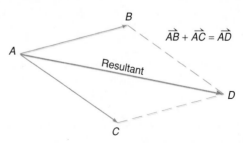

$$\overrightarrow{AB} + \overrightarrow{AC} = \overrightarrow{AD}$$

Resultant

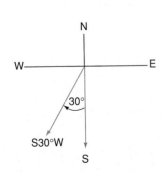

In certain applications involving moving objects, the direction of the motion is stated as a **bearing.** This is the direction given by the acute angle made with the north-south line. Thus, a bearing of North 30° East, abbreviated as N 30° E, means the object is traveling in a direction that is 30° east of the north-south line. (The letter N or S is written first.) There are two illustrations in the margin.

EXAMPLE 5   A motorboat starts at the south shore of a river and is *heading* due north to the opposite shore. The boat's speed (in still water) is 15 mph, and the river is flowing due east at 4 mph. What is the actual speed of the boat, and what is the final bearing?

Solution   Draw a diagram. Represent the boat's speed by $\overrightarrow{AB}$ and the river's speed by $\overrightarrow{AC}$. The resultant $\overrightarrow{AD} = \overrightarrow{AB} + \overrightarrow{DC}$ gives the final bearing, and its length is the actual speed.

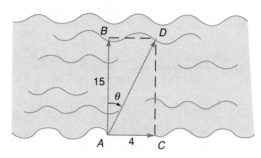

The Pythagorean theorem gives the magnitude.

$$|\overrightarrow{AD}|^2 = 15^2 + 4^2 = 241$$

Thus the actual speed is $\sqrt{241}$ or about 15.5 mph. Also,

$$\tan \theta = \frac{4}{15} = .2667$$

Then $\theta = 15°$ to the nearest degree, and the direction of the boat is given by the bearing N 15°E.   ∎

EXAMPLE 6   An airplane has an airspeed (the speed in still air) of 240 mph with a bearing of S 30°W. If a wind is blowing due west at 40 mph, find the final bearing of the plane and the ground speed (the speed relative to the ground).

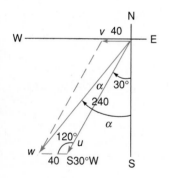

Solution   Draw a diagram in which **u** represents the velocity of the plane and **v** represents the velocity of the wind. Then the magnitude of the resultant $\mathbf{w} = \mathbf{u} + \mathbf{v}$ is the ground speed. Use the Law of Cosines to find $|\mathbf{w}|$.

$$|\mathbf{w}|^2 = 240^2 + 40^2 - 2(40)(240) \cos 120$$
$$= 68,800$$

Then the ground speed $|\mathbf{w}|$ is approximately 262 mph. Also, to find θ, we first find α using the Law of Sines.

$$\frac{40}{\sin \alpha} = \frac{262}{\sin 120°}$$

$$\sin \alpha = \frac{40 \sin 120°}{262} = .1322$$

Then $\alpha = 8°$ to the nearest degree, $\theta = 30° + 8° = 38°$ and the final bearing of the plane is S 38° W.   ∎

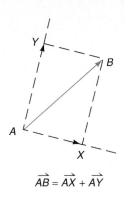

$$\overrightarrow{AB} = \overrightarrow{AX} + \overrightarrow{AY}$$

It is also possible to reverse the addition procedure for vectors. That is, a given vector can be *resolved* into the sum of two *component* vectors, whose sum is the given vector. The figure in the margin demonstrates how vector *AB* is resolved into the sum of two perpendicular components. This was done by first drawing perpendicular lines through *A* and then drawing lines through *B* to complete the parallelogram. The final example makes use of this procedure.

EXAMPLE 7   A 400-pound packing crate is sitting on a ramp that makes a 25° angle with the horizontal. Find the force required to hold the crate from sliding down the ramp. (Assume that there is no friction.)

Solution   In the diagram $\overrightarrow{PQ}$ is the force vector due to gravity. Thus $|\overrightarrow{PQ}| = 400$. Now resolve $\overrightarrow{PQ}$ into the sum of two *component* vectors $\overrightarrow{PR}$ and $\overrightarrow{PS}$ so that $\overrightarrow{PS}$ is parallel to the ramp, and $\overrightarrow{PR}$ is perpendicular to the ramp. Observe that $\overrightarrow{PS}$ is the

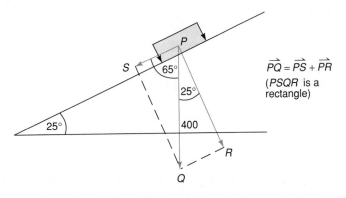

$$\overrightarrow{PQ} = \overrightarrow{PS} + \overrightarrow{PR}$$
(*PSQR* is a rectangle)

force vector that moves the crate down the ramp. Then a force equal to $|\overrightarrow{PS}|$, acting in the opposite direction, will hold the crate in place. (The magnitude of vector $\overrightarrow{PR}$ is the force that the crate exerts on the ramp, and is counteracted by the ramp itself.) From the diagram,

Observe that a force over 400 pounds is needed to lift the crate vertically, and just over 169 pounds is needed to move it upward on the ramp: an advantage of 400 − 169 = 231 pounds.

$$|\overrightarrow{PS}| = 400 \cos 65° = 169.04$$

Therefore, a force of 169 pounds (to the nearest pound) is needed to hold the crate in place.   ∎

EXERCISES 9.2

Use the Law of Sines to solve for the indicated parts of △*ABC.*
Do not use tables or a calculator.

1. Solve for *a* and *c* if $\alpha = 30°$, $\beta = 120°$, and $b = 54$.

2. Solve for *c* if $\gamma = 30°$, $\alpha = 135°$, and $a = 100$.

3. Solve for γ if $a = 12$, $c = 4\sqrt{3}$, and $\alpha = \dfrac{2\pi}{3}$.

4. Solve for β if $a = 5\sqrt{2}$, $b = 10$, and $\alpha = \dfrac{\pi}{6}$.

Use the Law of Sines to solve △ABC.

5. $\alpha = 25°$, $\gamma = 55°$, $b = 12$ 6. $\gamma = 110°$, $\beta = 28°$, $a = 8$

7. $\alpha = 62.2°$, $\beta = 50°$, $b = 5$ 8. $\alpha = 155°$, $\beta = 15.5°$, $c = 20$

Determine the number of triangles possible with the given parts.

9. $\alpha = 32°$, $a = 5.1$, $b = 10$ 10. $\alpha = 32°$, $a = 6$, $b = 10$

11. $\alpha = 30°$, $a = 9.5$, $b = 19$ 12. $\alpha = 50°$, $a = 15$, $b = 14.3$

13. $\alpha = 126°$, $a = 20$, $b = 25$ 14. $\alpha = 77°$, $a = 49$, $b = 50$

Use the Law of Sines to solve for the indicated parts of △ABC whenever possible.

15. Solve for β and c if $\alpha = 53°$, $a = 12$, and $b = 15$.

16. Solve for γ and a if $\beta = 122°$, $b = 20$, and $c = 8$.

17. Solve for β if $b = 25$, $a = 7$, and $\alpha = 75°$.

18. Solve for β if $\alpha = 44°$, $a = 9$, and $b = 12$.

19. Solve for γ if $\beta = 22.7°$, $b = 25$, and $c = 30$.

20. Solve for γ if $\beta = 22.7°$, $b = 8$, and $c = 30$.

21. Solve for the larger angle of the given parallelogram.

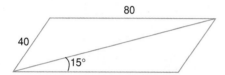

22. Two guy wires support a telephone pole. They are attached to the top of the pole and are anchored into the ground on opposite sides of the pole at points A and B. If $AB = 120$ feet and the angles of elevation at A and B are $72°$ and $56°$, respectively, find the length of the wires to the nearest tenth of a foot.

23. An airplane is flying in a straight line toward an airfield at a fixed altitude. At one point the angle of depression to the airfield is $32°$. After flying 2 more miles the angle of depression is $74°$. What is the distance between the airplane and the airfield when the angle of depression is $74°$? Give your answer to the nearest tenth of a mile.

24. Use the Law of Sines to find AD to the nearest unit.

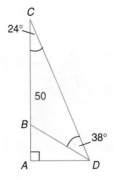

25. From the top of a 250-foot hill, the angles of depression of two cottages A and B on the shore of a lake are $15.5°$ and $29.2°$, respectively. If the cottages are due north of the observation point, find the distance between them to the nearest foot.

26. A 45-foot tower standing vertically on a hillside casts a shadow down the hillside that is 72 feet long. The angle at the tip of the shadow S, subtended by the tower, is 28°. Find the angle of elevation of the sun at S.

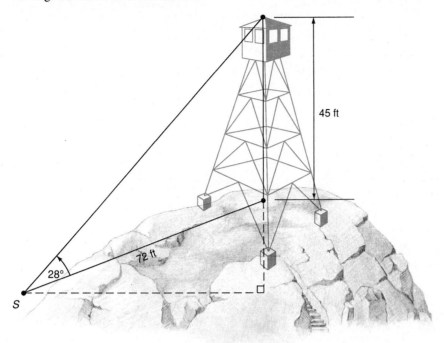

45 ft

72 ft

28°

S

27. In $\triangle ABC$ let D be a point between A and C such that $\angle BDA = 58°$. If $\angle A = 110°$, $\angle C = 43°$, and $DC = 20$ cm, find AD to the nearest tenth of a centimeter.

28. (a) Use the Law of Sines to prove that any two sides of a triangle are in the same ratio as the sines of their opposite angles.

   (b) Two of the angles of a triangle are 105° and 45°. The side opposite the smallest angle of the triangle has length 10. Use the result in (a) to find the length of the side opposite the 45° angle without using tables or a calculator.

29. In the figure $AB = 12$, $BC = 4$, $AD = 10$, and the measure of $\angle B$ is 20°. Solve for CD to the nearest unit. (*Hint:* Begin with the Law of Cosines.)

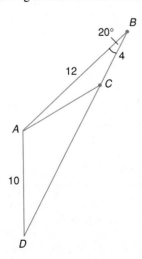

B

20°

4

12

C

A

10

D

Show that if $\alpha = 90°$, the Law of Sines is consistent with the definition of the sine of an acute angle.

31. In the figure, A and B are two points on one side of a canyon with $AB = 300$ meters. Use a calculator to find the distance between the points C and D on the opposite side of the canyon.

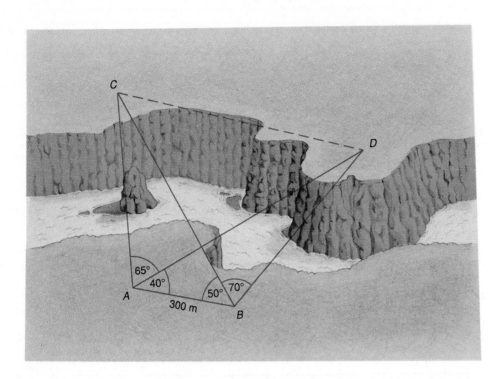

32. Two forces are acting on an object. An 80-pound force is pulling to the right, and a 60-pound force is pulling upward. Find the resultant force and the angle it makes with the 80-pound force. (Give the angle to the nearest degree.)

33. Two forces, one of 24 pounds and the other of 10 pounds, are pulling on an object at right angles to one another. Find the resultant force and the angle it makes with the 24-pound force. (Give the angle to the nearest degree.)

34. Two forces acting on an object form an angle of 60° with each other. One force is 24 pounds and the other is 7 pounds. Find the resultant force and the angle it makes with the smaller force. (Give the force to the nearest pound and the angle to the nearest degree.)

35. Answer the questions in Exercise 34 if the forces are 38 pounds and 12 pounds and the angle between them is 80°.

36. A boat is heading across a river from east to west. The boat's speed (in still water) is 10 mph, and the river is flowing north at 3 mph. What is the actual speed of the boat, and what is its final bearing? (Round the speed to the nearest tenth of an mph and the bearing to the nearest degree.)

37. An airplane has an airspeed of 210 mph with a bearing of N 30° E. A wind is blowing due west at 30 mph. Find the final bearing of the plane, and its ground speed. (Round the speed to the nearest mph and the bearing to the nearest degree.)

38. In Example 7, page 465, find the force that the crate exerts on the ramp. (Give the answer to the nearest pound.)

39. Find the force required to keep a vehicle weighing 1800 pounds from rolling down a driveway that makes an angle of 5° with the horizontal. (Round the force to the nearest pound.)

40. Find the force required to push a crate weighing 120 pounds up a ramp that makes an angle of 12° with the horizontal. (Round the force to the nearest pound.)

41. A force of 200 pounds is required to push a vehicle up a driveway that makes a 10° angle with the horizontal. Find the weight of the vehicle to the nearest 10 pounds.

42. A 36-pound force is required to stop a 160-pound skier from sliding down a hill. To the nearest degree, find the angle the hill makes with the horizontal.

Written Assignment

(a) Make a list of the various cases of solving triangles that can be done with the Law of Cosines and those that can be done with the Law of Sines. (Use abbreviations such as AAS for the given parts angle-angle-side.)

(b) Discuss the four possibilities of SSA for an acute ∠A in which parts a, α, and b are given.

EXPLORATIONS
Think Critically

1. Prove that the Pythagorean theorem follows from the Law of Cosines. In what sense is this result due to circular reasoning?

2. In △ABC, $\alpha = 60°$ and $b = 3c$. Use the Law of Cosines to express a in terms of c. For what value of c does $a = 7$? Is there a value c for which △ABC becomes a right triangle? Explain.

3. ABCD is a parallelogram having diagonals AC and BD with $AC > BD$. Let $AD = a$ and $AB = b$. Observe that the Law of Cosines gives the length of each diagonal in terms of a, b, and the cosine of an appropriate angle. Explain why $AC = \sqrt{a^2 + b^2 + 2ab \cos A}$.

4. As noted in the margin on page 459, the Law of Sines is an abbreviation of three equations. What do these three equations become when the Law of Sines is applied to a right △ABC with right angle at C?

5. In △ADC, ∠C = 90° and ∠A = 23°. Point B is between A and C so that $AB = .25$ miles and ∠DBC = 43°. How can the Law of Sines be used to find CD?

9.3
TRIGONOMETRIC
FORM
OF COMPLEX
NUMBERS

Complex numbers were introduced in Section 1.10 and then used later to solve equations having imaginary roots in Sections 3.3 and 4.7. Now you will learn how these numbers can be given both a geometric and a trigonometric interpretation that will result in new, and sometimes more efficient, methods for doing some basic computations.

An ordered pair of real numbers (x, y) determines the unique complex number $x + yi$, and vice versa. This correspondence between the complex numbers and the ordered pairs of real numbers allows us to give a geometric interpretation of the complex numbers by using the **complex plane**. This plane is a rectangular coordinate system in which the horizontal axis is called the **real axis** and the vertical axis is the **imaginary axis**. The following figure illustrates how complex numbers are plotted (graphed) in the complex plane.

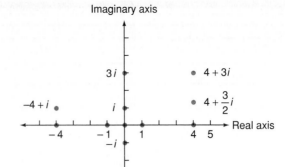

There is a geometric interpretation for the sum of two complex numbers. First locate $z = a + bi$ and $w = c + di$ and use line segments to connect these points to the origin. Then complete the parallelogram as shown. The tip of the diagonal of the parallelogram that passes through the origin will represent the sum $z + w$ (see Exercise 10). This graphic procedure for finding sums is known as the **parallelogram rule**.

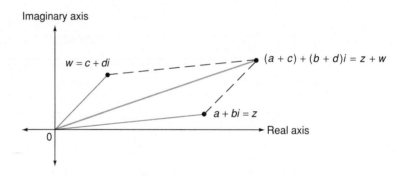

EXAMPLE 1 Find the sum $(3 + 2i) + (2 - 4i)$ using the parallelogram rule, and verify using the definition for addition of complex numbers.

Solution The parallelogram rule is used in the figure in the margin and computed algebraically as follows:

$$(3 + 2i) + (2 - 4i) = (3 + 2) + (2 - 4)i$$
$$= 5 - 2i \qquad \blacksquare$$

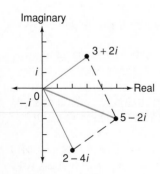

EXAMPLE 2 Describe the diagonal formed when finding the sum of $z = 3 + 2i$ and its conjugate $\bar{z}$ by use of the parallelogram rule.

Solution The diagonal is six units in length on the x-axis, from $(0, 0)$ to $(6, 0)$, as shown in the next figure. The sum is $6 + 0i = 6$.

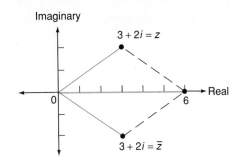

Complex numbers can also be subtracted graphically. This is done in Example 3 using a parallelogram after changing the difference $z - w$ into the sum $z + (-w)$.

EXAMPLE 3 Use a parallelogram to find $(3 + 2i) - (2 - 4i)$.

Solution First locate $3 + 2i$ and $-(2 - 4i) = -2 + 4i$. Then add using the parallelogram rule to get the difference, $1 + 6i$.

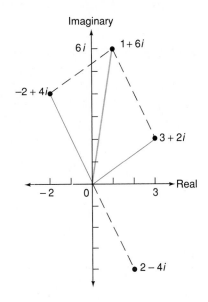

For z not on an axis, the triangle indicates why $r = \sqrt{x^2 + y^2}$. You can verify that this definition also applies for z on an axis. On the x-axis this is consistent with our earlier definition of absolute value.

Any complex number z determines a unique point in the plane. Its distance from the origin, $r \geq 0$, is called the **modulus** or **absolute value** of z and is also denoted by $|z|$.

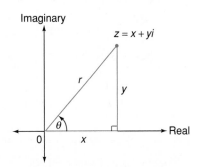

$$r = |x + yi| = \sqrt{x^2 + y^2}$$

The line segment connecting z to the origin and the positive part of the real axis determine an angle of measure θ where $0 \leq \theta < 360°$ (or $0 \leq \theta < 2\pi$). θ is called the **principal argument** of z, which is also referred to simply as the argument of z.

(The choice of principal argument can vary. Another interval that is used for this is $-180° \leq \theta < 180°$.) From the preceding figure we have

$$x = r \cos \theta \quad \text{and} \quad y = r \sin \theta$$

You can verify that these formulas also apply when z is on an axis.

Consequently,

$$z = x + iy = r \cos \theta + i(r \sin \theta)$$

or, when factored, produces the **trigonometric form**

$$z = r(\cos \theta + i \sin \theta)$$

For $z = 0 + 0i = 0$, $r = 0$ and any θ can be used.

Any angle coterminal with θ will give the same result and may also be referred to as an argument of z. However, we will work primarily with the principal argument.

EXAMPLE 4   Convert $\sqrt{3} + i$ and $-1 - i$ into trigonometric form.

Solution   Graph the points and observe that $\sqrt{3} + i$ determines a 30°–60°–90° triangle and $-1 - i$ determines a 45°–45°–90° triangle.

r is the modulus and θ is the argument.

$$r = \sqrt{(\sqrt{3})^2 + 1^2} = 2$$

$$\theta = 30° = \frac{\pi}{6}$$

$$\sqrt{3} + i = 2(\cos 30° + i \sin 30°)$$

$$= 2\left(\cos \frac{\pi}{6} + i \sin \frac{\pi}{6}\right)$$

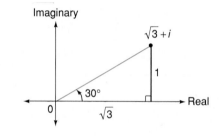

$$r = \sqrt{(-1)^2 + (-1)^2} = \sqrt{2}$$

$$\theta = 225° = \frac{5\pi}{4}$$

$$-1 - i = \sqrt{2}(\cos 225° + i \sin 225°)$$

$$= \sqrt{2}\left(\cos \frac{5\pi}{4} + i \sin \frac{5\pi}{4}\right)$$

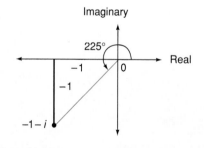

In contrast to the trigonometric form $z = r(\cos \theta + i \sin \theta)$, we call $z = x + yi$ the **rectangular form** of z.

EXAMPLE 5   Use Table VI or a calculator to convert $3(\cos 50° + i \sin 50°)$ into rectangular form $x + yi$.

Solution

$$3(\cos 50° + i \sin 50°) = 3(.6428 + .7660i)$$

$$= 1.9284 + 2.2980i$$

472   CHAPTER 9: Additional Applications of Trigonometry

The multiplication and division of complex numbers are sometimes simpler to do using the trigonometric forms. To see how this is done, we begin with

$$z_1 = r_1(\cos \theta_1 + i \sin \theta_1) \quad \text{and} \quad z_2 = r_2(\cos \theta_2 + i \sin \theta_2)$$

and multiply.

$$z_1 z_2 = [r_1(\cos \theta_1 + i \sin \theta_1)][r_2(\cos \theta_2 + i \sin \theta_2)]$$
$$= r_1 r_2[(\cos \theta_1 \cos \theta_2 - \sin \theta_1 \sin \theta_2) + i(\sin \theta_1 \cos \theta_2 + \cos \theta_1 \sin \theta_2)]$$

Now use addition formulas for sines and cosines to obtain

If $\theta_1 + \theta_2$ is not in $[0, 2\pi)$, replace it with the principal argument, which is the smallest nonnegative coterminal angle.

$$z_1 z_2 = r_1 r_2[\cos (\theta_1 + \theta_2) + i \sin(\theta_1 + \theta_2)]$$

To multiply two complex numbers in trigonometric form, multiply their moduli and add their arguments.

EXAMPLE 6 Let $z_1 = 2 + 2\sqrt{3}i$ and $z_2 = -1 - \sqrt{3}i$.

(a) Evaluate $z_1 z_2$ by using the rectangular forms.

(b) Evaluate $z_1 z_2$ by using trigonometric forms and verify that the result is the same as in part (a).

Solution
(a) $z_1 z_2 = (2 + 2\sqrt{3}i)(-1 - \sqrt{3}i)$
$$= (-2 + 6) + (-2\sqrt{3} - 2\sqrt{3})i = 4 - 4\sqrt{3}i$$

(b) Converting z_1 and z_2 to trigonometric form, we get

$$z_1 = 4\left(\cos \frac{\pi}{3} + i \sin \frac{\pi}{3}\right) \text{ and } z_2 = 2\left(\cos \frac{4\pi}{3} + i \sin \frac{4\pi}{3}\right)$$

Then

$$z_1 z_2 = \left[4\left(\cos \frac{\pi}{3} + i \sin \frac{4\pi}{3}\right)\right]\left[2\left(\cos \frac{4\pi}{3} + i \sin \frac{4\pi}{3}\right)\right]$$

$$= 4 \cdot 2\left[\cos \left(\frac{\pi}{3} + \frac{4\pi}{3}\right) + i \sin \left(\frac{\pi}{3} + \frac{4\pi}{3}\right)\right]$$

$$= 8\left(\cos \frac{5\pi}{3} + i \sin \frac{5\pi}{3}\right)$$

$$= 8\left[\frac{1}{2} + i\left(-\frac{\sqrt{3}}{2}\right)\right]$$

$$= 4 - 4\sqrt{3}i \qquad \blacksquare$$

Let $z_1 = -2 + 2i$ and $z_2 = 3\sqrt{3} - 3i$.
Convert to trigonometric form and evaluate $z_1 z_2$.

If $\theta_1 - \theta_2$ *is not in* $[0, 2\pi)$,
*replace it with the principal
argument, which is the
smallest nonnegative coterminal angle.*

By similar reasoning as for the product (see Exercise 46), the result for division is the following:

$$\frac{z_1}{z_2} = \frac{r_1}{r_2}[\cos(\theta_1 - \theta_2) + i \sin(\theta_1 - \theta_2)] \qquad z_2 \neq 0$$

*To divide two complex numbers in trigonometric form, divide their moduli and
subtract their arguments.*

EXAMPLE 7   For z_1, z_2 in Example 6, find:

(a) $\dfrac{z_1}{z_2}$ using rectangular forms.

(b) $\dfrac{z_1}{z_2}$ using trigonometric forms and verify that the result is the same as in part (a).

Solution

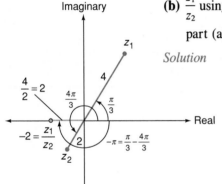

(a)
$$\frac{2 + 2\sqrt{3}i}{-1 - \sqrt{3}i} = \frac{2 + 2\sqrt{3}i}{-1 - \sqrt{3}i} \cdot \frac{-1 + \sqrt{3}i}{-1 + \sqrt{3}i}$$

$$= \frac{(-2 - 6) + (2\sqrt{3} - 2\sqrt{3})i}{1 + 3}$$

$$= \frac{-8}{4} = -2$$

(b)
$$\frac{4\left(\cos\dfrac{\pi}{3} + i \sin\dfrac{\pi}{3}\right)}{2\left(\cos\dfrac{4\pi}{3} + i \sin\dfrac{4\pi}{3}\right)} = 2\left[\cos\left(\dfrac{\pi}{3} - \dfrac{4\pi}{3}\right) + i \sin\left(\dfrac{\pi}{3} - \dfrac{4\pi}{3}\right)\right]$$

$$= 2[\cos(-\pi) + i \sin(-\pi)]$$

$$= 2(\cos \pi + i \sin \pi) \qquad (\pi \text{ is the principal}$$
$$\qquad\qquad\qquad\qquad\qquad \text{argument.)}$$

$$= 2(-1 + 0)$$

$$= -2 \qquad\qquad\qquad\qquad\qquad \blacksquare$$

EXERCISES 9.3

Find $z + w$ using the parallelogram rule and verify using the definition of addition for complex numbers.

1. $z = 3 + 4i$, $w = 4 + 3i$ 2. $z = 3 - 4i$, $w = 4 - 3i$
3. $z = -3 + 2i$, $w = -2 - 3i$ 4. $z = 1 + 3i$, $w = -3 - i$

Find z − w using parallelograms and verify using the definition of subtraction for complex numbers.

5. $z = 3 + 4i$, $w = 4 + 3i$ 6. $z = 3 - 4i$, $w = 4 - 3i$
7. $z = -3 + 2i$, $w = -2 - 3i$ 8. $z = 1 + 3i$, $w = -3 - i$
9. Find the sum graphically on a complex plane.

$$[(2 + 5i) + (2 - 2i)] + (3 - 5i)$$

10. Verify the parallelogram rule by showing that $x = a + c$ and $y = b + d$.

11. Use the parallelogram rule to show that the sum of a complex number and its conjugate is a real number.
12. Let $z = a + bi$ and $w = c + di$. What must be true so that $z + w$ has the form xi where x is a real number?

Convert each of the following into trigonometric form and plot each point in a plane.

13. $1 + i$ 14. $3 + 3i$ 15. $-1 + \sqrt{3}i$ 16. $-2 + 2\sqrt{3}i$ 17. $\dfrac{\sqrt{3}}{2} - \dfrac{1}{2}i$

18. 5 19. -10 20. $-4i$ 21. $-10 - 10i$

Express each of the following complex numbers in the rectangular form $x + yi$. Use tables only if necessary.

22. $2(\cos 30° + i \sin 30°)$ 23. $3(\cos 120° + i \sin 120°)$ 24. $\dfrac{1}{2}\left(\cos \dfrac{\pi}{2} + i \sin \dfrac{\pi}{2}\right)$

25. $5\left(\cos \dfrac{3\pi}{2} + i \sin \dfrac{3\pi}{2}\right)$ 26. $9(\cos 0° + i \sin 0°)$ 27. $\sqrt{2}(\cos 60° + i \sin 60°)$

28. $1(\cos \pi + i \sin \pi)$ 29. $4(\cos 315° + i \sin 315°)$ 30. $3(\cos 68° + i \sin 68°)$
31. $\cos 350° + i \sin 350°$

Find zw, $\dfrac{z}{w}$, and $\dfrac{w}{z}$. Give the answers in trigonometric form and convert to rectangular form. Use tables or a calculator only when necessary.

32. $z = 3(\cos 41° + i \sin 41°)$ 33. $z = \frac{1}{2}(\cos 100° + i \sin 100°)$
$\quad w = 2(\cos 20° + i \sin 20°)$ $\quad w = 10(\cos 50° + i \sin 50°)$

34. $z = \cos \dfrac{\pi}{3} + i \sin \dfrac{\pi}{3}$ 35. $z = \sqrt{2}\left(\cos \dfrac{5\pi}{4} + i \sin \dfrac{5\pi}{4}\right)$

$\quad w = 6\left(\cos \dfrac{\pi}{4} + i \sin \dfrac{\pi}{4}\right)$ $\quad w = 8\left(\cos \dfrac{7\pi}{4} + i \sin \dfrac{7\pi}{4}\right)$

Evaluate zw and $\dfrac{z}{w}$ using trigonometric forms and convert the answers to rectangular form.

36. $z = 1 + i$
$w = -2 - 2i$

37. $z = 7i$
$w = \sqrt{3} + i$

38. $z = -5i$
$w = 7$

39. $z = -2 + 2\sqrt{3}i$
$w = -2\sqrt{3} - 2i$

40. $z = 10 - 10i$
$w = -4\sqrt{2} + 4\sqrt{2}i$

41. $z = -1 + i$
$w = 1 - i$

Multiply as indicated and express the answer in rectangular form.
Use Table VI or a calculator only if necessary.

42. $[4(\cos 23° + i \sin 23°)][2(\cos 37° + i \sin 37°)]$

43. $[\frac{1}{2}(\cos 20° + i \sin 20°)][\sqrt{2}(\cos 70° + i \sin 70°)]$

44. $[15(\cos 25° + i \sin 25°)][3(\cos 205° + i \sin 205°)]$

45. $[\frac{2}{3}(\cos 122° + i \sin 122°)][\frac{3}{4}(\cos 77° + i \sin 77°)]$

46. Derive the formula for quotients in trigonometric form on page 474 (*Hint:* Use the conjugate of $\cos \theta_2 + i \sin \theta_2$ and formulas for sines and cosines.

Prove the following using $z = a + bi$ and $w = c + di$.

47. $|z| = |\bar{z}|$

48. $z\bar{z} = |z|^2$

49. $|zw| = |z||w|$

50. $\left|\dfrac{z}{w}\right| = \dfrac{|z|}{|w|}$

When the equation $|z| = 2$, for $z = x + yi$ is squared, we obtain $x^2 + y^2 = 4$. Thus $|z| = 2$ is the equation of a circle of radius 2 and center $(0, 0)$. Identify the curve defined by each equation.

51. $|z - 1| = 2$

52. $|z - 2i| = 3$

53. $|z - (2 + 3i)| = 1$

54. $3|z - 1| = |z + i|$

55. $|z + 2| = 8 - |z - 2|$

CHALLENGE
Think Creatively

Use the result of Exercise 48 and properties of conjugates (see Exercises 65–68, page 248) to prove that $|zw| = |z||w|$, without converting to the $a + bi$ forms.

9.4
DE MOIVRE'S
THEOREM

The formula for the product of two complex numbers in trigonometric form can be used to derive a special formula for the nth power of a complex number z. We begin with the case $n = 2$. First, let

$$z_1 = z_2 = z = r(\cos \theta + i \sin \theta)$$

Now substitute into

$$z_1 z_2 = r_1 r_2[\cos(\theta_1 + \theta_2) + i \sin(\theta_1 + \theta_2)]$$

Thus

$$z^2 = r^2[\cos 2\theta + i \sin 2\theta]$$

For $n = 3$, write $z^3 = z^2 z$ and let $z_1 = z^2$ and $z_2 = z$ in the formula for $z_1 z_2$ to obtain

$$z^3 = z^2 z = [r^2(\cos 2\theta + i \sin 2\theta)][r(\cos \theta + i \sin \theta)]$$
$$= r^3(\cos 3\theta + i \sin 3\theta)$$

In this manner it can be shown that $z^n = r^n(\cos n\theta + i \sin n\theta)$ for each positive integer n. Since $z = r(\cos \theta + i \sin \theta)$, we get the following result:

DE MOIVRE'S THEOREM

For any positive integer n:

$$[r(\cos \theta + i \sin \theta)]^n = r^n(\cos n\theta + i \sin n\theta)$$

To take an nth power of a complex number in trigonometric form, take the nth power of the modulus and n times the argument.

EXAMPLE 1   Use De Moivre's theorem to evaluate $(1 - i)^8$.

Solution   Converting $z = 1 - i$ to trigonometric form, we obtain

$$z = 2^{1/2}\left(\cos \frac{7\pi}{4} + i \sin \frac{7\pi}{4}\right)$$

Now use De Moivre's theorem with $r = 2^{1/2}$ and $\theta = \dfrac{7\pi}{4}$.

$$z^8 = (2^{1/2})^8\left[\cos\left(8 \cdot \frac{7\pi}{4}\right) + i \sin\left(8 \cdot \frac{7\pi}{4}\right)\right]$$

$$= 2^4(\cos 14\pi + i \sin 14\pi)$$

$$= 16(\cos 0 + i \sin 0) \qquad \text{(0 is the principal argument.)}$$

$$= 16(1 + 0)$$

$$= 16 \qquad\qquad\qquad\qquad\qquad\qquad\qquad \blacksquare$$

**TEST YOUR UNDERSTANDING
Think Carefully**

(Answers: Page 502)

Use De Moivre's theorem to evaluate the following powers.

1. $(\cos 15° + i \sin 15°)^3$ 2. $[\frac{1}{2}(\cos 15° + i \sin 15°)]^6$
3. $(-\frac{1}{2} + \frac{1}{2}i)^4$ 4. $(-\sqrt{3} + i)^{10}$

Example 1 shows that $(1 - i)^8 = 16$. This says that $1 - i$ is an 8th root of 16. There are seven more 8th roots of 16. For the integers $n \geq 2$ every complex number has n distinct nth roots. De Moivre's theorem can be used to derive the following formula, which enables us to find such roots.

The formula can be restated for θ in degree measure. In this case, replace 2π by $360°$.

THE nth-ROOT FORMULA

If n is a positive integer and if $z = r(\cos \theta + i \sin \theta)$ is a nonzero complex number, the nth roots of z are given by

$$r^{1/n}\left(\cos \frac{\theta + 2k\pi}{n} + i \sin \frac{\theta + 2k\pi}{n}\right) \qquad \text{for } k = 0, 1, 2, \ldots, n - 1$$

When $k = 0, 1, 2, \ldots, n - 1$ the nth-root formula produces the n distinct roots. For larger values of k, say $k = n, n + 1, \ldots, 2n - 1$, the periodicity of sin and cos will produce the same roots. This is discussed in Exercise 33 at the end of this section, and also in Example 2, which is presented after the following verification of the nth-root formula.

One way to justify the nth-root formula is to take an nth root of z, raise it to the nth power, and obtain z. Thus, if $k = 0, 1, 2, \ldots, n - 1$,

$$\left[r^{1/n} \left(\cos \frac{\theta + 2k\pi}{n} + i \sin \frac{\theta + 2k\pi}{n} \right) \right]^n$$

$$= (r^{1/n})^n \left[\cos \left(n \cdot \frac{\theta + 2k\pi}{n} \right) + i \sin \left(n \cdot \frac{\theta + 2k\pi}{n} \right) \right] \qquad \text{(by De Moivre's theorem)}$$

$$= r[\cos(\theta + 2k\pi) + i \sin(\theta + 2k\pi)]$$

$$= r[\cos \theta + i \sin \theta] \qquad \text{(by the addition formulas for sin and cos)}$$

$$= z$$

EXAMPLE 2   Find the three cube roots of $z = 8i$.

Solution   First convert z into trigonometric form.

$$z = 8i = 0 + 8i = 8 \left(\cos \frac{\pi}{2} + i \sin \frac{\pi}{2} \right)$$

Now use the nth-root formula with $n = 3$, $r = 8$, and $\theta = \dfrac{\pi}{2}$. The values $k = 0$, 1, 2 produce the cube roots as follows:

$$k = 0: \qquad 8^{1/3} \left(\cos \frac{\frac{\pi}{2} + 0}{3} + i \sin \frac{\frac{\pi}{2} + 0}{3} \right) = 2 \left(\cos \frac{\pi}{6} + i \sin \frac{\pi}{6} \right)$$

$$= \sqrt{3} + i$$

Substitute $k = 3, 4, 5$ into the nth-root formula to see that the same three cube roots are obtained.

$$k = 1: \qquad 8^{1/3} \left(\cos \frac{\frac{\pi}{2} + 2\pi}{3} + i \sin \frac{\frac{\pi}{2} + 2\pi}{3} \right) = 2 \left(\cos \frac{5\pi}{6} + i \sin \frac{5\pi}{6} \right)$$

$$= -\sqrt{3} + i$$

$$k = 2: \qquad 8^{1/3} \left(\cos \frac{\frac{\pi}{2} + 4\pi}{3} + i \sin \frac{\frac{\pi}{2} + 4\pi}{3} \right) = 2 \left(\cos \frac{3\pi}{2} + i \sin \frac{3\pi}{2} \right)$$

$$= -2i \qquad \blacksquare$$

You can also check this root by applying De Moivre's theorem to the trigonometric form of $\sqrt{3} + i$.

The roots found in Example 2 can be checked by showing that their cubes equal $8i$. For example, here is a check for $\sqrt{3} + i$ using rectangular forms.

$$(\sqrt{3} + i)^3 = (\sqrt{3} + i)^2(\sqrt{3} + i) = (3 - 1 + 2\sqrt{3}i)(\sqrt{3} + i)$$
$$= (2 + 2\sqrt{3}i)(\sqrt{3} + i) = 2\sqrt{3} - 2\sqrt{3} + 2i + 6i$$
$$= 8i$$

The *n*th-root formula can also be used to solve equations of the form $z^n - c = 0$, where *n* is a positive integer and *c* is a constant.

EXAMPLE 3   Solve the equation $z^4 + 16 = 0$.

Solution $z^4 + 16 = 0$ can be written as $z^4 = -16$. Therefore, the solutions of the given equation are the 4th roots of -16. Writing -16 in trigonometric form, we have

$$-16 = 16(\cos \pi + i \sin \pi)$$

Now use $n = 4$, $r = 16$, and $\theta = \pi$ in the *n*th-root formula to get this general form of the roots.

$$z = 16^{1/4}\left(\cos \frac{\pi + 2k\pi}{4} + i \sin \frac{\pi + 2k\pi}{4}\right)$$

Specifically, the four roots of the equation are obtained by letting $k = 0, 1, 2, 3$ as follows:

$$k = 0: \quad z = 2\left(\cos \frac{\pi}{4} + i \sin \frac{\pi}{4}\right) = \sqrt{2} + \sqrt{2}i$$

$$k = 1: \quad z = 2\left(\cos \frac{3\pi}{4} + i \sin \frac{3\pi}{4}\right) = -\sqrt{2} + \sqrt{2}i$$

Check for $z = \sqrt{2} - \sqrt{2}i$:
$$(\sqrt{2} - \sqrt{2}i)^4$$
$$= [(\sqrt{2} - \sqrt{2}i)^2]^2$$
$$= [2 - 4i - 2]^2$$
$$= [-4i]^2$$
$$= 16i^2 = -16$$

$$k = 2: \quad z = 2\left(\cos \frac{5\pi}{4} + i \sin \frac{5\pi}{4}\right) = -\sqrt{2} - \sqrt{2}i$$

$$k = 3: \quad z = 2\left(\cos \frac{7\pi}{4} + i \sin \frac{7\pi}{4}\right) = \sqrt{2} - \sqrt{2}i \quad \blacksquare$$

The *n*th-root formula shows that each root has modulus $r^{1/n}$. Therefore, since the modulus of a complex number is its distance from the origin, the *n*th roots are on a circle with center at the origin having radius $r^{1/n}$. Furthermore, the arguments of a pair of successive roots differ by $\dfrac{2\pi}{n}$. This can be observed by using $\dfrac{\theta + 2k\pi}{n}$, which is the argument of the *k*th root. For instance, when $k = 2$ and $k = 3$ we have

Generalize this argument using $k + 1$ and k in place of 3 and 2.

$$\text{argument for } k = 3: \quad \frac{\theta + 6\pi}{n} = \frac{\theta}{n} + \frac{6\pi}{n}$$
$$\text{argument for } k = 2: \quad \frac{\theta + 4\pi}{n} = \frac{\theta}{n} + \frac{4\pi}{n}$$

$$\left.\begin{array}{c} \\ \\ \end{array}\right\} \quad \text{difference} = \frac{2\pi}{n}$$

This means that the nth roots are equally spaced around the circle. Here is the diagram for the cube roots of $8i$ (see Example 2).

$$\frac{2\pi}{n} = \frac{360°}{3} = 120°$$

$$\text{radius} = 2$$

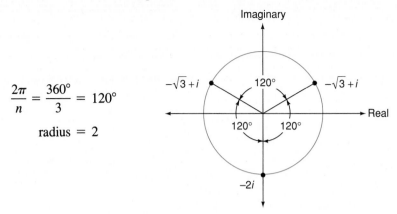

The trigonometric form of $z = 1$ is $1 = 1(\cos 0 + i \sin 0)$. Then, using $r = 1$ and $\theta = 0$ in the nth-root formula, we obtain the following:

> ### nth ROOTS OF UNITY
>
> The nth roots of 1 are given by
>
> $$\cos \frac{2k\pi}{n} + i \sin \frac{2k\pi}{n} \qquad \text{for } k = 0, 1, 2, \ldots, n - 1$$

These roots are uniformly spaced on the unit circle and determine a regular polygon.

EXAMPLE 4   Find the 6th roots of unity and sketch the regular hexagon.

Solution   Use $n = 6$ in the preceding formula.

k	$\cos\dfrac{k\pi}{3} + i\sin\dfrac{k\pi}{3} = x + yi$
0	$\cos 0 + i \sin 0 = 1$
1	$\cos\dfrac{\pi}{3} + i\sin\dfrac{\pi}{3} = \dfrac{1}{2} + \dfrac{\sqrt{3}}{2}i$
2	$\cos\dfrac{2\pi}{3} + i\sin\dfrac{2\pi}{3} = -\dfrac{1}{2} + \dfrac{\sqrt{3}}{2}i$
3	$\cos \pi + i \sin \pi = -1$
4	$\cos\dfrac{4\pi}{3} + i\sin\dfrac{4\pi}{3} = -\dfrac{1}{2} - \dfrac{\sqrt{3}}{2}i$
5	$\cos\dfrac{5\pi}{3} + i\sin\dfrac{5\pi}{3} = \dfrac{1}{2} - \dfrac{\sqrt{3}}{2}i$

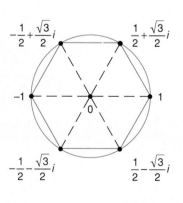

EXERCISES 9.4

Use De Moivre's theorem to evaluate the given powers and express the answers in the rectangular form x + yi. Use tables or a calculator only when necessary.

1. $(\cos 6° + i \sin 6°)^{10}$

2. $\left(\cos \dfrac{\pi}{5} + i \sin \dfrac{\pi}{5}\right)^{15}$

3. $(\cos 40° + i \sin 40°)^8$

4. $\left[2\left(\cos \dfrac{\pi}{9} + i \sin \dfrac{\pi}{9}\right)\right]^6$

5. $\left[\dfrac{1}{2}\left(\cos \dfrac{\pi}{8} + i \sin \dfrac{\pi}{8}\right)\right]^6$

6. $[3(\cos 15° + i \sin 15°)]^5$

7. $(-1 - i)^4$

8. $(1 - \sqrt{3}i)^8$

9. $(-\sqrt{3} + \sqrt{3}\,i)^6$

10. $\left(\dfrac{1}{2} + \dfrac{\sqrt{3}}{2}i\right)^6$

11. $\left(\dfrac{\sqrt{2}}{2} - \dfrac{\sqrt{2}}{2}i\right)^{10}$

12. $(2 - 2i)^4$

13. $\left(-\dfrac{\sqrt{3}}{2} + \dfrac{1}{2}i\right)^{12}$

14. $(2^{1/4} + 2^{1/4}i)^{12}$

Find the indicated roots and express them in trigonometric form.

15. The cube roots of $27\left(\cos \dfrac{\pi}{15} + i \sin \dfrac{\pi}{15}\right)$.

16. The cube roots of $-4\sqrt{2} - 4\sqrt{2}\,i$.

17. The 5th roots of $32\left(\cos \dfrac{\pi}{8} + i \sin \dfrac{\pi}{8}\right)$.

18. The 4th roots of $4 - 4\sqrt{3}\,i$.

Write the roots in rectangular form and graph them on the appropriate circle.

19. The 4th roots of 16.

20. The cube roots of 27.

21. The cube roots of -1.

22. The 6th roots of -1.

Find the following nth roots of unity for the given n, and sketch the regular polygon determined by the roots.

23. $n = 3$

24. $n = 4$

25. $n = 5$

26. $n = 8$

Solve the given equation.

27. $z^3 = -27$

28. $z^4 + 81 = 0$

29. $z^3 + 8i = 0$

30. $z^4 - \frac{1}{16}i = 0$

31. $z^3 + i = -1$

32. $z^4 + 1 - \sqrt{3}i = 0$

33. (a) Show that $k = 0$ and $k = n$ give the same nth roots in the nth-root formula.
   (b) Repeat part (a) for $k = 1$ and $k = n + 1$.
   (c) Repeat part (a) for k and $k + n$.

34. Prove that $(\cos \theta + i \sin \theta)^{-1} = \cos \theta - i \sin \theta$.

35. (a) Let n be a positive integer. Use the result in Exercise 34 and De Moivre's theorem to prove that $[r(\cos \theta + i \sin \theta)]^{-n} = r^{-n}(\cos n\theta - i \sin n\theta)$
   (b) Explain why $[r(\cos \theta + i \sin \theta)]^n = r^n(\cos n\theta + i \sin n\theta)$ holds for all integers n.

36. (a) Two complex numbers $a + bi$ and $c + di$ are equal if and only if $a = c$ and $b = d$. Use this criterion of equality in conjunction with De Moivre's theorem to derive the double-angle formulas for the sine and cosine. [*Hint:* Expand the right side of $\cos 2\theta + i \sin 2\theta = (\cos \theta + i \sin \theta)^2$.]
   (b) Follow the procedure that is described in part (a) to derive the formulas $\cos 3\theta = 4 \cos^3 \theta - 3 \cos \theta$ and $\sin 3\theta = 3 \sin \theta - 4 \sin^3 \theta$.

37. $z = \cos \dfrac{2\pi}{3} + i \sin \dfrac{2\pi}{3}$ is one of the cube roots of unity. Verify that $1 + z + z^2 = 0$.

Let $z = \cos \dfrac{2\pi}{n} + i \sin \dfrac{2\pi}{n}$ be an nth root of unity for $n \geq 2$.

Prove that $1 + z + z^2 + \cdots + z^{n-1} = 0$. (*Hint:* Consider the fact that $z^n = 1$.)

9.5 POLAR COORDINATES

When a complex number is given in trigonometric form, it can be located in the complex plane using its argument θ and modulus r. A similar method, using **polar coordinates**, can be used for locating points in the plane without using complex numbers. In a **polar coordinate system** a point in the plane is located in reference to a fixed point 0, called the **pole**, and a fixed ray called the **polar axis**. The pole is the endpoint of the polar axis, which is usually horizontal and extends endlessly to the right.

Any point $P \neq 0$ is on some ray that forms an angle of measure θ with the polar axis as its initial side and is some distance r from the pole. The ordered pair (r, θ) are the polar coordinates of P. Consequently, when the polar coordinates (r, θ) of a point P are given, P can be located as follows:

1. Start at the polar axis and rotate through an angle of measure θ to determine the ray θ.

*The terminal side of θ is also referred to as the **ray θ**. The pole 0 is the initial point of ray θ.*

2. On the ray θ move r units from the pole 0 to locate P.

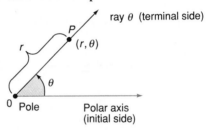

EXAMPLE 1 Plot the points with these polar coordinates:

$$\left(1, \frac{\pi}{6}\right), \qquad \left(2, \frac{2\pi}{3}\right), \qquad \left(\frac{3}{2}, 255°\right), \qquad \left(\frac{2}{3}, 330°\right)$$

Solution

Observe that for positive angles we rotate counterclockwise.

A ray for an angle θ may be extended through the pole 0 in the opposite or backward direction. Points on this opposite side have polar coordinates (r, θ), where $r < 0$.

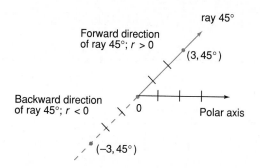

The opposite or backward extension of a ray θ is the same as the ray $\theta + \pi$. Therefore, (r, θ) and $(-r, \theta + \pi)$ are polar coordinates of the same point.

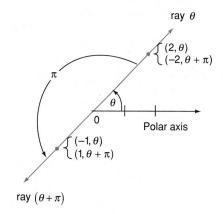

The angle θ may also be negative. In this case we rotate clockwise from the polar axis to determine the ray θ.

EXAMPLE 2   Plot the points with these polar coordinates:

$$\left(2, -\frac{\pi}{3}\right), \qquad \left(-2, -\frac{\pi}{3}\right)$$

Solution

A pair of polar coordinates (r, θ) determine exactly one point. However, a point in the plane has more than one set of polar coordinates. For example, the point P with coordinates $(2, 60°)$ also has these polar coordinates:

$$(2, 420°), \qquad (2, -300°), \qquad (-2, 240°), \qquad (-2, -120°)$$

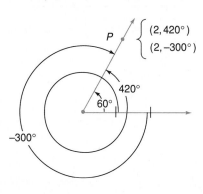 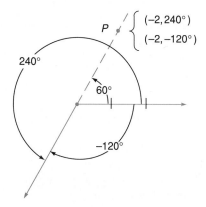

In fact, a point has an infinite number of polar coordinates. If (r, θ) are polar coordinates of point P, then

$$(r, \theta + 2k\pi) \quad \text{and} \quad (-r, \theta + (2k + 1)\pi) \qquad \text{for all integers } k$$

give all the polar coordinates of P. The pole 0 is assigned the polar coordinates $(0, \theta)$ for any angle θ.

EXAMPLE 3 Let $\left(3, \dfrac{\pi}{6}\right)$ be polar coordinates of point P. Find polar coordinates (r, θ) of P subject to these conditions:
(a) $r > 0, \ -2\pi < \theta < 0$ **(b)** $r < 0, \ 0 < \theta < 2\pi$

Solution
(a) 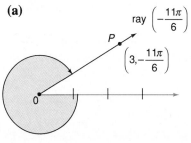 **(b)**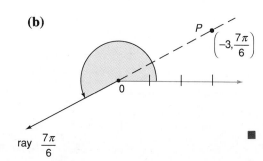

(Answers Page 502)

TEST YOUR UNDERSTANDING
Think Carefully

Plot the points having the given polar coordinates.

1. $(2, 45°), (-2, 45°), (2, -45°), (-2, -45°)$

2. $\left(-2, \dfrac{3\pi}{4}\right), \left(0, \dfrac{3\pi}{4}\right), \left(1, \dfrac{3\pi}{4}\right), \left(3, \dfrac{3\pi}{4}\right)$

3. If $\left(5, \dfrac{\pi}{12}\right)$ are polar coordinates of P, find the polar coordinates (r, θ) of P such that:
    (a) $r > 0, 2\pi < \theta < 4\pi$ **(b)** $r < 0, 0 < \theta < 2\pi$
    (c) $r > 0, -2\pi < \theta < 0$ **(d)** $r < 0, -2\pi < \theta < 0$

To see the relationship between polar and rectangular coordinates, we superimpose a rectangular system onto a polar system so that the pole coincides with the origin, the polar axis coincides with the positive part of the x-axis, and the ray $\frac{\pi}{2}$ coincides with the positive part of the y-axis.

$$x = r \cos \theta$$

$$y = r \sin \theta$$

$$x^2 + y^2 = r^2$$

$$\tan \theta = \frac{y}{x}$$

$r^2 = x^2 + y^2$ is equivalent to $r = \pm\sqrt{x^2 + y^2}$. Also, the signs of x and y together with $\tan \frac{y}{x}$ determine θ.

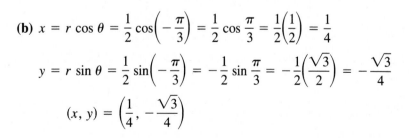

The equations listed next to the figure are easy to derive using the given triangle, which assumes that (x, y) is not on an axis, and $r > 0$. Furthermore, it can be shown that these equations hold when $r \leq 0$ or when (x, y) is on an axis (see Exercises 35, 36, and 37). The preceding equations are used to change from one type of coordinates into the other.

EXAMPLE 4   Convert from polar to rectangular coordinates.

(a) $\left(4, \frac{3\pi}{4}\right)$ **(b)** $\left(\frac{1}{2}, -\frac{\pi}{3}\right)$

Solution

(a) $x = r \cos \theta = 4 \cos\left(\frac{3\pi}{4}\right) = 4\left(-\frac{\sqrt{2}}{2}\right) = -2\sqrt{2}$

$y = r \sin \theta = 4 \sin\left(\frac{3\pi}{4}\right) = 4\left(\frac{\sqrt{2}}{2}\right) = 2\sqrt{2}$

$(x, y) = (-2\sqrt{2}, 2\sqrt{2})$

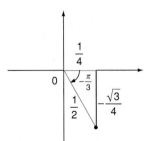

(b) $x = r \cos \theta = \frac{1}{2} \cos\left(-\frac{\pi}{3}\right) = \frac{1}{2} \cos \frac{\pi}{3} = \frac{1}{2}\left(\frac{1}{2}\right) = \frac{1}{4}$

$y = r \sin \theta = \frac{1}{2} \sin\left(-\frac{\pi}{3}\right) = -\frac{1}{2} \sin \frac{\pi}{3} = -\frac{1}{2}\left(\frac{\sqrt{3}}{2}\right) = -\frac{\sqrt{3}}{4}$

$(x, y) = \left(\frac{1}{4}, -\frac{\sqrt{3}}{4}\right)$ ■

EXAMPLE 5   Convert from rectangular to polar coordinate (r, θ) so that $r \geq 0$ and $0 \leq \theta < 2\pi$.

(a) $(3\sqrt{3}, -3)$ **(b)** $(-5, -5)$

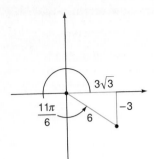

Solution

(a) $r = \sqrt{x^2 + y^2} = \sqrt{(3\sqrt{3})^2 + (-3)^2}$

$$= \sqrt{27 + 9}$$

$$= 6$$

$\tan \theta = \dfrac{-3}{3\sqrt{3}} = -\dfrac{\sqrt{3}}{3}$. Then $\theta = \dfrac{11\pi}{6}$, since (x, y) is in quadrant IV. Therefore,

$$(r, \theta) = \left(6, \frac{11\pi}{6}\right)$$

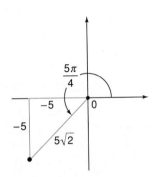

(b) $r = \sqrt{(-5)^2 + (-5)^2} = \sqrt{50} = 5\sqrt{2}$

$\tan \theta = \dfrac{-5}{-5} = 1$. Then $\theta = \dfrac{5\pi}{4}$, since (x, y) is in quadrant III. Therefore,

$$(r, \theta) = \left(5\sqrt{2}, \frac{5\pi}{4}\right)$$

■

EXERCISES 9.5

Plot the points in the same polar coordinate system.

1. $(1, 30°)$, $(-1, 30°)$, $(1, -30°)$, $(-1, -30°)$

2. $(2, 90°)$, $(-2, 90°)$, $(2, -90°)$, $(-2, -90°)$

3. $(4, 0°)$, $(-4, 0°)$

4. $\left(3, \dfrac{\pi}{4}\right)$, $\left(3, -\dfrac{\pi}{4}\right)$, $\left(-3, \dfrac{\pi}{4}\right)$, $\left(-3, -\dfrac{\pi}{4}\right)$

5. $(\tfrac{1}{2}, \pi)$, $(-\tfrac{1}{2}, \pi)$, $(\tfrac{1}{2}, -\pi)$, $(-\tfrac{1}{2}, -\pi)$

6. $\left(5, \dfrac{4\pi}{3}\right)$, $\left(-5, \dfrac{4\pi}{3}\right)$, $\left(5, -\dfrac{4\pi}{3}\right)$, $\left(-5, -\dfrac{4\pi}{3}\right)$

The polar coordinates of a point P are given. Find the polar coordinates (r, θ) of P subject to each of the following:

(a) $r > 0$ and $2\pi < \theta < 4\pi$ **(b)** $r > 0$ and $-2\pi < \theta < 0$

(c) $r < 0$ and $0 < \theta < 2\pi$ **(d)** $r < 0$ and $-2\pi < \theta < 0$

7. $(1, 45°)$ 8. $(\tfrac{1}{2}, 200°)$ 9. $\left(-3, \dfrac{\pi}{6}\right)$ 10. $\left(-\dfrac{3}{4}, \dfrac{5\pi}{4}\right)$

Convert the following polar coordinates into rectangular coordinates.

11. $\left(7, -\dfrac{5\pi}{2}\right)$ 12. $\left(-8, \dfrac{13\pi}{6}\right)$ 13. $\left(-\dfrac{3}{2}, \dfrac{5\pi}{2}\right)$

14. $(-1, -540°)$ 15. $\left(\dfrac{4}{5}, 405°\right)$ 16. $(-10, -420°)$

Convert the following rectangular coordinates into polar coordinates (r, θ) so that $r \geq 0$ and $0 \leq \theta < 2\pi$.

17. $(2, 0)$ 18. $(0, 2)$ 19. $(-2, 0)$

20. $(0, -2)$ 21. $(-6, -6)$ 22. $(3, -3\sqrt{3})$

23. $(2\sqrt{2}, 2\sqrt{2})$ 24. $(-2\sqrt{2}, 2\sqrt{2})$ 25. $(\sqrt{3}, -1)$

Convert the following rectangular coordinates into polar coordinates (r, θ) so that $r < 0$ and $0 \leq \theta < 2\pi$.

26. $(0, 7)$ **27.** $(7, 0)$ **28.** $(2, 2)$ **29.** $(-\sqrt{3}, 1)$ **30.** $(4, -4\sqrt{3})$ **31.** $(\sqrt{3}, \sqrt{3})$

Convert the following rectangular coordinates into polar coordinates (r, θ) so that $r > 0$ and $-180° \leq \theta < 180°$.

32. $(-1, -1)$ **33.** $(0, -10)$ **34.** $(4, -4\sqrt{3})$

The following exercises call for the verification of the formulas $x = r \cos \theta$ and $y = r \sin \theta$ in various cases. Assume that they hold when $r > 0$.

35. Verify the formulas for the origin whose polar coordinates are $(0, \theta)$, where θ is any angle measure.

36. Verify the formulas for a point whose rectangular coordinates are $(x, 0)$ with $x < 0$.

37. Verify the formulas for a point P with rectangular coordinates (x, y) and polar coordinates (r, θ) with $r < 0$. (*Hint:* Since P also has polar coordinates $(-r, \theta + \pi)$, you may apply the formulas since $-r > 0$.)

EXPLORATIONS
Think Critically

1. Let the real part of $z = x + yi$ be denoted by Re z, and the imaginary part by Im z, so that Re $z = x$ and Im $z = y$. Express Re z and Im z in terms of z and its conjugate $\bar{z}$.

2. Give a geometric justification of the triangle inequality $|z + w| \leq |z| + |w|$ for complex numbers z and w.

3. The cube roots of $8i$ are shown on page 480. Suppose a concentric circle of twice the radius is drawn and the lines through the cube roots of $8i$ are extended to obtain three points on the new circle. What would be the coordinates of these points, and what do they represent?

4. If (x, y) are the rectangular coordinates of the polar point (r, θ), verify that $(r, \theta + 2k\pi)$ and $(-r, \theta + (2k + 1)\pi)$ have the same rectangular coordinates for all integers k.

5. Find the rectangular coordinates of the point $(2, 15°)$ without tables or calculators. Express the coordinates in radical form and verify the results using a calculator.

9.6
GRAPHING
POLAR
EQUATIONS

The ray $\dfrac{\pi}{4}$ and its opposite ray form a straight line through the pole 0. This line contains all points with polar coordinates $\left(r, \dfrac{\pi}{4}\right)$ for all real numbers r. Thus the graph of the equation $\theta = \dfrac{\pi}{4}$ (for all r) is the line as shown.

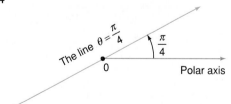

The graph of the equation

θ_0 designates a constant.

$$\theta = \theta_0$$

is a line through the pole forming an angle of measure θ_0 with the polar axis.

EXAMPLE 1 Write a polar equation of the line ℓ through the points with rectangular coordinates $(-2, 2)$ and $(3, -3)$.

Solution ℓ has equation $\theta = 135°$, since ℓ bisects quadrants II and IV.

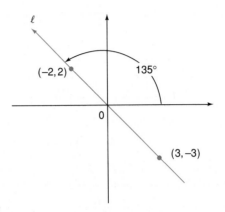

Some other equations for ℓ are: $\theta = \dfrac{3\pi}{4}$, $\theta = -\dfrac{5\pi}{4}$.

All points of the form $(4, \theta)$ for any value θ are four units from the pole and therefore determine a circle with center 0 and radius 4. Thus the graph of the polar equation $r = 4$ (for all θ) is the circle as shown in the following figure. Note that the points $(-4, \theta)$ are also on this circle. However, these are the same points as before, since $(-4, \theta)$ and $(4, \theta + \pi)$ represent the same point.

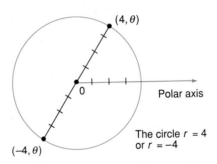

The graph of the equation

$$r = r_0$$

r_0 designates a nonzero constant.

is a circle with center 0 and radius $|r_0|$.

Equations written in terms of polar coordinates are called **polar equations**, and equations using rectangular coordinates may be referred to as **rectangular equations**. When you first learned to graph using rectangular equations, you became accustomed to locating points by moving horizontally for the x coordinate, and vertically for y. With polar coordinates you need to become accustomed to moving in a circular direction for the θ coordinate and forward or backward from the pole along the ray θ for the r coordinate.

EXAMPLE 2 Graph the polar equation $r = 2 \cos \theta$.

Solution As the values of θ increase from 0 to $\dfrac{\pi}{2}$ the corresponding values of r decrease from 2 to 0.

θ	0	$\dfrac{\pi}{6}$	$\dfrac{\pi}{4}$	$\dfrac{\pi}{3}$	$\dfrac{\pi}{2}$
$\cos \theta$	1	$\dfrac{\sqrt{3}}{2}$	$\dfrac{\sqrt{2}}{2}$	$\dfrac{1}{2}$	0
$r = 2 \cos \theta$	2	$\sqrt{3}$	$\sqrt{2}$	1	0

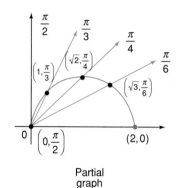

Partial
graph

Refer to the graph of the cosine function to see when the cosine is increasing or decreasing.

As the values of θ increase from $\dfrac{\pi}{2}$ to π, the values of r decrease from 0 to -2. Since the rays θ for $\dfrac{\pi}{2} < \theta < \pi$ are in quadrant II, and the r values are negative, the points are in quadrant IV. For values of $\theta = \pi$ to 2π the same points as before are obtained. For example, the pair $\left(\dfrac{5\pi}{4}, -\sqrt{2}\right)$ and $\left(\dfrac{\pi}{4}, \sqrt{2}\right)$ represent one point, as do the pair $\left(\dfrac{5\pi}{3}, 1\right)$ and $\left(\dfrac{2\pi}{3}, -1\right)$. Furthermore, the periodicity of the cosine will produce the same points for $\theta < 0$ or $\theta > 2\pi$. Therefore, the complete graph of $r = 2 \cos \theta$ is obtained for $\theta = 0$ to $\theta = \pi$, as shown.

The curve is traced out twice from $\theta = 0$ to $\theta = 2\pi$.

θ	$\dfrac{\pi}{2}$	$\dfrac{2\pi}{3}$	$\dfrac{3\pi}{4}$	$\dfrac{5\pi}{6}$	π
$\cos \theta$	0	$-\dfrac{1}{2}$	$-\dfrac{\sqrt{2}}{2}$	$-\dfrac{\sqrt{3}}{2}$	-1
$r = 2 \cos \theta$	0	-1	$-\sqrt{2}$	$-\sqrt{3}$	-2

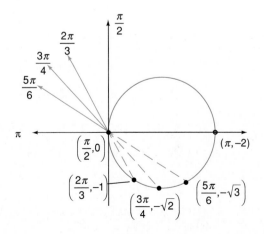

The graph in Example 2 is a circle, as are the graphs of the polar equations $r = 2a \cos \theta$, for $a > 0$. To prove this, consider the circle with radius a, center $(a, 0)$ and let P be a point on this circle with polar coordinates (r, θ). Now recall that an angle inscribed in a semicircle is a right angle. Then, from the right triangle,

Repeat the argument for P on the lower semicircle.

$$\cos \theta = \frac{\text{adjacent}}{\text{hypotenuse}} = \frac{r}{2a}, \text{ or } r = 2a \cos \theta.$$

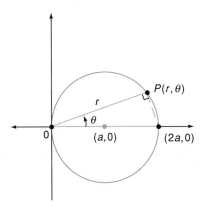

CIRCLE: $r = 2a \cos \theta, \; a > 0$

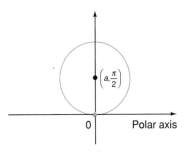

CIRCLE: $r = -2a \cos \theta, \; a > 0$

Similarly, the graph of $r = -2a \cos \theta$, $a > 0$, is the circle with radius a whose center has polar coordinates (a, π) as shown in the margin. In this case, note that

$$\cos \theta = -\cos(\pi - \theta) = -\frac{r}{2a}, \text{ or } r = -2a \cos \theta.$$

In the exercises you will be asked to verify that $r = \pm 2a \sin \theta$, $a > 0$, are polar equations of circles with radius a, as shown here.

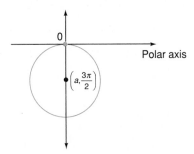

CIRCLE: $r = 2\,a \sin\theta, \; a > 0$

CIRCLE: $r = -2\,a \sin\theta, \; a > 0$

EXAMPLE 3   Graph the polar equations.

(a) $r = -4 \cos \theta$ **(b)** $r = \frac{1}{2} \sin \theta$

Solution
(a) $r = -4 \cos \theta$
$= -2(2) \cos \theta$

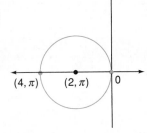

This is the form $-2a \cos \theta$, $a = 2$.

(b) $r = \frac{1}{2} \sin \theta$
$\quad = 2(\frac{1}{4}) \sin \theta$

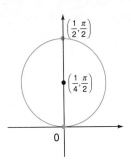

This is the form $2a \sin \theta$, $a = \frac{1}{4}$.   ∎

Another way to show that the graph of $r = 2a \cos \theta$ is a circle is to convert the equation into rectangular form. To do this, begin by multiplying the equation by r.

$$r = 2a \cos \theta$$
$$r^2 = 2a(r \cos \theta) \qquad \text{(multiplying by } r\text{)}$$

The purpose of multiplying by r is to obtain the factor r cos θ on the right and r² on the left, which easily convert into rectangular forms.

Since $r^2 = x^2 + y^2$ and $r \cos \theta = x$, we get

$$x^2 + y^2 = 2ax$$

Now complete the square in x.

$$x^2 - 2ax + y^2 = 0$$
$$x^2 - 2ax + a^2 + y^2 = a^2$$
$$(x - a)^2 + y^2 = a^2 \quad \longleftarrow \quad \text{a circle with radius } a \text{ and center } (a, 0)$$

The preceding demonstrates that converting from polar to rectangular coordinates can be helpful in identifying polar graphs.

EXAMPLE 4   Show that $r = \dfrac{2}{1 - \sin \theta}$ is the equation of a parabola by converting to rectangular coordinates.

Solution

$$r = \frac{2}{1 - \sin \theta}$$

$$r - r \sin \theta = 2$$

$$r = r \sin \theta + 2$$

$$r = y + 2 \qquad \longleftarrow \quad y = r \sin \theta$$

$$r^2 = y^2 + 4y + 4 \qquad \longleftarrow \quad \text{squaring}$$

$$x^2 + y^2 = y^2 + 4y + 4 \qquad \longleftarrow \quad r^2 = x^2 + y^2$$

$$4y = x^2 - 4$$

$$y = \tfrac{1}{4}x^2 - 1 \qquad \longleftarrow \quad \text{a parabola with vertex } (0, -1),$$
$$\text{opening upward} \qquad \blacksquare$$

EXAMPLE 5   Graph the polar curve $r = 2(1 + \sin \theta)$.

Solution   Since the sine has period 2π, it is necessary to consider only values of θ from 0 to 2π. We form a table of selected values for θ (in intervals of 30°), plot the points on a *polar grid system,* and connect the points with a smooth curve.

θ	0°	30°	60°	90°	120°	150°	180°	210°	240°	270°	300°	330°	360°
$\sin \theta$	0	.5	.87	1	.87	.5	0	−.5	−.87	−1	−.87	−.5	0
$r = 2(1 + \sin \theta)$	2	3	3.7	4	3.7	3	2	1	.3	0	.3	1	2

*This heart-shaped curve is a **cardioid**. Cardioids are given by these polar equations:*

$$r = a(1 \pm \sin \theta)$$
$$r = a(1 \pm \cos \theta)$$

If $\sin \theta$ is involved then $\theta = \dfrac{\pi}{2}$ is the axis of symmetry, and if $\cos \theta$ is involved, then the symmetry is around the polar axis.

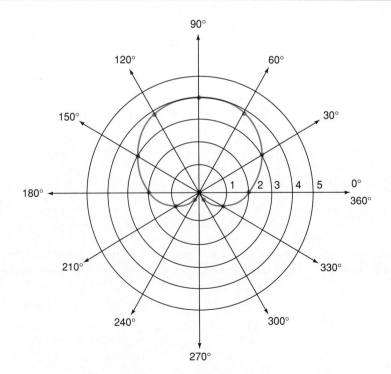

EXAMPLE 6 Sketch the polar curve $r = \sin 2\theta$.

Solution As θ increases from 0 to $\dfrac{\pi}{4}$, 2θ increases from 0 to $\dfrac{\pi}{2}$, and therefore $r = \sin 2\theta$ increases from 0 to 1. Also, as θ increases from $\dfrac{\pi}{4}$ to $\dfrac{\pi}{2}$, 2θ increases from $\dfrac{\pi}{2}$ to π; therefore r decreases from 1 to 0. This produces the curve for $\theta = 0$ to $\theta = \dfrac{\pi}{2}$.

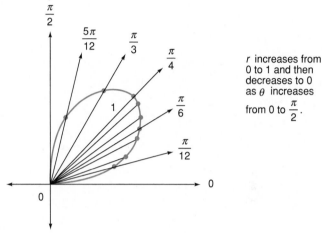

r increases from 0 to 1 and then decreases to 0 as θ increases from 0 to $\dfrac{\pi}{2}$.

Refer to the graph of the sine function to see when the sine is increasing or decreasing.

As θ increases from $\dfrac{\pi}{2}$ to $\dfrac{3\pi}{4}$, 2θ increases from π to $\dfrac{3\pi}{2}$ and r decreases from 0 to -1. These negative r-values produce points in the fourth quadrant, since the rays are in the second quadrant. Similarly, as θ increases from $\dfrac{3\pi}{4}$ to π, r increases from -1, to 0. This completes the loop for $\theta = \dfrac{\pi}{2}$ to $\theta = \pi$.

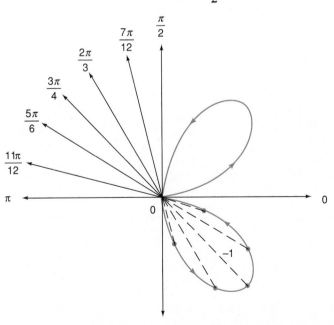

Rose curves are given by

$$r = a \sin n\theta$$

$$r = a \cos n\theta$$

There are n petals if n is odd and 2n petals if n is even.

By similar analysis we find that there is such a loop in each of the remaining quadrants. The complete graph has four loops, as shown in the following figure.

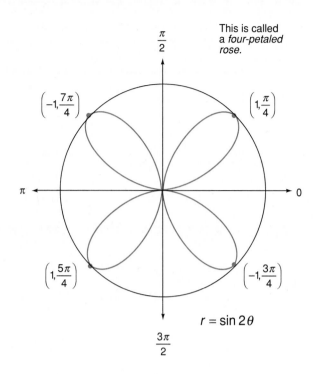

This is called a *four-petaled rose*.

$r = \sin 2\theta$

EXAMPLE 7 Graph the polar curve $r = \theta$ for $\theta \geq 0$.

Solution Form a table of values and note that as θ increases, so does r.

θ	0	$\dfrac{\pi}{4}$	$\dfrac{\pi}{2}$	π	$\dfrac{3\pi}{2}$	2π	$\dfrac{9\pi}{4}$
$r = \theta$	0	$\dfrac{\pi}{4}$	$\dfrac{\pi}{2}$	π	$\dfrac{3\pi}{2}$	2π	$\dfrac{9\pi}{4}$
(Approximations) $\longrightarrow$		.8	1.6	3.1	4.7	6.3	7.1

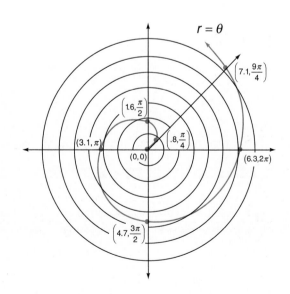

$r = \theta$

The curve is an endless, ever widening spiral called the **spiral of Archimedes.** ■

494   CHAPTER 9: Additional Applications of Trigonometry

EXERCISES 9.6

Sketch the graphs of the given polar equations in the same polar system.

1. $\theta = 30°, \theta = 210°, \theta = -150°, \theta = -30°$ 2. $\theta = \dfrac{2\pi}{3}, \theta = \dfrac{5\pi}{3}, \theta = -\dfrac{\pi}{4}, \theta = \dfrac{15\pi}{4}$

3. $\theta = 0, \theta = \dfrac{\pi}{2}, \theta = \pi, \theta = -\dfrac{\pi}{2}$ 4. $r = 1, r = \frac{3}{2}, r = 3$

5. $r = -2, r = 4, r = -4$

Write a polar equation of the line through the given points.

6. $(3, 150°), (2, -30°)$ 7. $(0, 0°), (10, -6°)$
8. $(1, -1), (-1, 1)$; (rectangular coordinates)   9. $(\sqrt{3}, 1), (-\sqrt{6}, -\sqrt{2})$; (rectangular coordinates)

Graph the polar equation and convert it into rectangular form.

10. $r = \cos\theta$ 11. $r = \sin\theta$ 12. $r = -3\cos\theta$
13. $r = 5\sin\theta$ 14. $r = -5\sin\theta$ 15. $r = -\frac{1}{3}\cos\theta$

Convert the polar equations into rectangular equations.

16. (a) $\theta = 45°$ (b) $\theta = \dfrac{\pi}{3}$ (c) $\theta = \dfrac{2\pi}{3}$ 17. (a) $r = 2$ (b) $r = -2$ (c) $r = \sqrt{5}$

Sketch each cardioid.

18. $r = 1 + \cos\theta$ 19. $r = 3 + 3\cos\theta$ 20. $r = 4 - 4\sin\theta$ 21. $r = 2(1 - \cos\theta)$

Graph the polar curves.

22. $r\cos\theta = 2$ 23. $r\sin\theta = -2$
24. $\tan\theta = 3$ 25. $\cot\theta = -3$
26. $r = \frac{1}{2}\theta, \theta \geq 0$ 27. $r = \theta, \theta \leq 0$
28. $r = 2\sin 2\theta$ 29. $r = \cos 2\theta$
   (four-petal rose)   (four-petal rose)
30. $r = \cos 3\theta$ 31. $r = 2\sin 3\theta$
   (three-petal rose)   (three-petal rose)
32. $r^2 = \sin 2\theta$ 33. $r^2 = 4\cos 2\theta$
   (lemniscate; two loops)   (lemniscate; two loops)
34. $r = 1 + 2\sin\theta$ 35. $r = 1 - 2\cos\theta$
   (limaçon; like a cardioid but with an inner loop)   (limaçon; like a cardioid but with an inner loop)

Convert the polar equation into rectangular form and identify the curve.

36. $r = \dfrac{1}{1 - \sin\theta}$ 37. $r = \dfrac{4}{1 + \cos\theta}$ 38. $r = 2\csc\theta$
39. $r = -5\sec\theta$ 40. $r = 2\cos\theta + 2\sin\theta$ 41. $r = 6\cos\theta + 8\sin\theta$
42. $r^2 = \dfrac{2}{\sin 2\theta}$ 43. $r(2\cos\theta + 3\sin\theta) = 3$ 44. $r(8\cos\theta - 4\sin\theta) = 20$
45. $r\cos\left(\theta - \dfrac{\pi}{3}\right) = \sqrt{3}$

46. (a) Show that the polar equation of the given circle is $r = 2a \sin \theta$, $a > 0$.

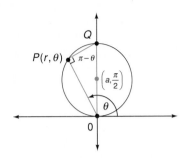 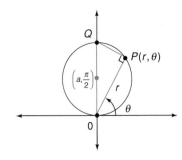

(b) Write the rectangular equation of the circle with center $(0, -a)$ and radius $a > 0$, and convert to the polar form $r = -2a \sin \theta$.

47. Since $x = r \cos \theta$, the equation of the vertical line $x = c$ becomes $r \cos \theta = c$ in polar form. Draw a line $x = c$ and use trigonometry to explain this result.

48. Since $y = r \sin \theta$, the equation of the horizontal line $y = c$ becomes $r \sin \theta = c$ in polar form. Draw a line $y = c$ and use trigonometry to explain this result.

Find the polar coordinates of the points of intersection of the given curves for the indicated interval of θ. (Note: A point of intersection must have the same polar coordinates for each curve.)

49. $r = 3 \cos \theta$, $r = 1 + \cos \theta$; $0 \le \theta < 2\pi$

50. $r = 1$, $r = 1 - \sin \theta$; $0 \le \theta \le 2\pi$

51. $r = \sin 2\theta$, $r = \cos \theta$, $0 \le \theta \le \pi$

52. $r = \sin \theta$, $r = \cos \theta$; $0 \le \theta \le \dfrac{\pi}{2}$

53. $r \cos \theta = \frac{1}{2}$, $2r \sin \theta = -\sqrt{3}$; $-\dfrac{\pi}{2} \le \theta \le 0$

 Written Assignment: The maximum distance that the polar curve $r = 4 \cos \theta$ is from the pole 0 is 4. Explain why this is so. Make a chart or list that gives the maximum distance that each of the following curves is from the pole, and include a θ value when this occurs in each case.

(a) $r = \pm 2a \cos \theta$ **(b)** $r = a(1 \pm \cos \theta)$ **(c)** $r = a \cos n\theta$
       $r = \pm 2a \sin \theta$ $r = a(1 \pm \sin \theta)$ $r = a \sin n\theta$

KEY TERMS

*Review these **key terms** so that you are able to define or describe them. A clear understanding of these terms will be very helpful when reviewing the developments of this chapter.*

Law of cosines • Heron's formula • Law of sines • Ambiguous case • Vector • Initial and terminal point • Parallelogram law • Resultant • Bearing • Heading • Component vectors • Complex plane • Real and imaginary axes • Parallelogram rule • Modulus • Principal argument • Trigonometric form • Rectangular form • De Moivres' theorem • nth root formula • nth roots of unity • Polar coordinate system • The pole and polar axis • Polar coordinates • Polar equation • Cardioid • Rose curve • Spiral of Archimedes

REVIEW EXERCISES

The solutions to the following exercises can be found within the text of Chapter 9.
Try to answer each question before referring to the text.

Section 9.1

1. Write the Law of Cosines for a triangle with sides a, b, c and angle measures α, β, γ.
2. Solve for c in $\triangle ABC$ if $a = 4$, $b = 7$, and $\gamma = 130°$.
3. Solve for α and β in $\triangle ABC$ given in Exercise 2.
4. An offshore lighthouse is 2 kilometers from a Coast Guard station C and 2.5 kilometers from a hospital H near the shoreline. If the angle formed by light beams to C and H is 143°, what is the distance CH between the Coast Guard station and the hospital?
5. Find the area of $\triangle ABC$ where $AB = 8$ centimeters, $AC = 12$ centimeters, and $\angle A = 110°$.
6. State Heron's formula for the area of a triangle with sides of measure a, b, and c.
7. Find the area of $\triangle ABC$ where $a = 10$ centimeters, $b = 9$ centimeters, and $c = 17$ centimeters, using Heron's formula.

Section 9.2

8. State the Law of Sines for a triangle with sides a, b, c and angle measures α, β, γ.
9. Solve $\triangle ABC$ if $\beta = 75°$, $a = 5$, and $\gamma = 41°$.
10. Solve for c in $\triangle ABC$ in which $\beta = 108°$, $\gamma = 39°$, and $b = 950$ meters.
11. Solve $\triangle ABC$ if $a = 50$, $b = 65$, and $\alpha = 57°$.
12. Approximate the remaining parts of $\triangle ABC$ if $a = 10$, $b = 13$, and $\alpha = 25°$.
13. A motorboat starts at the south shore of a river and is heading due north to the opposite shore. The boat's speed (in still water) is 15 mph, and the river is flowing due east at 4 mph. What is the actual speed of the boat, and what is the final bearing?
14. An airplane has an airspeed (the speed in still air) of 240 mph with a bearing of S 30° W. If a wind is blowing due west at 40 mph, find the final bearing of the plane and the ground speed (the speed relative to the ground).
15. A 400-pound packing crate is sitting on a ramp that makes a 25° angle with the horizontal. Find the force required to hold the crate from sliding down the ramp. (Assume that there is no friction.)

Section 9.3

For $z = 3 + 2i$ and $w = 2 - 4i$, evaluate the following using the parallelogram rule.
16. $z + w$ 17. $z + \bar{z}$ 18. $z - w$
19. What is the modulus of $z = x + yi$?

Convert the complex numbers into trigonometric form.
20. $\sqrt{3} + i$ 21. $-1 - i$
22. Use Table VI or a calculator to convert $3(\cos 50° + i \sin 50°)$ into rectangular form.
23. Use trigonometric forms to evaluate the product $(2 + 2\sqrt{3}i)(-1 - \sqrt{3}i)$. State the answer in both trigonometric and rectangular forms.
24. Use trigonometric forms to evaluate the quotient $\dfrac{2 + 2\sqrt{3}i}{-1 - \sqrt{3}i}$. State the answer in both trigonometric and rectangular forms.

Section 9.4

25. Complete this formula for De Moivre's theorem: $[r(\cos\theta + i\sin\theta)]^n = ?$

26. Use De Moivre's theorem to evaluate $(1 - i)^8$.

27. If $z = r(\cos\theta + i\sin\theta)$ is a nonzero complex number, write the form of the nth roots of z for $k = 0, 1, 2, \ldots, n - 1$.

28. What is the angular difference between two consecutive nth roots of a complex number?

29. Find the three cube roots of $8i$ and display them as points on an appropriate circle.

30. Solve the equation $z^4 + 16 = 0$.

31. Write the form of the nth roots of unity for $k = 0, 1, 2, \ldots n - 1$.

32. Find the 6th roots of unity and display them as points on an appropriate circle.

Section 9.5

33. Plot the points with these polar coordinates:

$$\left(1, \frac{\pi}{6}\right), \left(2, \frac{2\pi}{3}\right), \left(\frac{3}{2}, 225°\right), \left(\frac{2}{3}, 330°\right)$$

34. Plot the points with these polar coordinates:

$$\left(2, -\frac{\pi}{3}\right), \left(-2, -\frac{\pi}{3}\right)$$

35. Let P have polar coordinates $(2, 60°)$. Find the value of r in each case so that the resulting pair becomes a set of polar coordinates for P.

$$(r, 420°), (r, -300°), (r, 240°), (r, -120°)$$

36. Let P have polar coordinates $\left(3, \frac{\pi}{6}\right)$. Find polar coordinates of (r, θ) subject to these conditions:

    (a) $r > 0, -2\pi < \theta < 0$ (b) $r < 0, 0 < \theta < 2\pi$

37. Convert $\left(4, \frac{3\pi}{4}\right)$ and $\left(\frac{1}{2}, -\frac{\pi}{3}\right)$ into rectangular coordinates.

38. Convert the rectangular coordinates $(3\sqrt{3}, -3)$ and $(-5, -5)$ into polar coordinates so that $r \geq 0$ and $0 \leq \theta < 2\pi$.

Section 9.6

39. Write a polar equation for the line through points with rectangular coordinates $(-2, 2)$ and $(3, -3)$.

40. Write a polar equation of the circle with radius 4 and center at the pole.

Sketch the graphs of the polar equation.

41. $r = 2\cos\theta$ 42. $r = -4\cos\theta$ 43. $r = \frac{1}{2}\sin\theta$

44. How many times is the complete graph of $r = 2\cos\theta$ traced out as θ varies from $\theta = 0$ to $\theta = 2\pi$?

45. Convert the polar equation $r = 2a\cos\theta$ into a rectangular equation.

46. Show that $r = \dfrac{2}{1 - \sin\theta}$ is the equation of a parabola by converting to rectangular coordinates.

Sketch the graphs of the polar equations.

47. $r = 2(1 + \sin\theta)$ 48. $r = \sin 2\theta$ 49. $r = \theta$ for $\theta \geq 0$

Use these questions to test your knowledge of the basic skills and concepts of Chapter 9. Then check your answers with those given at the end of the book.

1. **(a)** Solve for a if $\angle A = 60°$, $b = 8$, and $c = 12$. Give $\angle A$ in radians
   (b) Write an expression for sin B in terms of other known parts of triangle ABC given in part (a).

2. Determine the number of triangles possible with the given parts.
   (a) $\alpha = 30°$, $a = 5.8$, $b = 12$ **(b)** $\alpha = 30°$, $a = 4.1$, $b = 8$

3. Find the exact area of $\triangle ABC$ where $\alpha = 60°$, $b = 20$ cm, and $c = 12$ cm, without using trigonometric tables or a calculator.

4. Solve for a in the figure.

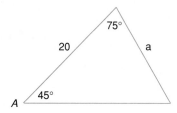

5. In $\triangle ABC$, $\alpha = 18°$, $b = 20$ centimeters, and $c = 34$ centimeters. Find the area of the triangle to the nearest square centimeter.

6. A triangle has sides that measure 7 centimeters, 9 centimeters, and 12 centimeters. Find the area to the nearest square centimeter using Heron's formula.

7. *ABCD* is a parallelogram. Solve for x and give the exact answer.

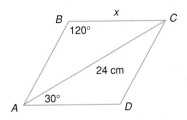

8. Triangle ABC has $a = 9$, $b = 13$, and $c = 6$. Find α and β to the nearest tenth of a degree.

9. Find the remaining parts β, γ, and c of all possible triangles ABC if $a = 10$, $b = 16$, and $\alpha = 30°$. (Round the answers to the nearest tenths.)

10. Find the force required to hold a 2,000-pound car from rolling down a ramp that makes a 9° angle with the horizontal. (Round the answer to the nearest pound.)

A boat that can travel at a speed of 18 mph in still water is heading in the direction N 40° W. A current is moving due east at 6 mph.

11. Find the boat's actual speed to the nearest mph.

12. Find the final bearing to the nearest degree.

13. Multiply as indicated and express the answer in rectangular form:

$$[5(\cos 250° + i \sin 250°)][\sqrt{2}(\cos 245° + i \sin 245°)]$$

For Exercises 14 and 15, let $z = \dfrac{\sqrt{3}}{4} + \dfrac{1}{4}i$ and $w = -1 + \sqrt{3}i$.

14. Convert z and w into trigonometric form and evaluate the quotient $\dfrac{z}{w}$.

15. Use De Moivre's theorem to evaluate z^5 and express the answer in rectangular form.

16. Find the three cube roots of 8 and locate them graphically.

17. Find the four fourth roots of $z = -8\sqrt{2} + 8\sqrt{2}i$ and express them in trigonometric form.

18. Point P has polar coordinates $(-3, 60°)$. Find the polar coordinates (r, θ) of P so that
    (a) $r < 0$ and $-360° < \theta < 0°$
    (b) $r > 0$ and $0 < \theta < 360°$

19. (a) Convert the polar coordinates $\left(\dfrac{1}{3}, -\dfrac{\pi}{6}\right)$ into rectangular coordinates.

    (b) Convert the rectangular coordinates $(-7\sqrt{2}, 7\sqrt{2})$ into the polar coordinates.

20. Write the polar equation of the circle with radius 3 and center having polar coordinates $(3, \pi)$.

Convert the polar equation into a rectangular equation and identify the curve.
21. $r = 2 \sin \theta$ 22. $r = 4 \sin \theta - 3 \cos \theta$

Sketch the graphs of the polar equations.
23. $2r \sin \theta = 6$ 24. $r = 2 + 2 \cos \theta$ 25. $r = 4 \sin 2\theta$

CHAPTER 9 TEST: MULTIPLE CHOICE

Do not use tables or a calculator for this test.

1. Which of the following formulas can be used to solve for side c in $\triangle ABC$ with angles A, B, and C and sides a, b, and c?
   I. $c^2 = a^2 + b^2 - 2ac \cos C$
   II. $c^2 = a^2 - b^2 - 2ab \cos C$
   III. $c^2 = a^2 - b^2 + 2ab \cos C$
   (a) Only I   (b) Only II   (c) Only III   (d) I, II, and III   (e) None of the preceding

2. In $\triangle ABC$, $a = 4$, $b = \sqrt{106}$, and $c = 5\sqrt{2}$. Using the Law of Cosines (without the aid of the trigonometric tables or a calculator), the measure of $\angle B$ is found to be
   (a) $-\dfrac{\pi}{4}$ (b) $\dfrac{\pi}{4}$ (c) $\dfrac{2\pi}{3}$ (d) $\dfrac{3\pi}{4}$ (e) None of the preceding

3. In $\triangle ABC$, $a = 8$ cm, $b = 8$ cm, and $c = 12$ cm. Using Heron's formula, the area of the triangle is which of the following?
   (a) $280\sqrt{2}$ cm^2 (b) $12\sqrt{7}$ cm^2 (c) $2\sqrt{7}$ cm^2 (d) $\frac{1}{2}(8)(12)(\sin 30°)$ cm^2 (e) None of the preceding

4. In $\triangle ABC$, $\angle A = 75°$, $\angle B = 40°$, and $b = 40$. Then side $c =$
   (a) $\dfrac{40 \sin 40°}{\sin 65°}$ (b) $\dfrac{40 \sin 65°}{\sin 40°}$ (c) $\dfrac{40 \sin 75°}{\sin 40°}$ (d) $\dfrac{40 \sin 40°}{\sin 75°}$ (e) None of the preceding

5. Determine the number of triangles possible having the parts $a = 4.2$, $b = 7$, and $\angle A = 30°$.
   (a) None   (b) One   (c) Two   (d) More than two   (e) None of the preceding

6. What is the force required to keep a 300-pound crate from sliding down a ramp that makes an angle of 15° with the horizontal?

   (a) 300 cos 75°   **(b)** 300 cos 15°   **(c)** $\dfrac{300}{\cos 15°}$ **(d)** $\dfrac{\cos 75°}{300}$ **(e)** None of the preceding

7. Convert $1 - i$ into trigonometric form:

   (a) $\sqrt{2}\left(\cos \dfrac{\pi}{4} + i \sin \dfrac{\pi}{4}\right)$ **(b)** $\sqrt{2} \cos \dfrac{7\pi}{4} + i \sin \dfrac{\pi}{4}$

   (c) $\sqrt{2}\left(\cos \dfrac{7\pi}{4} + i \sin \dfrac{7\pi}{4}\right)$ **(d)** $\sqrt{2} \cos \dfrac{\pi}{4} + i \sin \dfrac{\pi}{4}$ **(e)** None of the preceding

8. Let $z = 2(\cos 20° + i \sin 20°)$. Then $z^3 =$

   (a) $8(\cos^3 20° - i \sin^3 20°)$ **(b)** $3 + \dfrac{3\sqrt{3}}{2}i$ **(c)** $4\sqrt{3} + 4i$ **(d)** $4 + 4\sqrt{3}\,i$ **(e)** None of the preceding

9. Let $z = 8\left(\cos \dfrac{4\pi}{3} + i \sin \dfrac{4\pi}{3}\right)$ and $w = 24\left(\cos \dfrac{11\pi}{6} + i \sin \dfrac{11\pi}{6}\right)$. Then $\dfrac{z}{w} =$

   (a) $\dfrac{1}{3}\left(\cos \dfrac{\pi}{2} + i \sin \dfrac{\pi}{2}\right)$ **(b)** $3\left(\cos \dfrac{19\pi}{6} + i \sin \dfrac{19\pi}{6}\right)$

   (c) $\dfrac{1}{3}\left(\cos \dfrac{3\pi}{2} + i \sin \dfrac{3\pi}{2}\right)$ **(d)** $3\left(\cos\left(-\dfrac{\pi}{2}\right) + i \sin\left(-\dfrac{\pi}{2}\right)\right)$

   (e) None of the preceding

10. Convert the polar coordinates $\left(-\dfrac{1}{2}, \dfrac{\pi}{3}\right)$ to rectangular coordinates:

    (a) $\left(-\dfrac{1}{4}, -\dfrac{\sqrt{3}}{4}\right)$ **(b)** $\left(-\dfrac{1}{4}, \dfrac{\sqrt{3}}{4}\right)$ **(c)** $\left(-\dfrac{\sqrt{3}}{4}, -\dfrac{1}{4}\right)$ **(d)** $\left(-\dfrac{\sqrt{3}}{4}, \dfrac{1}{4}\right)$ **(e)** None of the preceding

11. Convert the rectangular coordinates $(-3, 3\sqrt{3})$ to polar coordinates (r, θ) so that $r \geq 0$ and $0 \leq \theta < 2\pi$.

    (a) $\left(6, \dfrac{5\pi}{3}\right)$ **(b)** $\left(6, \dfrac{2\pi}{3}\right)$ **(c)** $\left(-6, \dfrac{2\pi}{3}\right)$ **(d)** $\left(6, \dfrac{5\pi}{6}\right)$ **(e)** None of the preceding

12. Which of the following is a cube root of -64?
    (a) $-4(\cos \pi + i \sin \pi)$ **(b)** $2\sqrt{3} + 2i$ **(c)** $-2 + 2\sqrt{3}\,i$ **(d)** $2 - 2\sqrt{3}\,i$ **(e)** None of the preceding

13. Which of the following polar equations have circles as graphs?
    I. $r = -5$
    II. $r = -\frac{1}{2} \sin \theta$
    III. $r \cos \theta = 6$
    (a) Only I   **(b)** Only II   **(c)** Only I and II   **(d)** I, II, and III   **(e)** None of the preceding

14. The graph of which of the following is a cardioid?
    (a) $r = 3 - 3 \cos \theta$ **(b)** $r = 3 \sin 2\theta$ **(c)** $r = 2 \cos 3\theta$
    (d) $r \cos \theta + r \sin \theta = 1$ **(e)** None of the preceding

15. A rectangular form of the polar equation $r = -8 \cos \theta$ is
    (a) $x^2 + (y - 4)^2 = 16$ **(b)** $(x + 4)^2 + y^2 = 16$
    (c) $\sqrt{x^2 + y^2} = -8x$ **(d)** $x^2 + y^2 = 64 \cos^2 \theta$ **(e)** None of the preceding

Page 453

1. $2\sqrt{7}$ **2.** $5\sqrt{5}$ **3.** $\sqrt{399}$ **4.** $\dfrac{\pi}{3}$ **5.** $\dfrac{\pi}{4}$ **6.** $\dfrac{2\pi}{3}$

Page 461

1. $4\sqrt{2}$ **2.** $12\sqrt{3}$ **3.** $\frac{15}{2}\sqrt{6}$ **4.** $4\sqrt{2}$

Page 474

$z_1 = 2\sqrt{2}\left(\cos\dfrac{3\pi}{4} + i\sin\dfrac{3\pi}{4}\right)$ $z_1 z_2 = 12\sqrt{2}\left(\cos\dfrac{31\pi}{12} + i\sin\dfrac{31\pi}{12}\right)$

$z_2 = 6\left(\cos\dfrac{11\pi}{6} + i\sin\dfrac{11\pi}{6}\right)$ $= 12\sqrt{2}\left(\cos\dfrac{7\pi}{12} + i\sin\dfrac{7\pi}{12}\right)$

Page 477

1. $\dfrac{\sqrt{2}}{2} + \dfrac{\sqrt{2}}{2}i$ **2.** $\frac{1}{64}i$ **3.** $-\frac{1}{4}$ **4.** $512 + 512\sqrt{3}i$

Page 484

1.

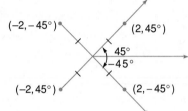

2.

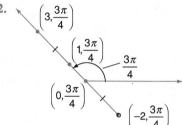

3. **(a)** $\left(5, \dfrac{25\pi}{12}\right)$ **(b)** $\left(-5, \dfrac{13\pi}{12}\right)$ **(c)** $\left(5, -\dfrac{23\pi}{12}\right)$ **(d)** $\left(-5, -\dfrac{11\pi}{12}\right)$

Page 491

1. $\theta = 30°$ **2.** $\theta = 140°$ **3.** $\theta = \dfrac{\pi}{2}$ **4.** $\theta = \dfrac{\pi}{4}$ **5.** $\theta = \dfrac{\pi}{2}$

6. $\theta = \dfrac{5\pi}{6}$ **7.** **(a)** $r = 4$ **(b)** $r = 5$ **(c)** $r = \sqrt{2}$

8. $r = 4\cos\theta$ **9.** $r = -3\cos\theta$ **10.** $r = -\frac{4}{3}\sin\theta$

LINEAR SYSTEMS, MATRICES, AND DETERMINANTS

10.1 SOLVING LINEAR SYSTEMS USING MATRICES

Linear systems of equations were solved in Section 2.5, using the substitution and multiplication-addition (or subtraction) methods. In this chapter three more procedures for solving such systems will be developed. The first, using matrices, is an efficient method that extends easily to systems having more than three variables.

We begin by forming the rectangular array of numbers, called a *matrix,* consisting of the coefficients and constants of the system. For example, the following linear system:

$$(1) \quad 2x + 5y + 8z = 11$$
$$(2) \quad x + 4y + 7z = 10$$
$$(3) \quad 3x + 6y + 12z = 15$$

is replaced by this corresponding matrix:

Coefficients

$$
\begin{array}{c}
 \quad x \quad y \quad z \\
 \quad \downarrow \quad \downarrow \quad \downarrow \\
\begin{array}{c}
\text{Row 1} \\
\text{Row 2} \\
\text{Row 3}
\end{array}
\left[
\begin{array}{ccc|c}
2 & 5 & 8 & 11 \\
1 & 4 & 7 & 10 \\
3 & 6 & 12 & 15
\end{array}
\right]
\end{array}
$$

The dashed vertical line serves as a reminder that the coefficients of the variables are to the left. The numbers to the right are the constants on the right-hand side of the equal signs in the given system.

Compare the steps in the two columns that follow. Note that the objective is to transform the given linear system into the form reached in step 5.

Working with the Equations	Working with the Matrices
Step 1 Write the system. $$2x + 5y + 8y = 11$$ $$x + 4y + 7z = 10$$ $$3x + 6y + 12z = 15$$	**Step 1** Write the corresponding matrix. $$\begin{bmatrix} 2 & 5 & 8 & 11 \\ 1 & 4 & 7 & 10 \\ 3 & 6 & 12 & 15 \end{bmatrix}$$
Step 2 Interchange the first two equations. $$x + 4y + 7z = 10$$ $$2x + 5y + 8z = 11$$ $$3x + 6y + 12z = 15$$	**Step 2** Interchange the first two rows. $$\begin{bmatrix} 1 & 4 & 7 & 10 \\ 2 & 5 & 8 & 11 \\ 3 & 6 & 12 & 15 \end{bmatrix}$$
Step 3 Add -2 times the first equation to the second, and -3 times the first to the third. $$x + 4y + 7z = 10$$ $$-3y - 6z = -9$$ $$-6y - 9z = -15$$	**Step 3** Add -2 times the first row to the second, and -3 times the first to the third. $$\begin{bmatrix} 1 & 4 & 7 & 10 \\ 0 & -3 & -6 & -9 \\ 0 & -6 & -9 & -15 \end{bmatrix}$$
Step 4 Multiply the second equation by $-\frac{1}{3}$. $$x + 4y + 7z = 10$$ $$y + 2z = 3$$ $$-6y - 9z = -15$$	**Step 4** Multiply row 2 by $-\frac{1}{3}$. $$\begin{bmatrix} 1 & 4 & 7 & 10 \\ 0 & 1 & 2 & 3 \\ 0 & -6 & -9 & -15 \end{bmatrix}$$
Step 5 Add 6 times the second equation to the third. $$x + 4y + 7z = 10$$ $$y + 2z = 3$$ $$3z = 3$$	**Step 5** Add 6 times row 2 to row 3. $$\begin{bmatrix} 1 & 4 & 7 & 10 \\ 0 & 1 & 2 & 3 \\ 0 & 0 & 3 & 3 \end{bmatrix}$$

Observe that the instructions stated in Steps 2, 3, 4, and 5 refer to what has been done to the preceding system or matrix.

The operations on the equations in the left column produced **equivalent systems** of equations. Equivalent systems have the same solutions. Since each of these systems has its corresponding matrix produced by comparable operations on the rows of the matrices, we say that the matrices are **row-equivalent**.

In step 5, we reached a row-equivalent matrix whose corresponding linear system is said to be in **triangular form**. From this form we solve for the variables using **back-substitution**. That is, we find z from the last equation, then substitute back into the second to find y, and finally substitute back into the first to find x.

When a different sequence of operations on the rows is used, it will most likely result in a different, but equivalent, triangular form.

$$x + 4y + 7z = 10$$
$$y + 2z = 3$$
$$3z = 3$$

$$3z = 3 \longrightarrow z = 1$$
$$y + 2(1) = 3 \longrightarrow y = 1$$
$$x + 4(1) + 7(1) = 10 \longrightarrow x = -1$$

The solution is the ordered triple $(-1, 1, 1)$.

Our new matrix method for solving a linear system has two major parts:

Part A: Use the following **fundamental row operations** to transform the initial matrix corresponding to the linear system into a row-equivalent matrix that translates into a linear system in triangular form.

1. Interchange two rows.
2. Multiply a row by a nonzero constant.
3. Add a multiple of a row to another row.

Sometimes, as in Example 2, an incomplete triangular form will be reached.

Part A can be completed by using a variety of row operations. The steps of a solution are therefore not unique, but the final solution will be the same.

Part B: Convert the matrix obtained in Part A into a linear system equivalent to the original system; solve for the variables by back-substitution.

EXAMPLE 1   Solve the linear system using row-equivalent matrices.

$$2x + 14y - 4z = -2$$
$$-4x - 3y + z = 8$$
$$3x - 5y + 6z = 7$$

Solution   Begin by writing the matrix corresponding to the linear system and apply the fundamental row operations.

$$\begin{bmatrix} 2 & 14 & -4 & \vdots & -2 \\ -4 & -3 & 1 & \vdots & 8 \\ 3 & -5 & 6 & \vdots & 7 \end{bmatrix}$$ This is the matrix for the given system.

*The explanations pointing to the rows state that row operations were applied to the rows of the **preceding matrix** to obtain the designated row.*

Getting a 1 in the circled position will make it easier to get zeros *below* this 1 in the next step using row operation (3).

$$\begin{bmatrix} ① & 7 & -2 & \vdots & -1 \\ -4 & -3 & 1 & \vdots & 8 \\ 3 & -5 & 6 & \vdots & 7 \end{bmatrix}$$ $\longleftarrow \frac{1}{2} \times (\text{row } 1)$

$$\begin{bmatrix} 1 & 7 & -2 & \vdots & -1 \\ 0 & 25 & -7 & \vdots & 4 \\ 0 & -26 & 12 & \vdots & 10 \end{bmatrix}$$ $\longleftarrow 4 \times (\text{row } 1) + \text{row } 2$
$\longleftarrow -3 \times (\text{row } 1) + \text{row } 3$

$$\begin{bmatrix} 1 & 7 & -2 & \vdots & -1 \\ 0 & 25 & -7 & \vdots & 4 \\ 0 & -1 & 5 & \vdots & 14 \end{bmatrix}$$ $\longleftarrow \text{Row } 2 + \text{row } 3$

Getting the −1 in the circled position will make it easier to get zero *below* this −1 in the next step.

$$\begin{bmatrix} 1 & 7 & -2 & \vdots & -1 \\ 0 & -① & 5 & \vdots & 14 \\ 0 & 25 & -7 & \vdots & 4 \end{bmatrix}$$ $\longleftarrow \text{Interchange rows } 2$
$\longleftarrow \text{and } 3.$

$$\begin{bmatrix} 1 & 7 & -2 & \vdots & -1 \\ 0 & -1 & 5 & \vdots & 14 \\ 0 & 0 & 118 & \vdots & 354 \end{bmatrix}$$ $\longleftarrow 25 \times (\text{row } 2) + \text{row } 3$

Now, convert to the corresponding linear system as follows:

Solve for the variables using back-substitution.

$$\begin{aligned} x + 7y - 2z &= -1 \\ -y + 5z &= 14 \\ 118z &= 354 \end{aligned}$$

$$118z = 354 \longrightarrow z = 3$$
$$-y + 5(3) = 14 \longrightarrow y = 1$$
$$x + 7(1) - 2(3) = -1 \longrightarrow x = -2$$

Thus the solution is the ordered triple $(-2, 1, 3)$. ∎

This matrix procedure also reveals when a linear system has no solutions, that is, when it is an inconsistent system. In such a case we will obtain a row in a matrix of the form

$$0 \quad 0 \cdots 0 \; \vdots \; p$$

where $p \neq 0$. But when this row is converted to an equation, we get the false statement $0 = p$. For example, solving the system

$$3x - 6y = 9$$
$$-2x + 4y = -8$$

we have

$$\begin{bmatrix} 3 & -6 & \vdots & 9 \\ -2 & 4 & \vdots & -8 \end{bmatrix}$$

$$\begin{bmatrix} 1 & -2 & \vdots & 3 \\ -2 & 4 & \vdots & -8 \end{bmatrix} \longleftarrow \tfrac{1}{3} \times (\text{row 1})$$

$$\begin{bmatrix} 1 & -2 & \vdots & 3 \\ 0 & 0 & \vdots & -2 \end{bmatrix} \longleftarrow 2 \times (\text{row 1}) + \text{row 2}$$

The last row gives the false equation $0 = -2$. Therefore, the system is inconsistent; there are no solutions.

The next example demonstrates how the matrix method can be used to solve a system that has infinitely many solutions, that is, a dependent system.

EXAMPLE 2 Solve the system

$$\begin{aligned} x + 2y - z &= 1 \\ 2x - y + 3z &= 4 \\ 5x + 5z &= 9 \end{aligned}$$

CAUTION
Note that there is no y-term in the third equation. You should think of this as 0y and record the 0 in the y-position in the third row of the matrix as shown.

Solution

$$\begin{bmatrix} 1 & 2 & -1 & \vdots & 1 \\ 2 & -1 & 3 & \vdots & 4 \\ 5 & 0 & 5 & \vdots & 9 \end{bmatrix}$$

$$\begin{bmatrix} 1 & 2 & -1 & \vdots & 1 \\ 0 & -5 & 5 & \vdots & 2 \\ 0 & -10 & 10 & \vdots & 4 \end{bmatrix} \quad \begin{array}{l} \longleftarrow -2 \times (\text{row 1}) + \text{row 2} \\ \longleftarrow -5 \times (\text{row 1}) + \text{row 3} \end{array}$$

$$\begin{bmatrix} 1 & 2 & -1 & \vdots & 1 \\ 0 & -5 & 5 & \vdots & 2 \\ 0 & 0 & 0 & \vdots & 0 \end{bmatrix} \quad \longleftarrow -2 \times (\text{row 2}) + \text{row 3}$$

Since the last row contains all zeros, we have an equivalent linear system of two equations in three variables.

$$\begin{aligned} x + 2y - z &= 1 \\ -5y + 5z &= 2 \end{aligned}$$

This system has an *incomplete* triangular form.

Again we use back-substitution. First use the last equation to solve for y in terms of z to get $y = -\frac{2}{5} + z$. Now let $z = c$ represent any number, giving $y = -\frac{2}{5} + c$, and substitute back into the first equation.

$$x + 2y - z = x + 2(-\tfrac{2}{5} + c) - c = 1 \quad \longrightarrow \quad x = \tfrac{9}{5} - c$$

The solutions are

Find the specific solutions of the system for the values $c = 0$, $c = 1$, $c = \frac{2}{5}$, and $c = -2$.

$$(\tfrac{9}{5} - c, \ -\tfrac{2}{5} + c, \ c) \qquad \text{for any number } c$$

This result can be checked in the original system as follows.

$$\begin{aligned} x + 2y - \ z &= \tfrac{9}{5} - c + 2(-\tfrac{2}{5} + c) - c = 1 \\ 2x - \ y + 3z &= 2(\tfrac{9}{5} - c) - (-\tfrac{2}{5} + c) + 3c = 4 \\ 5x \quad\ + 5z &= 5(\tfrac{9}{5} - c) + 5c = 9 \end{aligned}$$

$\blacksquare$

The solutions for the system in Example 2 are stated in terms of $z = c$, where c (or z) represents any number. There are other ways in which these solutions can be stated. For example, when c is replaced by z in the given solutions we have

$$x = \tfrac{9}{5} - z \quad \text{or} \quad z = \tfrac{9}{5} - x \quad \text{and} \quad y = -\tfrac{2}{5} + z$$

Now let $x = d$ be any number. Then

$$z = \tfrac{9}{5} - x = \tfrac{9}{5} - d$$

and

$$y = -\tfrac{2}{5} + z = -\tfrac{2}{5} + (\tfrac{9}{5} - d) = \tfrac{7}{5} - d$$

The solutions can be stated in this form:

You should also be able to obtain the solutions in this form
$(\frac{7}{5} - e, e, \frac{2}{5} + e)$
for any number e.

$$\left(d, \tfrac{7}{5} - d, \tfrac{9}{5} - d\right) \qquad \text{for any number } d$$

When using the matrix method, keep the following observations in mind.

1. If you reach a linear system in triangular form, as in Example 1, the system has a unique solution that can be found by back-substitution.

2. If you reach a linear system that has an incomplete triangular form, and there are no false equations, as in Example 2, then the system has many solutions that can be found by back-substitution.

3. If you reach a linear system having a false equation, then the system has no solutions.

EXERCISES 10.1

Use matrices and fundamental row operations to solve each system.

1. $\begin{aligned} x + 5y &= -9 \\ 4x - 3y &= -13 \end{aligned}$ 2. $\begin{aligned} 4x - y &= 6 \\ 2x + 3y &= 10 \end{aligned}$ 3. $\begin{aligned} 3x + 2y &= 18 \\ 6x + 5y &= 45 \end{aligned}$ 4. $\begin{aligned} 2x + 4y &= 24 \\ -3x + 5y &= -25 \end{aligned}$

5. $\begin{aligned} 4x - 5y &= -2 \\ 16x + 2y &= 3 \end{aligned}$ 6. $\begin{aligned} x &= y - 7 \\ 3y &= 2x + 16 \end{aligned}$ 7. $\begin{aligned} 2x &= -8y + 2 \\ 4y &= x - 1 \end{aligned}$ 8. $\begin{aligned} 2x - 5y &= 4 \\ -10x + 25y &= -20 \end{aligned}$

9. $\begin{aligned} 30x + 45y &= 60 \\ 4x + 6y &= 8 \end{aligned}$ 10. $\begin{aligned} x - y &= 3 \\ -\tfrac{1}{3}x + \tfrac{1}{3}y &= 1 \end{aligned}$ 11. $\begin{aligned} 2x &= 8 - 3y \\ 6x + 9y &= 14 \end{aligned}$ 12. $\begin{aligned} -10x + 5y &= 8 \\ 15x - 10y &= -4 \end{aligned}$

13. $\begin{aligned} x - 2y + 3z &= -2 \\ -4x + 10y + 2z &= -2 \\ 3x + y + 10z &= 7 \end{aligned}$ 14. $\begin{aligned} 2x + 4y + 8z &= 14 \\ 4x - 2y + 2z &= 6 \\ -5x + 3y - z &= -4 \end{aligned}$ 15. $\begin{aligned} -x + 2y + 3z &= 11 \\ 2x - 3y &= -6 \\ 3x - 3y + 3z &= 3 \end{aligned}$

16. $\begin{aligned} -2x + y + 2z &= 14 \\ 5x + z &= -10 \\ x - 2y - 3z &= -14 \end{aligned}$ 17. $\begin{aligned} x - 2z &= 5 \\ 3y + 4z &= -2 \\ -2x + 3y + 8z &= 4 \end{aligned}$ 18. $\begin{aligned} 2x + y &= 3 \\ 4x + 5z &= 6 \\ -2y + 5z &= -4 \end{aligned}$

19. $\begin{aligned} x - 2y + z &= 1 \\ -6x + y + 2z &= -2 \\ -4x - 3y + 4z &= 0 \end{aligned}$ 20. $\begin{aligned} 4x - 3y + z &= 0 \\ -3x + y + 2z &= 0 \\ -2x - y + 5z &= 0 \end{aligned}$

21. $\begin{aligned} w - x + 2y + 2z &= 0 \\ 2w - y - 3z &= 0 \\ 4x - 3y + z &= -2 \\ -3w + 2x + 4z &= 1 \end{aligned}$ 22. $\begin{aligned} 4w - 5x + 2z &= 0 \\ -2w + 10x + y - 3z &= 8 \\ 5x - 2y + 4z &= -16 \\ 6w + 3z &= 0 \end{aligned}$

23. $\begin{aligned} v + 2w - x + 3z &= 4 \\ -w + 5x - y - z &= 3 \\ 3v + 6x + 2z &= 1 \\ 2v + 3w + 3x - y + 5z &= 10 \\ v + 9x - 2y + z &= 5 \end{aligned}$ 24. $\begin{aligned} w + 3y &= 10 \\ -x + 2y + 6z &= 2 \\ -3w - 2x - 4z &= 2 \\ 4x - y &= 8 \end{aligned}$

Written Assignment: Continue using row operations after Step 5 on page 504 and reach the form

$$\begin{bmatrix} 1 & 0 & 0 & \vdots & -1 \\ 0 & 1 & 0 & \vdots & 1 \\ 0 & 0 & 1 & \vdots & 1 \end{bmatrix}$$

Describe the steps you have used to obtain this form, and explain why nothing more needs to be done.

The matrix method for solving linear systems of Section 10.1 uses row operations to transform a matrix into a row-equivalent matrix. That method, however, does not combine two or more matrices in any way. In this section we learn how to add and multiply matrices and also learn that with these operations matrices have many, but not all, of the basic properties of the real numbers stated in Section 1.1.

A matrix A with m rows and n columns is said to have dimensions m and n and can be written as

$$A = \begin{bmatrix} a_{11} & a_{12} & \cdots & a_{1n} \\ a_{21} & a_{22} & \cdots & a_{2n} \\ \vdots & \vdots & \vdots & \vdots \\ a_{m1} & a_{m2} & \cdots & a_{mn} \end{bmatrix} \begin{matrix} \longleftarrow \text{1st row} \\ \longleftarrow \text{2nd row} \\ \\ \longleftarrow \text{mth row} \end{matrix}$$

$$\begin{matrix} \uparrow & \uparrow & & \uparrow \\ \text{1st} & \text{2nd} & & \text{nth} \\ \text{column} & \text{column} & & \text{column} \end{matrix}$$

Instead of the symbol A we sometimes use $A_{m \times n}$ to emphasize the dimensions of A, in which $m \times n$ means m rows by n columns. The preceding rectangular display can be abbreviated by writing

$$A_{m \times n} = [a_{ij}] \quad \longleftarrow \quad a_{ij} \text{ is the element in the } i\text{th row and } j\text{th column}$$

in which a_{ij} represents *each* element in the matrix for $1 \le i \le m$ and $1 \le j \le n$.

Equality of Matrices

Two m by n matrices are equal if and only if they have precisely the same elements in the same positions. In symbols using $A = [a_{ij}]$ and $B = [b_{ij}]$, we have:

$$A = B \text{ if and only if } a_{ij} = b_{ij} \text{ for each } i \text{ and } j$$

To add two m by n matrices, we add corresponding elements. For example, here is the computation for the sum of two 3 by 2 matrices.

$$\begin{bmatrix} 2 & -5 \\ 3 & 7 \\ -1 & 0 \end{bmatrix} + \begin{bmatrix} -4 & 1 \\ 0 & 9 \\ 2 & -2 \end{bmatrix} = \begin{bmatrix} 2 + (-4) & -5 + 1 \\ 3 + 0 & 7 + 9 \\ -1 + 2 & 0 + (-2) \end{bmatrix} = \begin{bmatrix} -2 & -4 \\ 3 & 16 \\ 1 & -2 \end{bmatrix}$$

Addition of Matrices

In general, the sum of $A_{m \times n} = [a_{ij}]$ and $B_{m \times n} = [b_{ij}]$ can be expressed as

$$A + B = [a_{ij}] + [b_{ij}] = [a_{ij} + b_{ij}] \text{ for each } i \text{ and } j$$

It is important to note that matrices can be added only when they have the same dimensions. Thus

$$\begin{bmatrix} 1 & 2 & 3 \\ -2 & 0 & 6 \end{bmatrix} \quad \text{and} \quad \begin{bmatrix} 6 & -5 \\ 3 & 3 \end{bmatrix}$$

cannot be added.

Matrix addition is both commutative and associative. That is, for matrices having the same dimensions:

$$A + B = B + A$$

$$(A + B) + C = A + (B + C)$$

We will not present formal proofs of these and other properties. However, you can see that they make sense by verifying the properties for specific examples.

EXAMPLE 1 Let $A = \begin{bmatrix} 3 & 1 & -2 \\ -1 & 3 & 0 \end{bmatrix}$, $B = \begin{bmatrix} -5 & 5 & 6 \\ -2 & -2 & -2 \end{bmatrix}$, and $C = \begin{bmatrix} 0 & 1 & 0 \\ 1 & 0 & 1 \end{bmatrix}$. Show that $(A + B) + C = A + (B + C)$.

Solution

$$(A + B) + C = \left(\begin{bmatrix} 3 & 1 & -2 \\ -1 & 3 & 0 \end{bmatrix} + \begin{bmatrix} -5 & 5 & 6 \\ -2 & -2 & -2 \end{bmatrix} \right) + \begin{bmatrix} 0 & 1 & 0 \\ 1 & 0 & 1 \end{bmatrix}$$

$$= \begin{bmatrix} -2 & 6 & 4 \\ -3 & 1 & -2 \end{bmatrix} + \begin{bmatrix} 0 & 1 & 0 \\ 1 & 0 & 1 \end{bmatrix} = \begin{bmatrix} -2 & 7 & 4 \\ -2 & 1 & -1 \end{bmatrix}$$

$$A + (B + C) = \begin{bmatrix} 3 & 1 & -2 \\ -1 & 3 & 0 \end{bmatrix} + \left(\begin{bmatrix} -5 & 5 & 6 \\ -2 & -2 & -2 \end{bmatrix} + \begin{bmatrix} 0 & 1 & 0 \\ 1 & 0 & 1 \end{bmatrix} \right)$$

$$= \begin{bmatrix} 3 & 1 & -2 \\ -1 & 3 & 0 \end{bmatrix} + \begin{bmatrix} -5 & 6 & 6 \\ -1 & -2 & -1 \end{bmatrix} = \begin{bmatrix} -2 & 7 & 4 \\ -2 & 1 & -1 \end{bmatrix}$$

Thus

You should also verify that
$A + B = B + A$.

$$(A + B) + C = A + (B + C)$$ ∎

The m by n zero matrix, denoted by O, is the matrix all of whose entries are 0. For example, the 2 by 3 zero matrix is

$$O_{2\times3} = \begin{bmatrix} 0 & 0 & 0 \\ 0 & 0 & 0 \end{bmatrix}$$

The next rule is easy to verify:

$$A_{m\times n} + O_{m\times n} = A_{m\times n}$$

Negative of a Matrix

The negative of an m by n matrix $A = [a_{ij}]$ is the matrix denoted by $-A$ and is defined in this way:

$$-A = [-a_{ij}]$$

Additive Inverse Property

In other words, the negative of a matrix is formed by replacing each number in A by its opposite. Consequently, since $a_{ij} + (-a_{ij}) = 0$,

$$A + (-A) = O$$

For example, if $A = \begin{bmatrix} 8 & -5 \\ -4 & 9 \\ 0 & 2 \end{bmatrix}$, then $-A = \begin{bmatrix} -8 & 5 \\ 4 & -9 \\ 0 & -2 \end{bmatrix}$ and

$$A + (-A) = \begin{bmatrix} 8 + (-8) & -5 + 5 \\ -4 + 4 & 9 + (-9) \\ 0 + 0 & 2 + (-2) \end{bmatrix} = \begin{bmatrix} 0 & 0 \\ 0 & 0 \\ 0 & 0 \end{bmatrix} = O$$

Subtraction of Matrices

Subtraction of m by n matrices can be defined using the negative of a matrix.

$$A - B = A + (-B)$$

Scalar Multiplication

A matrix A can be multiplied by a real number c by multiplying each element in A by c. This is referred to as **scalar multiplication**.

$$cA = c[a_{ij}] = [ca_{ij}]$$

It is customary to write the scalar c to the left of the matrix, thus we do not use the notation Ac for scalar multiplication.

For example,

$$7\begin{bmatrix} 8 & -5 \\ -4 & 9 \\ 0 & 2 \end{bmatrix} = \begin{bmatrix} 56 & -35 \\ -28 & 63 \\ 0 & 14 \end{bmatrix}$$

Using this property, the negative of a matrix A may be written as $-A = -1A$.

Here is a list of scalar multiplication properties:

$$(cd)A = c(dA)$$
$$(c + d)A = cA + dA$$
$$c(A + B) = cA + cB$$

A, B have the same dimension and c, d are real numbers.

EXAMPLE 2 Verify $(cd)A = c(dA)$ using

$$c = -2, \quad d = \tfrac{1}{3}, \quad \text{and} \quad A = \begin{bmatrix} 6 & -9 \\ 1 & 5 \end{bmatrix}.$$

Solution

$$(cd)A = (-2 \cdot \tfrac{1}{3})\begin{bmatrix} 6 & -9 \\ 1 & 5 \end{bmatrix} = -\tfrac{2}{3}\begin{bmatrix} 6 & -9 \\ 1 & 5 \end{bmatrix} = \begin{bmatrix} -4 & 6 \\ -\tfrac{2}{3} & -\tfrac{10}{3} \end{bmatrix}$$

$$c(dA) = -2\left(\tfrac{1}{3}\begin{bmatrix} 6 & -9 \\ 1 & 5 \end{bmatrix}\right) = -2\begin{bmatrix} 2 & -3 \\ \tfrac{1}{3} & \tfrac{5}{3} \end{bmatrix} = \begin{bmatrix} -4 & 6 \\ -\tfrac{2}{3} & -\tfrac{10}{3} \end{bmatrix} \quad \blacksquare$$

TEST YOUR UNDERSTANDING
Think Carefully

Let $A = \begin{bmatrix} 1 & 4 & -3 \\ 2 & 5 & 0 \\ 0 & 1 & 1 \end{bmatrix}$, $B = \begin{bmatrix} -2 & 2 \\ 3 & -3 \\ 1 & 4 \end{bmatrix}$, $C = \begin{bmatrix} 5 & 1 \\ -1 & 6 \\ 8 & 2 \end{bmatrix}$

Evaluate each of the following, if possible.

1. $A + B$ 2. $B + C$ 3. $C + (B + C)$

4. $B - C$ 5. $C - B$ 6. $2A + 0A$

(Answers: Page 557)   7. $2A + 3C$ 8. $(-5)(B + C)$ 9. $4((-\tfrac{1}{2})C)$

Multiplication of matrices is a more complicated operation than addition. First, let us consider the product of

$$A_{2\times 3} = \begin{bmatrix} -4 & -1 & 2 \\ 3 & 8 & 1 \end{bmatrix} \quad \text{and} \quad B_{3\times 2} = \begin{bmatrix} -2 & 1 \\ 7 & 0 \\ 4 & -6 \end{bmatrix}$$

Here is the completed product followed by the explanation of how it was done.

$$\overset{A}{\begin{bmatrix} -4 & -1 & 2 \\ 3 & 8 & 1 \end{bmatrix}}\overset{B}{\begin{bmatrix} -2 & 1 \\ 7 & 0 \\ 4 & -6 \end{bmatrix}} = \overset{C}{\begin{bmatrix} 9 & -16 \\ 54 & -3 \end{bmatrix}} = \overset{C}{\begin{bmatrix} c_{11} & c_{12} \\ c_{21} & c_{22} \end{bmatrix}}$$

Row 1 of A and *column 1 of B* produce 9, the element in row 1, column 1 of *C*:

$$(-4)(-2) + (-1)(7) + (2)(4) = 9 = c_{11}$$

Row 1 of A and *column 2 of B* produce -16, the element in row 1, column 2 of *C*:

$$(-4)(1) + (-1)(0) + (2)(-6) = -16 = c_{12}$$

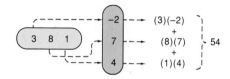

Row 2 of A and *column 1 of B* produce 54, the element in row 2, column 1 of *C*:

$$(3)(-2) + (8)(7) + (1)(4) = 54 = c_{21}$$

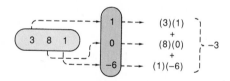

Row 2 of A and *column 2 of B* produce -3, the element in row 2, column 2 of *C*:

$$(3)(1) + (8)(0) + (1)(-6) = -3 = c_{22}$$

As you can observe in the preceding computations, an element in $AB = C$ is the sum of the products of numbers in a row of A by the numbers in a column B. Furthermore, the *i*th row of A together with the *j*th column of B produces the element in row *i*, column *j* of C.

Here is a diagram that displays how the row 2, column 3 element in a product is completed.

$$\begin{bmatrix} 2 & 1 \\ -6 & 3 \\ 1 & 0 \end{bmatrix} \quad \begin{bmatrix} 0 & -2 & 2 & 1 \\ 5 & 0 & 7 & 6 \end{bmatrix} = \begin{bmatrix} 5 & -4 & 11 & 8 \\ 15 & 12 & 9 & 12 \\ 0 & -2 & 2 & 1 \end{bmatrix}$$

Row 2   Column 3   Row 2, column 3 element

$\begin{matrix} 2 & \to & (-6)(2) \\ -6 & 3 & \\ 7 & \to & (3)(7) \end{matrix}$ add $\longrightarrow 9$

In each of the matrix products that have been found, you can see that it is possible to compute product AB only when *the number of columns of A equals the number of rows of B*. Furthermore, if A is m by n and B is n by p, then AB is m by p. For example,

$$(A_{3\times5})(B_{5\times7}) = C_{3\times7}$$

These must · number · number of
be the same. · of rows · columns
· in A · in B

TEST YOUR UNDERSTANDING Think Carefully

Let $A = \begin{bmatrix} 1 & 2 \\ 3 & -5 \end{bmatrix}$, $B = \begin{bmatrix} 6 & -4 & 9 \\ 1 & 2 & 0 \end{bmatrix}$, $C = \begin{bmatrix} 3 & 7 \\ -4 & 6 \\ 2 & -9 \end{bmatrix}$, $D = \begin{bmatrix} 8 & -8 \\ -7 & -7 \end{bmatrix}$

Find each product if possible.

1. AD 2. AB 3. AC 4. CA
5. BD 6. BC 7. CB 8. $(AB)C$

(Answers: Page 557)

Using A and D from the preceding set of exercises, we have

$$AD = \begin{bmatrix} 1 & 2 \\ 3 & -5 \end{bmatrix}\begin{bmatrix} 8 & -8 \\ -7 & -7 \end{bmatrix} = \begin{bmatrix} -6 & -22 \\ 59 & 11 \end{bmatrix}$$

and

$$DA = \begin{bmatrix} 8 & -8 \\ -7 & -7 \end{bmatrix}\begin{bmatrix} 1 & 2 \\ 3 & -5 \end{bmatrix} = \begin{bmatrix} -16 & 56 \\ -28 & 21 \end{bmatrix}$$

Therefore, $AD \neq DA$, which proves that matrix multiplication is not commutative. We do have the following properties, though, assuming that the operations are possible.

Associative Property

Distributive Properties

$$(AB)C = A(BC)$$
$$A(B + C) = AB + AC$$
$$(B + C)A = BA + CA$$

In the exercises you will be asked to verify these properties for specific cases.

Scalar multiplication can also be combined with the product of matrices according to this result, in which c represents any real number.

$$c(AB) = (cA)B = A(cB)$$

There is also a multiplicative identity for **square matrices**, matrices that are n by n. This is the matrix whose main diagonal consists of 1's and all other entries are 0. We use I_n for this matrix.

In particular, when $n = 3$,

$$I_3 = \begin{bmatrix} 1 & 0 & 0 \\ 0 & 1 & 0 \\ 0 & 0 & 1 \end{bmatrix} \longleftarrow \text{ main diagonal}$$

Using $A = \begin{bmatrix} a & b & c \\ d & e & f \\ g & h & i \end{bmatrix}$, we have

$$I_3 A = \begin{bmatrix} 1 & 0 & 0 \\ 0 & 1 & 0 \\ 0 & 0 & 1 \end{bmatrix}\begin{bmatrix} a & b & c \\ d & e & f \\ g & h & i \end{bmatrix} = \begin{bmatrix} a & b & c \\ d & e & f \\ g & h & i \end{bmatrix} = A$$

Similarly, $AI_3 = A$. In general, if A is n by n, then

Multiplication Identity Property

$$AI_n = I_n A = A$$

I_n is the multiplicative identity for the n by n matrices. It is also called the **nth-order identity matrix**.

EXERCISES 10.2

Evaluate the given matrix expression, if possible, using these matrices.

$$A = \begin{bmatrix} 1 \\ -2 \end{bmatrix} \quad B = [3 \;\; 5] \quad C = \begin{bmatrix} 5 & -1 \\ -3 & 4 \end{bmatrix} \quad D = \begin{bmatrix} 1 & 2 \\ 3 & 4 \end{bmatrix} \quad E = \begin{bmatrix} 1 & -3 & 2 \\ -5 & 2 & 0 \\ 0 & -1 & 3 \end{bmatrix}$$

$$F = \begin{bmatrix} 1 & 1 & 1 \\ -1 & 1 & 1 \\ -1 & -1 & 1 \end{bmatrix} \quad G = \begin{bmatrix} 1 & -4 & 0 \\ 2 & -3 & 5 \end{bmatrix} \quad H = \begin{bmatrix} 5 & 2 & -3 & 0 \\ 1 & 0 & 3 & -1 \\ 4 & -2 & 0 & 1 \end{bmatrix} \quad J = \begin{bmatrix} 1 & 0 \\ 0 & -1 \\ 1 & 1 \\ 0 & 0 \end{bmatrix}$$

1. $C + D$
2. $D - C$
3. $A + B$
4. $C + G$
5. $E + 2F$
6. $-2E + 5E$
7. BA
8. AB
9. CD
10. DC
11. CA
12. BC
13. GE
14. EG
15. GJ
16. EF
17. HJ
18. $(DA)C$

19. GH
20. $(\frac{1}{2}D)J$
21. $F\begin{bmatrix} 0 \\ 0 \\ 0 \end{bmatrix}$
22. $\begin{bmatrix} 0 \\ 0 \\ 0 \end{bmatrix}F$
23. $I_3 H$
24. EI_3

Verify the matrix equation using these matrices.

$$A = \begin{bmatrix} 2 & 1 \\ -4 & 1 \end{bmatrix} \quad B = \begin{bmatrix} 0 & 6 \\ 5 & -4 \end{bmatrix} \quad C = \begin{bmatrix} 3 & 7 \\ -1 & -2 \end{bmatrix}$$

25. $B + C = C + B$
26. $3(AB) = (3A)B = A(3B)$

27. $A + (B + C) = (A + B) + C$

28. $A(B + C) = AB + AC$

29. $(A + B)C = AC + BC$

30. $(AB)C = A(BC)$

31. Verify that $(AB)C = A(BC)$ using

$$A = \begin{bmatrix} 1 & -2 \\ 3 & 3 \\ 0 & 4 \end{bmatrix} \qquad B = \begin{bmatrix} -1 & 6 & 8 \\ 2 & 1 & 3 \end{bmatrix} \qquad C = \begin{bmatrix} 5 \\ -2 \\ 1 \end{bmatrix}$$

32. Arrange the four matrices with the indicated dimensions so that a product of all four can be formed. What will be the dimensions of this product?

$$A_{4\times 1}, \quad B_{3\times 5}, \quad C_{1\times 3}, \quad D_{2\times 4}$$

33. Let $E = \begin{bmatrix} 0 & 1 & 0 \\ 1 & 0 & 0 \\ 0 & 0 & 1 \end{bmatrix}$ and $A = \begin{bmatrix} a & b & c \\ d & e & f \\ g & h & i \end{bmatrix}$

   (a) Evaluate EA.

   (b) Compare the answer in part (a) to matrix A and find a matrix F so that FA is the same as A except that the first and third rows are interchanged.

34. **(a)** Use E and A in Exercise 33 and evaluate AE.

   (b) Find a matrix F so that AF is the same as A except that the second and third columns are interchanged.

35. Find matrices E and F so that for matrix A in Exercise 33,

$$EA = \begin{bmatrix} a & b & c \\ d & e & f \\ g + 3a & h + 3b & i + 3c \end{bmatrix} \quad \text{and} \quad AF = \begin{bmatrix} a & b - c & c \\ d & e - f & f \\ g & h - i & i \end{bmatrix}$$

36. Find matrices E and F so that for matrix A in Exercise 33,

$$EA = \begin{bmatrix} 2a & 2b & 2c \\ d & e & f \\ g & h & i \end{bmatrix} \quad \text{and} \quad AF = \begin{bmatrix} a & 2b & c \\ d & 2e & f \\ g & 2h & i \end{bmatrix}$$

37. **(a)** Find x, y so that

$$\begin{bmatrix} 1 & 1 \\ 0 & 0 \end{bmatrix} \begin{bmatrix} 2 & 3 \\ x & y \end{bmatrix} = \begin{bmatrix} 0 & 0 \\ 0 & 0 \end{bmatrix}$$

   (b) Explain why it is not possible to find x, y so that

$$\begin{bmatrix} 2 & 3 \\ x & y \end{bmatrix} \begin{bmatrix} 1 & 1 \\ 0 & 0 \end{bmatrix} = \begin{bmatrix} 0 & 0 \\ 0 & 0 \end{bmatrix}$$

38. Find $a, b, c,$ and d so that $\begin{bmatrix} 1 & 1 \\ 0 & 1 \end{bmatrix} \begin{bmatrix} a & b \\ c & d \end{bmatrix} = I_2$ and verify that $\begin{bmatrix} a & b \\ c & d \end{bmatrix} \begin{bmatrix} 1 & 1 \\ 0 & 1 \end{bmatrix} = I_2$.

39. **(a)** Let $A = \begin{bmatrix} x & 0 & 0 \\ 0 & y & 0 \\ 0 & 0 & z \end{bmatrix}$ and evaluate $A^2 = AA$.

(b) Use the result in part (a) to evaluate $A^3 = AAA$.

(c) If n is a positive integer, find A^n.

40. **(a)** Use $A = \begin{bmatrix} 3 & -1 \\ 2 & 1 \end{bmatrix}$, $B = \begin{bmatrix} 0 & -2 \\ 1 & 4 \end{bmatrix}$

Show that $(A + B)^2 \neq A^2 + 2AB + B^2$.

(b) Give a reason for each numbered step.

$$(A + B)^2 = (A + B)(A + B)$$
$$= (A + B)A + (A + B)B \qquad \text{(i)}$$
$$= A^2 + BA + AB + B^2 \qquad \text{(ii)}$$

(c) Verify the result of part (b) using matrices A and B in part (a).

 Written Assignment: Consider the real number properties for addition and multiplication given in the boxed display on page 3. Some of these properties apply to addition and multiplication of the 2 by 2 matrices. Make lists of those that do apply and those that don't. For those that don't, give counterexamples.

CHALLENGE
Think Creatively

For $A = \begin{bmatrix} 2 & 0 \\ -4 & 2 \end{bmatrix}$ find matrices E_1, E_2 so that $E_2E_1A = \begin{bmatrix} 1 & 0 \\ 0 & 1 \end{bmatrix}$. (*Hint:* Consider the ideas similar to those contained in Exercises 33–36.)

**10.3
SOLVING
LINEAR
SYSTEMS
USING
INVERSES**

You may have wondered why nothing was said in the preceding section about division of matrices. The reason is simple. Division of matrices is undefined. There is, however, a process possible for some matrices that resembles the division of real numbers when the division is converted to multiplication using reciprocals as in

$$2 \div 5 = 2 \cdot \frac{1}{5} = 2 \cdot 5^{-1}$$

Inverse Matrices

Similarly, some *square* matrices have inverses, denoted by A^{-1}, which makes it possible to compute products of the form BA^{-1}.

An n by n square matrix B is said to be the **inverse** of an n by n square matrix A if and only if we have the following:

$$AB = BA = I_n, \qquad \text{where } I_n \text{ is the identity matrix}$$

This compares to
$$x \cdot \frac{1}{x} = \frac{1}{x} \cdot x = 1 \text{ for real}$$
numbers $x \neq 0$.

In such a case matrix A is also the inverse of B; A and B are inverses of one another. The notation A^{-1} is frequently used for the inverse of a matrix A, so that

$$AA^{-1} = A^{-1}A = I_n$$

When a matrix has an inverse, it has only one inverse. That is, the inverse of a matrix is unique (see Exercise 37).

Does $A = \begin{bmatrix} 1 & -1 \\ 0 & 2 \end{bmatrix}$ have an inverse? If so, then there are numbers v, w, x, and y such that

$$\begin{bmatrix} 1 & -1 \\ 0 & 2 \end{bmatrix}\begin{bmatrix} v & w \\ x & y \end{bmatrix} = \begin{bmatrix} 1 & 0 \\ 0 & 1 \end{bmatrix}$$

Multiply to get

$$\begin{bmatrix} v - x & w - y \\ 2x & 2y \end{bmatrix} = \begin{bmatrix} 1 & 0 \\ 0 & 1 \end{bmatrix}$$

Since two matrices are equal if and only if their elements in the same positions are equal, we get

$$v - x = 1$$
$$w - y = 0$$
$$2x = 0 \quad \longrightarrow \quad x = 0$$
$$2y = 1 \quad \longrightarrow \quad y = \tfrac{1}{2}$$
$$v - x = v - 0 = 1 \quad \longrightarrow \quad v = 1$$
$$w - y = w - \tfrac{1}{2} = 0 \quad \longrightarrow \quad w = \tfrac{1}{2}$$

Therefore,

$$\begin{bmatrix} v & w \\ x & y \end{bmatrix} = \begin{bmatrix} 1 & \tfrac{1}{2} \\ 0 & \tfrac{1}{2} \end{bmatrix}$$

Now check by multiplying:

This shows that $AA^{-1} = I_2$.

$$\begin{bmatrix} 1 & -1 \\ 0 & 2 \end{bmatrix}\begin{bmatrix} 1 & \tfrac{1}{2} \\ 0 & \tfrac{1}{2} \end{bmatrix} = \begin{bmatrix} 1 + 0 & \tfrac{1}{2} - \tfrac{1}{2} \\ 0 + 0 & 0 + 1 \end{bmatrix} = \begin{bmatrix} 1 & 0 \\ 0 & 1 \end{bmatrix} = I_2$$

Also,

This shows that $A^{-1}A = I_2$.

$$\begin{bmatrix} 1 & \tfrac{1}{2} \\ 0 & \tfrac{1}{2} \end{bmatrix}\begin{bmatrix} 1 & -1 \\ 0 & 2 \end{bmatrix} = \begin{bmatrix} 1 & 0 \\ 0 & 1 \end{bmatrix} = I_2$$

Thus

$$A^{-1} = \begin{bmatrix} 1 & -1 \\ 0 & 2 \end{bmatrix}^{-1} = \begin{bmatrix} 1 & \tfrac{1}{2} \\ 0 & \tfrac{1}{2} \end{bmatrix}$$

Not all square matrices have inverses. For example, the matrix $\begin{bmatrix} 0 & 1 \\ 0 & 1 \end{bmatrix}$ has no inverse because if it did there would have to be v, w, x, and y so that

$$\begin{bmatrix} 0 & 1 \\ 0 & 1 \end{bmatrix}\begin{bmatrix} v & w \\ x & y \end{bmatrix} = \begin{bmatrix} 1 & 0 \\ 0 & 1 \end{bmatrix}$$

$$\begin{bmatrix} x & y \\ x & y \end{bmatrix} = \begin{bmatrix} 1 & 0 \\ 0 & 1 \end{bmatrix}$$

Equating the elements gives

$$x = 1 \quad \text{and} \quad x = 0$$

which is not possible. Consequently, $\begin{bmatrix} 0 & 1 \\ 0 & 1 \end{bmatrix}$ has no inverse.

The method of finding an inverse used in the preceding discussion is not practical for larger matrices. As you have seen, when $n = 2$ we have to solve 4 equations. With $n = 3$, there would be 9 equations, then 16 equations for $n = 4$, and so on.

Fortunately, there are other more efficient methods for finding inverses. We will now present such a procedure and first demonstrate how it works using the matrix $A = \begin{bmatrix} 1 & -1 \\ 0 & 2 \end{bmatrix}$, for which we already know the inverse.

Begin by constructing a 2 by 4 matrix that contains both A and the identity matrix I_2.

$$\begin{array}{cc} A & I_2 \\ \downarrow & \downarrow \end{array}$$

$$\left[\begin{array}{cc|cc} 1 & -1 & 1 & 0 \\ 0 & 2 & 0 & 1 \end{array}\right]$$

Put A to the left and I_2 to the right of the dashed line.

Now apply fundamental row operations to the rows of this 2 by 4 matrix until the left half has been transformed into I_2. At this step the right half will be A^{-1}. Here are the details.

$$\begin{array}{cc} A & I_2 \\ \downarrow & \downarrow \end{array}$$

$$\left[\begin{array}{cc|cc} 1 & -1 & 1 & 0 \\ 0 & 2 & 0 & 1 \end{array}\right]$$

$\frac{1}{2} \times (\text{row } 2) \longrightarrow \left[\begin{array}{cc|cc} 1 & -1 & 1 & 0 \\ 0 & 1 & 0 & \frac{1}{2} \end{array}\right]$

$\text{row } 2 + \text{row } 1 \longrightarrow \left[\begin{array}{cc|cc} 1 & 0 & 1 & \frac{1}{2} \\ 0 & 1 & 0 & \frac{1}{2} \end{array}\right]$

$$\begin{array}{cc} \uparrow & \uparrow \\ I_2 & A^{-1} \end{array}$$

From our earlier work we know that $\begin{bmatrix} 1 & \frac{1}{2} \\ 0 & \frac{1}{2} \end{bmatrix}$ is the inverse of A.

This method can also be used to find out that a matrix does not have an inverse. The fundamental row operations are applied until we discover that it is not possible to reach I on the left. When this happens there is no inverse.

EXAMPLE 1 Show that $A = \begin{bmatrix} 2 & -4 \\ -1 & 2 \end{bmatrix}$ has no inverse.

Solution

$$\begin{array}{cc} A & I_2 \\ \downarrow & \downarrow \end{array}$$

$$\left[\begin{array}{rr:rr} 2 & -4 & 1 & 0 \\ -1 & 2 & 0 & 1 \end{array}\right]$$

$$\frac{1}{2} \times (\text{row } 1) \longrightarrow \left[\begin{array}{rr:rr} 1 & -2 & \frac{1}{2} & 0 \\ -1 & 2 & 0 & 1 \end{array}\right]$$

$$\text{row } 1 + \text{row } 2 \longrightarrow \left[\begin{array}{rr:rr} 1 & -2 & \frac{1}{2} & 0 \\ 0 & 0 & \frac{1}{2} & 1 \end{array}\right]$$

Because of the row of zeros to the left it is not possible to use row operations to obtain $1 \ 0$ in the first row on the left. Then I_2 cannot be reached on the left, and therefore there is no inverse of A. ∎

The procedure used for finding the inverse of a 2×2 matrix also applies to larger square matrices and can be summarized for an $n \times n$ matrix A as follows.

Let $[A \ \vdots \ I_n]$ be transformed by fundamental row operations.

1. If the form $[I_n \ \vdots \ B]$ is obtained, then the $n \times n$ matrix B is the inverse of A.
2. If all zeros are obtained in one or more rows to the left of the dashed vertical line, then A does not have an inverse.

The next example illustrates this procedure for a 3×3 matrix. This will be considerably more work than for the 2×2 case; however, it is usually more efficient than the method of solving a system of nine linear equations that was mentioned on page 519.

EXAMPLE 2 Let $A = \begin{bmatrix} 2 & 0 & -1 \\ -1 & 2 & 1 \\ 3 & -2 & -4 \end{bmatrix}$. Find A^{-1}.

Solution

Our first step will be to interchange the first two rows to get the -1 in the second row into the top left position.

$$
\begin{array}{cc}
A & I_3 \\
\downarrow & \downarrow
\end{array}
$$

$$
\left[\begin{array}{ccc|ccc}
2 & 0 & -1 & 1 & 0 & 0 \\
-1 & 2 & 1 & 0 & 1 & 0 \\
3 & -2 & -4 & 0 & 0 & 1
\end{array}\right]
$$

Interchange $\longrightarrow$
R_1 and R_2 $\longrightarrow$
$$
\left[\begin{array}{ccc|ccc}
-1 & 2 & 1 & 0 & 1 & 0 \\
2 & 0 & -1 & 1 & 0 & 0 \\
3 & -2 & -4 & 0 & 0 & 1
\end{array}\right]
$$

$2 \times R_1 + R_2 \longrightarrow$
$3 \times R_1 + R_3 \longrightarrow$
$$
\left[\begin{array}{ccc|ccc}
-1 & 2 & 1 & 0 & 1 & 0 \\
0 & 4 & 1 & 1 & 2 & 0 \\
0 & 4 & -1 & 0 & 3 & 1
\end{array}\right]
$$

The explanations of the row operations at the left use the abbreviation R_2 in place of row 2, etc. Recall that these notes describe what was done in the preceding matrix to get the indicated rows.

$R_3 + (-R_2) \longrightarrow$
$$
\left[\begin{array}{ccc|ccc}
-1 & 2 & 1 & 0 & 1 & 0 \\
0 & 4 & 1 & 1 & 2 & 0 \\
0 & 0 & -2 & -1 & 1 & 1
\end{array}\right]
$$

$-\frac{1}{2} \times R_3 \longrightarrow$
$$
\left[\begin{array}{ccc|ccc}
-1 & 2 & 1 & 0 & 1 & 0 \\
0 & 4 & 1 & 1 & 2 & 0 \\
0 & 0 & 1 & \frac{1}{2} & -\frac{1}{2} & -\frac{1}{2}
\end{array}\right]
$$

$-1 \times R_3 + R_1 \longrightarrow$
$-1 \times R_3 + R_2 \longrightarrow$
$$
\left[\begin{array}{ccc|ccc}
-1 & 2 & 0 & -\frac{1}{2} & \frac{3}{2} & \frac{1}{2} \\
0 & 4 & 0 & \frac{1}{2} & \frac{5}{2} & \frac{1}{2} \\
0 & 0 & 1 & \frac{1}{2} & -\frac{1}{2} & -\frac{1}{2}
\end{array}\right]
$$

$-1 \times R_1 \longrightarrow$
$\frac{1}{4} \times R_2 \longrightarrow$
$$
\left[\begin{array}{ccc|ccc}
1 & -2 & 0 & \frac{1}{2} & -\frac{3}{2} & -\frac{1}{2} \\
0 & 1 & 0 & \frac{1}{8} & \frac{5}{8} & \frac{1}{8} \\
0 & 0 & 1 & \frac{1}{2} & -\frac{1}{2} & -\frac{1}{2}
\end{array}\right]
$$

$2 \times R_2 + R_1 \longrightarrow$
$$
\left[\begin{array}{ccc|ccc}
1 & 0 & 0 & \frac{3}{4} & -\frac{1}{4} & -\frac{1}{4} \\
0 & 1 & 0 & \frac{1}{8} & \frac{5}{8} & \frac{1}{8} \\
0 & 0 & 1 & \frac{1}{2} & -\frac{1}{2} & -\frac{1}{2}
\end{array}\right]
$$

$$
\begin{array}{cc}
\uparrow & \uparrow \\
I_3 & A^{-1}
\end{array}
$$

You should now verify that the indicated matrix A^{-1} satisfies

$$AA^{-1} = A^{-1}A = I_3$$

■

One application of inverses is to the solution of linear systems in which the number of variables is the same as the number of equations. For such a system the coefficients of the variables form a square matrix A. Then, as you will see, if A^{-1} exists the system can be solved using this inverse.

The first step is to convert the linear system into a matrix equation. For example, consider this system:

$$
\begin{aligned}
2x \quad - \quad z &= 2 \\
-x + 2y + \quad z &= 0 \\
3x - 2y - 4z &= 10
\end{aligned}
$$

In terms of matrices it can be written as

Observe that after the left side has been multiplied we get
$$
\begin{bmatrix} 2x + 0y - z \\ -x + 2y + z \\ 3x - 2y + 4z \end{bmatrix} = \begin{bmatrix} 2 \\ 0 \\ 10 \end{bmatrix}
$$
Now equate the elements to obtain the original system.

$$
\begin{bmatrix} 2 & 0 & -1 \\ -1 & 2 & 1 \\ 3 & -2 & -4 \end{bmatrix} \begin{bmatrix} x \\ y \\ z \end{bmatrix} = \begin{bmatrix} 2 \\ 0 \\ 10 \end{bmatrix}
$$

matrix of   matrix   matrix
coefficients   of   of
            variables constants

Using A for the matrix of coefficients, X for the matrix of variables, and C for the matrix of constants, we have

$$
AX = C
$$

Now assume that A^{-1} exists, and multiply the preceding equation by A^{-1} on the left. Then

$$
\begin{aligned}
A^{-1}(AX) &= A^{-1}C \\
(A^{-1}A)X &= A^{-1}C \quad \text{(Matrix multiplication is associative.)} \\
I_3X &= A^{-1}C \quad (A^{-1}A = I_3) \\
X &= A^{-1}C \quad \text{(Note that } I_3 X = X.)
\end{aligned}
$$

The solution can now be found by equating the entries in $X = \begin{bmatrix} x \\ y \\ z \end{bmatrix}$ with the three

$A^{-1}C$ has dimensions 3 by 1.

numbers in $A^{-1}C$. The preceding result can be generalized as follows.

Verify that if B is n by m, $m \neq n$, then $I_nB = B$, but BI_n is undefined.

Let $AX = C$ be the matrix form of a linear system with n equations and n variables, where A is the $n \times n$ matrix of coefficients, X is the $n \times 1$ matrix of variables, and C is the $n \times 1$ matrix of constants. Then if matrix A has an inverse, the solution for the system is given by

$$
X = A^{-1}C
$$

EXAMPLE 3 Solve the linear system using the inverse of the matrix of coefficients.

$$3x + 4y = 5$$
$$x + 2y = 3$$

Verify that $AA^{-1} = I_2$ and $A^{-1}A = I_2$

Solution The matrix of coefficients is $A = \begin{bmatrix} 3 & 4 \\ 1 & 2 \end{bmatrix}$. Its inverse is found to be $A^{-1} = \begin{bmatrix} 1 & -2 \\ -\frac{1}{2} & \frac{3}{2} \end{bmatrix}$. Then, using $X = A^{-1}C$, we have

$$\begin{bmatrix} x \\ y \end{bmatrix} = \begin{bmatrix} 1 & -2 \\ -\frac{1}{2} & \frac{3}{2} \end{bmatrix}\begin{bmatrix} 5 \\ 3 \end{bmatrix} = \begin{bmatrix} -1 \\ 2 \end{bmatrix}$$

Therefore, $x = -1$, $y = 2$. Check this in the original system. ∎

EXAMPLE 4 Solve the linear system using A^{-1}, where A is the matrix of coefficients.

$$2x \quad\quad - z = 2$$
$$-x + 2y + z = 0$$
$$3x - 2y - 4z = 10$$

Solution The matrix of coefficients A is given by

$$A = \begin{bmatrix} 2 & 0 & -1 \\ -1 & 2 & 1 \\ 3 & -2 & -4 \end{bmatrix}$$

which is the same matrix as in Example 2. Then, using A^{-1} from Example 2 and the result $X = A^{-1}C$, we get

$$\begin{bmatrix} x \\ y \\ z \end{bmatrix} = \begin{bmatrix} \frac{3}{4} & -\frac{1}{4} & -\frac{1}{4} \\ \frac{1}{8} & \frac{5}{8} & \frac{1}{8} \\ \frac{1}{2} & -\frac{1}{2} & -\frac{1}{2} \end{bmatrix}\begin{bmatrix} 2 \\ 0 \\ 10 \end{bmatrix} = \begin{bmatrix} -1 \\ \frac{3}{2} \\ -4 \end{bmatrix}$$

Therefore, $x = -1$, $y = \frac{3}{2}$, $z = -4$. ∎

The system in Example 4 was solved with little effort because A^{-1} was already available. In fact, any system having the same A as its matrix of coefficients can be solved just as quickly, regardless of the constants. Therefore, this inverse method is particularly useful in solving more than one linear system having the same matrix of coefficients.

Find A^{-1}, if it exists, and verify that $AA^{-1} = A^{-1}A = I_n$.

1. $A = \begin{bmatrix} 4 & -1 \\ 2 & 0 \end{bmatrix}$

2. $A = \begin{bmatrix} -\frac{1}{5} & -\frac{2}{5} \\ 0 & \frac{1}{4} \end{bmatrix}$

3. $A = \begin{bmatrix} \frac{1}{3} & -\frac{4}{3} \\ -2 & 8 \end{bmatrix}$

4. $A = \begin{bmatrix} -\frac{3}{2} & \frac{5}{3} \\ \frac{9}{4} & -\frac{5}{2} \end{bmatrix}$

5. $A = \begin{bmatrix} 2 & -5 \\ -3 & 4 \end{bmatrix}$

6. $A = \begin{bmatrix} 10 & 15 \\ -5 & -1 \end{bmatrix}$

7. $A = \begin{bmatrix} \frac{1}{3} & \frac{2}{3} \\ \frac{2}{3} & \frac{1}{3} \end{bmatrix}$

8. $A = \begin{bmatrix} -3 & 6 \\ 6 & -3 \end{bmatrix}$

9. $A = \begin{bmatrix} 0 & a \\ b & 0 \end{bmatrix}$, $\quad ab \neq 0$

10. $A = \begin{bmatrix} 2 & 1 & 0 \\ 0 & 0 & -2 \\ 4 & 4 & 0 \end{bmatrix}$

11. $A = \begin{bmatrix} 0 & 1 & 0 \\ 1 & 0 & 0 \\ 0 & 0 & 1 \end{bmatrix}$

12. $A = \begin{bmatrix} 1 & 2 & -1 \\ -1 & 3 & 4 \\ 1 & 7 & 2 \end{bmatrix}$

13. $A = \begin{bmatrix} 1 & 0 & 2 \\ 2 & -1 & 0 \\ 0 & 3 & 4 \end{bmatrix}$

14. $A = \begin{bmatrix} 1 & 1 & -1 \\ 1 & -1 & -1 \\ -1 & -1 & -1 \end{bmatrix}$

15. $A = \begin{bmatrix} 4 & -3 & 1 \\ 0 & -1 & 9 \\ -2 & 1 & 4 \end{bmatrix}$

16. $A = \begin{bmatrix} 8 & -13 & 2 \\ -4 & 7 & -1 \\ 3 & -5 & 1 \end{bmatrix}$

17. $A = \begin{bmatrix} -11 & 2 & 2 \\ -4 & 0 & 1 \\ 6 & -1 & -1 \end{bmatrix}$

18. $A = \begin{bmatrix} 1 & 2 & 0 \\ -1 & 1 & 4 \\ 2 & 3 & -1 \end{bmatrix}$

19. $A = \begin{bmatrix} 1 & 1 & 0 & 2 \\ -1 & 0 & 2 & -1 \\ 0 & 2 & 0 & -2 \\ 2 & 0 & 0 & 5 \end{bmatrix}$

20. $A = \begin{bmatrix} 2 & -3 & 0 & 4 \\ -4 & 1 & -1 & 0 \\ -2 & 7 & -2 & 12 \\ 10 & 0 & 3 & -4 \end{bmatrix}$

Write the system of linear equations obtained from the matrix equation $AX = C$ for the given matrices.

21. $A = \begin{bmatrix} 3 & 1 \\ 2 & -2 \end{bmatrix}$, $X = \begin{bmatrix} x \\ y \end{bmatrix}$, $C = \begin{bmatrix} 9 \\ 14 \end{bmatrix}$

22. $A = \begin{bmatrix} 5 & -2 & 3 \\ 0 & 4 & 1 \\ 2 & -1 & 6 \end{bmatrix}$, $X = \begin{bmatrix} x \\ y \\ z \end{bmatrix}$, $C = \begin{bmatrix} -2 \\ 7 \\ 0 \end{bmatrix}$

Solve the linear system using the inverse of the matrix of coefficients. Observe that each matrix of coefficients is one of the matrices given in Exercises 1–20.

23. $\begin{aligned} 2x - 5y &= 7 \\ -3x + 4y &= -14 \end{aligned}$

24. $\begin{aligned} 2x - 5y &= -21 \\ -3x + 4y &= -7 \end{aligned}$

25. $\begin{aligned} \tfrac{1}{3}x + \tfrac{2}{3}y &= -8 \\ \tfrac{2}{3}x + \tfrac{1}{3}y &= 5 \end{aligned}$

26. $\begin{aligned} \tfrac{1}{3}x + \tfrac{2}{3}y &= 0 \\ \tfrac{2}{3}x + \tfrac{1}{3}y &= 1 \end{aligned}$

27. $\begin{aligned} x \quad\;\; + 2z &= 4 \\ 2x - y \quad\;\; &= -8 \\ 3y + 4z &= 0 \end{aligned}$

28. $\begin{aligned} x \quad\;\; + 2z &= -2 \\ 2x - y \quad\;\; &= 2 \\ 3y + 4z &= 1 \end{aligned}$

29. $\begin{aligned} x + y - z &= 1 \\ x - y - z &= 2 \\ -x - y - z &= 3 \end{aligned}$

30. $\begin{aligned} x + y - z &= -4 \\ x - y - z &= 6 \\ -x - y - z &= 10 \end{aligned}$

31. $\begin{aligned} 8x - 13y + 2z &= 1 \\ -4x + 7y - z &= 3 \\ 3x - 5y + z &= -2 \end{aligned}$

32. $\begin{aligned} -11x + 2y + 2z &= 0 \\ -4x \quad\;\; + z &= 5 \\ 6x - y - z &= -1 \end{aligned}$

33. $\begin{aligned} w + x \quad\;\; + 2z &= 2 \\ -w \quad\;\; + 2y - z &= -6 \\ 2x \quad\;\; - 2z &= 0 \\ 2w \quad\;\; + 5z &= 8 \end{aligned}$

34. $\begin{aligned} w + x \quad\;\; + 2z &= 1 \\ -w \quad\;\; + 2y - z &= 3 \\ 2x \quad\;\; - 2z &= 2 \\ 2w \quad\;\; + 5z &= -1 \end{aligned}$

35. For $A = \begin{bmatrix} 2 & 1 \\ 0 & -4 \end{bmatrix}$ and $B = \begin{bmatrix} 0 & 2 \\ 1 & -2 \end{bmatrix}$ show that $(AB)^{-1} = B^{-1}A^{-1}$.

36. For matrix A in Exercise 35, verify that $(A^{-1})^2 = (A^2)^{-1}$

37. (a) Assume that the matrix A has an inverse and let B and C be any matrices such that

$$AB = BA = I_n \quad \text{and} \quad AC = CA = I_n$$

Begin with the equation $AB = I_n$ and prove that $B = C$.

(b) What has been proved about the inverse of a matrix in part (a)?

38. (a) Let A and B be n by n matrices having inverses. Simplify the products $(AB)(B^{-1}A^{-1})$ and $(B^{-1}A^{-1})(AB)$.

(b) As a result of part (a), what can you say about the matrix $B^{-1}A^{-1}$?

39. (a) Let A be an invertible square matrix. Use the result in Exercise 38(b) to prove that $(A^2)^{-1} = (A^{-1})^2$.

(b) Write a generalization of the result in part (a) for any positive integer power of A.

 Written Assignment: Study Exercise 38 and state the result as a mathematical theorem about two invertible matrices of the same dimensions. Do not use any symbolism whatever in your statement.

**CHALLENGE
Think Creatively**

Let A be an n by n matrix. Show that if A satisfies the equation $A^2 + 5A - I = 0$, where $I = I_n$ and 0 is the n by n zero matrix, then A has an inverse. Find the inverse.

**EXPLORATIONS
Think Critically**

1. Why is it necessary to verify that both the left and right distributive properties hold for matrices, yet for the real numbers it is only necessary to begin with the left distributive property $a(b + c) = ab + ac$?

2. Let $A = [a_{ij}]$ and $B = [b_{ij}]$ be m by n matrices. The definition of addition states that for all i and j, $A + B = [a_{ij} + b_{ij}]$ and $B + A = [b_{ij} + a_{ij}]$. How does the commutative property for addition of matrices follow from this?

3. Suppose for matrices A, B, and C we have $AB = AC$. Under what conditions does it follow that $B = C$? Justify.

4. If A is an invertible matrix, is it correct to say $(5A)^{-1} = \frac{1}{5}A^{-1}$? Explain.

5. For $A = \begin{bmatrix} a & 0 \\ 0 & b \end{bmatrix}$, where $ab \neq 0$, find $(A^n)^{-1}$ for any positive integer n. Justify your result.

**10.4
INTRODUCTION
TO
DETERMINANTS**

The general linear equation has the form $ax + by = c$. Since we are now dealing with two linear equations, it is appropriate to use subscripts to distinguish the constants in the first equation from those in the second. Let

$$a_1 x + b_1 y = c_1$$
$$a_2 x + b_2 y = c_2$$

represent any system of linear equations, and refer to the system by the letter S.

Suppose that system S is a consistent system. Then S has a unique solution that can be found by the multiplication-addition method.

Multiply the first equation by b_2 and the second by $-b_1$.

$$a_1 b_2 x + b_1 b_2 y = c_1 b_2$$
$$-a_2 b_1 x - b_1 b_2 y = -c_2 b_1$$

Add to eliminate y.

$$a_1 b_2 x - a_2 b_1 x = c_1 b_2 - c_2 b_1$$

Factor.

$$(a_1 b_2 - a_2 b_1)x = c_1 b_2 - c_2 b_1$$

To solve for x it must be the case that $a_1 b_2 - a_2 b_1 \neq 0$. Then

$$x = \frac{c_1 b_2 - c_2 b_1}{a_1 b_2 - a_2 b_1}$$

Solve

$$8x - 20y = 3$$
$$4x + 10y = \tfrac{3}{2}$$

by substituting into the general results for x and y.

Similarly, multiplying the first and second equations of S by $-a_2$ and a_1, respectively, will produce this solution for y.

$$y = \frac{a_1 c_2 - a_2 c_1}{a_1 b_2 - a_2 b_1}$$

We have found that if $a_1 b_2 - a_2 b_1 \neq 0$, then system S has the common solution given above. The situation $a_1 b_2 - a_2 b_1 = 0$ will be taken up at the end of this section.

It is somewhat clumsy to apply this general solution, since it is necessary to keep track of the position of each constant. (Try this with the example suggested in the margin.) However, this process can be simplified with the introduction of some special symbolism.

First notice the denominator for both x and y is the same value $a_1 b_2 - a_2 b_1$. This value is called the **determinant** of the matrix $A = \begin{bmatrix} a_1 & b_1 \\ a_2 & b_2 \end{bmatrix}$ and is symbolized by replacing the brackets of the matrix by vertical bars as in the following definition.

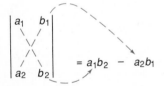

You can use this diagram to help remember the definition.

DEFINITION OF A 2 BY 2 DETERMINANT

For a matrix $A = \begin{bmatrix} a_1 & b_1 \\ a_2 & b_2 \end{bmatrix}$ the determinant of A is symbolized by

$$|A| = \begin{vmatrix} a_1 & b_1 \\ a_2 & b_2 \end{vmatrix} \text{ and is defined by } |A| = \begin{vmatrix} a_1 & b_1 \\ a_2 & b_2 \end{vmatrix} = a_1 b_2 - a_2 b_1$$

Observe that the same four numbers, placed in different positions, can result in unequal determinants, as shown on the next page.

$$\begin{vmatrix} 10 & -20 \\ 8 & 4 \end{vmatrix} = (10)(4) - (8)(-20) = 200$$

$$\begin{vmatrix} -20 & 8 \\ 10 & 4 \end{vmatrix} = (-20)(4) - (10)(8) = -160$$

Evaluate each of the determinants.

1. $\begin{vmatrix} 1 & 2 \\ 3 & 4 \end{vmatrix}$
2. $\begin{vmatrix} 1 & 3 \\ 2 & 4 \end{vmatrix}$
3. $\begin{vmatrix} 2 & 1 \\ 4 & 3 \end{vmatrix}$
4. $\begin{vmatrix} 1 & 3 \\ 4 & 2 \end{vmatrix}$

5. $\begin{vmatrix} 2 & 3 \\ 1 & 4 \end{vmatrix}$
6. $\begin{vmatrix} 4 & 2 \\ 1 & 3 \end{vmatrix}$
7. $\begin{vmatrix} 10 & -5 \\ 2 & 1 \end{vmatrix}$
8. $\begin{vmatrix} \frac{1}{2} & 6 \\ 0 & 4 \end{vmatrix}$

9. $\begin{vmatrix} -8 & -4 \\ -7 & -3 \end{vmatrix}$
10. $\begin{vmatrix} -1 & 0 \\ 0 & -1 \end{vmatrix}$
11. $\begin{vmatrix} 0 & 2 \\ 2 & 0 \end{vmatrix}$
12. $\begin{vmatrix} \frac{1}{3} & -\frac{1}{4} \\ 8 & -6 \end{vmatrix}$

The numerator in the general form for x is $c_1 b_2 - c_2 b_1$. Using our new symbolism, this number is the **second-order determinant**:

$$\begin{vmatrix} c_1 & b_1 \\ c_2 & b_2 \end{vmatrix} = c_1 b_2 - c_2 b_1$$

Similarly, the numerator for y is

$$\begin{vmatrix} a_1 & c_1 \\ a_2 & c_2 \end{vmatrix} = a_1 c_2 - a_2 c_1$$

In summary, we have the following, known as *Cramer's rule*, for solving such a system of linear equations.

CRAMER'S RULE

The system of equations S

$$a_1 x + b_1 y = c_1$$
$$a_2 x + b_2 y = c_2$$

has the unique solution

$$x = \frac{\begin{vmatrix} c_1 & b_1 \\ c_2 & b_2 \end{vmatrix}}{\begin{vmatrix} a_1 & b_1 \\ a_2 & b_2 \end{vmatrix}} \quad \text{and} \quad y = \frac{\begin{vmatrix} a_1 & c_1 \\ a_2 & c_2 \end{vmatrix}}{\begin{vmatrix} a_1 & b_1 \\ a_2 & b_2 \end{vmatrix}}$$

provided that $a_1 b_2 - a_2 b_1 \neq 0$.

Another way of stating that $a_1 b_2 - a_2 b_1 \neq 0$ is to say that the determinant of the matrix of coefficients is not zero.

Cramer's rule becomes easier to apply after making a few observations. First, the determinant

$$\begin{vmatrix} a_1 & b_1 \\ a_2 & b_2 \end{vmatrix}$$

is the denominator for each fraction. The first column of the determinant consists of the coefficients of the x-terms in S, and the second column contains the coefficients of the y-terms. To write the fraction for x, first record the denominator. Then, to get the determinant in the numerator, simply remove the first column (the coefficients of x) and replace it with constants c_1 and c_2, as shown in the following figure. Similarly, the replacement of the second column by c_1, c_2 gives the numerator in the fraction for y.

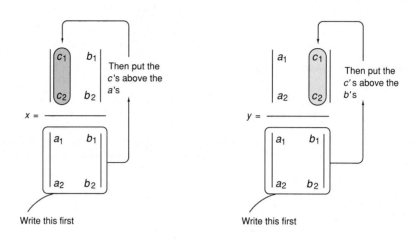

EXAMPLE 1 Solve the following system using Cramer's rule.

$$5x - 9y = 7$$
$$-8x + 10y = 2$$

Solution

$$x = \frac{\begin{vmatrix} 7 & -9 \\ 2 & 10 \end{vmatrix}}{\begin{vmatrix} 5 & -9 \\ -8 & 10 \end{vmatrix}} = \frac{70 - (-18)}{50 - 72} = \frac{88}{-22} = -4$$

Check the ordered pair $(-4, -3)$ in the given system.

$$y = \frac{\begin{vmatrix} 5 & 7 \\ -8 & 2 \end{vmatrix}}{\begin{vmatrix} 5 & -9 \\ -8 & 10 \end{vmatrix}} = \frac{10 - (-56)}{-22} = -3$$ ∎

EXAMPLE 2 Use determinants to solve the given system.

$$3x = 2y + 22$$
$$2(x + y) = x - 2y - 2$$

It is important to have both equations in the standard form

$$ax + by = c$$

since Cramer's rule is based on this form.

Solution   First write the system in standard form.

$$3x - 2y = 22$$
$$x + 4y = -2$$

$$x = \frac{\begin{vmatrix} 22 & -2 \\ -2 & 4 \end{vmatrix}}{\begin{vmatrix} 3 & -2 \\ 1 & 4 \end{vmatrix}} = \frac{84}{14} = 6 \qquad y = \frac{\begin{vmatrix} 3 & 22 \\ 1 & -2 \end{vmatrix}}{14} = \frac{-28}{14} = -2$$

■

The system

$$a_1 x + b_1 y = c_1$$
$$a_2 x + b_2 y = c_2$$

Recall that a system of two linear equations in two variables is dependent when the two equations are equivalent. Inconsistent means that there are no (common) solutions.

is either dependent or inconsistent when the determinant of the coefficients is zero, that is, when

$$\begin{vmatrix} a_1 & b_1 \\ a_2 & b_2 \end{vmatrix} = 0$$

For example, consider these two systems:

(a) $2x - 3y = 5$ (b) $2x - 3y = 8$
         $-10x + 15y = 8$ $-10x + 15y = -40$

In each case we have the following:

$$\begin{vmatrix} a_1 & b_1 \\ a_2 & b_2 \end{vmatrix} = \begin{vmatrix} 2 & -3 \\ -10 & 15 \end{vmatrix} = 0$$

Verify these conclusions; see Section 2.5.

System (a) turns out to be inconsistent and (b) is dependent.

EXERCISES 10.4

Evaluate each determinant.

1. $\begin{vmatrix} 5 & -1 \\ -3 & 4 \end{vmatrix}$ 2. $\begin{vmatrix} 1 & 2 \\ 3 & 4 \end{vmatrix}$ 3. $\begin{vmatrix} 17 & -3 \\ 20 & 2 \end{vmatrix}$ 4. $\begin{vmatrix} -7 & 9 \\ -5 & 5 \end{vmatrix}$

5. $\begin{vmatrix} 10 & 5 \\ 6 & -3 \end{vmatrix}$ 6. $\begin{vmatrix} 6 & 11 \\ 0 & -9 \end{vmatrix}$ 7. $\begin{vmatrix} 16 & 0 \\ -9 & 0 \end{vmatrix}$ 8. $\begin{vmatrix} a & b \\ 3a & 3b \end{vmatrix}$

Solve each system using Cramer's rule.

9. $3x + 9y = 15$ 10. $x - y = 7$ 11. $-4x + 10y = 8$
    $6x + 12y = 18$ $-2x + 5y = -8$ $11x - 9y = 15$

12. $7x + 4y = 5$ 13. $5x + 2y = 3$ 14. $3x + 3y = 6$
    $-x + 2y = -2$ $2x + 3y = -1$ $4x - 2y = -1$

15. $\frac{1}{3}x + \frac{3}{8}y = \quad 13$
    $x - \frac{9}{4}y = -42$

16. $\quad x - 3y = 7$
    $-\frac{1}{2}x + \frac{1}{4}y = 1$

17. $3x + y = 20$
    $\quad y = x$

18. $3x + 2y = 5$
    $2x + 3y = 0$

19. $\frac{1}{2}x - \frac{2}{7}y = -\frac{1}{2}$
    $-\frac{1}{3}x - \frac{1}{2}y = \quad \frac{31}{6}$

20. $-4x + 3y = -20$
    $\quad 2x + 6y = -15$

21. $9x - 12 = 4y$
    $3x + 2y = 3$

22. $\dfrac{x - y}{3} - \dfrac{y}{6} = \dfrac{2}{3}$
    $22x + 9(y - 2x) = 8$

Verify that the determinant of the coefficients is zero for each of the following. Then decide whether the system is dependent or inconsistent.

23. $\quad 5x - 2y = \quad 3$
    $-15x + 6y = -4$

24. $2x - 3y = 5$
    $10 - 4x = -6y$

25. $16x - 4y = 20$
    $12x - 3y = 15$

26. $\quad 2x - 6y = -12$
    $-3x + 9y = \quad 18$

27. $3x = 5y - 10$
    $6x - 10y = -25$

28. $10y = 2x - 4$
    $\quad x - 5y = 2$

When variables are used for some of the entries in the symbolism of a determinant, the determinant itself can be used to state equations. Solve for x.

29. $\begin{vmatrix} x & 2 \\ 5 & 3 \end{vmatrix} = 8$

30. $\begin{vmatrix} 7 & 3 \\ 4 & x \end{vmatrix} = 15$

31. $\begin{vmatrix} -2 & 4 \\ x & 3 \end{vmatrix} = -1$

Solve each system.

32. $\begin{vmatrix} x & y \\ 3 & 2 \end{vmatrix} = 2$
    $\begin{vmatrix} x & -1 \\ y & 3 \end{vmatrix} = 14$

33. $\begin{vmatrix} x & y \\ 2 & 4 \end{vmatrix} = 5$
    $\begin{vmatrix} 1 & y \\ -1 & x \end{vmatrix} = -\frac{1}{2}$

34. $\begin{vmatrix} 3 & x \\ 2 & y \end{vmatrix} = 13$
    $\begin{vmatrix} 3 & 2 \\ y & x \end{vmatrix} = -12$

35. Show that if the rows and columns of a second-order determinant are interchanged, the value of the determinant remains the same.

36. Show that if one of the rows of $\begin{vmatrix} a_1 & b_1 \\ a_2 & b_2 \end{vmatrix}$ is a multiple of the other, then the determinant is zero. (*Hint:* Let $a_2 = ka_1$ and $b_2 = kb_1$.)

37. Use Exercises 35 and 36 to demonstrate that the determinant is zero if one column is a multiple of the other.

38. If a common factor k is factored out of each element of a row or column of a two-by-two matrix A, resulting in a matrix B, then $|A| = k|B|$.

39. Make repeated use of the result in Exercise 38 to show the following:

$$\begin{vmatrix} 27 & 3 \\ 105 & -75 \end{vmatrix} = (45)\begin{vmatrix} 9 & 1 \\ 7 & -5 \end{vmatrix} \quad \text{or} \quad \begin{vmatrix} 27 & 3 \\ 105 & -75 \end{vmatrix} = (45)\begin{vmatrix} 3 & 1 \\ 7 & -15 \end{vmatrix}$$

Then evaluate each side to check.

40. Prove: $\begin{vmatrix} a_1 + t_1 & b_1 \\ a_2 + t_2 & b_2 \end{vmatrix} = \begin{vmatrix} a_1 & b_1 \\ a_2 & b_2 \end{vmatrix} + \begin{vmatrix} t_1 & b_1 \\ t_2 & b_2 \end{vmatrix}$

41. Prove that if to each element of a row (or column) of a second-order determinant we add k times the corresponding element of another row (or column), then the value of the new determinant is the same as that of the original determinant.

42. (a) Evaluate $\begin{vmatrix} 3 & 5 \\ -6 & -1 \end{vmatrix}$ by definition.

    (b) Evaluate the same determinant using the result of Exercise 41 by adding 2 times row one to row two.

(c) Evaluate the same determinant using the result of Exercise 41 by adding -6 times column two to column one.

Use the results of Exercises 38 and 41 to evaluate each determinant.

43. $\begin{vmatrix} 12 & -42 \\ -6 & 27 \end{vmatrix}$ **44.** $\begin{vmatrix} 45 & 75 \\ 40 & -25 \end{vmatrix}$

10.5 SOLVING LINEAR SYSTEMS USING THIRD-ORDER DETERMINANTS

Just as a system of two linear equations in two variables can be solved using second-order determinants, so can a system of three linear equations in three variables be solved by using *third-order* determinants.

A **third-order determinant** may be defined in terms of second-order determinants as follows:

> ### DEFINITION OF A THIRD-ORDER DETERMINANT
>
> $$\begin{vmatrix} a_1 & b_1 & c_1 \\ a_2 & b_2 & c_2 \\ a_3 & b_3 & c_3 \end{vmatrix} = a_1 \begin{vmatrix} b_2 & c_2 \\ b_3 & c_3 \end{vmatrix} - a_2 \begin{vmatrix} b_1 & c_1 \\ b_3 & c_3 \end{vmatrix} + a_3 \begin{vmatrix} b_1 & c_1 \\ b_2 & c_2 \end{vmatrix}$$

Note that the first term on the right is the product of a_1 times a second-order determinant. This determinant, also called the **minor of a_1**, can be found by eliminating the row and column that a_1 is in. Thus

$$\begin{vmatrix} \cancel{a_1} & \cancel{b_1} & \cancel{c_1} \\ a_2 & b_2 & c_2 \\ a_3 & b_3 & c_3 \end{vmatrix} \longrightarrow \begin{vmatrix} b_2 & c_2 \\ b_3 & c_3 \end{vmatrix} \qquad (\text{minor of } a_1)$$

Similar schemes can be used to obtain the minors for a_2 and a_3.

$$\begin{vmatrix} a_1 & b_1 & c_1 \\ \cancel{a_2} & \cancel{b_2} & \cancel{c_2} \\ a_3 & b_3 & c_3 \end{vmatrix} \longrightarrow \begin{vmatrix} b_1 & c_1 \\ b_3 & c_3 \end{vmatrix} \qquad (\text{minor of } a_2)$$

$$\begin{vmatrix} a_1 & b_1 & c_1 \\ a_2 & b_2 & c_2 \\ \cancel{a_3} & \cancel{b_3} & \cancel{c_3} \end{vmatrix} \longrightarrow \begin{vmatrix} b_1 & c_1 \\ b_2 & c_2 \end{vmatrix} \qquad (\text{minor of } a_3)$$

EXAMPLE 1 Evaluate: $\begin{vmatrix} 2 & -2 & 2 \\ 3 & 1 & 1 \\ 2 & -1 & 1 \end{vmatrix}$

Solution

$$\begin{vmatrix} 2 & -2 & 2 \\ 3 & 1 & 0 \\ 2 & -1 & 1 \end{vmatrix} = 2 \begin{vmatrix} 1 & 0 \\ -1 & 1 \end{vmatrix} - 3 \begin{vmatrix} -2 & 2 \\ -1 & 1 \end{vmatrix} + 2 \begin{vmatrix} -2 & 2 \\ 1 & 0 \end{vmatrix}$$

$$= 2(1 - 0) - 3(-2 + 2) + 2(0 - 2)$$
$$= 2 - 0 - 4$$
$$= -2 \qquad \blacksquare$$

A third-order determinant can be evaluated and simplified as follows:

Simplified Form of a Third-Order Determinant

$$\begin{vmatrix} a_1 & b_1 & c_1 \\ a_2 & b_2 & c_2 \\ a_3 & b_3 & c_3 \end{vmatrix} = a_1(b_2c_3 - b_3c_2) - a_2(b_1c_3 - b_3c_1) + a_3(b_1c_2 - b_2c_1)$$

$$= a_1b_2c_3 + a_2b_3c_1 + a_3b_1c_2 - a_1b_3c_2 - a_2b_1c_3 - a_3b_2c_1$$

The given definition of a third-order determinant can be described as an *expansion by minors* along the first column. It turns out that there are six such expansions, one for each row and column, all giving the same result. Here is the expansion by minors along the first row.

Simplify this expansion to see that it agrees with the preceding simplification of the expansion along the first column.

$$\begin{vmatrix} a_1 & b_1 & c_1 \\ a_2 & b_2 & c_2 \\ a_3 & b_3 & c_3 \end{vmatrix} = \boxed{+\, a_1} \begin{vmatrix} b_2 & c_2 \\ b_3 & c_3 \end{vmatrix} \boxed{-\, b_1} \begin{vmatrix} a_2 & c_2 \\ a_3 & c_3 \end{vmatrix} \boxed{+\, c_1} \begin{vmatrix} a_2 & b_2 \\ a_3 & b_3 \end{vmatrix}$$

When the expansion by minors is done along the second column we have

$$\begin{vmatrix} a_1 & b_1 & c_1 \\ a_2 & b_2 & c_2 \\ a_3 & b_3 & c_3 \end{vmatrix} = \boxed{-\, b_1} \begin{vmatrix} a_2 & c_2 \\ a_3 & c_3 \end{vmatrix} \boxed{+\, b_2} \begin{vmatrix} a_1 & c_1 \\ a_3 & c_3 \end{vmatrix} \boxed{-\, b_3} \begin{vmatrix} a_1 & c_1 \\ a_2 & c_2 \end{vmatrix}$$

$$= -b_1(a_2c_3 - a_3c_3) + b_2(a_1c_3 - a_3c_1) - b_3(a_1c_2 - a_2c_1)$$

$$= a_1b_2c_3 + a_2b_3c_1 + a_3b_1c_2 - a_1b_3c_2 - a_2b_1c_3 - a_3b_2c_1$$

For any expansion along a row or column keep the following display of signs in mind. It gives the signs preceding the row or column elements in the expansion.

Observe how the signs in the second column have been used in the preceding expansion.

$$\begin{vmatrix} + & - & + \\ - & + & - \\ + & - & + \end{vmatrix}$$

EXAMPLE 2 For $A = \begin{bmatrix} 2 & 1 & -3 \\ -4 & 0 & 2 \\ 5 & -1 & 6 \end{bmatrix}$ evaluate $|A|$ using these methods:

(a) Expand by minors along the third column.
(b) Expand by minors along the second row.

Solution

(a) $|A| = \begin{vmatrix} 2 & 1 & -3 \\ -4 & 0 & 2 \\ 5 & -1 & 6 \end{vmatrix} = -3 \begin{vmatrix} -4 & 0 \\ 5 & -1 \end{vmatrix} - 2 \begin{vmatrix} 2 & 1 \\ 5 & -1 \end{vmatrix} + 6 \begin{vmatrix} 2 & 1 \\ -4 & 0 \end{vmatrix}$

$$= -3(4 - 0) - 2(-2 - 5) + 6(0 + 4)$$
$$= -12 + 14 + 24$$
$$= 26$$

(b) $|A| = \begin{vmatrix} 2 & 1 & -3 \\ -4 & 0 & 2 \\ 5 & -1 & 6 \end{vmatrix} = -(-4)\begin{vmatrix} 1 & -3 \\ -1 & 6 \end{vmatrix} + 0\begin{vmatrix} 2 & -3 \\ 5 & 6 \end{vmatrix} - 2\begin{vmatrix} 2 & 1 \\ 5 & -1 \end{vmatrix}$

$$= 4(3) + 0 - 2(-7)$$
$$= 26 \qquad \blacksquare$$

Here is another procedure that can be used to evaluate a third-order determinant. Rewrite the first two columns at the right as shown. Follow the arrows pointing downward to get the three products having a plus sign, and the arrows pointing upward give the three products having a negative sign.

Note that this process does not work for higher-order determinants.

$$\begin{vmatrix} a_1 & b_1 & c_1 \\ a_2 & b_2 & c_2 \\ a_3 & b_3 & c_3 \end{vmatrix}\begin{matrix} a_1 & b_1 \\ a_2 & b_2 \\ a_3 & b_3 \end{matrix} = a_1b_2c_3 + b_1c_2a_3 + c_1a_2b_3 - a_3b_2c_1 - b_3c_2a_1 - c_3a_2b_1$$

EXAMPLE 3   Evaluate by rewriting the first two columns: $\begin{vmatrix} -1 & 3 & 4 \\ 2 & 1 & 2 \\ 5 & 1 & 3 \end{vmatrix}$

Solution

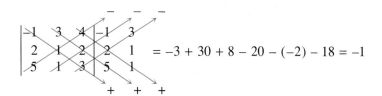

$$= -3 + 30 + 8 - 20 - (-2) - 18 = -1 \qquad \blacksquare$$

TEST YOUR UNDERSTANDING
Think Carefully

Evaluate each determinant using each of these methods:
(a) Expand by minors along one of the rows.
(b) Expand by minors along one of the columns.
(c) Rewrite the first two columns.

1. $\begin{vmatrix} 2 & 1 & 0 \\ -1 & 0 & 3 \\ 0 & 4 & -5 \end{vmatrix}$ 2. $\begin{vmatrix} 3 & 6 & 9 \\ 0 & 0 & 0 \\ -2 & -4 & -6 \end{vmatrix}$ 3. $\begin{vmatrix} 3 & -2 & 7 \\ 0 & 3 & -5 \\ 0 & 0 & 3 \end{vmatrix}$

4. $\begin{vmatrix} -1 & 4 & -2 \\ 6 & -6 & 1 \\ 3 & 3 & 2 \end{vmatrix}$ 5. $\begin{vmatrix} 1 & -2 & 3 \\ -4 & 5 & -4 \\ 3 & -2 & 1 \end{vmatrix}$ 6. $\begin{vmatrix} 2 & -1 & 9 \\ -7 & 3 & -4 \\ 2 & -1 & 9 \end{vmatrix}$

(Answers: Page 557)

In Exercises 22–27, a number of determinant properties are considered. The following two properties can be particularly helpful when evaluating determinants.

1. If a common factor is factored out of each element in a row (or column) of a square matrix A, resulting in a new matrix B, then $|A| = k|B|$. (See Exercise 26.)
2. If to each element of a row (or column) of a square matrix we add k times the elements of another row (or column), then the resulting matrix has the same determinant as the original matrix. (See Exercise 27.)

EXAMPLE 4 Evaluate $|A|$ using properties (1) and (2) where

$$A = \begin{bmatrix} 4 & 6 & -3 \\ -2 & -18 & 7 \\ 5 & 12 & -8 \end{bmatrix}$$

Solution

Observe that property (2) has been used twice to produce one row containing two zeros. This resulted in only one nonzero term when expanding by minors along that row.

$$|A| = \begin{vmatrix} 4 & 6 & -3 \\ -2 & -18 & 7 \\ 5 & 12 & -8 \end{vmatrix} = 6\begin{vmatrix} 4 & 1 & -3 \\ -2 & -3 & 7 \\ 5 & 2 & -8 \end{vmatrix} \qquad \text{6 has been factored out of column 2.}$$

$$= 6\begin{vmatrix} 0 & 1 & 0 \\ 10 & -3 & -2 \\ -3 & 2 & -2 \end{vmatrix} \qquad \begin{array}{l}-4 \times (\text{col. 2}) + \text{col. 1} \\ \text{and } 3 \times (\text{col. 2}) + \text{col. 3}\end{array}$$

$$= 6(-1)\begin{vmatrix} 10 & -2 \\ -3 & -2 \end{vmatrix} \qquad \begin{array}{l}\text{expanding by minors along} \\ \text{row 1}\end{array}$$

$$= -6(-20 - 6) = 156 \qquad \blacksquare$$

Third-order determinants can be used to extend Cramer's rule for the solution of three linear equations in three variables. Assume the following general system has a unique solution.

$$a_1x + b_1y + c_1z = d_1$$
$$a_2x + b_2y + c_2z = d_2$$
$$a_3x + b_3y + c_3z = d_3$$

By completing the tedious computations involved, it can be shown that the unique solution for this system is the following. (Also, see Exercises 37 and 38 for a proof of this result that is based on special properties of determinants.)

Note that D is the determinant of coefficients of the variables, in order. Then the numerators for the solutions for x, y, and z consist of the coefficients, but in each case the constants are used to replace the coefficients of the variable under consideration.

$$x = \frac{\begin{vmatrix} d_1 & b_1 & c_1 \\ d_2 & b_2 & c_2 \\ d_3 & b_3 & c_3 \end{vmatrix}}{D} \qquad y = \frac{\begin{vmatrix} a_1 & d_1 & c_1 \\ a_2 & d_2 & c_2 \\ a_3 & d_3 & c_3 \end{vmatrix}}{D} \qquad z = \frac{\begin{vmatrix} a_1 & b_1 & d_1 \\ a_2 & b_2 & d_2 \\ a_3 & b_3 & d_3 \end{vmatrix}}{D}$$

where

$$D = \begin{vmatrix} a_1 & b_1 & c_1 \\ a_2 & b_2 & c_2 \\ a_3 & b_3 & c_3 \end{vmatrix} \quad \text{and} \quad D \neq 0$$

If we let the determinants in the numerators be denoted by D_x, D_y, and D_z so that

$$D_x = \begin{vmatrix} d_1 & b_1 & c_1 \\ d_2 & b_2 & c_2 \\ d_3 & b_3 & c_3 \end{vmatrix} \qquad D_y = \begin{vmatrix} a_1 & d_1 & c_1 \\ a_2 & d_2 & c_2 \\ a_3 & d_3 & c_3 \end{vmatrix} \qquad D_z = \begin{vmatrix} a_1 & b_1 & d_1 \\ a_2 & b_2 & d_2 \\ a_3 & b_3 & d_3 \end{vmatrix}$$

then the solution of the system in Cramer's rule can be stated in the following condensed form.

$$x = \frac{D_x}{D} \qquad y = \frac{D_y}{D} \qquad z = \frac{D_z}{D}, \qquad \text{where } D \neq 0$$

EXAMPLE 5 Use Cramer's rule to solve this system.

$$\begin{aligned} x + 2y + z &= 3 \\ 2x - y - z &= 4 \\ -x - y + 2z &= -5 \end{aligned}$$

Solution First we find D and note that $D \neq 0$.

$$D = \begin{vmatrix} 1 & 2 & 1 \\ 2 & -1 & -1 \\ -1 & -1 & 2 \end{vmatrix} = \begin{vmatrix} 1 & 2 & 1 \\ 0 & -5 & -3 \\ 0 & 1 & 3 \end{vmatrix} \qquad \begin{aligned} & -2 \times (\text{row 1}) + \text{row 2} \\ & \text{and} \\ & 1 \times (\text{row 1}) + \text{row 3} \end{aligned}$$

$$= (1) \begin{vmatrix} -5 & -3 \\ 1 & 3 \end{vmatrix} \qquad \text{expanding along column 1}$$

$$= -12$$

Verify each of the following computations.

$$x = \frac{D_x}{D} = \frac{\begin{vmatrix} 3 & 2 & 1 \\ 4 & -1 & -1 \\ -5 & -1 & 2 \end{vmatrix}}{D} = \frac{-24}{-12} = 2$$

$$y = \frac{D_y}{D} = \frac{\begin{vmatrix} 1 & 3 & 1 \\ 2 & 4 & -1 \\ -1 & -5 & 2 \end{vmatrix}}{D} = \frac{-12}{-12} = 1$$

$$z = \frac{D_z}{D} = \frac{\begin{vmatrix} 1 & 2 & 3 \\ 2 & -1 & 4 \\ -1 & -1 & -5 \end{vmatrix}}{D} = \frac{12}{-12} = -1$$

Check this solution in the original system.

Thus $x = 2$, $y = 1$, and $z = -1$. ∎

EXERCISES 10.5

1. For $A = \begin{bmatrix} 6 & -2 & -1 \\ 0 & -9 & 4 \\ -3 & 5 & 1 \end{bmatrix}$ evaluate $|A|$ using each of these methods.

   (a) Expand by minors along the first column.
   (b) Expand by minors along the third row.
   (c) Rewrite the first two columns.

2. For $A = \begin{bmatrix} -8 & -1 & 0 \\ 4 & 7 & -5 \\ 3 & 0 & 2 \end{bmatrix}$ evaluate $|A|$ using each of these methods.

   (a) Expand by minors along the second column.
   (b) Expand by minors along the second row.
   (c) Rewrite the first two columns.

Evaluate each determinant.

3. $\begin{vmatrix} 2 & 2 & -1 \\ -1 & 3 & -3 \\ 1 & 2 & 3 \end{vmatrix}$

4. $\begin{vmatrix} 2 & 0 & -1 \\ 3 & -2 & 1 \\ -3 & 0 & 4 \end{vmatrix}$

5. $\begin{vmatrix} 1 & -3 & 2 \\ -5 & 2 & 0 \\ 4 & -1 & 3 \end{vmatrix}$

6. $\begin{vmatrix} 1 & 2 & 3 \\ 4 & 5 & 6 \\ 7 & 8 & 9 \end{vmatrix}$

7. $\begin{vmatrix} 1 & 1 & 1 \\ -1 & 1 & 1 \\ -1 & -1 & 1 \end{vmatrix}$

8. $\begin{vmatrix} 1 & 1 & 4 \\ 2 & 2 & -5 \\ 3 & 3 & 6 \end{vmatrix}$

Solve for x.

9. $\begin{vmatrix} -1 & x & -1 \\ x & -3 & 0 \\ -3 & 5 & -1 \end{vmatrix} = 0$

10. $\begin{vmatrix} x & 5 & 2x \\ 2x & 0 & x^2 \\ 1 & -1 & 2 \end{vmatrix} = 0$

11. $\begin{vmatrix} 5-x & 0 & -2 \\ 4 & -1-x & 3 \\ 2 & 0 & 1-x \end{vmatrix} = 0$

Evaluate the following for this system:

$$3x - y + 4z = 2$$
$$-5x + 3y - 7z = 0$$
$$7x - 4y + 4z = 12$$

12. D 13. D_x 14. $\dfrac{D_x}{D}$ 15. $\dfrac{D_y}{D}$

Use Cramer's rule to solve each system.

16. $\begin{aligned} x + y + z &= 12 \\ x - y + 3z &= 12 \\ 2x + 5y + 2z &= -2 \end{aligned}$

17. $\begin{aligned} x + 2y + 3z &= 5 \\ 3x - y &= -3 \\ -4x + z &= 6 \end{aligned}$

18. $\begin{aligned} x - 8y - 2z &= 12 \\ -3x + 3y + z &= -10 \\ 4x + y + 5z &= 2 \end{aligned}$

19. $\begin{aligned} 2x + y &= 5 \\ 3x - 2z &= -7 \\ -3y + 8z &= -5 \end{aligned}$

20. $\begin{aligned} 4x - 2y - z &= 1 \\ 2x + y + 2z &= 9 \\ x - 3y - z &= \tfrac{3}{2} \end{aligned}$

21. $\begin{aligned} 6x + 3y - 4z &= 5 \\ \tfrac{3}{2}x + y - 4z &= 0 \\ 3x - y + 8z &= 5 \end{aligned}$

A general property of determinants is stated in each exercise. Prove this property for the indicated special case.

22. If a square matrix A contains a row of zeros or a column of zeros, then $|A| = 0$. Prove this case:

$$\begin{vmatrix} a_1 & b_1 & c_1 \\ 0 & 0 & 0 \\ a_3 & b_3 & c_3 \end{vmatrix} = 0$$

23. If the rows and columns of a square matrix are interchanged to obtain the matrix A' (called the transpose of A), then $|A'| = |A|$. Prove this case:

$$\begin{vmatrix} a_1 & b_1 & c_1 \\ a_2 & b_2 & c_2 \\ a_3 & b_3 & c_3 \end{vmatrix} = \begin{vmatrix} a_1 & a_2 & a_3 \\ b_1 & b_2 & b_3 \\ c_1 & c_2 & c_3 \end{vmatrix}$$

24. **(a)** If one row of a square matrix A is a multiple of another row, then $|A| = 0$. Prove this case:

$$\begin{vmatrix} a & b & c \\ ka & kb & kc \\ d & e & f \end{vmatrix} = 0$$

(b) Use the property in part (a) and the property in Exercise 23 to prove that *if one column in A, is a multiple of another column, then $|A| = 0$*.

25. Interchanging any two rows or any two columns changes the sign of the determinant. Prove this case:

$$\begin{vmatrix} c_1 & b_1 & a_1 \\ c_2 & b_2 & a_2 \\ c_3 & b_3 & a_3 \end{vmatrix} = - \begin{vmatrix} a_1 & b_1 & c_1 \\ a_2 & b_2 & c_2 \\ a_3 & b_3 & c_3 \end{vmatrix}$$

26. If a common factor k is factored out of each element of a row (or column) of a square matrix A, resulting in a matrix B, then $|A| = k|B|$.
Prove the following case in which the elements in the first row of A may be written as $a_1 = ka_1'$, $b_1 = kb_1'$, and $c_1 = kc_1'$:

$$|A| = \begin{vmatrix} ka_1' & kb_1' & kc_1' \\ a_2 & b_2 & c_2 \\ a_3 & b_3 & c_3 \end{vmatrix} = k \begin{vmatrix} a_1' & b_1' & c_1' \\ a_2 & b_2 & c_2 \\ a_3 & b_3 & c_3 \end{vmatrix} = k|B|$$

27. *If to each element of a row (or column) of a square matrix we add k times the elements of another row (or column), then the resulting matrix has the same determinant as the original matrix.*
Prove this case:

$$\begin{vmatrix} a_1 + kb_1 & b_1 & c_1 \\ a_2 + kb_2 & b_2 & c_2 \\ a_3 + kb_3 & b_3 & c_3 \end{vmatrix} = \begin{vmatrix} a_1 & b_1 & c_1 \\ a_2 & b_2 & c_2 \\ a_3 & b_3 & c_3 \end{vmatrix}$$

28. Use the property in Exercise 24(a) to evaluate $\begin{vmatrix} 3 & 5 & -5 \\ 1 & 2 & 3 \\ -4 & -8 & -12 \end{vmatrix}$

29. Use the property in Exercise 24(b) to evaluate $\begin{vmatrix} 7 & 2 & -14 \\ 0 & 6 & 0 \\ -3 & 1 & 6 \end{vmatrix}$

30. Use the property in Exercise 26 to explain each step.

$$\begin{vmatrix} 8 & -10 & 2 \\ 4 & 25 & -1 \\ 2 & 10 & 0 \end{vmatrix} = 2\begin{vmatrix} 4 & -10 & 2 \\ 2 & 25 & -1 \\ 1 & 10 & 0 \end{vmatrix} = 10\begin{vmatrix} 4 & -2 & 2 \\ 2 & 5 & -1 \\ 1 & 2 & 0 \end{vmatrix} = 20\begin{vmatrix} 2 & -1 & 1 \\ 2 & 5 & -1 \\ 1 & 2 & 0 \end{vmatrix}$$

$\qquad\qquad$ (i) $\qquad\qquad\qquad$ (ii) $\qquad\qquad\qquad$ (iii)

31. Evaluate the determinant given on the left side of (i) in Exercise 30, and also evaluate the result given in (iii).

32. You can verify that $\begin{vmatrix} 1 & 2 & 3 \\ -4 & -1 & 5 \\ 3 & 1 & 7 \end{vmatrix} = 71$

It follows directly from this result that $\begin{vmatrix} 1 & -4 & 3 \\ 2 & -1 & 1 \\ 3 & 5 & 7 \end{vmatrix} = 71$

Which of the preceding properties (see Exercises 22–27) justifies this conclusion without doing any further calculations?

33. Evaluate $\begin{vmatrix} 5 & -4 & 3 \\ -6 & 6 & 2 \\ -7 & 3 & 4 \end{vmatrix}$ by first making repeated use of the property in Exercise 27 to obtain either a row or a column that has two zeros.

34. Show that $\begin{vmatrix} a^2 & b^2 & c^2 \\ a & b & c \\ 1 & 1 & 1 \end{vmatrix} = (a - b)(a - c)(b - c).$

35. A fourth-order determinant can be evaluated by expanding by minors along a row or column as was done for third-order determinants, and a similar pattern of signs is used in the expansion as shown at the left below. Write the expansion by minors along the first row for the determinant at the right, and evaluate the determinant.

$$\begin{vmatrix} + & - & + & - \\ - & + & - & + \\ + & - & + & - \\ - & + & - & + \end{vmatrix} \qquad \begin{vmatrix} 2 & -1 & 3 & 0 \\ 1 & 0 & 5 & -3 \\ 0 & 2 & -4 & 6 \\ -5 & 3 & 0 & 1 \end{vmatrix}$$

36. Let $A = \begin{bmatrix} -3 & 4 & 0 & 1 \\ 9 & -1 & 2 & 0 \\ 0 & 0 & -6 & 0 \\ 0 & 3 & 0 & -2 \end{bmatrix}$

Along which row or column would you expand to evaluate $|A|$ in the most efficient way? Use this expansion to find $|A|$ and check by using a different expansion.

37. Below is a proof that $x = \dfrac{D_x}{D}$ for the general system on page 534, assuming that $D \neq 0$.

Give reasons for the steps that are numbered at the right. Since $D = \begin{vmatrix} a_1 & b_1 & c_1 \\ a_2 & b_2 & c_2 \\ a_3 & b_3 & c_3 \end{vmatrix}$, then

$$xD = \begin{vmatrix} a_1 x & b_1 & c_1 \\ a_2 x & b_2 & c_2 \\ a_3 x & b_3 & c_3 \end{vmatrix} \qquad (i)$$

$$xD = \begin{vmatrix} a_1 x + b_1 y & b_1 & c_1 \\ a_2 x + b_2 y & b_2 & c_2 \\ a_3 x + b_3 y & b_3 & c_3 \end{vmatrix} \qquad (ii)$$

$$xD = \begin{vmatrix} a_1 x + b_1 y + c_1 z & b_1 & c_1 \\ a_2 x + b_2 y + c_2 z & b_2 & c_2 \\ a_3 x + b_3 y + c_3 z & b_3 & c_3 \end{vmatrix} \qquad (iii)$$

$$xD = \begin{vmatrix} d_1 & b_1 & c_1 \\ d_2 & b_2 & c_2 \\ d_3 & b_3 & c_3 \end{vmatrix} \qquad (iv)$$

$$xD = D_x$$

$$x = \frac{D_x}{D} \qquad (v)$$

38. Use arguments like those in Exercise 37 to prove that $y = \dfrac{D_y}{D}$ and $z = \dfrac{D_z}{D}$ for the general system on page 534.

**CHALLENGE
Think Creatively**

The straight line through the two points (a, b) and (c, d), where $a \neq c$, is given by this equation:

$$\begin{vmatrix} 1 & 1 & 1 \\ x & a & c \\ y & b & d \end{vmatrix} = 0$$

Show how to convert this equation into the point-slope form for this line. (*Hint:* Apply the result in Exercise 27 to the columns in the determinant.)

**10.6
SYSTEMS
OF LINEAR
INEQUALITIES**

The graphing of linear inequalities in two variables was introduced in Section 2.4. (You should review that material at this time.) There you learned that the graph of an inequality such as $3x - 5y < 10$ consists of all points in the plane on one side of the line $3x - 5y = 10$. To determine which side of the line is correct, two procedures were used.

Method I	Method II
(a) Draw the line $3x - 5y = 10$	(a) Draw the line $3x - 5y = 10$
(b) Solve the inequality for y.	(b) Select a convenient point, such as $(0, 0)$, on one side of the line and substitute the coordinates into the inequality

$$3x - 5y < 10$$

$$-5y < -3x + 10$$

$$y > \tfrac{3}{5}x - 2$$

$$3(0) - 5(0) = 0 < 10$$

(c) Since $y > \frac{3}{5}x - 2$, the graph consists of all points above the line indicated by the shading.

(c) Since $0 < 10$ is a true statement, $(0, 0)$ is on the correct side of the line, and the graph consists of all points on the same side of the line as the point $(0, 0)$, indicated by the shading.

The dashed line means that the points on the line $3x - 5y = 10$ are not part of the graph. The graph of $3x - 5y \le 10$ would include the points on the line, and a solid line would be used to indicate this.

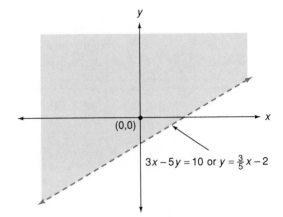

The methods of graphing a linear inequality can be extended to graphing systems of linear inequalities. This is demonstrated in Example 1.

EXAMPLE 1   Graph the system of linear inequalities:

$$2x + y \le 6$$
$$3x - 4y \ge 12$$

Solution   Draw the lines $2x + y = 6$ and $3x - 4y = 12$. Shade the region $2x + y \le 6$ vertically and the region $3x - 4y \ge 12$ horizontally. Since the coordinates of a point $P(x, y)$ must satisfy *both* conditions, the graph consists of all points shaded in both directions, as shown. The parts of the two lines above their point of intersection are dashed since they are not included.

Observe that the unshaded region consists of the points that satisfy this system:
$$2x + y > 6$$
$$3x - 4y < 12$$
What is the system for the part shaded only vertically? (See Exercise 29.)

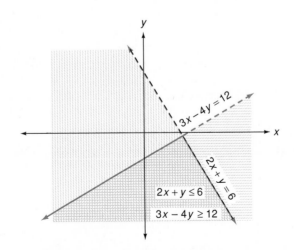

A system of linear inequalities may have more than two inequalities. Example 2 demonstrates how to graph a system of three linear inequalities.

EXAMPLE 2 Graph the system:

$$x + 3y \geq 12$$

$$-2x + y \leq 4$$

$$8x + 3y \leq 54$$

Solution Draw the lines ℓ_1: $x + 3y = 12$; ℓ_2: $-2x + y = 4$; and ℓ_3: $8x + 3y = 54$. Solve each inequality for y to obtain the equivalent system

$$y \geq -\frac{1}{3}x + 4$$

$$y \leq 2x + 4$$

$$y \leq -\frac{8}{3}x + 18$$

The first inequality gives the points on and above ℓ_1, and the last two inequalities give the points on and below ℓ_2 and above ℓ_3. Since a point (x, y) belongs to the graph of the system when it satisfies all three inequalities, the graph is the triangular region including the sides of the triangle as shown.

As an alternative procedure, draw the three lines and use a test point from each of the 7 regions determined by the lines (six regions are outside the triangle and one is inside). When a point from a region satisfies all 3 inequalities, then that region is part of the graph.

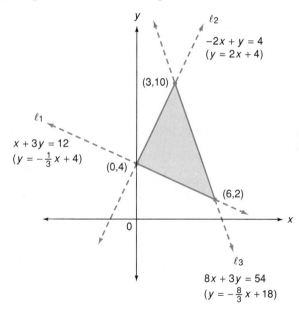

The final example shows how absolute values can be used to write and graph systems of linear inequalities.

EXAMPLE 3 Graph the system:

$$|x| \geq 3$$

$$|y| \leq 5$$

Such inequalities were introduced in Section 1.4

Solution The inequality $|x| \geq 3$ is equivalent to the compound inequality $x \leq -3$ or $x \geq 3$. Thus the points in the plane that satisfy $|x| \geq 3$ are all points to the left or on the vertical line $x = -3$, as well as all points to the right or on the line $x = 3$.

Similarly, since $|y| \leq 5$ is equivalent to $-5 \leq y \leq 5$, the points for this inequality are those on or between the horizontal lines $y = \pm 5$.

Since a point is on the graph of the given system provided *both* inequalities of the system are satisfied, the graph of the given system is the shaded region below.

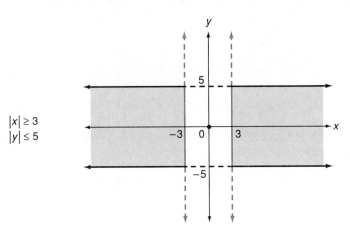

$|x| \geq 3$
$|y| \leq 5$

EXERCISES 10.6

Graph the regions satisfying the given inequalities.

1. $y \geq x - 2$
2. $y \leq x + 1$
3. $y > x$
4. $y \geq -\frac{1}{3}x - 2$
5. $x + y \leq 4$
6. $3x - y < 6$
7. $|x| \leq 3$
8. $|x| \geq 3$
9. $|x| < 2$
10. $|y| > 1$
11. $|y| \leq 2$
12. $|y| \geq 2$

Graph each system of inequalities.

13. $y \geq x + 2$
    $y \leq -x + 1$

14. $y \geq -4x + 1$
    $y \leq 4x + 1$

15. $y \leq -4x + 1$
    $y \geq 4x + 1$

16. $2x - y + 1 \geq 0$
    $x - 2y + 2 \leq 0$

17. $3x + y < 6$
    $x > 1$

18. $2x - 3y > 6$
    $y > -2$

19. $x + 2y \leq 10$
    $3x + 2y \leq 18$
    $x \geq 0$
    $y \geq 0$

20. $x + 2y \geq 10$
    $3x + 2y \leq 18$
    $x \geq 0$

21. $x + 2y \leq 10$
    $3x + 2y \geq 18$
    $y \geq 0$

22. $x + 2y \geq 10$
    $3x + 2y \geq 18$
    $x \geq 0$
    $y \geq 0$

23. $x - y \leq -1$
    $2x + y \geq 7$
    $4x - y \leq 11$

24. $x - y \geq -1$
    $2x + y \geq 7$
    $4x - y \leq 11$

25. $x - y \geq -1$
    $2x + y \leq 7$
    $4x - y \leq 11$
    $x \geq 0$
    $y \geq 0$

26. $|x| > 2$
    $|y| > 1$

27. $|x - 2| \leq 1$
    $|y + 1| \geq 2$

28. $|x| \leq 2$
    $|y| \leq 1$

29. Find the system of linear inequalities whose graph is the region having only vertical shading in the figure for Example 1, page 540.

30. Find the system of linear inequalities whose graph is the region having only horizontal shading in the figure for Example 1, page 540.

Find the system of linear inequalities whose graph is indicated by the Roman numeral. Note, for example, that region I includes all three sides, whereas region III includes one side and excludes the other two.

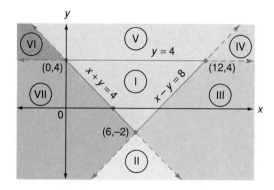

31. I   32. II   33. III   34. IV   35. V   36. VI   37. VII

1. What must be true for the determinant of the coefficients of a linear system having two variables and two equations in order for the system to have a unique solution?

2. If two lines are parallel, what can you say about the determinant of the matrix of coefficients for the corresponding linear system?

3. Consider the procedure for finding $|A|$ where A is a 3×3 matrix in which the first two columns are repeated to the right. Find similar procedures by repeating parts of A to the left of $|A|$, also by repeating parts above $|A|$, and also by repeating parts below $|A|$.

4. Suppose a linear system in three variables and three equations has a nonzero determinant for the matrix of coefficients. If each of the constants c_1, c_2, c_3 is zero, what can you say about the solution of the system? Justify.

5. Write the system of linear inequalities whose graph consists of all points inside and on the square whose vertices are on the coordinate axes and whose diagonals have length d.

10.7 LINEAR PROGRAMMING

In this chapter, you have learned additional procedures for finding the solution of linear systems of equations, and have graphed systems of linear inequalities. We will now apply this knowledge as we consider **linear programming**, a mathematical procedure for solving problems related to the logistics of decision making. Applications of this procedure can be found in numerous areas, including business and industry, agriculture, the field of nutrition, and the military.

This introduction to linear programming will primarily be a geometric approach based on the graphing of linear systems in two variables. We begin with the following system and use it to develop some of the basic concepts.

$$x \geq 0$$
$$y \geq 0$$
$$\tfrac{3}{4}x + \tfrac{1}{2}y \leq 6$$
$$\tfrac{1}{2}x + y \leq 6$$

The graph of this system is a region R consisting of all points (x, y) that satisfy *each* of the inequalities in the system. First note that the conditions $x \geq 0$ and $y \geq 0$ require the points to be in quadrant I, or on the nonnegative parts of each axis. Now graph the lines $\tfrac{3}{4}x + \tfrac{1}{2}y = 6$, $\tfrac{1}{2}x + y = 6$ and determine their point of intersection $(6, 3)$. Then, by the procedure used in Section 10.6, the (x, y) that satisfy both inequalities $\tfrac{3}{4}x + \tfrac{1}{2}y \leq 6$ and $\tfrac{1}{2}x + y \leq 6$ are found to be below or on these lines. The completed region R looks like this.

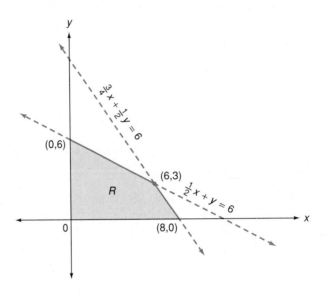

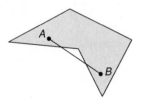

The region R is said to be a **convex set of points**, because any two points inside R can be joined by a line segment that is totally inside R. Roughly speaking, the boundary of R "bends outward." In the margin is a figure of a region that is *not* convex; note, in this figure, that points A and B cannot be connected by a line segment that is totally inside the region.

Draw the line ℓ_1: $x + y = 4$ in the same coordinate system as region R. (See the next figure.) All points (x, y) in R that are on this line have coordinates whose sum is 4. Now draw lines ℓ_2 and ℓ_3 parallel to ℓ_1 and also having equations $x + y = 2$ and $x + y = 8$, respectively. Obviously, all points in R on ℓ_2 or on ℓ_3 have coordinates whose sum is *not* 4. These three lines all have the form $x + y = k$, where k is some constant. There is an infinite number of such parallel lines, depending on the constant k. They all have slope -1. Some of these lines intersect R and others do not; we will be interested only in those that do.

Can you guess which line of this form will intersect R and have the largest possible value k? (See the following figure.)

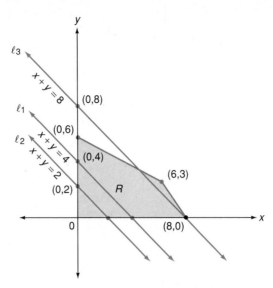

It is not difficult to see the answer, because all these lines are parallel; and the "higher" the line, the larger will be the value k. As a matter of fact, the line $x + y = k$ has k as its y-intercept. So the line we are looking for will be the line that intersects the y-axis as high up as possible, has slope -1, and still meets region R. Because of the shape of R (it is convex), this will be the line through $(6, 3)$ with slope -1, as shown in the following figure. The equation of the line through $(6, 3)$ with slope -1 is $x + y = 9$. Any other parallel line with a larger k-value will be higher and cannot intersect R; and those others that do will be lower and have a smaller k-value.

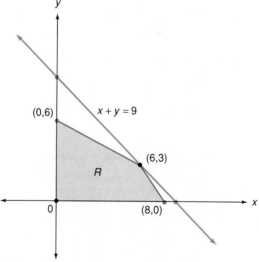

To sum up, we can say that of all the lines with form $x + y = k$ that intersect region R, the one that has the largest k-value is the line $x + y = 9$. Putting it another way, we can say that of all the points in R, the point $(6, 3)$ is the point whose coordinates give the maximum value for the quantity $k = x + y$.

Suppose that we now look for the point in R that produces the maximum value for k, where $k = 2x + 3y$. First we draw a few parallel lines, each with equation of the form $k = 2x + 3y$. The following figure shows such lines for k taking on the values 6, 10, and 18.

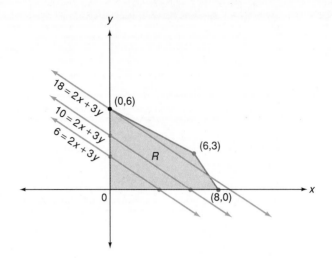

We see that the higher the line, the larger will be the value of k. Because of the convex shape of R, the highest such parallel line (intersecting R) must pass through the vertex $(6, 3)$. For this line we get $2(6) + 3(3) = 21 = k$. This is the largest value of $k = 2x + 3y$ for the (x, y) in R.

The preceding observations should be convincing evidence for the following result:

> Whenever we have a convex-shaped region R, then the point in R that produces the maximum value of a quantity of the form $k = ax + by$, where a and b are positive, will be at a vertex of R.

Because of this result, it is no longer necessary to draw lines through R. All that needs to be done to find the maximum of $k = ax + by$ is to graph the region R, find the coordinates of all vertices on the boundary, and see which one produces the largest value for k.

EXAMPLE 1 Maximize the quantity $k = 4x + 5y$ for (x, y) in the region S given by this system of four inequalities:

$$-x + 2y \leq 2$$
$$3x + 2y \leq 10$$
$$x + 6y \geq 6$$
$$3x + 2y \geq 6$$

Solution First graph the corresponding four lines and shade in the required region. Next find the four vertices of the region S by solving the appropriate pairs of equations. Since $k = 4x + 5y$ will be a maximum only at a vertex, the listing in the following table shows that 18 is the maximum value.

If $k > 18$, then the line given by $4x + 5y = k$ will not intersect S so that $4x + 5y = 18$ is the highest such line that does intersect S, and any parallel line $4x + 5y = k$ below $4x + 5y = 18$ will have $k < 18$.

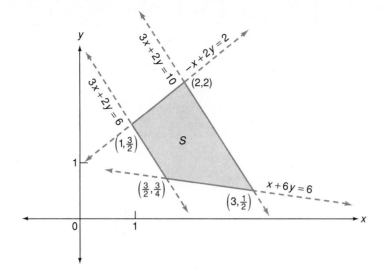

Vertex	$k = 4x + 5y$
$(1, \frac{3}{2})$	$4(1) + 5(\frac{3}{2}) = 11\frac{1}{2}$
$(\frac{3}{2}, \frac{3}{4})$	$4(\frac{3}{2}) + 5(\frac{3}{4}) = 9\frac{3}{4}$
$(2, 2)$	$4(2) + 5(2) = 18$
$(3, \frac{1}{2})$	$4(3) + 5(\frac{1}{2}) = 14\frac{1}{2}$

Instead of asking for the maximum of $k = 4x + 5y$, as in Example 1, it is also possible to find the minimum value of $k = 4x + 5y$ for the same region S. The earlier discussions explaining why a vertex of a convex region gives the maximum can easily be adjusted to show that a vertex will also give the minimum. Essentially, it becomes a matter of finding the line with correct slope, that has the lowest possible y-intercept, and that still intersects S. Because of this, the table in the solution to the preceding example shows that $9\frac{3}{4}$ is the minimum value of $k = 4x + 5y$. It occurs at the vertex $(\frac{3}{2}, \frac{3}{4})$.

Suppose we wish to maximize $k = 3x + 2y$ in Example 1. Then all points on the side from $(2, 2)$ to $(3, \frac{1}{2})$ will give the same maximum value for k.

TEST YOUR UNDERSTANDING
Think Carefully

Find the required values for the region R, where R is the graph of the system.

$$5x + 4y \geq 40 \qquad x + 4y \leq 40 \qquad 7x + 4y \leq 112$$

1. Maximum of $k = x + y$ 2. Minimum of $k = x + y$
3. Maximum of $k = x + 2y$ 4. Minimum of $k = x + 2y$
5. Maximum of $k = x + 5y$ 6. Minimum of $k = x + 5y$
7. Maximum of $k = 2x + y$ 8. Minimum of $k = 2x + y$
9. Maximum of $k = 3x + 20y$ 10. Minimum of $k = 3x + 20y$

(Answers: Page 557)

The preceding method of finding the maximum or minimum of a quantity $k = ax + by$, relative to a convex region, can be applied to a variety of applied situations. Here is a typical problem.

Before studying the solution, read this problem several times. Then list the given information and try to write the algebraic expression that needs to be maximized.

EXAMPLE 2 A store sells two kinds of bicycles, model A and model B. The store buys them unassembled from a wholesaler. Two employees are responsible for assembling the bicycles, and they are permitted to work no more than 6 hours each per week to do this job. Working together to assemble model A, employee I works $\frac{3}{4}$ hour and employee II works $\frac{1}{2}$ hour. Model B requires $\frac{1}{2}$ hour's work by employee I as well as 1 hour's work by employee II. There is a $55 profit on each model A sold and $48 on each model B. Because of the popularity of the sport, the store is able to sell as many bicycles as they decide to assemble. How many bicycles of each model should they assemble in order to get the maximum profit?

Solution   Let x be the number of model A bicycles assembled per week and y the number of model B's per week. Then $55x$ is the profit earned for model A, and $48y$ is the profit for model B. The total profit p is given by $p = 55x + 48y$. It is this quantity we need to maximize.

Next we find the time each employee works. For employee I, $\frac{3}{4}x + \frac{1}{2}y$ will be the number of hours worked per week, because each model A (there are x of these) requires $\frac{3}{4}$ hour, and each model B (there are y of these) requires $\frac{1}{2}$ hour. But each employee works *no more than* 6 hours per week. Thus the total time for this worker satisfies

$$\tfrac{3}{4}x + \tfrac{1}{2}y \le 6$$

By very similar reasoning, the total time for employee II is $\frac{1}{2}x + y$, which satisfies

$$\tfrac{1}{2}x + y \le 6$$

It is also known that $x \ge 0$ and $y \ge 0$ because there cannot be a negative number of either model.

Collecting the preceding conditions, we have that x and y must satisfy this system:

$$x \ge 0$$
$$y \ge 0$$
$$\tfrac{3}{4}x + \tfrac{1}{2}y \le 6$$
$$\tfrac{1}{2}x + y \le 6$$

We want to find the (x, y) for this system so that $p = 55x + 48y$ is a maximum.

The graph of this system is the same region R given at the beginning of this section. The points (x, y) in R represent all the possibilities because for all such points the numbers x and y satisfy *all* the conditions of the stated problem, and any (x, y) not in R cannot be a possibility because the x- and y-values would not satisfy *all* the stated conditions.

The table in the margin shows that the vertex (6, 3) gives the answer. Therefore, a maximum weekly profit is realized by assembling 6 bicycles of model A and 3 of model B.   ■

Vertex	p = 55x + 48y
(0, 0)	0
(0, 6)	288
(6, 3)	474
(8, 0)	440

Problems such as this are called **linear programming** problems. The inequalities that give the region R are sometimes called the **constraints**. The region R is called the set of **feasible points**, and (6, 3) is referred to as the **optimal point**.

EXERCISES 10.7

1. **(a)** Graph the region given by this system of constraints.

$$x + 3y \le 6, \qquad x \ge 0, \qquad y \ge 0$$

(b) Find all the vertices.   **(c)** Find the maximum and minimum of $p = x + y$.
(d) Find the maximum and minimum of $q = 6x + 10y$.
(e) Find the maximum and minimum of $r = 2x + 9y$.

Follow the instructions in Exercise 1 for each of these systems.

2. $x + 4y \le 72$
$5x + 4y \le 120$
$x \ge 0$
$y \ge 0$

3. $x + 6y \le 96$
$4x + 5y \le 118$
$3x + y \le 72$
$x \ge 0$
$y \ge 0$

4. $x - 2y \ge -10$
$2x + 7y \le 57$
$5x + 6y \le 85$
$5x + 2y \le 75$
$x \ge 0$
$y \ge 0$

5. $y - x \le 3$
$x + 4y \le 22$
$2x + y \le 16$
$x - 3y \le 1$
$x + 4y \ge 8$
$x \ge 0$

Find the maximum and minimum values of p, q, and r for the given constraints.

6. $y \ge 0$
$8x + 7y \le 56$
$8x + 3y \ge 24$

$p = 2x + y$
$q = 5x + 4y$
$r = 3x + 3y$

7. $x \ge 0$
$6x + 5y \le 60$
$2x + 5y \ge 20$

$p = x + 3y$
$q = 4x + 3y$
$r = \frac{1}{2}x + \frac{1}{4}y$

8. $x \ge 4$
$y \ge 2$
$x + 2y \le 20$

$p = x + 3y$
$q = 2x + 3y$
$r = 2x + 5y$

9. $x \le 7$
$0 \le y \le 5$
$2x + y \ge 8$

$p = 2x + 2y$
$q = 3x + y$
$r = x + \frac{1}{5}y$

The regions for the systems in Exercises 10 and 11 are "open-ended"; that is, their borders do not form a closed polygon and the graph extends endlessly. These regions are still convex in the sense discussed in the text.

10. Graph the region given by the system of constraints.

$$4x - y \ge 20, \quad 2x - 3y \ge 0, \quad x - 4y \ge -20, \quad y \ge 0$$

and evaluate the maximum and minimum (when they exist) of the following:
(a) $p = x + y$ **(b)** $q = 6x + 10y$ **(c)** $r = 2x + y$

11. Follow the instructions of Exercise 10 for this system.

$$5x + 2y \ge 22, \quad 6x + 7y \ge 54, \quad 2x + 11y \ge 44, \quad 4x - 5y \le 34, \quad x \ge 0$$

12. A publisher prints and sells both hardcover and paperback copies of the same book. Two machines, M_1 and M_2, are needed jointly to manufacture these books. To produce one hardcover copy, machine M_1 works $\frac{1}{6}$ hour and machine M_2 works $\frac{1}{12}$ hour. For a paperback copy, machines M_1 and M_2 work $\frac{1}{15}$ hour and $\frac{1}{10}$ hour, respectively. Each machine may be operated no more than 12 hours per day. If the profit is $2 on a hardcover copy and $1 on a paperback copy, how many of each type should be made per day to earn the maximum profit?

13. A manufacturer produces two models of a certain product: model I and model II. There is a $5 profit on model I and an $8 profit on model II. Three machines, M_1, M_2, and M_3, are used jointly to manufacture these models. The number of hours that each machine operates to produce 1 unit of each model is given in the table:

	Model I	Model II
Machine M_1	$1\frac{1}{2}$	1
Machine M_2	$\frac{3}{4}$	$1\frac{1}{2}$
Machine M_3	$1\frac{1}{3}$	$1\frac{1}{3}$

No machine is in operation more than 12 hours per day.
(a) If x is the number of model I made per day, and y the number of model II per day, show that x and y satisfy the following constraints.

$$x \ge 0, \quad y \ge 0, \quad \tfrac{3}{2}x + y \le 12, \quad \tfrac{3}{4}x + \tfrac{3}{2}y \le 12, \quad \tfrac{4}{3}x + \tfrac{4}{3}y \le 12$$

(b) Express the daily profit p in terms of x and y.

(c) Graph the feasible region given by the constraints in part (a) and find the coordinates of the vertices.

(d) What is the maximum profit, and how many of each model are produced daily to realize it?

(e) Find the maximum profit possible for the constraints stated in part (a) if the unit profits are $8 and $5 for models I and II, respectively.

14. A farmer buys two varieties of animal feed. Type A contains 8 ounces of corn and 4 ounces of oats per pound; type B contains 6 ounces of corn and 8 ounces of oats per pound. The farmer wants to combine the two feeds so that the resulting mixture has at least 60 pounds of corn and at least 50 pounds of oats. Feed A costs him 5¢ per pound and feed B costs 6¢ per pound. How many pounds of each type should the farmer buy to minimize the cost?

15. An appliance manufacturer makes two kinds of refrigerators: model A, which earns $100 profit, and model B, which earns $120 profit. Each month the manufacturer can produce up to 600 units of model A and up to 500 units of model B. If there are only enough man-hours available to produce no more than a total of 900 refrigerators per month, how many of each kind should be produced to obtain the maximum profit?

16. Two dog foods, A and B, each contain three types of ingredients, I_1, I_2, I_3. The number of ounces of these ingredients in each pound of a dog food is given in the table.

	I_1	I_2	I_3
A	8	3	1
B	8	1	3

A mixture of the two dog foods is to be formed to contain at least 1600 ounces of I_1, at least 300 ounces of I_2, and at least 360 ounces of I_3. If a pound of dog food A costs 8¢ and a pound of dog food B costs 6¢, how many pounds of each dog food should be used so that the resulting mixture meets all the requirements at the least cost?

Written Assignment: Give a geometric explanation as to why 18 is the maximum value in Example 1.

REVIEW EXERCISES

The solutions of the following exercises can be found within the text of Chapter 10.
Try to answer each question before referring to the text.

Section 10.1

Solve each system by using the row-equivalent matrix procedure.

1. $2x + 5y + 8z = 11$
    $x + 4y - 7z = 10$
    $3x + 6y + 12z = 15$

2. $2x + 14y - 4z = -2$
    $-4x - 3y + z = 8$
    $3x - 5y + 6z = 7$

3. $3x - 6y = 9$
    $-2x + 4y = -8$

4. $x + 2y - z = 1$
    $2x - y + 3z = 4$
    $5x + 5z = 9$

Section 10.2

5. Add: $\begin{bmatrix} 2 & -5 \\ 3 & 7 \\ -1 & 0 \end{bmatrix} + \begin{bmatrix} -4 & 1 \\ 0 & 9 \\ 2 & -2 \end{bmatrix}$

6. Verify $(A + B) + C = A + (B + C)$ using

$$A = \begin{bmatrix} 3 & 1 & -2 \\ -1 & 3 & 0 \end{bmatrix}, \quad B = \begin{bmatrix} -5 & 5 & 6 \\ -2 & -2 & -2 \end{bmatrix}, \quad \text{and } C = \begin{bmatrix} 0 & 1 & 0 \\ 1 & 0 & 1 \end{bmatrix}$$

7. Verify $A + (-A) = 0$ using $A = \begin{bmatrix} 8 & -5 \\ -4 & 9 \\ 0 & 2 \end{bmatrix}$.

8. Find $7A$ using A in Exercise 7.

9. Verify $(cd)A = c(dA)$ using $c = -2$, $d = \frac{1}{3}$, and $A = \begin{bmatrix} 6 & -9 \\ 1 & 5 \end{bmatrix}$.

10. Find AB for $A = \begin{bmatrix} -4 & -1 & 2 \\ 3 & 8 & 1 \end{bmatrix}$ and $B = \begin{bmatrix} -2 & 1 \\ 7 & 0 \\ 4 & -6 \end{bmatrix}$.

11. Multiply: $\begin{bmatrix} 2 & 1 \\ -6 & 3 \\ 1 & 0 \end{bmatrix} \begin{bmatrix} 0 & -2 & 2 & 1 \\ 5 & 0 & 7 & 6 \end{bmatrix}$

12. Verify $AD \neq DA$ using $A = \begin{bmatrix} 1 & 2 \\ 3 & -5 \end{bmatrix}$ and $D = \begin{bmatrix} 8 & -8 \\ -7 & -7 \end{bmatrix}$

13. Find $I_3 A$ for $A = \begin{bmatrix} a & b & c \\ d & e & f \\ g & h & i \end{bmatrix}$.

Section 10.3

Find A^{-1}, if it exists.

14. $A = \begin{bmatrix} 1 & -1 \\ 0 & 2 \end{bmatrix}$

15. $A = \begin{bmatrix} 2 & -4 \\ -1 & 2 \end{bmatrix}$

16. $A = \begin{bmatrix} 2 & 0 & -1 \\ -1 & 2 & 1 \\ 3 & -2 & -4 \end{bmatrix}$

17. Write the linear system as a matrix equation.

$$2x \quad - z = 2$$
$$-x + 2y + z = 0$$
$$3x - 2y - 4z = 10$$

18. Solve the system using the inverse of the matrix of coefficients.

$$3x + 4y = 5$$
$$x + 2y = 3$$

19. Solve the system in Exercise 17 using the inverse of the matrix of coefficients.

20. If A^{-1} exists, prove that the solution to the matrix equation $AX = C$ is given by $X = A^{-1}C$.

Section 10.4

Evaluate the determinants.

21. $\begin{vmatrix} a_1 & b_1 \\ a_2 & b_2 \end{vmatrix}$
  22. $\begin{vmatrix} -20 & 8 \\ 10 & 4 \end{vmatrix}$
  23. $\begin{vmatrix} 10 & -20 \\ 8 & 4 \end{vmatrix}$

Solve the given system using Cramer's rule.

24. $\begin{aligned} 5x - 9y &= 7 \\ -8x + 10y &= 2 \end{aligned}$
  25. $\begin{aligned} 3x &= 2y + 22 \\ 2(x + y) &= x - 2y - 2 \end{aligned}$

Section 10.5

Evaluate the determinant.

26. $\begin{vmatrix} 2 & -2 & 2 \\ 3 & 1 & 0 \\ 2 & -1 & 1 \end{vmatrix}$
  27. $\begin{vmatrix} 4 & 6 & -3 \\ -2 & -18 & 7 \\ 5 & 12 & -8 \end{vmatrix}$

28. Evaluate the determinant $\begin{vmatrix} a_1 & b_1 & c_1 \\ a_2 & b_2 & c_2 \\ a_3 & b_3 & c_3 \end{vmatrix}$

by expanding along the second column and simplify.

29. For $A = \begin{vmatrix} 2 & 1 & -3 \\ -4 & 0 & 2 \\ 5 & -1 & 6 \end{vmatrix}$ evaluate $|A|$ using each of these methods.

   (a) Expand by minors along the third column.
   (b) Expand by minors along the second row.

30. Evaluate by rewriting the first two columns: $\begin{vmatrix} -1 & 3 & 4 \\ 2 & 1 & 2 \\ 5 & 1 & 3 \end{vmatrix}$

31. Solve the system using Cramer's rule.

$$\begin{aligned} x + 2y + z &= 3 \\ 2x - y - z &= 4 \\ -x - y + 2z &= -5 \end{aligned}$$

Section 10.6

32. Graph the region satisfying the linear inequality $3x - 5y < 10$.

Graph the system of inequalities.

33. $\begin{aligned} 2x + y &\le 6 \\ 3x - 4y &\ge 12 \end{aligned}$
  34. $\begin{aligned} x + 3y &\ge 12 \\ -2x + y &\le 4 \\ 8x + 3y &\le 54 \end{aligned}$

35. Graph the system: $\begin{aligned} |x| &\ge 3 \\ |y| &\le 5 \end{aligned}$

36. Maximize the quantity $k = 4x + 5y$ for (x, y) in the region S given by this system:

$$-x + 2y \le 2, \qquad 3x + 2y \le 10, \qquad x + 6y \ge 6, \qquad 3x + 2y \ge 6$$

37. Find the minimum of $k = 4x + 5y$ for the system in Exercise 36.

38. A store sells two kinds of bicycles, model A and model B. The store buys them unassembled from a wholesaler. Two employees are responsible for assembling the bicycles, and they are permitted to work no more than 6 hours each per week to do this job. Working together to assemble model A, employee I works $\frac{3}{4}$ hour and employee II works $\frac{1}{2}$ hour. Model B requires $\frac{1}{2}$ hour's work by employee I as well as 1 hour's work by employee II. There is a $55 profit on each model A sold and $48 on each model B. Because of the popularity of the sport, the store is able to sell as many bicycles as they decide to assemble. How many bicycles of each model should they assemble in order to get the maximum profit?

CHAPTER 10 TEST: STANDARD ANSWER

Use the questions to test your knowledge of the basic skills and concepts of Chapter 10. Then check your answers with those given at the back of the book.

Solve the linear system using row-equivalent matrices.

1.
$$\begin{aligned} -6x + 3y &= 9 \\ 10x - 5y &= -12 \end{aligned}$$

2.
$$\begin{aligned} 2x - y + 2z &= -3 \\ x + 4y - 3z &= 18 \\ -4x + 2y - 3z &= 0 \end{aligned}$$

3.
$$\begin{aligned} x + y + 2z &= 3 \\ 3x \quad\ + 2z &= 2 \\ -x + 2y + 2z &= 4 \end{aligned}$$

4. (a) Write the system in Question 1 as a matrix equation.
   (b) Write the system in Question 2 as a matrix equation.

Evaluate whenever possible, using these matrices.

$$A = \begin{bmatrix} 2 & 0 & 3 \\ -1 & 4 & 9 \end{bmatrix} \qquad B = \begin{bmatrix} 1 & 0 & -1 \\ 0 & -1 & 1 \\ -1 & 0 & 11 \end{bmatrix} \qquad C = \begin{bmatrix} 1 & 2 & 3 \\ 4 & 5 & 6 \\ 7 & 8 & 9 \\ 1 & 0 & 1 \end{bmatrix}$$

$$D = \begin{bmatrix} -1 & -1 \\ 2 & 2 \\ -3 & -3 \end{bmatrix} \qquad E = \begin{bmatrix} 3 & 0 \\ 1 & 1 \end{bmatrix} \qquad F = \begin{bmatrix} -5 & 7 \\ 4 & -9 \end{bmatrix}$$

$$G = \begin{bmatrix} 0 & -1 & 1 \\ 6 & 2 & 3 \\ -5 & -5 & -8 \end{bmatrix} \qquad H = \begin{bmatrix} -1 \\ 1 \\ 2 \end{bmatrix} \qquad J = \begin{bmatrix} 2 & -2 & 4 \end{bmatrix}$$

5. (a) $B + G$ **(b)** $A - B$ **6.** $5E - 2F$ **7.** AB

8. (a) m_{32} in the product $CB = M$.   **(b)** $(|E|)(|F|)$

9. AD **10.** $A(EF)$ **11.** E^3 (means EEE)   **12.** HJ

13. (a) The dimension of $CBDA$.
   (b) The dimension of $DBAE$.

14. Evaluate: $(A + \frac{1}{2}B)(C - 2D)$ for:

$$A = [11, 7, 2] \qquad B = [-2, 6, -2] \qquad C = \begin{bmatrix} -5 \\ -2 \\ 4 \end{bmatrix} \qquad D = \begin{bmatrix} -8 \\ 5 \\ 1 \end{bmatrix}$$

15. Find the inverse of $A = \begin{bmatrix} 3 & 1 \\ 2 & -2 \end{bmatrix}$.

16. Use the result of Question 15 to solve this system: $3x + y = 9$
$$2x - 2y = 14$$

17. Find the inverse of $A = \begin{bmatrix} 1 & 0 & -2 \\ 0 & 1 & 0 \\ 3 & 2 & 0 \end{bmatrix}$.

18. The matrix $A = \begin{bmatrix} -11 & 2 & 2 \\ -4 & 0 & 1 \\ 6 & -1 & -1 \end{bmatrix}$ has inverse $A^{-1} = \begin{bmatrix} 1 & 0 & 2 \\ 2 & -1 & 3 \\ 4 & 1 & 8 \end{bmatrix}$. Use this informa-

tion to solve this linear system: $\begin{aligned} -11x + 2y + 2z &= 3 \\ -4x + z &= -1 \\ 6x - y - z &= -2 \end{aligned}$

Evaluate each determinant.

19. (a) $\begin{vmatrix} 4a & b \\ -a & 3b \end{vmatrix}$ (b) $\begin{vmatrix} -2 & 4 & 3 \\ 6 & -1 & -4 \\ 5 & 2 & 0 \end{vmatrix}$

Use Cramer's rule to solve the system.

20. $\begin{aligned} 4x - 2y &= 15 \\ 3x + 2y &= -8 \end{aligned}$ 21. $\begin{aligned} 2x + y - z &= -3 \\ x - 2y + z &= 8 \\ 3x - y - 2z &= -1 \end{aligned}$

22. Graph this system of inequalities:

$$x + y - 2 \le 0, \qquad 2x - y + 2 \ge 0$$

23. Graph the system

$$x \ge 0, \qquad y \ge 0, \qquad x + 2y \le 8, \qquad x + y \le 5, \qquad 2x + y \le 8$$

24. Find the maximum value of $k = 3x + 2y$ for the region in Question 23.

25. A manufacturer of television sets makes two kinds of 19-inch color TV sets: model A, which earns \$80 profit, and model B, which earns \$100 profit. Each month the manufacturer can produce up to 400 units of model A and up to 300 units of model B but no more than 600 sets altogether. How many of each should be produced to obtain the maximum profit?

CHAPTER 10 TEST: MULTIPLE CHOICE

1. Which of the following are permissible fundamental row operations when using matrices to solve a linear system?
   I. Interchange two rows.
   II. Multiply a row by a nonzero constant.
   III. Add a multiple of a row to another row.
   (a) Only I   **(b)** Only II   **(c)** Only III   **(d)** I, II, and III   **(e)** None of the preceding

2. Which of the following is equal to the determinant $\begin{vmatrix} 2 & 5 \\ 7 & 3 \end{vmatrix}$?

   (a) $\begin{vmatrix} 5 & 7 \\ 3 & 2 \end{vmatrix}$ **(b)** $\begin{vmatrix} 7 & 3 \\ 2 & 5 \end{vmatrix}$ **(c)** $\begin{vmatrix} 2 & 5 \\ 3 & 7 \end{vmatrix}$ **(d)** $\begin{vmatrix} -3 & -5 \\ 7 & 2 \end{vmatrix}$ **(e)** None of the preceding.

3. For the matrix A shown at the right, which of the following represents $9A - 2A$? $A = \begin{bmatrix} 3 & -5 \\ -2 & 6 \\ 0 & 4 \end{bmatrix}$

(a) $\begin{bmatrix} 21 & -35 \\ -2 & 6 \\ 0 & 4 \end{bmatrix}$ (b) $\begin{bmatrix} 21 & -5 \\ -14 & 6 \\ 0 & 4 \end{bmatrix}$ (c) $\begin{bmatrix} 21 & -35 \\ -14 & 42 \\ 0 & 28 \end{bmatrix}$ (d) $\begin{bmatrix} 21 & -55 \\ -22 & 42 \\ 0 & 28 \end{bmatrix}$ (e) None of the preceding

4. Which of the following is equal to AB for matrices A and B below?

$$A = \begin{bmatrix} -2 & 1 & 5 \\ 3 & 0 & 2 \end{bmatrix} \quad B = \begin{bmatrix} 4 & -2 \\ 3 & 4 \\ 0 & -1 \end{bmatrix}$$

(a) $\begin{bmatrix} -5 & 3 \\ 12 & -8 \end{bmatrix}$ (b) $\begin{bmatrix} -5 & 12 \\ 3 & -8 \end{bmatrix}$ (c) $\begin{bmatrix} -6 & 4 & 5 \\ 1 & 4 & 1 \end{bmatrix}$ (d) AB is impossible to compute.

(e) None of the preceding.

5. Suppose a linear system of three equations in three variables is being solved by making use of row-equivalent matrices. What does it mean when we arrive at the following matrix?

$$\begin{bmatrix} 1 & -3 & 6 & | & 4 \\ 0 & 4 & 3 & | & -1 \\ 0 & 0 & 0 & | & 0 \end{bmatrix}$$

(a) $(4, -1, 0)$ is a solution of the original system.

(b) The original system of equations has more than one solution.

(c) The original system of equations has exactly one solution.

(d) The original system of equations has no solutions.

(e) None of the preceding

6. Which of the following is the inverse of the matrix shown at the right? $\begin{bmatrix} -1 & 2 \\ 2 & -3 \end{bmatrix}$

(a) $\begin{bmatrix} -3 & 2 \\ -1 & 1 \end{bmatrix}$ (b) $\begin{bmatrix} -1 & 2 \\ 0 & 1 \end{bmatrix}$ (c) $\begin{bmatrix} 2 & 3 \\ 1 & 2 \end{bmatrix}$ (d) $\begin{bmatrix} 3 & 2 \\ 2 & 1 \end{bmatrix}$ (e) None of the preceding

7. Matrix A has dimensions 3 by 5, matrix B has dimensions 7 by 4, and matrix C has dimensions 5 by 7. Which of the following matrix products is possible?

(a) ABC (b) BCA (c) CBA (d) ACB (e) None of the preceding

8. If $\begin{bmatrix} -2 & 4 \\ 6 & -8 \end{bmatrix} \begin{bmatrix} w & x \\ y & z \end{bmatrix} = I_2$, then $Z =$

(a) $-\frac{1}{2}$ (b) 0 (c) $\frac{1}{4}$ (d) 1 (e) None of the preceding

9. Let $AX = C$ be the matrix equation of a system of linear equations, where A is the 3 by 3 matrix of coefficients, X is the 3 by 1 matrix of variables, and C is the 3 by 1 matrix of constants. If A has an inverse, which of the following is true about the solution X?

(a) $X = CA^{-1}$ (b) $X = A^{-1}C$ (c) $X = \frac{1}{A}C$ (d) No solution is possible. (e) None of the preceding

10. if for the system $\begin{matrix} a_1x + b_1y = c_1 \\ a_2x + b_2y = c_2 \end{matrix}$ we have $\begin{vmatrix} a_1 & b_1 \\ a_2 & b_2 \end{vmatrix} \neq 0$, then which of the following is true?

(a) The system has a unique solution.

(b) The system has no solution.

(c) The system has many solutions.

(d) The value of x is equal to the determinant $\begin{vmatrix} a_1 & c_1 \\ a_2 & c_2 \end{vmatrix}$ divided by $a_1 b_2 - a_2 b_1$.

(e) None of the preceding

11. Evaluate: $\begin{vmatrix} 1 & -2 & 3 \\ -1 & 0 & 1 \\ -2 & 3 & -1 \end{vmatrix}$

(a) 8   **(b)** -14 **(c)** -6 **(d)** 0   **(e)** None of the preceding

12. Which response below is true regarding the graph of this system?

$$2x + y \le 6$$
$$x + y \ge 4$$
$$y \ge 0$$
$$x \ge 0$$

(a) The graph consists of all points inside and on the border of just one triangle.

(b) The graph consists of all points inside and on the border of two triangles.

(c) The graph is a four-sided figure.

(d) All points on the positive part of the y-axis are in the graph.

(e) None of the preceding

13. When Cramer's rule is used to solve for y for the system shown below, which of the following becomes the numerator of the fraction?

$$2x + y - 3z = 5$$
$$-x + 2y + z = 1$$
$$3x - y + 2z = -3$$

(a) $\begin{vmatrix} 2 & 1 & -3 \\ -1 & 2 & 1 \\ 3 & -1 & 2 \end{vmatrix}$ **(b)** $\begin{vmatrix} 2 & -3 & 5 \\ -1 & 1 & 1 \\ 3 & 2 & -3 \end{vmatrix}$ **(c)** $\begin{vmatrix} 5 & 2 & -3 \\ 1 & -1 & 1 \\ -3 & 3 & 2 \end{vmatrix}$

(d) $\begin{vmatrix} 5 & 2 & 1 \\ 1 & -1 & 2 \\ -3 & 3 & -1 \end{vmatrix}$ **(e)** None of the preceding

14. Which of the following is the second row of the inverse of matrix A shown below?

$$A = \begin{bmatrix} 1 & 0 & 2 \\ 0 & 2 & -1 \\ 2 & 5 & 2 \end{bmatrix}$$

(a) 0   1 $-\frac{1}{2}$ **(b)** -2 -2 1   **(c)** 0 -2 1   **(d)** 0 $\frac{1}{2}$ 0   **(e)** None of the preceding

15. What is the maximum of $k = 4x + 3y$ for the region given by the system shown below?

$$x \ge 0$$
$$y \ge 0$$
$$x + 2y \le 22$$
$$3x + 2y \le 30$$

(a) 88   **(b)** 44   **(c)** 43   **(d)** 33   **(e)** None of the preceding

Page 512

1. Not possible.

2. $\begin{bmatrix} 3 & 3 \\ 2 & 3 \\ 9 & 6 \end{bmatrix}$

3. $\begin{bmatrix} 8 & 4 \\ 1 & 9 \\ 17 & 8 \end{bmatrix}$

4. $\begin{bmatrix} -7 & 1 \\ 4 & -9 \\ -7 & 2 \end{bmatrix}$

5. $\begin{bmatrix} 7 & -1 \\ -4 & 9 \\ 7 & -2 \end{bmatrix}$

6. $\begin{bmatrix} 2 & 8 & -6 \\ 4 & 10 & 0 \\ 0 & 2 & 2 \end{bmatrix}$

7. Not possible

8. $\begin{bmatrix} -15 & -15 \\ -10 & -15 \\ -45 & -30 \end{bmatrix}$

9. $\begin{bmatrix} -10 & -2 \\ 2 & -12 \\ -16 & -4 \end{bmatrix}$

Page 514

1. $\begin{bmatrix} -6 & -22 \\ -59 & 11 \end{bmatrix}$

2. $\begin{bmatrix} 8 & 0 & 9 \\ 13 & -22 & 27 \end{bmatrix}$

3. Not possible.

4. $\begin{bmatrix} 24 & -29 \\ 14 & -38 \\ -25 & 49 \end{bmatrix}$

5. Not possible.

6. $\begin{bmatrix} 52 & -63 \\ -5 & 19 \end{bmatrix}$

7. $\begin{bmatrix} 25 & 2 & 27 \\ -18 & 28 & -36 \\ 3 & -26 & 18 \end{bmatrix}$

8. $\begin{bmatrix} 42 & -25 \\ 181 & -284 \end{bmatrix}$

Page 520

1. $\begin{bmatrix} \frac{5}{3} & 2 \\ \frac{2}{3} & 1 \end{bmatrix}$

2. Does not exist.

3. $\begin{bmatrix} 0 & -1 \\ -\frac{1}{2} & \frac{3}{2} \end{bmatrix}$

4. $\begin{bmatrix} 4 & 3 \\ -4 & -6 \end{bmatrix}$

5. Does not exist

6. $\begin{bmatrix} \frac{1}{110} & \frac{1}{110} \\ \frac{1}{110} & -\frac{1}{1100} \end{bmatrix}$

Page 527

1. -2 2. -2 3. 2 4. -10 5. 5 6. 10
7. 20 8. 2 9. -4 10. 1 11. -4 12. 0

Page 533

1. -29 2. 0 3. 27 4. -93 5. -8 6. 0

Page 547

1. 19 2. 1 3. 26 4. -34 5. 50 6. -139
7. 37 8. 10 9. 200 10. -592

11 SEQUENCES AND SERIES

11.1 SEQUENCES

The same equation can be used to define a variety of functions by changing the domain. For example, below are the graphs of three functions all of whose range values are given by the equation $y = x^2$ for the indicated domains.

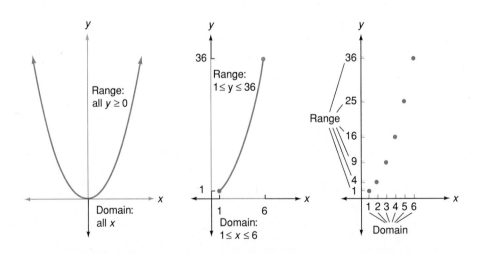

The type of function that is studied in this chapter is illustrated by the preceding graph at the right, where the domain consists of the consecutive integers 1, 2, 3, 4, 5, 6. This kind of function is called a **sequence**.

DEFINITION OF A SEQUENCE

A sequence is a function whose domain is a set of consecutive positive integers.

Instead of using the variable x, letters, such as n, k, i are normally used for the domain variable of a sequence. Frequently, sequences (functions) will be denoted by the lowercase letters, such as a, and the range values by a_n, which are also called the **terms** of the sequence.

Sequences are often given by stating their **general** or **nth terms**. Thus the general term of the sequence, previously given by $y = x^2$, becomes $a_n = n^2$.

EXAMPLE 1 Find the range values of the sequence given by $a_n = \dfrac{1}{n}$ for the domain $\{1, 2, 3, 4, 5\}$ and graph.

Solution The range values and graph are as follows.

This is an example of a finite sequence since the domain is finite. That is, the domain is a set of positive integers having a last element.

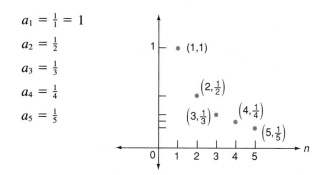

$$a_1 = \tfrac{1}{1} = 1$$

$$a_2 = \tfrac{1}{2}$$

$$a_3 = \tfrac{1}{3}$$

$$a_4 = \tfrac{1}{4}$$

$$a_5 = \tfrac{1}{5}$$

EXAMPLE 2 List the first six terms of the sequence given by $b_k = \dfrac{(-1)^k}{k^2}$.

Solution

$$b_1 = \frac{(-1)^1}{1^2} = -1 \qquad b_4 = \frac{(-1)^4}{4^2} = \frac{1}{16}$$

$$b_2 = \frac{(-1)^2}{2^2} = \frac{1}{4} \qquad b_5 = \frac{(-1)^5}{5^2} = -\frac{1}{25}$$

$$b_3 = \frac{(-1)^3}{3^2} = -\frac{1}{9} \qquad b_6 = \frac{(-1)^6}{6^2} = \frac{1}{36}$$

TEST YOUR UNDERSTANDING
Think Carefully

Write the first five terms of the given sequence.

1. $a_n = 2n + 1$ 2. $a_n = -2n$ 3. $a_n = -2n + 2$

4. $b_k = \dfrac{(-1)^k}{k}$ 5. $b_k = \dfrac{1}{k^2}$ 6. $b_k = \dfrac{-3}{k(k + 1)}$

7. $c_n = \dfrac{3}{n(2n - 1)}$ 8. $c_n = \left(\dfrac{1}{3}\right)^n$ 9. $c_n = 1 - (-1)^n$

(Answers: Page 600)

Sometimes a sequence is given by a verbal description. If, for example, we ask for the increasing sequence of odd integers beginning with -3, then this implies the **infinite sequence** whose first few terms are

$$-3, -1, 1, \ldots$$

A sequence can also be given by presenting a listing of its first few terms, possibly including the general term. Thus the preceding sequence is

$$-3, -1, 1, \ldots, 2n - 5, \ldots$$

EXAMPLE 3 Find the tenth term of the sequence

*This is an example of an **infinite** sequence since the domain is infinite. That is, the domain consists of all the positive integers.*

$$-3, 4, \frac{5}{3}, \ldots, \frac{n + 2}{2n - 3}, \ldots$$

Solution Since the first term, -3, is obtained by letting $n = 1$ in the general term $\frac{n + 2}{2n - 3}$, the tenth term is

$$a_{10} = \frac{10 + 2}{2(10) - 3} = \frac{12}{17}$$

∎

EXAMPLE 4 Write the first four terms of the sequence given by $a_n = \left(1 + \frac{1}{n}\right)^n$. Round off to two decimal places when appropriate.

Solution

$$a_1 = \left(1 + \frac{1}{1}\right)^1 = 2$$
$$a_2 = \left(1 + \frac{1}{2}\right)^2 = \left(\frac{3}{2}\right)^2 = \frac{9}{4} = 2.25$$
$$a_3 = \left(1 + \frac{1}{3}\right)^3 = \left(\frac{4}{3}\right)^3 = \frac{64}{27} = 2.37$$
$$a_4 = \left(1 + \frac{1}{4}\right)^4 = \left(\frac{5}{4}\right)^4 = \frac{625}{256} = 2.44$$

∎

The terms of the sequence in Example 4 are getting successively larger. But the increase from term to term is getting smaller. That is, the differences between successive terms are decreasing:

$$a_2 - a_1 = 0.25$$
$$a_3 - a_2 = 0.12$$
$$a_4 - a_3 = 0.07$$

If more terms of $a_n = \left(1 + \frac{1}{n}\right)^n$ were computed, you would see that while the terms keep on increasing, the amount by which each new term increases keeps getting smaller.

Use a calculator to verify these table entries to four decimal places.

n	$a_n = \left(1 + \dfrac{1}{n}\right)^n$
10	2.5937
50	2.6916
100	2.7048
500	2.7156
1000	2.7169
5000	2.7180
10,000	2.7181

It turns out that no matter how large n is, the value of $\left(1 + \dfrac{1}{n}\right)^n$ is never more than 2.72. In fact, the larger the n that is taken, the closer $\left(1 + \dfrac{1}{n}\right)^n$ gets to the irrational value $e = 2.71828\ldots$. This is the number that was introduced in Chapter 6 in reference to natural logarithms and exponential functions.

Sequences can also be defined **recursively**, which means that the first term is given and the nth term a_n is defined in terms of the preceding term a_{n-1}. This is illustrated in the next example.

EXAMPLE 5 Let a_n be the nth term of a sequence defined by

$$a_1 = 6$$
$$a_n = 3a_{n-1} - 7 \qquad \text{for } n > 1$$

Find the first five terms of this sequence.

Solution

$$a_1 = 6$$
$$a_2 = 3a_1 - 7 = 3 \cdot 6 - 7 = 11$$
$$a_3 = 3a_2 - 7 = 3 \cdot 11 - 7 = 26$$
$$a_4 = 3a_3 - 7 = 3 \cdot 26 - 7 = 71$$
$$a_5 = 3a_4 - 7 = 3 \cdot 71 - 7 = 206$$

∎

EXERCISES 11.1

The domain of the sequence in each exercise consists of the integers 1, 2, 3, 4, 5. Write the corresponding range values.

1. $a_n = 2n - 1$ 2. $a_n = 10 - n^2$ 3. $a_k = (-1)^k$ 4. $b_k = -\dfrac{6}{k}$ 5. $b_i = 8(-\tfrac{1}{2})^i$ 6. $b_i = (\tfrac{1}{2})^{i-3}$

Write the first four terms of the sequence given by the formula in each exercise.

7. $c_k = (-1)^k k^2$

8. $c_j = 3(\tfrac{1}{10})^{j-1}$

9. $c_j = 3(\tfrac{1}{10})^j$

10. $a_j = 3(\tfrac{1}{10})^{j+1}$

11. $a_j = 3(\tfrac{1}{10})^{2j}$

12. $a_n = \dfrac{(-1)^{n+1}}{n + 3}$

13. $c_n = \dfrac{1}{n} - \dfrac{1}{n + 1}$

14. $a_n = \dfrac{n^2 - 4}{n + 2}$

15. $a_k = (2k - 10)^2$

16. $a_k = 1 + (-1)^k$

17. $a_n = -2 + (n - 1)(3)$

18. $a_n = a_1 + (n - 1)(d)$

19. $b_i = \dfrac{i - 1}{i + 1}$

20. $b_i = 64^{1/i}$

21. $b_n = \left(1 + \dfrac{1}{n}\right)^{n-1}$

22. $u_n = \dfrac{1}{2^n}$

23. $u_n = -2(\tfrac{3}{4})^{n-1}$

24. $u_k = a_1 r^{k-1}$

25. $x_k = \dfrac{k}{2^k}$

26. $x_n = \dfrac{(-1)^n}{n} + n$

27. $x_k = \dfrac{k}{k + 1} - \dfrac{k + 1}{k}$

28. $y_n = \left(1 + \dfrac{1}{n + 1}\right)^n$

29. $y_n = 4$

30. $y_n = \dfrac{n}{(n + 1)(n + 2)}$

31. Find the sixth term of $1, 2, 5, \ldots, \frac{1}{2}(1 + 3^{n-1}), \ldots$.

32. Find the ninth and tenth terms of $0, 4, 0, \ldots, \dfrac{2^n + (-2)^n}{n}, \ldots$.

33. Find the seventh term of $a_k = 3(0.1)^{k-1}$.

34. Find the twentieth term of $a_n = (-1)^{n-1}$.

35. Find the twelfth term of $a_i = i$.

36. Find the twelfth term of $a_i = (i - 1)^2$.

37. Find the twelfth term of $a_i = (1 - i)^3$.

38. Find the hundredth term of $a_n = \dfrac{n + 1}{n^2 + 5n + 4}$.

39. Write the first four terms of the sequence of even increasing integers beginning with 4.

40. Write the first four terms of the sequence of decreasing odd integers beginning with 3.

41. Write the first five positive multiples of 5 and find the formula for the nth term.

42. Write the first five powers of 5 and find the formula for the nth term.

43. Write the first five powers of -5 and find the formula for the nth term.

44. Write the first five terms of the sequence of reciprocals of the negative integers and find the formula for the nth term.

45. The numbers 1, 3, 6, and 10 are called **triangular numbers** because they correspond to the number of dots in the triangular arrays. Find the next three triangular numbers.

46. When an investment earns **simple interest** it means the interest is earned only on the original investment. For example, if P dollars are invested in a bank that pays simple interest at the annual rate of r percent, then the interest for the first year is Pr, and the amount in the bank at the end of the year is $P + Pr$. For the second year, the interest is again Pr; the amount now would be $(P + Pr) + Pr = P + 2Pr$.

  (a) What is the amount after n years?

  (b) What is the amount in the bank if an investment of \$750 has been earning simple interest for 5 years at the annual rate of 12%?

  (c) If the amount in the bank is \$5395 after 12 years, what was the original investment P if it has been earning simple interest at the annual rate of $12\frac{1}{2}\%$?

47. Find the first eight terms of the sequence defined recursively by $a_1 = 12$ and $a_n = -\frac{1}{2}a_{n-1} + 2$ for $n > 1$.

48. Find the first six terms of the sequence defined recursively by $a_1 = 6$ and $a_n = \dfrac{3}{a_{n-1}}$ for $n > 1$.

49. Write the first eight terms of $a_n = \dfrac{1 + (-1)^{n+1}}{2i^{n-1}}$. (See page 66 for the powers of $i = \sqrt{-1}$.)

50. **(a)** Write the first seven terms of $a_n = n!$ where $n!$ is read as "n factorial" and is defined by $n! = n(n - 1) \cdot (n - 2) \cdots 3 \cdot 2 \cdot 1$

  (b) If $a_n = \dfrac{(2n)!}{(n!)^2}$, then show that $\dfrac{a_{n+1}}{a_n} = 2\left(\dfrac{2n + 1}{n + 1}\right)$.

51. Write the first four terms of

$$a_n = \frac{3 \cdot 5 \cdots (2n - 1)(2n + 1)}{2 \cdot 4 \cdots (2n - 2)(2n)}$$

Find a formula for the *n*th triangular number. (*Hint:* See Exercise 45 and consider the additional number of dots in each new figure.)

11.2
SUMS OF
FINITE
SEQUENCES

How long would it take you to add up the integers from 1 to 1000? Here is a quick way. List the sequence displaying the first few and last few terms.

$$1, 2, 3, \ldots, 998, 999, 1000$$

Add them in pairs, the first and last, the second and second from last, and so on.

It is told that Carl Friedrich Gauss (1777–1855) discovered how to compute such sums when he was 10 years old.

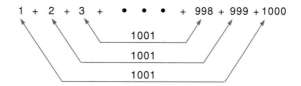

Since there are 500 such pairs to be added, the total is

$$500(1001) = 500,500$$

For any finite sequence we can add up all its terms and say that we have found the *sum of the sequence*. The sum of a sequence is called a **series**. For example, the sequence

$$1, \quad 3, \quad 5, \quad 7, \quad 9, \quad 11$$

can be associated with the series

$$1 + 3 + 5 + 7 + 9 + 11$$

The sum of the terms in this series can easily be found, by adding, to be 36. As another example, the sequence $a_n = \dfrac{1}{n}$ for $n = 1, 2, 3, 4, 5$ has the sum

$$1 + \frac{1}{2} + \frac{1}{3} + \frac{1}{4} + \frac{1}{5} = \frac{60 + 30 + 20 + 15 + 12}{60} = \frac{137}{60}$$

EXAMPLE 1 Find the sum of the first seven terms of $a_k = 2k$.

Solution

$$a_1 + a_2 + a_3 + a_4 + a_5 + a_6 + a_7 = 2 + 4 + 6 + 8 + 10 + 12 + 14 = 56$$

∎

Just think of sigma as a command to add.

There is a very handy notational device available for expressing the sum of a sequence. The Greek letter Σ (capital sigma) is used for this purpose. Referring to Example 1, the sum of the seven terms is expressed by the symbol $\Sigma_{k=1}^{7} a_k$; that is,

$$\sum_{k=1}^{7} a_k = a_1 + a_2 + a_3 + a_4 + a_5 + a_6 + a_7$$

Add the terms a_k for consecutive values of k, starting with $k = 1$ up to and including $k = 7$. With this symbolism, the question in Example 1 can now be stated by asking for the value of $\Sigma_{k=1}^{7} a_k$, where $a_k = 2k$, or by asking for the value of $\Sigma_{k=1}^{7} 2k$.

EXAMPLE 2   Find $\displaystyle\sum_{n=1}^{5} b_n$, where $b_n = \dfrac{n}{n + 1}$.

Solution

$$\sum_{n=1}^{5} b_n = b_1 + b_2 + b_3 + b_4 + b_5$$

Now replace each b_n by its numerical value:

$$\sum_{n=1}^{5} b_n = \frac{1}{2} + \frac{2}{3} + \frac{3}{4} + \frac{4}{5} + \frac{5}{6}$$

$$= \frac{30 + 40 + 45 + 48 + 50}{60}$$

$$= \frac{213}{60} = \frac{71}{20} \qquad \blacksquare$$

EXAMPLE 3   Evaluate $\displaystyle\sum_{k=1}^{4} (2k + 1)$.

Solution   It is understood here that we are to find the sum of the first four terms of the sequence whose general term is $a_k = 2k + 1$.

$$\sum_{k=1}^{4} (2k + 1) = (2 \cdot 1 + 1) + (2 \cdot 2 + 1) + (2 \cdot 3 + 1) + (2 \cdot 4 + 1)$$

$$= 3 + 5 + 7 + 9$$

$$= 24 \qquad \blacksquare$$

*The letter k that appears on the bottom of the sigma, and takes on the values from 1 to 4, is called the **index of summation**. The letter n is used as the index of summation in Example 2.*

EXAMPLE 4   Find $\displaystyle\sum_{i=1}^{5} x_i$, where $x_i = (-1)^i(i + 1)$.

Solution

$$\sum_{i=1}^{5} x_i = x_1 + x_2 + x_3 + x_4 + x_5$$

$$= (-1)^1(1 + 1) + (-1)^2(2 + 1) + (-1)^3(3 + 1)$$
$$+ (-1)^4(4 + 1) + (-1)^5(5 + 1)$$

$$= -2 + 3 - 4 + 5 - 6$$

$$= -4 \qquad \blacksquare$$

Evaluate each of the following.

1. $\displaystyle\sum_{k=1}^{5}(4k)$ **2.** $\displaystyle\sum_{k=1}^{5}(2k-1)$ **3.** $\displaystyle\sum_{k=1}^{5}(k^2-k)$

4. $\displaystyle\sum_{n=1}^{6}(-1)^n$ **5.** $\displaystyle\sum_{n=1}^{4}(2n^2-n)$ **6.** $\displaystyle\sum_{n=1}^{4}2(-\tfrac{1}{2})^{n+1}$

When the general term of a given series can be found, then the series can be rewritten using the sigma notation. For example, since the series

$$3 + 6 + 9 + 12 + 15 + 18 + 21$$

Note: In some of the exercises the summation begins with values of i, n, or k other than 1. For example, see Exercises 8, 22, 24, and 25.

is the sum of the first seven multiples of 3, it can be written as $\sum_{k=1}^{7} 3k$.

EXERCISES 11.2

Find the sum of the first five terms of the sequence given by the formula in each exercise.

1. $a_n = 3n$ **2.** $a_k = (-1)^k\dfrac{1}{k}$ **3.** $a_i = i^2$ **4.** $b_i = i^3$ **5.** $b_k = \dfrac{3}{10^k}$ **6.** $b_n = -6 + 2(n-1)$

7. Find $\sum_{n=1}^{8} t_n$ where $t_n = 2^n$.   **8.** Find $\sum_{n=0}^{8} x_n$ where $x_n = \dfrac{1}{2^n}$.   **9.** Find $\sum_{k=1}^{20} y_k$ where $y_k = 3$.

Find each of the following sums for n = 7.

10. $2 + 4 + \cdots + 2n$ **11.** $2 + 4 + \cdots + 2^n$

12. $-7 + 2 + \cdots + (9n - 16)$ **13.** $3 + \frac{3}{2} + \cdots + 3(\frac{1}{2})^{n-1}$

Evaluate each of the following.

14. $\displaystyle\sum_{k=1}^{6}(5k)$ **15.** $5\left(\displaystyle\sum_{k=1}^{6}k\right)$ **16.** $\displaystyle\sum_{n=1}^{4}(n^2+n)$

17. $\displaystyle\sum_{n=1}^{4}n^2 + \displaystyle\sum_{n=1}^{4}n$ **18.** $\displaystyle\sum_{t=1}^{8}(i - 2i^2)$ **19.** $\displaystyle\sum_{k=1}^{4}\dfrac{k}{2^k}$

20. $\displaystyle\sum_{k=1}^{7}(-1)^k$ **21.** $\displaystyle\sum_{k=1}^{8}(-1)^k$ **22.** $\displaystyle\sum_{k=3}^{7}(2k - 5)$

23. $\displaystyle\sum_{j=1}^{6}[-3 + (j - 1)5]$ **24.** $\displaystyle\sum_{k=-3}^{3}10^k$ **25.** $\displaystyle\sum_{k=-3}^{3}\dfrac{1}{10^k}$

26. $\displaystyle\sum_{k=1}^{5}4(-\tfrac{1}{2})^{k-1}$ **27.** $\displaystyle\sum_{t=1}^{4}(-1)^i 3^i$ **28.** $\displaystyle\sum_{n=1}^{3}\left(\dfrac{n+1}{n} - \dfrac{n}{n+1}\right)$

29. $\displaystyle\sum_{n=1}^{3}\dfrac{n+1}{n} - \displaystyle\sum_{n=1}^{3}\dfrac{n}{n+1}$ **30.** $\displaystyle\sum_{k=1}^{8}\dfrac{1 + (-1)^k}{2}$ **31.** $\displaystyle\sum_{k=1}^{3}(0.1)^{2k}$

Rewrite each series using sigma notation.

32. $4 + 8 + 12 + 16 + 20 + 24$ **33.** $5 + 10 + 15 + 20 + \cdots + 50$

34. $-4 - 2 + 0 + 2 + 4 + 6 + 8$ **35.** $-9 - 6 - 3 + 0 + 3 + \cdots + 24$

36. Read the discussion at the beginning of this section, where we found the sum of the first 1000 positive integers, and find a formula for the sum of the first n positive integers for n even.

37. **(a)** Find $\sum_{k=1}^{n}(2k - 1)$ for each of the following values of n: 2, 3, 4, 5, 6.

   (b) On the basis of the results in part (a), find a formula for the sum of the first n odd numbers.

38. The sequence 1, 1, 2, 3, 5, 8, 13, . . . is called the *Fibonacci sequence.* Its first two terms are ones, and each term thereafter is computed by adding the preceding two terms.

   (a) Write the next seven terms of this sequence.

   (b) Let $u_1, u_2, u_3, \ldots, u_n, \ldots$ be the Fibonacci sequence. Evaluate $S_n = \sum_{k=1}^{n} u_k$ for these values of n: 1, 2, 3, 4, 5, 6, 7, 8.

   (c) Note that $u_1 = u_3 - u_2$, $u_2 = u_4 - u_3$, $u_3 = u_5 - u_4$, and so on. Use this form for the first n numbers to derive a formula for the sum of the first n Fibonacci numbers.

39. Let a_n be a sequence with $a_1 = 2$ and $a_m + a_n = a_{m+n}$, where m and n are any positive integers. Show that $a_n = 2n$ for any n.

40. Show that $\sum_{k=1}^{9} \log_{10} \dfrac{k + 1}{k} = 1$. $\left(\textit{Hint: } \log_{10} \dfrac{a}{b} = \log_{10} a - \log_{10} b. \right)$

41. Prove: $\sum_{k=1}^{n} s_k + \sum_{k=1}^{n} t_k = \sum_{k=1}^{n} (s_k + t_k)$.

42. Prove: $\sum_{k=1}^{n} cs_k = c \sum_{k=1}^{n} s_k$, c a constant.

43. Prove: $\sum_{k=1}^{n} (s_k + c) = \left(\sum_{k=1}^{n} s_k \right) + nc$, c a constant.

44. **(a)** Evaluate $\sum_{k=1}^{10} \dfrac{1}{k(k + 1)}$ using the result $\dfrac{1}{k(k + 1)} = \dfrac{1}{k} - \dfrac{1}{k + 1}$.

   (b) Use the identity given in part (a) to prove that $\dfrac{1}{1 \cdot 2} + \dfrac{1}{2 \cdot 3} + \dfrac{1}{3 \cdot 4} + \cdots + \dfrac{1}{98 \cdot 99} + \dfrac{1}{99 \cdot 100} = \dfrac{99}{100}$.

45. Use the identity $\dfrac{2}{k(k + 2)} = \dfrac{1}{k} - \dfrac{1}{k + 2}$ to show that $\sum_{k=1}^{10} \dfrac{2}{k(k + 2)} = \dfrac{175}{132}$.

CHALLENGE
Think Creatively

This question refers to matrix multiplication developed in Section 10.2. Let $A = [a_{ij}]$ be m by n and $B = [b_{ij}]$ be n by p so that $AB = C = [c_{ij}]$ is m by p. The computation that produces an element c_{ij} in the product involves the ith row of A and jth column of B. Express this computation using the sigma notation.

11.3
ARITHMETIC
SEQUENCES
AND SERIES

Here are the first five terms of the sequence whose general term is $a_k = 7k - 2$:

$$5, 12, 19, 26, 33$$

Do you notice any special pattern? It does not take long to observe that each term, after the first, is 7 more than the preceding term. This sequence is an example of an *arithmetic sequence.*

DEFINITION OF AN ARITHMETIC SEQUENCE

A sequence is said to be *arithmetic* if each term, after the first, is obtained from the preceding term by adding a common value.

Let us consider the first four terms of two different arithmetic sequences:

$$2, 4, 6, 8, \ldots$$

$$-\tfrac{1}{2}, -1, -\tfrac{3}{2}, -2, \ldots$$

For the first sequence, the common value (or difference) that is added to each term to get the next is 2. Thus it is easy to see that 10, 12, and 14 are the next three terms. You might guess that the *n*th term is $a_n = 2n$.

The second sequence has the common difference $-\tfrac{1}{2}$. This can be found by subtracting the first term from the second, or the second from the third, and so forth. The *n*th term is $a_n = -\tfrac{1}{2}n$.

Unlike the preceding sequences, it is not always easy to see what the *n*th term of a specific arithmetic sequence is. Therefore, we will now develop a general form that makes it possible to write the *n*th term of any such sequence.

Let a_n be the *n*th term of an arithmetic sequence, and let *d* be the **common difference**. Then the first four terms are:

$$a_1$$

$$a_2 = a_1 + d$$

$$a_3 = a_2 + d = (a_1 + d) + d = a_1 + 2d$$

$$a_4 = a_3 + d = (a_1 + 2d) + d = a_1 + 3d$$

The pattern is clear. Without further computation we see that

$$a_5 = a_1 + 4d \quad \text{and} \quad a_6 = a_1 + 5d$$

Since the coefficient of *d* is always 1 less than the number of the term, the *n*th term is given as follows.

GENERAL TERM OF AN ARITHMETIC SEQUENCE

The *n*th term of an arithmetic sequence is

$$a_n = a_1 + (n - 1)d$$

where a_1 is the first term and *d* is the common difference.

By substituting the values $n = 1, 2, 3, 4, 5, 6$, you can check that this formula gives the preceding terms a_1 through a_6.

EXAMPLE 1   Find the *n*th term of the arithmetic sequence 11, 2, −7, ...

Solution   Use the formula for a_n with $a_1 = 11$ and $d = a_2 - a_1 = 2 - 11 = -9$. Thus

$$a_n = a_1 + (n - 1)d$$
$$= 11 + (n - 1)(-9) = -9n + 20 \qquad \blacksquare$$

Each of the following gives the first few terms of an arithmetic sequence. Find the nth term in each case.

1. 5, 10, 15, . . .   **2.** 6, 2, −2, . . .   **3.** $\frac{1}{10}, \frac{1}{5}, \frac{3}{10}, \ldots$

4. −5, −13, −21, . . .   **5.** 1, 2, 3, . . .   **6.** −3, −2, −1, . . .

Find the nth term a_n of the arithmetic sequence with the given values for the first term and the common difference.

7. $a_1 = \frac{2}{3}; d = \frac{2}{3}$ **8.** $a_1 = 53; d = -12$

9. $a_1 = 0; d = \frac{1}{5}$ **10.** $a_1 = 2; d = 1$

EXAMPLE 2   The first term of an arithmetic sequence is −15 and the fifth term is 13. Find the fortieth term.

Solution   Since $a_5 = 13$ use $n = 5$ in the formula $a_n = a_1 + (n - 1)d$ in order to solve for d.

$$a_5 = a_1 + (5 - 1)d$$
$$13 = -15 + 4d$$
$$28 = 4d$$
$$7 = d$$

Then $a_{40} = -15 + (39)7 = 258.$ \qquad $\blacksquare$

Adding the terms of a finite sequence may not be much work when the number of terms to be added is small. When many terms are to be added, however, the amount of time and effort needed can be overwhelming. For example, to add the first 10,000 terms of the arithmetic sequence beginning with

$$246, \quad 261, \quad 276, \ldots$$

would call for an enormous effort, unless some shortcut could be found. Fortunately, there is an easy way to find such sums. This method (in disguise) was already used in the question at the start of Section 11.2. Let us look at the general situation. Let S_n denote the sum of the first *n* terms of the arithmetic sequence give by $a_k = a_1 + (k - 1)d$:

*The sum of an arithmetic sequence is called an **arithmetic series**.*

$$S_n = \sum_{k=1}^{n} [a_1 + (k - 1)d]$$
$$= a_1 + [a_1 + d] + [a_1 + 2d] + \cdots + [a_1 + (n - 1)d]$$

Put this sum in reverse order and write the two equalities together as follows:

$$S_n = a_1 + [a_1 + d] + \cdots + [a_1 + (n - 2)d] + [a_1 + (n - 1)d]$$

$$\updownarrow \qquad \updownarrow \qquad \qquad \updownarrow \qquad \qquad \updownarrow$$

$$S_n = [a_1 + (n - 1)d] + [a_1 + (n - 2)d] + \cdots + [a_1 + d] + a_1$$

Now add to get

$$2S_n = [2a_1 + (n - 1)d] + [2a_1 + (n - 1)d] + \cdots + [2a_1 + (n - 1)d] + [2a_1 + (n - 1)d]$$

On the right-hand side of this equation there are n terms, each of the form $2a_1 + (n - 1)d$. Therefore,

$$2S_n = n[2a_1 + (n - 1)d]$$

Divide by 2 to solve for S_n:

$$S_n = \frac{n}{2}[2a_1 + (n - 1)d]$$

Returning to the sigma notation, we can summarize our results this way:

ARITHMETIC SERIES

$$\sum_{k=1}^{n} [a_1 + (k - 1)d] = \frac{n}{2}[2a_1 + (n - 1)d]$$

EXAMPLE 3 Find S_{20} for the arithmetic sequence whose first term is $a_1 = 3$ and whose common difference is $d = 5$.

Solution Substituting $a_1 = 3$, $d = 5$, and $n = 20$ into the formula for S_n, we have

$$S_{20} = \frac{20}{2}[2(3) + (20 - 1)5]$$

$$= 10(6 + 95)$$

$$= 1010 \qquad \blacksquare$$

EXAMPLE 4 Find the sum of the first 10,000 terms of the arithmetic sequence beginning with 246, 261, 276, ...

Solution Since $a_1 = 246$ and $d = 15$,

$$S_{10,000} = \frac{10,000}{2}[2(246) + (10,000 - 1)15]$$

$$= 5000(150,477)$$

$$= 752,385,000 \qquad \blacksquare$$

EXAMPLE 5 Find the sum of the first n positive integers.

Solution   First observe that the problem calls for the sum of the sequence $a_k = k$ for $k = 1, 2, \ldots, n$. This is an arithmetic sequence with $a_1 = 1$ and $d = 1$. Therefore,

$$\sum_{k=1}^{n} k = \frac{n}{2}[2(1) + (n - 1)1] = \frac{n(n + 1)}{2} \qquad \blacksquare$$

With the result of Example 5 we are able to check the answer for the sum of the first 1000 positive integers, found at the beginning of Section 11.2, as follows:

$$\sum_{k=1}^{1000} k = \frac{1000(1001)}{2} = 500,500$$

The form $a_k = a_1 + (k - 1)d$ for the general term of an arithmetic sequence easily converts to $a_k = dk + (a_1 - d)$. It is this latter form that is ordinarily used when the general term of a *specific* arithmetic sequence is given. For example, we would usually begin with the form $a_k = 3k + 5$ instead of $a_k = 8 + (k - 1)3$. The important thing to notice in the form $a_k = dk + (a_1 - d)$ is that the common difference is the coefficient of k.

EXAMPLE 6   Evaluate: $\displaystyle\sum_{k=1}^{50}(-6k + 10)$

Solution   First note that $a_k = -6k + 10$ is an arithmetic sequence with $d = -6$ and with $a_1 = 4$.

$$\sum_{k=1}^{50}(-6k + 10) = \tfrac{50}{2}[2(4) + (50 - 1)(-6)]$$

$$= -7150 \qquad \blacksquare$$

The formula for an arithmetic series can be converted into another useful form when a_n is substituted for $a_1 + (n - 1)d$. Thus

When this form is rewritten as

$$S_n = n\left(\frac{a_1 + a_n}{2}\right)$$

the sum can be viewed as n times the average of the first and nth terms.

$$S_n = \frac{n}{2}[2a_1 + (n - 1)d]$$

$$= \frac{n}{2}[a_1 + a_1 + (n - 1)d]$$

$$= \frac{n}{2}[a_1 + a_n]$$

ARITHMETIC SERIES (Alternative Form)

$$S_n = \frac{n}{2}(a_1 + a_n)$$

In order to decide when a number N is divisible by 3, use the fact that N is divisible by 3 only when the sum of its digits is divisible by 3. Thus 261 is divisible by 3 because 2 + 6 + 1 = 9 is divisible by 3.

EXAMPLE 7 Find the sum of all the multiples of 3 between 4 and 262.

Solution The first multiple of 3 after 4 is $6 = a_1$ and the one preceding 262 is $261 = a_n$. Now find n.

$$a_n = a_1 + (n - 1)d$$

$$261 = 6 + (n - 1)3$$

$$86 = n$$

Then

$$S_{86} = \tfrac{86}{2}[a_1 + a_{86}] = \tfrac{86}{2}[6 + 261] = 11{,}481 \quad \blacksquare$$

EXAMPLE 8 Suppose you take out a short-term loan of $6000 that you agree to pay back in 12 equal monthly payments, plus 3% interest per month on the monthly balance. Each month's payment is made during the first week of the new month.

(a) Write the general term of the sequence that gives the monthly balance.

(b) How much is the interest for each of the first 3 months? Write the general term of the sequence that gives the amount of the monthly interest.

(c) What is the total interest payment for the year, and what percent is this total of the $6000 loan?

Solution

(a) The equal monthly payments on the loan are $6000 \div 12 = 500$ dollars. Since these payments are made during the first week of each month, the monthly balance is given by

$$a_k = 6000 - 500(k - 1)$$

At the end of the first month your balance is
$6000 - 500(0) = 6000$,
and for $k = 12$,
$6000 - 500(11) = 500$,
which is the last payment made during the first week of the thirteenth month.

for $k = 1, 2, \ldots , 12$.

(b) The first month's interest is 3% of $6000, or $6000(.03) = 180$ dollars. You also pay back $500, so your second month's interest is $5500(.03) = 165$ dollars. Similarly, the interest for the third month is $5000(.03) = 150$ dollars. In general, the monthly interest payment is

This is an arithmetic sequence with $a_1 = 180$ and $d = -15$.

$$\overset{\left(\substack{\text{monthly}\\\text{balance}}\right)}{\downarrow} \qquad \overset{\left(\substack{\text{monthly}\\\text{rate}}\right)}{\downarrow}$$

$$[6000 - 500(k - 1)](.03) = 180 - 15(k - 1)$$

$$= 195 - 15k$$

(c) The total interest is the sum of the 12 interest payments. Thus

$$\sum_{k=1}^{12}(195 - 15k) = \frac{12}{2}(2 \cdot 180 + 11(-15))$$

$$= 6(195)$$

$$= 1170$$

The annual rate is $\dfrac{1170}{6000} = 0.195 = 19.5\%$. $\quad \blacksquare$

EXERCISES 11.3

Each of the following gives the first two terms of an arithmetic sequence. Write the next three terms; find the nth term; and find the sum of the first 20 terms.

1. 1, 3, . . .
2. 2, 4, . . .
3. 2, −4, . . .
4. 1, −3, . . .
5. $\frac{15}{2}$, 8, . . .
6. $-\frac{4}{3}$, $-\frac{11}{3}$, . . .
7. $\frac{2}{5}$, $-\frac{1}{5}$, . . .
8. $-\frac{1}{2}$, $\frac{1}{4}$, . . .
9. 50, 100, . . .
10. −27, −2, . . .
11. −10, 10, . . .
12. 225, 163, . . .

Find the indicated sum by using ordinary addition; also find the sum by using the formula for the sum of an arithmetic sequence.

13. 5 + 10 + 15 + 20 + 25 + 30 + 35 + 40 + 45 + 50 + 55 + 60 + 65
14. −33 − 25 − 17 − 9 − 1 + 7 + 15 + 23 + 31 + 39
15. $\frac{3}{4} + 1 + \frac{5}{4} + \frac{3}{2} + \frac{7}{4} + 2 + \frac{9}{4} + \frac{5}{2} + \frac{11}{4}$
16. 128 + 71 + 14 − 43 − 100 − 157
17. Find a_{30} for the arithmetic sequence having $a_1 = -30$ and $a_{10} = 69$.
18. Find a_{51} for the arithmetic sequence having $a_1 = 9$ and $a_8 = -19$.

Find S_{100} for the arithmetic sequence with the given values for a_1 and d.

19. $a_1 = 3; d = 3$
20. $a_1 = 1; d = 8$
21. $a_1 = -91; d = 21$
22. $a_1 = -7; d = -10$
23. $a_1 = \frac{1}{7}; d = 5$
24. $a_1 = \frac{2}{5}; d = -4$
25. $a_1 = 725; d = 100$
26. $a_1 = 0.1; d = 10$

27. Find S_{28} for the sequence $-8, 8, \ldots, 16n - 24, \ldots$.
28. Find S_{25} for the sequence $96, 100, \ldots, 4n + 92, \ldots$.
29. Find the sum of the first 50 positive multiples of 12.
30. (a) Find the sum of the first 100 positive even numbers.
    (b) Find the sum of the first n positive even numbers.
31. (a) Find the sum of the first 100 positive odd numbers.
    (b) Find the sum of the first n positive odd numbers.

Evaluate the series in each exercise.

32. $\sum_{k=1}^{12}[3 + (k - 1)9]$
33. $\sum_{k=1}^{9}[-6 + (k - 1)\frac{1}{2}]$
34. $\sum_{k=1}^{20}(4k - 15)$
35. $\sum_{k=1}^{30}(10k - 1)$
36. $\sum_{k=1}^{40}(-\frac{1}{3}k + 2)$
37. $\sum_{k=1}^{49}(\frac{3}{4}k - \frac{1}{2})$
38. $\sum_{k=1}^{20}5k$
39. $\sum_{k=1}^{n}5k$

40. Find u such that $7, u, 19$ is an arithmetic sequence.
41. Find u such that $-7, u, \frac{5}{2}$ is an arithmetic sequence.
42. Find the twenty-third term of the arithmetic sequence $6, -4, \ldots$.
43. Find the thirty-fifth term of the arithmetic sequence $-\frac{2}{3}, -\frac{1}{5}, \ldots$.
44. An object is dropped from an airplane and falls 32 feet during the first second. During each successive second it falls 48 feet more than in the preceding second. How many feet does it travel during the first 10 seconds? How far does it fall during the tenth second?
45. Suppose you save $10 one week and that each week thereafter you save 50¢ more than the preceding week. How much will you have saved by the end of 1 year?
46. Suppose a 100-pound bag of grain has a small hole in the bottom that is steadily getting larger. The first minute $\frac{1}{3}$ ounce of grain leaks out, and each successive minute $\frac{1}{3}$ ounce more grain leaks out than during the preceding minute. How many pounds of grain remain in the bag after one hour?

47. A $12,000 loan is paid back in 12 equal monthly payments, made during the first week of the new month. The interest rate is 2% per month on the monthly balance. (This is the amount at the end of a month before the 1000 dollar payment for that month is made.)

(a) Find the monthly interest payment for each of the first 3 months.

(b) Write the general term of the sequences that give the monthly loan balance and the monthly interest.

(c) Find the total interest paid for the year and the annual interest rate.

48. A pyramid of blocks has 26 blocks in the bottom row and 2 fewer blocks in each successive row thereafter. How many blocks are there in the pyramid?

49. Evaluate $\sum_{n=6}^{20}(5n - 3)$.

Use the form $S_n = \dfrac{n}{2}(a_1 + a_n)$ to find S_{80} for the sequences in Exercises 50 and 51.

50. $a_k = \frac{1}{2}k + 10$ **51.** $a_k = 3k - 8$

52. Find the sum of all the even numbers between 33 and 427.

53. Find the sum of all odd numbers from 33 to 427.

54. Find the sum of all multiples of 4 from -100 to 56.

55. If $\sum_{k=1}^{30}[a_1 + (k - 1)d] = -5865$ and $\sum_{k=1}^{20}[a_1 + (k - 1)d] = -2610$, find a_1 and d.

56. Listing the first few terms of a sequence like 2, 4, 6, . . . , without stating its general term or describing just what kind of sequence it is makes it impossible to predict the next term. Show that both sequences $t_n = 2n$ and $u_n = 2n + (n - 1)(n - 2)(n - 3)$ produce these first three terms but that their fourth terms are different.

57. Find u and v such that 3, u, v, 10 is an arithmetic sequence.

58. What is the connection between arithmetic sequences and linear functions?

59. The function f defined by $f(x) = 3x + 7$ is a linear function. Evaluate the series $\sum_{k=1}^{16} f_k$, where $f_k = f(k)$ is the arithmetic sequence associated with f.

EXPLORATIONS
Think Critically

1. Infinite sequences may or may not have an infinite number of distinct terms. Give examples of general terms for infinite sequences having the following number of distinct terms: (a) one, (b) two, (c) three.

2. If the addend $2k - 1$ in $\sum_{k=1}^{n}(2k - 1)$ is changed to $2k + 1$, then the value of the summation will remain the same provided that the proper change is made with the index of summation. Make this adjustment by completing the equation $\sum_{k=1}^{n}(2k - 1) = \sum(2k + 1)$. Now complete the equality $\sum_{k=3}^{n+1}(k + 1)^2 = \sum_{k=1}^{n-1} \ldots$, in which the index of summation has been modified, by adjusting the addend.

3. The procedure used in Exercises 44 and 45 on page 566 avoids tedious computations in evaluating certain sums. Use the method that was presented in Section 4.8 to convert $a_k = \dfrac{4}{(k + 1)(k + 3)}$ into the difference of two fractions and then evaluate $\sum_{k=1}^{10} a_k$.

4. In Exercises 30(b) and 31(b) on page 572, the sum of the first n odd integers and the sum of the first n even integers are called for. When these two answers are added, the result represents the sum of which consecutive integers beginning with 1? Use the formula of an arithmetic series to verify your result.

Suppose that a ball is dropped from a height of 4 feet and bounces straight up and down, always bouncing up exactly one-half the distance it just came down. How far will the ball have traveled if you catch if after it reaches the top of the fifth bounce? The following figure will help you to answer this question. For the sake of clarity, the bounces have been separated in the figure.

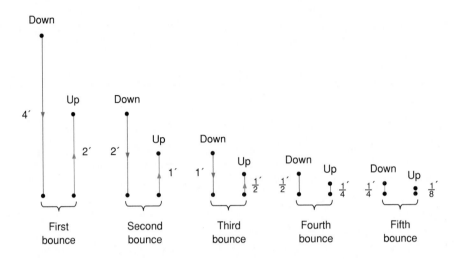

From this diagram we can determine how far the ball has traveled on each bounce. On the first bounce it goes 4 feet down and 2 feet up, for a total of 6 feet; on the second bounce the total distance is $2 + 1 = 3$ feet; and so on. These distances from the following sequence of five terms (one for each bounce):

$$6, \quad 3, \quad \frac{3}{2}, \quad \frac{3}{4}, \quad \frac{3}{8}$$

This sequence has the special property that, after the first term, each successive term can be obtained by multipying the preceding term by $\frac{1}{2}$; that is, the second term, 3, is half the first, 6, and so on. This is an example of a **geometric sequence.** Later we will develop a formula for finding the sum of such a sequence; in the meantime, we can find the total distance the ball has traveled during the five bounces by adding the first five terms as follows:

$$6 + 3 + \frac{3}{2} + \frac{3}{4} + \frac{3}{8} = \frac{48 + 24 + 12 + 6 + 3}{8} = 11\frac{5}{8}$$

*A geometric sequence is also referred to as a **geometric progression**.*

DEFINITION OF A GEOMETRIC SEQUENCE

A sequence is said to be *geometric* if each term, after the first, is obtained by multiplying the preceding term by a common value.

Here are the first four terms of a geometric sequence

$$2, \, -4, \, 8, \, -16, \, \ldots$$

By inspection, you can determine that the common multiplier for this sequence is -2. To find the nth term we first derive the formula for the nth term of any geometric sequence.

Let a_n be the nth term of a geometric sequence, and let a_1 be its first term. The common multiplier, which is also called the **common ratio**, is denoted by r. Here are the first four terms:

$$a_1$$

$$a_2 = a_1 r$$

$$a_3 = a_2 r = (a_1 r)r = a_1 r^2$$

$$a_4 = a_3 r = (a_1 r^2)r = a_1 r^3$$

Notice that the exponent of r is 1 less than the number of the term. This observation allows us to write the nth term as follows:

GENERAL TERM OF A GEOMETRIC SEQUENCE

This formula says that the nth term of a geometric sequence is completely determined by its first term a₁ and common ratio r.

The nth term of a geometric sequence is

$$a_n = a_1 r^{n-1}$$

where a_1 is the first term and r is the common ratio.

With this result, the first four terms and the nth terms of the geometric sequence given previously are as follows.

$$2, -4, 8, -16, \ldots, 2(-2)^{n-1} \qquad (r = -2)$$

Here are two more illustrations:

$$1, \frac{1}{3}, \frac{1}{9}, \frac{1}{27}, \ldots, 1\left(\frac{1}{3}\right)^{n-1} \qquad (r = \tfrac{1}{3})$$

$$5, -5, 5, -5, \ldots, 5(-1)^{n-1} \qquad (r = -1)$$

You can substitute the values $n = 1, 2, 3$, and 4 into the forms for the nth terms and see that the given first four terms are obtained in each case.

EXAMPLE 1 Find the hundredth term of the geometric sequence having $r = \tfrac{1}{2}$ and $a_1 = \tfrac{1}{2}$.

Solution The nth term of this sequence is given by

$$a_n = \frac{1}{2}\left(\frac{1}{2}\right)^{n-1} = \frac{1}{2}\left(\frac{1}{2^{n-1}}\right) = \frac{1}{2^n}$$

Thus $a_{100} = \dfrac{1}{2^{100}}$.

■

The reason r is called the common ratio of a geometric sequence $a_n = a_1 r^{n-1}$ is that for each n the ratio of the $(n + 1)$st term to the nth term equals r. Thus

$$\frac{a_{n+1}}{a_n} = \frac{a_1 r^n}{a_1 r^{n-1}} = r$$

Note that r can also be found using a_2 and a_3.

$$\frac{a_3}{a_2} = \frac{\frac{27}{2}}{9} = \frac{27}{2 \cdot 9} = \frac{3}{2}$$

EXAMPLE 2   Find the nth term of the geometric sequence $6, 9, \frac{27}{2}, \ldots$ and find the seventh term.

Solution   First find r.

$$r = \frac{a_2}{a_1} = \frac{9}{6} = \frac{3}{2}$$

Then the nth term is

$$a_n = 6\left(\frac{3}{2}\right)^{n-1}$$

Let $n = 7$ to get $a_7 = 6\left(\frac{3}{2}\right)^{7-1} = 6\left(\frac{3}{2}\right)^6$

$$= (3 \cdot 2) \cdot \frac{3^6}{2^6} = \frac{3^7}{2^5} = \frac{2187}{32} \qquad \blacksquare$$

EXAMPLE 3   Write the kth term of the geometric sequence $a_k = \left(\frac{1}{2}\right)^{2k}$ in the form $a_1 r^{k-1}$ and find the value of a_1 and r.

Solution

$$a_k = \left(\tfrac{1}{2}\right)^{2k} = \left[\left(\tfrac{1}{2}\right)^2\right]^k = \left(\tfrac{1}{4}\right)^k$$

$$= \tfrac{1}{4}\left(\tfrac{1}{4}\right)^{k-1} \quad \longleftarrow \quad \text{This is now in the form } a_1 r^{n-1}.$$

The first term a_1 can also be found by substituting 1 into the given formula for a_k; then find a_2 and evaluate

$$r = \frac{a_2}{a_1}.$$

Then $a_1 = \frac{1}{4}$ and $r = \frac{1}{4}$. $\qquad \blacksquare$

TEST YOUR UNDERSTANDING
Think Carefully

Write the first five terms of the geometric sequences with the given general term. Also write the nth term in the form $a_1 r^{n-1}$ and find r.

1. $a_n = \left(\tfrac{1}{2}\right)^{n-1}$ **2.** $a_n = \left(\tfrac{1}{2}\right)^{n+1}$ **3.** $a_n = \left(-\tfrac{1}{2}\right)^n$ **4.** $a_n = \left(-\tfrac{1}{3}\right)^{3n}$

Find r and the nth term of the geometric sequence with the given first two terms.

(Answers: Page 600)

5. $\frac{1}{5}, 2$ **6.** $27, -12$

EXAMPLE 4   A geometric sequence consisting of positive numbers has $a_1 = 18$ and $a_5 = \frac{32}{9}$. Find r.

Solution   Use $n = 5$ in $a_n = a_1 r^{n-1}$.

$$\frac{32}{9} = 18r^4$$

$$r^4 = \frac{32}{9 \cdot 18} = \frac{16}{81}$$

$$r = \pm\sqrt[4]{\frac{16}{81}} = \pm\frac{2}{3}$$

Check this result by writing $a_1 = 18$ and finding $a_2, a_3, a_4,$ and a_5 using $a_n = a_{n-1}r$.

Since the terms are positive, $r = \frac{2}{3}$. $\qquad \blacksquare$

Let us return to the original problem of this section. We found that the total distance the ball traveled was $11\frac{5}{8}$ feet. This is the sum of the first five terms of the geometric sequence whose nth term is $6(\frac{1}{2})^{n-1}$. Adding these five terms was easy. But what about adding the first 100 terms? There is a formula for the sum of a geometric sequence that will enable us to find such answers efficiently.

The sum of a geometric sequence is called a **geometric series.** Just as with arithmetic series, there is a formula for finding such sums. To discover this formula, let $a_k = a_1 r^{k-1}$ be a geometric sequence and denote the sum of the first n terms by $S_n = \sum_{k=1}^{n} a_1 r^{k-1}$. Then

$$S_n = a_1 + a_1 r + a_1 r^2 + \cdots + a_1 r^{n-2} + a_1 r^{n-1}$$

Multiplying this equation by r gives

$$rS_n = a_1 r + a_1 r^2 + \cdots + a_1 r^{n-1} + a_1 r^n$$

Now consider these two equations:

$$S_n = a_1 + a_1 r + a_1 r^2 + \cdots + a_1 r^{n-2} + a_1 r^{n-1}$$
$$rS_n = \qquad a_1 r + a_1 r^2 + \cdots + a_1 r^{n-2} + a_1 r^{n-1} + a_1 r^n$$

Subtract and factor:

$$S_n - rS_n = a_1 - a_1 r^n$$
$$(1 - r)S_n = a_1(1 - r^n)$$

Divide by $1 - r$ to solve for S_n:

Here $r \neq 1$. However, when $r = 1$, $a_k = a_1 r^{k-1} = a_1$, which is also an arithmetic sequence having $d = 0$.

$$S_n = \frac{a_1(1 - r^n)}{1 - r}$$

Returning to sigma notation, we can summarize our results this way:

GEOMETRIC SERIES

$$\sum_{k=1}^{n} a_1 r^{k-1} = \frac{a_1(1 - r^n)}{1 - r}$$

The preceding formula can be used to verify the earlier result for the bouncing ball:

$$\sum_{k=1}^{5} 6\left(\frac{1}{2}\right)^{k-1} = \frac{6[1 - (\frac{1}{2})^5]}{1 - \frac{1}{2}}$$

$$= \frac{6(1 - \frac{1}{32})}{\frac{1}{2}}$$

$$= \frac{93}{8} = 11\frac{5}{8}$$

EXAMPLE 5 Find the sum of the first 100 terms of the geometric sequence given by $a_k = 6(\frac{1}{2})^{k-1}$ and show that the answer is very close to 12.

Solution

$$S_{100} = \frac{6\left(1 - \dfrac{1}{2^{100}}\right)}{1 - \frac{1}{2}}$$

$$= 12\left(1 - \frac{1}{2^{100}}\right)$$

Next observe that the fraction $\dfrac{1}{2^{100}}$ is so small that $1 - \dfrac{1}{2^{100}}$ is very nearly equal to 1, and therefore S_{100} is very close to 12. ∎

Another way to find a_1 and r is to write the first few terms as follows:

$$\frac{3}{100} + \frac{3}{1000} + \frac{3}{10,000} + \cdots$$

$$= \frac{3}{100} + \frac{3}{100}\left(\frac{1}{10}\right)$$

$$+ \frac{3}{100}\left(\frac{1}{10}\right)^2 + \cdots$$

Then $a_1 = 0.03$ and $r = 0.1$.

EXAMPLE 6 Evaluate: $\displaystyle\sum_{k=1}^{8} 3\left(\frac{1}{10}\right)^{k+1}$

Solution

$$3\left(\frac{1}{10}\right)^{k+1} = \frac{3}{100}\left(\frac{1}{10}\right)^{k-1}$$

Then $a_1 = 0.03$, $r = 0.1$, and

$$S_8 = \frac{0.03[1 - (0.1)^8]}{1 - 0.1}$$

$$= \frac{0.03(1 - 0.00000001)}{0.9}$$

$$= 0.033333333 \qquad\qquad ∎$$

Geometric sequences have many applications, as illustrated by Examples 7 and 8. You will find others in the exercises at the end of this section.

EXAMPLE 7 Suppose that you save $128 in January and that each month thereafter you only manage to save half of what you saved the previous month. How much do you save in the tenth month, and what are your total savings after 10 months?

Solution The amounts saved each month form a geometric sequence with $a_1 = 128$ and $r = \frac{1}{2}$. Then $a_n = 128(\frac{1}{2})^{n-1}$ and

$$a_{10} = 128\left(\frac{1}{2}\right)^9 = \frac{2^7}{2^9} = \frac{1}{4} = 0.25$$

This means that you saved 25¢ in the tenth month. Your total savings are:

$$S_{10} = \frac{128\left(1 - \dfrac{1}{2^{10}}\right)}{1 - \frac{1}{2}} \qquad = 256\left(1 - \frac{1}{2^{10}}\right)$$

$$= 256 - \frac{256}{2^{10}}$$

$$= 256 - \frac{2^8}{2^{10}}$$

$$= 255.75$$

The total savings is $255.75. ■

EXAMPLE 8 A roll of wire is 625 feet long. If $\frac{1}{5}$ of the wire is cut off repeatedly, what is the general term of the sequence for the length of wire remaining? Use a calculator and the general term to determine the length of wire remaining after 7 cuts.

Solution Since $\frac{1}{5}$ is cut off, $\frac{4}{5} = 0.8$ must remain. Thus, $625(0.8) = 500$ feet remain after one cut, $625(0.8)(0.8) = 500(0.8) = 400$ feet remain after two cuts, and after n cuts are made, the length of wire remaining is $625(0.8)^n$ feet. Using a calculator, we have

$$625(0.8)^7 = 131.072$$

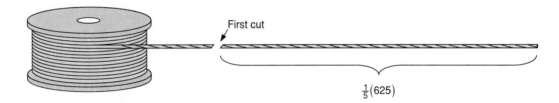

Therefore, to the nearest tenth of a foot, 131.1 feet of wire remains after 7 cuts. ■

EXERCISES 11.4

The first three terms of a geometric sequence are given. Write the next three terms and also find the formula for the nth term.

1. 2, 4, 8, . . .   2. 2, −4, 8, . . .   3. 1, 3, 9, . . .   4. 2, −2, 2, . . .

5. −3, 1, −$\frac{1}{3}$, . . .   6. 100, 10, 1, . . .   7. −1, −5, −25, . . .   8. 12, −6, 3, . . .

9. −6, −4, −$\frac{8}{3}$, . . .   10. −64, 16, −4, . . .   11. $\frac{1}{1000}$, $\frac{1}{10}$, 10, . . .   12. $\frac{27}{8}$, $\frac{3}{2}$, $\frac{2}{3}$, . . .

Find the sum of the first six terms of the indicated sequence by using ordinary addition and also by using the formula for a geometric series.

13. The sequence in Exercise 1.   14. The sequence in Exercise 5.

15. The sequence in Exercise 9.

16. Find the tenth term of the geometric sequence 2, 4, 8, ...

17. Find the fourteenth term of the geometric sequence $\frac{1}{8}$, $\frac{1}{4}$, $\frac{1}{2}$, ...

18. Find the fifteenth term of the geometric sequence $\dfrac{1}{100,000}$, $\dfrac{1}{10,000}$, $\dfrac{1}{1000}$, ...

19. What is the one-hundred-first term of the geometric sequence having $a_1 = 3$ and $r = -1$?

20. For the geometric sequence with $a_1 = 100$ and $r = \frac{1}{10}$, use the formula $a_n = a_1 r^{n-1}$ to find which term is equal to $\dfrac{1}{10^{10}}$.

21. Find r for the geometric series having $a_1 = 20$ and $a_6 = -\frac{5}{8}$.

22. Find r for the geometric series having $a_1 = -25$ and $a_5 = -3.24$.

Evaluate the series.

23. $\displaystyle\sum_{k=1}^{10} 2^{k-1}$

24. $\displaystyle\sum_{j=1}^{10} 2^{j+2}$

25. $\displaystyle\sum_{k=1}^{n} 2^{k-1}$

26. $\displaystyle\sum_{k=1}^{8} 3(\tfrac{1}{10})^{k-1}$

27. $\displaystyle\sum_{k=1}^{5} 3^{k-4}$

28. $\displaystyle\sum_{k=1}^{6} (-3)^{k-2}$

29. $\displaystyle\sum_{j=1}^{5} (\tfrac{2}{3})^{j-2}$

30. $\displaystyle\sum_{k=1}^{8} 16(\tfrac{1}{2})^{k+2}$

31. $\displaystyle\sum_{k=1}^{8} 16(-\tfrac{1}{2})^{k+2}$

32. Find $u > 0$ such that 2, u, 98 forms a geometric sequence.

33. Find $u < 0$ such that $\frac{1}{7}$, u, $\frac{25}{63}$ forms a geometric sequence.

34. Find a sequence whose first term is 5 that is both geometric and arithmetic. What are r and d?

35. Suppose someone offered you a job that pays 1¢ the first day, 2¢ the second day, 4¢ the third day, etc., each day earning double what was earned the preceding day. How many dollars would you earn for 30 days work?

36. Suppose that the amount you save in any given month is twice the amount you saved in the previous month. How much will you have saved at the end of 1 year if you save $1 in January? How much if you saved 25¢ in January?

37. A certain bacterial culture doubles in number every day. If there were 1000 bacteria at the end of the first day, how many will there be after 10 days? How many after n days?

38. A radioactive substance is decaying so that at the end of each month there is only one-third as much as there was at the beginning of the month. If there were 75 grams of the substance at the beginning of the year, how much is left at midyear?

39. Suppose that an automobile depreciates 10% in value each year for the first 5 years. What is it worth after 5 years if its original cost was $5280? (*Hint:* Use $a_1 = 5280$ and $n = 6$.)

40. In the compound-interest formula $A_t = P(1 + r)^t$, developed in Section 6.5, P is the initial investment, r is the annual interest rate, and t is the number of years during which the interest has been compounded annually to obtain the total value A_t. Explain how this formula may be viewed as the general term of a geometric sequence.

41. A sum of $800 is invested at 11% interest compounded annually.

    (a) What is the amount after n years?

    (b) What is the amount after 5 years? (Use common logarithms or a calculator to get an approximation.)

42. How much money must be invested at the interest rate of 12%, compounded annually, so that after 3 years the amount is $1000?

43. Use a calculator to find the amount of money that an investment of $1500 earns at the interest rate of 8% compounded annually for 5 years.

44. (a) If $\frac{3}{5}$ of the wire in Example 8, page 000, is cut off repeatedly, what is the general form of the sequence for the length of the remaining wire?

    (b) What length remains after six cuts have been made? Give the answer to the nearest tenth of a foot.

    (c) What is the general form of the total length of wire that has been cut off after n cuts have been made?

45. (a) A set of containers are decreasing in size so that the second container is $\frac{1}{2}$ the volume of the first, the third is $\frac{1}{2}$ the volume of the second, etc. If the first container is empty and the other five are filled with water, can all the full ones be emptied into the first without the water spilling over? Explain.

(b) Answer the question in part (a) assuming that each container, after the first, is $\frac{2}{3}$ the volume of the one preceding it.

CHALLENGE	Suppose you snap your fingers, wait 1 minute, and snap them again. Then you snap them
Think Creatively	after 2 more minutes, then again after 4 more minutes, again after 8 more minutes, etc., each time waiting twice as long as you waited for the preceding snap. First *guess* how many times you would snap your fingers if you continued this process for one year. Following this process, use a calculator to compute how long it would take you to snap your fingers (a) 10 times, (b) 15 times, and (c) 20 times.

11.5
INFINITE
GEOMETRIC
SERIES

In decimal form the fraction $\frac{3}{4}$ becomes 0.75, which means $\frac{75}{100}$. This can also be written as $\frac{7}{10} + \frac{5}{100}$. What about $\frac{1}{3}$? As a decimal we can write

$$\frac{1}{3} = 0.333 \ldots$$

where the dots mean that the 3 repeats endlessly. We can express this decimal as the sum of fractions whose denominators are powers of 10:

$$\frac{1}{3} = \frac{3}{10} + \frac{3}{100} + \frac{3}{1000} + \cdots$$

*The sum of an infinite sequence is an **infinite series**.*

The numbers being added here are the terms of the *infinite geometric sequence* with first term $a_1 = \frac{3}{10}$ and common ratio $r = \frac{1}{10}$. Thus the nth term is

$$a_1 r^{n-1} = \frac{3}{10}\left(\frac{1}{10}\right)^{n-1} = 3\left(\frac{1}{10}\right)\left(\frac{1}{10}\right)^{n-1}$$

$$= 3\left(\frac{1}{10}\right)^n$$

$$= \frac{3}{10^n}$$

The sum of the first n terms is found by using the formula

$$S_n = \sum_{k=1}^{n} a_1 r^{k-1} = \frac{a_1(1 - r^n)}{1 - r}$$

Here are some cases:

$$S_1 = \frac{\frac{3}{10}\left(1 - \frac{1}{10}\right)}{1 - \frac{1}{10}} = \frac{1}{3}\left(1 - \frac{1}{10}\right) = 0.3$$

$$S_2 = \frac{\frac{3}{10}\left(1 - \frac{1}{10^2}\right)}{1 - \frac{1}{10}} = \frac{1}{3}\left(1 - \frac{1}{10^2}\right) = 0.33$$

$$S_{10} = \frac{\frac{3}{10}\left(1 - \frac{1}{10^{10}}\right)}{1 - \frac{1}{10}} = \frac{1}{3}\left(1 - \frac{1}{10^{10}}\right) = 0.\underbrace{3333333333}_{10 \text{ places}}$$

$$S_n = \frac{\frac{3}{10}\left(1 - \frac{1}{10^n}\right)}{1 - \frac{1}{10}} = \frac{1}{3}\left(1 - \frac{1}{10^n}\right) = 0.\underbrace{333 \ldots 3}_{n \text{ places}}$$

You can see that as more and more terms are added, the closer and closer the answer gets to $\frac{1}{3}$. This can be seen by studying the form for the sum of the first n terms:

$$S_n = \frac{1}{3}\left(1 - \frac{1}{10^n}\right)$$

It is clear that the bigger n is, the closer $\frac{1}{10^n}$ is to zero, the closer $1 - \frac{1}{10^n}$ is to 1, and, finally, the closer S_n is to $\frac{1}{3}$. Although it is true that S_n is never exactly equal to $\frac{1}{3}$, for very large n the difference between S_n and $\frac{1}{3}$ is very small. Saying this another way:

By taking n large enough, S_n can be made as close to $\frac{1}{3}$ as we like.

This is what we mean when we say that the sum of all the terms is $\frac{1}{3}$.

$$\frac{3}{10} + \frac{3}{10^2} + \frac{3}{10^3} + \cdots + \frac{3}{10^n} + \cdots = \frac{1}{3}$$

The summation symbol, Σ, can also be used here after an adjustment in notation is made. Traditionally, the symbol ∞ has been used to suggest an infinite number of objects. So we use this and make the transition from the sum of a finite number of terms

$$S_n = \sum_{k=1}^{n} \frac{3}{10^k} = \frac{3}{10} + \frac{3}{10^2} + \cdots + \frac{3}{10^n} = \frac{1}{3}\left(1 - \frac{1}{10^n}\right)$$

In calculus the symbol S_∞ is replaced by $\lim\limits_{n \to \infty} S_n = \frac{1}{3}$, which is read as "the limit of S_n as n gets arbitrarily large is $\frac{1}{3}$."

to the sum of an infinite number of terms:

$$S_\infty = \sum_{k=1}^{\infty} \frac{3}{10^k} = \frac{3}{10} + \frac{3}{10^2} + \cdots + \frac{3}{10^n} + \cdots = \frac{1}{3}$$

Not every geometric sequence produces an infinite geometric series that has a finite sum. For instance, the sequence

$$2, 4, 8, \ldots, 2^n, \ldots$$

is geometric, but the corresponding geometric series

$$2 + 4 + 8 + \cdots + 2^n + \cdots$$

cannot have a finite sum.

By now you might suspect that the common ratio r determines whether or not an infinite geometric sequence can be added. This turns out to be true. To see how this works, the general case will be considered next.

Let

$$a_1, \qquad a_1 r, \qquad a_1 r^2, \ldots, a_1 r^{n-1}, \ldots$$

be an infinite geometric sequence.

Then the sum of the first n terms is

$$S_n = \frac{a_1(1 - r^n)}{1 - r}$$

Rewrite in this form:

$$S_n = \frac{a_1}{1 - r}(1 - r^n)$$

Use a calculator to verify the powers of $r = 0.9$ and $r = 1.1$ to the indicated decimal places.

$(0.9)^1$	$= 0.9$
$(0.9)^{10}$	$= 0.35$
$(0.9)^{20}$	$= 0.12$
$(0.9)^{40}$	$= 0.015$
$(0.9)^{80}$	$= 0.0002$
$(0.9)^{100}$	$= 0.00003$

$\downarrow$

getting close to 0

At this point the importance of r^n becomes clear. If, as n gets larger, r^n gets very large, then the infinite geometric series will not have a finite sum. But if r^n gets arbitrarily close to zero as n gets larger, then $1 - r^n$ gets close to 1 and S_n gets closer and closer to $\dfrac{a_1}{1 - r}$.

The values of r for which r^n gets arbitrarily close to zero are precisely those values between -1 and 1; that is, $|r| < 1$. For instance, $\frac{3}{5}$, $-\frac{1}{10}$, and 0.09 are values of r for which r^n gets close to zero; and 1.01, -2, and $\frac{3}{2}$ are values for which the series does not have a finite sum.

To sum up, we have the following useful result:

$(1.1)^1$	$= 1.1$
$(1.1)^5$	$= 1.6$
$(1.1)^{10}$	$= 2.6$
$(1.1)^{20}$	$= 6.7$
$(1.1)^{50}$	$= 117.4$
$(1.1)^{100}$	$= 13780.6$

$\downarrow$

getting very large

SUM OF AN INFINITE GEOMETRIC SERIES

If $|r| < 1$, then $\displaystyle\sum_{k=1}^{\infty} a_1 r^{k-1} = \frac{a_1}{1 - r}$. For other values of r the series has no finite sum.

EXAMPLE 1   Find the sum of the infinite geometric series

$$27 + 3 + \frac{1}{3} + \cdots$$

Solution   Since $r = \frac{3}{27} = \frac{1}{9}$ and $a_1 = 27$, the preceding result gives

$$27 + 3 + \frac{1}{3} + \cdots = \frac{27}{1 - \frac{1}{9}} = \frac{27}{\frac{8}{9}} = \frac{243}{8}$$ ∎

EXAMPLE 2   Why does the infinite geometric series $\sum\limits_{k=1}^{\infty} 5(\frac{4}{3})^{k-1}$ have no finite sum?

Solution   The series has no finite sum because the common ratio $r = \frac{4}{3}$ is not between -1 and 1.   ∎

Another way to find a_1 is to let $k = 1$ in $\dfrac{7}{10^{k+1}}$:

$$a_1 = \frac{7}{10^2} = \frac{7}{100}$$

Also, r can be found by taking the ratio of the second term to the first term.

$$r = \frac{\frac{7}{10^3}}{\frac{7}{10^2}} = \frac{1}{10}$$

EXAMPLE 3   Find: $\sum\limits_{k=1}^{\infty} \dfrac{7}{10^{k+1}}$.

Solution   Since $\dfrac{7}{10^{k+1}} = 7\left(\dfrac{1}{10^{k+1}}\right) = 7\left(\dfrac{1}{10^2}\right)\dfrac{1}{10^{k-1}} = \dfrac{7}{100}\left(\dfrac{1}{10}\right)^{k-1}$, it follows that $a_1 = \frac{7}{100}$ and $r = \frac{1}{10}$. Therefore, by the formula for the sum of an infinite geometric series we have

$$S_\infty = \sum_{k=1}^{\infty} \frac{7}{10^{k+1}} = \frac{\frac{7}{100}}{1 - \frac{1}{10}} = \frac{7}{100 - 10} = \frac{7}{90}$$ ∎

TEST YOUR UNDERSTANDING
Think Carefully

Find the common ratio r, and then find the sum if the given infinite geometric series has one.

1. $10 + 1 + \frac{1}{10} + \cdots$ 2. $\frac{1}{64} + \frac{1}{16} + \frac{1}{4} + \cdots$

3. $36 - 6 + 1 - \cdots$ 4. $-16 - 4 - 1 - \cdots$

5. $\sum\limits_{k=1}^{\infty} (\frac{4}{3})^{k-1}$ 6. $\sum\limits_{k=1}^{\infty} 3(0.01)^k$

7. $\sum\limits_{i=1}^{\infty} (-1)^i 3^i$ 8. $\sum\limits_{n=1}^{\infty} 100(-\frac{9}{10})^{n+1}$

9. $101 - 102.01 + 103.0301 - \cdots$

(Answers: Page 600)

The introduction to this section indicated how the endless repeating decimal 0.333. . . can be regarded as an infinite geometric series. The next example illustrates how such decimal fractions can be written in the rational form $\dfrac{a}{b}$ (the ratio of two integers) by using the formula for the sum of an infinite geometric series.

Compare the method shown in Example 4 with that developed in Exercise 57 of Section 1.1.

EXAMPLE 4 Express the repeating decimal 0.242424. . . in rational form.

Solution First write

$$0.242424 \ldots = \frac{24}{100} + \frac{24}{10,000} + \frac{24}{1,000,000} + \cdots$$

$$= \frac{24}{10^2} + \frac{24}{10^4} + \frac{24}{10^6} + \cdots + \frac{24}{10^{2k}} + \cdots$$

$$= \frac{24}{100} + \frac{24}{100}\left(\frac{1}{100}\right) + \frac{24}{100}\left(\frac{1}{100}\right)^2 + \cdots + \frac{24}{100}\left(\frac{1}{100}\right)^{k-1} + \cdots$$

Observe that

$$\frac{24}{10^{2k}} = 24\left(\frac{1}{10^{2k}}\right)$$

$$= 24\left(\frac{1}{10^2}\right)^k = 24\left(\frac{1}{100}\right)^k$$

$$= \frac{24}{100}\left(\frac{1}{100}\right)^{k-1}$$

Then $a_1 = \frac{24}{100}$, $r = \frac{1}{100}$, and

$$0.242424 \ldots = \sum_{k=1}^{\infty} \frac{24}{100}\left(\frac{1}{100}\right)^{k-1}$$

$$= \frac{\frac{24}{100}}{1 - \frac{1}{100}}$$

$$= \frac{24}{99} = \frac{8}{33} \qquad \text{(Check this result by dividing 33 into 8.)} \qquad \blacksquare$$

It seems as if the horse cannot finish the race this way. But read on to see that there is really no contradiction with this interpretation.

EXAMPLE 5 A racehorse running at the constant rate of 30 miles per hour will finish a 1-mile race in 2 minutes. Now consider the race broken down into the following parts: before the racehorse can finish the 1-mile race it must first reach the halfway mark; having done that, the horse must next reach the quarter pole; then it must reach the eighth pole; and so on. That is, it must always cover half the distance remaining before it can cover the whole distance. Show that the sum of the infinite number of time intervals is also 2 minutes.

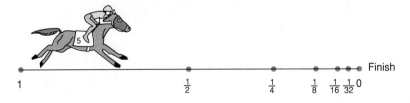

Solution For the first $\frac{1}{2}$ mile the time will be $\frac{\frac{1}{2}}{\frac{1}{2}} = 1$ minute; for the next $\frac{1}{4}$ mile the

$$T = \frac{D}{R}\left(time = \frac{distance}{rate}\right)$$

time will be $\frac{\frac{1}{4}}{\frac{1}{2}} = \frac{1}{2}$ minute; for the next $\frac{1}{8}$ mile the time will be $\frac{\frac{1}{8}}{\frac{1}{2}} = \frac{1}{4}$ minute; and

for the *n*th distance, which is $\frac{1}{2^n}$ miles, the time will be $\frac{\frac{1}{2^n}}{\frac{1}{2}} = \frac{1}{2^{n-1}}$.

Thus the total time is given by this series:

$$\sum_{k=1}^{\infty} \frac{1}{2^{k-1}} = 1 + \frac{1}{2} + \frac{1}{4} + \cdots + \frac{1}{2^{n-1}} + \cdots$$

This is an infinite geometric series having $a_1 = 1$ and $r = \frac{1}{2}$. Thus

$$\sum_{k=1}^{\infty} 1\left(\frac{1}{2}\right)^{k-1} = \frac{1}{1 - \frac{1}{2}} = 2$$

which is the same result as before. ∎

EXAMPLE 6 Rectangle $ABCD$ has dimensions 1 by 2. The next rectangle $PQRS$ has dimensions $\frac{1}{2}$ by 1. In like manner, each inner rectangle has dimensions half the size of the preceding rectangle. If this sequence of rectangles continues endlessly, what is the sum of the areas of all the rectangles?

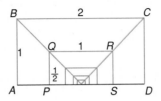

Solution The area of rectangle $ABCD$ is $1 \cdot 2$, the area of rectangle $PQRS$ is $\frac{1}{2} \cdot 1$, the next has area $\frac{1}{4} \cdot \frac{1}{2}$, and so on. The sum of all the areas is this infinite geometric series.

$$1 \cdot 2 + \tfrac{1}{2} \cdot 1 + \tfrac{1}{4} \cdot \tfrac{1}{2} + \tfrac{1}{8} \cdot \tfrac{1}{4} + \cdots$$
$$= 2 + \tfrac{1}{2} + \tfrac{1}{8} + \tfrac{1}{32} + \cdots$$
$$= 2 + \tfrac{1}{2} + \left(\tfrac{1}{2}\right)^3 + \left(\tfrac{1}{2}\right)^5 + \cdots$$

Since $a_1 = 2$ and $r = \frac{1}{4}$, the sum equals

$$\frac{a_1}{1 - r} = \frac{2}{1 - \frac{1}{4}} = \frac{8}{3} = 2\tfrac{2}{3}$$ ∎

CAUTION: Learn to Avoid These Mistakes	
WRONG	**RIGHT**
$\displaystyle\sum_{k=1}^{\infty} \left(\frac{1}{3}\right)^{n+1} = \dfrac{1}{1 - \dfrac{1}{3}}$	$\displaystyle\sum_{n=1}^{\infty} \left(\frac{1}{3}\right)^{n+1} = \sum_{n=1}^{\infty} \frac{1}{9}\left(\frac{1}{3}\right)^{n-1}$ $= \dfrac{\dfrac{1}{9}}{1 - \dfrac{1}{3}}$
$\displaystyle\sum_{n=1}^{\infty} \left(-\frac{1}{2}\right)^{n-1} = \dfrac{1}{1 - \dfrac{1}{2}}$	$\displaystyle\sum_{n=1}^{\infty} \left(-\frac{1}{2}\right)^{n-1} = \dfrac{1}{1 - \left(-\dfrac{1}{2}\right)}$
$\displaystyle\sum_{n=1}^{\infty} 2(1.03)^{n-1} = \dfrac{2}{1 - 1.03}$	$\displaystyle\sum_{n=1}^{\infty} 2(1.03)^{n-1}$ is not a finite sum since $r = 1.03 > 1$

EXERCISES 11.5

Find the sum, if it exists, of each infinite geometric series.

1. $2 + 1 + \frac{1}{2} + \cdots$

2. $8 + 4 + 2 + \cdots$

3. $25 + 5 + 1 + \cdots$

4. $1 + \frac{4}{3} + \frac{16}{9} + \cdots$

5. $1 - \frac{1}{2} + \frac{1}{4} - \cdots$

6. $100 - 1 + \frac{1}{100} - \cdots$

7. $1 + 0.1 + 0.01 + \cdots$

8. $52 + 0.52 + 0.0052 + \cdots$

9. $-2 - \frac{1}{4} - \frac{1}{32} - \cdots$

10. $-729 + 81 - 9 + \cdots$

*Explain what is **wrong** with each of these statements.*

11. $\displaystyle\sum_{n=1}^{\infty} \left(\frac{1}{2}\right)^{n+1} = \frac{1}{1 - \frac{1}{2}}$

12. $\displaystyle\sum_{n=1}^{\infty} (-1)^{n-1} = \frac{1}{1 - (-1)} = \frac{1}{2}$

13. $\displaystyle\sum_{n=1}^{\infty} \left(-\frac{1}{3}\right)^{n-1} = \frac{1}{1 - \frac{1}{3}}$

14. $\displaystyle\sum_{n=1}^{\infty} 3(1.02)^{n-1} = \frac{3}{1 - 1.02}$

Decide whether or not the given infinite geometric series has a sum. If it does, find it using

$$S_\infty = \frac{a_1}{1 - r}.$$

15. $\displaystyle\sum_{k=1}^{\infty} \left(\frac{1}{3}\right)^{k-1}$

16. $\displaystyle\sum_{k=1}^{\infty} \left(\frac{1}{3}\right)^{k}$

17. $\displaystyle\sum_{k=1}^{\infty} \left(\frac{1}{3}\right)^{k+1}$

18. $\displaystyle\sum_{n=1}^{\infty} \frac{1}{2^{n+1}}$

19. $\displaystyle\sum_{n=1}^{\infty} \frac{1}{2^{n-2}}$

20. $\displaystyle\sum_{k=1}^{\infty} \left(\frac{1}{10}\right)^{k-1}$

21. $\displaystyle\sum_{k=1}^{\infty} 2(0.1)^{k-1}$

22. $\displaystyle\sum_{k=1}^{\infty} \left(-\frac{1}{2}\right)^{k-1}$

23. $\displaystyle\sum_{n=1}^{\infty} \left(\frac{3}{2}\right)^{n-1}$

24. $\displaystyle\sum_{n=1}^{\infty} \left(-\frac{1}{3}\right)^{n+2}$

25. $\displaystyle\sum_{k=1}^{\infty} (0.7)^{k-1}$

26. $\displaystyle\sum_{k=1}^{\infty} 5(0.7)^{k}$

27. $\displaystyle\sum_{k=1}^{\infty} 5(1.01)^{k}$

28. $\displaystyle\sum_{k=1}^{\infty} \left(\frac{1}{10}\right)^{k-4}$

29. $\displaystyle\sum_{k=1}^{\infty} 10\left(\frac{2}{3}\right)^{k-1}$

30. $\displaystyle\sum_{k=1}^{\infty} (-1)^{k}$

31. $\displaystyle\sum_{k=1}^{\infty} (0.45)^{k-1}$

32. $\displaystyle\sum_{k=1}^{\infty} (-0.9)^{k+1}$

33. $\displaystyle\sum_{n=1}^{\infty} 7\left(-\frac{3}{4}\right)^{n-1}$

34. $\displaystyle\sum_{k=1}^{\infty} (0.1)^{2k}$

35. $\displaystyle\sum_{k=1}^{\infty} \left(-\frac{2}{5}\right)^{2k}$

Find a rational form for each of the following repeating decimals in a manner similar to that in Example 4. Check your answers.

36. $0.444 \ldots$

37. $0.777 \ldots$

38. $7.777 \ldots$

39. $0.131313 \ldots$

40. $13.131313 \ldots$

41. $0.0131313 \ldots$

42. $0.050505 \ldots$

43. $0.999 \ldots$

44. $0.125125125 \ldots$

45. Suppose that a 1-mile distance a racehorse must run is divided into an infinite number of parts, obtained by always considering $\frac{2}{3}$ of the remaining distance to be covered. Then the lengths of these parts form the sequence

$$\frac{2}{3}, \frac{2}{9}, \frac{2}{27}, \ldots, \frac{2}{3^n}, \ldots$$

(a) Find the sequence of times corresponding to these distances. (Assume that the horse is moving at a rate of $\frac{1}{2}$ mile per minute.)

(b) Show that the sum of the times in part (a) is 2 minutes.

46. A certain ball always rebounds $\frac{1}{3}$ of the distance it falls. If the ball is dropped from a height of 9 feet, how far does it travel before coming to rest? (See the similar situation at the beginning of Section 11.4.)

47. A substance initially weighing 64 grams is decaying at a rate such that after 4 hours there are only 32 grams left. In another 2 hours only 16 grams remain; in another 1 hour after

that only 8 grams remain; and so on so that the time intervals and amounts remaining form geometric sequences. How long does it take altogether until nothing of the substance is left?

48. After it is set in motion, each swing in either direction of a particular pendulum is 40% as long as the preceding swing. What is the total distance that the end of the pendulum travels before coming to rest if the first swing is 30 inches long?

49. Assume that a racehorse takes 1 minute to go the first $\frac{1}{2}$ mile of a 1-mile race. After that, the horse's speed is no longer constant: for the next $\frac{1}{4}$ mile it takes $\frac{2}{3}$ minute; for the next $\frac{1}{8}$ mile it takes $\frac{4}{9}$ minute; for the next $\frac{1}{16}$ mile it takes $\frac{40}{81}$ minute; and so on, so that the time intervals form a geometric sequence. Why can't the horse finish the race?

50. **(a)** *ABCD* is a square whose sides measure 8 units. *PQRS* is a square whose sides are $\frac{1}{2}$ the length of the sides of square *ABCD*. The next square has sides $\frac{1}{2}$ the length of those of square *PQRS*. In like manner each inner square has sides $\frac{1}{2}$ the length of the preceding square. If this sequence of squares continues endlessly, what is the sum of all the areas of the squares in the sequence?

   (b) What is the sum of all the perimeters?

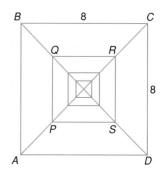

51. *ABC* is an isosceles right triangle with right angle at *C*. P_1 is the midpoint of the hypotenuse *AB* so that CP_1 divides triangle *ABC* into two congruent triangles. P_2 is the midpoint of *BC* so that P_1P_2 divides triangle CBP_1 into two congruent triangles. This process continues endlessly.

   (a) If $AC = CB = 4$, what do you expect the sum of the area of all the triangles labeled 1, 2, 3, . . . to be equal to?

   (b) Verify the result in part (a) by using an infinite geometric series.

   (c) Find the sum of all the triangles labeled with odd numbers and also find the sum of all the triangles labeled with even numbers. What is the sum of the two sums?

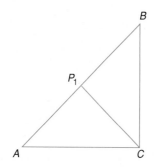

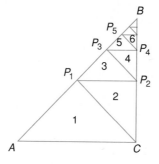

52. The largest circle has radius $A_1B = 1$. The next circle has radius $A_2B = \frac{1}{2}A_1B$, the one after that has radius $A_3B = \frac{1}{2}A_2B$, and so on. If these circles continue endlessly in this manner, what is the sum of the areas of all the circles?

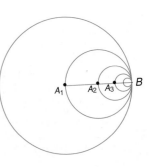

53. Triangle AB_1C_1 has a right angle at C_1. $AC_1 = 9$ and $B_1C_1 = 3$. Points $C_2, C_3, C_4, \ldots$ are chosen so that $AC_2 = \frac{2}{3}AC_1$, $AC_3 = \frac{2}{3}AC_2$, and so on. Find the sum of the areas of all the right triangles labeled AB_kC_k for $k = 1, 2, 3, \ldots$.

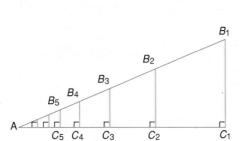

54. A marble is released on a semicircular track and travels 8 feet before it reverses direction. Then it travels 6 feet before reversing a second time. This movement with reversals continues indefinitely so that the distances traveled between reversals forms an infinite geometric sequence. Find the total distance traveled before coming to rest.

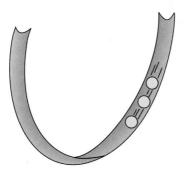

55. (a) Suppose you have $2000 in a special savings account and spend 60% on various products. Next, assume that the owners of the stores from which the products were purchased also spend 60% of what you paid them. If this process continues indefinitely, the total amount spent forms an infinite geometric series whose first term is 1200. Find the total spent on all the purchases made.

(b) Assume that at each step in part (a) the amount not spent is put into savings. Use an infinite geometric series, having first term 800, to find the total savings.

56. In the study of calculus it is shown that the number $e = 2.718281 \ldots$ (see Section 6.4)

is given by the infinite series $e = 1 + 1 + \dfrac{1}{2!} + \dfrac{1}{3!} + \cdots + \dfrac{1}{k!} + \cdots$

where $k! = k(k - 1)(k - 2) \cdots 3 \cdot 2 \cdot 1$. Approximate e by adding the first six terms of the series and compare to the decimal form of e.

CHALLENGE
Think Creatively

Use the identity $\dfrac{1}{k(k + 1)} = \dfrac{1}{k} - \dfrac{1}{(k + 1)}$ to find the sum $\displaystyle\sum_{k=1}^{\infty} \dfrac{1}{k(k + 1)}$. (*Hint:* Consider

the sum $S_n = \displaystyle\sum_{k=1}^{n} \dfrac{1}{k(k + 1)}$ as n gets arbitrarily large.)

EXPLORATIONS
Think Critically

1. Which is larger, the sum of the first n powers of 2, starting with 2^0, or the single power 2^n? Justify your answer.

2. Observe that $\displaystyle\sum_{k=1}^{2n} [2^{k-1} + (-2)^{k-1}] = \sum_{k=1}^{2n} 2^{k-1} + \sum_{k=1}^{2n} (-2)^{k-1}$. Therefore, the series at the left can be evaluated by evaluating each series at the right and adding the results. Do this computation and use the answer to show that the series on the left can be written in the form $\displaystyle\sum_{k=1}^{n} a_1 r^{k-1}$.

3. For each x such that $|x| < 1$, $\displaystyle\sum_{k=1}^{\infty} x^{k-1}$ is an infinite geometric series having a finite sum. Thus we may define the function f by $f(x) = \displaystyle\sum_{k=1}^{\infty} x^{k-1}$ for $|x| < 1$. Sketch the graph of this function.

4. Decide if the series $\displaystyle\sum_{k=1}^{\infty} \dfrac{3^{k+1}}{5^{k-1}}$ is geometric having a ratio r, where $-1 < r < 1$. If it is, then find the sum. If it isn't, give a reason.

5. For which values of x does the geometric series $\displaystyle\sum_{k=1}^{\infty} (3x - 4)^{k-1}$ have a finite sum? Find the sum for such x.

Study these statements:

$$1 = 1^2$$
$$1 + 3 = 2^2$$
$$1 + 3 + 5 = 3^2$$
$$1 + 3 + 5 + 7 = 4^2$$
$$1 + 3 + 5 + 7 + 9 = 5^2$$

Do you see the pattern? The last statement shows that the sum of the first five positive odd integers is 5^2. What about the sum of the first six positive odd integers? The pattern is the same;

$$1 + 3 + 5 + 7 + 9 + 11 = 6^2$$

It would be reasonable to guess that the sum of the first n positive odd integers is n^2. That is,

$$1 + 3 + 5 + \cdots + (2n - 1) = n^2$$

But a guess is not a proof. It is our objective in this section to learn how to prove a statement that involves an infinite number of cases.

Let us refer to the nth statement above as S_n. Thus S_1, S_2, S_3, S_4, S_5, and S_6 are the first six cases of S_n that we know are true.

Does the truth of the first six cases allow us to conclude that S_n is true for all positive integers n? No! We cannot assume that a few special cases guarantee an infinite number of cases. If we allowed "proving by a finite number of cases," then the following example is such a "proof" that all positive even integers are less than 100.

The first positive even integer is 2, and we know that $2 < 100$.

The second is 4, and we know that $4 < 100$.

The third is 6, and $6 < 100$.

Therefore, since $2n < 100$ for a finite number of cases, we might conclude that $2n < 100$ for all n.

This false result should convince you that in trying to prove a collection of statements S_n for all positive integers $n = 1, 2, 3, \ldots$, we need to do more than just check it out for a finite number of cases. We need to call on a type of proof known as **mathematical induction**.

Suppose that we had a long (endless) row of dominoes each 2 inches long all standing up in a straight row so that the distance between any two of them is $1\frac{1}{2}$ inches. How can you make them all fall down with the least effort?

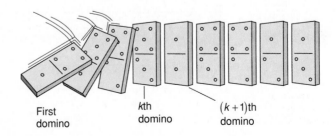

First
domino

*k*th
domino

$(k+1)$th
domino

The answer is obvious. Push the first domino down toward the second. Since the first one must fall, and because the space between each pair is less than the length of a domino, they will all (eventually) fall down. The first knocks down the second, the second knocks down the third, and, in general, the kth domino knocks down the $(k + 1)$st. Two things guaranteed this "chain reaction":

1. The first domino will fall.
2. If any domino falls, then so will the next.

These two conditions are guidelines in forming the **principle of mathematical induction**.

THE PRINCIPLE OF MATHEMATICAL INDUCTION

Condition 1 *starts the "chain reaction" and condition* 2 *keeps it going.*

Let S_n be a statement for each positive integer n. Suppose that the following two conditions hold:

1. S_1 is true.
2. If S_k is true, then S_{k+1} is true, where k is any positive integer.

Then S_n is a true statement for all n.

Note that we are not proving this principle; rather, it is a basic principle that we accept and use to construct proofs. It is very important to realize that in condition 2 we are not proving S_k to be true; rather, we must prove this proposition:

If S_k is true, then S_{k+1} is true.

Consequently, a proof by mathematical induction *includes* a proof of the proposition that S_k implies S_{k+1}, a proof within a proof. Within that inner proof, we are allowed to *assume* and use S_k.

At the beginning of this section, we guessed at the formula for the sum of the first n odd integers. Now, using mathematical induction, this formula is proved in Example 1.

EXAMPLE 1 Prove by mathematical induction that S_n is true for all positive integers n, where S_n is the statement

$$1 + 3 + 5 + \cdots + (2n - 1) = n^2$$

Proof Both conditions 1 and 2 of the principle of mathematical induction must be satisfied. We begin with the first.

Proving S_1 starts the chain. The first domino has fallen.

1. S_1 is true because $1 = 1^2$.

We assume S_k to be true to see what effect it has on the next case, S_{k+1}. This is comparable to considering what happens when the kth domino falls.

2. Suppose that S_k is true, where k is a positive integer. That is, we *assume*

$$1 + 3 + 5 + \cdots + (2k - 1) = k^2$$

We want to prove that S_{k+1} follows from this. To do so note that the next odd number after $2k - 1$ is $2(k + 1) - 1 = 2k + 1$, which is added to the preceding equation:

$$1 + 3 + 5 + \cdots + (2k - 1) \qquad\qquad = k^2$$
$$\underline{\qquad\qquad\qquad\qquad\qquad\qquad 2k + 1 = 2k + 1}$$
$$1 + 3 + 5 + \cdots + (2k - 1) + (2k + 1) = k^2 + 2k + 1$$

Now factor $k^2 + 2k + 1$:

$$1 + 3 + 5 + \cdots + (2k - 1) + (2k + 1) = (k + 1)^2$$

Now that S_k implies S_{k+1}; the chain reaction keeps going.

This is the statement S_{k+1}. Therefore, we have shown that if S_k is given, then S_{k+1} must follow. This, together with the fact that S_1 is true, allows us to say that S_n is true for all n by the principle of mathematical induction. ∎

EXAMPLE 2 Prove that the sum of the squares of the first n consecutive positive integers is given by $\dfrac{n(n + 1)(2n + 1)}{6}$.

Proof Let S_n be the statement

$$\underbrace{1^2 + 2^2 + 3^2 + \cdots + n^2}_{\text{(sum of first } n \text{ squares)}} = \frac{n(n + 1)(2n + 1)}{6}.$$

for any positive integer n.

1. S_1 is true because $1^2 = \dfrac{1(1 + 1)(2 \cdot 1 + 1)}{6}$.

2. Suppose that S_k is true for any k. That is, we *assume*

$$1^2 + 2^2 + 3^2 + \cdots + k^2 = \frac{k(k + 1)(2k + 1)}{6}$$

We want to prove that S_{k+1} follows from this. To do so, add the next square, $(k + 1)^2$, to both sides.

(A) $1^2 + 2^2 + 3^2 + \cdots + k^2 + (k + 1)^2 = \dfrac{k(k + 1)(2k + 1)}{6} + (k + 1)^2$

Combine the right side.

$$\frac{k(k + 1)(2k + 1)}{6} + (k + 1)^2 = \frac{k(k + 1)(2k + 1) + 6(k + 1)^2}{6}$$

$$= \frac{(k + 1)[k(2k + 1) + 6(k + 1)]}{6}$$

$$= \frac{(k + 1)(k + 2)(2k + 3)}{6}$$

Substitute back into equation (A).

$$1^2 + 2^2 + 3^2 + \cdots + (k + 1)^2 = \frac{(k + 1)(k + 2)(2k + 3)}{6}$$

To see that this is S_{k+1}, rewrite the right side:

Observe that on the left the first k + 1 squares are added, and on the right we have the required form in terms of k + 1.

$$1^2 + 2^2 + 3^2 + \cdots + (k + 1)^2 = \frac{(k + 1)[(k + 1) + 1][2(k + 1) + 1]}{6}$$

Now both conditions of the principle of mathematical induction have been satisfied, and it follows that

$$1^2 + 2^2 + 3^3 + \cdots + n^2 = \frac{n(n + 1)(2n + 1)}{6}$$

is true for all integers $n \geq 1$. ∎

Examples 1 and 2 involved equations that were established by mathematical induction. However, this principle is also used for other types of mathematical situations. The next example demonstrates the application of this principle when an inequality is involved. Also note that in Example 3 the proof begins by establishing S_2 rather than S_1. This is an acceptable use of the principle of induction as long as we also establish part (2). In fact, if S_1 in a proof by induction is replaced by S_a, where a is any (fixed) positive integer, and part (2) proves that S_k implies S_{k+1} for all $k \geq a$, then S_n holds for all $n \geq a$.

Try a few specific cases. Use a calculator to verify these:

$(1.02)^2 > 1 + 2(0.02)$

$(1.001)^2 > 1 + 2(0.001)$

$(1.00054)^2 > 1 + 2(0.00054)$

EXAMPLE 3 Let $t > 0$ and use mathematical induction to prove that $(1 + t)^n > 1 + nt$ for all positive integers $n \geq 2$.

Proof Let S_n be the statement $(1 + t)^n > 1 + nt$, where $t > 0$ and n is any integer where $n \geq 2$.

1. For $n = 2$, $(1 + t)^2 = 1 + 2t + t^2$. Since $t^2 > 0$, we get

$$1 + 2t + t^2 > 1 + 2t$$

because

$$(1 + 2t + t^2) - (1 + 2t) = t^2 > 0.$$

Our objective here is to prove $(1 + t)^{k+1} > 1 + (k + 1)t$.

2. Suppose that for $k \geq 2$ we have $(1 + t)^k > 1 + kt$. Multiply both sides by the positive number $(1 + t)$.

$$(1 + t)^k(1 + t) > (1 + kt)(1 + t)$$

Then

$$(1 + t)^{k+1} > 1 + (k + 1)t + kt^2$$

But $1 + (k + 1)t + kt^2 > 1 + (k + 1)t$. Therefore, by transitivity of $>$,

$$(1 + t)^{k+1} > 1 + (k + 1)t$$

By (1) and (2) above, the principle of mathematical induction imples that $(1 + t)^n > 1 + nt$ for all integers $n \geq 2$. ∎

EXERCISES 11.6

Use mathematical induction to prove the following statements for all positive integers n.

1. $1 + 2 + 3 + \cdots + n = \dfrac{n(n + 1)}{2}$

2. $2 + 4 + 6 + \cdots + 2n = n(n + 1)$

3. $\displaystyle\sum_{i=1}^{n} 3i = \dfrac{3n(n + 1)}{2}$

4. $1 + 4 + 7 + \cdots + (3n - 2) = \dfrac{n(3n - 1)}{2}$

5. $\dfrac{5}{3} + \dfrac{4}{3} + 1 + \cdots + \left(-\dfrac{1}{3}n + 2\right) = \dfrac{n(11 - n)}{6}$

6. $1 \cdot 2 + 2 \cdot 3 + 3 \cdot 4 + \cdots + n(n + 1) = \dfrac{n(n + 1)(n + 2)}{3}$

7. $\dfrac{1}{1 \cdot 2} + \dfrac{1}{2 \cdot 3} + \dfrac{1}{3 \cdot 4} + \cdots + \dfrac{1}{n(n + 1)} = \dfrac{n}{n + 1}$

8. $3 + 3^2 + 3^3 + \cdots + 3^n = \dfrac{3^{n+1} - 3}{2}$

9. $-2 - 4 - 6 - \cdots - 2n = -n - n^2$

10. $1 + \dfrac{1}{2} + \dfrac{1}{2^2} + \cdots + \dfrac{1}{2^{n-1}} = 2\left(1 - \dfrac{1}{2^n}\right)$

11. $1 + \frac{2}{5} + \frac{4}{25} + \cdots + (\frac{2}{5})^{n-1} = \frac{5}{3}[1 - (\frac{2}{5})^n]$

12. $1 - \frac{1}{3} + \frac{1}{9} - \cdots + (-\frac{1}{3})^{n-1} = \frac{3}{4}[1 - (-\frac{1}{3})^n]$

13. $1^3 + 2^3 + 3^3 + \cdots + n^3 = \dfrac{n^2(n + 1)^2}{4}$

14. Which of Exercises 1–13 can also be proved using the formula for an arithmetic series? Which ones can be proved using the formula for a geometric series?

For Exercises 15 and 16 use mathematical induction to prove the following for all positive integers n ≥ 1.

15. $\displaystyle\sum_{i=1}^{n} ar^{i-1} = \dfrac{a(1 - r^n)}{1 - r}, r \neq 1$

16. $\displaystyle\sum_{i=1}^{n} [a + (i - 1)d] = \dfrac{n}{2}[2a + (n - 1)d]$

17. Use mathematical induction to prove $2^n > 4n$ for $n \geq 5$.
18. Use mathematical induction to prove $(\frac{3}{4})^n < \frac{3}{4}$ for $n \geq 2$.
19. Use mathematical induction to prove $a^n < 1$, where $0 < a < 1$, for $n \geq 1$.
20. Let $0 < a < 1$. Use mathematical induction to prove that $a^n < a$ for all integers $n \geq 2$.
21. Let a and b be real numbers. Then use mathematical induction to prove that $(ab)^n = a^n b^n$ for all positive integers n.
22. Use mathematical induction to prove the generalized distributive property $a(b_0 + b_1 + \cdots + b_n) = ab_0 + ab_1 + \cdots + ab_n$ for all positive integers n, where a and b_i are real numbers. (Assume that parentheses may be inserted into or extracted from an indicated sum of real numbers).
23. Give an inductive proof of $|a_0 + a_1 + \cdots + a_n| \leq |a_0| + |a_1| + \cdots + |a_n|$ for all positive integers n, where the a_i are real numbers.
24. Use induction to prove that if $a_0 a_1 \cdots a_n = 0$, then at least one of the factors is zero for all positive integers n. (Assume that parentheses may be inserted into or extracted from an indicated product of real numbers.)
25. (a) Prove by induction that $\dfrac{a^n - b^n}{a - b} = a^{n-1} + a^{n-2}b + \cdots + ab^{n-2} + b^{n-1}$ for all integers $n \geq 2$.

$\left(\textit{Hint:}\quad \text{Consider}\quad \dfrac{a^{n+1} - b^{n+1}}{a - b} = \dfrac{a^n a - b^n b}{a - b} = \dfrac{a^n a - b^n a + b^n a - b^n b}{a - b}\right)$

(b) How does the result give the factorization of the difference of two nth powers?

26. Let u_n be the sequence such that $u_1 = 1$, $u_2 = 1$, and $u_{n+1} = u_{n-1} + u_n$ for $n \geq 2$. (This is the Fibonacci sequence (see Exercise 38, page 566.) Use mathematical induction to prove that for any positive integer n, $\sum_{i=1}^{n} u_i = u_{n+2} - 1$.

For Exercises 27 and 28, first investigate a pattern leading to a formula and then prove the formula by mathematical induction.

27. (a) Complete the statements

$$1 \qquad\qquad\qquad =$$
$$1 + 2 + 1 \qquad\qquad\quad =$$
$$1 + 2 + 3 + 2 + 1 \qquad\quad =$$
$$1 + 2 + 3 + 4 + 3 + 2 + 1 \quad =$$
$$1 + 2 + 3 + 4 + 5 + 4 + 3 + 2 + 1 =$$

(b) Use the results in part (a) to *guess* the sum in the general case.

$$1 + 2 + 3 + \cdots + (n - 1) + n + (n - 1) + \cdots + 3 + 2 + 1 =$$

(c) Prove part (b) by mathematical induction.

28. (a) The left-hand column contains a number of points in a plane, no three of which are collinear. The right-hand column contains the number of distinct lines that the points determine. Complete this information for the last three cases.

Number of Points	Number of Lines
2	1
3	3
4	
5	
6	

(b) Observe that when there are 4 points, there are $6 = \dfrac{4(3)}{2}$ lines. Write similar statements when there are 2, 3, 5, or 6 points.

(c) Conjecture (guess) what the number of distinct lines is for n points, no three of which are collinear. Prove the conjecture by mathematical induction.

29. Use mathematical induction to prove that $7^n - 1$ is divisible by 6 for $n \geq 1$. (*Hint:* A number divisible by 6 can be written in the form $6b$, where b is an integer.)

30. Use mathematical induction to prove that $7^n - 4$ is divisible by 3 for $n \geq 1$. (*Hint:* A number divisible by 3 can be written in the form $3b$, where b is an integer.)

**CHALLENGE
Think Creatively**   In Exercise 29 you were asked to prove that $7^n - 1$ is divisible by 6 for $n \geq 1$. Prove this result *without* using mathematical induction.

KEY TERMS

Review these **key terms** *so that you are able to define or describe them. A clear understanding of these terms will be very helpful when reviewing the developments of this chapter.*

Sequence • Terms of a sequence • General term of a sequence • Infinite sequence • Recursive sequence • Triangular numbers • Series • Sigma notation • Fibonacci sequence • Arithmetic sequence • Common difference • Arithmetic series • Geometric sequence • Common ratio • Geometric series • Infinite geometric series • Sum of an infinite geometric series • Principle of mathematical induction • Proof by mathematical induction

REVIEW EXERCISES

The solutions to the following exercises can be found within the text of Chapter 11. Try to answer each question before referring to the text.

Section 11.1

1. State the definition of a sequence.

2. Find the range values of the sequence $a_n = \dfrac{1}{n}$ for the domain $\{1, 2, 3, 4, 5,\}$ and graph.

3. List the first six terms of the sequence $b_k = \dfrac{(-1)^k}{k^2}$.

4. Find the tenth term of the sequence $a_n = \dfrac{n + 2}{2n - 3}$.

5. Write the first four terms of the sequence $a_n = \left(1 + \dfrac{1}{n}\right)^n$ and round off the terms to two decimal places.

6. Let $a_1 = 6$ and $a_n = 3a_{n-1} - 7$ for $n > 1$. Find the first five terms of this sequence.

Section 11.2

7. Find the sum of the first seven terms of $a_k = 2k$.

8. Find $\Sigma_{n=1}^{5} b_n$, where $b_n = \dfrac{n}{n + 1}$.

9. Evaluate $\Sigma_{k=1}^{4}(2k + 1)$.   10. Evaluate $\Sigma_{i=1}^{5}(-1)^i(i + 1)$.

11. Rewrite the series using the sigma notation: $3 + 6 + 9 + 12 + 15 + 18 + 21$.

Section 11.3

12. State the definition of an arithmetic sequence.

13. What is the *n*th term of an arithmetic sequence whose first term is a_1 and whose common differerence is d?

14. Find the *n*th term of the arithmetic sequence $11, 2, -7, \ldots$.

15. Find a_{40} for an arithmetic sequence having $a_1 = -15$ and $a_5 = 13$.

16. Write the formula for the sum S_n of the arithmetic sequence $a_k = a_1 + (k - 1)d$.

17. Find S_{20} for the arithmetic sequence with $a_1 = 3$ and $d = 5$.

18. Find the sum of the first 10,000 terms of the arithmetic sequence $246, 261, 276, \ldots$.

19. Find the sum of the first n positive integers.

20. Evaluate $\Sigma_{k=1}^{50}(-6k + 10)$.

21. Find the sum of all multiples of 3 between 4 and 262.

22. Suppose you take out a short-term loan of $6000 that you agree to pay back in 12 equal monthly payments, plus 3% interest per month on the monthly balance. Each month's payment is made during the first week of the new month. (a) Write the general term of the sequence that gives the monthly balance. (b) How much is the interest for each of the first 3 months? Write the general term for the monthly balance. (c) What is the total interest payment for the year, and what percent is this total of the $6000 loan?

Section 11.4

23. What is a geometric sequence?

Write the nth term of the geometric sequence.

24. $2, -4, 8, \ldots$ 25. $5, -5, 5, \ldots$ 26. $6, 9, \frac{27}{2}, \ldots$

27. Find the hundredth term of the geometric sequence having $r = \frac{1}{2}$ and $a_1 = \frac{1}{2}$.
28. Write the kth term of the geometric sequence $a_k = (\frac{1}{2})^{2k}$ in the form $a_1 r^{k-1}$ and find the values of a_1 and r.
29. A geometric sequence of positive terms has $a_1 = 18$ and $a_5 = \frac{32}{9}$. Find r.
30. Write the formula for the sum S_n of a geometric sequence $a_k = a_1 r^{k-1}$.
31. Find the sum of the first 100 terms of the geometric sequence $a_k = 6(\frac{1}{2})^{k-1}$.
32. Evaluate $\sum_{k=1}^{8} 3\left(\dfrac{1}{10}\right)^{k+1}$.

33. Suppose that you save $128 in January and that each month thereafter you manage to save only half of what you saved the previous month. How much do you save in the tenth month? What are your total savings after 10 months?
34. A roll of wire is 625 feet long. If $\frac{1}{5}$ of the wire is cut off repeatedly, what is the general term of the sequence for the length of wire remaining? Use a calculator and the general term to determine the length of wire remaining after 7 cuts.

Section 11.5

35. For which values of r does $\sum_{k=1}^{\infty} a_1 r^{k-1} = \dfrac{a_1}{1-r}$?
36. Find the sum of the infinite geometric series $27 + 3 + \frac{1}{3} + \cdots$.
37. Why does the infinite series $\sum_{k=1}^{\infty} 5(\frac{4}{3})^{k-1}$ have no finite sum?
38. Evaluate $\sum_{k=1}^{\infty} \dfrac{7}{10^{k+1}}$.

39. Express the repeating decimal $0.242424 \ldots$ in rational form (the ratio of two integers).
40. A racehorse running at the constant rate of 30 miles per hour will finish a 1-mile race in 2 minutes. Now consider the race broken down into the following parts. Before the racehorse can finish the 1-mile race it must first reach the halfway mark; having done that, the horse must next reach the quarter pole; then it must reach the eighth pole; and so on. That is, it must always cover half the distance before it can cover the whole distance. Show that the sum of the infinite number of time intervals is also 2 minutes.

Section 11.6

41. To prove that a statement S_n is true for each positive integer n, what two conditions must be established according to the principle of mathematical induction?

42. Prove by mathematical induction:

$$1 + 3 + 5 + \cdots + (2n - 1) = n^2$$

43. Use mathematical induction to prove:

$$1^2 + 2^2 + 3^2 + \cdots + n^2 = \frac{n(n+1)(2n+1)}{6}$$

44. Let $t > 0$ and use mathematical induction to prove that $(1 + t)^n > 1 + nt$ for all positive integers $n \geq 2$.

CHAPTER 11 TEST: STANDARD ANSWER

Use these questions to test your knowledge of the basic skills and concepts of Chapter 11. Then check your answers with those given at the back of the book.

1. Find the first four terms and the fortieth term of the sequence given by $a_n = \dfrac{n^2}{6 - 5n}$.

2. Write the hundredth term of $a_n = \dfrac{n + 2}{3n^2 + 6n}$ in simplified form.

3. Write the first four terms of the sequence given by $b_n = (-1)^n + n$.

4. Find the tenth term of $u_k = \dfrac{(1 - k)^4}{(-3)^{k-1}}$ and simplify.

5. Find a_5 for the sequence where $a_1 = -\frac{2}{3}$ and $a_n = (-1)^n a_{n-1} + \frac{4}{3}$ for $n > 1$.

6. Find $\displaystyle\sum_{n=1}^{5} c_n$, where $c_n = \dfrac{(-2)^n}{n}$.

In questions 7 and 8, an arithmetic sequence has $a_1 = -3$ and $d = \frac{1}{2}$.

7. Find the forty-ninth term.

8. What is the sum of the first 20 terms?

9. Write the next three terms and the nth term of the geometric sequence $-768, 192, -48, \ldots$.

10. Find the sum of the first 50 positive multiples of 7.

11. For an arithmetic sequence $a_1 = -\frac{2}{3}$ and $a_9 = 10$, find a_{21}.

12. Find the sum of all multiples of 3 between 89 and 301.

13. Use the formula for the sum of a finite geometric sequence to show that

$$\sum_{k=1}^{4} 8\left(\tfrac{1}{2}\right)^k = \tfrac{15}{2}$$

14. Evaluate $\displaystyle\sum_{j=1}^{101} (4j - 50)$.   15. Evaluate $\displaystyle\sum_{k=1}^{8} 12\left(\tfrac{1}{2}\right)^{k-1}$.

16. A geometric sequence has $a_1 = -24$ and $a_6 = \frac{3}{4}$. Find r.

Decide whether each of the given infinite geometric series has a sum. Find the sum if it exists; otherwise, give a reason why there is no sum.

17. $\displaystyle\sum_{k=1}^{\infty} 8\left(\tfrac{3}{4}\right)^{k+1}$ 18. $1 + \frac{3}{2} + \frac{9}{4} + \cdots$ 19. $0.06 - 0.009 + 0.00135 - \cdots$

20. Change the repeating decimal $0.363636\ldots$ into rational form.

21. Suppose you save $10 one week and that each week thereafter you save 10¢ more than the week before. How much will you have saved after 1 year?

22. A promotional display of canned peaches in a supermarket is in the form of a pyramid. A sign that gives the discount price for the peaches is standing on the top row, which consists of 3 cans of peaches. The second row has 4 cans, the third row has 5, etc. If the pyramid has 20 rows, how many cans are there in the pyramid?

23. An object is moving along a straight line such that each minute it travels one-third as far as it did during the preceding minute. How far will the object have moved before coming to rest if it moves 24 feet during the first minute?

24. Prove by mathematical induction that for all $n \geq 1$

$$5 + 10 + 15 + \cdots + 5n = \frac{5n(n + 1)}{2}$$

25. Prove by mathematical induction that for all $n \geq 1$

$$\frac{1}{1 \cdot 3} + \frac{1}{3 \cdot 5} + \frac{1}{5 \cdot 7} + \cdots + \frac{1}{(2n - 1)(2n + 1)} = \frac{n}{2n + 1}$$

CHAPTER 11 TEST: MULTIPLE CHOICE

1. What is the tenth term of the sequence $1, \frac{1}{3}, \ldots, \frac{n + 2}{2n^2 + 3n - 2}, \ldots$?

(a) $\frac{1}{7}$ (b) $\frac{1}{19}$ (c) $\frac{3}{107}$ (d) $\frac{2}{201}$ (e) None of the preceding

2. List the first four terms of the sequence given by the formula $a_n = 1 + (-1)^n$.

(a) $0, 2, 0, 2$ (b) $2, 0, 2, 0$ (c) $0, -2, 0, -2,$ (d) $-2, 0, -2, 0$ (e) None of the preceding

3. Find $\sum\limits_{n=1}^{4} c_n$ where $c_n = 3^{k-1}(n - 1)$.

(a) 38   (b) 102   (c) 103   (d) 306   (e) None of the preceding

4. What is the fiftieth term of the arithmetic sequence having $a_1 = -2$ and $d = 5$?

(a) 243   (b) 245   (c) 248   (d) 252   (e) None of the preceding

5. Find S_{20} for the arithmetic sequence whose first term is $a_1 = 2$ and whose common difference is $d = -3$.

(a) 640   (b) 610   (c) -690 (d) -530 (e) None of the preceding

6. Find $\sum\limits_{k=1}^{50}(-2k + 3)$.

(a) 2550   (b) 2500   (c) -2300 (d) -2400 (e) None of the preceding

7. A slow leak in a water pipe develops in such a way that the first day of the leak one ounce of water drips out. Each day thereafter the amount of water lost is one-half ounce more than the day before. How many ounces of water will leak out in 60 days?

(a) $30\frac{1}{2}$ (b) 915   (c) 945   (d) 960   (e) None of the preceding

8. What is the hundredth term of the geometric sequence having $r = \frac{1}{2}$ and $a_1 = -\frac{1}{2}$?

(a) 2^{-100} (b) $-\frac{1}{4^{99}}$ (c) $-\frac{1}{2^{100}}$ (d) $-\frac{1}{2^{98}}$ (e) None of the preceding

9. Find $\sum\limits_{k=1}^{100} 3\left(\frac{1}{3}\right)^k$.

(a) $\frac{3}{2}\left(1 - \frac{1}{3^{100}}\right)$ (b) $1 - \frac{1}{3^{100}}$ (c) $\frac{2}{3}\left(1 - \frac{1}{3^{100}}\right)$ (d) $1 - \left(\frac{1}{3}\right)^{99}$ (e) None of the preceding

10. Suppose you save \$512 in January and then each month thereafter you save only half as much as you saved the preceding month. How much money will you have saved after one year?

(a) \$1023   (b) \$1023.25   (c) \$1023.50   (d) 1023.75   (e) None of the preceding

11. Find the sum of the infinite geometric sequence $-27, -9, -3, \ldots$.

(a) $\frac{81}{4}$ (b) $-\frac{81}{4}$ (c) $\frac{81}{2}$ (d) -81 (e) None of the preceding

12. Compute: $\sum\limits_{n=1}^{\infty} 100(\frac{7}{100})^{n+1}$.

   (a) $\frac{49}{93}$ (b) $\frac{490}{3}$ (c) $\frac{1000}{3}$ (d) No finite sum   (e) None of the preceding

13. Compute: $\sum\limits_{n=1}^{\infty} (-1)^n 2^n$.

   (a) $-\frac{1}{2}$ (b) $\frac{1}{2}$ (c) 2   (d) No finite sum   (e) None of the preceding

14. The first triangle in an infinite sequence of triangles has vertices $A(0, 16)$, $B(-4, 0)$, and $C(4, 0)$. Then each triangle after the first has coordinates that are one-half the coordinates of the preceding triangle. What is the sum of the areas of all the triangles?

   (a) $\frac{128}{3}$ (b) $\frac{256}{3}$ (c) $\frac{512}{3}$ (d) 96   (e) 128

15. Which of the following is correct regarding the proof, by mathematical induction, of a statement S_n for all positive integers n?

   I. In part (1) of the proof we prove that S_1 is true.

   II. In part (2) of the proof we assume that S_k is true for all integers $k \geq 1$, and then prove that S_k implies S_{k+1}.

   III. In part (2) of the proof we first prove S_k is true for all integers $k \geq 1$ and then use this to prove that S_{k+1} is true.

   (a) Only I   (b) Only III   (c) Only I and II   (d) Only I and III   (e) None of the preceding

---

ANSWERS TO THE TEST YOUR UNDERSTANDING EXERCISES

Page 559

1. 3, 5, 7, 9, 11   2. $-2, -4, -6, -8, -10$ 3. $0, -2, -4, -6, -8$

4. $-1, \frac{1}{2}, -\frac{1}{3}, \frac{1}{4}, -\frac{1}{5}$ 5. $1, \frac{1}{4}, \frac{1}{9}, \frac{1}{16}, \frac{1}{25}$ 6. $-\frac{3}{2}, -\frac{1}{2}, -\frac{1}{4}, -\frac{3}{20}, -\frac{1}{10}$

7. $3, \frac{1}{2}, \frac{1}{5}, \frac{3}{28}, \frac{1}{15}$ 8. $\frac{1}{3}, \frac{1}{9}, \frac{1}{27}, \frac{1}{81}, \frac{1}{243}$ 9. 2, 0, 2, 0, 2

Page 565

1. 60   2. 25   3. 40   4. 0   5. 50   6. $\frac{5}{16}$

Page 568

1. $5n$ 2. $-4n + 10$ 3. $\frac{1}{10}n$ 4. $-8n + 3$ 5. n

6. $n - 4$ 7. $\frac{2}{3}n$ 8. $-12n + 65$ 9. $\frac{1}{5}n - \frac{1}{5}$ 10. $n + 1$

Page 576

1. $1, \frac{1}{2}, \frac{1}{4}, \frac{1}{8}, \frac{1}{16}, 1(\frac{1}{2})^{n-1}; r = \frac{1}{2}$ 2. $\frac{1}{4}, \frac{1}{8}, \frac{1}{16}, \frac{1}{32}, \frac{1}{64}, \frac{1}{4}(\frac{1}{2})^{n-1}; r = \frac{1}{2}$

3. $-\frac{1}{2}, \frac{1}{4}, -\frac{1}{8}, \frac{1}{16}, -\frac{1}{32}; -\frac{1}{2}(-\frac{1}{2})^{n-1}; r = -\frac{1}{2}$ 4. $-\frac{1}{27}, \frac{1}{27^2}, -\frac{1}{27^3}, \frac{1}{27^4}, -\frac{1}{27^5}; -\frac{1}{27}(-\frac{1}{27})^{n-1}; r = -\frac{1}{27}$

5. $r = 10; \frac{1}{5}(10)^{n-1}$ 6. $r = -\frac{4}{9}; 27(-\frac{4}{9})^{n-1}$

Page 584

1. $r = \frac{1}{10}; S_\infty = 11\frac{1}{9}$ 2. $r = 4$; no finite sum.   3. $r = -\frac{1}{6}; S_\infty = 30\frac{6}{7}$

4. $r = \frac{1}{4}; S_\infty = -21\frac{1}{3}$ 5. $r = \frac{4}{3}$; no finite sum.   6. $r = 0.01; S_\infty = \frac{1}{33}$

7. $r = -3$; no finite sum.   8. $r = -\frac{9}{10}; S_\infty = \frac{810}{19}$ 9. $r = -1.01$; no finite sum.

12.1 PERMUTATIONS

You will be able to answer this question using the concept of combinations in Section 12.2. See Exercise 15, page 614.

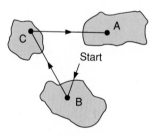

This map shows the trip where the first city is B, then C, and then A.

Suppose that there are 30 students in your class, and you all decide to become acquainted by shaking hands. Each person shakes hands with every other person. How many handshakes take place? Although this problem may not be an important or realistic one, it does suggest types of problems we may solve using counting procedures that will be studied in this chapter.

As a start, let us assume that you are planning a trip that will consist of visiting three cities, A, B, and C. You have your choice as to the order in which you are to visit the cities. How many different trips are possible? A trip begins with a stop at any one of the three cities, the second stop will be at any one of the remaining two cities, and the trip is completed by stopping at the remaining city. One way to answer this question is to sketch the possible trips, one of which is shown in the margin. However, it can get clumsy or tedious to find all trips using such diagrams, especially if more cities are involved.

A better way to obtain the solution is to draw a **tree diagram** that illustrates all possible routes. From the diagram we can read the six possible trips. The arrangement ABC means that the trip begins with city A, goes to B, and ends at C. The other arrangements have similar interpretations.

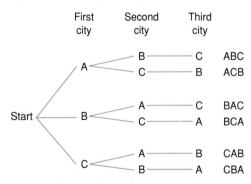

Tree diagrams can be useful in solving such counting problems when the number of possibilities is relatively small. However, when large numbers of possibilities are involved, such tree diagrams are not practical. A more efficient method is

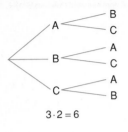

$3 \cdot 2 = 6$

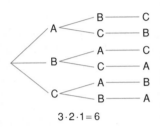

$3 \cdot 2 \cdot 1 = 6$

needed. There is such a method whose underlying idea can be observed using the preceding tree diagram.

After the start, the tree branches into 3 paths A, B, and C. Then, from each of these 3 points there are 2 new branches, giving 6 paths thus far. In other words, using the concept of multiplication, we have 3 groups each containing 2 paths for a total of $3 \cdot 2 = 6$ possibilities for the first two cities. Then the last choice consists of only the 1 remaining city. So now we have $3 \cdot 2 = 6$ groups, each of which contains 1 path, for a final total of $(3 \cdot 2) \cdot 1 = 6$ possible trips.

When the preceding observations are generalized, we obtain the following important principle of counting.

FUNDAMENTAL PRINCIPLE OF COUNTING

Suppose that a first task can be completed in m_1 ways, a second task in m_2 ways, and so on, until we reach the rth task that can be done in m_r ways; then the total number of ways in which these tasks can be completed together is the product

$$m_1 m_2 \cdots m_r$$

The following examples illustrate the use of this counting principle. Because of the large number of possibilities involved, these examples also demonstrate the advantage this principle has over the construction of tree diagrams.

EXAMPLE 1　A club consists of 15 boys and 20 girls. They wish to elect officers consisting of a girl as president and a boy as vice-president. They also wish to elect a treasurer and a secretary, who may be of either sex. How many sets of officers are possible?

We must assume here that no person may hold two offices at the same time.

Solution　There are 20 choices for president and 15 choices for vice-president. Thereafter, since two club members have been chosen and the remaining positions can be filled by either a boy or a girl, 33 members are left for the post of treasurer, and then 32 choices for secretary. Then by the fundamental principle of counting, the total number of choices is

$$20 \cdot 15 \cdot 33 \cdot 32 = 316,800 \qquad \blacksquare$$

EXAMPLE 2
(a) How many three-digit whole numbers can be formed if zero is not an acceptable digit in the hundreds place and repetitions of digits are allowed?
(b) How many if repetitions are not allowed?

Solution
(a) Imagine that you must place a digit in each of three positions, as in the display below.

____ ____ ____　　　　　　　　　Digits to use:

0 cannot be used here ⤴　　　　　0, 1, 2, 3, 4, 5, 6, 7, 8, 9

A tree diagram would begin this way:

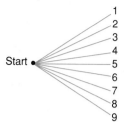

If the tree were continued what nine numbers would follow the 4? What eight numbers would follow the path 47?

There are 9 choices for the first position and 10 for each of the others. By the fundamental principle of counting, the solution is

$$9 \cdot 10 \cdot 10 = 900$$

(b) There are 9 choices for the hundreds place since zero is not allowed here. Once a choice is made, there are still 9 choices available for the tens digit since zero is permissible here. Finally, there are only 8 choices available for the units digits. The final solution is the product $9 \cdot 9 \cdot 8 = 648$. ∎

In our first illustration, where the six trips to the three cities were listed, the order of the elements A, B, C was crucial; trip ABC is different from trip BAC. We say that each of the six arrangements is a *permutation* of three objects taken three at a time. In general, for n elements (n a positive integer) and r a positive integer, $r \leq n$, we have this definition.

When considering permutations, the order of the elements is important. Thus 213 and 312 are two different permutations of the digits 1, 2, 3.

> ### DEFINITION OF PERMUTATION
>
> A **permutation** of n elements taken r at a time is an ordered arrangement, without repetitions, of r of the n elements. The number of permutations of n elements taken r at a time is denoted by $_nP_r$.

Note that a "word" here is to be interpreted as any collection of three letters. That is, "nlm" is considered to be a word in this example.

EXAMPLE 3 How many three-letter "words" composed from the 26 letters of the alphabet are possible? No duplication of letters is permitted.

Solution Since duplication of letters is *not* permitted, once a letter is chosen, it may not be selected again. Therefore, the first letter may be any one of the 26 letters of the alphabet, the second may be any one of the remaining 25, and the third is chosen from the remaining 24. Thus the total number of different "words" is $26 \cdot 25 \cdot 24 = 15,600$. Since we have taken 3 elements out of 26, without repetitions and in all possible orders, we may say that there are 15,600 permutations; that is

$$_{26}P_3 = 26 \cdot 25 \cdot 24 = 15,600$$ ∎

Note that $_{26}P_3 = 26 \cdot 25 \cdot 24$ has three factors beginning with 26 and with each successive factor decreasing by 1. In general, for $_nP_r$ there will be r factors beginning with n, as follows.

$$_nP_r = n(n - 1)(n - 2)(n - 3) \cdots [n - (r - 1)]$$

Thus

$$_nP_r = n(n - 1)(n - 2)(n - 3) \cdots (n - r + 1)$$

When applying this formula, begin with n and use a total of r factors successively decreasing by 1.

Illustration:

$$\underset{\underset{\textstyle \raisebox{1ex}{$\scriptstyle r = 8 \text{ factors successively decreasing by 1}$}}{}}{\overset{\overset{\textstyle n = 20}{}}{_{20}P_8}} = \underbrace{20 \cdot 19 \cdot 18 \cdot 17 \cdot 16 \cdot 15 \cdot 14 \cdot 13}$$

A specific application of this formula occurs when $n = r$. In this case we have the permutation of n elements taken n at a time, and the product has n factors.

$$_nP_n = n(n - 1)(n - 2)(n - 3) \cdots 3 \cdot 2 \cdot 1$$

We may abbreviate this formula by using **factorial notation**.

$$_nP_n = n!$$

In the notation n! the ! is not used as a typical exclamation mark. Rather, it means to multiply all the positive integers from n down to 1 as shown.

For example:

$$_3P_3 = 3! = 3 \cdot 2 \cdot 1 = 6$$

$$_4P_4 = 4! = 4 \cdot 3 \cdot 2 \cdot 1 = 24$$

$$_5P_5 = 5! = 5 \cdot 4 \cdot 3 \cdot 2 \cdot 1 = 120$$

For future consistency we find it convenient to define 0! as equal to 1.

$$0! = 1$$

EXAMPLE 4 How many different ways can the four letters of the word MATH be arranged using each letter only once in each arrangement?

Solution Here we have the permutations of four elements taken four at a time. Thus $_4P_4 = 4! = 4 \cdot 3 \cdot 2 \cdot 1 = 24$. This includes such arrangements as MATH, AMTH, TMAH, and HMAT. Can you list all 24 possibilities? Complete the a tree diagram in the margin to help you find all possible cases. ∎

First
letter

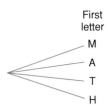

M
A
T
H

TEST YOUR UNDERSTANDING
Think Carefully

Evaluate.

1. $_{10}P_4$ **2.** $_8P_3$ **3.** $_6P_6$ **4.** $\dfrac{10!}{8!}$ **5.** $\dfrac{12!}{9! \, 3!}$

6. How many four-letter "words" from the 26 letters of the alphabet are possible? No duplication of letters is permitted.

7. Answer Exercise 6 if duplications are permitted.

8. How many different ways can the letters of the word EAT be arranged using each letter once in each arrangement? List all the possibilities.

9. Draw a tree diagram that shows all the three-digit whole numbers that can be formed using the digits 2, 5, 8 so that each of the digits is used once in each number.

10. How many four-digit whole numbers greater than 5000 can be formed using each of the digits 3, 5, 6, 8 once in each number?

(Answers: Page 634)

You have seen that $n!$ consists of n factors, beginning with n and successively decreasing to 1. At times it will be useful to display only some of the specific factors in $n!$ For example:

$$n! = n(n - 1)! = n(n - 1)(n - 2)! = n(n - 1)(n - 2)(n - 3)!$$

In particular,

$$5! = 5 \cdot 4! = 5 \cdot 4 \cdot 3! = 5 \cdot 4 \cdot 3 \cdot 2!$$

We can use the formula for $_nP_n$ to obtain a different form for $_nP_r$. To do so we write the formula for $_nP_r$ and then multiply numerator and denominator by $(n - r)!$ as follows.

$$_nP_r = \frac{n(n - 1)(n - 2) \cdots [n - (r - 1)]}{1} \cdot \frac{(n - r)!}{(n - r)!}$$

Now observe that, after multiplying, the numerator can be written as $n!$ to produce another useful formula for the permutation of n elements taken r at a time.

When applying these formulas there are three things to keep in mind. First, the n elements or objects must all be different. Second, no element is repeated in a permutation. And third, order of the elements is important.

PERMUTATION FORMULAS

For n distinct elements taken r at a time, where $1 \leq r \leq n$:

$$_nP_r = n(n - 1)(n - 2) \cdots (n - r + 1)$$

$$_nP_r = \frac{n!}{(n - r)!}$$

Illustrations:

Using the first formula, $_7P_4 = 7 \cdot 6 \cdot 5 \cdot 4 = 840$

Using the second formula, $_7P_4 = \frac{7!}{3!} = \frac{7 \cdot 6 \cdot 5 \cdot 4 \cdot 3!}{3!} = 840$

EXAMPLE 5 A club contains 10 members. They wish to elect officers consisting of a president, vice-president, and secretary-treasurer. How many sets of officers are possible?

Solution We may think of the three offices to be filled in terms of a first office (president), a second office (vice-president), and a third office (secretary-treasurer). Therefore we need to select 3 out of 10 members and arrange them in all possible orders; we need to find the permutations of 10 elements taken 3 at a time.

$$_{10}P_3 = \frac{10!}{7!} = \frac{10 \cdot 9 \cdot 8 \cdot 7!}{7!} = 720$$

■

This example makes use of both permutations and the fundamental counting principle.

EXAMPLE 6 A family of 5 consisting of the parents and 3 children are going to be arranged in a row by a photographer. If the parents are to be next to each other, how many arrangements are possible?

Solution Suppose the parents occupy the first two positions and the children the last three. In this case we have

$$\begin{pmatrix} \text{Parents occupy} \\ \text{these two positions} \\ \text{in } _2P_2 \text{ ways} \\ \underline{2} \cdot \underline{1} = 2 \end{pmatrix} \cdot \begin{pmatrix} \text{Children occupy} \\ \text{these three positions} \\ \text{in } _3P_3 \text{ ways} \\ \underline{3} \cdot \underline{2} \cdot \underline{1} = 6 \end{pmatrix}$$

Since the parents can occupy the first two positions in 2 ways and the children occupy the remaining three positions in 6 ways, the fundamental counting principle gives $2 \cdot 6 = 12$ ways. But 4 adjacent positions are possible for the parents, and each of these gives 12 arrangements. Therefore, the total number of arrangement is $4(12) = 48$. ∎

These are the 4 adjacent positions for the parents:

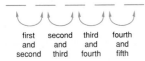

first
and
second

second
and
third

third
and
fourth

fourth
and
fifth

Permutations can also be formed using collections of objects not all of which are distinct from one another. For example, in the word ELEMENT the 3 E's are not distinguishable. How many ways can these 7 letters be arranged to form all _distinguishable permutations?_

One such permutation is TENEMEL. If the E's were momentarily made distinguishable, by using subscripts, and if the consonants remain fixed, then TENEMEL produces these $3! = 6$ permutations:

_E_1, E_2, E_3 can be put into 3 positions in $3! = 6$ ways._

$$\text{TE}_1\text{NE}_2\text{ME}_3\text{L} \qquad \text{TE}_2\text{NE}_1\text{ME}_3\text{L} \qquad \text{TE}_3\text{NE}_1\text{ME}_2\text{L}$$

$$\text{TE}_1\text{NE}_3\text{ME}_2\text{L} \qquad \text{TE}_2\text{NE}_3\text{ME}_1\text{L} \qquad \text{TE}_3\text{NE}_2\text{ME}_1\text{L}$$

Similarly, every distinguishable permutation of the letters in ELEMENT produces 3! permutations of the letters in $E_1LE_2ME_3NT$, of which there are $7! = 5040$. Then, letting P be the number of distinguishable permutations of the letters in ELEMENT, we have $6P = 5040$, or $P = \dfrac{5040}{6} = \dfrac{7!}{3!}$.

This type of reasoning can be extended to produce the following result:

> The number of distinguishable permutations of n objects of which n_1 are of one kind, n_2 are of another kind, . . . , n_k are of another kind is given by
>
> $$\frac{n!}{n_1! \, n_2! \cdots n_k!}$$

These are the permutations for part (a):

BKOO   OBKO
BOKO   OKBO
BOOK   OBOK
KBOO   OKOB
KOBO   OOBK
KOOB   OOKB

EXAMPLE 7 Find the number of distinguishable permutations using all of the letters in each word.

(a) BOOK **(b)** REFERRED **(c)** BEGINNING

Solution

(a) Use $n_1 = 2$, since there are two O's, and $n = 4$ to get $\dfrac{4!}{2!} = 12$.

(b) Use $n_1 = 3$, $n_2 = 3$, and $n = 8$ to get $\dfrac{8!}{3!\,3!} = 1120$

(c) Use $n_1 = 2$, $n_2 = 2$, $n_3 = 3$, and $n = 9$ to get

$$\frac{9!}{2!\,2!\,3!} = \frac{9 \cdot 8 \cdot 7 \cdot 6 \cdot 5 \cdot 4 \cdot 3!}{2 \cdot 2 \cdot 3!}$$

$$= 9 \cdot 8 \cdot 7 \cdot 6 \cdot 5 = 15{,}120 \qquad \blacksquare$$

EXERCISES 12.1

Evaluate.

1. $\dfrac{7!}{6!}$ **2.** $\dfrac{12!}{10!}$ **3.** $\dfrac{12!}{2!\,10!}$ **4.** $\dfrac{15!}{10!\,5!}$ **5.** $_5P_4$ **6.** $_5P_5$ **7.** $_4P_1$ **8.** $_8P_5$

9. Write $_nP_{n-3}$ in factorial notation.   **10.** Show that $_nP_n = {_nP_{n-1}}$.

11. How many ways can the manager of a baseball team select a pitcher and a catcher for a game if there are 5 pitchers and 3 catchers on the team?

12. How many different outfits can Laura wear if she is able to match any one of five blouses, four skirts, and three pairs of shoes?

Consider three-letter "words" to be formed by using the vowels a, e, i, o, and u.

13. How many different three-letter words can be formed if repetitions are not allowed?

14. How many different words can be formed if repetitions are allowed?

15. How many different words without repetitions can be formed whose middle letter is *o*?

16. How many different words without repetitions can be formed whose first letter is *e*?

17. How many different words without repetitions can be formed whose letters at the ends are *u* and *i*?

18. If repetitions are allowed, how many different words can be formed whose middle letter is *a*?

19. How many different words can be formed containing the letter *a* and two other letters?

20. How many different words can be formed containing the letters *a* and *e* so that these letters are not next to each other?

For Exercises 21–26, consider three-digit numbers to be formed using the digits 1, 2, 3, 4, 5, 6, 7, 8, and 9. Also assume that repetition of digits is not allowed unless specified otherwise.

21. How many three-digit whole numbers can be formed?

22. How many three-digit whole numbers can be formed if repetition of digits is allowed?

23. How many three-digit whole numbers can be formed that are even?

24. How many three-digit whole numbers can be formed that are divisible by 5?

25. How many three-digit whole numbers can be formed that are greater than 600?

26. How many three-digit whole numbers can be formed that are less than 400 and are divisible by 5?

27. **(a)** In how many different ways can the letters of STUDY be arranged using each letter only once in each arrangement?

     (b) How many arrangements are there if the S and T are in the first two positions?

     (c) How many arrangements are there if S and T are next to each other?

     (d) How many arrangements are there if the S and T are not next to each other?

28. A class consists of 20 members. In how many different ways can the class select a set of officers consisting of a president, a vice-president, a secretary, and treasurer?

29. A baseball team consists of nine players. How many different batting orders are possible? How many are possible if the pitcher bats last?

30. (a) Each question in a multiple-choice exam has the four choices indicated by the letters a, b, c, and d. If there are eight questions, how many ways can the test be answered? (*Hint:* Use the fundamental counting principle.)

   (b) How many ways are there if no two consecutive questions can have the same answer?

31. When people are seated at a circular table, we consider only their positions relative to each other and are not concerned with the particular seat that a person occupies. How many arrangements are there for seven people to seat themselves around a circular table? (*Hint:* Consider one person's position as fixed.)

32. Review Exercise 31 and conjecture a formula for the number of different permutations of n distinct objects placed around a circle.

33. A license plate is formed by listing two letters of the alphabet followed by three digits. How many different license plates are possible:

   (a) If repetitions of letters and digits are not allowed?

   (b) If repetitions are allowed?

34. Write an expression that gives the number of arrangements of n objects taken r at a time if repetitions are allowed.

35. Solve for n: **(a)** $_nP_1 = 10$; **(b)** $_nP_2 = 12$.

36. Show that $2(_nP_{n-2}) = {}_nP_{n-1}$.

37. A pair of dice is rolled. Each die has six faces on each of which is one of the numbers 1 through 6. One possible outcome is $(3, 5)$, where the first digit shows the outcome of one die and the second digit shows the outcome of the other die. How many different outcomes are possible? List all possible outcomes as pairs of numbers (x, y).

38. To avoid electronic detection, a ship can send coded messages to neighboring ships by displaying a sequence of signal flags having different shapes. If 12 different-shaped flags are available, how many messages can be displayed using a 4-flag sequence?

39. (a) A social security number is a sequence of nine digits. How many different social security numbers are possible?

   (b) How many are there in which no digits repeat?

   (c) How many are there in which there are some repetitions of digits?

   (d) How many are there in which a digit appears exactly three times in succession and no other digits repeat?

40. A local telephone number consists of 7 digits, the first 3 of which are called the telephone exchange, such as

$$627{-}4195$$

$$\underbrace{}$$

(telephone exchange)

(a) For a given exchange, how many different telephone numbers are possible?

(b) Suppose a city has 73,500 telephones. What is the minimum number of exchanges needed to accommodate the city's phones?

41. A student has room for 6 books on a shelf near her study area. The books consist of a dictionary and textbooks in the areas of chemistry, English, history, mathematics, and philosophy.

(a) In how many ways can the books be arranged on the shelf?

(b) In how many ways can they be arranged if the dictionary is put into the first position?

(c) How many arrangements are possible in which the mathematics and philosophy books are next to each other?

42. Solve for n:

(a) $_nP_6 = 15(_nP_5)$ (b) $_nP_6 = 90(_nP_4)$

43. Solve for r: $_{12}P_r = 8(_{12}P_{r-1})$

How many distinguishable permutations can be formed using all the letters in each word?

44. (a) SEVEN   (b) INNING   (c) ORDERED

45. (a) DELEGATE   (b) COLLEGE   (c) STATEMENTS

46. List the distinguishable permutations using all the letters in ERIE. (*Hint:* use a tree diagram.)

**CHALLENGE
Think Creatively**

How many ways can you choose the letters to form the word PYRAMID in the given diagram if

P

Y   Y

R   R   R

A   A   A   A

M   M   M   M   M

I   I   I   I   I   I

D   D   D   D   D   D   D

(a) each letter in the word PYRAMID can be any one of that particular letter listed?

(b) the letter being chosen, other than P, is below and to the immediate left or right of the preceding letter?

(c) The letters are chosen according to part (b) except that the D must be the middle D in the last row.

**12.2
COMBINATIONS**

A permutation may be regarded as an *ordered* collection of elements. For instance, a visit to cities A, B, and C in the order ABC is different from a visit in the order ACB. On the other hand, there are times when we need to consider situations where the order of the elements is not essential. For example, suppose that on a mathematics test you are given the choice of answering any three of five given questions denoted by Q_1, Q_2, Q_3, Q_4, Q_5. If you choose questions Q_2, Q_3, Q_4, it makes no difference in which order you answer the three questions, so this is not a permutation. Rather, we say that the set $\{Q_2, Q_3, Q_4\}$ is a *combination* of 3 things taken out of 5, according to the following definition in which n and r are integers with $0 \leq r \leq n$.

*Recall that a set A is a **subset** of a set B provided that each element in A is also in B.*

DEFINITION OF COMBINATION

A **combination** is a subset of r distinct elements selected out of n elements without regard to order. The number of combinations of n elements taken r at a time is denoted by either of these symbols:

$$_nC_r \quad \text{or} \quad \binom{n}{r}$$

As an example, suppose that a class of 10 members wishes to elect a committee consisting of three of its members. These three members are not to be designated as holding any special office. Then a committee consisting of members David, Ellen, and Robert is the same regardless of the order in which they are selected. In other words, using D, E, and R respectively, as abbreviations for their names, the combination $\{D, E, R\}$ gives rise to the following six permutations:

$$\text{DER} \qquad \text{DRE} \qquad \text{EDR} \qquad \text{ERD} \qquad \text{RDE} \qquad \text{RED}$$

Each combination of three members in this illustration actually gives rise to $3! = 6$ permutations. To find the number of possible committees we need to find the combinations of 10 elements taken 3 at a time. The *number* of such combinations is expressed as $_{10}C_3$ and is read as "the number of combinations of 10 elements taken 3 at a time." Since each of these combinations produces $3! = 6$ permutations, it follows that

The 120 combinations give rise to 720 permutations.

$$3!(_{10}C_3) = \,_{10}P_3 \quad \text{or} \quad _{10}C_3 = \frac{_{10}P_3}{3!} = \frac{10 \cdot 9 \cdot 8}{3 \cdot 2 \cdot 1} = 120$$

In general, using $_nP_r = \dfrac{n!}{(n-r)!}$, we obtain

$$_nC_r = \frac{_nP_r}{r!} = \frac{\dfrac{n!}{(n-r)!}}{r!} = \frac{n}{r!(n-r)!}$$

COMBINATION FORMULAS

*When applying these formulas there are three things to keep in mind. First, the n elements or objects must all be different. Second, no element is used more than once in a combination. And third, order of the elements is **not** important.*

For n elements taken r at a time, when $0 \leq r \leq n$:

$$_nC_r = \frac{_nP_r}{r!}$$

$$_nC_r = \frac{n!}{r!(n-r)!}$$

Recall that we have defined
$0! = 1$. *Do you see here why
this definition was made?*

As noted, the symbol $\binom{n}{r}$ can be used in place of $_nC_r$. Thus $\binom{n}{0} = \dfrac{n!}{0!\,n!} = 1$; also $\binom{n}{n} = \dfrac{n!}{n!\,0!} = 1$. Can you prove that $\binom{n}{r} = \binom{n}{n-r}$?

Illustrations:

$$\binom{10}{2} = \frac{10!}{2!\,8!} = \frac{10 \cdot 9 \cdot 8!}{2 \cdot 8!} = 45$$

$$_{10}C_9 = \frac{10!}{9!\,1!} = \frac{10 \cdot 9!}{9!} = 10$$

The question of order is the essential ingredient that determines whether a problem involves permutations or combinations. Examples 1 and 2 below will illustrate this distinction.

EXAMPLE 1   Using the digits 1 through 9, how many different four-digit whole numbers can be formed if repetition of digits is not allowed?

Solution   Order is important here; thus 4923 is a different number from 9432. Therefore we need to find the *permutations* of nine elements taken four at a time.

$$_9P_4 = \frac{9!}{5!} = 9 \cdot 8 \cdot 7 \cdot 6 = 3024$$

■

EXAMPLE 2   A student has a penny, a nickel, a dime, a quarter, and a half-dollar and wishes to leave a tip consisting of exactly three coins. How many different amounts as tips are possible?

*Here are four of the 10 tips
possible:*

$1 + 5 + 10 = 16$

$1 + 5 + 25 = 31$

$1 + 5 + 50 = 56$

$1 + 10 + 25 = 36$

*Complete this list to find the
other 6 possibilities.*

Solution   Order is not important here; a tip of $5¢ + 10¢ + 25¢$ is the same as one of $25¢ + 10¢ + 5¢$. Therefore, we need to find the *combinations* of five things taken three at a time.

$$\binom{5}{3} = {_5C_3} = \frac{5!}{3!\,2!} = \frac{5 \cdot 4}{2} = 10$$

■

The next example illustrates how the fundamental principle of counting is used in a problem involving combinations.

EXAMPLE 3   A class consists of 10 boys and 15 girls. How many committees of five can be selected if each committee is to consist of two boys and three girls?

Solution   The order is not essential here since the committee members do not hold any special offices. Thus the problem involves combinations.

To select two boys: $_{10}C_2 = \dfrac{10!}{2!\,8!} = 45$

To select three girls: $_{15}C_3 = \dfrac{15!}{3!\,12!} = 455$

Since there are 45 pairs of boys possible, and since each of these pairs can be matched with any of the possible 455 triples of girls, the fundamental principle of counting gives

$$45 \cdot 455 = 20,475$$

as the total number of committees that can be formed. ■

TEST YOUR UNDERSTANDING
Think Carefully

Evaluate.

1. $_{10}C_4$ 2. $_5C_5$ 3. $_8C_0$ 4. $\binom{12}{3}$ 5. $\binom{8}{5}$

6. Show that $_{10}C_3 = {_{10}C_7}$.
7. How many different ways can a committee of 4 members be selected from a group of 12 students?
8. A supermarket carries 6 brands of canned peas and 8 brands of canned corn. A shopper wants to try 2 different brands of peas and 3 different brands of corn. How many ways can the shopper select the 5 items?
9. How many lines are determined by eight points in a plane if no three points are on the same line?
10. How many triangles can be formed using five points in a plane no three of which are on the same line?

(Answers: Page 634)

In the next example an ordinary deck of playing cards consisting of 52 different cards is used. These are divided into four suits: spades, hearts, diamonds, and clubs. There are 13 cards in each suit: from 1 (ace) through 10, jack, queen, and king.

EXAMPLE 4 A "poker hand" consists of 5 cards. How many different hands can be dealt from a deck of 52 cards?

Solution The order of the 5 cards dealt is not important, so that this becomes a problem involving combinations rather than permutations. We wish to find the number of combinations of 52 elements taken 5 at a time.

Use a calculator to solve problems with extensive computations such as shown here. First simplify the fraction.

$$_{52}C_5 = \frac{52!}{5!\,47!} = \frac{52 \cdot 51 \cdot 50 \cdot 49 \cdot 48 \cdot \cancel{47!}}{5 \cdot 4 \cdot 3 \cdot 2 \cdot 1 \cdot \cancel{47!}} = 2,598,960$$ ■

EXAMPLE 5 An ice cream parlor advertises that you may have your choice of five different toppings, and you may choose none, one, two, three, four, or all five toppings. How many choices are there in all?

Solution There are several ways to approach this problem. From one point of view you may consider yourself on a cafeteria line with five stations. At each one you have two choices, to accept the topping or not to accept it. Thus, by the fundamental principal of counting, the total number of choices is $2 \cdot 2 \cdot 2 \cdot 2 \cdot 2 = 32$. From another point of view, the solution is the number of different ways that we can select none, one, two, three, four, or five elements from a total of five possibilities; that is,

This example illustrates how the fundamental counting principle is used in a problem involving combinations.

$$\binom{5}{0} + \binom{5}{1} + \binom{5}{2} + \binom{5}{3} + \binom{5}{4} + \binom{5}{5}$$

Since, for example, choosing one topping and choosing two toppings are not done together, we add the number of possibilities rather than multiply.

Show that this sum is also equal to 32. ∎

EXAMPLE 6 How many subsets, each consisting of 3 elements, does the set $S = \{a, b, c, d, e\}$ have? List these subsets.

Solution Since a subset is a combination, the number of required subsets is given by

$$_5C_3 = \frac{5!}{3!\,2!} = 10$$

The subsets are

How many times does each of the 5 elements of S appear in the 10 subsets?

$$\{a, b, c\}, \quad \{a, b, d\}, \quad (a, b, e), \quad \{a, c, d\}, \quad \{a, c, e\}$$
$$\{a, d, e\}, \quad \{b, c, d\}, \quad \{b, c, e\}, \quad \{b, d, e\}, \quad \{c, d, e\}$$ ∎

EXAMPLE 7 A dinner party of 10 people arrives at a restaurant that has only two tables available. One table seats 6 and the other 4. If the seating arrangement at either table is not taken into account, how many ways can the 10 people divide themselves to be seated at these two tables?

Solution Each time 6 people sit at the table for 6, the remaining 4 will be at the other table. Therefore, it is necessary to find the number of ways we can select subsets of 6 out of 10 people. Thus

If the seating arrangements at each table are taken into account, how many ways can the people be seated? (See Exercise 41.)

$$\binom{10}{6} = \frac{10!}{6!\,4!} = \frac{10 \cdot 9 \cdot 8 \cdot 7 \cdot 6!}{6! \cdot 4 \cdot 3 \cdot 2 \cdot 1} = 210$$

The dinner party can split into the two tables in 210 ways. ∎

EXERCISES 12.2

Evaluate.

1. $_5C_2$ 2. $_{10}C_1$ 3. $_{10}C_0$ 4. $_4C_3$ 5. $\binom{15}{15}$ 6. $\binom{30}{3}$ 7. $\binom{30}{27}$ 8. $\binom{n}{3}$

9. A class consists of 20 members. In how many different ways can the class select **(a)** a committee of 4? **(b)** a set of 4 officers?

10. On a test a student must select 8 questions out of a total of 10. In how many different ways can this be done?

11. How many different ways can six people be split up into two teams of three each?

12. How many straight lines are determined by five points, no three of which are collinear?

13. There are 15 women on a basketball team. In how many different ways can a coach field a team of 5 players?

14. Answer Exercise 13 if two of the players can only play center and the others can play any of the remaining positions. (Assume that exactly one center is in a game at one time.)

15. Answer the question stated at the beginning of Section 12.1: How many handshakes take place when each person in a group of 30 shakes hands with every other person?

16. Box A contains 8 balls and box B contains 10 balls. In how many different ways can 5 balls be selected from these boxes if 2 are to be taken from box A and 3 from box B?

17. A class consists of 12 women and 10 men. A committee is to be selected consisting of 3 women and 4 men. How many different committees are possible?

18. Solve for n: **(a)** $_nC_1 = 6$; **(b)** $_nC_2 = 6$.

19. Convert to fraction form and simplify: **(a)** $_nC_{n-1}$; **(b)** $_nC_{n-2}$.

20. Prove: $\binom{n}{r} = \binom{n}{n-r}$.   21. Evaluate $_nC_4$ given that $_nP_4 = 1680$.

22. Solve for n: $5\binom{n}{2} = 2\binom{n+2}{2}$

23. Consider this expression: $\binom{n}{0} + \binom{n}{1} + \binom{n}{2} + \cdots + \binom{n}{n-1} + \binom{n}{n}$.

Evaluate for: **(a)** $n = 2$ **(b)** $n = 3$ **(c)** $n = 4$ **(d)** $n = 5$

24. Use the results of Exercise 23 and conjecture the value of the expression for any positive integer n.

25. Explain the equality in Exercise 20 in terms of subsets.

26. Give subset interpretations of the results $_nC_0 = 1$ and and $_nC_n = 1$.

27. Interpret the result in Exercise 24 in terms of subsets of a set.

28. Ten points are marked on a circle. How many different triangles do these points determine so that the vertices of each triangle are marked points on the circle?

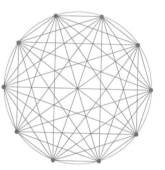

29. How many ways can 5-card hands be selected out of a deck of 52 cards so that all 5 cards are in the same suit?

30. How many ways can 5-card hands be selected out of a deck of 52 cards such that 4 of the cards have the same face value? (Four 10's and some fifth card is one such hand.)

31. How many ways can 5-card hands be selected out of a deck of 52 cards so that the 5 cards consist of a pair and three of a kind? (Two kings and three 7's is one such hand.)

32. A student wants to form a schedule consisting of 2 mathematics courses, 2 history courses, and 1 art course. The student can make these selections from 6 mathematics courses, 10 history courses, and 5 art courses. Assuming that there are no time conflicts, how many ways can the student select the 5 courses?

33. Suppose that in the U.S. Senate 25 Republicans and 19 Democrats are eligible for membership on a new committee. If this new committee is to consist of 9 senators, how many committees would be possible if the committee contained:

(a) 5 Republicans and 4 Democrats?

(b) 5 Democrats and 4 Republicans?

34. In Exercise 33, how many committees are possible if the chairperson of the committee is an eligible Republican senator appointed by the Vice-President, and the rest of the committee is evenly divided between Democrats and Republicans?

35. A basketball squad has 12 players consisting of 3 centers, 5 forwards, and 4 guards. How many ways can the coach field a team having 1 center, 2 forwards, and 2 guards?

36. The Green Lawn Tennis Club has scheduled a round-robin tennis tournament in which each player plays one match against every other player. If 12 players are signed up for the tournament, how many matches have been scheduled?

37. A college club has 18 members of which 10 are men and 8 are women. The chairperson of the club is one of the men. A committee of 5 club members is to be formed that must include the chairperson of the club.

(a) How many committees can be formed consisting of 2 women and 3 men?

(b) How many committees can be formed containing no less than 1 woman?

38. Solve for n: $_nC_6 = 3(_{n-2}C_4)$

39. Suppose that a 9-member committee is to vote on an amendment. How many different ways can the votes be cast so that the amendment passes by a simple majority? (A simple majority means 5 or more yes votes.)

40. In Exercise 39, how many different favorable ways can the votes be cast if the amendment needs at least a $\frac{2}{3}$ majority?

41. If the dinner tables in Example 7 of this section are round, how many seating arrangements are possible if the arrangement at each table is taken into account.

 Written Assignment: Use specific examples of your own to explain the distinction between a permutation and a combination of n elements taken r at a time.

**CHALLENGE
Think Creatively**   Prove: $\left(\begin{array}{c} n \\ r-1 \end{array}\right) + \left(\begin{array}{c} n \\ r \end{array}\right) = \left(\begin{array}{c} n+1 \\ r \end{array}\right)$

**12.3
THE BINOMIAL
EXPANSION**

The factored form of the trinomial square $a^2 + 2ab + b^2$ is $(a + b)^2$. Turning this around, we say that the *expanded form* of $(a + b)^2$ is $a^2 + 2ab + b^2$. And if $(a + b)^2$ is multiplied by $a + b$, we get the expansion of $(a + b)^3$. Here is a list of the expansions of the first five powers of the binomial $a + b$.

You can verify these results by multiplying the expansion in each row by $a + b$ to get the expansion in the next row.

$$(a + b)^1 = a + b$$

$$(a + b)^2 = a^2 + 2ab + b^2$$

$$(a + b)^3 = a^3 + 3a^2b + 3ab^2 + b^3$$

$$(a + b)^4 = a^4 + 4a^3b + 6a^2b^2 + 4ab^3 + b^4$$

$$(a + b)^5 = a^5 + 5a^4b + 10a^3b^2 + 10a^2b^3 + 5ab^4 + b^5$$

Our objective here is to learn how to find such expansions directly without having to multiply. That is, we want to be able to expand $(a + b)^n$, especially for larger values of n, without having to multiply $a + b$ by itself repeatedly.

Let n represent a positive integer. As seen in the preceding display, each expansion begins with a^n and ends with b^n. Moreover, each expansion, has $n + 1$ terms that are all preceded by plus signs. Now look at the case for $n = 5$. Replace the first term a^5 by a^5b^0 and use a^0b^5 in place of b^5. Then

$$(a + b)^5 = a^5b^0 + 5a^4b + 10a^3b^2 + 10a^2b^3 + 5ab^4 + a^0b^5$$

In this form it becomes clear that (from left to right) the exponents of a successively decrease by 1, beginning with 5 and ending with zero. At the same time, the exponents of b increase from zero to 5. Also note that the sum of the exponents for each term is 5. Verify that similar patterns also hold for the other cases shown.

Using the preceding observations, we would *expect* the expansion of $(a + b)^6$ to have seven terms that, except for the unknown coefficients, look like this:

$$a^6 + \underline{\hspace{1em}} a^5b + \underline{\hspace{1em}} a^4b^2 + \underline{\hspace{1em}} a^3b^3 + \underline{\hspace{1em}} a^2b^4 + \underline{\hspace{1em}} ab^5 + b^6$$

Our list of expansions reveals that the second coefficient, as well as the coefficient of the next to the last term, is the number n. Filling in these coefficients for the case $n = 6$ gives

$$a^6 + 6a^5b + \underline{\qquad}a^4b^2 + \underline{\qquad}a^3b^3 + \underline{\qquad}a^2b^4 + 6ab^5 + b^6$$

To get the remaining coefficients we return to the case $n = 5$ and learn how such coefficients can be generated. Look at the second and third terms.

$$\underset{\text{2nd term}}{\underbrace{(5)a^{(4)}b}} \qquad \underset{\text{3rd term}}{\underbrace{10a^3b^{(2)}}}$$

If the exponent 4 of a in the *second* term is multiplied by the coefficient 5 of the *second* term and then divided by the exponent 2 of b in the *third* term, the result is 10, the coefficient of the third term.

$$\text{coefficient of third term} = \frac{4(5)}{2} = 10$$

exponent of a in 2nd term ↘   ↗ coefficient of 2nd term

↑ exponent of b in 3rd term

Verify that this procedure words for the next coefficient.

On the basis of the evidence, we expect the missing coefficients for the case $n = 6$ to be obtainable in the same way. Here are the computations:

Use $(6)\ a^{(5)}b + \underline{\ }a^4b^{(2)}$: 3rd coefficient $= \dfrac{5(6)}{2} = 15$

Use $(15)\ a^{(4)}b^2 + \underline{\ }a^3b^{(3)}$: 4th coefficient $= \dfrac{4(15)}{3} = 20$

Use $(20)\ a^{(3)}b^3 + \underline{\ }a^2b^{(4)}$: 5th coefficient $= \dfrac{3(20)}{4} = 15$

Finally, we may write the following expansion:

You can verify that this equality is correct by multiplying the expansion for $(a + b)^5$ by $a + b$.

$$(a + b)^6 = a^6 + 6a^5b + 15a^4b^2 + 20a^3b^3 + 15a^2b^4 + 6ab^5 + b^6$$

More labor can be saved by observing the symmetry in the expansions of $(a + b)^n$. For instance, when $n = 6$ the coefficients around the middle term are symmetric. Similarly, when $n = 5$ the coefficients around the two middle terms are symmetric.

To get an expansion of the binomial $a - b$, write $a - b = a + (-b)$ and substitute into the previous form. For example, with $n = 6$,

Be certain that you recognize the difference between the expansion of $(a - b)^n$ and the factorization of $a^n - b^n$ (see page 50).

$$(a - b)^6 = [a + (-b)]^6 = a^6 + 6a^5(-b) + 15a^4(-b)^2 + 20a^3(-b)^3$$
$$+ 15a^2(-b)^4 + 6a(-b)^5 + (-b)^6$$

$$= a^6 - 6a^5b + 15a^4b^2 - 20a^3b^3 + 15a^2b^4 - 6ab^5 + b^6$$

This result indicates that the expansion of $(a - b)^n$ is the same as the expansion of $(a + b)^n$ except that the signs alternate, beginning with plus.

EXAMPLE 1   Expand: **(a)** $(x + 2)^7$ **(b)** $(x - 2)^7$

Solution

(a) Let x and 2 play the role of a and b in $(a + b)^7$, respectively.

$$(x + 2)^7 = x^7 + 7x^62 + \_\_x^52^2 + \_\_x^42^3 + \_\_x^32^4 + \_\_x^22^5 + 7x2^6 + 2^7$$

Now find the missing coefficients:

$$\text{3rd coefficient} = \frac{6(7)}{2} = 21 = \text{6th coefficient}$$

$$\text{4th coefficient} = \frac{5(21)}{3} = 35 = \text{5th coefficient}$$

The completed expansion may now be given as follows:

$$(x + 2)^7 = x^7 + 7x^62 + 21x^52^2 + 35x^42^3 + 35x^32^4 + 21x^22^5 + 7x2^6 + 2^7$$

$$= x^7 + 14x^6 + 84x^5 + 280x^4 + 560x^3 + 672x^2 + 448x + 128$$

Note that since x − 2 = x + (−2), you can set a = x and b = −2. Then each odd power of b is negative, and the signs alternate.

(b) The expansion of $(x - 2)^7$ may be obtained from the expansion of $(x + 2)^7$ by alternating the signs. Thus,

$$(x - 2)^7 = x^7 - 14x^6 + 84x^5 - 280x^4 + 560x^3 - 672x^2 + 448x - 128 \quad \blacksquare$$

The preceding work can be used to obtain the expansion of the general form $(a + b)^n$. Begin by writing the variable parts of the first few terms.

$$a^n + \_\_\_\_a^{n-1}b^1 + \_\_\_\_a^{n-2}b^2 + \_\_\_\_a^{n-3}b^3 + \cdots$$

As before, to get the second coefficient multiply 1 by n and divide by 1.

$$a^n + \frac{n}{1}a^{n-1}b^1 + \_\_\_\_a^{n-2}b^2 + \_\_\_\_a^{n-3}b^3 + \cdots$$

To get the third coefficient multiply $\frac{n}{1}$ by $n - 1$ and divide by 2.

$$a^n + \frac{n}{1}a^{n-1}b^1 + \frac{n(n - 1)}{1 \cdot 2}a^{n-2}b^2 + \_\_\_\_a^{n-3}b^3 + \cdots$$

The next coefficient is $\frac{n(n - 1)}{1 \cdot 2}$ times $n - 2$ divided by 3, and we now have

$$a^n + \frac{n}{1}a^{n-1}b^1 + \frac{n(n - 1)}{1 \cdot 2}a^{n-2}b^2 + \frac{n(n - 1)(n - 2)}{1 \cdot 2 \cdot 3}a^{n-3}b^3 + \cdots$$

Proceeding in this manner and noting the symmetry of the coefficients, we obtain the following result.

BINOMIAL FORMULA

$$(a + b)^n = a^n + \frac{n}{1}a^{n-1}b + \frac{n(n-1)}{1 \cdot 2}a^{n-2}b^2$$

$$+ \frac{n(n-1)(n-2)}{1 \cdot 2 \cdot 3}a^{n-3}b^3 + \cdots + \frac{n}{1}ab^{n-1} + b^n$$

The term having the factor b^r is the $(r + 1)$st term and can be written as

$$\frac{n(n-1)(n-2) \cdots (n-r+1)}{r!}a^{n-r}b^r$$

When expanding a binomial you may find it easier to follow the steps that precede the statement of the binomial formula rather than substitute directly into it.

EXAMPLE 2 Expand $(2x - y)^5$ and simplify.

Solution Use $a = 2x$, $b = y$, and $n = 5$ in the binomial formula with alternating signs.

$$(2x - y)^5 = (2x)^5 - \frac{5}{1}(2x)^4y + \frac{5 \cdot 4}{1 \cdot 2}(2x)^3y^2 - \frac{5 \cdot 4 \cdot 3}{1 \cdot 2 \cdot 3}(2x)^2y^3$$

$$+ \frac{5 \cdot 4 \cdot 3 \cdot 2}{1 \cdot 2 \cdot 3 \cdot 4}(2x)y^4 - \frac{5 \cdot 4 \cdot 3 \cdot 2 \cdot 1}{1 \cdot 2 \cdot 3 \cdot 4 \cdot 5}y^5$$

$$= 32x^5 - 80x^4y + 80x^3y^2 - 40x^2y^3 + 10xy^4 - y^5 \qquad \blacksquare$$

The coefficients in the expansion of $(a + b)^7$ were found in Example 1.

EXAMPLE 3 Evaluate 2^7 by expanding $(1 + 1)^7$.

Solution Since all powers of 1 equal 1, the expansion of $(1 + 1)^7$ is the sum of the coefficients in the expansion of $(a + b)^7$. Thus

$$2^7 = (1 + 1)^7 = 1 + 7 + 21 + 35 + 35 + 21 + 7 + 1 = 128 \qquad \blacksquare$$

Another way to develop the binomial formula is to make use of our knowledge of combinations. Let us consider the expansion of $(a + b)^5$ again from a different point of view.

$$(a + b)^5 = \underbrace{(a + b)(a + b)(a + b)(a + b)(a + b)}_{5 \text{ factors}}$$

To expand $(a + b)^5$, consider each term as follows.

First term: Multiply all the a's together to obtain a^5.

Second term: We need to combine all terms of the form a^4b. How are these terms formed in the multiplication process that gives the expansion? One of these terms is formed by multiplying the a's in the first four factors times the b in the last factor.

$$(\widehat{a} + b)(\widehat{a} + b)(\widehat{a} + b)(\widehat{a} + b)(a + \boxed{b})$$

$$\text{Product} = a^4 b$$

Another of these terms is formed like this:

$$(\widehat{a} + b)(\widehat{a} + b)(\widehat{a} + b)(a + \boxed{b})(\widehat{a} + b)$$

$$\text{Product} = a^4 b$$

Now you can see that the number of such terms is the same as the number of ways we can select just one of the b's from the five factors. This can be done in five ways, which can be expressed as $_5C_1$ or as $\binom{5}{1}$, the coefficient of $a^4 b$.

Third term: Search for all terms of the form $a^3 b^2$. The number of ways of selecting two b's from five terms, that is, $_5C_2$ or $\binom{5}{2}$.

Fourth term: The number of terms of the form $a^2 b^3$ is the number of ways of selecting three b's from five terms, that is, $_5C_3$ or $\binom{5}{3}$.

Fifth term: The number of ways of selecting four b's from the five terms is $_5C_4$ or $\binom{5}{4}$, the coefficient of ab^4.

Sixth term: Multiply all the b's together to obtain b^5.

Thus we may write the expansion of $(a + b)^5$ in this form:

$$a^5 + \binom{5}{1}a^4 b + \binom{5}{2}a^3 b^2 + \binom{5}{3}a^2 b^3 + \binom{5}{4}ab^4 + b^5$$

For consistency of form, we may write the coefficient of a^5 as $\binom{5}{0}$ and that of b^5 as $\binom{5}{5}$. In each case note that $\binom{5}{0} = \binom{5}{5} = 1$.

A similar argument can be made for each of the terms of the expansion of $(a + b)^n$. For example, to find the coefficient of the term that has the factor $a^{n-r}b^r$, we need to find the number of different ways of selecting r b's from n terms. This can be expressed as $_nC_r$ or $\binom{n}{r}$. Now we are ready to generalize and write this second form of the *binomial formula:*

This expansion can be written in sigma notation as

$$(a + b)^n = \sum_{r=0}^{n} \binom{n}{r} a^{n-r} b^r$$

$$(a + b)^n = \binom{n}{0}a^n b^0 + \binom{n}{1}a^{n-1}b^1 + \binom{n}{2}a^{n-2}b^2 + \cdots$$

$$+ \binom{n}{r}a^{n-r}b^r + \cdots + \binom{n}{n-1}a^1 b^{n-1} + \binom{n}{n}a^0 b^n$$

*The numbers $\binom{n}{r}$ are referred to as **binominal coefficients**.*

Observe that the $(r + 1)$st term here and in the formula on page 618 are the same:

$$\binom{n}{r}a^{n-r}b^r = \frac{{}_nP_r}{r!}a^{n-r}b^r = \frac{n(n-1)(n-2)\cdots(n-r+1)}{r!}a^{n-r}b^r$$

EXAMPLE 4 Find the sixth term in the expansion of $(a + b)^8$.

Recall that the sum of the exponents in each term is equal to n.

Solution Note that for any term in the expansion the exponent r of b is one less than the number of the term. Then, since we need the sixth term, $r = 5$. Also, since the sum of the exponents is 8, the sixth term is

$$\binom{8}{5}a^3b^5 = 56a^3b^5 \qquad \blacksquare$$

Example 4 can also be solved by using the method in the next example.

EXAMPLE 5 Find the fourth term in the expansion of $(x - 2y)^{10}$.

Note that in Example 5 we may think of $(x - 2y)^{10}$ as $[x + (-2y)]^{10}$ so that we may apply the binomial formula for $(a + b)^n$.

Solution Use the general term $\binom{n}{r}a^{n-r}b^r$, which is the $(r + 1)$st term. Then $r + 1 = 4$ and $r = 3$, $n = 10$, and $n - r = 7$. Thus the fourth term is

$$\binom{10}{3}x^7(-2y)^3 = 120x^7(-8y^3) = -960x^7y^3 \qquad \blacksquare$$

EXERCISES 12.3

Expand and simplify.

1. $(x + 1)^5$ **2.** $(x - 1)^6$ **3.** $(x + 1)^7$ **4.** $(x - 1)^8$

5. $(a - b)^4$ **6.** $(3x - 2)^4$ **7.** $(3x - y)^5$ **8.** $(x + y)^8$

9. $(a^2 + 1)^5$ **10.** $(2 + h)^9$ **11.** $(1 - h)^{10}$ **12.** $(-2 + x)^7$

13. $\left(\frac{1}{2} - a\right)^4$ **14.** $\left(\frac{x}{2} + \frac{2}{y}\right)^5$ **15.** $\left(\frac{1}{x} - x^2\right)^6$ **16.** $\left(2a - \frac{1}{a^2}\right)^6$

$\triangle$ *Simplify.*

17. $\dfrac{(1 + h)^3 - 1}{h}$ **18.** $\dfrac{(3 + h)^4 - 81}{h}$ **19.** $\dfrac{(c + h)^3 - c^3}{h}$

20. $\dfrac{(x + h)^6 - x^6}{h}$ **21.** $\dfrac{2(x + h)^5 - 2x^5}{h}$ **22.** $\dfrac{\dfrac{1}{(2 + h)^2} - \dfrac{1}{4}}{h}$

23. Evaluate 2^{10} by expanding $(1 + 1)^{10}$.

24. Write the first five terms in the expansion of $(x + 1)^{15}$. What are the last five terms?

25. Write the first five terms and the last five terms in the expansion of $(c + h)^{20}$.

26. Write the first four terms and the last four terms in the expansion of $(a - 1)^{30}$.

27. Study this triangular array of numbers and discover the connection with the expansions of $(a + b)^n$, where $n = 1, 2, 3, 4, 5, 6$.

$$
\begin{array}{ccccccccccccc}
 & & & & & 1 & & 1 & & & & & \\
 & & & & 1 & & 2 & & 1 & & & & \\
 & & & 1 & & 3 & & 3 & & 1 & & & \\
 & & 1 & & 4 & & 6 & & 4 & & 1 & & \\
 & 1 & & 5 & & 10 & & 10 & & 5 & & 1 & \\
1 & & 6 & & 15 & & 20 & & 15 & & 6 & & 1 \\
\end{array}
$$

This triangular array of numbers is called **Pascal's triangle,** named after French mathematician Blaise Pascal (1623–1662). However, the triangle appeared in Chinese writings as early as 1303.

In Exercise 27 you learned that the nth row in Pascal's triangle contains the coefficients in the expansion of $(a + b)^n$. Use this result to expand the following.

28. $(x + 1)^5$ 29. $(a + 2)^6$ 30. $(2x - 3)^3$ 31. $(3p + 2q)^4$

32. Discover how the 6th row of the triangle in Exercise 27 can be obtained from the 5th row by studying the connection between the 4th and 5th rows indicated by this scheme.

$$
\begin{array}{ccccccccccc}
1 & & 4 & & 6 & & 4 & & 1 & & \\
 & + & & + & & + & & + & & \\
1 & & 5 & & 10 & & 10 & & 5 & & 1 \\
\end{array}
$$

33. Using the result of Exercise 32, write the 7th, 8th, 9th, and 10th rows of the triangle.
34. Use the 9th row found in Exercise 33 to expand $(x + h)^9$.
35. Use the 10th row found in Exercise 33 to expand $(x - h)^{10}$.
36. Why does the sum of all the numbers in one line of Pascal's triangle equal twice the sum of the numbers in the preceding line?
37. Find the sixth term in the expansion of $(a + 2b)^{10}$.   38. Find the fifth term in the expansion of $(2x - y)^8$.
39. Find the fourth term in the expansion of $\left(\dfrac{1}{x} + \sqrt{x}\right)^7$.
40. Find the eighth term in the expansion of $\left(\dfrac{a}{2} + \dfrac{b^2}{3}\right)^{10}$.
41. Find the term that contains x^5 in the expansion of $(2x + 3y)^8$.
42. Find the term that contains y^{10} in the expansion of $(x - 2y^2)^8$.
43. Write the middle term of the expansion of $\left(3a - \dfrac{b}{2}\right)^{10}$.
44. Write the last three terms of the expansion of $(a^2 - 2b^3)^7$.
45. Evaluate $(2.1)^4$ by expanding $(2 + 0.1)^4$.   46. Evaluate $(1.9)^4$ by expanding $(2 - 0.1)^4$.
47. Evaluate $(3.98)^3$ by expanding an appropriate binomial.   48. Evaluate $(1.2)^5$ by expanding an appropriate binomial.

CHALLENGE
Think Creatively

1. For x such that $|x| < 1$, find the infinite binomial expansion of $(1 - x)^{-1}$ and $(1 + x)^{-1}$. (*Hint:* Consider infinite geometric series.)
2. Prove that the sum of all the numbers in any two successive rows of Pascal's triangle is divisible by 6.

1. True or false: The permutation formula $_nP_r = n(n-1)(n-2)\cdots(n-r+1)$ is based directly on the fundamental principle of counting. Justify your answer.

2. From Exercise 27, page 614, the number of subsets of a set of n elements is 2^n. Arrive at this result by making use of the fundamental principle of counting. (*Hint:* Use a tree diagram applied to a set of three elements and generalize for n elements.)

3. In order to win a certain state lottery, a player must have chosen the same six integers that are selected at random, from the integers 1 through 40, by the lottery agency. Is the number of groups of six numbers a matter of permutations or combinations? How many such groups are there?

4. What can be done to the formula $_nP_r = \dfrac{n!}{(n-r)!}$ in order to produce the formula for $_{n+1}P_r$? On the basis of the preceding observation explain why $_{n+1}P_r \geq {_nP_r}$.

5. For any positive integer n evaluate the expression

$$\binom{n}{0} - \binom{n}{1} + \binom{n}{2} - \binom{n}{3} + \cdots + (-1)^n\binom{n}{n}$$

On the basis of the preceding result, what can you conclude about the number of subsets formed from a set of n elements that have an even number of elements and the number of subsets that have an odd number of elements?

12.4
PROBABILITY

Concepts of probability are encountered frequently in daily life, such as a weather forecaster's statement that there is "a 20% chance of rain." Actually, the formal study of probability started in the seventeenth century, when two famous mathematicians, Pascal and Fermat, considered the following problem that was posed to them by a gambler. Two people are involved in a game of chance and are forced to quit before either one has won. The number of points needed to win the game is known, and the number of points that each player has at the time is known. The problem was to determine how the stakes should be divided.

From this beginning mathematicians developed the theory of probability that has had far-reaching effects in many fields of endeavor. In this section we explore some basic aspects of probability, and the counting procedures studied earlier in this chapter will be useful in solving a variety of probability questions.

Let us begin by considering the situation of tossing two coins. What is the probability that both coins will be heads? We can approach this problem by forming a list of all possible outcomes. (A tree diagram is helpful in identifying all the possibilities.)

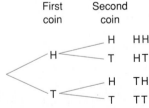

Notice that there are four possible outcomes {HH, HT, TH, TT} and that only one of these gives the required two heads. Thus we say that the probability that both coins will be heads is $\frac{1}{4}$. In symbols we may write $P(HH) = \frac{1}{4}$. This reflects what happens *in the long run*. If we continue to toss two coins repeatedly, we would ex-

$P(HH) = \frac{1}{4}$ is read as "the probability of two heads is $\frac{1}{4}$."

pect that on the average one out of four tosses will show two heads. Of course, there may be times when we toss double heads several times in a row; but if the experiment were to be repeated 1000 times, we can expect to have *about* 250 cases of double heads.

First coin	Second coin		Event	Probability
Heads	Heads		2 heads	$\frac{1}{4}$
Heads	Tails		1 head	$\frac{2}{4}$
Tails	Heads			
Tails	Tails		0 heads	$\frac{1}{4}$

Let us assume that an experiment can have n different outcomes, each *equally likely,* and that s of these outcomes produce the event E. Then the **probability** that the event E will occur, $P(E)$, is given as

Note: This definition assumes that all events are equally likely to occur; that is, each event has the same chance of happening as any other event.

$$P(E) = \frac{s}{n} = \frac{\text{number of successful outcomes}}{\text{total number of outcomes}}$$

EXAMPLE 1   A die is tossed. What is the probability of tossing a 5?

Solution   There are six possible outcomes $\{1, 2, 3, 4, 5, 6\}$. Each has the same chance of occurring as the others. There is only one way to succeed, namely by tossing a 5. Thus $P(5) = \frac{s}{n} = \frac{1}{6}$.   ■

EXAMPLE 2   To win the jackpot of a state lottery, the six numbers chosen by a person, from the numbers 1 through 54, must be the same as the six numbers selected at random by the state lottery system. On each lottery ticket purchased, there are two separate selections for the six numbers. What is the probability of winning the jackpot with one lottery ticket?

Solution   Since the order of the six numbers chosen does not matter, the number of ways of selecting six numbers out of 54 is $_{54}C_6$.

$$_{54}C_6 = \frac{54 \cdot 53 \cdot 52 \cdot 51 \cdot 50 \cdot 49 \cdot 48!}{6 \cdot 5 \cdot 4 \cdot 3 \cdot 2 \cdot 1 \cdot 48!}$$

$$= 25,827,165$$

The chances of winning the jackpot is about 1 in 13 million, an extremely small possibility.

Then, since there are two chances on one ticket,

$$P(\text{jackpot}) = \frac{2}{25,827,165} \approx \frac{1}{12,913,583}$$ ■

We can phrase the result of Example 3 in another way: $P(7 \text{ cannot occur}) = 1$.

EXAMPLE 3   A die is tossed. What is the probability of tossing a 7?

Solution   None of the outcomes is 7. Thus $P(7) = \frac{0}{6} = 0$.   ■

From Example 3 we see that the probability of an event that *cannot* occur is 0. Furthermore, the probability for an event that will *always* occur is 1. For example, the probability of tossing a number less than 7 on a single toss of a die is $\frac{6}{6} = 1$ since all numbers are less than 7. This leads to the following observation for the probability that an event E will occur:

$$0 \le P(E) \le 1$$

As an extension of this idea, we note that every event will either occur or fail to occur. That is, $P(E) + P(\text{not } E) = 1$. Therefore,

$$P(\text{not } E) = 1 - P(E)$$

EXAMPLE 4   Two cards are drawn simultaneously from a deck of playing cards. What is the probability that both cards are not spades?

The advantage of the formula $P(not\ E) = 1 - P(E)$ is demonstrated by this example. It is much more difficult to solve otherwise. Try to explain this solution:

$P(not\ 2\ spades)$

$= \dfrac{_{39}C_2 + {}_{39}C_1 \cdot {}_{13}C_1}{_{52}C_2}$

$= \dfrac{16}{17}.$

Solution   Two cards can be selected out of 52 in $_{52}C_2$ ways. Also, since there are 13 spades, $_{13}C_2$ is the number of ways of selecting 2 spades. Then

$$P(2 \text{ spades}) = \frac{_{13}C_2}{_{52}C_2} = \frac{1}{17}$$

Now use the formula $P(\text{not } E) = 1 - P(E)$.

$$P(\text{not } 2 \text{ spades}) = 1 - P(2 \text{ spades}) = 1 - \frac{1}{16} = \frac{16}{17} \qquad \blacksquare$$

We need to be careful when adding probabilities, since we may do this only when events are *mutually exclusive,* that is, when they cannot both happen at the same time. For example, consider the probability of drawing a king or a queen when a single card is drawn from a deck of cards.

$$P(\text{king}) = \frac{4}{52} \qquad P(\text{queen}) = \frac{4}{52}$$

$$P(\text{king or queen}) = \frac{4}{52} + \frac{4}{52} = \frac{8}{52} = \frac{2}{13}$$

This seems to agree with our intuition, since there are 8 cards in a deck of 52 cards that meet the necessary conditions. These conditions are mutually exclusive because a card cannot be a king and a queen at the same time. Note, however, the difference in the conditions of Example 5.

For mutually exclusive events E or F,

$$P(E \text{ or } F) = P(E) + P(F)$$

---

*In Example 5 the events are **not** mutually exclusive because it is possible for a card to be both a queen and a spade.*

EXAMPLE 5 A single card is drawn from a deck of cards. What is the probability that the card is either a queen or a spade?

Solution The probability of drawing a queen is $\frac{4}{52}$, and the probability of drawing a spade is $\frac{13}{52}$. However there is one card that is being counted twice in these probabilities, namely the queen of spades. Thus we account for this as follows:

$$P(\text{queen or spade}) = \frac{13}{52} + \frac{4}{52} - \frac{1}{52} = \frac{16}{52} = \frac{4}{13} \qquad \blacksquare$$

In general, for events E and F,

$$P(E \text{ or } F) = P(E) + P(F) - P(E \text{ and } F)$$

TEST YOUR UNDERSTANDING
Think Carefully

A pair of coins is tossed. Find the probability of each outcome.

1. Both are tails.　　　　　　**2.** One is heads and one is tails.
3. At least one is heads.　　　**4.** Both are heads or both are tails.

A single die is tossed. Find the probability of each outcome.

5. An even number comes up.　　　**6.** A number less than 5 comes up.
7. An even number or a number greater than 3 comes up.

A single card is drawn from a deck of cards. Find the probability of each outcome.

8. A red card is drawn.　　　　　**9.** A heart or a spade is drawn.
10. An ace or a heart is drawn.　　**11.** A picture card is drawn.
12. A picture card or an ace is drawn.

(Answers: Page 634)

We consider the picture cards to be the jacks, queens, and kings.

Sometimes a probability example can be solved using multiplication. For example, consider the probability of drawing two aces when two cards are selected from a deck of cards. In all such cases, unless stated otherwise, we shall assume that a card is drawn and *not* replaced in the deck before the second card is drawn.

Probability that the first card is an ace $= \frac{4}{52}$.

Assume that an ace was drawn. Then there are only 51 cards left in the deck, of which 3 are aces. Thus the probability that the second card is an ace $= \frac{3}{51}$.

We now make use of the following principle. Suppose that $P(E)$ represents the probability that an event E will occur, and $P(F \text{ given } E)$ is the probability that event F occurs after E has occurred. Then

$$P(E \text{ and } F) = P(E) \times P(F \text{ given } E)$$

Thus to find the probability that both cards are aces in the preceding illustration, we must multiply:

$$\frac{4}{52} \cdot \frac{3}{51} = \frac{1}{221} \approx 0.0045$$

This example can also be expressed through the use of combinations. Thus the total number of ways to select two cards from the deck is $_{52}C_2$. Furthermore, the number of ways of selecting two aces from the four aces in a deck of cards is $_4C_2$. Therefore,

Show that $\dfrac{_4C_2}{_{52}C_2} = \dfrac{1}{221}$.

$$\text{Probability of selecting two aces} = \frac{_4C_2}{_{52}C_2} = \frac{1}{221}$$

When each of two events can occur so that neither one affects the occurrence of the other, we say that the events are **independent**. For example, if E is the event that a head will come up when tossing a coin and F is the event that a 4 comes up when rolling a die, then neither outcome affects the other and therefore they are independent events. Also, since $P(E) = \frac{1}{2}$ and $P(F) = \frac{1}{6}$, the probability of both events occurring is found by multiplying. Thus

This can also be found by noting that there are 2 outcomes when a coin is tossed and 6 when a die is rolled. Then there are $2 \cdot 6 = 12$ ways for both events to occur together. Since only 1 of these 12 consists of heads on the coin and 4 on the die, $P(E \text{ and } F) = \frac{1}{12}$.

$$P(E \text{ and } F) = \frac{1}{2} \cdot \frac{1}{6} = \frac{1}{12}$$

In general,

> For independent events E and F,
>
> $$P(E \text{ and } F) = P(E) \cdot P(F)$$

EXAMPLE 6   A single card is drawn from a deck of cards. It is then replaced and a second card is drawn. What is the probability that both cards are aces?

Solution   Since the card is replaced after the first drawing, the two events are independent; neither outcome depends on the other. In each case the probability is $\frac{4}{52}$. The probability that both are aces is $\frac{4}{52} \cdot \frac{4}{52} = \frac{1}{169}$.   ■

*In Example 7 we assume that we are to draw **exactly** two aces. A more difficult problem is to find the probability of drawing **at least** two aces. Can you solve that problem?*

The final two illustrative examples demonstrate the use of combinations in conjunction with the fundamental counting principle to solve probability problems.

EXAMPLE 7   Five cards are dealt from a deck of 52 cards. What is the probability that exactly two of the cards are aces?

Solution   The number of ways of selecting two aces from the four aces in the deck is $_4C_2$. However, we also need to select three additional cards from the remaining 48 cards in the deck that are *not* aces; this can be done in $_{48}C_3$ ways. Thus, by the fundamental counting principle, the total number of ways of drawing two aces (and three other cards) is the product $(_4C_2) \cdot (_{48}C_3)$.

$$\text{Probability of drawing two aces in five cards} = \frac{(_4C_2) \cdot (_{48}C_3)}{_{52}C_5}$$

Show that this ratio reduces to $\dfrac{2162}{54{,}145} \approx 0.04$ ■

*If the members are selected **at random**, then each one has an equally likely chance of being selected. For example, a random selection could involve having all 18 names placed in a hat and 4 names drawn as in a lottery.*

EXAMPLE 8 A class consists of 10 men and 8 women. Four members are to be selected *at random* to represent the class. What is the probability that the selection will consist of two men and two women?

Solution Number of ways of selecting two men $= {}_{10}C_2$.
Number of ways of selecting two women $= {}_8C_2$.
Number of ways of selecting four members of the class $= {}_{18}C_4$.
Thus the probability that the committee will consist of two men and two women is given as

$$\frac{({}_{10}C_2) \cdot ({}_8C_2)}{{}_{18}C_4} = \frac{7}{17}$$

$\blacksquare$

EXERCISES 12.4

Use the tree diagram, showing the results of tossing three coins, to find the probability of each outcome.

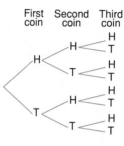

1. All three coins are heads.
2. Exactly one coin is heads.
3. At least one coin is heads.
4. None of the coins is heads.
5. At most one coin is heads.
6. At least two coins are heads.

Assume that after spinning the pointer, its chance of stopping in any one region is just as great as its chance of stopping in any of the other regions and that it does not stop on a line. Find the probabilities.

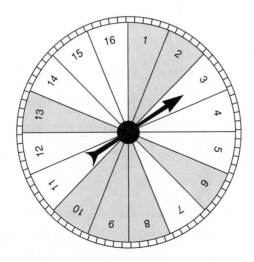

7. P(odd number)
8. P(multiple of 3)
9. P(white)
10. P(7 or 11)
11. P(even and red)
12. P(no more than 7)
13. P(prime)
14. P(red and between 3 and 10)

There are 36 different ways that a pair of dice can land, as shown in the diagram. Each outcome is shown as a pair of numbers showing the outcome on each die. The first number in each pair shows the outcome of tossing one die; the other number is the outcome of the second die. Use this diagram to find the probability of each outcome when a pair of dice are tossed.

$$(1, 1) \quad (1, 2) \quad (1, 3) \quad (1, 4) \quad (1, 5) \quad (1, 6)$$
$$(2, 1) \quad (2, 2) \quad (2, 3) \quad (2, 4) \quad (2, 5) \quad (2, 6)$$
$$(3, 1) \quad (3, 2) \quad (3, 3) \quad (3, 4) \quad (3, 5) \quad (3, 6)$$
$$(4, 1) \quad (4, 2) \quad (4, 3) \quad (4, 4) \quad (4, 5) \quad (4, 6)$$
$$(5, 1) \quad (5, 2) \quad (5, 3) \quad (5, 4) \quad (5, 5) \quad (5, 6)$$
$$(6, 1) \quad (6, 2) \quad (6, 3) \quad (6, 4) \quad (6, 5) \quad (6, 6)$$

15. Both dice show the same number.

16. The sum is 11.

17. The sum is 7.

18. The sum is 7 or 11.

19. The sum is 2, 3, or 12.

20. The sum is 6 or 8.

21. The sum is an odd number.

22. The sum is not 7.

Two cards are drawn from a deck of 52 playing cards without replacement. Find the probability of each outcome.

23. Both cards are red.

24. Both cards are spades.

25. Both cards are the ace of hearts.

26. Both cards are the same suit.

(*Hint for Ex. 26:* A successful outcome is to have both cards spades *or* both hearts *or* both diamonds *or* both clubs. The sum of these probabilities gives the solution.)

Two cards are drawn from a deck of 52 playing cards with the first card replaced before the second card is drawn. Find the probability of each outcome.

27. Both cards are black or both are red.

28. Both cards are hearts.

29. Both cards are the ace of hearts.

30. Both cards are picture cards.

31. Neither card is an ace.

32. Neither card is a spade.

33. The first card is an ace and the second card is a king.

34. The first card is an ace and the second card is not an ace.

A bag of marbles contains 8 red marbles and 5 green marbles. Three marbles are drawn at random at the same time. Find the probability of each outcome.

35. All are red.

36. All are green.

37. Two are red and one is green.

38. One is red and two are green.

39. A student takes a true-false test consisting of 10 questions by just guessing at each answer. Find the probability that the student's score will be:
   (a) 100%   **(b)** 0%   **(c)** 80% or better

40. A die is tossed three times in succession. Find the probability that:
   (a) All three tosses show 5.   **(b)** Exactly one of the tosses shows 5.
   (c) At least one of the tosses shows 5.

In Exercises 41 and 42 use $P(not\ E) = 1 - P(E)$

41. Two cards are drawn simultaneously from a deck of playing cards. Find P (not two red cards).

42. Three cards are drawn simultaneously from a deck of playing cards. Find P (not three hearts).

43. Five cards are dealt from a deck of 52 cards. Find the probability of obtaining:

(a) Four aces

(b) Four of a kind (that is, four aces or four twos or four threes, etc.).

(c) A flush (that is, all five cards of the same suit).

(d) A royal straight flush (10, jack, queen, king, ace, all in the same suit).

44. You are given 10 red marbles and 10 green marbles to distribute into two boxes. You will then be blindfolded and asked to select one of the boxes and then draw one marble from that box. You win $10,000 if the marble you select is red. Finally, you are allowed to distribute the 20 marbles into the two boxes in any way that you wish before you begin to make the selection. Try to determine what is the best strategy for distributing the marbles; that is, how many of each color should you place in each box?

The odds in favor of an event occurring is the ratio of the probability that it will occur to the probability that it will not occur. For example, the odds in favor of tossing two heads in two tosses of a coin are $\dfrac{1/4}{3/4} = \dfrac{1}{3}$ or 1 to 3. The odds against this event are 3 to 1. Exercises 45–52 involve computation of odds.

45. A card is drawn from a deck of cards. What are the odds in favor of obtaining an ace? What are the odds against this event occurring?

46. What are the odds in favor of tossing three successive heads with a coin?

47. What are the odds against tossing a 7 or an 11 in a single throw of a pair of dice?

48. Two cards are drawn simultaneously from a deck of cards. What are the odds in favor of both cards being hearts?

49. What are the odds for getting 2 heads when tossing 3 coins?

50. Suppose you have four playing cards in your hand whose face values are 5, 6, 7, 8. What are the odds of selecting a fifth card from the remaining 48 that has a face value of either 4 or 9?

51. Show that if the odds for an event E are a to b, then $P(E) = \dfrac{a}{a + b}$.

52. Suppose you are given 1 to 3 odds that event E occurs and 3 to 5 odds that event F occurs. If E and F cannot occur together, what are the odds that E or F will occur?

**CHALLENGE
Think Creatively**

Three cards are placed in a hat. One is red on each side, one is green on each side, and the third is red on one side and green on the other side. A card is drawn at random and placed on the table. Assume that the color showing is red. What is the probability that the other side is also red? Note that the answer, suprisingly, is *not* $\frac{1}{2}$. (*Hint:* Consider making a list of all possibilities using R_1, R_2, and R_3 for the red sides and G_1, G_2, and G_3 for the green sides.)

KEY TERMS

*Review these **key terms** so that you are able to define or describe them. A clear understanding of these terms will be very helpful when reviewing the developments of this chapter.*

Tree diagram • Fundamental principle of counting • Permutation • Factorial notation • Permutation formulas • Distinguishable permutations • Combination • Combination formulas • Expanded form of a binomial • Binomial formula • Binomial coefficients • Probability of an event • Mutually exclusive events • Probability of mutually exclusive events • Independent events • Probability of independent events • Odds for an event

REVIEW EXERCISES

The solutions to the following exercises can be found within the text of Chapter 12. Try to answer each question before referring to the text.

Section 12.1

1. Suppose you are planning a trip that will consist of visiting three cities, A, B, and C. You have your choice as to the order in which you are to visit the cities. List all possible different trips.

2. A club consists of 15 boys and 20 girls. They wish to elect officers consisting of a girl as president and a boy as vice-president. They also wish to elect a treasurer and a secretary who may be of either sex. How many sets of officers are possible?

3. How many three-digit whole numbers can be formed if zero is not an acceptable digit in the hundreds place and repetitions are allowed?

4. Repeat Exercise 3 if repetitions are not allowed.

5. How many three-letter "words" from the 26 letters of the alphabet are possible? No duplication of letters is permitted.

6. How many different ways can the four letters of the word MATH be arranged using each letter only once in each arrangement?

7. Evaluate $_7P_4$ by using each of the formulas for $_nP_r$.

8. A club contains 10 members. They wish to elect officers consisting of a president, vice-president, and secretary-treasurer. How many sets of officers are possible?

9. A family of 5 consisting of the parents and 3 children are going to be arranged in a row by a photographer. If the parents are to be next to each other, how many arrangements are possible?

10. Find the number of distinguishable permutations using all of the letters in each word.
    (a) BOOK   **(b)** REFERRED   **(c)** BEGINNING

Section 12.2

11. Evaluate:
    (a) $\binom{10}{2}$ **(b)** $_{52}C_5$.

12. Using the digits 1 through 9, how many different four-digit whole numbers can be formed if repetition of digits is not allowed?

13. A student has a penny, a nickel, a dime, a quarter, and a half-dollar and wishes to leave a tip consisting of exactly three coins. How many different amounts as tips are possible?

14. A class consists of 10 boys and 15 girls. How many committees of five can be selected if each committee is to consist of two boys and three girls?

15. An ice cream parlor advertises that you may have your choice of five different toppings, and you may choose none, one, two, three, four, or all five toppings. How many choices are there in all?

16. A dinner party of 10 people arrives at a restaurant that has only two tables available. One table seats 6 and the other 4. If the seating arrangement at either table is not taken into account, how many ways can the 10 people divide themselves to be seated at these two tables?

Section 12.3

17. Write the expansion of $(x + 2)^7$.

18. Expand $(x - 2)^7$ and simplify.

19. Expand $(2x - y)^5$ and simplify.

20. Evaluate 2^7 by expanding $(1 + 1)^7$.

21. Write the $(r + 1)$st term in the expansion of $(a + b)^n$.

22. Find the sixth term in the expansion of $(a + b)^8$.

23. Find the fourth term in the expansion of $(x - 2y)^{10}$.

Section 12.4

24. A die is tossed.

    (a) What is the probability of tossing a 5?

    (b) What is the probability of tossing a 7?

25. To win the jackpot of a state lottery, the six numbers chosen by a person, from the numbers 1 through 54, must be the same as the six numbers selected at random by the state lottery system. On each lottery ticket purchased, there are two separate selections for the six numbers. What is the probability of winning the jackpot with one lottery ticket?

26. Two cards are drawn simultaneously from a deck of playing cards. What is the probability that both cards are not spades?

27. A single card is drawn from a deck of cards. What is the probability that the card is either a queen or a king?

28. A single card is drawn from a deck of cards. What is the probability that the card is either a queen or a spade?

29. What is the probability of drawing two aces when two cards are selected from a deck of cards?

30. Two cards are drawn simultaneously from a deck of playing cards. What is the probability that both cards are not spades?

31. A single card is drawn from a deck of cards. It is then replaced and a second card is drawn. What is the probability that both cards are aces?

32. Five cards are dealt from a deck of 52 cards. What is the probability that exactly two of the cards are aces?

33. A class consists of 10 men and 8 women. Four members are to be selected at random to represent the class. What is the probability that the selection will consist of two men and two women?

CHAPTER 12 TEST: STANDARD ANSWER

Use these questions to test your knowledge of the basic skills and concepts of Chapter 12. Then check your answers with those given at the back of the book.

1. Evaluate: (a) $_{10}P_3$ (b) $_{10}C_3$.

2. How many different ways can the letters of the word TODAY be arranged so that each arrangement uses each of the letters once?

3. How many three-digit whole numbers can be formed if zero is not an acceptable hundreds digit?

    (a) Repetitions are allowed.   (b) Repetitions are not allowed.

4. In how many different ways can a class of 15 students select a committee consisting of three students?

5. How many 3-digit even numbers can be formed if repetition of digits is allowed?

6. How many different four-letter "words" can be formed from the letters m, o, n, d, a, y if no letter may be used more than once?

7. How many distinguishable permutations can be made from all of the letters in the word MINIMUM?

8. A class consists of 12 boys and 10 girls. How many committees of four can be selected if each committee is to consist of two boys and two girls?.

9. Solve for n: $7\binom{n}{2} = 3\binom{n+1}{3}$

10. A student takes a five-question true-false test just by guessing at each answer.
    (a) How many different sets of answers are possible?
    (b) What is the probability of obtaining a score of 100% on the test?
    (c) What is the probability that the score will be 0%?

11. A single die is tossed. What is the probability that it will show an even number or a number less than 3?

12. A pair of dice is tossed. What is the probability that the sum will be:
    (a) 12   (c) Not 12   (c) Less than 5?

13. Two cards are drawn from a deck of 52 playing cards. What is the probability that they are both aces or both kings if:
    (a) The first card is replaced before the second card is drawn?
    (b) No replacements are made?

14. Five cards are dealt from a deck of 52 playing cards. What is the probability that exactly four of the cards are picture cards?

15. A box contains 20 red chips and 15 green chips. Five chips are selected from the box at random. What is the probability that exactly four of the chips will be red?

16. Four coins are tossed. Find the probability that:
    (a) All four coins will land showing heads.
    (b) None of the coins will show heads.

17. Two integers from 0 through 10 are selected simultaneously and at random. What is the probability that they will both be odd?

18. Box A contains 5 green and 3 red chips. Box B contains 4 green and 6 red chips. You are to select one chip from each box. What is the probability that both chips will be red?

19. In Exercise 15, suppose that you are blindfolded and told to select just one chip from one of the boxes. What is the probability that the chip you select will be red?

20. A box contains 8 red marbles, 13 blue marbles, and 6 green marbles. If two marbles are drawn simultaneously from the box, what is the probability that neither marble is blue?

21. Expand: $(x - 2y)^5$.

22. Write the first four terms of the expansion of $\left(\dfrac{1}{2a} - b\right)^{10}$.

23. Write the seventh term in the expansion of $(3a + b)^{11}$.

24. Write the middle term in the expansion of $(2x - y)^{16}$.

25. Evaluate $(3.1)^4$ by using the expansion of $(a + b)^n$ for appropriate values of n, a, and b.

CHAPTER 12 TEST: MULTIPLE CHOICE

1. How many four-digit whole numbers can be formed if zero is not an acceptable digit in the thousands place and repetition of digits is not allowed?
    (a) 3360   (b) 4032   (c) 4536   (d) 5040   (e) None of the preceding

2. Which of the following are correct?

    I. $_nP_r = \dfrac{n!}{(n-r)!}$ II. $_nP_n = n!$ III. $_nC_r = \dfrac{n!}{r!(n-r)!}$

    (a) Only I   (b) Only II   (c) Only III   (d) I, II, and III   (e) None of the preceding

3. How many different ways can the five letters of the word EIGHT be arranged using each letter only once in each arrangement?

   (a) $_8P_5$ (b) $_5C_5$ (c) $5!$ (d) 5^5 (e) None of the preceding

4. A group consists of 5 boys and 6 girls. How many committees of five can be selected if each committee is to consist of 2 boys and 3 girls?

   (a) 150   (b) 200   (c) 1800   (d) 2400   (e) None of the preceding

5. How many distinguishable permutations can be formed using all of the letters in the word PRESSURE?

   (a) 8   (b) 5040   (c) 6720   (d) 56   (e) None of the preceding

6. Which of the following are true?

   I. $_{10}C_3 = {_{10}C_7}$ **II.** $_{10}C_3 = \dfrac{_{10}P_3}{3!}$ **III.** $_{10}P_3 = {_{10}P_7}$

   (a) Only I   (b) Only II   (c) Only III   (d) I, II, and III   (e) None of the preceding

7. How many triangles can be formed using six points in a plane no three of which are on the same straight line?

   (a) 10   (b) 12   (c) 18   (d) 20   (e) None of the preceding

8. What is the coefficient of x^3 in the expansion of $(2 + x)^5$?

   (a) 40   (b) 20   (c) 10   (d) 80   (e) None of the preceding

9. What is the fifth term in the expansion of $(3a - b)^6$?

   (a) $-135a^2b^4$ (b) $135a^2b^4$ (c) $-540a^3b^3$ (d) $-18ab^5$ (e) None of the preceding

10. What is the middle term in the expansion of $(x - 2)^6$?

    (a) $-20x^3$ (b) $60x^4$ (c) $-160x^3$ (d) $240x^2$ (e) None of the preceding

11. Which of the following expressions represents the ninth term in the expansion of $(a + b)^n$?

    (a) $\dbinom{n}{8}a^{n-8}b^8$ (b) $\dbinom{n}{8}a^8b^{n-8}$ (c) $\dbinom{n}{9}a^{n-9}b^9$ (d) $\dbinom{n}{9}a^9b^{n-9}$ (e) None of the preceding

12. A coin is tossed three times. What is the probability that not all three tosses are the same?

    (a) $\frac{1}{8}$ (b) $\frac{3}{8}$ (c) $\frac{1}{4}$ (d) $\frac{3}{4}$ (e) None of the preceding

13. Five cards are dealt from a deck of 52 cards. Which of the following shows the probability that four aces will be dealt?

    (a) $\dfrac{_4C_4}{_{12}C_5}$ (b) $\dfrac{_{52}C_4}{_{52}C_5}$ (c) $\dfrac{(_4C_4)(_{48}C_1)}{_{52}C_5}$ (d) $\dfrac{(_4C_4)(_{52}C_1)}{_{52}C_5}$ (e) None of the preceding.

14. Two cards are drawn from a deck of 52 cards with the first card replaced before the second card is drawn. What is the probability that neither card is a spade?

    (a) $\frac{9}{16}$ (b) $\frac{3}{4}$ (c) $\frac{1}{16}$ (d) $\frac{19}{34}$ (e) None of the preceding

15. A pair of dice is tossed. What is the probability that the sum of the faces showing on top is 10?

    (a) $\frac{2}{9}$ (b) $\frac{1}{12}$ (c) $\frac{1}{9}$ (d) $\frac{1}{6}$ (e) None of the preceding

Page 604

1. 5040   2. 336   3. 720   4. 90   5. 220
6. 358,800   7. 456,976   8. 6; EAT, ETA, AET, ATE, TEA, TAE
9.   10. 18.

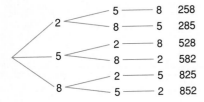

5 —— 8	258	
2 < 8 —— 5	285	
5 < 2 —— 8	528	
8 —— 2	582	
8 < 2 —— 5	825	
5 —— 2	852	

Page 612

1. 210   2. 1   3. 1   4. 220   5. 56

6. $_{10}C_3 = \dfrac{10!}{3!\,7!} = \dfrac{10!}{7!\,3!} = {}_{10}C_7$ 7. 495   8. 840   9. 28   10. 10

Page 625

1. $\frac{1}{4}$ 2. $\frac{1}{2}$ 3. $\frac{3}{4}$ 4. $\frac{1}{2}$ 5. $\frac{1}{2}$ 6. $\frac{2}{3}$ 7. $\frac{2}{3}$ 8. $\frac{1}{2}$ 9. $\frac{1}{2}$ 10. $\frac{4}{13}$ 11. $\frac{3}{13}$ 12. $\frac{4}{13}$

TABLES

N	$\sqrt{N}$	$\sqrt[3]{N}$	N	$\sqrt{N}$	$\sqrt[3]{N}$	N	$\sqrt{N}$	$\sqrt[3]{N}$	N	$\sqrt{N}$	$\sqrt[3]{N}$
1	1.000	1.000	51	7.141	3.708	101	10.050	4.657	151	12.288	5.325
2	1.414	1.260	52	7.211	3.733	102	10.100	4.672	152	12.329	5.337
3	1.732	1.442	53	7.280	3.756	103	10.149	4.688	153	12.369	5.348
4	2.000	1.587	54	7.348	3.780	104	10.198	4.703	154	12.410	5.360
5	2.236	1.710	55	7.416	3.803	105	10.247	4.718	155	12.450	5.372
6	2.449	1.817	56	7.483	3.826	106	10.296	4.733	156	12.490	5.383
7	2.646	1.913	57	7.550	3.849	107	10.344	4.747	157	12.530	5.395
8	2.828	2.000	58	7.616	3.871	108	10.392	4.762	158	12.570	5.406
9	3.000	2.080	59	7.681	3.893	109	10.440	4.777	159	12.610	5.418
10	3.162	2.154	60	7.746	3.915	110	10.488	4.791	160	12.649	5.429
11	3.317	2.224	61	7.810	3.936	111	10.536	4.806	161	12.689	5.440
12	3.464	2.289	62	7.874	3.958	112	10.583	4.820	162	12.728	5.451
13	3.606	2.351	63	7.937	3.979	113	10.630	4.835	163	12.767	5.463
14	3.742	2.410	64	8.000	4.000	114	10.677	4.849	164	12.806	5.474
15	3.873	2.466	65	8.062	4.021	115	10.724	4.863	165	12.845	5.485
16	4.000	2.520	66	8.124	4.041	116	10.770	4.877	166	12.884	5.496
17	4.123	2.571	67	8.185	4.062	117	10.817	4.891	167	12.923	5.507
18	4.243	2.621	68	8.246	4.082	118	10.863	4.905	168	12.961	5.518
19	4.359	2.668	69	8.307	4.102	119	10.909	4.919	169	13.000	5.529
20	4.472	2.714	70	8.367	4.121	120	10.954	4.932	170	13.038	5.540
21	4.583	2.759	71	8.426	4.141	121	11.000	4.946	171	13.077	5.550
22	4.690	2.802	72	8.485	4.160	122	11.045	4.960	172	13.115	5.561
23	4.796	2.844	73	8.544	4.179	123	11.091	4.973	173	13.153	5.572
24	4.899	2.884	74	8.602	4.198	124	11.136	4.987	174	13.191	5.583
25	5.000	2.924	75	8.660	4.217	125	11.180	5.000	175	13.229	5.593
26	5.099	2.962	76	8.718	4.236	126	11.225	5.013	176	13.267	5.604
27	5.196	3.000	77	8.775	4.254	127	11.269	5.027	177	13.304	5.615
28	5.292	3.037	78	8.832	4.273	128	11.314	5.040	178	13.342	5.625
29	5.385	3.072	79	8.888	4.291	129	11.358	5.053	179	13.379	5.636
30	5.477	3.107	80	8.944	4.309	130	11.402	5.066	180	13.416	5.646
31	5.568	3.141	81	9.000	4.327	131	11.446	5.079	181	13.454	5.657
32	5.657	3.175	82	9.055	4.344	132	11.489	5.092	182	13.491	5.667
33	5.745	3.208	83	9.110	4.362	133	11.533	5.104	183	13.528	5.677
34	5.831	3.240	84	9.165	4.380	134	11.576	5.117	184	13.565	5.688
35	5.916	3.271	85	9.220	4.397	135	11.619	5.130	185	13.601	5.698
36	6.000	3.302	86	9.274	4.414	136	11.662	5.143	186	13.638	5.708
37	6.083	3.332	87	9.327	4.431	137	11.705	5.155	187	13.675	5.718
38	6.164	3.362	88	9.381	4.448	138	11.747	5.168	188	13.711	5.729
39	6.245	3.391	89	9.434	4.465	139	11.790	5.180	189	13.748	5.739
40	6.325	3.420	90	9.487	4.481	140	11.832	5.192	190	13.784	5.749
41	6.403	3.448	91	9.539	4.498	141	11.874	5.205	191	13.820	5.759
42	6.481	3.476	92	9.592	4.514	142	11.916	5.217	192	13.856	5.769
43	6.557	3.503	93	9.644	4.531	143	11.958	5.229	193	13.892	5.779
44	6.633	3.530	94	9.695	4.547	144	12.000	5.241	194	13.928	5.789
45	6.708	3.557	95	9.747	4.563	145	12.042	5.254	195	13.964	5.799
46	6.782	3.583	96	9.798	4.579	146	12.083	5.266	196	14.000	5.809
47	6.856	3.609	97	9.849	4.595	147	12.124	5.278	197	14.036	5.819
48	6.928	3.634	98	9.899	4.610	148	12.166	5.290	198	14.071	5.828
49	7.000	3.659	99	9.950	4.626	149	12.207	5.301	199	14.107	5.838
50	7.071	3.684	100	10.000	4.642	150	12.247	5.313	200	14.142	5.848

TABLE II: EXPONENTIAL FUNCTIONS

x	e^x	e^{-x}	x	e^x	e^{-x}
0.0	1.00	1.000	3.1	22.2	0.045
0.1	1.11	0.905	3.2	24.5	0.041
0.2	1.22	0.819	3.3	27.1	0.037
0.3	1.35	0.741	3.4	30.0	0.033
0.4	1.49	0.670	3.5	33.1	0.030
0.5	1.65	0.607	3.6	36.6	0.027
0.6	1.82	0.549	3.7	40.4	0.025
0.7	2.01	0.497	3.8	44.7	0.022
0.8	2.23	0.449	3.9	49.4	0.020
0.9	2.46	0.407	4.0	54.6	0.018
1.0	2.72	0.368	4.1	60.3	0.017
1.1	3.00	0.333	4.2	66.7	0.015
1.2	3.32	0.301	4.3	73.7	0.014
1.3	3.67	0.273	4.4	81.5	0.012
1.4	4.06	0.247	4.5	90.0	0.011
1.5	4.48	0.223	4.6	99.5	0.010
1.6	4.95	0.202	4.7	110	0.0091
1.7	5.47	0.183	4.8	122	0.0082
1.8	6.05	0.165	4.9	134	0.0074
1.9	6.69	0.150	5.0	148	0.0067
2.0	7.39	0.135	5.5	245	0.0041
2.1	8.17	0.122	6.0	403	0.0025
2.2	9.02	0.111	6.5	665	0.0015
2.3	9.97	0.100	7.0	1097	0.00091
2.4	11.0	0.091	7.5	1808	0.00055
2.5	12.2	0.082	8.0	2981	0.00034
2.6	13.5	0.074	8.5	4915	0.00020
2.7	14.9	0.067	9.0	8103	0.00012
2.8	16.4	0.061	9.5	13360	0.00075
2.9	18.2	0.055	10.0	22026	0.000045
3.0	20.1	0.050			

TABLE III: NATURAL LOGARITHMS (BASE *e*)

x	ln *x*	*x*	ln *x*	*x*	ln *x*
0.0		3.4	1.224	6.8	1.917
0.1	−2.303	3.5	1.253	6.9	1.932
0.2	−1.609	3.6	1.281	7.0	1.946
0.3	−1.204	3.7	1.308	7.1	1.960
0.4	−0.916	3.8	1.335	7.2	1.974
0.5	−0.693	3.9	1.361	7.3	1.988
0.6	−0.511	4.0	1.386	7.4	2.001
0.7	−0.357	4.1	1.411	7.5	2.015
0.8	−0.223	4.2	1.435	7.6	2.028
0.9	−0.105	4.3	1.459	7.7	2.041
1.0	0.000	4.4	1.482	7.8	2.054
1.1	0.095	4.5	1.504	7.9	2.067
1.2	0.182	4.6	1.526	8.0	2.079
1.3	0.262	4.7	1.548	8.1	2.092
1.4	0.336	4.8	1.569	8.2	2.104
1.5	0.405	4.9	1.589	8.3	2.116
1.6	0.470	5.0	1.609	8.4	2.128
1.7	0.531	5.1	1.629	8.5	2.140
1.8	0.588	5.2	1.649	8.6	2.152
1.9	0.642	5.3	1.668	8.7	2.163
2.0	0.693	5.4	1.686	8.8	2.175
2.1	0.742	5.5	1.705	8.9	2.186
2.2	0.788	5.6	1.723	9.0	2.197
2.3	0.833	5.7	1.740	9.1	2.208
2.4	0.875	5.8	1.758	9.2	2.219
2.5	0.916	5.9	1.775	9.3	2.230
2.6	0.956	6.0	1.792	9.4	2.241
2.7	0.993	6.1	1.808	9.5	2.251
2.8	1.030	6.2	1.825	9.6	2.262
2.9	1.065	6.3	1.841	9.7	2.272
3.0	1.099	6.4	1.856	9.8	2.282
3.1	1.131	6.5	1.872	9.9	2.293
3.2	1.163	6.6	1.887	10.0	2.303
3.3	1.194	6.7	1.902		

TABLE IV: FOUR-PLACE COMMON LOGARITHMS (BASE 10)

x	0	1	2	3	4	5	6	7	8	9
1.0	.0000	.0043	.0086	.0128	.0170	.0212	.0253	.0294	.0334	.0374
1.1	.0414	.0453	.0492	.0531	.0569	.0607	.0645	.0682	.0719	.0755
1.2	.0792	.0828	.0864	.0899	.0934	.0969	.1004	.1038	.1072	.1106
1.3	.1139	.1173	.1206	.1239	.1271	.1303	.1335	.1367	.1399	.1430
1.4	.1461	.1492	.1523	.1553	.1584	.1614	.1644	.1673	.1703	.1732
1.5	.1761	.1790	.1818	.1847	.1875	.1903	.1931	.1959	.1987	.2014
1.6	.2041	.2068	.2095	.2122	.2148	.2175	.2201	.2227	.2253	.2279
1.7	.2304	.2330	.2355	.2380	.2405	.2430	.2455	.2480	.2504	.2529
1.8	.2553	.2577	.2601	.2625	.2648	.2672	.2695	.2718	.2742	.2765
1.9	.2788	.2810	.2833	.2856	.2878	.2900	.2923	.2945	.2967	.2989
2.0	.3010	.3032	.3054	.3075	.3096	.3118	.3139	.3160	.3181	.3201
2.1	.3222	.3243	.3263	.3284	.3304	.3324	.3345	.3365	.3385	.3404
2.2	.3424	.3444	.3464	.3483	.3502	.3522	.3541	.3560	.3579	.3598
2.3	.3617	.3636	.3655	.3674	.3692	.3711	.3729	.3747	.3766	.3784
2.4	.3802	.3820	.3838	.3856	.3874	.3892	.3909	.3927	.3945	.3962
2.5	.3979	.3997	.4014	.4031	.4048	.4065	.4082	.4099	.4116	.4133
2.6	.4150	.4166	.4183	.4200	.4216	.4232	.4249	.4265	.4281	.4298
2.7	.4314	.4330	.4346	.4362	.4378	.4393	.4409	.4425	.4440	.4456
2.8	.4472	.4487	.4502	.4518	.4533	.4548	.4564	.4579	.4594	.4609
2.9	.4624	.4639	.4654	.4669	.4683	.4698	.4713	.4728	.4742	.4757
3.0	.4771	.4786	.4800	.4814	.4829	.4843	.4857	.4871	.4886	.4900
3.1	.4914	.4928	.4942	.4955	.4969	.4983	.4997	.5011	.5024	.5038
3.2	.5051	.5065	.5079	.5092	.5105	.5119	.5132	.5145	.5159	.5172
3.3	.5185	.5198	.5211	.5224	.5237	.5250	.5263	.5276	.5289	.5302
3.4	.5315	.5328	.5340	.5353	.5366	.5378	.5391	.5403	.5416	.5428
3.5	.5441	.5453	.5465	.5478	.5490	.5502	.5514	.5527	.5539	.5551
3.6	.5563	.5575	.5587	.5599	.5611	.5623	.5635	.5647	.5658	.5670
3.7	.5682	.5694	.5705	.5717	.5729	.5740	.5752	.5763	.5775	.5786
3.8	.5798	.5809	.5821	.5832	.5843	.5855	.5866	.5877	.5888	.5899
3.9	.5911	.5922	.5933	.5944	.5955	.5966	.5977	.5988	.5999	.6010
4.0	.6021	.6031	.6042	.6053	.6064	.6075	.6085	.6096	.6107	.6117
4.1	.6128	.6138	.6149	.6160	.6170	.6180	.6191	.6201	.6212	.6222
4.2	.6232	.6243	.6253	.6263	.6274	.6284	.6294	.6304	.6314	.6325
4.3	.6335	.6345	.6355	.6365	.6375	.6385	.6395	.6405	.6415	.6425
4.4	.6435	.6444	.6454	.6464	.6474	.6484	.6493	.6503	.6513	.6522
4.5	.6532	.6542	.6551	.6561	.6571	.6580	.6590	.6599	.6609	.6618
4.6	.6628	.6637	.6646	.6656	.6665	.6675	.6684	.6693	.6702	.6712
4.7	.6721	.6730	.6739	.6749	.6758	.6767	.6776	.6785	.6794	.6803
4.8	.6812	.6821	.6830	.6839	.6848	.6857	.6866	.6875	.6884	.6893
4.9	.6902	.6911	.6920	.6928	.6937	.6946	.6955	.6964	.6972	.6981
5.0	.6990	.6998	.7007	.7016	.7024	.7033	.7042	.7050	.7059	.7067
5.1	.7076	.7084	.7093	.7101	.7110	.7118	.7126	.7135	.7143	.7152
5.2	.7160	.7168	.7177	.7185	.7193	.7202	.7210	.7218	.7226	.7235
5.3	.7243	.7251	.7259	.7267	.7275	.7284	.7292	.7300	.7308	.7316
5.4	.7324	.7332	.7340	.7348	.7356	.7364	.7372	.7380	.7388	.7396
N	0	1	2	3	4	5	6	7	8	9

x	0	1	2	3	4	5	6	7	8	9
5.5	.7404	.7412	.7419	.7427	.7435	.7443	.7451	.7459	.7466	.7474
5.6	.7482	.7490	.7497	.7505	.7513	.7520	.7528	.7536	.7543	.7551
5.7	.7559	.7566	.7574	.7582	.7589	.7597	.7604	.7612	.7619	.7627
5.8	.7634	.7642	.7649	.7657	.7664	.7672	.7679	.7686	.7694	.7701
5.9	.7709	.7716	.7723	.7731	.7738	.7745	.7752	.7760	.7767	.7774
6.0	.7782	.7789	.7796	.7803	.7810	.7818	.7825	.7832	.7839	.7846
6.1	.7853	.7860	.7868	.7875	.7882	.7889	.7896	.7903	.7910	.7917
6.2	.7924	.7931	.7938	.7945	.7952	.7959	.7966	.7973	.7980	.7987
6.3	.7993	.8000	.8007	.8014	.8021	.8028	.8035	.8041	.8048	.8055
6.4	.8062	.8069	.8075	.8082	.8089	.8096	.8102	.8109	.8116	.8122
6.5	.8129	.8136	.8142	.8149	.8156	.8162	.8169	.8176	.8182	.8189
6.6	.8195	.8202	.8209	.8215	.8222	.8228	.8235	.8241	.8248	.8254
6.7	.8261	.8267	.8274	.8280	.8287	.8293	.8299	.8306	.8312	.8319
6.8	.8325	.8331	.8338	.8344	.8351	.8357	.8363	.8370	.8376	.8382
6.9	.8388	.8395	.8401	.8407	.8414	.8420	.8426	.8432	.8439	.8445
7.0	.8451	.8457	.8463	.8470	.8476	.8482	.8488	.8494	.8500	.8506
7.1	.8513	.8519	.8525	.8531	.8537	.8543	.8549	.8555	.8561	.8567
7.2	.8573	.8579	.8585	.8591	.8597	.8603	.8609	.8615	.8621	.8627
7.3	.8633	.8639	.8645	.8651	.8657	.8663	.8669	.8675	.8681	.8686
7.4	.8692	.8698	.8704	.8710	.8716	.8722	.8727	.8733	.8739	.8745
7.5	.8751	.8756	.8762	.8768	.8774	.8779	.8785	.8791	.8797	.8802
7.6	.8808	.8814	.8820	.8825	.8831	.8837	.8842	.8848	.8854	.8859
7.7	.8865	.8871	.8876	.8882	.8887	.8893	.8899	.8904	.8910	.8915
7.8	.8921	.8927	.8932	.8938	.8943	.8949	.8954	.8960	.8965	.8971
7.9	.8976	.8982	.8987	.8993	.8998	.9004	.9009	.9015	.9020	.9025
8.0	.9031	.9036	.9042	.9047	.9053	.9058	.9063	.9069	.9074	.9079
8.1	.9085	.9090	.9096	.9101	.9106	.9112	.9117	.9122	.9128	.9133
8.2	.9138	.9143	.9149	.9154	.9159	.9165	.9170	.9175	.9180	.9186
8.3	.9191	.9196	.9201	.9206	.9212	.9217	.9222	.9227	.9232	.9238
8.4	.9243	.9248	.9253	.9258	.9263	.9269	.9274	.9279	.9284	.9289
8.5	.9294	.9299	.9304	.9309	.9315	.9320	.9325	.9330	.9335	.9340
8.6	.9345	.9350	.9355	.9360	.9365	.9370	.9375	.9380	.9385	.9390
8.7	.9395	.9400	.9405	.9410	.9415	.9420	.9425	.9430	.9435	.9440
8.8	.9445	.9450	.9455	.9460	.9465	.9469	.9474	.9479	.9484	.9489
8.9	.9494	.9499	.9504	.9509	.9513	.9518	.9523	.9528	.9533	.9538
9.0	.9542	.9547	.9552	.9557	.9562	.9566	.9571	.9576	.9581	.9586
9.1	.9590	.9595	.9600	.9605	.9609	.9614	.9619	.9624	.9628	.9633
9.2	.9638	.9643	.9647	.9652	.9657	.9661	.9666	.9671	.9675	.9680
9.3	.9685	.9689	.9694	.9699	.9703	.9708	.9713	.9717	.9722	.9727
9.4	.9731	.9736	.9741	.9745	.9750	.9754	.9759	.9763	.9768	.9773
9.5	.9777	.9782	.9786	.9791	.9795	.9800	.9805	.9809	.9814	.9818
9.6	.9823	.9827	.9832	.9836	.9841	.9845	.9850	.9854	.9859	.9863
9.7	.9868	.9872	.9877	.9881	.9886	.9890	.9894	.9899	.9903	.9908
9.8	.9912	.9917	.9921	.9926	.9930	.9934	.9939	.9943	.9948	.9952
9.9	.9956	.9961	.9965	.9969	.9974	.9978	.9983	.9987	.9991	.9996
N	0	1	2	3	4	5	6	7	8	9

TABLE V: TRIGONOMETRIC FUNCTION VALUES—RADIANS

Radians	sin	cos	tan	cot	sec	csc
.00	.0000	1.0000	.0000		1.000	
.01	.0100	1.0000	.0100	99.997	1.000	100.00
.02	.0200	.9998	.0200	49.993	1.000	50.00
.03	.0300	.9996	.0300	33.323	1.000	33.34
.04	.0400	.9992	.0400	24.987	1.001	25.01
.05	.0500	.9988	.0500	19.983	1.001	20.01
.06	.0600	.9982	.0601	16.647	1.002	16.68
.07	.0699	.9976	.0701	14.262	1.002	14.30
.08	.0799	.9968	.0802	12.473	1.003	12.51
.09	.0899	.9960	.0902	11.081	1.004	11.13
.10	.0998	.9950	.1003	9.967	1.005	10.02
.11	.1098	.9940	.1104	9.054	1.006	9.109
.12	.1197	.9928	.1206	8.293	1.007	8.353
.13	.1296	.9916	.1307	7.649	1.009	7.714
.14	.1395	.9902	.1409	7.096	1.010	7.166
.15	.1494	.9888	.1511	6.617	1.011	6.692
.16	.1593	.9872	.1614	6.197	1.013	6.277
.17	.1692	.9856	.1717	5.826	1.015	5.911
.18	.1790	.9838	.1820	5.495	1.016	5.586
.19	.1889	.9820	.1923	5.200	1.018	5.295
.20	.1987	.9801	.2027	4.933	1.020	5.033
.21	.2085	.9780	.2131	4.692	1.022	4.797
.22	.2182	.9759	.2236	4.472	1.025	4.582
.23	.2280	.9737	.2341	4.271	1.027	4.386
.24	.2377	.9713	.2447	4.086	1.030	4.207
.25	.2474	.9689	.2553	3.916	1.032	4.042
.26	.2571	.9664	.2660	3.759	1.035	3.890
.27	.2667	.9638	.2768	3.613	1.038	3.749
.28	.2764	.9611	.2876	3.478	1.041	3.619
.29	.2860	.9582	.2984	3.351	1.044	3.497
.30	.2955	.9553	.3093	3.233	1.047	3.384
.31	.3051	.9523	.3203	3.122	1.050	3.278
.32	.3146	.9492	.3314	3.018	1.053	3.179
.33	.3240	.9460	.3425	2.920	1.057	3.086
.34	.3335	.9428	.3537	2.827	1.061	2.999
.35	.3429	.9394	.3650	2.740	1.065	2.916
.36	.3523	.9359	.3764	2.657	1.068	2.839
.37	.3616	.9323	.3879	2.578	1.073	2.765
.38	.3709	.9287	.3994	2.504	1.077	2.696
.39	.3802	.9249	.4111	2.433	1.081	2.630
.40	.3894	.9211	.4228	2.365	1.086	2.568
.41	.3986	.9171	.4346	2.301	1.090	2.509
.42	.4078	.9131	.4466	2.239	1.095	2.452
.43	.4169	.9090	.4586	2.180	1.100	2.399
.44	.4259	.9048	.4708	2.124	1.105	2.348
	sin	cos	tan	cot	sec	csc

Radians	sin	cos	tan	cot	sec	csc
.45	.4350	.9004	.4831	2.070	1.111	2.299
.46	.4439	.8961	.4954	2.018	1.116	2.253
.47	.4529	.8916	.5080	1.969	1.122	2.208
.48	.4618	.8870	.5206	1.921	1.127	2.166
.49	.4706	.8823	.5334	1.875	1.133	2.125
.50	.4794	.8776	.5463	1.830	1.139	2.086
.51	.4882	.8727	.5594	1.788	1.146	2.048
.52	.4969	.8678	.5726	1.747	1.152	2.013
.53	.5055	.8628	.5859	1.707	1.159	1.978
.54	.5141	.8577	.5994	1.668	1.166	1.945
.55	.5227	.8525	.6131	1.631	1.173	1.913
.56	.5312	.8473	.6269	1.595	1.180	1.883
.57	.5396	.8419	.6310	1.560	1.188	1.853
.58	.5480	.8365	.6552	1.526	1.196	1.825
.59	.5564	.8309	.6696	1.494	1.203	1.797
.60	.5646	.8253	.6841	1.462	1.212	1.771
.61	.5729	.8196	.6989	1.431	1.220	1.746
.62	.5810	.8139	.7139	1.401	1.229	1.721
.63	.5891	.8080	.7291	1.372	1.238	1.697
.64	.5972	.8021	.7445	1.343	1.247	1.674
.65	.6052	.7961	.7602	1.315	1.256	1.652
.66	.6131	.7900	.7761	1.288	1.266	1.631
.67	.6210	.7838	.7923	1.262	1.276	1.610
.68	.6288	.7776	.8087	1.237	1.286	1.590
.69	.6365	.7712	.8253	1.212	1.297	1.571
.70	.6442	.7648	.8423	1.187	1.307	1.552
.71	.6518	.7584	.8595	1.163	1.319	1.534
.72	.6594	.7518	.8771	1.140	1.330	1.517
.73	.6669	.7452	.8949	1.117	1.342	1.500
.74	.6743	.7385	.9131	1.095	1.354	1.483
.75	.6816	.7317	.9316	1.073	1.367	1.467
.76	.6889	.7248	.9505	1.052	1.380	1.452
.77	.6961	.7179	.9697	1.031	1.393	1.437
.78	.7033	.7109	.9893	1.011	1.407	1.422
.79	.7104	.7038	1.009	.9908	1.421	1.408
.80	.7174	.6967	1.030	.9712	1.435	1.394
.81	.7243	.6895	1.050	.9520	1.450	1.381
.82	.7311	.6822	1.072	.9331	1.466	1.368
.83	.7379	.6749	1.093	.9146	1.482	1.355
.84	.7446	.6675	1.116	.8964	1.498	1.343
.85	.7513	.6600	1.138	.8785	1.515	1.331
.86	.7578	.6524	1.162	.8609	1.533	1.320
.87	.7643	.6448	1.185	.8437	1.551	1.308
.88	.7707	.6372	1.210	.8267	1.569	1.297
.89	.7771	.6294	1.235	.8100	1.589	1.287
	sin	cos	tan	cot	sec	csc

$\frac{\pi}{6} \rightarrow$ (at row .50)

$\frac{\pi}{4} \rightarrow$ (at row .79)

Radians	sin	cos	tan	cot	sec	csc
.90	.7833	.6216	1.260	.7936	1.609	1.277
.91	.7895	.6137	1.286	.7774	1.629	1.267
.92	.7956	.6058	1.313	.7615	1.651	1.257
.93	.8016	.5978	1.341	.7458	1.673	1.247
.94	.8076	.5898	1.369	.7303	1.696	1.238
.95	.8134	.5817	1.398	.7151	1.719	1.229
.96	.8192	.5735	1.428	.7001	1.744	1.221
.97	.8249	.5653	1.459	.6853	1.769	1.212
.98	.8305	.5570	1.491	.6707	1.795	1.204
.99	.8360	.5487	1.524	.6563	1.823	1.196
1.00	.8415	.5403	1.557	.6421	1.851	1.188
1.01	.8468	.5319	1.592	.6281	1.880	1.181
1.02	.8521	.5234	1.628	.6142	1.911	1.174
1.03	.8573	.5148	1.665	.6005	1.942	1.166
1.04	.8624	.5062	1.704	.5870	1.975	1.160
1.05	.8674	.4976	1.743	.5736	2.010	1.153
1.06	.8724	.4889	1.784	.5604	2.046	1.146
1.07	.8772	.4801	1.827	.5473	2.083	1.140
1.08	.8820	.4713	1.871	.5344	2.122	1.134
1.09	.8866	.4625	1.917	.5216	2.162	1.128
1.10	.8912	.4536	1.965	.5090	2.205	1.122
1.11	.8957	.4447	2.014	.4964	2.249	1.116
1.12	.9001	.4357	2.066	.4840	2.295	1.111
1.13	.9044	.4267	2.120	.4718	2.344	1.106
1.14	.9086	.4176	2.176	.4596	2.395	1.101
1.15	.9128	.4085	2.234	.4475	2.448	1.096
1.16	.9168	.3993	2.296	.4356	2.504	1.091
1.17	.9208	.3902	2.360	.4237	2.563	1.086
1.18	.9246	.3809	2.427	.4120	2.625	1.082
1.19	.9284	.3717	2.498	.4003	2.691	1.077
1.20	.9320	.3624	2.572	.3888	2.760	1.073
1.21	.9356	.3530	2.650	.3773	2.833	1.069
1.22	.9391	.3436	2.733	.3659	2.910	1.065
1.23	.9425	.3342	2.820	.3546	2.992	1.061
1.24	.9458	.3248	2.912	.3434	3.079	1.057
1.25	.9490	.3153	3.010	.3323	3.171	1.054
1.26	.9521	.3058	3.113	.3212	3.270	1.050
1.27	.9551	.2963	3.224	.3102	3.375	1.047
1.28	.9580	.2867	3.341	.2993	3.488	1.044
1.29	.9608	.2771	3.467	.2884	3.609	1.041
1.30	.9636	.2675	3.602	.2776	3.738	1.038
1.31	.9662	.2579	3.747	.2669	3.878	1.035
1.32	.9687	.2482	3.903	.2562	4.029	1.032
1.33	.9711	.2385	4.072	.2456	4.193	1.030
1.34	.9735	.2288	4.256	.2350	4.372	1.027
	sin	cos	tan	cot	sec	csc

$\frac{\pi}{3} \rightarrow$ (points to row 1.05)

Radians	sin	cos	tan	cot	sec	csc
1.35	.9757	.2190	4.455	.2245	4.566	1.025
1.36	.9779	.2092	4.673	.2140	4.779	1.023
1.37	.9799	.1994	4.913	.2035	5.014	1.021
1.38	.9819	.1896	5.177	.1931	5.273	1.018
1.39	.9837	.1798	5.471	.1828	5.561	1.017
1.40	.9854	.1700	5.798	.1725	5.883	1.015
1.41	.9871	.1601	6.165	.1622	6.246	1.013
1.42	.9887	.1502	6.581	.1519	6.657	1.011
1.43	.9901	.1403	7.055	.1417	7.126	1.010
1.44	.9915	.1304	7.602	.1315	7.667	1.009
1.45	.9927	.1205	8.238	.1214	8.299	1.007
1.46	.9939	.1106	8.989	.1113	9.044	1.006
1.47	.9949	.1006	9.887	.1011	9.938	1.005
1.48	.9959	.0907	10.983	.0910	11.029	1.004
1.49	.9967	.0807	12.350	.0810	12.390	1.003
1.50	.9975	.0707	14.101	.0709	14.137	1.003
1.51	.9982	.0608	16.428	.0609	16.458	1.002
1.52	.9987	.0508	19.670	.0508	19.965	1.001
1.53	.9992	.0408	24.498	.0408	24.519	1.001
1.54	.9995	.0308	32.461	.0308	32.477	1.000
1.55	.9998	.0208	48.078	.0208	48.089	1.000
1.56	.9999	.0108	92.620	.0108	92.626	1.000
$\frac{\pi}{2} \to$ 1.57	1.0000	.0008	1255.8	.0008	1255.8	1.000
	sin	cos	tan	cot	sec	csc

Table V contains the trigonometric function values for the radians, to two decimal places, from .00 to $1.57 \approx \frac{\pi}{2}$.

For radian values of 1.58 and larger, or -1.58 and smaller, the trigonometric function values can be found by locating the terminal side of the angle and using the reference angles. This figure can be useful in this regard.

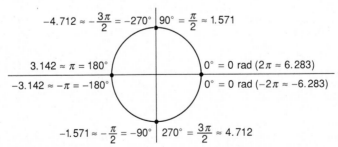

Illustrations:
The terminal side for $\Theta = 2.95$ is in quadrant II, since $1.571 < 2.95 < 3.142$. The reference angle $\phi = 3.14 - 2.95 = .19$. Thus, $\tan 2.95 = -\tan .19 = -.1923$.

For $\Theta = -9.05$ the terminal side is in quadrant III, since -9.05 is coterminal with $-9.05 + 6.28 = -2.77$. The reference angle $\phi = 3.14 - 2.77 = .37$. Thus, $\sin(-9.05) = -\sin(.37) = -.3616$.

Note: These results will differ slightly from the more accurate results obtained with a calculator.

TABLE VI: TRIGONOMETRIC FUNCTION VALUES—DEGREES

Degrees	sin	cos	tan	cot	sec	csc	
.0	.0000	1.0000	.0000	undef	1.0000	undef	**90.0**
.1	.0017	1.0000	.0017	572.96	1.0000	572.96	89.9
.2	.0035	1.0000	.0035	286.48	1.0000	286.48	89.8
.3	.0052	1.0000	.0052	190.98	1.0000	190.99	89.7
.4	.0070	1.0000	.0070	143.24	1.0000	143.24	89.6
.5	.0087	1.0000	.0087	114.59	1.0000	114.59	**89.5**
.6	.0105	.9999	.0105	95.489	1.0001	95.495	89.4
.7	.0122	.9999	.0122	81.847	1.0001	81.853	89.3
.8	.0140	.9999	.0140	71.615	1.0001	71.622	89.2
.9	.0157	.9999	.0157	63.657	1.0001	63.665	89.1
1.0	.0175	.9998	.0175	57.290	1.0002	57.299	**89.0**
1.1	.0192	.9998	.0192	52.081	1.0002	52.090	88.9
1.2	.0209	.9998	.0209	47.740	1.0002	47.750	88.8
1.3	.0227	.9997	.0227	44.066	1.0003	44.077	88.7
1.4	.0244	.9997	.0244	40.917	1.0003	40.930	88.6
1.5	.0262	.9997	.0262	38.188	1.0003	38.202	**88.5**
1.6	.0279	.9996	.0279	35.801	1.0004	35.815	88.4
1.7	.0297	.9996	.0297	33.694	1.0004	33.708	88.3
1.8	.0314	.9995	.0314	31.821	1.0005	31.836	88.2
1.9	.0332	.9995	.0332	30.145	1.0006	30.161	88.1
2.0	.0349	.9994	.0349	28.636	1.0006	28.654	**88.0**
2.1	.0366	.9993	.0367	27.271	1.0007	27.290	87.9
2.2	.0384	.9993	.0384	26.031	1.0007	26.050	87.8
2.3	.0401	.9992	.0402	24.898	1.0008	24.918	87.7
2.4	.0419	.9991	.0419	23.859	1.0009	23.880	87.6
2.5	.0436	.9990	.0437	22.904	1.0010	22.926	**87.5**
2.6	.0454	.9990	.0454	22.022	1.0010	22.044	87.4
2.7	.0471	.9989	.0472	21.205	1.0011	21.229	87.3
2.8	.0488	.9988	.0489	20.446	1.0012	20.471	87.2
2.9	.0506	.9987	.0507	19.740	1.0013	19.766	87.1
3.0	.0523	.9986	.0524	19.081	1.0014	19.107	**87.0**
3.1	.0541	.9985	.0542	18.464	1.0015	18.492	86.9
3.2	.0558	.9984	.0559	17.886	1.0016	17.914	86.8
3.3	.0576	.9983	.0577	17.343	1.0017	17.372	86.7
3.4	.0593	.9982	.0594	16.832	1.0018	16.862	86.6
3.5	.0610	.9981	.0612	16.350	1.0019	16.380	**86.5**
3.6	.0628	.9980	.0629	15.895	1.0020	15.926	86.4
3.7	.0645	.9979	.0647	15.464	1.0021	15.496	86.3
3.8	.0663	.9978	.0664	15.056	1.0022	15.089	86.2
3.9	.0680	.9977	.0682	14.669	1.0023	14.703	86.1
4.0	.0698	.9976	.0699	14.301	1.0024	14.336	**86.0**
4.1	.0715	.9974	.0717	13.951	1.0026	13.987	85.9
4.2	.0732	.9973	.0734	13.617	1.0027	13.654	85.8
4.3	.0750	.9972	.0752	13.300	1.0028	13.337	85.7
4.4	.0767	.9971	.0769	12.996	1.0030	13.305	85.6
4.5	.0785	.9969	.0787	12.706	1.0031	12.746	**85.5**
4.6	.0802	.9968	.0805	12.429	1.0032	12.469	85.4
4.7	.0819	.9966	.0822	12.163	1.0034	12.204	85.3
4.8	.0837	.9965	.0840	11.909	1.0035	11.951	85.2
4.9	.0854	.9963	.0857	11.665	1.0037	11.707	85.1
5.0	.0872	.9962	.0875	11.430	1.0038	11.474	**85.0**
	cos	sin	cot	tan	csc	sec	Degrees

Degrees	sin	cos	tan	cot	sec	csc	
5.0	.0872	.9962	.0875	11.430	1.0038	11.474	**85.0**
5.1	.0889	.9960	.0892	11.205	1.0040	11.249	84.9
5.2	.0906	.9959	.0910	10.988	1.0041	11.034	84.8
5.3	.0924	.9957	.0928	10.780	1.0043	10.826	84.7
5.4	.0941	.9956	.0945	10.579	1.0045	10.626	84.6
5.5	.0958	.9954	.0963	10.385	1.0046	10.433	**84.5**
5.6	.0976	.9952	.0981	10.199	1.0048	10.248	84.4
5.7	.0993	.9951	.0998	10.019	1.0050	10.069	84.3
5.8	.1011	.9949	.1016	9.8448	1.0051	9.8955	84.2
5.9	.1028	.9947	.1033	9.6768	1.0053	9.7283	84.1
6.0	.1045	.9945	.1051	9.5144	1.0055	9.5668	**84.0**
6.1	.1063	.9943	.1069	9.3572	1.0057	9.4105	83.9
6.2	.1080	.9942	.1086	9.2052	1.0059	9.2593	83.8
6.3	.1097	.9940	.1104	9.0579	1.0061	9.1129	83.7
6.4	.1115	.9938	.1122	8.9152	1.0063	8.9711	83.6
6.5	.1132	.9936	.1139	8.7769	1.0065	8.8337	**83.5**
6.6	.1149	.9934	.1157	8.6427	1.0067	8.7004	83.4
6.7	.1167	.9932	.1175	8.5126	1.0069	8.5711	83.3
6.8	.1184	.9930	.1192	8.3863	1.0071	8.4457	83.2
6.9	.1201	.9928	.1210	8.2636	1.0073	8.3238	83.1
7.0	.1219	.9925	.1228	8.1443	1.0075	8.2055	**83.0**
7.1	.1236	.9923	.1246	8.0285	1.0077	8.0905	82.9
7.2	.1253	.9921	.1263	7.9158	1.0079	7.9787	82.8
7.3	.1271	.9919	.1281	7.8062	1.0082	7.8700	82.7
7.4	.1288	.9917	.1299	7.6996	1.0084	7.7642	82.6
7.5	.1305	.9914	.1317	7.5958	1.0086	7.6613	**82.5**
7.6	.1323	.9912	.1334	7.4947	1.0089	7.5611	82.4
7.7	.1340	.9910	.1352	7.3962	1.0091	7.4635	82.3
7.8	.1357	.9907	.1370	7.3002	1.0093	7.3684	82.2
7.9	.1374	.9905	.1388	7.2066	1.0096	7.2757	82.1
8.0	.1392	.9903	.1405	7.1154	1.0098	7.1853	**82.0**
8.1	.1409	.9900	.1423	7.0264	1.0101	7.0972	81.9
8.2	.1426	.9898	.1441	6.9395	1.0103	7.0112	81.8
8.3	.1444	.9895	.1459	6.8548	1.0106	6.9273	81.7
8.4	.1461	.9893	.1477	6.7720	1.0108	6.8454	81.6
8.5	.1478	.9890	.1495	6.6912	1.0111	6.7655	**81.5**
8.6	.1495	.9888	.1512	6.6122	1.0114	6.6874	81.4
8.7	.1513	.9885	.1530	6.5350	1.0116	6.6111	81.3
8.8	.1530	.9882	.1548	6.4596	1.0119	6.5366	81.2
8.9	.1547	.9880	.1566	6.3859	1.0122	6.4637	81.1
9.0	.1564	.9877	.1584	6.3138	1.0125	6.3925	**81.0**
9.1	.1582	.9874	.1602	6.2432	1.0127	6.3228	80.9
9.2	.1599	.9871	.1620	6.1742	1.0130	6.2546	80.8
9.3	.1616	.9869	.1638	6.1066	1.0133	6.1880	80.7
9.4	.1633	.9866	.1655	6.0405	1.0136	6.1227	80.6
9.5	.1650	.9863	.1673	5.9758	1.0139	6.0589	**80.5**
9.6	.1668	.9860	.1691	5.9124	1.0142	5.9963	80.4
9.7	.1685	.9857	.1709	5.8502	1.0145	5.9351	80.3
9.8	.1702	.9854	.1727	5.7894	1.0148	5.8751	80.2
9.9	.1719	.9851	.1745	5.7297	1.0151	5.8164	80.1
10.0	.1736	.9848	.1763	5.6713	1.0154	5.7588	**80.0**
	cos	sin	cot	tan	csc	sec	Degrees

Degrees	sin	cos	tan	cot	sec	csc	
10.0	.1736	.9848	.1763	5.6713	1.0154	5.7588	**80.0**
10.1	.1754	.9845	.1781	5.6140	1.0157	5.7023	79.9
10.2	.1771	.9842	.1799	5.5578	1.0161	5.6470	79.8
10.3	.1788	.9839	.1817	5.5026	1.0164	5.5928	79.7
10.4	.1805	.9836	.1835	5.4486	1.0167	5.5396	79.6
10.5	.1822	.9833	.1853	5.3955	1.0170	5.4874	**79.5**
10.6	.1840	.9829	.1871	5.3435	1.0174	5.4362	79.4
10.7	.1857	.9826	.1890	5.2923	1.0177	5.3860	79.3
10.8	.1874	.9823	.1908	5.2422	1.0180	5.3367	79.2
10.9	.1891	.9820	.1926	5.1929	1.0184	5.2883	79.1
11.0	.1908	.9816	.1944	5.1446	1.0187	5.2408	**79.0**
11.1	.1925	.9813	.1962	5.0970	1.0191	5.1942	78.9
11.2	.1942	.9810	.1980	5.0504	1.0194	5.1484	78.8
11.3	.1959	.9806	.1998	5.0045	1.0198	5.1034	78.7
11.4	.1977	.9803	.2016	4.9594	1.0201	5.0593	78.6
11.5	.1994	.9799	.2035	4.9152	1.0205	5.0159	**78.5**
11.6	.2011	.9796	.2053	4.8716	1.0209	4.9732	78.4
11.7	.2028	.9792	.2071	4.8288	1.0212	4.9313	78.3
11.8	.2045	.9789	.2089	4.7867	1.0216	4.8901	78.2
11.9	.2062	.9785	.2107	4.7453	1.0220	4.8496	78.1
12.0	.2079	.9781	.2126	4.7046	1.0223	4.8097	**78.0**
12.1	.2096	.9778	.2144	4.6646	1.0227	4.7706	77.9
12.2	.2113	.9774	.2162	4.6252	1.0231	4.7321	77.8
12.3	.2130	.9770	.2180	4.5864	1.0235	4.6942	77.7
12.4	.2147	.9767	.2199	4.5483	1.0239	4.6569	77.6
12.5	.2164	.9763	.2217	4.5107	1.0243	4.6202	**77.5**
12.6	.2181	.9759	.2235	4.4737	1.0247	4.5841	77.4
12.7	.2198	.9755	.2254	4.4374	1.0251	4.5486	77.3
12.8	.2215	.9751	.2272	4.4015	1.0255	4.5137	77.2
12.9	.2233	.9748	.2290	4.3662	1.0259	4.4793	77.1
13.0	.2250	.9744	.2309	4.3315	1.0263	4.4454	**77.0**
13.1	.2267	.9740	.2327	4.2972	1.0267	4.4121	76.9
13.2	.2284	.9736	.2345	4.2635	1.0271	4.3792	76.8
13.3	.2300	.9732	.2364	4.2303	1.0276	4.3469	76.7
13.4	.2317	.9728	.2382	4.1976	1.0280	4.3150	76.6
13.5	.2334	.9724	.2401	4.1653	1.0284	4.2837	**76.5**
13.6	.2351	.9720	.2419	4.1335	1.0288	4.2527	76.4
13.7	.2368	.9715	.2438	4.1022	1.0293	4.2223	76.3
13.8	.2385	.9711	.2456	4.0713	1.0297	4.1923	76.2
13.9	.2402	.9707	.2475	4.0408	1.0302	4.1627	76.1
14.0	.2419	.9703	.2493	4.0108	1.0306	4.1336	**76.0**
14.1	.2436	.9699	.2512	3.9812	1.0311	4.1048	75.9
14.2	.2453	.9694	.2530	3.9520	1.0315	4.0765	75.8
14.3	.2470	.9690	.2549	3.9232	1.0320	4.0486	75.7
14.4	.2487	.9686	.2568	3.8947	1.0324	4.0211	75.6
14.5	.2504	.9681	.2586	3.8667	1.0329	3.9939	**75.5**
14.6	.2521	.9677	.2605	3.8391	1.0334	3.9672	75.4
14.7	.2538	.9673	.2623	3.8118	1.0338	3.9408	75.3
14.8	.2554	.9668	.2642	3.7848	1.0343	3.9147	75.2
14.9	.2571	.9664	.2661	3.7583	1.0348	3.8890	75.1
15.0	.2588	.9659	.2679	3.7321	1.0353	3.8637	**75.0**
	cos	sin	cot	tan	csc	sec	Degrees

Degrees	sin	cos	tan	cot	sec	csc	
15.0	.2588	.9659	.2679	3.7321	1.0353	3.8637	**75.0**
15.1	.2605	.9655	.2698	3.7062	1.0358	3.8387	74.9
15.2	.2622	.9650	.2717	3.6806	1.0363	3.8140	74.8
15.3	.2639	.9646	.2736	3.6554	1.0367	3.7897	74.7
15.4	.2656	.9641	.2754	3.6305	1.0372	3.7657	74.6
15.5	.2672	.9636	.2773	3.6059	1.0377	3.7420	**74.5**
15.6	.2689	.9632	.2792	3.5816	1.0382	3.7186	74.4
15.7	.2706	.9627	.2811	3.5576	1.0388	3.6955	74.3
15.8	.2723	.9622	.2830	3.5339	1.0393	3.6727	74.2
15.9	.2740	.9617	.2849	3.5105	1.0398	3.6502	74.1
16.0	.2756	.9613	.2867	3.4874	1.0403	3.6280	**74.0**
16.1	.2773	.9608	.2886	3.4646	1.0408	3.6060	73.9
16.2	.2790	.9603	.2905	3.4420	1.0413	3.5843	73.8
16.3	.2807	.9598	.2924	3.4197	1.0419	3.5629	73.7
16.4	.2823	.9593	.2943	3.3977	1.0424	3.5418	73.6
16.5	.2840	.9588	.2962	3.3759	1.0429	3.5209	**73.5**
16.6	.2857	.9583	.2981	3.3544	1.0435	3.5003	73.4
16.7	.2874	.9578	.3000	3.3332	1.0440	3.4799	73.3
16.8	.2890	.9573	.3019	3.3122	1.0446	3.4598	73.2
16.9	.2907	.9568	.3038	3.2914	1.0451	3.4399	73.1
17.0	.2924	.9563	.3057	3.2709	1.0457	3.4203	**73.0**
17.1	.2940	.9558	.3076	3.2506	1.0463	3.4009	72.9
17.2	.2957	.9553	.3096	3.2305	1.0468	3.3817	72.8
17.3	.2974	.9548	.3115	3.2106	1.0474	3.3628	72.7
17.4	.2990	.9542	.3134	3.1910	1.0480	3.3440	72.6
17.5	.3007	.9537	.3153	3.1716	1.0485	3.3255	**72.5**
17.6	.3024	.9532	.3172	3.1524	1.0491	3.3072	72.4
17.7	.3040	.9527	.3191	3.1334	1.0497	3.2891	72.3
17.8	.3057	.9521	.3211	3.1146	1.0503	3.2712	72.2
17.9	.3074	.9516	.3230	3.0961	1.0509	3.2536	72.1
18.0	.3090	.9511	.3249	3.0777	1.0515	3.2361	**72.0**
18.1	.3107	.9505	.3269	3.0595	1.0521	3.2188	71.9
18.2	.3123	.9500	.3288	3.0415	1.0527	3.2017	71.8
18.3	.3140	.9494	.3307	3.0237	1.0533	3.1848	71.7
18.4	.3156	.9489	.3327	3.0061	1.0539	3.1681	71.6
18.5	.3173	.9483	.3346	2.9887	1.0545	3.1515	**71.5**
18.6	.3190	.9478	.3365	2.9714	1.0551	3.1352	71.4
18.7	.3206	.9472	.3385	2.9544	1.0557	3.1190	71.3
18.8	.3223	.9466	.3404	2.9375	1.0564	3.1030	71.2
18.9	.3239	.9461	.3424	2.9208	1.0570	3.0872	71.1
19.0	.3256	.9455	.3443	2.9042	1.0576	3.0716	**71.0**
19.1	.3272	.9449	.3463	2.8878	1.0583	3.0561	70.9
19.2	.3289	.9444	.3482	2.8716	1.0589	3.0407	70.8
19.3	.3305	.9438	.3502	2.8556	1.0595	3.0256	70.7
19.4	.3322	.9432	.3522	2.8397	1.0602	3.0106	70.6
19.5	.3338	.9426	.3541	2.8239	1.0608	2.9957	**70.5**
19.6	.3355	.9421	.3561	2.8083	1.0615	2.9811	70.4
19.7	.3371	.9415	.3581	2.7929	1.0622	2.9665	70.3
19.8	.3387	.9409	.3600	2.7776	1.0628	2.9521	70.2
19.9	.3404	.9403	.3620	2.7625	1.0636	2.9379	70.1
20.0	.3420	.9397	.3640	2.7475	1.0642	2.9238	**70.0**
	cos	sin	cot	tan	csc	sec	Degrees

TABLE VI: TRIGONOMETRIC FUNCTION VALUES—DEGREES
(continued)

Degrees	sin	cos	tan	cot	sec	csc	
20.0	.3420	.9397	.3640	2.7475	1.0642	2.9238	**70.0**
20.1	.3437	.9391	.3659	2.7326	1.0649	2.9099	69.9
20.2	.3453	.9385	.3679	2.7179	1.0655	2.8960	69.8
20.3	.3469	.9379	.3699	2.7034	1.0662	2.8824	69.7
20.4	.3486	.9373	.3719	2.6889	1.0669	2.8688	69.6
20.5	.3502	.9367	.3739	2.6746	1.0676	2.8555	**69.5**
20.6	.3518	.9361	.3759	2.6605	1.0683	2.8422	69.4
20.7	.3535	.9354	.3779	2.6464	1.0690	2.8291	69.3
20.8	.3551	.9348	.3799	2.6325	1.0697	2.8161	69.2
20.9	.3567	.9342	.3819	2.6187	1.0704	2.8032	69.1
21.0	.3584	.9336	.3839	2.6051	1.0711	2.7904	**69.0**
21.1	.3600	.9330	.3859	2.5916	1.0719	2.7778	68.9
21.2	.3616	.9323	.3879	2.5782	1.0726	2.7653	68.8
21.3	.3633	.9317	.3899	2.5649	1.0733	2.7529	68.7
21.4	.3649	.9311	.3919	2.5517	1.0740	2.7407	68.6
21.5	.3665	.9304	.3939	2.5386	1.0748	2.7285	**68.5**
21.6	.3681	.9298	.3959	2.5257	1.0755	2.7165	68.4
21.7	.3697	.9291	.3979	2.5129	1.0763	2.7046	68.3
21.8	.3714	.9285	.4000	2.5002	1.0770	2.6927	68.2
21.9	.3730	.9278	.4020	2.4876	1.0778	2.6811	68.1
22.0	.3746	.9272	.4040	2.4751	1.0785	2.6695	**68.0**
22.1	.3762	.9265	.4061	2.4627	1.0793	2.6580	67.9
22.2	.3778	.9259	.4081	2.4504	1.0801	2.6466	67.8
22.3	.3795	.9252	.4101	2.4383	1.0808	2.6354	67.7
22.4	.3811	.9245	.4122	2.4262	1.0816	2.6242	67.6
22.5	.3827	.9239	.4142	2.4142	1.0824	2.6131	**67.5**
22.6	.3843	.9232	.4163	2.4023	1.0832	2.6022	67.4
22.7	.3859	.9225	.4183	2.3906	1.0840	2.5913	67.3
22.8	.3875	.9219	.4204	2.3789	1.0848	2.5805	67.2
22.9	.3891	.9212	.4224	2.3673	1.0856	2.5699	67.1
23.0	.3907	.9205	.4245	2.3559	1.0864	2.5593	**67.0**
23.1	.3923	.9198	.4265	2.3445	1.0872	2.5488	66.9
23.2	.3939	.9191	.4286	2.3332	1.0880	2.5384	66.8
23.3	.3955	.9184	.4307	2.3220	1.0888	2.5282	66.7
23.4	.3971	.9178	.4327	2.3109	1.0896	2.5180	66.6
23.5	.3987	.9171	.4348	2.2998	1.0904	2.5078	**66.5**
23.6	.4003	.9164	.4369	2.2889	1.0913	2.4978	66.4
23.7	.4019	.9157	.4390	2.2781	1.0921	2.4879	66.3
23.8	.4035	.9150	.4411	2.2673	1.0929	2.4780	66.2
23.9	.4051	.9143	.4431	2.2566	1.0938	2.4683	66.1
24.0	.4067	.9135	.4452	2.2460	1.0946	2.4586	**66.0**
24.1	.4083	.9128	.4473	2.2355	1.0955	2.4490	65.9
24.2	.4099	.9121	.4494	2.2251	1.0963	2.4395	65.8
24.3	.4115	.9114	.4515	2.2148	1.0972	2.4300	65.7
24.4	.4131	.9107	.4536	2.2045	1.0981	2.4207	65.6
24.5	.4147	.9100	.4557	2.1943	1.0989	2.4114	**65.5**
24.6	.4163	.9092	.4578	2.1842	1.0998	2.4022	65.4
24.7	.4179	.9085	.4599	2.1742	1.1007	2.3931	65.3
24.8	.4195	.9078	.4621	2.1642	1.1016	2.3841	65.2
24.9	.4210	.9070	.4642	2.1543	1.1025	2.3751	65.1
25.0	.4226	.9063	.4663	2.1445	1.1034	2.3662	**65.0**
	cos	sin	cot	tan	csc	sec	Degrees

Degrees	sin	cos	tan	cot	sec	csc	
25.0	.4226	.9063	.4663	2.1445	1.1034	2.3662	**65.0**
25.1	.4242	.9056	.4684	2.1348	1.1043	2.3574	64.9
25.2	.4258	.9048	.4706	2.1251	1.1052	2.3486	64.8
25.3	.4274	.9041	.4727	2.1155	1.1061	2.3400	64.7
25.4	.4289	.9033	.4748	2.1060	1.1070	2.3314	64.6
25.5	.4305	.9026	.4770	2.0965	1.1079	2.3228	**64.5**
25.6	.4321	.9018	.4791	2.0872	1.1089	2.3144	64.4
25.7	.4337	.9011	.4813	2.0778	1.1098	2.3060	64.3
25.8	.4352	.9003	.4834	2.0686	1.1107	2.2976	64.2
25.9	.4368	.8996	.4856	2.0594	1.1117	2.2894	64.1
26.0	.4384	.8988	.4877	2.0503	1.1126	2.2812	**64.0**
26.1	.4399	.8980	.4899	2.0413	1.1136	2.2730	63.9
26.2	.4415	.8973	.4921	2.0323	1.1145	2.2650	63.8
26.3	.4431	.8965	.4942	2.0233	1.1155	2.2570	63.7
26.4	.4446	.8957	.4964	2.0145	1.1164	2.2490	63.6
26.5	.4462	.8949	.4986	2.0057	1.1174	2.2412	**63.5**
26.6	.4478	.8942	.5008	1.9970	1.1184	2.2333	63.4
26.7	.4493	.8934	.5029	1.9883	1.1194	2.2256	63.3
26.8	.4509	.8926	.5051	1.9797	1.1203	2.2179	63.2
26.9	.4524	.8918	.5073	1.9711	1.1213	2.2103	63.1
27.0	.4540	.8910	.5095	1.9626	1.1223	2.2027	**63.0**
27.1	.4555	.8902	.5117	1.9542	1.1233	2.1952	62.9
27.2	.4571	.8894	.5139	1.9458	1.1243	2.1877	62.8
27.3	.4586	.8886	.5161	1.9375	1.1253	2.1803	62.7
27.4	.4602	.8878	.5184	1.9292	1.1264	2.1730	62.6
27.5	.4617	.8870	.5206	1.9210	1.1274	2.1657	**62.5**
27.6	.4633	.8862	.5228	1.9128	1.1284	2.1584	62.4
27.7	.4648	.8854	.5250	1.9047	1.1294	2.1513	62.3
27.8	.4664	.8846	.5272	1.8967	1.1305	2.1441	62.2
27.9	.4679	.8838	.5295	1.8887	1.1315	2.1371	62.1
28.0	.4695	.8829	.5317	1.8807	1.1326	2.1301	**62.0**
28.1	.4710	.8821	.5340	1.8728	1.1336	2.1231	61.9
28.2	.4726	.8813	.5362	1.8650	1.1347	2.1162	61.8
28.3	.4741	.8805	.5384	1.8572	1.1357	2.1093	61.7
28.4	.4756	.8796	.5407	1.8495	1.1368	2.1025	61.6
28.5	.4772	.8788	.5430	1.8418	1.1379	2.0957	**61.5**
28.6	.4787	.8780	.5452	1.8341	1.1390	2.0890	61.4
28.7	.4802	.8771	.5475	1.8265	1.1401	2.0824	61.3
28.8	.4818	.8763	.5498	1.8190	1.1412	2.0757	61.2
28.9	.4833	.8755	.5520	1.8115	1.1423	2.0692	61.1
29.0	.4848	.8746	.5543	1.8040	1.1434	2.0627	**61.0**
29.1	.4863	.8738	.5566	1.7966	1.1445	2.0562	60.9
29.2	.4879	.8729	.5589	1.7893	1.1456	2.0498	60.8
29.3	.4894	.8721	.5612	1.7820	1.1467	2.0434	60.7
29.4	.4909	.8712	.5635	1.7747	1.1478	2.0371	60.6
29.5	.4924	.8704	.5658	1.7675	1.1490	2.0308	**60.5**
29.6	.4939	.8695	.5681	1.7603	1.1501	2.0245	60.4
29.7	.4955	.8686	.5704	1.7532	1.1512	2.0183	60.3
29.8	.4970	.8678	.5727	1.7461	1.1524	2.0122	60.2
29.9	.4985	.8669	.5750	1.7391	1.1535	2.0061	60.1
30.0	.5000	.8660	.5774	1.7321	1.1547	2.0000	**60.0**
	cos	sin	cot	tan	csc	sec	Degrees

Degrees	sin	cos	tan	cot	sec	csc	
30.0	.5000	.8660	.5774	1.7321	1.1547	2.0000	**60.0**
30.1	.5015	.8652	.5797	1.7251	1.1559	1.9940	59.9
30.2	.5030	.8643	.5820	1.7182	1.1570	1.9880	59.8
30.3	.5045	.8634	.5844	1.7113	1.1582	1.9821	59.7
30.4	.5060	.8625	.5867	1.7045	1.1594	1.9762	59.6
30.5	.5075	.8616	.5890	1.6977	1.1606	1.9703	**59.5**
30.6	.5090	.8607	.5914	1.6909	1.1618	1.9645	59.4
30.7	.5105	.8599	.5938	1.6842	1.1630	1.9587	59.3
30.8	.5120	.8590	.5961	1.6775	1.1642	1.9530	59.2
30.9	.5135	.8581	.5985	1.6709	1.1654	1.9473	59.1
31.0	.5150	.8572	.6009	1.6643	1.1666	1.9416	**59.0**
31.1	.5165	.8563	.6032	1.6577	1.1679	1.9360	58.9
31.2	.5180	.8554	.6056	1.6512	1.1691	1.9304	58.8
31.3	.5195	.8545	.6080	1.6447	1.1703	1.9249	58.7
31.4	.5210	.8536	.6104	1.6383	1.1716	1.9194	58.6
31.5	.5225	.8526	.6128	1.6319	1.1728	1.9139	**58.5**
31.6	.5240	.8517	.6152	1.6255	1.1741	1.9084	58.4
31.7	.5255	.8508	.6176	1.6191	1.1753	1.9031	58.3
31.8	.5270	.8499	.6200	1.6128	1.1766	1.8977	58.2
31.9	.5284	.8490	.6224	1.6066	1.1779	1.8924	58.1
32.0	.5299	.8480	.6249	1.6003	1.1792	1.8871	**58.0**
32.1	.5314	.8471	.6273	1.5941	1.1805	1.8818	57.9
32.2	.5329	.8462	.6297	1.5880	1.1818	1.8766	57.8
32.3	.5344	.8453	.6322	1.5818	1.1831	1.8714	57.7
32.4	.5358	.8443	.6346	1.5757	1.1844	1.8663	57.6
32.5	.5373	.8434	.6371	1.5697	1.1857	1.8612	**57.5**
32.6	.5388	.8425	.6395	1.5637	1.1870	1.8561	57.4
32.7	.5402	.8415	.6420	1.5577	1.1883	1.8510	57.3
32.8	.5417	.8406	.6445	1.5517	1.1897	1.8460	57.2
32.9	.5432	.8396	.6469	1.5458	1.1910	1.8410	57.1
33.0	.5446	.8387	.6494	1.5399	1.1924	1.8361	**57.0**
33.1	.5461	.8377	.6519	1.5340	1.1937	1.8312	56.9
33.2	.5476	.8368	.6544	1.5282	1.1951	1.8263	56.8
33.3	.5490	.8358	.6569	1.5224	1.1964	1.8214	56.7
33.4	.5505	.8348	.6594	1.5166	1.1978	1.8166	56.6
33.5	.5519	.8339	.6619	1.5108	1.1992	1.8118	**56.5**
33.6	.5534	.8329	.6644	1.5051	1.2006	1.8070	56.4
33.7	.5548	.8320	.6669	1.4994	1.2020	1.8023	56.3
33.8	.5563	.8310	.6694	1.4938	1.2034	1.7976	56.2
33.9	.5577	.8300	.6720	1.4882	1.2048	1.7929	56.1
34.0	.5592	.8290	.6745	1.4826	1.2062	1.7883	**56.0**
34.1	.5606	.8281	.6771	1.4770	1.2076	1.7837	55.9
34.2	.5621	.8271	.6796	1.4715	1.2091	1.7791	55.8
34.3	.5635	.8261	.6822	1.4659	1.2105	1.7745	55.7
34.4	.5650	.8251	.6847	1.4605	1.2120	1.7700	55.6
34.5	.5664	.8241	.6873	1.4550	1.2134	1.7655	**55.5**
34.6	.5678	.8231	.6899	1.4496	1.2149	1.7610	55.4
34.7	.5693	.8221	.6924	1.4442	1.2163	1.7566	55.3
34.8	.5707	.8211	.6950	1.4388	1.2178	1.7522	55.2
34.9	.5721	.8202	.6976	1.4335	1.2193	1.7478	55.1
35.0	.5736	.8192	.7002	1.4281	1.2208	1.7434	**55.0**
	cos	sin	cot	tan	csc	sec	Degrees

Degrees	sin	cos	tan	cot	sec	csc	
35.0	.5736	.8192	.7002	1.4281	1.2208	1.7434	**55.0**
35.1	.5750	.8181	.7028	1.4229	1.2223	1.7391	54.9
35.2	.5764	.8171	.7054	1.4176	1.2238	1.7348	54.8
35.3	.5779	.8161	.7080	1.4124	1.2253	1.7305	54.7
35.4	.5793	.8151	.7107	1.4071	1.2268	1.7263	54.6
35.5	.5807	.8141	.7133	1.4019	1.2283	1.7221	**54.5**
35.6	.5821	.8131	.7159	1.3968	1.2299	1.7179	54.4
35.7	.5835	.8121	.7186	1.3916	1.2314	1.7137	54.3
35.8	.5850	.8111	.7212	1.3865	1.2329	1.7095	54.2
35.9	.5864	.8100	.7239	1.3814	1.2345	1.7054	54.1
36.0	.5878	.8090	.7265	1.3764	1.2361	1.7013	**54.0**
36.1	.5892	.8080	.7292	1.3713	1.2376	1.6972	53.9
36.2	.5906	.8070	.7319	1.3663	1.2392	1.6932	53.8
36.3	.5920	.8059	.7346	1.3613	1.2408	1.6892	53.7
36.4	.5934	.8049	.7373	1.3564	1.2424	1.6852	53.6
36.5	.5948	.8039	.7400	1.3514	1.2440	1.6812	**53.5**
36.6	.5962	.8028	.7427	1.3465	1.2456	1.6772	53.4
36.7	.5976	.8018	.7454	1.3416	1.2472	1.6733	53.3
36.8	.5990	.8007	.7481	1.3367	1.2489	1.6694	53.2
36.9	.6004	.7997	.7508	1.3319	1.2505	1.6655	53.1
37.0	.6018	.7986	.7536	1.3270	1.2521	1.6616	**53.0**
37.1	.6032	.7976	.7563	1.3222	1.2538	1.6578	52.9
37.2	.6046	.7965	.7590	1.3175	1.2554	1.6540	52.8
37.3	.6060	.7955	.7618	1.3127	1.2571	1.6502	52.7
37.4	.6074	.7944	.7646	1.3079	1.2588	1.6464	52.6
37.5	.6088	.7934	.7673	1.3032	1.2605	1.6427	**52.5**
37.6	.6101	.7923	.7701	1.2985	1.2622	1.6390	52.4
37.7	.6115	.7912	.7729	1.2938	1.2639	1.6353	52.3
37.8	.6129	.7902	.7757	1.2892	1.2656	1.6316	52.2
37.9	.6143	.7891	.7785	1.2846	1.2673	1.6279	52.1
38.0	.6157	.7880	.7813	1.2799	1.2690	1.6243	**52.0**
38.1	.6170	.7869	.7841	1.2753	1.2708	1.6207	51.9
38.2	.6184	.7859	.7869	1.2708	1.2725	1.6171	51.8
38.3	.6198	.7848	.7898	1.2662	1.2742	1.6135	51.7
38.4	.6211	.7837	.7926	1.2617	1.2760	1.6099	51.6
38.5	.6225	.7825	.7954	1.2572	1.2778	1.6064	**51.5**
38.6	.6239	.7815	.7983	1.2527	1.2796	1.6029	51.4
38.7	.6252	.7804	.8012	1.2482	1.2813	1.5994	51.3
38.8	.6266	.7793	.8040	1.2437	1.2831	1.5959	51.2
38.9	.6280	.7782	.8069	1.2393	1.2849	1.5925	51.1
39.0	.6293	.7771	.8098	1.2349	1.2868	1.5890	**51.0**
39.1	.6307	.7760	.8127	1.2305	1.2886	1.5856	50.9
39.2	.6320	.7749	.8156	1.2261	1.2904	1.5822	50.8
39.3	.6334	.7738	.8185	1.2218	1.2923	1.5788	50.7
39.4	.6347	.7727	.8214	1.2174	1.2941	1.5755	50.6
39.5	.6361	.7716	.8243	1.2131	1.2960	1.5721	**50.5**
39.6	.6374	.7705	.8273	1.2088	1.2978	1.5688	50.4
39.7	.6388	.7694	.8302	1.2045	1.2997	1.5655	50.3
39.8	.6401	.7683	.8332	1.2002	1.3016	1.5622	50.2
39.9	.6414	.7672	.8361	1.1960	1.3035	1.5590	50.1
40.0	.6428	.7660	.8391	1.1918	1.3054	1.5557	**50.0**
	cos	sin	cot	tan	csc	sec	Degrees

Degrees	sin	cos	tan	cot	sec	csc	
40.0	.6428	.7660	.8391	1.1918	1.3054	1.5557	**50.0**
40.1	.6441	.7649	.8421	1.1875	1.3073	1.5525	49.9
40.2	.6455	.7638	.8451	1.1833	1.3093	1.5493	49.8
40.3	.6468	.7627	.8481	1.1792	1.3112	1.5461	49.7
40.4	.6481	.7615	.8511	1.1750	1.3131	1.5429	49.6
40.5	.6494	.7604	.8541	1.1708	1.3151	1.5398	**49.5**
40.6	.6508	.7593	.8571	1.1667	1.3171	1.5366	49.4
40.7	.6521	.7581	.8601	1.1626	1.3190	1.5335	49.3
40.8	.6534	.7570	.8632	1.1585	1.3210	1.5304	49.2
40.9	.6547	.7559	.8662	1.1544	1.3230	1.5273	49.1
41.0	.6561	.7547	.8693	1.1504	1.3250	1.5243	**49.0**
41.1	.6574	.7536	.8724	1.1463	1.3270	1.5212	48.9
41.2	.6587	.7524	.8754	1.1423	1.3291	1.5182	48.8
41.3	.6600	.7513	.8785	1.1383	1.3311	1.5151	48.7
41.4	.6613	.7501	.8816	1.1343	1.3331	1.5121	48.6
41.5	.6626	.7490	.8847	1.1303	1.3352	1.5092	**48.5**
41.6	.6639	.7478	.8878	1.1263	1.3373	1.5062	48.4
41.7	.6652	.7466	.8910	1.1224	1.3393	1.5032	48.3
41.8	.6665	.7455	.8941	1.1184	1.3414	1.5003	48.2
41.9	.6678	.7443	.8972	1.1145	1.3435	1.4974	48.1
42.0	.6691	.7431	.9004	1.1106	1.3456	1.4945	**48.0**
42.1	.6704	.7420	.9036	1.1067	1.3478	1.4916	47.9
42.2	.6717	.7408	.9067	1.1028	1.3499	1.4887	47.8
42.3	.6730	.7396	.9099	1.0990	1.3520	1.4859	47.7
42.4	.6743	.7385	.9131	1.0951	1.3542	1.4830	47.6
42.5	.6756	.7373	.9163	1.0913	1.3563	1.4802	**47.5**
42.6	.6769	.7361	.9195	1.0875	1.3585	1.4774	47.4
42.7	.6782	.7349	.9228	1.0837	1.3607	1.4746	47.3
42.8	.6794	.7337	.9260	1.0799	1.3629	1.4718	47.2
42.9	.6807	.7325	.9293	1.0761	1.3651	1.4690	47.1
43.0	.6820	.7314	.9325	1.0724	1.3673	1.4663	**47.0**
43.1	.6833	.7302	.9358	1.0686	1.3696	1.4635	46.9
43.2	.6845	.7290	.9391	1.0649	1.3718	1.4608	46.8
43.3	.6858	.7278	.9424	1.0612	1.3741	1.4581	46.7
43.4	.6871	.7266	.9457	1.0575	1.3763	1.4554	46.6
43.5	.6884	.7254	.9490	1.0538	1.3786	1.4527	**46.5**
43.6	.6896	.7242	.9523	1.0501	1.3809	1.4501	46.4
43.7	.6909	.7230	.9556	1.0464	1.3832	1.4474	46.3
43.8	.6921	.7218	.9590	1.0428	1.3855	1.4448	46.2
43.9	.6934	.7206	.9623	1.0392	1.3878	1.4422	46.1
44.0	.6947	.7193	.9657	1.0355	1.3902	1.4396	**46.0**
44.1	.6959	.7181	.9691	1.0319	1.3925	1.4370	45.9
44.2	.6972	.7169	.9725	1.0283	1.3949	1.4344	45.8
44.3	.6984	.7157	.9759	1.0247	1.3973	1.4318	45.7
44.4	.6997	.7145	.9793	1.0212	1.3996	1.4293	45.6
44.5	.7009	.7133	.9827	1.0176	1.4020	1.4267	**45.5**
44.6	.7022	.7120	.9861	1.0141	1.4044	1.4242	45.4
44.7	.7034	.7108	.9896	1.0105	1.4069	1.4217	45.3
44.8	.7046	.7096	.9930	1.0070	1.4093	1.4192	45.2
44.9	.7059	.7083	.9965	1.0035	1.4118	1.4167	45.1
45.0	.7071	.7071	1.0000	1.0000	1.4142	1.4142	**45.0**
	cos	sin	cot	tan	csc	sec	Degrees

Answers to Odd-Numbered Exercises and Chapter Test Questions

CHAPTER 1: FUNDAMENTALS OF ALGEBRA

1.1 Real Numbers and Their Properties (page 5)

1. $\{1, 2, 3, 4\}$ **3.** $\{3, 4, 5, 6\}$ **5.** $\{-2, -1\}$ **7.** $\{\ldots, -3, -2, -1, 0\}$ **9.** There are none.   **11.** True
13. False; $7 \div 3$ is not an integer.   **15.** False; $8 - 2 \neq 2 - 8$ **17.** True   **19.** True   **21.** (c), (d), (f)   **23.** (e), (f)
25. (a), (b), (c), (d), (f)   **27.** (b), (c), (d), (f)   **29.** (e), (f)   **31.** Closure property for addition
33. Inverse property for addition   **35.** Commutative property for multiplication   **37.** Commutative property for addition
39. Identity property for addition   **41.** Inverse property for addition   **43.** Multiplication property of zero
45. Distributive property   **47.** 3   **49.** 5   **51.** 7
53. If $ab = 0$, then at least one of a or b is zero. Since $5 \neq 0$, $x - 2 = 0$ which implies $x = 2$.
55. No; as a counterexample $2^3 \neq 3^2$.   **57.** (a) $\frac{5}{11}$; (b) $\frac{37}{99}$; (c) $\frac{26}{111}$
59. Let $\frac{0}{0} = x$, where x is some number. Then the definition of division gives $0 \cdot x = 0$. Since any number x will work, the answer to $\frac{0}{0}$ is not unique; therefore, $\frac{0}{0}$ is undefined.

1.2 Introduction to Equations and Problem Solving (page 12)

1. $x = 4$ **3.** $x = -4$ **5.** $x = -4$ **7.** $x = -8$ **9.** $x = \frac{9}{2}$ **11.** $x = -9$ **13.** $x = \frac{20}{3}$ **15.** $x = 15$
17. $x = -10$ **19.** $x = -\frac{22}{3}$ **21.** $x = 5$ **23.** $x = 3$ **25.** $x = 90$ **27.** $w = \dfrac{P - 2\ell}{2}$ **29.** $t = \dfrac{N - u}{10}$
31. $r = \dfrac{C}{2\pi}$ **33.** $v = \dfrac{w - 7}{4}$ **35.** $w = 7, l = 21$ **37.** 10 at 10¢, 13 at 20¢, 5 at 25¢   **39.** $3\frac{1}{2}$ hours

41. Let $x + (x + 2) + (x + 4) = 180$. Then $3x = 174$ and $x = 58$. Therefore, the integers must be even.
43. $w = 5, l = 14$ **45.** 77, 79, 81   **47.** 36 min.   **49.** \$7000 at 9%; \$9700 at 12%   **51.** \$8500
53. $x, x + 2, 3x + 6, 3x + 15, 6x + 30, x + 5, (x + 5) - x = 5$

1.3 Statements of Inequality and Their Graphs (page 20)

1. False; $1 < 2$ and 1 is not negative.   **3.** True   **5.** False; $\frac{1}{2} > 0$ but $\frac{1}{4} < \frac{1}{2}$.   **7.** True   **9.** $\{x \mid x > 12\}$
11. $\{x \mid x \geq 4\}$ **13.** $\{x \mid x < 10\}$ **15.** $\{x \mid x > -7\}$ **17.** $\{x \mid x > -10\}$ **19.** $\{x \mid x \leq 5\}$ **21.** $\{x \mid x \leq -\frac{1}{4}\}$
23. $\{x \mid x > 2\}$ **25.** $\{x \mid x < -3\}$ **27.** $\{x \mid x > 32\}$ **29.** $\{x \mid x > -65\}$ **31.** $\{x \mid x < 0\}$
33. $\{x \mid x < 1 \text{ or } x > 1\}$ **35.** $\{x \mid x \geq 1\}$ **37.** $\{x \mid x \geq -\frac{13}{2}\}$ **39.** $\{x \mid x \leq 3\}$ **41.**
43.   **45.**   **47.**   **49.** $[-5, 2]$ **51.** $[-6, 0)$
53. $(-10, 10)$ **55.** $(-\infty, 5)$ **57.** $[-2, \infty)$ **59.** $(-\infty, -1]$ **61.** $-1 \leq x < 3; [-1, 3)$ **63.** $-1 < x < 3; (-1, 3)$
65. $x < 1; (-\infty, 1)$ **67.** $\{x \mid x < 4\}$ **69.** $\{x \mid x \leq -2\}$
71. $\{x \mid x > -4\}$ **73.** (10, 25), (11, 28), (12, 31), (13, 34), (14, 37)
75. At least 83 but less than 93   **77.** Between 77°F and 86°F   **79.** **(a)** $8 \leq g \leq 68$ **(b)** $2 \leq h \leq 16$
81. Two earn between \$74 and \$88, and the third earns between \$62 and \$76.

1.4 Absolute Value (page 25)

1. False   **3.** False   **5.** True   **7.** True   **9.** False   **11.** **(a)** 6 **(b)** -5 **(c)** -6 **(d)** $\frac{3}{2}$ **13.** $x = \pm \frac{1}{2}$
15. $x = -2, x = 4$ **17.** $x = -2, x = 5$ **19.** $x = 1, x = 7$ **21.** $x = -\frac{20}{3}, x = 4$ **23.** $x > 0$
25. $x = -4$ or $x = 2$:   **27.** $x \leq -2$ or $x \geq 4$:
29. $-5 \leq x \leq 1$:   **31.** $x = -5$ or $x = 5$:
33. $x \leq -5$ or $x \geq 5$:   **35.** $2 \leq x \leq 8$:
37. $2.9 < x < 3.1$:   **39.** $-3 < x < 4$:
41. $2 < x < 6$:   **43.** $3 \leq x \leq 5$:
45. All $x \neq 3$:

47. **(a)** $|5 - 2| \geq \big||5| - |2|\big|; 3 \geq 3$ **(b)** $|3 + 4| \leq |3| + |4|; 7 \leq 7$
$|5 - (-2)| \geq \big||5| - |-2|\big|; 7 \geq 3$ $|3 + (-4)| \leq |3| + |-4|; 1 \leq 7$
$|-5 - 2| \geq \big||-5| - |2|\big|; 7 \geq 3$ $|-3 + 4| \leq |-3| + |4|; 1 \leq 7$
$|-5 - (-2)| \geq \big||-5| - |-2|\big|; 3 \geq 3$ $|-3 + (-4)| \leq |-3| + |-4|; 7 \leq 7$

1.5 Integral Exponents (page 32)

1. False; 3^6 **3.** False; 2^7 **5.** True   **7.** True   **9.** False; $2 \cdot 3^4$ **11.** False; 1   **13.** False; 2^3 **15.** False; 2^{-8}

17. 7   **19.** $\frac{5}{3}$ **21.** 1   **23.** 4   **25.** 24   **27.** 16   **29.** $\frac{15}{4}$ **31.** $\frac{1}{x^6}$ **33.** x^{12} **35.** $72a^5$ **37.** $16a^8$ **39.** x^6y^2

41. $\frac{x^8}{y^6}$ **43.** $\frac{y^7}{x^5}$ **45.** $5x$ **47.** $\frac{b^2}{9}$ **49.** $(a + b)^6$ **51.** $-3x^3y^3$ **53.** $\frac{1}{(a + 3b)^{22}}$ **55.** $\frac{1}{x^2} + \frac{1}{y^2}$ **57.** x^2y^3

59. $\frac{10}{(4 - 5x)^3}$ **61.** 9   **63.** 8   **65.** -3 **67.** 5 grams; $640(\frac{1}{2})^n$ grams   **69.** 32 feet; $243(\frac{2}{3})^n$ feet   **71.** \$1331

73. $3^4 \neq 4^3; (2^3)^4 = 2^{12} \neq 2^{81} = 2^{(3^4)}$ **75.** $\dfrac{1}{a^{-n}} = \dfrac{1}{\frac{1}{a^n}} = 1 \cdot \dfrac{a^n}{1} = a^n$

1.6 Radicals and Rational Exponents (page 41)

1. $\frac{1}{9}$ 3. $\frac{1}{16}$ 5. 25   7. -3 9. $\frac{1}{2}$ 11. 9   13. 50   15. -2 17. 13   19. $\sqrt[3]{35}/6$ 21. $\frac{148}{135}$ 23. $-\frac{11}{4}$
25. $4\sqrt{2}$ 27. $6\sqrt{2}$ 29. $17\sqrt{5}$ 31. $10\sqrt{2}$ 33. $6\sqrt[3]{2}$ 35. $14\sqrt{2}$ 37. $|x|\sqrt{2}$ 39. $5\sqrt{10}$ 41. $7\sqrt{5}$
43. $12|x|$ 45. $|x|\sqrt{y} + 12|x|\sqrt{2y}$ 47. $4x\sqrt{2}$ 49. $\sqrt{2}/6$ 51. $8\sqrt{3}/|x|$ 53. $4\sqrt[3]{4}$ 55. 1   57. $16a^4b^6$
59. a^2/b 61. $\dfrac{3x}{(3x^2 + 2)^{1/2}}$ 63. $\dfrac{x + 2}{(x^2 + 4x)^{1/2}}$ 65. (a) 25 cm (b) 18.9 cm   67. $\dfrac{\sqrt[3]{4x}}{x}$ 69. $\dfrac{1}{xy^n}$ 71. x

73. Let $x = \sqrt[n]{a}$, $y = \sqrt[n]{b}$. Then $x^n = a$ and $y^n = b$. Now we get $\dfrac{a}{b} = \dfrac{x^n}{y^n} = \left(\dfrac{x}{y}\right)^n$. Thus, by definition, $\sqrt[n]{\dfrac{a}{b}} = \dfrac{x}{y}$. But

   $x = \sqrt[n]{a}$ and $y = \sqrt[n]{b}$. Therefore, $\sqrt[n]{\dfrac{a}{b}} = \dfrac{\sqrt[n]{a}}{\sqrt[n]{b}}$.   75. $|xy| = \sqrt{(xy)^2} = \sqrt{x^2y^2} = \sqrt{x^2}\sqrt{y^2} = |x||y|$

1.7 Fundamental Operations with Polynomials (page 47)

1. $8x^2 - 2x + 7$ 3. $4x^3 - 9x^2 + 3$ 5. $2x^3 + x^2 + 11x - 28$ 7. $x^3 - 7x^2 - 10x + 8$
9. $4x^3 - x^2 - 5x - 22$ 11. $3x + 1$ 13. $6y + 6$ 15. $x^3 + 2x^2 + x$ 17. $7y + 8$ 19. $-10xy + 2x^2y^2$
21. $-20x^4 + 4x^3 + 2x^2$ 23. $x^2 + 2x + 1$ 25. $4x^2 + 26x - 14$ 27. $-6x^2 - 3x + 18$ 29. $-6x^2 + 3x + 18$
31. $\frac{4}{9}x^2 + 8x + 36$ 33. $-63 + 55x - 12x^2$ 35. $\frac{1}{25}x^2 - \frac{1}{10}x + \frac{1}{16}$ 37. $x - 100$ 39. $x - 2$
41. $x^4 - 2x^3 + 2x^2 - 31x - 36$ 43. $x^3 - 8$ 45. $x^5 - 32$ 47. $x^{4n} - x^{2n} - 2$ 49. $3x - 6x^2 + 3x^3$
51. $-6x^3 + 19x^2 - x - 6$ 53. $5x^4 - 12x^2 + 2x + 4$ 55. $6x^5 + 15x^4 - 12x^3 - 27x^2 + 10x + 15$
57. $x^3 - 3x^2 + 3x - 1$ 59. $a^4 + 4a^3b + 6a^2b^2 + 4ab^3 + b^4$ 61. $8x^3 + 36x^2 + 54x + 27$
63. $\frac{1}{27}x^3 + x^2 + 9x + 27$ 65. $6(\sqrt{5} + \sqrt{3})$ 67. $-2(\sqrt{2} + 3)$ 69. $\dfrac{x + 2\sqrt{xy} + y}{x - y}$ 71. $\dfrac{-4}{5 - 3\sqrt{5}}$
73. $\dfrac{x - y}{x - 2\sqrt{xy} + y}$
75. Multiply numerator and denominator of the first fraction by $\sqrt{4 + h} + 2$ and simplify.

1.8 Factoring Polynomials (page 55)

1. $(2x + 3)(2x - 3)$ 3. $(a + 11b)(a - 11b)$ 5. $(x + 4)(x^2 - 4x + 16)$ 7. $(5x - 4)(25x^2 + 20x + 16)$
9. $(2x + 7y)(4x^2 - 14xy + 49y^2)$ 11. $(\sqrt{3} + 2x)(\sqrt{3} - 2x)$ 13. $(\sqrt{x} + 6)(\sqrt{x} - 6)$
15. $(2\sqrt{2} + \sqrt{3x})(2\sqrt{2} - \sqrt{3x})$ 17. $(\sqrt[3]{7} + a)[(\sqrt[3]{7})^2 - a\sqrt[3]{7} + a^2]$ 19. $(3\sqrt[3]{x} + 1)(9\sqrt[3]{x^2} - 3\sqrt[3]{x} + 1)$
21. $(\sqrt[3]{3x} - \sqrt[3]{4})(\sqrt[3]{9x^2} + \sqrt[3]{12x} + \sqrt[3]{16})$ 23. $(x - 1)(x + y)$ 25. $(x - 1)(y - 1)$ 27. $(2 - y^2)(1 + x)$
29. $7(x + h)(x^2 - hx + h^2)$ 31. $(a^4 + b^4)(a^2 + b^2)(a + b)(a - b)$ 33. $(a - 2)(a^4 + 2a^3 + 4a^2 + 8a + 16)$
35. $(x - y)(a + b)(a^2 - ab + b^2)$ 37. $(x^2 + 4y^2)(x + 2y)(x - 2y)^2$ 39. $(5a - 1)(4a - 1)$ 41. $(3x + 1)^2$
43. $(7x + 1)(2x + 5)$ 45. $(8x - 1)(x - 1)$ 47. $2(2x - 3)(2x - 1)$ 49. $(4a - 3)(3a - 4)$
51. $(2x - 1)(2x + 3)$ 53. $(8a + 3b)(3a + 2b)$ 55. Not factorable   57. $2(3x - 5)(x + 2)$
59. $2(b + 2)(b + 4)$ 61. $ab(a - b)^2$ 63. $8(2x - 1)(x - 1)$ 65. $25(a + b)^2$ 67. $(a - 1)^2(a^2 + a + 1)^2$
69. $3xy(x^2 + 2y)(2x^2 - 5y)$ 71. $(x + 2)^2(8x + 7)$
73. $x(x + 1)^2(x^2 - x + 1)^2(11x^3 - 9x + 2)$ or, in complete factored form, $x(x + 1)^3(x^2 - x + 1)^2(11x^2 - 11x + 2)$.
75. (a) $(x + 2)(x^4 - 2x^3 + 4x^2 - 8x + 16)$ (b) $x(2x + y)(64x^6 - 32x^5y + 16x^4y^2 - 8x^3y^3 + 4x^2y^4 - 2xy^5 + y^6)$
77. $(x^2 - x + 1)(x^2 + x + 1)$ 79. $20(2x + 1)^9$
81. $\sqrt{1 + y^2} = \sqrt{1 + x^2(x^2 + 2)} = \sqrt{x^4 + 2x^2 + 1} = \sqrt{(x^2 + 1)^2} = x^2 + 1$

1.9 Fundamental Operations with Rational Expressions (page 61)

1. False; $\frac{1}{21}$ 3. False; $\dfrac{ax}{2} - \dfrac{5b}{6}$ 5. True   7. $\dfrac{2x}{3z}$ 9. $3x + 1$ 11. $3x^2 + 2x + 1$ 13. $a + 1 - ab$

15. $3 - 4a^2x^4$ 17. $-x$ 19. $\dfrac{n + 1}{n^2 + 1}$ 21. $\dfrac{3(x - 1)}{2(x + 1)}$ 23. $\dfrac{2x + 3}{2x - 3}$ 25. $\dfrac{a - 4b}{a^2 - 4ab + 16b^2}$ 27. $\dfrac{2y}{x}$ 29. $\dfrac{2}{a^2}$

31. $\dfrac{x}{y}$ 33. $\dfrac{-3a - 8b}{6}$ 35. $\dfrac{x^2 + 1}{3(x + 1)}$ 37. -1 39. $\dfrac{2 - xy}{x}$ 41. $\dfrac{5y^2 - y}{y^2 - 1}$ 43. $\dfrac{3x}{x + 1}$ 45. $\dfrac{8 - x}{x^2 - 4}$

47. $\dfrac{3(x + 9)}{(x - 3)(x + 3)^2}$ 49. $\dfrac{2}{x + 5}$ 51. $x - 3$ 53. $\dfrac{2(n^2 + 4)}{n + 2}$ 55. $\dfrac{1}{2(x + 2)}$ 57. $-\dfrac{1}{4(4 + h)}$

59. $-\dfrac{1}{3(x + 3)}$ 61. $\dfrac{4 - x}{16x^2}$ 63. $\dfrac{xy}{y + x}$ 65. $\dfrac{-4x}{(1 + x^2)^2}$ 67. $-\dfrac{4}{x^2}$ 69. $-\dfrac{1}{ab}$ 71. $\dfrac{y^2 - x^2}{x^3 y^3}$

73. **(a)** $\dfrac{\dfrac{AD}{B} + C}{D} = \dfrac{AD + BC}{BD} = \dfrac{AD}{BD} + \dfrac{BC}{BD} = \dfrac{A}{B} + \dfrac{C}{D}$

(b) $\left[\dfrac{\left(\dfrac{AB}{D} + C\right)D}{F} + E \right] \cdot F = \left[\dfrac{AB + CD}{F} + E \right] \cdot F = AB + CD + EF$

75. $\dfrac{2xy^2 - 2x^2 y\left(\dfrac{x^2}{y^2}\right)}{y^4} = \dfrac{2xy^2 - \dfrac{2x^4}{y}}{y^4} = \dfrac{2xy^3 - 2x^4}{y^5} = \dfrac{2x(x^3 + 8) - 2x^4}{y^5} = \dfrac{16x}{y^5}$

77. $\sqrt{1 + y^2} = \sqrt{1 + \dfrac{x^4}{64} - \dfrac{1}{2} + \dfrac{4}{x^4}} = \sqrt{\dfrac{x^4}{64} + \dfrac{1}{2} + \dfrac{4}{x^4}} = \sqrt{\left(\dfrac{x^2}{8} + \dfrac{2}{x^2}\right)^2} = \dfrac{x^2}{8} + \dfrac{2}{x^2}$

1.10 Introduction to Complex Numbers (page 67)

1. True 3. True 5. False 7. $5 + 2i$ 9. $-5 + 0i$ 11. $4i$ 13. $12i$ 15. $\frac{3}{4}i$ 17. $-\sqrt{5}\,i$ 19. -27

21. $-\sqrt{6}$ 23. $15i$ 25. $12i$ 27. $5i\sqrt{2}$ 29. $(3 - \sqrt{3})i$ 31. $10 + 7i$ 33. $5 - 3i$ 35. $10 + 2i$

37. $-10 + 6i$ 39. $13i$ 41. $23 + 14i$ 43. $5 - 3i$ 45. $\frac{13}{5} + \frac{1}{5}i$ 47. $\frac{4}{5} - \frac{3}{5}i$ 49. $-3i$ 51. $-2i$ 53. 4

55. $\frac{3}{13} - \frac{2}{13}i$ 57. $\sqrt{(-4)(-9)} = \sqrt{36} = 6;\ \sqrt{-4}\sqrt{-9} = (2i)(3i) = 6i^2 = -6$ 59. $\dfrac{ac + bd}{c^2 + d^2} + \dfrac{bc - ad}{c^2 + d^2}i$

61. $[(3 + i)(3 - i)](4 + 3i) = (9 - i^2)(4 + 3i) = 10(4 + 3i) = 40 + 30i$

$(3 + i)[(3 - i)(4 + 3i)] = (3 + i)(15 + 5i) = 40 + 30i$

63. $4 - 3i$ 65. $\frac{53}{13} - \frac{21}{13}i$ 67. $(x + i)(x - i)$ 69. $3(x + 5i)(x - 5i)$

CHAPTER 1 TEST: STANDARD ANSWER (page 72)

1. **(a)** False; **(b)** true; **(c)** false; **(d)** true; **(e)** false; **(f)** false; **(g)** true; **(h)** false 2. 3 3. $x = -3$

4. Width $= 7$ inches; length $= 19$ inches 5. 2:27 P.M. 6. **(a)** **(b)**

7. $x < -2$ 8. $-3 < x < -1$

9. $x < \frac{2}{9}$: 10. $x \le -1$ or $x \ge 2$:

11. **(a)** False; **(b)** true; **(c)** false; **(d)** false; **(e)** true; **(f)** false 12. **(a)** $\dfrac{4x^8}{y^7}$; **(b)** $\dfrac{4y^7}{3x^6}$

13. **(a)** $4\sqrt{3}$; **(b)** $3\sqrt{5}$; **(c)** $-3x^2\sqrt[3]{3}$ 14. **(a)** $10\sqrt{2}$; **(b)** $6\sqrt{3}$ 15. $-|x|\sqrt{3}$ 16. $5x^4 + 12x^3 - 2x - 3$

17. $(4 - 3b)(16 + 12b + 9b^2)$ 18. $(2x - 3)(3x + 1)$ 19. $(x - 3y)(2x + y^2)$ 20. $\dfrac{2(x + 3)}{(x + 2)(x + 5)}$ 21. x

22. $-\dfrac{x + 7}{49x^2}$ 23. $\dfrac{3x^2 - 18x + 25}{(x - 3)(x + 3)(2x - 5)}$ 24. $43 + 23i$ 25. $-\frac{13}{41} + \frac{47}{41}i$

CHAPTER 1 TEST: MULTIPLE CHOICE (page 73)

1. (c) 2. (c) 3. (d) 4. (b) 5. (a) 6. (d) 7. (b) 8. (c) 9. (b) 10. (a) 11. (c) 12. (b) 13. (d)

14. (e) 15. (d)

CHAPTER 2: LINEAR FUNCTIONS AND EQUATIONS

2.1 Introduction to the Function Concept (page 81)

1. Function: all reals   **3.** Function: all $x > 0$ **5.** Not a function   **7.** Function: all $x \neq -1$ **9.** Not a function
11. True   **13.** False: -3 **15.** True   **17.** False: $-x^2 + 16$ **19.** True   **21.** (a) -3; (b) -1; (c) 0
23. (a) 1; (b) 0; (c) $\frac{1}{4}$ **25.** (a) -2; (b) -1; (c) $-\frac{7}{8}$ **27.** (a) 2; (b) 0; (c) $\frac{5}{16}$ **29.** (a) $-\frac{1}{2}$; (b) -1; (c) -2
31. (a) -1; (b) does not exist; (c) $\sqrt[3]{2}$ **33.** (a) 81; (b) 5; (c) $\frac{25}{36}$; (d) $\frac{1}{36}$ **35.** $3h(2) = 24 \neq 48 = h(6)$
37. (a) -11 (b) -5; (c) not defined; (d) not defined; (e) -9 **39.** $x + 3$ **41.** $-\dfrac{1}{3x}$ **43.** 2 **45.** 1
47. $-4 - h$ **49.** $-\dfrac{4 + h}{4(2 + h)^2}$

2.2 Graphing Lines in the Rectangular Coordinate System (page 90)

1.

x	-3	-2	-1	0	1	2
y	-5	-4	-3	-2	-1	0

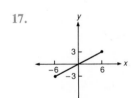

3.

x	-2	-1	0	1	2
y	-8	-6	-4	-2	0

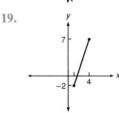

5.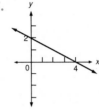

7.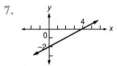

9.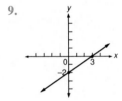

11.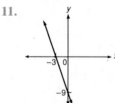

13.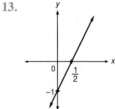

15.

17.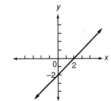

19.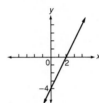

21. Each quotient gives the same slope, $-\frac{1}{2}$.

(a) $\dfrac{3 - 1}{-2 - 2}$; (b) $\dfrac{2 - 0}{0 - 4}$; (c) $\dfrac{1 - 0}{2 - 4}$; (d) $\dfrac{3 - (-1)}{-2 - 6}$; (e) $\dfrac{2 - (-1)}{0 - 6}$; (f) $\dfrac{1 - (-1)}{2 - 6}$

23. $\frac{1}{9}$ **25.** 0 **27.** $-\frac{17}{28}$

29.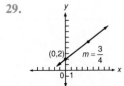

31.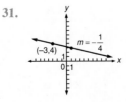

33.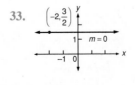

35.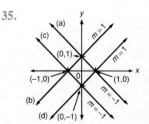

37. Both have the same slope of $-\frac{3}{2}$.

39. The slopes of PQ and RS are each $-\frac{5}{12}$; the slopes of PS and QR are each $\frac{12}{5}$. Thus the sides of the figure are perpendicular to each other and form right angles. Also, the diagonals are perpendicular because the slope of PR is $\frac{7}{17}$ and the slope of QS is the negative reciprocal, $-\frac{17}{7}$.

41. Horizontal lines have slope 0, which does not have a reciprocal. Also, vertical lines have no slope, and therefore a slope comparison cannot be made.

43. $\frac{49}{2}$

45. (a) Slope $\ell_1 = m_1 = \dfrac{DA}{CD} = \dfrac{DA}{1} = DA$; **(b)** slope $\ell_2 = m_2 = \dfrac{DB}{CD} = \dfrac{DB}{1} = DB$; $m_2 < 0$;

(c) $\dfrac{DA}{CD} = \dfrac{CD}{BD}$ or $\dfrac{m_1}{1} = \dfrac{1}{-m_2}$, since $BD = -DB = -m_2$.

47. \$135; 1625

2.3 Algebraic Forms of Linear Functions (page 99)

1. $y = 2x + 3$ **3.** $y = x + 1$ **5.** $y = 5$ **7.** $y = \frac{1}{2}x + 3$ **9.** $y = \frac{1}{4}x - 2$ **11.** $y - 3 = x - 2$

13. $y - 3 = 4(x + 2)$ **15.** $y - 5 = 0$ **17.** $y - 1 = \frac{1}{2}(x - 2)$ **19.** $y = 5x$ **21.** $y + \sqrt{2} = 10(x - \sqrt{2})$

23. $y = -3x + 4$; $m = -3$; $b = 4$ **25.** $y = 2x - \frac{1}{3}$; $m = 2$; $b = -\frac{1}{3}$ **27.** $y = \frac{5}{3}$; $m = 0$; $b = \frac{5}{3}$

29. $y = \frac{4}{3}x - \frac{7}{3}$; $m = \frac{4}{3}$; $b = -\frac{7}{3}$ **31.** $y = \frac{1}{2}x - 2$; $m = \frac{1}{2}$; $b = -2$ **33.** $x + y = 5$ **35.** $x - y = 3$

37. $2x + y = -15$ **39.** $y = 27$ **41.** $x = 5, y = -7$ **43.** $y = -\frac{2}{3}x - \frac{1}{3}$ **45.** $y - 7 = -\frac{1}{3}(x - 4)$

47. $y - 1 = -\frac{1}{2}(x + 5)$ **49.** $y = 3x + 2$; $y = -\frac{1}{2}x + 4$; $y = -\frac{5}{3}x + \frac{14}{3}$ **51.** They are negative reciprocals.

53.

55.

57.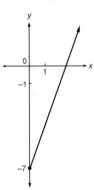

59.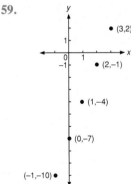

61. $x = -3$; not a function   **63.** $y = x + 2$; Domain: $x \geq -2$; Range: $y \geq 0$

65. $y = -1$ for $-1 \leq x < 0$; $y = 3$ for $0 \leq x < 3$; $y = 1$ for $3 \leq x \leq 4$; Domain: $-1 \leq x \leq 4$; Range: $\{-1, 1, 3\}$

67. (a) If $c = 0$ then $(0, 0)$ fits the equation and the line would then pass through the origin.

(b) $\dfrac{c}{b} = y$-intercept; $\dfrac{c}{a} = x$-intercept

(c) $ax + by = c$; $\dfrac{ax}{c} + \dfrac{by}{c} = 1$; $\dfrac{x}{\left(\dfrac{c}{a}\right)} = \dfrac{y}{\left(\dfrac{c}{b}\right)} = 1$; $\dfrac{x}{q} + \dfrac{y}{p} = 1$ **(d)** $\dfrac{x}{\frac{3}{2}} + \dfrac{y}{-5} = 1$ or $10x - 3y = 15$

(e) $m = \dfrac{0 - (-5)}{\frac{3}{2} - (0)} = \dfrac{5}{\frac{3}{2}} = \dfrac{10}{3}$; $y = \dfrac{10}{3}x - 5$

69. $y - 3 = -\frac{5}{7}(x + 2)$ or $y + 2 = -\frac{5}{7}(x - 5)$

1. 99   **3.** −100   **5.** −4   **7.** 1

9. Domain: all reals
Range: $y \geq 0$

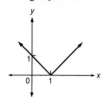

11. Domain: all reals
Range: $y \geq 0$

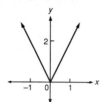

13. Domain: all reals
Range: $y \geq 0$

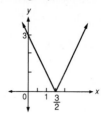

15. Domain: $x \geq -1$
Range: $y \leq 3$

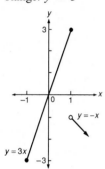

17. Domain: $-2 < x \leq 3$
Range: $-2 < y \leq 4$

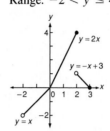

19.

21.

23.

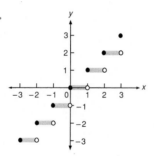

25.

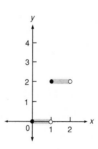

27.

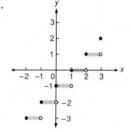

29. $P(x) = \frac{1}{10}(8 + [x])$ for $0 < x < 10$ **31.**

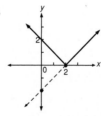

33. $y \leq 2x$ **35.** $y \leq 1$

37.

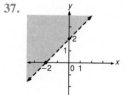

39.

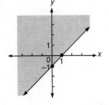

41.

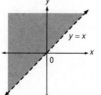

43.

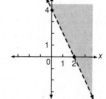

45.

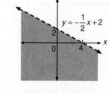

2.5 Systems of Linear Equations (page 112)

1. $(-2, -6)$ 3. $(4, -10)$ 5. $(-4, -1)$ 7. $(-26, -62)$ 9. $(-16, -38)$ 11. $(\frac{21}{88}, -\frac{9}{22})$ 13. $(5, 1)$
15. $(8, 5)$ 17. $(3, -11)$ 19. $(1, 0)$ 21. $(1, \frac{1}{2})$ 23. $(1, 0)$ 25. $(\frac{17}{18}, \frac{7}{18})$ 27. Inconsistent   29. Inconsistent
31. Consistent   33. Consistent   35. Dependent   37. Inconsistent   39. $(-1, 0, 2)$ 41. $(1, 2, -3)$
43. $(3, -2, -4)$ 45. $(\frac{1}{2}, -2, 5)$ 47. $(1, 0, 3)$ 49. Inconsistent; no solution   51. $(4, -3, 2)$ 53. $(-\frac{3}{2}c, \frac{9}{2}c)$
55. $(-20, -20)$ 57. $a = 4, b = -7$ 59. $\left(\dfrac{ce - bf}{ae - bd}, \dfrac{af - cd}{ae - bd}\right)$

2.6 Applications of Linear Systems (page 118)

1. 40 field goals; 16 free throws   3. $\ell = 21$ cm, $w = 9$ cm   5. 8 lb of potatoes and $1\frac{1}{2}$ lb of string beans
7. \$3200 room and board; \$5200 tuition   9. 14 quarters; 8 nickels   11. 344
13. $2\frac{3}{4}$ hr at 5 kph; $1\frac{3}{4}$ hr at 3 kph; total time $= 4\frac{1}{2}$ hr   15. Speed of plane $= 410$ mph; wind velocity $= 10$ mph
17. 550 at 25¢ and 360 at 45¢   19. \$2800 at 8%; \$3200 at $7\frac{1}{2}$%   21. 16 milliliters of each
23. Cost $= \$18$; profit $= \$13$ 25. 6 miles by car; 72 miles by train   27. An infinite number of answers
29. The clerk was wrong because if there were common unit prices, say $x = $ cost per orange and $y = $ cost per tangerine, then $6x + 12y = 234$, $2x + 4y = 77$ which is an inconsistent system. (*Note:* The smaller bag is a better buy since 2 (oranges) + 4 (tangerines) $= \frac{1}{3}$(6 oranges + 12 tangerines) $= \frac{1}{3}(2.34) = 0.78$, and 77¢ is cheaper.)
31. \$120   33. 90   35. $(90, 1800)$ 37. 115   39. 25 ones, 30 fives, 40 tens   41. $40°, 50°, 90°$
43. 15 lb peanuts, 10 lb pecans, 25 lb brazil nuts   45. 10 grams of A, 15 grams of B, 20 grams of C

2.7 Extracting Functions from Geometric Figures (page 124)

1. $h(x) = \frac{5}{2}x$ 3. $w(h) = \frac{20}{9}h$ 5. $s(x) = \frac{1}{3}x$ 7. $A(x) = \frac{5}{2}x(4 - x)$ 9. $V(x) = x(50 - 2x)^2$
11. $A(x) = \pi r^2 + 2r(200 - \pi r)$ 13. $V(x) = \frac{4}{3}x(30 - x^2)$ 15. $A(x) = \frac{7}{5}x(10 - 2x)$
17. $A(x) = 2x\sqrt{144 - x^2}$ 19. $A(h) = h(10 + \sqrt{25 - h^2})$

CHAPTER 2 TEST: STANDARD ANSWER (page 129)

1. All $x \neq -2$ 2. (a) $\dfrac{3}{2 + x}$; (b) $\dfrac{3}{2} + \dfrac{3}{x} = \dfrac{3x + 6}{2x}$ 3. $x + 8$ 4.   5. (a) $-\dfrac{7}{3}$ (b) 0

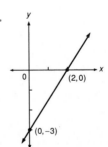

6. Domain: all x
   Range: all y

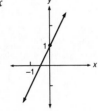

7. (a) $x = 3$; (b) $y = -2$

8. $y = \frac{1}{2}x - 3$ 9. $y = \frac{2}{3}x - \frac{5}{3}; \frac{2}{3}; -\frac{5}{3}$ 10. $y = -\frac{9}{5}x + \frac{2}{5}$ 11. $y - 8 = \frac{5}{2}(x - 2)$

12. 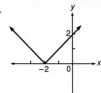 **13.** 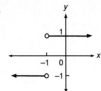 **14.** 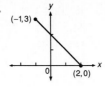 **15.** 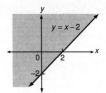 **16.**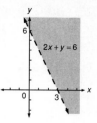

17. $V(x) = x(40 - 2x)^2$ **18.** $x = 3$, $y = -1$ **19.** $x = -1$, $y = 2$, $z = 5$ **20.** Dependent

21. Consistent; $(3, -5)$ **22.** Inconsistent   **23.** 200 kph   **24.** $(750, 2325)$ **25.** 6 oz of I; 8 oz of II; 3 oz of III

CHAPTER 2 TEST: MULTIPLE CHOICE (page 131)

1. (b)   **2.** (a)   **3.** (d)   **4.** (a)   **5.** (c)   **6.** (b)   **7.** (b)   **8.** (d)   **9.** (e)   **10.** (a)   **11.** (b)   **12.** (d)
13. (c)   **14.** (d)   **15.** (d)

CHAPTER 3: QUADRATIC FUNCTIONS AND THE CONIC SECTIONS, WITH APPLICATIONS

3.1 Graphing Quadratic Functions (page 141)

1. 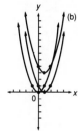   **3.** 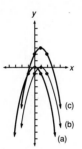   **5.** 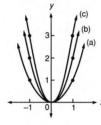   **7.** 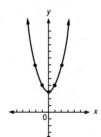   **9.**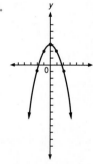

11. 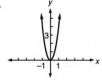   **13.** 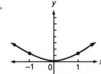   **15.** 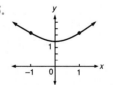   **17.**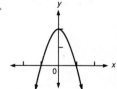

19. Decreasing on $(-\infty, 1]$; increasing on $[1, \infty)$; concave up

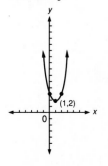

21. Increasing on $(-\infty, -1]$; decreasing on $[-1, \infty)$; concave down

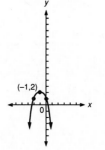

23. Decreasing on $(-\infty, 3]$; increasing on $[3, \infty)$; concave up

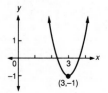

25. (a) $(3, 5)$; **(b)** $x = 3$; **(c)** set of real numbers; **(d)** $y \geq 5$

27. (a) $(3, 5)$; **(b)** $x = 3$; **(c)** set of real numbers; **(d)** $y \leq 5$

29. (a) $(-1, -3)$; **(b)** $x = -1$; **(c)** set of real numbers; **(d)** $y \geq -3$

31. (a) $(1, 2)$; **(b)** $x = 1$; **(c)** set of real numbers; **(d)** $y \leq 2$

33. (a) $(-2, -4)$; **(b)** $x = -2$; **(c)** set of real numbers; **(d)** $y \geq -4$

35.

37. Not a function because for $x > 0$ there are two corresponding y-values.

39. The graph of $x^2 - 4 > 0$ consists of all points on the number line where $x < -2$ or where $x > 2$. On the coordinate plane, the graph of $y = x^2 - 4$ is positive (above the x-axis) for $x < -2$ or $x > 2$.

41. $a = -2$ **43.** $k = 3$

45.

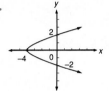

47.

49.

51.

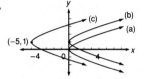

53. $y = (x - 4)^2 + 2$ **55.** $x = (y - 2)^2 - 5$ **57.** $y = \frac{1}{2}(x - 3)^2 - 2$ **59.**

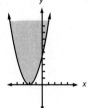

61.

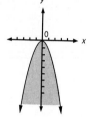

63.

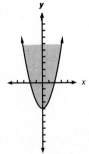

65.

67.

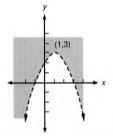

69. Yes   **71.** Yes   **73.** No   **75.** Yes

3.2 Completing the Square (page 148)

1. 9   **3.** 4   **5.** $\frac{9}{4}$ **7.** $(x + 1)^2 - 6$ **9.** $-(x + 3)^2 + 11$ **11.** $-(x - \frac{3}{2})^2 - \frac{7}{4}$ **13.** $(x + \frac{5}{2})^2 - \frac{33}{4}$

15. $2(x + 1)^2 - 5$ **17.** $3(x + 1)^2 - 8$ **19.** $(x - \frac{1}{4})^2 + \frac{15}{16}$ **21.** $\frac{3}{4}(x - \frac{2}{3})^2 - \frac{2}{3}$ **23.** $-5(x + \frac{1}{5})^2 + 1$

25. $(x - \frac{1}{2})^2 - 3$ **27.** $h = -\dfrac{b}{2a}$; $k = \dfrac{4ac - b^2}{4a}$ **29.** $y = (x - 2)^2 + 3$; $(2, 3)$; $x = 2$; $y = 7$

31. $y = (x - 3)^2 - 4$; $(3, -4)$; $x = 3$; $y = 5$ **33.** $y = 2(x - 1)^2 - 6$; $(1, -6)$; $x = 1$; $y = -4$

35.

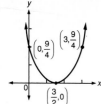

37.

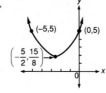

39.

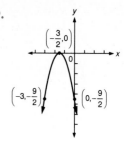

41.

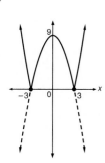

43.

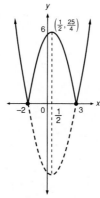

45.

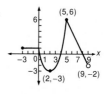

Decreasing on $(-\infty, -3]$ and on $[0, 3]$; increasing on $[-3, 0]$ and on $[3, \infty)$. Concave up on $(-\infty, -3)$ and on $(3, \infty)$; Concave down on $(-3, 3)$.

Decreasing on $(-\infty, -2]$ and on $[\frac{1}{2}, 3]$; increasing on $[-2, \frac{1}{2}]$ and on $[3, \infty)$. Concave up on $(-\infty, -2)$ and on $(3, \infty)$; concave down on $(-2, 3)$.

47. $a > 0$; $k = 0$.   **49.** $a > 0$; $k = 2$.

3.3 The Quadratic Formula (page 154)

1. $x = 2, x = 3$ **3.** $x = 0, x = 10$ **5.** $x = -\frac{1}{5}, x = \frac{3}{2}$ **7.** $x = 4$ **9.** $x = 0, x = 5$ **11.** $x = 2 \pm \sqrt{3}$

13. $x = -2$; $x = 5$ **15.** $x = 1 - \sqrt{5} = -1.24$; $x = 1 + \sqrt{5} = 3.24$ **17.** $x = -2$; $x = -\frac{1}{3}$

19. $x = 3 + \sqrt{3} = 4.73$; $x = 3 - \sqrt{3} = 1.27$ **21.** $x = 3 \pm i\sqrt{5}$

23. $x = \dfrac{3 + \sqrt{17}}{4} = 1.78$; $x = \dfrac{3 - \sqrt{17}}{4} = -0.28$ **25.** $x = \dfrac{-3 \pm i\sqrt{7}}{2}$ **27.** $x = \dfrac{4 \pm 2i\sqrt{6}}{5}$ **29.** 3; $-\frac{1}{2}$

31. None   **33.** $\frac{3}{2}$; 2 **35.** None   **37.** (a)   **39.** (b)   **41.** (d)   **43.** (c)   **45.** (c)   **47.** (d)

49. Once; **(a)** $(2, 0)$; **(b)** 4; **(c)** 2 **51.** Once; **(a)** $(\frac{1}{3}, 0)$; **(b)** 1; **(c)** $\frac{1}{3}$ **53.** None; **(a)** $(1, 1)$; **(b)** 3; **(c)** none

55. -6 or 6 **57.** $\pm 2\sqrt{7}$ **59.** $k > -4$ **61.** $k > -\frac{1}{4}$ **63.** $t > 9$ **65.** $t < -\frac{1}{4}$

67. Sum: $\dfrac{-b + \sqrt{b^2 - 4ac}}{2a} + \dfrac{-b - \sqrt{b^2 - 4ac}}{2a} = \dfrac{-2b}{2a} = -\dfrac{b}{a}$

Product: $\dfrac{-b + \sqrt{b^2 - 4ac}}{2a} \cdot \dfrac{-b - \sqrt{b^2 - 4ac}}{2a} = \dfrac{b^2 - (b^2 - 4ac)}{4a^2} = \dfrac{4ac}{4a^2} = \dfrac{c}{a}$

69. Sum $= -\frac{5}{6}$ **71.** Sum $= \frac{26}{3}$ **73.** Sum $= -3$ **75.** $-1 < x < 3$ **77.** $x \leq -5$ or $x \geq 2$

Product $= -\frac{2}{3}$ Product $= \frac{35}{3}$ Product $= -\frac{9}{2}$

79. No solution.   **81.**

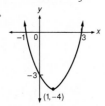

83.

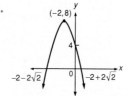

85.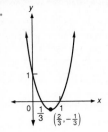

87.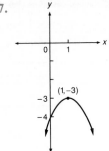

89. $x = \pm\sqrt{7}; x = \pm\dfrac{\sqrt{2}}{2}i$ **91.** $x = -3; x = -2; x = 2$ **93.** $x = -\frac{3}{2}, x = \frac{1}{4}$

3.4 Applications of Quadratic Functions (page 161)

1. 14, 15 **3.** 5, 8 **5.** 2 feet **7.** Width = 3 cm; length = 5 cm **9.** $-7 + \sqrt{62}$ meters **11.** $n = 15$
13. 20 mph; 1.5 hours downstream; 2.5 hours upstream **15.** Maximum = 7 at $x = 5$ **17.** Minimum = 36 at $x = 2$
19. Minimum = 0 at $x = \frac{7}{2}$ **21.** Maximum = $\frac{1}{9}$ at $x = -\frac{1}{3}$ **23.** 60; \$600 **25.** 6, 6
27. -11 and 11; product = -121 **29.** 10 feet by 20 feet **31.** 4 seconds **33.** 2 seconds and 6 seconds
35. \$9.75 **37.** (1, 2) **39.** $(\frac{3}{2}, \frac{9}{2})$; minimum value = $\frac{9}{2}$ **41.** $(-\frac{1}{3}, \frac{19}{18})$; maximum value = $\frac{19}{18}$
43. (0, 9); maximum value = 9 **45.** **(a)** 116 ft; **(b)** 197 ft; **(c)** 262 ft **47.** \$15

3.5 Conic Sections: The Circle (page 170)

1.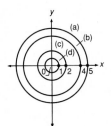

3. $(x - 2)^2 + y^2 = 25, C(2, 0), r = 5$

5. $(x - 1)^2 + (y - 3)^2 = 1; C(1, 3), r = 1$ **7.** $(x - 2)^2 + (y - 5)^2 = 1; C(2, 5), r = 1$
9. $(x - 4)^2 + (y - 0)^2 = 2; C(4, 0), r = \sqrt{2}$ **11.** $(x - 10)^2 + (y + 10)^2 = 100; C(10, -10), r = 10$
13. $(x + \frac{3}{4})^2 + (y - 1)^2 = 9; C(-\frac{3}{4}, 1), r = 3$ **15.** $(x - 2)^2 + y^2 = 4$ **17.** $(x + 3)^2 + (y - 3)^2 = 7$

19.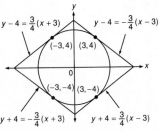

21. $x = 2; x = -2$

23.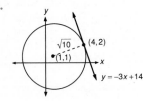

25.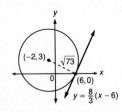

27.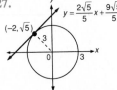

29. $(\frac{1}{2}, -2)$ **31.** $(-\frac{5}{2}, \frac{1}{2})$

33. $y = -\frac{1}{2}x + 10$ **35.** $5x + 12y = 26$ **37. (a)** $y(x) = \sqrt{25 + 24x - x^2} - 5$; **(b)** $y(7) = 7$

39. $r = \frac{1}{2}\sqrt{a^2 + b^2}$ for semicircle on AB, and area $= \frac{1}{2}\pi\left(\frac{1}{2}\sqrt{a^2 + b^2}\right)^2 = \frac{\pi}{8}(a^2 + b^2)$.

$r = \frac{a}{2}$ for semicircle on OA, and area $= \frac{1}{2}\pi\left(\frac{a}{2}\right)^2 = \frac{1}{8}\pi a^2$.

$r = \frac{b}{2}$ for semicircle on OB, and area $= \frac{1}{2}\pi\left(\frac{b}{2}\right)^2 = \frac{1}{8}\pi b^2$.

Then, $\frac{1}{8}\pi a^2 + \frac{1}{8}\pi b^2 = \frac{1}{8}\pi(a^2 + b^2)$

41. $\left(\frac{2}{5}, \frac{16}{5}\right)$

3.6 Conic Sections: The Ellipse (page 177)

1.

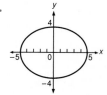

Center: $(0, 0)$
Vertices: $(\pm 5, 0)$
Foci: $(\pm 3, 0)$

3.

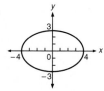

Center: $(0, 0)$
Vertices: $(\pm 4, 0)$
Foci: $(\pm\sqrt{7}, 0)$

5.

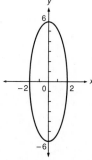

Center: $(0, 0)$
Vertices: $(0, \pm 6)$
Foci: $(0, \pm 4\sqrt{2})$

7.

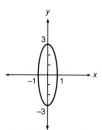

Center: $(0, 0)$
Vertices: $(0, \pm 3)$
Foci: $(0, \pm 2\sqrt{2})$

9.

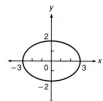

Center: $(0, 0)$
Vertices: $(\pm 3, 0)$
Foci: $(\pm\sqrt{5}, 0)$

11.

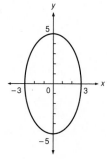

Center: $(0, 0)$
Vertices: $(0, \pm 5)$
Foci: $(0, \pm 4)$

13.

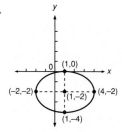

Center: $(1, -2)$
Vertices: $(-2, -2), (4, -2)$
Foci: $(1 \pm \sqrt{5}, -2)$

15.

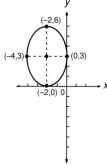

Center: $(-2, 3)$
Vertices: $(-2, 6), (-2, 0)$
Foci: $(-2, 3 \pm \sqrt{5})$

17. $\dfrac{x^2}{25} + \dfrac{y^2}{9} = 1$ 19. $\dfrac{(x-2)^2}{32} + \dfrac{(y-3)^2}{16} = 1$ 21. $\dfrac{x^2}{16} + \dfrac{y^2}{25} = 1$ 23. $\dfrac{x^2}{64} + \dfrac{y^2}{16} = 1$

25. $\dfrac{(x-6)^2}{9} + \dfrac{(y-1)^2}{4} = 1$

27. $x^2 + \dfrac{(y-6)^2}{25} = 1$; center: (0, 6); vertices: (0, 1), (0, 11); foci: $(0,\ 6 \pm 2\sqrt{6})$

29. $\dfrac{(x+3)^2}{13} + \dfrac{(y-1)^2}{4} = 1$; center: (−3, 1); vertices; $(-3 \pm \sqrt{13},\ 1)$; foci: (−6, 1), (0, 1)

31. (a) $\dfrac{x^2}{b^2} + \dfrac{y^2}{(4355)^2} = 1$ where $b^2 = (4355)^2 - (225)^2$ (b) 4349 miles; $\dfrac{x^2}{(4349)^2} + \dfrac{y^2}{(4355)^2} = 1$

33. (a) $y = \frac{2}{3}\sqrt{900 - 400} = \frac{20}{3}\sqrt{5}$ (b) 14.9 ft

3.7 Conic Sections: The Hyperbola (page 183)

1.

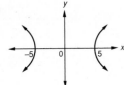

Center: (0, 0)
Vertices: $(\pm 5, 0)$
Foci: $(\pm\sqrt{34}, 0)$
Asymptotes: $y = \pm\frac{3}{5}x$

3.

Center: (0, 0)
Vertices: $(\pm 4, 0)$
Foci: $(\pm\sqrt{41}, 0)$
Asymptotes: $y = \pm\frac{5}{4}x$

5.

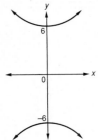

Center: (0, 0)
Vertices: $(0, \pm 6)$
Foci: $(0, \pm 3\sqrt{5})$
Asymptotes: $y = \pm 2x$

7.

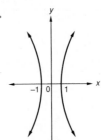

Center: (0, 0)
Vertices: $(\pm 1, 0)$
Foci: $(\pm\sqrt{10}, 0)$
Asymptotes: $y = \pm 3x$

9.

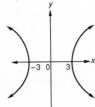

Center: (0, 0)
Vertices: $(\pm 3, 0)$
Foci: $(\pm\sqrt{13}, 0)$
Asymptotes: $y = \pm\frac{2}{3}x$

11.

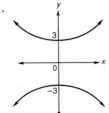

Center: (0, 0)
Vertices: $(0, \pm 3)$
Foci: $(0, \pm\sqrt{34})$
Asymptotes: $y = \pm\frac{3}{5}x$

13.

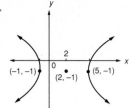

15.

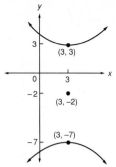

Center: $(2, -1)$
Vertices: $(-1, -1)$, $(5, -1)$
Foci: $(2 \pm \sqrt{13}, -1)$
Asymptotes: $y + 1 = \pm\frac{2}{3}(x - 2)$

Center: $(3, -2)$
Vertices: $(3, 3)$, $(3, -7)$
Foci: $(3, -2 \pm \sqrt{41})$
Asymptotes: $y + 2 = \pm\frac{5}{4}(x - 3)$

17. $\dfrac{x^2}{16} - \dfrac{y^2}{20} = 1$ **19.** $\dfrac{(y - 3)^2}{9} - \dfrac{(x + 2)^2}{7} = 1$ **21.** $\dfrac{x^2}{16} - \dfrac{y^2}{4} = 1$ **23.** $\dfrac{y^2}{64} - \dfrac{x^2}{225} = 1$

25. Center: $(-1, 3)$; vertices: $(-5, 3)$, $(3, 3)$; foci: $(-6, 3)$, $(4, 3)$

27. Center: $(3, -1)$; vertices: $(3, -9)$, $(3, 7)$; foci: $(3, -11)$, $(3, 9)$

29. $\dfrac{(x - 1)^2}{9} - \dfrac{(y + 2)^2}{4} = 1$; Center: $(1, -2)$; vertices: $(-2, -2)$, $(4, -2)$; foci: $(1 \pm \sqrt{13}, -2)$

31. $\dfrac{(y + 2)^2}{4} - \dfrac{(x - 1)^2}{1} = 1$; Center: $(1, -2)$; vertices: $(1, -4)$, $(1, 0)$; foci: $(1, -2 \pm \sqrt{5})$

3.8 Conic Sections: The Parabola (page 188)

1. Focus: $(0, \frac{1}{16})$; directrix: $y = -\frac{1}{16}$ **3.** Focus: $(0, \frac{1}{8})$; directrix: $y = -\frac{1}{8}$ **5.** Focus: $(\frac{1}{2}, 0)$; directrix: $x = -\frac{1}{2}$

7. Focus: $(0, -\frac{1}{16})$; directrix: $y = \frac{1}{16}$ **9.** Focus: $(-\frac{3}{8}, 0)$; directrix: $x = \frac{3}{8}$ **11.** Focus: $(-\frac{1}{12}, 0)$; directrix: $x = \frac{1}{12}$

13. $x^2 = -12y$

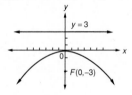

15. $x^2 = \frac{8}{3}y$

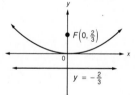

17. $y^2 = -3x$

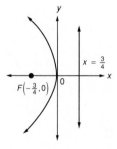

19. $y^2 = 3x$

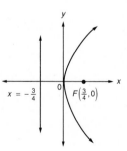

21. $(x - 2)^2 = 8(y + 5)$; $y = \frac{1}{8}(x - 2)^2 - 5$; $y = \frac{1}{8}x^2 - \frac{1}{2}x - \frac{9}{2}$

23. Vertex: $(2, -5)$; axis: $x = 2$; focus: $(2, -4)$; directrix: $y = -6$

25. $(x - 2)^2 = 4(1)(y - 3)$; vertex: $(2, 3)$; focus: $(2, 4)$; directrix: $y = 2$; axis of symmetry: $x = 2$

27. $(y - k)^2 = 4p(x - h)$ **29.** $(y + 2)^2 = 6(x + 3)$; focus: $(-\frac{3}{2}, -2)$; axis: $y = -2$

31. Vertex: $(4, 0)$; axis: $y = 0$; focus: $(\frac{7}{4}, 0)$; directrix: $x = \frac{25}{4}$ **33.** $\frac{9}{8}$ ft; 8π ft   **35.** 90 ft

37. Ellipse; center $(0, 0)$, vertices $(\pm 5, 0)$, foci $(\pm 3, 0)$

39. Hyperbola; center $(0, 0)$, vertices $(\pm 6, 0)$, foci $(\pm\sqrt{61}, 0)$ asymptotes $y = \pm\frac{5}{6}x$

41. Circle; center $(0, 0)$, radius 4

43. Hyperbola:

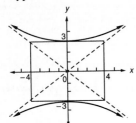

45. Ellipse:

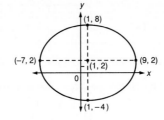

47. Hyperbola:

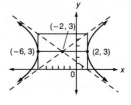

49. $(x - 1)^2 + (y + 2)^2 = 4$; circle

51. $\dfrac{(x + 1)^2}{4} + \dfrac{y^2}{1} = 1$; ellipse

53. $\dfrac{(x + 1)^2}{16} - \dfrac{(y - 3)^2}{9} = 1$; hyperbola

55. $(y + 5)^2 = 6(x + 4)$; parabola

3.9 Solving Nonlinear Systems **(page 195)**

1.

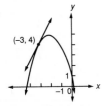

3.

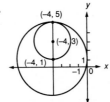

5.

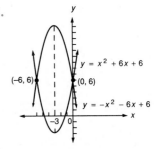

7. $(0, 1)$ **9.** No solutions **11.** No solutions **13.** $(-1, -1), (-2, -2)$ **15.** $(2, 1), (2, -1)$ **17.** $(1, 0), (0, -2)$
19. No solutions **21.** $(2, \sqrt{3}), (2, -\sqrt{3}), (-2, \sqrt{3}), (-2, -\sqrt{3})$ **23.** $(0, 0), (2, 0)$ **25.** $(0, 0), (1, 1)$
27. $(3, -3); (3 + \sqrt{3}, -2), (3 - \sqrt{3}, -2)$

29.

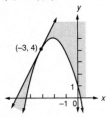

31.

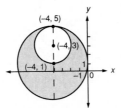

33.

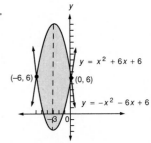

35.

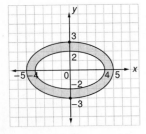

37.

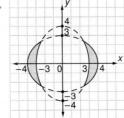

CHAPTER 3 TEST: STANDARD ANSWER (page 198)

1.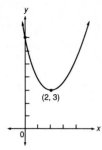

2.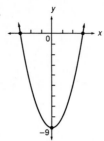

3. (a) $y = -5(x - 2)^2 + 19$; (b) $(2, 19)$; (c) $x = 2$; (d) domain: all real numbers; range: all $y \leq 19$

4. $x = -\frac{1}{3}$, $x = 3$ **5.** $2 - \sqrt{11}, 2 + \sqrt{11}$ **6.** 5; two irrational numbers **7.** 169; two rational numbers

8. Maximum $= 20$ at $x = -6$ **9.** 6 feet **10.** 64 feet **11.** (a)   (b) $y = -\frac{3}{4}x + \frac{9}{2}$

12. $C(-\frac{1}{2}, 7)$; $r = 5$ **13.** $\sqrt{106}$ **14.** $(-1, -1)$; $\sqrt{20}$; $(x + 1)^2 + (y + 1)^2 = 20$

15. $\dfrac{(x + 2)^2}{1} - \dfrac{(y - 3)^2}{4} = 1$; a hyperbola with center at $(-2, 3)$; vertices: $(-3, 3)$, $(-1, 3)$; foci: $(-2 \pm \sqrt{5}, 3)$; asymptotes: $y - 3 = \pm 2(x + 2)$

16. $\dfrac{(x - 1)^2}{4} + \dfrac{(y + 2)^2}{9} = 1$; an ellipse with center at $(1, -2)$, vertices: $(1, 1)$, $(1, -5)$; foci: $(1, -2 \pm \sqrt{5})$

17. $\dfrac{x^2}{16} + \dfrac{y^2}{9} = 1$ **18.** $\dfrac{x^2}{36} - \dfrac{y^2}{28} = 1$ **19.** Focus: $(0, -2)$; directrix: $y = 2$ **20.** $y^2 = -\frac{8}{3}x$; focus: $(-\frac{2}{3}, 0)$

21. $(x + 3)^2 = 8(y - 1)$; directrix: $y = -1$; axis: $x = -3$ **22.** $(0, 2)$; $(4, 2)$; $(2 + \sqrt{3}, 1)$; $(2 - \sqrt{3}, 1)$

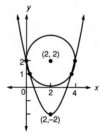

23. $(3, 1)$; $(3, -1)$; $(-3, 1)$; $(-3, -1)$ **24.** No solution

25.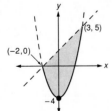

CHAPTER 3 TEST: MULTIPLE CHOICE (page 200)

1. (c) **2.** (e) **3.** (a) **4.** (b) **5.** (a) **6.** (d) **7.** (c) **8.** (a) **9.** (d) **10.** (c) **11.** (b) **12.** (a)
13. (c) **14.** (d) **15.** (d)

CHAPTER 4: POLYNOMIAL AND RATIONAL FUNCTIONS

4.1 Hints for Graphing (page 207)

1.

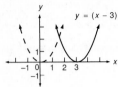

For $y = (x - 3)^2$:
Domain: all reals
Range: all $y \geq 0$
Decreasing on $(-\infty, 3]$
Increasing on $[3, \infty)$
Concave up on $(-\infty, \infty)$

3.

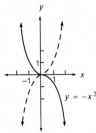

For $y = -x^3$:
Domain: all reals
Range: all reals
Decreasing on $(-\infty, \infty)$
Concave up on $(-\infty, 0)$
Concave down on $(0, \infty)$

5.

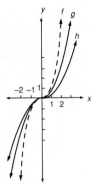

For $h(x) = \frac{1}{4}x^3$:

Domain: all reals
Range: all reals
Increasing on $(-\infty, \infty)$
Concave down on $(-\infty, 0)$
Concave up on $(0, \infty)$

7.

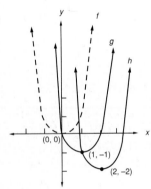

For $h(x) = (x - 2)^4 - 2$:
Domain: all reals
Range: all $y \geq -2$
Decreasing on $(-\infty, 2]$
Increasing on $[2, \infty)$
Concave up on $(-\infty, \infty)$

9.

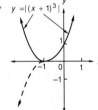

$y = |(x + 1)^3|$

11.

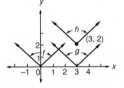

13. Translate the graph of $y = x^3$ one unit to the right and 2 units up.

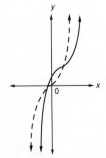

15. First sketch $y = 2x^3$ by multiplying the ordinates of $y = x^3$ by 2. Then translate $y = 2x^3$ three units left, and 3 units down.

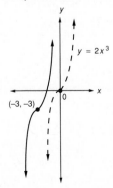

$y = 2x^3$

$(-3, -3)$

17. Reflect the graph of $y = x^3$ in the x-axis, and then translate 1 unit to the left and 1 unit down.

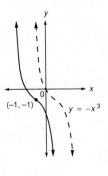

$(-1, -1)$

$y = -x^3$

19. $y = (x - 3)^4 + 2$ **21.** $y = |x + \frac{3}{4}|$ **23.** $y = |x^3 - 1|$

25.
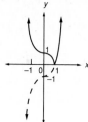

27.
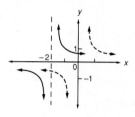
$y = (x + 1)^3$

29.
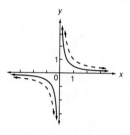
$y = -(x - 1)^3$

31. $x^2 + 3x + 9$ **33.** $4 + 6h + 4h^2 + h^3$

4.2 Graphing Some Special Rational Functions (page 213)

1.
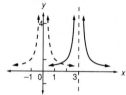

3.
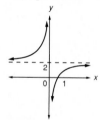

5.
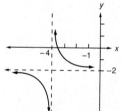

7.
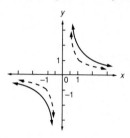

9. Asymptotes: $x = 0$, $y = 2$

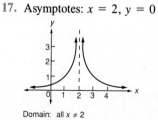

Domain: all $x \neq 0$
Range: all $y \neq 2$
Increasing and
   concave up on $(-\infty, 0)$
Increasing and
   concave down on $(0, \infty)$

11. Asymptotes $x = -4$, $y = -2$

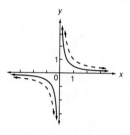

Domain: all $x \neq -4$
Range: all $y \neq -2$
Decreasing and
   concave down on $(-\infty, -4)$
Decreasing and
   concave up on $(-4, \infty)$

13. Asymptotes: $x = 2$, $y = 1$

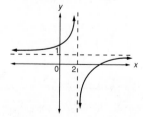

Domain: all $x \neq 2$
Range: all $y \neq 1$
Increasing and
   concave up on $(-\infty, 2)$
Increasing and
   concave down on $(2, \infty)$

15. Asymptotes: $x = -1$, $y = -2$

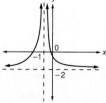

Domain: all $x \neq -1$
Range: all $y > -2$
Increasing and
   concave up on $(-\infty, -1)$
Decreasing and
   concave up on $(-1, \infty)$

17. Asymptotes: $x = 2$, $y = 0$

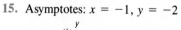

Domain: all $x \neq 2$
Range: all $y > 0$
Increasing and
   concave up on $(-\infty, 2)$
Decreasing and
   concave up on $(2, \infty)$

19. Asymptotes: $x = 0$, $y = 0$

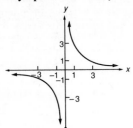

21. Asymptotes: $x = 1$, $y = 0$

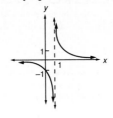

23. $f(x) = x + 3$, $x \neq 3$

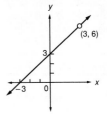

25. $f(x) = x + 2$, $x \neq 3$

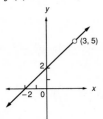

27. $f(x) = \dfrac{1}{x - 1}$, $x \neq -1$

Asymptotes: $x = 1$, $y = 0$

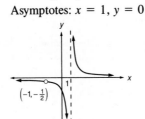

29. $f(x) = x^2 + 2x + 4$, $x \neq 2$

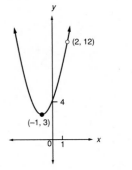

31.

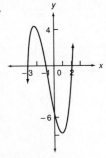

33. $-\dfrac{1}{3x}$

4.3 Polynomial and Rational Functions (page 219)

1. $f(x) < 0$ on both $(-\infty, 1)$ and $(2, 3)$; $f(x) > 0$ on both $(1, 2)$ and $(3, \infty)$
3. $f(x) < 0$ on $(-\infty, -4)$ and $(\frac{1}{3}, 2)$; $f(x) > 0$ on $(-4, 0)$, $(0, \frac{1}{3})$, and $(2, \infty)$ 5. $x < -1$ or $x > 4$
7. $x < -1$ or $\frac{1}{5} < x < 10$

9.

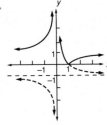

11.

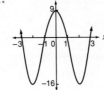

13.

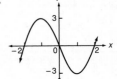

15.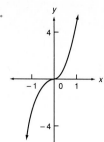

17.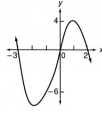

19.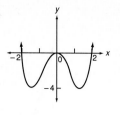

21.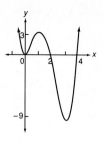

23. Asymptotes: $x = -2$; $x = 1$; $y = 0$

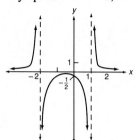

25. Asymptotes: $x = -2$, $x = 2$; $y = 0$

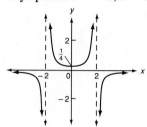

27. Asymptotes: $x = -1$; $x = 1$; $y = 0$

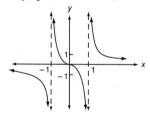

29. Asymptote: $y = 0$

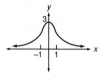

31. Asymptotes: $x = 2$; $y = 1$

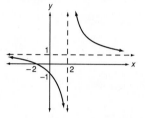

33. Asymptotes: $x = -3$; $y = 1$

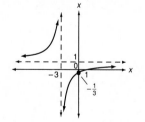

35. Asymptotes: $x = -4$; $x = 3$; $y = 1$

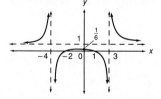

37. Horizontal asymptote: $y = 0$

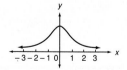

39. Horizontal asymptote: $y = 2$

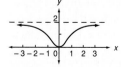

41. $g(x) = 6(x - 2)(x - 1)$; $g(x) > 0$ on both $(-\infty, 1)$ and $(2, \infty)$; $g(x) < 0$ on $(1, 2)$

43. $g(x) = -\dfrac{x + 4}{x^3}$; $g(x) > 0$ on $(-4, 0)$; $g(x) < 0$ on both $(-\infty, -4)$ and $(0, \infty)$

45. $g(x) = \dfrac{14(x - 5)}{(x + 2)^3}$; $g(x) > 0$ on both $(-\infty, -2)$ and $(5, \infty)$; $g(x) < 0$ on $(-2, 5)$

4.4 Equations and Inequalities with Fractions (page 227)

1. $x = 20$ **3.** $x = 4$ **5.** $x = \frac{17}{4}$ **7.** $x = 2$ **9.** $x = 12$ **11.** No solutions   **13.** $x = -\frac{5}{2}; x = 2$

15. $x = 1; x = 2$ **17.** $x = 20$ **19.** $x = -3; x = 3$ **21.** $-\frac{1}{2}, 3$; domain: all $x \neq -1$ or 0

23. $\frac{10}{7}, 5$; domain: all $x \neq -5$ or 0 **25.** $-5, 7$; domain: all $x \neq -\frac{1}{2}$ or 2 **27.** $m = \dfrac{2gK}{v^2}$ **29.** $r_1 = \dfrac{S}{\pi s} - r_2$

31. $s = a + (n - 1)d$ **33.** $m = \dfrac{fp}{p - f}$ **35.** $x \leq 30$ **37.** $x < -1$ **39.** $x > 10$ **41.** $-3 < x < 0$

43. $x < -5$ or $x > -2$ **45.** $x < -5$ or $x > 2$ **47.** 3 **49.** $1\frac{5}{7}$ hours   **51.** $\frac{2}{9}, \frac{4}{9}$ **53.** 60 feet   **55.** 19 **57.** $\frac{2}{3}$

59. $\dfrac{a}{b} = \dfrac{c}{d}$ implies $ad = bc$. Then, $ad + bd = bc + bd$; $d(a + b) = b(c + d)$; therefore $\dfrac{a + b}{b} = \dfrac{c + d}{d}$.

61. 110 paperbacks and 55 hardcover copies   **63.** 8000

65. Use $\dfrac{4(4) + 4(3) + 3(3) + 1(2) + x(3)}{15} = 3.4$ to get $x = 4$; grade must be A.   **67.** 5000 ohms   **69.** 10 cm

71.

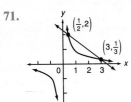

4.5 Variation (page 234)

1. $P = 4s; k = 4$ **3.** $A = 5l; k = 5$ **5.** $z = kxy^3$ **7.** $z = \dfrac{kx}{y^3}$ **9.** $w = \dfrac{kx^2}{yz}$ **11.** $\frac{1}{2}$ **13.** $\frac{1}{5}$ **15.** $\frac{2}{5}$

17. $\frac{243}{8}$ **19.** 27 feet   **21.** 4 pounds   **23.** 75 pounds per square inch   **25.** 784 feet   **27.** $\dfrac{32\pi}{3}$ cubic inches

29. 5.12 ohms; 512 ohms   **31.** **(a)** $s(y) = ky(\frac{25}{4} - y^2)$; **(b)** $s(x) = \frac{1}{2}kx^2\sqrt{25 - 4x^2}$; **(c)** $5.52k$

4.6 Division of Polynomials and Synthetic Division (page 239)

1. $x^2 - 5x + 2; r = 0$ **3.** $x^2; r = 7$ **5.** $x^2 + 7x + 7; r = 22$ **7.** $4x + 3; r = 7x - 7$ **9.** $x + 2; r = 0$
11. $x^2 + x - 2; r = 0$ **13.** $2x^2 - x - 2; r = 9$ **15.** $x^2 + 7x + 7; r = 22$ **17.** $x^3 - 5x^2 + 17x - 36; r = 73$
19. $2x^3 + 2x^2 - x + 3; r = 1$ **21.** $x^2 + 3x + 9; r = 0$ **23.** $x^2 - 3x + 9; r = 0$ **25.** $x^3 + 2x^2 + 4x + 8; r = 0$
27. $x^3 - 2x^2 + 4x - 8; r = 32$ **29.** $x^3 + \frac{1}{2}x^2 + \frac{5}{6}x + \frac{7}{12}; r = \frac{47}{60}$ **31.** $3x^2 - x - 2; r = 2$
33. $3x^2 - x + 2; r = 2$ **35.** $3x^6 + x^5 + 2x^4 + 5x^3 + 2x^2 - x - 2; r = 0$

4.7 The Remainder, Factor, and Rational Root Theorems (page 247)

1. 8 **3.** 59 **5.** 0 **7.** 1 **9.** 67 **11.** -4

For Exercises 13–23, use synthetic division to obtain $p(c) = 0$, showing $x - c$ to be a factor of $p(x)$. The remaining factors of $p(x)$ are obtained by factoring the quotient obtained in the synthetic division.

13. $(x + 1)(x + 2)(x + 3)$ **15.** $(x - 2)(x + 3)(x + 4)$ **17.** $-(x + 2)(x - 3)(x + 1)$
19. $(x - 5)(3x + 4)(2x - 1)$ **21.** $(x + 2)^2(x + 1)(x - 1)$ **23.** $x^2(x + 3)^2(x + 1)(x - 1)$ **25.** -3 (a double root), 5
27. $-\frac{2}{3}, -5, 5$ **29.** -1 (a triple root), 0 **31.** $-5, -\sqrt{3}, -1, \sqrt{3}$ **33.** $-3, 3$ **35.** $-\sqrt{3}, 0, \sqrt{3}, 2, 3$

37. $-1, \dfrac{1 - \sqrt{7}}{2}, \dfrac{1 + \sqrt{7}}{2}$ **39.** $\frac{2}{3}, 4$ **41.** $-(x + 4)^2(x - 5)$ **43.** $3(x + 2)(2x - 1)(x^2 + 1)$

45. $-1, 2, 1 - 2i, 1 + 2i$ **47.** $2, -\dfrac{1}{6} \pm \dfrac{\sqrt{47}}{6}i$ **49.** 1 (a triple root), $3 + i, 3 - i$ **51.** $2, 3$ **53.** 2

55. $(x + 2)(x + 1)(x - 1)(x^2 - x + 3)$ **57.** $(-2, -27); (2, 1); (3, 8)$ **59.** $(-3, -161); (5, 335); (\frac{1}{3}, \frac{253}{27})$

61. $\overline{z + w} = \overline{(-6 + 8i) + (\frac{1}{2} + \frac{1}{2}i)} = \overline{-\frac{11}{2} + \frac{17}{2}i} = -\frac{11}{2} - \frac{17}{2}i$

$\overline{z} + \overline{w} = \overline{-6 + 8i} + \overline{\frac{1}{2} + \frac{1}{2}i} = (-6 - 8i) + (\frac{1}{2} - \frac{1}{2}i) = -\frac{11}{2} - \frac{17}{2}i$

63. $\overline{zw} = \overline{(-6 + 8i)(\frac{1}{2} + \frac{1}{2}i)} = \overline{-7 + i} = -7 - i;\ \overline{z}\,\overline{w} = \overline{(-6 + 8i)}\ \overline{(\frac{1}{2} + \frac{1}{2}i)} = (-6 - 8i)(\frac{1}{2} - \frac{1}{2}i) = -7 - i$

65. $\overline{z + w} = \overline{(a + bi) + (c + di)} = \overline{(a + c) + (b + d)i} = (a + c) - (b + d)i$

$\overline{z} + \overline{w} = \overline{a + bi} + \overline{c + di} = (a - bi) + (c - di) = (a + c) - (b + d)i$

67. $\overline{\left(\dfrac{z}{w}\right)} = \overline{\left(\dfrac{a + bi}{c + di}\right)} = \overline{\dfrac{ac + bd}{c^2 + d^2} + \dfrac{bc - ad}{c^2 + d^2}i} = \dfrac{ac + bd}{c^2 + d^2} - \dfrac{bc - ad}{c^2 + d^2}i$

$\dfrac{\overline{z}}{\overline{w}} = \dfrac{\overline{a + bi}}{\overline{c + di}} = \dfrac{a - bi}{c - di} = \dfrac{a - bi}{c - di} \cdot \dfrac{c + di}{c + di} = \dfrac{ac + bd}{c^2 + d^2} - \dfrac{bc - ad}{c^2 + d^2}i$

69. The following proof makes repeated use of the properties in Exercises 65 and 68. Also used is the observation that the conjugate of a real number is itself.

Since $0 = p(z), \overline{0} = \overline{p(z)}$. Then, since $\overline{0} = 0$, we have

$$0 = \overline{p(z)} = \overline{a_n z^n + a_{n-1} z^{n-1} + \cdots + a_1 z + a_0}$$

$$= \overline{a_n z^n} + \overline{a_{n-1} z^{n-1}} + \cdots + \overline{a_1 z} + \overline{a_0} \qquad \text{(Ex. 65)}$$

$$= \overline{a_n}\,\overline{z^n} + \overline{a_{n-1}}\,\overline{z^{n-1}} + \cdots + \overline{a_1}\,\overline{z} + \overline{a_0} \qquad \text{(Ex. 68)}$$

$$= a_n \overline{z^n} + a_{n-1} \overline{z^{n-1}} + \cdots + a_1 \overline{z} + a_0 \qquad \text{(The } a_i \text{ are real numbers)}$$

$$= a_n (\overline{z})^n + a_{n-1}(\overline{z})^{n-1} + \cdots + a_1(\overline{z}) + a_0 \qquad \text{(Ex. 68)}$$

$$= p(\overline{z})$$

Thus, $p(\overline{z}) = 0$

4.8 Decomposing Rational Functions (page 254)

1. $\dfrac{1}{x + 1} + \dfrac{1}{x - 1}$ **3.** $\dfrac{2}{x - 3} - \dfrac{1}{x + 2}$ **5.** $\dfrac{6}{x - 4} + \dfrac{3}{x - 2} - \dfrac{4}{x + 1}$ **7.** $\dfrac{3}{x - 2} + \dfrac{3}{(x - 2)^2}$

9. $\dfrac{6}{5x + 2} - \dfrac{3}{3x - 4}$ **11.** $-\dfrac{1}{x} + \dfrac{2}{x + 2} - \dfrac{1}{(x + 2)^2}$ **13.** $x + \dfrac{1}{x - 1} - \dfrac{1}{x + 1}$ **15.** $3x^2 + 1 + \dfrac{1}{2x - 1} - \dfrac{3}{(2x - 1)^2}$

17. $10x - 5 + \dfrac{6}{x - 3} + \dfrac{14}{x + 2}$ **19.** $\dfrac{4}{x + 7} - \dfrac{2}{x - 5} + \dfrac{3}{x + 2}$ **21.** $\dfrac{x - 3}{x^2 + 3} - \dfrac{1}{x - 1}$ **23.** $\dfrac{3}{x - 1} + \dfrac{x - 2}{x^2 + x + 1}$

CHAPTER 4 TEST: STANDARD ANSWER (page 257)

1. No asymptotes

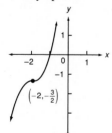

$\left(-2, -\frac{3}{2}\right)$

2. No asymptotes

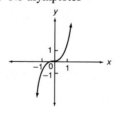

3. Asymptotes: $x = 2, y = 0$

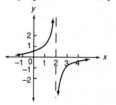

4. Asymptotes: $x = -1, x = 2, y = 0$

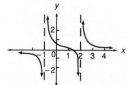

5.

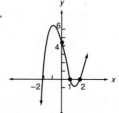

6. $f(x) < 0$ on both $(-\infty, -3)$ and $(0, 2)$;
   $f(x) > 0$ on both $(-3, 0)$ and $(2, \infty)$

7.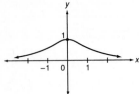

8.

9. $a = \dfrac{S}{n} - \dfrac{1}{2}(n - 1)d$ **10.** $-\dfrac{3}{2}; 2$ **11.** $x < -\dfrac{14}{3}$ **12.** 1   **13.** $5\dfrac{5}{7}$ feet   **14.** 90 minutes   **15.** $z = 2$

16. $V = 288\pi$ cubic inches

17. $-3\vert\,2 + 5 + 0 - \;1 - 21 + \;7$ quotient: $2x^4 - x^3 + 3x^2 - 10x + 9$
$\underline{\qquad - 6 + 3 - \;9 + 30 - 27}$ remainder: -20
$\qquad 2 - 1 + 3 - 10 + \;9\vert - 20$

18. When $p(x)$ is divided by $x - \frac{1}{3}$, the remainder is $\frac{2}{3}$. Then, by the remainder theorem, we have $p(\frac{1}{3}) = \frac{2}{3}$.

19. Since $p(\frac{1}{3}) \neq 0$, the factor theorem says that $x - \frac{1}{3}$ is not a factor of $p(x)$.

20. $2\vert\,1 - 4 + 7 - 12 + 12$
$\underline{\qquad + 2 - 4 + \;\;6 - 12}$
$\quad 1 - 2 + 3 - \;\;6\vert + \;\;0 = r$

Since $r = 0$, $x - 2$ is a factor of $p(x)$, and we get
$p(x) = (x - 2)(x^3 - 2x^2 + 3x - 6) = (x - 2)[x^2(x - 2) + 3(x - 2)] = (x - 2)(x^2 + 3)(x - 2)$
$\qquad = (x - 2)^2(x^2 + 3)$

21. $(x + 3)^2(x^2 - x + 1)$ **22.** -1 (a double root), $-3, 2$

23. $-1, 3, \pm i$ **24.** $\dfrac{2}{x + 5} - \dfrac{1}{x - 5}$

25. $-\dfrac{1}{x - 1} + \dfrac{5}{x - 3} + \dfrac{2}{x + 2}$

CHAPTER 4 TEST: MULTIPLE CHOICE (page 258)

1. (d)   **2.** (b)   **3.** (a)   **4.** (c)   **5.** (b)   **6.** (e)   **7.** (d)   **8.** (a)   **9.** (b)   **10.** (a)   **11.** (d)   **12.** (c)
13. (b)   **14.** (c)   **15.** (a)

CHAPTER 5: THE ALGEBRA OF FUNCTIONS

5.1 Graphing Some Special Radical Functions (page 265)

1.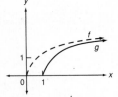

3.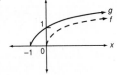

5.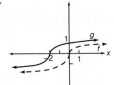

7.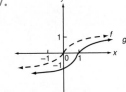

9. Domain: $x \geq -2$
Increasing for $x \geq -2$
concave down for $x > -2$

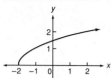

11. Domain: $x \geq 3$
Increasing for $x \geq 3$;
concave down for $x > 3$

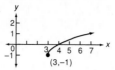

13. Domain: $x \leq 0$
Decreasing for $x \leq 0$;
concave down for $x < 0$

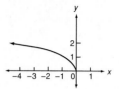

15. Domain: all real x
Increasing for all x
Concave up for $x < 0$
Concave down for $x > 0$

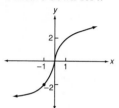

17. Domain: all real x
Decreasing for all x
Concave down for $x < 0$
Concave up for $x > 0$

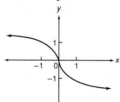

19. Domain: $x > 0$
Asymptotes: $x = 0$, $y = -1$
Decreasing and concave up
for $x > 0$

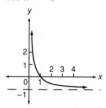

21. **(a)** $f(-x) = \dfrac{1}{\sqrt[3]{-x}} = \dfrac{1}{-\sqrt[3]{x}} = -\dfrac{1}{\sqrt[3]{x}} = -f(x)$

(b) all $x \neq 0$

(c)

x	$\frac{1}{27}$	$\frac{1}{8}$	1	8
y	3	2	1	$\frac{1}{2}$

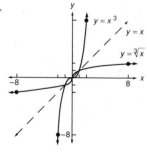

(d) $x = 0$; $y = 0$

23. $y = \sqrt[4]{x}$ is equivalent to $y^4 = x$ for $x \geq 0$.

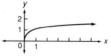

25.

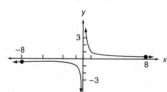

27.

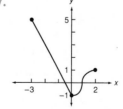

29. $\dfrac{\sqrt{4+h}-2}{h} = \dfrac{(\sqrt{4+h}-2)(\sqrt{4+h}+2)}{h(\sqrt{4+h}+2)} = \dfrac{4+h-4}{h(\sqrt{4+h}+2)} = \dfrac{1}{\sqrt{4+h}+2}$

31. **(a)** $d(x) = \sqrt{25+x^2} + \sqrt{100+(20-x)^2}$; **(b)** $t(x) = \dfrac{\sqrt{25+x^2}}{12} + \dfrac{\sqrt{100+(20-x)^2}}{10}$; **(c)** 2.4 hours

33. $d(x) = \sqrt{x^2 + \frac{1}{x}}$

5.2 Radical Equations (page 271)

1. 17; domain: $x \geq 1$ 3. $-2, 0$; domain; $x \leq -2$ or $x \geq 0$ 5. $-1, 6$; domain: $x \leq -1$ or $x \geq 6$ 7. $x = 10$
9. $x = 5$ 11. $x = \pm 10$ 13. No solution   15. $x = 2$ 17. $x = \frac{2}{3}$ 19. $x = \frac{5}{16}$ 21. $x = 9$ 23. $x = 2$
25. $x = 2$ 27. $x = 4$ 29. $x = 4$ 31. $x = 0, x = 8$ 33. $x = \frac{27}{8}$ 35. $x = 1$ 37. $x = 9$ 39. $x = 4$
41. $A = 4\pi r^2$; 16π 43. $h = \sqrt{s^2 - r^2}$; 14.72 cm   45. $S = \dfrac{10}{h} + 2\sqrt{5\pi h}$ 47. $(9, 3)$

49. 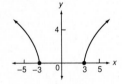   51. $y = \frac{5}{3}\sqrt{9 - x^2}$ 53. $f = \dfrac{y}{\sqrt{xy} - x}$

5.3 Determining the Signs of Radical Functions (page 276)

1. (a) All $x \geq 2$; (b) $f(x) > 0$ for all $x > 2$; (c) 2   3. (a) All real x; (b) $f(x) > 0$ for all $x \neq 2$; (c) 2
5. (a) All $x > 0$; (b) $f(x) > 0$ for $x > 0$; (c) none
7. (a) All $x \neq 0$; (b) $f(x) < 0$ on both $(-\infty, -4)$, $(0, 2)$; $f(x) > 0$ on both $(-4, 0)$ $(2, \infty)$; (c) $-4, 2$
9. (a) $-3 \leq x \leq 3$, $x \neq 0$; (b) $f(x) < 0$ on $(-3, 0)$; $f(x) > 0$ on $(0, 3)$; (c) $-3, 3$

11. $x^{1/2} + (x - 4)\frac{1}{2}x^{-1/2} = x^{1/2} + \dfrac{x - 4}{2x^{1/2}} = \dfrac{2x + x - 4}{2\sqrt{x}} = \dfrac{3x - 4}{2\sqrt{x}}$

13. $\dfrac{1}{2}(4 - x^2)^{-1/2}(-2x) = \dfrac{-2x}{2(4 - x^2)^{1/2}} = -\dfrac{x}{\sqrt{4 - x^2}}$

15. $\dfrac{x}{3(x - 1)^{2/3}} + (x - 1)^{1/3} = \dfrac{x + 3(x - 1)}{3(x - 1)^{2/3}} = \dfrac{4x - 3}{3(\sqrt[3]{x - 1})^2}$

17. $\dfrac{x}{(5x - 6)^{4/5}} + (5x - 6)^{1/5} = \dfrac{x + (5x - 6)}{(5x - 6)^{4/5}} = \dfrac{6(x - 1)}{(5x - 6)^{4/5}}$

19. $f(x) = \dfrac{x - 1}{\sqrt{x}}$; $f(x) < 0$ on $(0, 1)$, $f(x) > 0$ on $(1, \infty)$

21. $f(x) = \dfrac{3(x - 3)}{2\sqrt{x}}$; $f(x) < 0$ on $(0, 3)$, $f(x) > 0$ on $(3, \infty)$ 23. $f(x) = \dfrac{x(5x - 8)}{2\sqrt{x - 2}}$; $f(x) > 0$ on $(2, \infty)$

25. Domain: $x \leq -3$ or $x \geq 3$ 27. Domain: $-3 \leq x \leq 3$

(Note: This is the upper half of the hyperbola $x^2 - y^2 = 9$.)

(Note: This is the upper half of the circle $x^2 + y^2 = 9$.)

29. Domain: $x \geq 0$

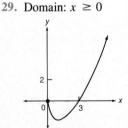

5.4 Combining and Decomposing Functions (page 283)

1. (a) $-1, 5, 4$; (b) $5x - 1$, all reals; (c) 4 **3.** (a) $-2, \frac{7}{2}, -7$; (b) $6x^2 - 5x - 6$, all reals; (c) -7

5. (a) $2, 1$; (b) $6x + 1$, all reals; (c) 1 **7.** (a) $x^2 + \sqrt{x}$; $x \geq 0$; (b) $x^{3/2}$; $x > 0$; (c) x; $x \geq 0$

9. (a) $x^3 - 1 + \dfrac{1}{x}$; all $x \neq 0$; (b) $x^4 - x$; all $x \neq 0$; (c) $\dfrac{1}{x^3} - 1$; all $x \neq 0$

11. (a) $x^2 + 6x + 8 + \sqrt{x - 2}$; $x \geq 2$; (b) $\dfrac{x^2 + 6x + 8}{\sqrt{x - 2}}$; $x > 2$; (c) $x + 6 + 6\sqrt{x - 2}$; $x \geq 2$

13. (a) $6x - 6$, all reals; (b) $-8x^2 + 22x - 5$, all reals; (c) $-8x + 19$, all reals

15. (a) $\dfrac{1}{2x} - 2x^2 + 1$, all $x \neq 0$; (b) $x - \dfrac{1}{2x}$, all $x \neq 0$; (c) $\dfrac{1}{4x^2 - 2}$, all $x \neq \pm\dfrac{1}{\sqrt{2}}$

17. $(f \circ g)(x) = (x - 1)^2$; $(g \circ f)(x) = x^2 - 1$ **19.** $(f \circ g)(x) = \dfrac{x + 3}{3 - x}$; $(g \circ f)(x) = 4 - \dfrac{6}{x}$

21. $(f \circ g)(x) = x^2$; $(g \circ f)x = x^2 + 2x$ **23.** $(f \circ g)(x) = 2$; $(g \circ f)(x) = 4$

25. (a) $\dfrac{1}{2\sqrt[3]{x} - 1}$; (b) $\dfrac{2}{\sqrt[3]{x}} - 1$; (c) $\dfrac{1}{\sqrt[3]{2x - 1}}$ **27.** $(f \circ f)(x) = x$; $(f \circ f \circ f)(x) = \dfrac{1}{x}$

(Other answers are possible for Exercises 29–52.)

29. $g(x) = 3x + 1; f(x) = x^2$ **31.** $g(x) = 1 - 4x; f(x) = \sqrt{x}$ **33.** $g(x) = \dfrac{x + 1}{x - 1}; f(x) = x^2$

35. $g(x) = 3x^2 - 1; f(x) = x^{-3}$ **37.** $g(x) = \dfrac{x}{x - 1}; f(x) = \sqrt{x}$ **39.** $g(x) = (x^2 - x - 1)^3; f(x) = \sqrt{x}$

41. $g(x) = 4 - x^2; f(x) = \dfrac{2}{\sqrt{x}}$ **43.** $f(x) = 2x + 1; g(x) = x^{1/2}; h(x) = x^3$

45. $f(x) = \dfrac{x}{x + 1}; g(x) = x^5; h(x) = x^{1/2}$ **47.** $f(x) = x^2 - 9; g(x) = x^2; h(x) = x^{1/3}$

49. $f(x) = x^2 - 4x + 7; g(x) = x^3; h(x) = -\sqrt{x}$ **51.** $f(x) = 2x - 11; g(x) = 1 + \sqrt{x}; h(x) = x^2$

53. $f(x) = (x + 1)^2$ **55.** $g(x) = \sqrt[3]{x}$ **57.** $x^2 - 2x$

5.5 Inverse Functions (page 288)

1. Not one-to-one   **3.** One-to-one   **5.** One-to-one

7. $(f \circ g)(x) = \frac{1}{3}(3x + 9) - 3 = x$

$(g \circ f)(x) = 3(\frac{1}{3}x - 3) + 9 = x$

9. $(f \circ g)(x) = (\sqrt[3]{x} - 1 + 1)^3 = x$

$(g \circ f)(x) = \sqrt[3]{(x + 1)^3} - 1 = x$

11. $(f \circ g)(x) = \dfrac{1}{\dfrac{1}{x} + 1 - 1} = x$

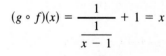

$(g \circ f)(x) = \dfrac{1}{\dfrac{1}{x - 1}} + 1 = x$

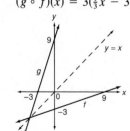

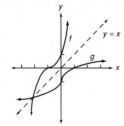

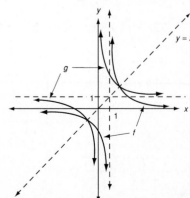

13. $g(x) = \sqrt[3]{x} + 5$ **15.** $g(x) = \frac{3}{2}x + \frac{3}{2}$ **17.** $g(x) = \sqrt[5]{x} + 1$ **19.** $g(x) = x^{5/3}$

21. $f^{-1}(x) = 2 + \frac{2}{x}; f(f^{-1}(x)) = f\left(2 + \frac{2}{x}\right) = \dfrac{2}{\left(2 + \frac{2}{x}\right) - 2} = \dfrac{2}{\frac{2}{x}} = x$

$f^{-1}(f(x)) = f^{-1}\left(\frac{2}{x - 2}\right) = 2 + \dfrac{2}{\frac{2}{x - 2}} = 2 + (x - 2) = x$

23. $f^{-1}(x) = \frac{3}{x} - 2; f(f^{-1}(x)) = f\left(\frac{3}{x} - 2\right) = \dfrac{3}{\left(\frac{3}{x} - 2\right) + 2} = \dfrac{3}{\frac{3}{x}} = x$

$f^{-1}(f(x)) = f^{-1}\left(\frac{3}{x + 2}\right) = \dfrac{3}{\frac{3}{x + 2}} - 2 = (x + 2) - 2 = x$

25. $f^{-1}(x) = x^{-1/5}; f(f^{-1}(x)) = f(x^{-1/5}) = (x^{-1/5})^{-5} = x; f^{-1}(f(x)) = f^{-1}(x^{-5}) = (x^{-5})^{-1/5} = x$

27. $(f \circ f)(x) = f(f(x)) = f\left(\frac{1}{x}\right) = \dfrac{1}{\frac{1}{x}} = x$

29. $(f \circ f)(x) = f(f(x)) = f\left(\frac{x}{x - 1}\right) = \dfrac{\frac{x}{x - 1}}{\frac{x}{x - 1} - 1} = \dfrac{x}{x - (x - 1)} = x$

31. $g(x) = \sqrt{x} - 1$

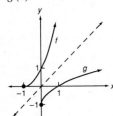

33. $g(x) = \dfrac{1}{x^2}$

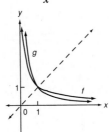

35. $g(x) = \sqrt{x + 4}$

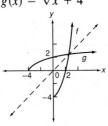

37. $y = mx + k$, where $m = -1$

CHAPTER 5 TEST: STANDARD ANSWER (page 291)

1. Domain: all reals

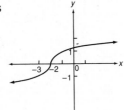

2. Domain: $x > 0$
Asymptotes: $y = 2, x = 0$

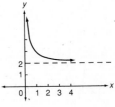

3. Domain: all reals

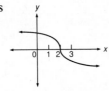

4. Domain: $x > 3$
Asymptotes: $x = 3, y = 0$

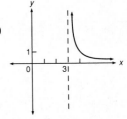

5.

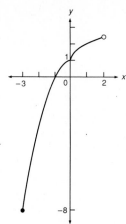

6. $\frac{2}{9}$ **7.** $-\frac{64}{27}; \frac{1}{8}$ **8.** $x = 16$

9. $(x + 1)^{1/2}(2x) + (x + 1)^{-1/2}\left(\dfrac{x^2}{2}\right) = 2x(x + 1)^{1/2} + \dfrac{x^2}{2(x + 1)^{1/2}} = \dfrac{4x(x + 1) + x^2}{2\sqrt{x + 1}} = \dfrac{5x^2 + 4x}{2\sqrt{x + 1}} = \dfrac{x(5x + 4)}{2\sqrt{x + 1}}$

10. $f(x) < 0$ on $(2, 4)$; $f(x) > 0$ on $(-\infty, 2)$ and $(4, \infty)$ **11.** $f(x) < 0$ on $(0, 4)$; $f(x) > 0$ on $(4, \infty)$

12. $-2, 1$; domain: $x \leq -2$ or $x \geq 1$

13. **(a)** $\dfrac{1}{x^2 - 1} + \sqrt{x + 2}$; all $x \geq -2$ and $x \neq \pm 1$ **(b)** $\dfrac{1}{x^2 - 1} - \sqrt{x + 2}$; all $x \geq -2$ and $x \neq \pm 1$

14. **(a)** $\dfrac{1}{(x^2 - 1)\sqrt{x + 2}}$; all $x > -2$ and $x \neq \pm 1$ **(b)** $\dfrac{\sqrt{x + 2}}{x^2 - 1}$; all $x \geq -2$ and $x \neq \pm 1$

15. Shift 4 units to the right and reflect in the x-axis.   **16.** $\frac{1}{17}; 4; \frac{4}{17}$ **17.** $\dfrac{1}{37}; \dfrac{2}{\sqrt{82}}$

18. $(f \circ g)(x) = \dfrac{1}{1 - x}$; all $x \geq 0$ and $x \neq 1$ $(g \circ f)(x) = \dfrac{1}{\sqrt{1 - x^2}}$; $-1 < x < 1$

(Other answers are possible for Exercises 19 and 20.)

19. $g(x) = x - 2; f(x) = x^{2/3}$ **20.** $h(x) = 2x - 1; g(x) = x^{3/2}; f(x) = \dfrac{1}{x}$ **21.** $f^{-1}(x) = \dfrac{x + 2}{3}$

22.

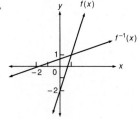

23. $g(x) = (x + 1)^3$; $(f \circ g)(x) = f(g(x)) = f((x + 1)^3) = \sqrt[3]{(x + 1)^3} - 1 = x$

24. All real numbers; $x = -3, x = 0$ **25.** $(2, 2), (10, 6)$

CHAPTER 5 TEST: MULTIPLE CHOICE (page 292)

1. (c)   **2.** (d)   **3.** (a)   **4.** (e)   **5.** (c)   **6.** (b)   **7.** (c)   **8.** (d)   **9.** (e)   **10.** (c)   **11.** (d)   **12.** (d)   **13.** (b)
14. (b)   **15.** (c)

CHAPTER 6: EXPONENTIAL AND LOGARITHMIC FUNCTIONS

6.1 Exponential Functions (page 300)

1.

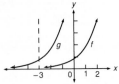

Horizontal asymptote: $y = 0$

3.

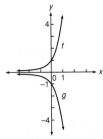

Horizontal asymptote: $y = 0$

5.

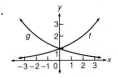

Horizontal asymptote: $y = 0$

7.

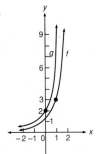

Horizontal asymptote: $y = 0$

9.

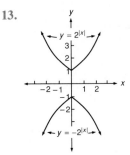

Horizontal asymptotes: $y = 0$, $y = -3$

11.

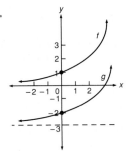

13.

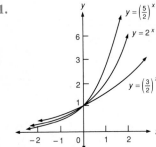

15. 6   **17.** ± 3 **19.** 1   **21.** $-2; 1$ **23.** -5 **25.** $\frac{1}{2}$ **27.** $\frac{3}{2}$ **29.** $-\frac{1}{2}$ **31.** $\frac{2}{3}$ **33.**

35. $x = \frac{1}{2}$

37. (a)

x	1.4	1.41	1.414	1.4142	1.41421
3^x	4.6555	4.7070	4.7277	4.7287	4.7288

estimate: 4.729; calculator: 4.728804…

(b)

x	1.7	1.73	1.732	1.7320	1.73205
3^x	6.4730	6.6899	6.7046	6.7046	6.7050

estimate: 6.705; calculator: 6.704991…

(c)

x	2.2	2.23	2.236	2.2360	2.23606
2^x	4.5948	4.6913	4.7109	4.7109	4.7111

estimate: 4.711; calculator: 4.711113…

(d)

x	3.1	3.14	3.141	3.1415	3.14159
4^x	77.5167	77.7085	77.8163	77.8702	77.8799

estimate: 77.880; calculator: 77.88023…

6.2 Logarithmic Functions (page 304)

1. $g(x) = \log_4 x$

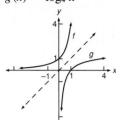

3. $g(x) = \log_{1/3} x$

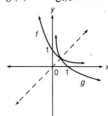

5. Shift 2 units left; $x > -2$; $x = -2$ **7.** Shift 2 units upward; $x > 0$; $x = 0$.

9. Domain: all $x > 0$

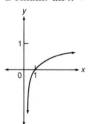

11. Domain: all $x > 0$

13. Domain: all $x \neq 0$

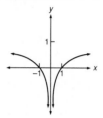

15. $\log_2 256 = 8$ **17.** $\log_{1/3} 3 = -1$ **19.** $\log_{17} 1 = 0$ **21.** $10^{-4} = 0.0001$ **23.** $(\sqrt{2})^2 = 2$ **25.** $12^{-3} = \frac{1}{1728}$
27. 4   **29.** -3 **31.** $\frac{1}{216}$ **33.** 5   **35.** 4   **37.** $\frac{1}{3}$ **39.** $\frac{2}{3}$ **41.** 9   **43.** -2 **45.** $-\frac{1}{3}$ **47.** 9   **49.** -1
51. $g(x) = 3^x - 3$, $(f \circ g)(x) = \log_3(3^x - 3 + 3) = \log_3 3^x = x$; $(g \circ f)(x) = 3^{\log_3(x+3)} - 3 = (x + 3) - 3 = x$

6.3 The Laws of Logarithms (page 309)

1. $\log_b 3 + \log_b x - \log_b(x + 1)$ **3.** $\frac{1}{2}\log_b(x^2 - 1) - \log_b x = \frac{1}{2}\log_b(x + 1) + \frac{1}{2}\log_b(x - 1) - \log_b x$

5. $-2\log_b x$ **7.** $\log_b \dfrac{x + 1}{x + 2}$ **9.** $\log_b \sqrt{\dfrac{x^2 - 1}{x^2 + 1}}$ **11.** $\log_b \dfrac{x^3}{2(x + 5)}$

13. $\log_b 27 + \log_b 3 = \log_b 81$ (Law 1)   **15.** $-2\log_b \frac{4}{9} = \log_b(\frac{4}{9})^{-2}$ (Law 3)
   $\log_b 243 - \log_b 3 = \log_b 81$ (Law 2) $= \log_b \frac{81}{16}$

17. (a) 0.6020; **(b)** 0.9030; **(c)** -0.3010 **19. (a)** 1.6811; **(b)** -0.1761; **(c)** 2.0970

21. (a) 0.2330; **(b)** 1.9515; **(c)** 1.4771   **23.** 20   **25.** $\frac{1}{20}$ **27.** 17   **29.** 8   **31.** 2   **33.** 7   **35.** 1.01   **37.** 5

39. Let $r = \log_b M$ and $s = \log_b N$. Then $b^r = M$ and $b^s = N$. Divide:
$\dfrac{M}{N} = \dfrac{b^r}{b^s} = b^{r-s}$. Convert to log form and substitute: $\log_b \dfrac{M}{N} = r - s = \log_b M - \log_b N$ **41.** $-1; 0$ **43.** 8

45. $B^{\log_B N} = N$
   $\log_b B^{\log_B N} = \log_b N$ (take log base b)
   $(\log_B N)(\log_b B) = \log_b N$ (Law 3)
   $\log_B N = \dfrac{\log_b N}{\log_b B}$ (divide by $\log_b B$)

6.4 The Base e (page 314)

1.

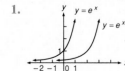

3.

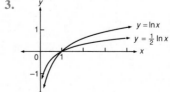

5.

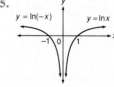

7.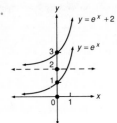

$y = e^x + 2$

$y = e^x$

9. $f(x) = -e^{-x}$
$g(x) = 1 - e^{-x}$

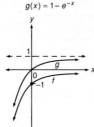

11. $t(x) = 1 - e^{\left(-\frac{1}{2}\right)x}$
$g(x) = 1 - e^{-x}$

13. Since $f(x) = 1 + \ln x$, shift 1 unit upward.   **15.** Since $f(x) = \frac{1}{2} \ln x$, multiply the ordinates by $\frac{1}{2}$.

17. Since $f(x) = \ln(x - 1)$, shift 1 unit to the right.   **19.** All $x > -2$ **21.** All $x > \frac{1}{2}$ **23.** All $x > 1$ except for $x = 2$

25. $\ln 5 + \ln x - \ln(x^2 - 4) = \ln 5 + \ln x - \ln(x + 2) - \ln(x - 2)$ **27.** $\ln(x - 1) + 2 \ln(x + 3) - \frac{1}{2}\ln(x^2 + 2)$

29. $\frac{3}{2} \ln x + \frac{1}{2} \ln(x + 1)$ **31.** $\ln \sqrt{x}(x^2 + 5)$ **33.** $\ln(x^2 - 1)^3$ **35.** $\ln \dfrac{\sqrt{x}}{(x - 1)^2 \sqrt[3]{x^2 + 1}}$ **37.** $\sqrt{x}$ **39.** $\dfrac{1}{x^2}$

41. 1   **43.** $x = -100 \ln 27$ **45.** $x = \frac{1}{3}$ **47.** $x = 0$ **49.** $x = 8$ **51.** $x = 3$ **53.** $x = 8$ **55.** $x = \dfrac{e^2}{1 + e^2}$

57.

n	2	10	100	500	1000	5000	10000	100000	1000000
$\left(1 + \dfrac{1}{n}\right)^n$	2.25000	2.59374	2.70481	2.71557	2.71692	2.71801	2.71815	2.71827	2.71828

(Other answers are possible for Exercises 59–67.)

59. Let $g(x) = -x^2 + x$ and $f(x) = e^x$. Then $(f \circ g)(x) = f(g(x)) = f(-x^2 + x) = e^{-x^2 + x} = h(x)$.

61. Let $g(x) = \dfrac{x}{x + 1}$ and $f(x) = \ln x$. Then $(f \circ g)(x) = f(g(x)) = f\left(\dfrac{x}{x + 1}\right) = \ln \dfrac{x}{x + 1} = h(x)$.

63. Let $g(x) = \ln x$ and $f(x) = \sqrt[3]{x}$. Then $(f \circ g)(x) = f(g(x)) = f(\ln x) = \sqrt[3]{\ln x} = h(x)$.

65. Let $h(x) = 3x - 1$, $g(x) = x^2$, and $f(x) = e^x$. Then

$$(f \circ g \circ h)(x) = f(g(h(x))) = f(g(3x - 1)) = f((3x - 1)^2) = e^{(3x-1)^2} = F(x)$$

67. Let $h(x) = e^x + 1$, $g(x) = \sqrt{x}$, and $f(x) = \ln x$. Then

$$(f \circ g \circ h)(x) = f(g(h(x))) = f(g(e^x + 1)) = f(\sqrt{e^x + 1}) = \ln \sqrt{e^x + 1} = F(x)$$

69. $f(x) > 0$ on the interval $(-\infty, \frac{1}{2})$; $f(x) < 0$ on the interval $(\frac{1}{2}, \infty)$

71. $f(x) > 0$ on the interval $\left(\dfrac{1}{e}, \infty\right)$; $f(x) < 0$ on the interval $\left(0, \dfrac{1}{e}\right)$

73. $\ln\left(\dfrac{x}{4} - \dfrac{\sqrt{x^2 - 4}}{4}\right) = \ln\left(\dfrac{x - \sqrt{x^2 - 4}}{4}\right) = \ln\left(\dfrac{x - \sqrt{x^2 - 4}}{4} \cdot \dfrac{x + \sqrt{x^2 - 4}}{x + \sqrt{x^2 - 4}}\right) = \ln\left(\dfrac{x^2 - (x^2 - 4)}{4(x + \sqrt{x^2 - 4})}\right)$

$= \ln\left(\dfrac{1}{x + \sqrt{x^2 - 4}}\right) = -\ln(x + \sqrt{x^2 - 4})$ **75.** $x = \ln(y + \sqrt{y^2 + 1})$

6.5 Exponential Growth and Decay (page 323)

1. $\dfrac{1.792}{0.693} = 2.586$ **3.** $\dfrac{2.079}{-1.609} = -1.292$

5. $\ln 15 = \ln(3)(5) = \ln 3 + \ln 5 = 1.099 + 1.609 = 2.708$; $\dfrac{2.708}{2(1.099)} = 1.232$

7. $\ln 100 = \ln 10^2 = 2 \ln 10$; $\dfrac{2 \ln 10}{-4(\ln 10)} = -0.5$

(The answers for questions 33–45 were found using a calculator. The remaining answers were found using Tables II and III. Using a calculator may result in slightly different answers in some cases.)

9. 2010 **11.** 27 **13.** $\frac{1}{2} \ln 100$ **15.** $\frac{1}{4} \ln \frac{1}{3}$ **17.** 667,000 **19.** 1.83 days **21.** 17.33 years
23. (a) $\frac{1}{5} \ln \frac{4}{5}$; **(b)** 6.4 grams; **(c)** 15.5 years **25.** 93.2 seconds **27.** 18.47 years **29.** 4200 years
31. 115,000 years **33. (a)** \$15,605 **(b)** \$15,657 **(c)** \$15,677 **(d)** \$15,682 **(e)** \$15,683
35. (a) \$13,655 **(b)** \$13,686 **(c)** \$13,699 **(d)** \$13,702 **(e)** \$13,703 **37.** 7.7 years; 5.8 years **39.** 13.86%
41. 8.69 years **43.** \$2914 **45.** \$11,009 **47.** $y = Ax^k$

6.6 Scientific Notation (page 327)

1. 4.68×10^3 **3.** 9.2×10^{-1} **5.** 7.583×10^6 **7.** 2.5×10^1 **9.** 5.55×10^{-7} **11.** 2.024×10^2
13. 78,900 **15.** 3000 **17.** 0.174 **19.** 17.4 **21.** 0.0906 **23.** 10^2 **25.** 10^0 **27.** 10^{17} **29.** 2000 **31.** 400
33. 3125 **35.** 0.000008 **37.** 500 seconds

6.7 Common Logarithms and Applications (page 331)

(These answers were found using Table IV. If a calculator is used for the common logarithms, then some of these answers will be slightly different.)

1. 2.6599 **3.** $9.6599 - 10$ **5.** 1.8627 **7.** 3.72 **9.** 68.1 **11.** 0.14 **13.** 43,000,000 **15.** 2770
17. 0.000125 **19.** 1.22 **21.** 0.00887 **23.** 6.58 **25.** \$1.21 per gallon **27.** 2.27 years **29.** \$4050
31. 0.23 gal **33.** 7220 cubic centimeters using $\frac{4}{3} = 1.33$ **35.** 2.15 seconds
37. For $1 \le x < 10$ we get $\log 1 \le \log x < \log 10$ because $f(x) = \log x$ is an increasing function. Substituting $0 = \log 1$ and $1 = \log 10$ into the preceding inequality gives $0 \le \log x < 1$.
39. (a) 2.825; **(b)** 9.273; **(c)** 1.378; **(d)** 5.489; **(e)** 4.495; **(f)** 7.497; **(g)** 1.021; **(h)** 9.508; **(i)** 2.263

CHAPTER 6 TEST: STANDARD ANSWER (page 337)

1. (i) a; **(ii)** c; **(iii)** e; **(iv)** h; **(v)** d; **(vi)** b **2. (a)** $5^3 = 125$; **(b)** $\log_{16} 8 = \frac{3}{4}$ **3. (a)** $x = \frac{1}{2}$; **(b)** $x = -2, x = 3$
4. $x = -7$ **5.** 10 **6. (a)** $b = \frac{2}{3}$; **(b)** $b = -2$
7. (a) Domain: all real x; range: $y > 0$; **(b)** increasing for all x; **(c)** concave up for all x; **(d)** y-intercept $= 1$; no x-intercept; **(e)** horizontal asymptote: $y = 0$
8.   **9.** All real x; $y = -4$ **10.** All $x > -4$; $x = -4$

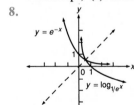

11. (a) $\log_b x + 10 \log_b(x^2 + 1)$; **(b)** $3 \ln x - \ln(x + 1) - \frac{1}{2} \ln(x^2 + 2)$ **12. (a)** $\log_7 10x^2$; **(b)** $\log_b \dfrac{x^{1/3}}{(x + 2)^2}$
13. $x = 5$ **14.** $x = \frac{10}{3}$ **15.** Shift up 2 units and to the right 1 unit **16.** $\dfrac{\ln 5}{2 \ln 4}$ **17.** $25 \ln 2$ **18.** $2000(1.02)^{24}$
19. $\dfrac{\ln 2}{4 \ln 1.02}$ **20.** \$12,060 (using Table II); \$12,083 (using a calculator) **21. (a)** 4.29×10^4; **(b)** 0.00325 **22.** 6
23. 0.11 **24.** $k = \frac{1}{3} \ln \frac{3}{2}$ **25.** $x = \dfrac{e - 2}{e + 2}$

CHAPTER 6 TEST: MULTIPLE CHOICE (page 338)

1. (d) **2. (a)** **3. (c)** **4. (e)** **5. (a)** **6. (c)** **7. (e)** **8. (d)** **9. (c)** **10. (b)** **11. (a)** **12. (b)** **13. (c)**
14. (b) **15. (d)**

7.1 Trigonometric Ratios (page 346)

	$\sin \theta$	$\cos \theta$	$\tan \theta$	$\cot \theta$	$\sec \theta$	$\csc \theta$
1.	$\dfrac{8}{17}$	$\dfrac{15}{17}$	$\dfrac{8}{15}$	$\dfrac{15}{8}$	$\dfrac{17}{15}$	$\dfrac{17}{8}$
3.	$\dfrac{\sqrt{3}}{2}$	$\dfrac{1}{2}$	$\sqrt{3}$	$\dfrac{1}{\sqrt{3}}$	2	$\dfrac{2}{\sqrt{3}}$
5.	$\dfrac{12}{13}$	$\dfrac{5}{13}$	$\dfrac{12}{5}$	$\dfrac{5}{12}$	$\dfrac{13}{5}$	$\dfrac{13}{12}$

7.   **9.**   **11.**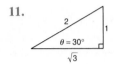

	$\sin A$	$\cos A$	$\tan A$	$\cot A$	$\sec A$	$\csc A$
13.	$\dfrac{3}{5}$	$\dfrac{4}{5}$	$\dfrac{3}{4}$	$\dfrac{4}{3}$	$\dfrac{5}{4}$	$\dfrac{5}{3}$
15.	$\dfrac{4}{\sqrt{41}}$	$\dfrac{5}{\sqrt{41}}$	$\dfrac{4}{5}$	$\dfrac{5}{4}$	$\dfrac{\sqrt{41}}{5}$	$\dfrac{\sqrt{41}}{4}$
17.	$\dfrac{1}{\sqrt{5}}$	$\dfrac{2}{\sqrt{5}}$	$\dfrac{1}{2}$	2	$\dfrac{\sqrt{5}}{2}$	$\sqrt{5}$
19.	$\dfrac{3}{4}$	$\dfrac{\sqrt{7}}{4}$	$\dfrac{3}{\sqrt{7}}$	$\dfrac{\sqrt{7}}{3}$	$\dfrac{4}{\sqrt{7}}$	$\dfrac{4}{3}$
21.	$\dfrac{\sqrt{21}}{5}$	$\dfrac{2}{5}$	$\dfrac{\sqrt{21}}{2}$	$\dfrac{2}{\sqrt{21}}$	$\dfrac{5}{2}$	$\dfrac{5}{\sqrt{21}}$
23.	$\dfrac{21}{29}$	$\dfrac{20}{29}$	$\dfrac{21}{20}$	$\dfrac{20}{21}$	$\dfrac{29}{20}$	$\dfrac{29}{21}$
	$\cos B$	$\sin B$	$\cot B$	$\tan B$	$\csc B$	$\sec\ B$

25. 1 **27.** $\dfrac{x^2}{z^2}$ **29.** 1 **31.** 1 **33.** $\cos A = \dfrac{\sqrt{7}}{4}$; $\tan A = \dfrac{3}{\sqrt{7}}$; $\cot A = \dfrac{\sqrt{7}}{3}$; $\sec A = \dfrac{4}{\sqrt{7}}$; $\csc A = \dfrac{4}{3}$

35. $\sin A = \dfrac{9}{41}$; $\cos A = \dfrac{40}{41}$, $\cot A = \dfrac{40}{9}$; $\sec A = \dfrac{41}{40}$; $\csc\ A = \dfrac{41}{9}$

37. $\sin B = \dfrac{3}{\sqrt{11}}$; $\cos B = \dfrac{\sqrt{2}}{\sqrt{11}}$; $\tan A = \dfrac{3}{\sqrt{2}}$; $\cot A = \dfrac{\sqrt{2}}{3}$; $\csc A = \dfrac{\sqrt{11}}{3}$

39. $(\sin A)(\cos A) = (\sqrt{1-x^2})(x) = x\sqrt{1-x^2}$ **41.** $\dfrac{4\sin^2 A}{\cos A} = \dfrac{4\left(\dfrac{x}{4}\right)^2}{\dfrac{\sqrt{16-x^2}}{4}} = \dfrac{x^2}{\sqrt{16-x^2}}$

43. $(\sin \theta)(\tan \theta) = \left(\dfrac{x}{2}\right)\left(\dfrac{x}{\sqrt{4 - x^2}}\right) = \dfrac{x^2}{2\sqrt{4 - x^2}}$ **45.** $(\tan^2 \theta)(\cos^2 \theta) = \left(\dfrac{x}{4}\right)^2\left(\dfrac{4}{\sqrt{x^2 + 16}}\right)^2 = \dfrac{x^2}{x^2 + 16}$

47. $A(s) = \dfrac{\sqrt{3}}{4}s^2$ **49.** $A(x) = 2(9 - x^2)$ **51.** $A(x) = \dfrac{\sqrt{3}}{36}x^2 + \dfrac{(100 - x)^2}{4\pi}$; domain: $0 < x < 100$

7.2 Angle Measure (page 354)

1. $\dfrac{\pi}{4}$ **3.** $\dfrac{\pi}{2}$ **5.** $\dfrac{3\pi}{2}$ **7.** $\dfrac{5\pi}{6}$ **9.** $\dfrac{5\pi}{4}$ **11.** $\dfrac{7\pi}{6}$ **13.** $\dfrac{11\pi}{6}$ **15.** $\dfrac{5\pi}{12}$ **17.** $\dfrac{5\pi}{9}$; 1.75   **19.** $\dfrac{17\pi}{9}$; 5.95   **21.** 180°

23. 360°   **25.** 100°   **27.** 120°   **29.** 225°   **31.** 300°   **33.** 50°   **35.** 12°; 12.0°   **37.** $\left(\dfrac{540}{\pi}\right)^\circ$; 171.9°   **39.** $\dfrac{\sqrt{2}}{2}$

41. $\dfrac{\sqrt{3}}{3}$ **43.** $s = 2\pi$ **45.** $\theta = \dfrac{\pi}{2}$ **47.** $\theta = \dfrac{4\pi}{3}$ **49.** 12π square centimeters   **51.** 54π square centimeters

53. 126π square centimeters   **55.** 8 square inches   **57.** $\dfrac{\pi}{2}$; 1.6   **59.** 11°   **61.** 150.7   **63.** $\dfrac{1225\pi}{6}$ square feet

65. (a) 15439 mph; 22644 ft/sec   **(b)** $\dfrac{8\pi}{7}$ rad/hr   **67. (a)** $\dfrac{\pi}{30}$ rad/min   **(b)** 2π rad/hr

69. (a) 440π ft./min   **(b)** approximately 3.8 min   **71.** 1.4 rad

7.3 Right Triangle Trigonometry (page 362)

1. $x = \dfrac{25}{\sqrt{3}}$ **3.** $x = 15\sqrt{2}$ **5.** $x = 20\sqrt{3}$ **7.** $x = \dfrac{8}{\sqrt{3}}$ **9.** $\theta = \dfrac{\pi}{4}$

11. $\sin \theta = .5779$, $\cos \theta = .8161$, $\tan \theta = .7080$, $\cot \theta = 1.4124$; $\sec \theta = 1.2253$; $\csc \theta = 1.7305$ **13.** .6691

15. .2126   **17.** 57.299   **19.** .0454   **21.** 1.0035   **23.** 1.0589   **25.** 81.0°   **27.** 65.3°   **29.** 6.3°   **31.** 12.7

33. 18.2   **35.** 71.6°   **37.** 21.1°   **39.** 35 meters   **41.** 274.7 feet   **43.** 1915 feet   **45.** 79.8°   **47.** 12.3

49. 1028 feet   **51.** 145 cm

7.4 Extending the Definitions (page 372)

1. (a) I; $\phi = 73°$; **(b)** $-287°$ **3. (a)** II; $\phi = 40°$; **(b)** $-220°$ **5. (a)** III; $\phi = \dfrac{\pi}{4}$; **(b)** $-\dfrac{3\pi}{4}$

7. (a) IV; $\phi = 73°$; **(b)** $287°$ **9. (a)** III; $\phi = 50°$; **(b)** $230°$ **11. (a)** I; $\phi = \dfrac{\pi}{3}$; **(b)** $\dfrac{\pi}{3}$

13. $\sin \dfrac{2\pi}{3} = \dfrac{\sqrt{3}}{2}$; $\csc \dfrac{2\pi}{3} = \dfrac{2}{\sqrt{3}}$ **15.** $\sin\left(-\dfrac{7\pi}{4}\right) = \dfrac{1}{\sqrt{2}}$; $\csc\left(-\dfrac{7\pi}{4}\right) = \sqrt{2}$ **17.** Ratios are the same as in Exercise 15.

$\cos \dfrac{2\pi}{3} = -\dfrac{1}{2}$; $\sec \dfrac{2\pi}{3} = -2$ $\cos\left(-\dfrac{7\pi}{4}\right) = \dfrac{1}{\sqrt{2}}$; $\sec\left(-\dfrac{7\pi}{4}\right) = \sqrt{2}$

$\tan \dfrac{2\pi}{3} = -\sqrt{3}$; $\cot \dfrac{2\pi}{3} = -\dfrac{1}{\sqrt{3}}$ $\tan\left(-\dfrac{7\pi}{4}\right) = 1$; $\cot\left(-\dfrac{7\pi}{4}\right) = 1$

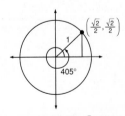
reference angle $= \dfrac{\pi}{4}$

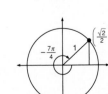
reference angle $= \dfrac{\pi}{3}$

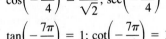
reference angle $= \dfrac{\pi}{4}$

19. $(0, -1)$; $\sin \theta = -1$; $\cos \theta = 0$; $\tan \theta$ is undefined; $\cot \theta = 0$; $\sec \theta$ is undefined; $\csc \theta = -1$.

21. $(0, 1)$; $\sin \theta = 1$; $\cos \theta = 0$; $\tan \theta$ is undefined; $\cot \theta = 0$; $\sec \theta$ is undefined; $\csc \theta = 1$.   **23.** .8391

25. $-.9877$ **27.** Undefined.   **29.** 0   **31.** Undefined   **33.** 0   **35.** .0872   **37.** 7.1154   **39.** .3007

41. $-\dfrac{\sqrt{3}}{3}$ **43.** -1.015 **45.** $-.9348$ **47.** $-\dfrac{\sqrt{2}}{2}$ **49.** $-\dfrac{2\sqrt{3}}{3}$

51. $P_1\left(\dfrac{\sqrt{3}}{2}, \dfrac{1}{2}\right)$; $P_2\left(\dfrac{\sqrt{2}}{2}, \dfrac{\sqrt{2}}{2}\right)$; $P_3\left(\dfrac{1}{2}, \dfrac{\sqrt{3}}{2}\right)$; $P_4\left(-\dfrac{\sqrt{3}}{2}, -\dfrac{1}{2}\right)$; $P_5\left(-\dfrac{\sqrt{2}}{2}, -\dfrac{\sqrt{2}}{2}\right)$; $P_6\left(-\dfrac{1}{2}, -\dfrac{\sqrt{3}}{2}\right)$

53. (a) $\left(\dfrac{2}{3}\right)^2 + \left(\dfrac{\sqrt{5}}{3}\right)^2 = \dfrac{4}{9} + \dfrac{5}{9} = \dfrac{9}{9} = 1$ **(b)**

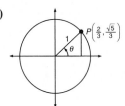

(c) $\sin \theta = \dfrac{\sqrt{5}}{3}$; $\cos \theta = \dfrac{2}{3}$; $\tan \theta = \dfrac{\sqrt{5}}{2}$; $\cot \theta = \dfrac{2\sqrt{5}}{5}$; $\sec \theta = \dfrac{3}{2}$; $\csc \theta = \dfrac{3\sqrt{5}}{5}$ **(d)** 48.2°

55. $y = -\dfrac{\sqrt{13}}{4} = -.9014$; $\theta = -64.3°$ **57.** $\dfrac{\pi}{2}$ **59.** 0; π **61.** $\dfrac{\pi}{6}; \dfrac{5\pi}{6}$ **63.** $\dfrac{3\pi}{4}; \dfrac{5\pi}{4}$

65. III; $\cos \theta = -\dfrac{\sqrt{2}}{2}$; $\cot \theta = 1$; $\sec \theta = -\sqrt{2}$; $\csc \theta = -\sqrt{2}$

67. IV; $\cos \theta = \dfrac{\sqrt{3}}{2}$; $\tan \theta = -\dfrac{\sqrt{3}}{3}$; $\sec \theta = \dfrac{2\sqrt{3}}{3}$; $\csc \theta = -2$ **69.** $-\dfrac{2\sqrt{2}}{3}$ **71.** $\dfrac{4\sqrt{15}}{15}$ **73.** $-\dfrac{\sqrt{5}}{5}$

75. $AB = \dfrac{AB}{1} = \dfrac{AB}{OA} = \dfrac{y}{x} = \tan \theta$; OB

7.5 Identities (page 379)

1. $\sin \theta$ **3.** 1   **5.** $\cos \theta$ **7.** 1   **9.** -1 **11.** $(\sin \theta)(\cot \theta)(\sec \theta) = (\sin \theta)\left(\dfrac{\cos \theta}{\sin \theta}\right)\left(\dfrac{1}{\cos \theta}\right) = 1$

13. $\cos^2 \theta(\tan^2 \theta + 1) = \cos^2 \theta\left(\dfrac{\sin^2 \theta}{\cos^2 \theta} + 1\right) = \sin^2 \theta + \cos^2 \theta = 1$ **15.** θ not coterminal with $\dfrac{\pi}{2}, \dfrac{3\pi}{2}$.

17. θ not a quadrantal angle   **19.** Same as Exercise 17.   **21.** $\dfrac{\sin^2 \theta}{\cos^2 \theta}$, θ not a quadrantal angle

23. $\dfrac{\sin \theta - 1}{\cos \theta}$; θ not a quadrantal angle   **25.** $\dfrac{1}{\sin \theta \cos \theta}$; θ not a quadrantal angle   **27.** $1 - 2\sin^2 \theta$

29. $\cos^2 \theta + \cos \theta - 1$ **31.** $\dfrac{\cos \theta}{\cot \theta} = (\cos \theta)\dfrac{\sin \theta}{\cos \theta} = \sin \theta$; θ not a quadrantal angle

33. $(\tan \theta - 1)^2 = \tan^2 \theta - 2\tan \theta + 1 = \sec^2 \theta - 2\tan \theta$; θ not coterminal with $\dfrac{\pi}{2}, \dfrac{3\pi}{2}$

35.

$\sec \theta - \cos \theta$	$\sin \theta \tan \theta$
$\dfrac{1}{\cos \theta} - \cos \theta$	$\sin \theta \dfrac{\sin \theta}{\cos \theta}$
$\dfrac{1 - \cos^2 \theta}{\cos \theta}$	$\dfrac{\sin^2 \theta}{\cos \theta}$
$\dfrac{\sin^2 \theta}{\cos \theta}$	

θ not coterminal with $\dfrac{\pi}{2}$ or $\dfrac{3\pi}{2}$

37.

$$\dfrac{\cot\theta-1}{1-\tan\theta} \quad \bigg| \quad \dfrac{\csc\theta}{\sec\theta}$$

$$\dfrac{\dfrac{\cos\theta}{\sin\theta}-1}{1-\dfrac{\sin\theta}{\cos\theta}} \quad \bigg| \quad \dfrac{\dfrac{1}{\sin\theta}}{\dfrac{1}{\cos\theta}}$$

$$\dfrac{\cos^2\theta-\sin\theta\cos\theta}{\sin\theta\cos\theta-\sin^2\theta} \quad \bigg| \quad \dfrac{\cos\theta}{\sin\theta}$$

$$\dfrac{\cos\theta\,(\cos\theta-\sin\theta)}{\sin\theta\,(\cos\theta-\sin\theta)}$$

$$\dfrac{\cos\theta}{\sin\theta}$$

θ not a quadrantal angle or not coterminal with $\dfrac{\pi}{4},\dfrac{5\pi}{4}$

39. $\tan\theta+\cot\theta=\dfrac{\sin\theta}{\cos\theta}+\dfrac{\cos\theta}{\sin\theta}=\dfrac{\sin^2\theta+\cos^2\theta}{\sin\theta\cos\theta}=\dfrac{1}{\sin\theta\cos\theta}$

41. $(\sec\theta+\tan\theta)(1-\sin\theta)=\sec\theta+\tan\theta-\sec\theta\sin\theta-\tan\theta\sin\theta=\sec\theta+\tan\theta-\dfrac{\sin\theta}{\cos\theta}-\dfrac{\sin^2\theta}{\cos\theta}$

$$=\sec\theta-\dfrac{\sin^2\theta}{\cos\theta}=\dfrac{1}{\cos\theta}-\dfrac{\sin^2\theta}{\cos\theta}=\dfrac{1-\sin^2\theta}{\cos\theta}=\dfrac{\cos^2\theta}{\cos\theta}=\cos\theta$$

43. $(\csc^2\theta-1)\sin^2\theta=\cot^2\theta\sin^2\theta=\dfrac{\cos^2\theta}{\sin^2\theta}\sin^2\theta=\cos^2\theta$

45. $\tan^2\theta-\sin^2\theta=\dfrac{\sin^2\theta}{\cos^2\theta}-\sin^2\theta=\sin^2\theta\left(\dfrac{1}{\cos^2\theta}-1\right)=\sin^2\theta\left(\dfrac{1-\cos^2\theta}{\cos^2\theta}\right)$

$$=\sin^2\theta\left(\dfrac{\sin^2\theta}{\cos^2\theta}\right)=\sin^2\theta\tan^2\theta$$

47. $\dfrac{1+\sec\theta}{\csc\theta}=\dfrac{1+\dfrac{1}{\cos\theta}}{\dfrac{1}{\sin\theta}}=\sin\theta+\dfrac{\sin\theta}{\cos\theta}=\sin\theta+\tan\theta$

49. $\dfrac{1}{1+\cos\theta}+\dfrac{1}{1-\cos\theta}=\dfrac{1-\cos\theta+1+\cos\theta}{(1+\cos\theta)(1-\cos\theta)}=\dfrac{2}{1-\cos^2\theta}=\dfrac{2}{\sin^2\theta}=2\csc^2\theta$

51. $\dfrac{1+\tan^2\theta}{1+\cot^2\theta}=\dfrac{\sec^2\theta}{\csc^2\theta}=\dfrac{\dfrac{1}{\cos^2\theta}}{\dfrac{1}{\sin^2\theta}}=\dfrac{\sin^2\theta}{\cos^2\theta}=\tan^2\theta=\sec^2\theta-1$

53. $\dfrac{\sec\theta+\csc\theta}{\sec\theta-\csc\theta}=\dfrac{\dfrac{1}{\cos\theta}+\dfrac{1}{\sin\theta}}{\dfrac{1}{\cos\theta}-\dfrac{1}{\sin\theta}}=\dfrac{\sin\theta+\cos\theta}{\sin\theta-\cos\theta}$

55. $(\csc\theta-\cot\theta)^2=\left(\dfrac{1}{\sin\theta}-\dfrac{\cos\theta}{\sin\theta}\right)^2$

$$=\dfrac{(1-\cos\theta)^2}{\sin^2\theta}=\dfrac{(1-\cos\theta)^2}{1-\cos^2\theta}$$

$$=\dfrac{(1-\cos\theta)^2}{(1-\cos\theta)(1+\cos\theta)}=\dfrac{1-\cos\theta}{1+\cos\theta}$$

57. $(\sin^2\theta+\cos^2\theta)^5=1^5=1$

59. $\dfrac{1}{\cos^2 \theta} + \dfrac{1}{\sin^2 \theta} = \dfrac{\sin^2 \theta + \cos^2 \theta}{\cos^2 \theta \sin^2 \theta} = \dfrac{1}{(1 - \sin^2 \theta)\sin^2 \theta} = \dfrac{1}{\sin^2 \theta - \sin^4 \theta}$

61. $\dfrac{\tan^2 \theta + 1}{\tan^2 \theta} = 1 + \dfrac{1}{\tan^2 \theta} = 1 + \cot^2 \theta = \csc^2 \theta$

63. $\begin{aligned}\dfrac{\tan \theta}{\sec \theta - 1} &= \dfrac{\tan \theta(\sec \theta + 1)}{(\sec \theta - 1)(\sec \theta + 1)} \\ &= \dfrac{\tan \theta(\sec \theta + 1)}{\sec^2 \theta - 1} = \dfrac{\tan \theta(\sec \theta + 1)}{\tan^2 \theta} \\ &= \dfrac{\sec \theta + 1}{\tan \theta}\end{aligned}$

65. $\begin{aligned}\sec^3 \theta + \dfrac{\sin^2 \theta}{\cos^3 \theta} &= \dfrac{1}{\cos^3 \theta} + \dfrac{\sin^2 \theta}{\cos^3 \theta} \\ &= \dfrac{1 + \sin^2 \theta}{\cos^3 \theta} \\ &= \dfrac{(\cos^2 \theta + \sin^2 \theta) + \sin^2 \theta}{\cos^3 \theta} \\ &= \dfrac{\cos^2 \theta + 2\sin^2 \theta}{\cos^3 \theta}\end{aligned}$

67. $\begin{aligned}\dfrac{1}{(\csc \theta - \sec \theta)^2} &= \dfrac{1}{\left(\dfrac{1}{\sin \theta} - \dfrac{1}{\cos \theta}\right)^2} = \dfrac{1}{\left(\dfrac{\cos \theta - \sin \theta}{\sin \theta \cos \theta}\right)^2} \\ &= \dfrac{\sin^2 \theta \cos^2 \theta}{(\cos \theta - \sin \theta)^2} = \dfrac{\sin^2 \theta \cos^2 \theta}{\cos^2 \theta - 2\cos \theta \sin \theta + \sin^2 \theta} \\ &= \dfrac{\sin^2 \theta \cos^2 \theta}{1 - 2\cos \theta \sin \theta} = \dfrac{\sin^2 \theta}{\dfrac{1}{\cos^2 \theta} - 2\dfrac{\sin \theta}{\cos \theta}} = \dfrac{\sin^2 \theta}{\sec^2 \theta - 2\tan \theta}\end{aligned}$

69. $\begin{aligned}\dfrac{\sec \theta}{\tan \theta - \sin \theta} &= \dfrac{\dfrac{1}{\cos \theta}}{\dfrac{\sin \theta}{\cos \theta} - \sin \theta} = \dfrac{1}{\sin \theta - \sin \theta \cos \theta} = \dfrac{1}{\sin \theta(1 - \cos \theta)} \\ &= \dfrac{1 + \cos \theta}{\sin \theta(1 - \cos^2 \theta)} = \dfrac{1 + \cos \theta}{\sin \theta \sin^2 \theta} = \dfrac{1 + \cos \theta}{\sin^3 \theta}\end{aligned}$

71. $-\ln|\cos x| = \ln|\cos x|^{-1} = \ln\dfrac{1}{|\cos x|} = \ln\left|\dfrac{1}{\cos x}\right| = \ln|\sec x|$

73. $\begin{aligned}-\ln|\sec x - \tan x| &= \ln|\sec x - \tan x|^{-1} = \ln\dfrac{1}{|\sec x - \tan x|} \\ &= \ln\left|\dfrac{1}{\sec x - \tan x}\right| = \ln\left|\dfrac{\sec x + \tan x}{\sec^2 x - \tan^2 x}\right| = \ln|\sec x + \tan x|\end{aligned}$

75. $\tan\left(-\dfrac{\pi}{4}\right) = -1 \neq 1 = \tan\dfrac{\pi}{4}$

77. $\sin\dfrac{\pi}{2} = 1 \neq -1 = -\sqrt{1 - \cos^2\dfrac{\pi}{2}}$

79. $\dfrac{\cot\dfrac{\pi}{3} - 1}{1 - \tan\dfrac{\pi}{3}} = \dfrac{\dfrac{1}{\sqrt{3}} - 1}{1 - \sqrt{3}} = \dfrac{1}{\sqrt{3}} \neq \sqrt{3} = \dfrac{2}{\dfrac{2}{\sqrt{3}}} = \dfrac{\sec\dfrac{\pi}{3}}{\csc\dfrac{\pi}{3}}$

81. (a)

$\begin{aligned}\ln(\sec \theta + \tan \theta) &= \ln\left(\dfrac{\sqrt{9 + x^2}}{3} + \dfrac{x}{3}\right) \\ &= \ln\left(\dfrac{\sqrt{9 + x^2} + x}{3}\right)\end{aligned}$

(b) $\ln 3$

1. (a) $\frac{15}{8}$; (b) $\frac{8}{17}$, (c) $\frac{17}{8}$ 2. $\sin A = \frac{\sqrt{5}}{3}$, $\cos A = \frac{2}{3}$; $\tan A = \frac{\sqrt{5}}{2}$.

3. $AB = \sqrt{9 + x^2}$; $3(\sin^2 A)(\sec A)$

$\quad = 3\left(\frac{x^2}{x^2 + 9}\right)\left(\frac{\sqrt{x^2 + 9}}{3}\right) = \frac{x^2}{\sqrt{x^2 + 9}}$

4. $\sin B = \frac{5}{7} \qquad \csc B = \frac{7}{5}$

$\quad \cos B = \frac{2\sqrt{6}}{7} \qquad \sec B = \frac{7}{2\sqrt{6}}$

$\quad \tan B = \frac{5}{2\sqrt{6}} \qquad \cot B = \frac{2\sqrt{6}}{5}$

5. $243°$; $\frac{8\pi}{9}$ 6. $\frac{5}{2}$; $\left(\frac{450}{\pi}\right)°$ 7. $\frac{20\pi}{3}$ 8. (a) $\frac{\sqrt{3}}{2}$; (b) 0; (c) $-\frac{\sqrt{2}}{2}$ 9. (a) $-\frac{\sqrt{3}}{3}$; (b) -1; (c) 2

	θ	$\sin \theta$	$\cos \theta$	$\tan \theta$	$\cot \theta$	$\sec \theta$	$\csc \theta$
10.	$-\dfrac{7\pi}{2}$	1	0	undefined	0	undefined	1
11.	$\dfrac{7\pi}{4}$	$-\dfrac{1}{\sqrt{2}}$	$\dfrac{1}{\sqrt{2}}$	-1	-1	$\sqrt{2}$	$-\sqrt{2}$

12. 132.3 meters   13. $20\sqrt{2}$ 14. 373 feet   15. $71.6°$ 16. $\dfrac{1}{\sin \theta}$

17. $\dfrac{1}{1 - \sin \theta} + \dfrac{1}{1 + \sin \theta} = \dfrac{1 + \sin \theta + 1 - \sin \theta}{1 - \sin^2 \theta} = \dfrac{2}{\cos^2 \theta} = 2 \sec^2 \theta$; θ not coterminal with $\dfrac{\pi}{2}$ or $\dfrac{3\pi}{2}$ 18. $\dfrac{3}{\sqrt{55}}$

19. $\cot^2 \theta - \cos^2 \theta = \dfrac{\cos^2 \theta}{\sin^2 \theta} - \cos^2 \theta$

$\qquad = \cos^2 \theta\left(\dfrac{1}{\sin^2 \theta} - 1\right) = \cos^2 \theta\left(\dfrac{1 - \sin^2 \theta}{\sin^2 \theta}\right)$

$\qquad = \cos^2 \theta\left(\dfrac{\cos^2 \theta}{\sin^2 \theta}\right) = \cos^2 \theta \cot^2 \theta$

20. $(\sec \theta - \tan \theta)^2 = \left(\dfrac{1}{\cos \theta} - \dfrac{\sin \theta}{\cos \theta}\right)^2$

$\qquad = \dfrac{(1 - \sin \theta)^2}{\cos^2 \theta} = \dfrac{(1 - \sin \theta)^2}{1 - \sin^2 \theta}$

$\qquad = \dfrac{(1 - \sin \theta)^2}{(1 - \sin \theta)(1 + \sin \theta)} = \dfrac{1 - \sin \theta}{1 + \sin \theta}$

21. $\dfrac{\tan \theta - \sin \theta}{\csc \theta - \cot \theta} = \dfrac{\dfrac{\sin \theta}{\cos \theta} - \sin \theta}{\dfrac{1}{\sin \theta} - \dfrac{\cos \theta}{\sin \theta}} = \dfrac{\sin^2 \theta - \sin^2 \theta \cos \theta}{\cos \theta - \cos^2 \theta}$

$\qquad = \dfrac{\sin^2 \theta(1 - \cos \theta)}{\cos \theta(1 - \cos \theta)} = \dfrac{\sin^2 \theta}{\cos \theta} = \dfrac{1 - \cos^2 \theta}{\cos \theta}$

$\qquad = \dfrac{1}{\cos \theta} - \cos \theta = \sec \theta - \cos \theta$

22. $\dfrac{\sqrt{5}}{5}$ 23. $2\sqrt{3} - \pi$ square units   24. -2 25. (a) 492π rad/min (b) 533π ft/min (c) $\dfrac{533\pi}{176}$ miles

1. (b)   2. (c)   3. (a)   4. (c)   5. (b)   6. (a)   7. (e)   8. (d)   9. (b)   10. (d)   11. (a)   12. (d)
13. (c)   14. (c)   15. (b)

8.1 The Sine and Cosine Functions (page 398)

1.

x	$-\pi$	$-\dfrac{5\pi}{6}$	$-\dfrac{2\pi}{3}$	$-\dfrac{\pi}{2}$	$-\dfrac{\pi}{3}$	$-\dfrac{\pi}{6}$	0
$y = \sin x$	0	$-\dfrac{1}{2}$	$-\dfrac{\sqrt{3}}{2}$	-1	$-\dfrac{\sqrt{3}}{2}$	$-\dfrac{1}{2}$	0

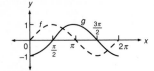

3. Shift the graph of f to the right $\dfrac{\pi}{2}$ units.

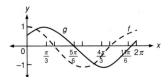

5. Shift the graph of f to the right $\dfrac{\pi}{3}$ units.

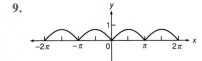

7. Shift the graph of f to the left π units and multiply the ordinates by 2.

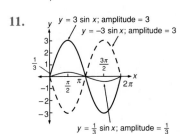

9.

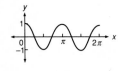

11.

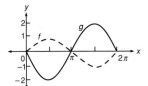

13. Amplitude = 1; period = π. **15.** Amplitude = $\dfrac{3}{2}$; period = $\dfrac{\pi}{2}$. **17.** Amplitude = 1; period = 4π.

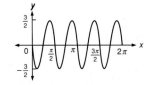

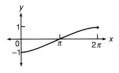

19. $p = 8\pi$

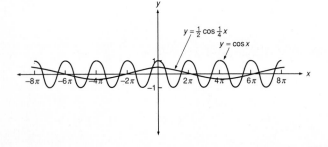

21. $p = 2$

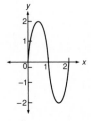

23. $a = 1$, $b = 3$; $y = \sin 3x$ **25.** $a = -3$, $b = \frac{1}{2}$; $y = -3 \sin \frac{1}{2}x$

	amplitude	period	phase shift
27.	5	$\dfrac{2\pi}{3}$	$\dfrac{\pi}{6}$
29.	2	π	$-\dfrac{\pi}{2}$
31.	$\dfrac{3}{2}$	8π	4

33. Amplitude $= 1$; period $= \dfrac{\pi}{2}$; phase shift $= \dfrac{\pi}{4}$.

35. Amplitude $= 2$; period $= \pi$; phase shift $= \dfrac{\pi}{4}$.

37. Amplitude $= \dfrac{5}{2}$; period $= 4\pi$; phase shift $= -\dfrac{\pi}{2}$.

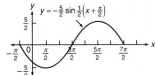

39. Shift the graph of $y = \sin \frac{1}{2}(x + \pi)$ three units upward.

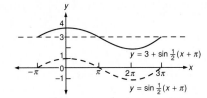

41.

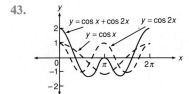

43.

45.

47. $(f \circ g)(x) = 2 \cos x + 5$; $(g \circ f)(x) = \cos(2x + 5)$ **49.** $g(x) = 5x^2$; $f(x) = \cos x$

51. $h(x) = 1 - 2x$; $g(x) = \sqrt[3]{x}$; $f(x) = \cos x$ **53.** $f(x + 2p) = f((x + p) + p) = f(x + p) = f(x)$

8.2 Graphing Other Trigonometric Functions (page 406)

1.

x	-1.4	-1.3	$-\dfrac{\pi}{3}$	$-\dfrac{\pi}{4}$	$-\dfrac{\pi}{6}$	0
$y = \tan x$	-5.8	-3.6	$-\sqrt{3}$	-1	$-\dfrac{\sqrt{3}}{3}$	0

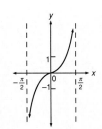

3. (a) $\cot(-x) = \dfrac{\cos(-x)}{\sin(-x)} = \dfrac{\cos x}{-\sin x} = -\dfrac{\cos x}{\sin x} = -\cot x$

(b) $\csc(-x) = \dfrac{1}{\sin(-x)} = \dfrac{1}{-\sin x} = -\dfrac{1}{\sin x} = -\csc x$

(c) $\csc(x + 2\pi) = \dfrac{1}{\sin(x + 2\pi)} = \dfrac{1}{\sin x} = \csc x$

5. Shift the graph of g to the left $\dfrac{\pi}{2}$ units.   **7.** Shift the graph of g to the right $\dfrac{\pi}{3}$ units.

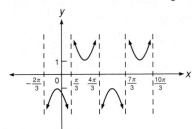

9. $p = \dfrac{\pi}{3}$

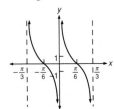

11. $p = 2\pi$

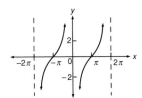

13. $p = \dfrac{\pi}{2}$

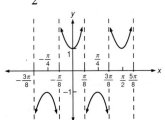

15. $p = 6\pi$

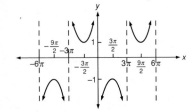

17. $p = \dfrac{4\pi}{3}$

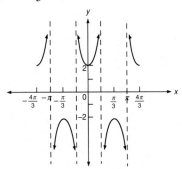

19. Period $= \dfrac{\pi}{2}$; phase shift $= -\dfrac{\pi}{4}$

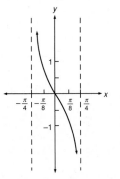

21. Period $= 4\pi$; phase shift $= \dfrac{\pi}{2}$

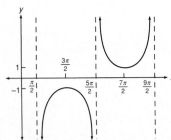

23.

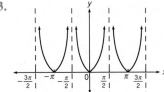

25. $(f \circ g)(x) = \tan^2 x;\ (g \circ f)(x) = \tan x^2$

27. $(f \circ g \circ h)(x) = e^{\sqrt{\sec x}};\ (g \circ f \circ h)(x) = \sqrt{e^{\sec x}};\ (h \circ g \circ f)(x) = \sec \sqrt{e^x}$

29. $h(x) = 2x + 1,\ g(x) = \tan x; f(x) = \sqrt{x}$ **31.** $0 < QP = \sin \theta < \overset{\frown}{AP} = \theta$; divide by θ to get $0 < \dfrac{\sin \theta}{\theta} < 1$

33.

θ	1	.5	.25	.1	.01
$\dfrac{\sin \theta}{\theta}$	.841471	.958851	.989616	.998334	.999983

$\dfrac{\sin \theta}{\theta}$ appears to be getting close to 1 as θ approaches 0.

8.3 The Addition and Subtraction Formulas (page 414)

1. $\cos 60° = \frac{1}{2}$ **3.** $\tan(-30°) = -\dfrac{1}{\sqrt{3}}$

5. $\sin 75° = \sin(45° + 30°) = \sin 45° \cos 30° + \cos 45° \sin 30° = \frac{1}{4}(\sqrt{6} + \sqrt{2})$
$\cos 75° = \cos(45° + 30°) = \cos 45° \cos 30° - \sin 45° \sin 30° = \frac{1}{4}(\sqrt{6} - \sqrt{2})$
$\tan 75° = \tan(45° + 30°) = \dfrac{\tan 45° + \tan 30°}{1 - \tan 45° \tan 30°} = \dfrac{\sqrt{3} + 1}{\sqrt{3} - 1}$

7. $\sin \dfrac{\pi}{12} = \sin\left(\dfrac{\pi}{3} - \dfrac{\pi}{4}\right) = \sin \dfrac{\pi}{3} \cos \dfrac{\pi}{4} - \cos \dfrac{\pi}{3} \sin \dfrac{\pi}{4} = \frac{1}{4}(\sqrt{6} - \sqrt{2})$

$\cos \dfrac{\pi}{12} = \cos\left(\dfrac{\pi}{3} - \dfrac{\pi}{4}\right) = \cos \dfrac{\pi}{3} \cos \dfrac{\pi}{4} + \sin \dfrac{\pi}{3} \sin \dfrac{\pi}{4} = \frac{1}{4}(\sqrt{2} + \sqrt{6})$

$\tan \dfrac{\pi}{12} = \tan\left(\dfrac{\pi}{3} - \dfrac{\pi}{4}\right) = \dfrac{\tan \dfrac{\pi}{3} - \tan \dfrac{\pi}{4}}{1 + \tan \dfrac{\pi}{3} \tan \dfrac{\pi}{4}} = \dfrac{\sqrt{3} - 1}{1 + \sqrt{3}}$

9. $\sin 165° = \sin(135° + 30°) = \sin 135° \cos 30° + \cos 135° \sin 30° = \frac{1}{4}(\sqrt{6} - \sqrt{2})$
$\cos 165° = \cos(135° + 30°) = \cos 135° \cos 30° - \sin 135° \sin 30° = -\frac{1}{4}(\sqrt{6} + \sqrt{2})$
$\tan 165° = \tan(135° + 30°) = \dfrac{\tan 135° + \tan 30°}{1 - \tan 135° \tan 30°} = \dfrac{-\sqrt{3} + 1}{\sqrt{3} + 1}$

11. $\sin \dfrac{17\pi}{12} = \sin\left(\dfrac{7\pi}{6} + \dfrac{\pi}{4}\right) = \sin \dfrac{7\pi}{6} \cos \dfrac{\pi}{4} + \cos \dfrac{7\pi}{6} \sin \dfrac{\pi}{4} = -\frac{1}{4}(\sqrt{2} + \sqrt{6})$

$\cos \dfrac{17\pi}{12} = \cos\left(\dfrac{7\pi}{6} + \dfrac{\pi}{4}\right) = \cos \dfrac{7\pi}{6} \cos \dfrac{\pi}{4} - \sin \dfrac{7\pi}{6} \sin \dfrac{\pi}{4} = \frac{1}{4}(\sqrt{2} - \sqrt{6})$

$\tan \dfrac{17\pi}{12} = \tan\left(\dfrac{7\pi}{6} + \dfrac{\pi}{4}\right) = \dfrac{\tan \dfrac{7\pi}{6} + \tan \dfrac{\pi}{4}}{1 - \tan \dfrac{7\pi}{6} \tan \dfrac{\pi}{4}} = \dfrac{\sqrt{3} + 1}{\sqrt{3} - 1}$

13. $\sin(\pi - \theta) = \sin \pi \cos \theta - \cos \pi \sin \theta = -(-1) \sin \theta = \sin \theta$

15. $\tan(\pi - \theta) = \dfrac{\tan \pi - \tan \theta}{1 + \tan \pi \tan \theta} = \dfrac{0 - \tan \theta}{1 + 0} = -\tan \theta$

17. $\cos(90° - \theta) = \cos 90° \cos \theta + \sin 90° \sin \theta$
$= 0 \cdot \cos \theta + 1 \cdot \sin \theta$
$= \sin \theta$

19. $\sin(\alpha - \beta) = \frac{16}{63};\ \tan(\alpha + \beta) = \frac{56}{33}$ **21.** $\sin(\alpha + \beta) = \frac{2}{15}(\sqrt{42} - 1);\ \cos(\alpha + \beta) = -\frac{1}{15}(\sqrt{21} + 4\sqrt{2})$

23. **(a)** $\cos 191° = \cos(180° + 11°) = -\cos 11° = -.9816;$ **(b)** $\sin 132° = \sin(90° + 42°) = \cos 42° = .7431;$
(c) $\tan 200° = \tan(180° + 20°) = \tan 20° = .3640;$ **(d)** $\cos 102° = \cos(90° + 12°) = -\sin 12° = -.2079$

25. $\cos\left(\theta - \dfrac{\pi}{4}\right) = \cos\theta\cos\dfrac{\pi}{4} + \sin\theta\sin\dfrac{\pi}{4} = \dfrac{\sqrt{2}}{2}(\cos\theta + \sin\theta)$

27. $\cos(\theta + 30°) + \cos(\theta - 30°) = \cos\theta\cos 30° - \sin\theta\sin 30° + \cos\theta\cos 30° + \sin\theta\sin 30°$

$$= 2\cos\theta\cos 30°$$
$$= \sqrt{3}\cos\theta$$

29. $\dfrac{\sin\left(\theta + \dfrac{\pi}{2}\right)}{\cos\left(\theta + \dfrac{\pi}{2}\right)} = \dfrac{\cos\theta}{-\sin\theta} = -\cot\theta$ **31.** $\csc(\pi - \theta) = \dfrac{1}{\sin(\pi - \theta)} = \dfrac{1}{\sin\theta} = \csc\theta$

33. $\sin(\alpha + \beta) - \sin(\alpha - \beta) = \sin\alpha\cos\beta + \cos\alpha\sin\beta - \sin\alpha\cos\beta + \cos\alpha\sin\beta$
$$= 2\cos\alpha\sin\beta$$

35. $\dfrac{\sin(\alpha + \beta)}{\sin(\alpha - \beta)} = \dfrac{\sin\alpha\cos\beta + \cos\alpha\sin\beta}{\sin\alpha\cos\beta - \cos\alpha\sin\beta} = \dfrac{\dfrac{\sin\alpha\cos\beta}{\cos\alpha\cos\beta} + \dfrac{\cos\alpha\sin\beta}{\cos\alpha\cos\beta}}{\dfrac{\sin\alpha\cos\beta}{\cos\alpha\cos\beta} - \dfrac{\cos\alpha\sin\beta}{\cos\alpha\cos\beta}} = \dfrac{\tan\alpha + \tan\beta}{\tan\alpha - \tan\beta}$

37. $2\sin\left(\dfrac{\pi}{4} - \theta\right)\sin\left(\dfrac{\pi}{4} + \theta\right) = 2\left[\sin\dfrac{\pi}{4}\cos\theta - \cos\dfrac{\pi}{4}\sin\theta\right]\left[\sin\dfrac{\pi}{4}\cos\theta + \cos\dfrac{\pi}{4}\sin\theta\right]$

$$= 2\left[\dfrac{\cos\theta}{\sqrt{2}} - \dfrac{\sin\theta}{\sqrt{2}}\right]\left[\dfrac{\cos\theta}{\sqrt{2}} + \dfrac{\sin\theta}{\sqrt{2}}\right]$$

$$= 2\left[\dfrac{\cos^2\theta}{2} - \dfrac{\sin^2\theta}{2}\right] = \cos^2\theta - \sin^2\theta$$

39. $\cos\left(\dfrac{\pi}{2} - x\right) = \sin x;$ **41.** $\sin(\alpha - \beta) = \sin[\alpha + (-\beta)]$

$\cos\left[\dfrac{\pi}{2} - \left(\dfrac{\pi}{2} - \theta\right)\right] = \sin\left(\dfrac{\pi}{2} - \theta\right);$ $\hspace{1.5cm} = \sin\alpha\cos(-\beta) + \cos\alpha\sin(-\beta)$

$\hspace{3cm} = \sin\alpha\cos\beta - \cos\alpha\sin\beta$

$\cos\theta = \sin\left(\dfrac{\pi}{2} - \theta\right)$

43. $\cot(\alpha + \beta) = \dfrac{\cos(\alpha + \beta)}{\sin(\alpha + \beta)}$

$$= \dfrac{\cos\alpha\cos\beta - \sin\alpha\sin\beta}{\sin\alpha\cos\beta + \cos\alpha\sin\beta}$$

$$= \dfrac{\dfrac{\cos\alpha\cos\beta}{\sin\alpha\sin\beta} - \dfrac{\sin\alpha\sin\beta}{\sin\alpha\sin\beta}}{\dfrac{\sin\alpha\cos\beta}{\sin\alpha\sin\beta} + \dfrac{\cos\alpha\sin\beta}{\sin\alpha\sin\beta}} = \dfrac{\cot\alpha\cot\beta - 1}{\cot\beta + \cot\alpha}$$

45. Since $\sin\theta = -\cos\left(\theta + \dfrac{\pi}{2}\right)$, shift $y = \cos\theta$ by $\dfrac{\pi}{2}$ units to the left and reflect through the θ-axis. **47.** $\frac{12}{37}$

49. (a) $\dfrac{f(x + h) - f(x)}{h} = \dfrac{\sin(x + h) - \sin x}{h}$ **(b)** $\dfrac{g(x + h) - g(x)}{h} = \dfrac{\cos(x + h) - \cos x}{h}$

$\hspace{1.2cm} = \dfrac{\sin x\cos h + \cos x\sin h - \sin x}{h}$ $\hspace{1cm} = \dfrac{\cos x\cos h - \sin x\sin h - \cos x}{h}$

$\hspace{1.2cm} = \dfrac{\sin x(\cos h - 1)}{h} + \dfrac{\cos x\sin h}{h}$ $\hspace{1cm} = \dfrac{\cos x(\cos h - 1)}{h} - \dfrac{\sin x\sin h}{h}$

$\hspace{1.2cm} = \left(\dfrac{\cos h - 1}{h}\right)\sin x + \left(\dfrac{\sin h}{h}\right)\cos x$ $\hspace{1cm} = \left(\dfrac{\cos h - 1}{h}\right)\cos x - \left(\dfrac{\sin h}{h}\right)\sin x$

51. $24.2°$ **53.** $121.8°$

8.4 The Double- and Half-Angle Formulas (page 424)

1. $\cos\dfrac{7\pi}{6} = -\dfrac{\sqrt{3}}{2}$ **3.** $3\tan 150° = -\sqrt{3}$ **5.** $\cos 75° = \cos\dfrac{150°}{2} = \sqrt{\dfrac{1 + \cos 150°}{2}}$

$$= \sqrt{\dfrac{1 - \dfrac{\sqrt{3}}{2}}{2}} = \dfrac{1}{2}\sqrt{2 - \sqrt{3}}$$

7. $\tan\dfrac{3\pi}{8} = \dfrac{1 - \cos\dfrac{3\pi}{4}}{\sin\dfrac{3\pi}{4}} = \dfrac{1 + \dfrac{1}{\sqrt{2}}}{\dfrac{1}{\sqrt{2}}} = \sqrt{2} + 1$

9. (a) $\tan 15° = \tan\dfrac{30°}{2} = \dfrac{1 - \cos 30°}{\sin 30°} = \dfrac{1 - \dfrac{\sqrt{3}}{2}}{\dfrac{1}{2}} = 2 - \sqrt{3}$

(b) $\tan 15° = 0.2679$ to four decimal places from Table VI or a calculator.
$2 - \sqrt{3} = 0.2679$ to four decimal places using a calculator.
$2 - \sqrt{3} = 0.268$ using Table I.

11. (a) $\tan 22.5° = \dfrac{1 - \cos 45°}{\sin 45°} = \dfrac{1 - \dfrac{1}{\sqrt{2}}}{\dfrac{1}{\sqrt{2}}} = \sqrt{2} - 1$

(b) $\tan 22.5° = 0.4142$ to four decimal places from Table VI or a calculator.
$\sqrt{2} - 1 = 0.4142$ to four decimal places using a calculator.
$\sqrt{2} - 1 = 0.414$ using Table I.

13. (a) $-\dfrac{240}{289}$; **(b)** $-\dfrac{161}{289}$; **(c)** $\dfrac{240}{161}$ **15.** $-\dfrac{12}{13}$

17. $9\sin 2\theta = 18\sin\theta\cos\theta = 18\left(\dfrac{x}{3}\right)\left(\dfrac{\sqrt{9 - x^2}}{3}\right) = 2x\sqrt{9 - x^2}$

19. $\dfrac{b}{10} = \sin 15° = \sin\dfrac{30°}{2} = \sqrt{\dfrac{1 - \cos 30°}{2}} = \dfrac{\sqrt{2 - \sqrt{3}}}{2}$; $b = 5\sqrt{2 - \sqrt{3}}$

21. Use $\tan 2\theta = \dfrac{2\tan\theta}{1 - \tan^2\theta}$, where $\tan 2\theta = \dfrac{h}{50}$ and $\tan\theta = \dfrac{h}{150}$ to find that $h = 50\sqrt{3}$.

23. $9\sqrt{7}$ **25.** $\cot x\sin 2x = \dfrac{\cos x}{\sin x}\cdot 2\sin x\cos x = 2\cos^2 x$ **27.** $\sec^2\dfrac{x}{2} = \dfrac{1}{\cos^2\dfrac{x}{2}} = \dfrac{1}{\dfrac{1 + \cos x}{2}} = \dfrac{2}{1 + \cos x}$

29. $\cos^2\dfrac{x}{2} = \dfrac{1 + \cos x}{2} = \dfrac{\tan x + \tan x\cos x}{2\tan x} = \dfrac{\tan x + \sin x}{2\tan x}$

31. $\cot\dfrac{x}{2} = \dfrac{1}{\tan\dfrac{x}{2}} = \dfrac{1}{\dfrac{\sin x}{1 + \cos x}} = \dfrac{1 + \cos x}{\sin x} = \dfrac{1}{\sin x} + \dfrac{\cos x}{\sin x} = \csc x + \cot x$

33. $(\sin x + \cos x)^2 = \sin^2 x + 2\sin x\cos x + \cos^2 x = 1 + 2\sin x\cos x = 1 + \sin 2x$

35. $\tan 3x = \tan(2x + x) = \dfrac{\tan 2x + \tan x}{1 - \tan 2x\tan x} = \dfrac{\dfrac{2\tan x}{1 - \tan^2 x} + \tan x}{1 - \dfrac{2\tan x}{1 - \tan^2 x}\cdot\tan x}$

$$= \dfrac{2\tan x + \tan x - \tan^3 x}{1 - \tan^2 x - 2\tan^2 x} = \dfrac{3\tan x - \tan^3 x}{1 - 3\tan^2 x}$$

37. $\cot 2x = \dfrac{1}{\tan 2x} = \dfrac{1}{\dfrac{2\tan x}{1 - \tan^2 x}} = \dfrac{1 - \tan^2 x}{2\tan x}$

$= \dfrac{\dfrac{1}{\tan^2 x} - \dfrac{\tan^2 x}{\tan^2 x}}{\dfrac{2\tan x}{\tan^2 x}} = \dfrac{\cot^2 x - 1}{2\cot x}$

39. $\csc 2x - \cot 2x = \dfrac{1}{\sin 2x} - \dfrac{\cos 2x}{\sin 2x} = \dfrac{1 - \cos 2x}{\sin 2x} = \tan \dfrac{2x}{2} = \tan x$

41. $\cos^4 x - \sin^4 x = (\cos^2 x - \sin^2 x)(\cos^2 x + \sin^2 x) = (\cos^2 x - \sin^2 x) = \cos 2x$

43. $\cos 4x = \cos 2(2x) = 2\cos^2 2x - 1$
$\quad = 2(2\cos^2 x - 1)^2 - 1$
$\quad = 2(4\cos^4 x - 4\cos^2 x + 1) - 1$
$\quad = 8\cos^4 x - 8\cos^2 x + 1$

45. $\cos^4 x = (\cos^2 x)^2 = \left(\dfrac{1 + \cos 2x}{2}\right)^2 = \dfrac{1}{4}(1 + 2\cos 2x + \cos^2 2x)$

$\qquad\qquad\qquad = \dfrac{1}{4}\left(1 + 2\cos 2x + \dfrac{1 + \cos 4x}{2}\right)$

$\qquad\qquad\qquad = \dfrac{1}{8}\cos 4x + \dfrac{1}{2}\cos 2x + \dfrac{3}{8}$

47. (a) $\cos\alpha\cos\beta - \sin\alpha\sin\beta = \cos(\alpha + \beta)$ **(b)** $\cos\alpha\cos\beta + \sin\alpha\sin\beta = \cos(\alpha - \beta)$

$\quad\underline{\cos\alpha\cos\beta + \sin\alpha\sin\beta = \cos(\alpha - \beta)}\qquad\qquad \underline{\cos\alpha\cos\beta - \sin\alpha\sin\beta = \cos(\alpha + \beta)}$

$\quad 2\cos\alpha\cos\beta = \cos(\alpha + \beta) + \cos(\alpha - \beta)\qquad\qquad 2\sin\alpha\sin\beta = \cos(\alpha - \beta) - \cos(\alpha + \beta)$

$\quad \cos\alpha\cos\beta = \tfrac{1}{2}[\cos(\alpha + \beta) + \cos(\alpha - \beta)]\qquad\qquad \sin\alpha\sin\beta = \tfrac{1}{2}[\cos(\alpha - \beta) - \cos(\alpha + \beta)]$

$\qquad\qquad\qquad\qquad\qquad\qquad\qquad\qquad\qquad\qquad \sin\alpha\cos\beta + \cos\alpha\sin\beta = \sin(\alpha + \beta)$

$\qquad\qquad\qquad\qquad\qquad\qquad\qquad\qquad\qquad\qquad \underline{\sin\alpha\cos\beta - \cos\alpha\sin\beta = \sin(\alpha - \beta)}$

$\qquad\qquad\qquad\qquad\qquad\qquad\qquad\qquad\qquad\qquad 2\sin\alpha\cos\beta = \sin(\alpha + \beta) + \sin(\alpha - \beta)$

$\qquad\qquad\qquad\qquad\qquad\qquad\qquad\qquad\qquad\qquad \sin\alpha\cos\beta = \tfrac{1}{2}[\sin(\alpha + \beta) + \sin(\alpha - \beta)]$

$\qquad\qquad\qquad\qquad\qquad\qquad\qquad\qquad\qquad\qquad \sin\alpha\cos\beta + \cos\alpha\sin\beta = \sin(\alpha + \beta)$

$\qquad\qquad\qquad\qquad\qquad\qquad\qquad\qquad\qquad\qquad \underline{\sin\alpha\cos\beta - \cos\alpha\sin\beta = \sin(\alpha - \beta)}$

$\qquad\qquad\qquad\qquad\qquad\qquad\qquad\qquad\qquad\qquad 2\cos\alpha\sin\beta = \sin(\alpha + \beta) - \sin(\alpha - \beta)$

$\qquad\qquad\qquad\qquad\qquad\qquad\qquad\qquad\qquad\qquad \cos\alpha\sin\beta = \tfrac{1}{2}[\sin(\alpha + \beta) - \sin(\alpha - \beta)]$

49. $2\cos\dfrac{u + v}{2}\cos\dfrac{u - v}{2} = \cos\left(\dfrac{u + v}{2} - \dfrac{u - v}{2}\right) + \cos\left(\dfrac{u + v}{2} + \dfrac{u - v}{2}\right) = \cos v + \cos u$

Similarly for the other formulas.

8.5 Trigonometric Equations (page 431)

In the following answers k represents any integer.

1. $2k\pi$ **3.** $\dfrac{\pi}{6} + 2k\pi; \dfrac{5\pi}{6} + 2k\pi$ **5.** $\pi + 2k\pi$ **7.** $\dfrac{7\pi}{12} + k\pi; \dfrac{11\pi}{12} + k\pi$ **9.** No solutions. **11.** $\dfrac{\pi}{6} + \dfrac{k\pi}{2}; \dfrac{2\pi}{3} + \dfrac{k\pi}{2}$

13. $\pi + 2k\pi - 1$ **15.** $\dfrac{2k\pi}{3}$ **17.** $60°, 120°$ **19.** $45°; 315°$ **21.** $90°; 270°$ **23.** $60°; 300°$ **25.** $30°; 150°; 270°$

27. $45°; 225°$ **29.** $0; \dfrac{7\pi}{6}; \dfrac{11\pi}{6}$ **31.** $\dfrac{\pi}{3}; \dfrac{4\pi}{3}$ **33.** $\dfrac{\pi}{2}; \dfrac{3\pi}{2}; \dfrac{\pi}{6}; \dfrac{11\pi}{6}$ **35.** $\dfrac{\pi}{2}$ **37.** $\dfrac{\pi}{6}; \dfrac{5\pi}{6}; \dfrac{\pi}{2}; \dfrac{3\pi}{2}$ **39.** No solutions.

41. $\dfrac{\pi}{3}; \dfrac{5\pi}{3}$ **43.** $0; \dfrac{\pi}{4}; \dfrac{3\pi}{4}; \dfrac{5\pi}{4}; \dfrac{7\pi}{4}; \pi$ **45.** $\dfrac{\pi}{4}; \dfrac{\pi}{2}; \dfrac{3\pi}{4}; \dfrac{5\pi}{4}; \dfrac{3\pi}{2}; \dfrac{7\pi}{4}$ **47.** $\dfrac{\pi}{4}; \dfrac{5\pi}{4}$ **49.** $\dfrac{\pi}{3}; \dfrac{5\pi}{3}$

51. $0; \dfrac{\pi}{6}; \dfrac{5\pi}{6}; \pi$ **53.** $\dfrac{\pi}{6}; \dfrac{5\pi}{6}; \dfrac{7\pi}{6}; \dfrac{11\pi}{6}$ **55.** $0; \dfrac{\pi}{3}; \dfrac{\pi}{2}; \dfrac{2\pi}{3}; \pi; \dfrac{4\pi}{3}; \dfrac{3\pi}{2}; \dfrac{5\pi}{3}$ **57.** $0; \dfrac{\pi}{2}$ **59.** 62.2°; 297.8°

61. 111.4°; 248.6° **63.** 33.7°; 213.7° **65.** 69.3°; 290.7°; 110.7°; 249.3°

67. (a) $\cos \dfrac{\pi}{4} = \dfrac{\sqrt{2}}{2} = .7071 \ldots; \cos \alpha = \dfrac{28}{35} = .8 > .7071 \ldots$. Then $\alpha < \dfrac{\pi}{4}$ since the cosine is decreasing for

$0 \le x \le \dfrac{\pi}{2}$.

(b) $\cos 2\alpha = 2\cos^2 \alpha - 1 = 2\left(\dfrac{4}{5}\right)^2 - 1 = \dfrac{7}{25} = \dfrac{28}{100} = \cos(\alpha + \beta)$. Then, since the cosine is one-to-one for

$0 \le x \le \dfrac{\pi}{2}, 2\alpha = \alpha + \beta$, or $\alpha = \beta$.

69. $\left(\dfrac{3\pi}{4}, \dfrac{\sqrt{2}}{2}\right), \left(\dfrac{7\pi}{4}, -\dfrac{\sqrt{2}}{2}\right)$

8.6 Inverse Trigonometric Functions (page 441)

1. 0 **3.** $-\dfrac{\pi}{2}$ **5.** $-\dfrac{\pi}{4}$ **7.** $\dfrac{3\pi}{4}$ **9.** 1.56 **11.** .59 **13.** 2.23 **15.** $\dfrac{1}{\sqrt{145}}$ **17.** 0 **19.** $\dfrac{\sqrt{3}}{2}$

21. undefined **23.** 0 **25.** Sin and arcsin are inverse functions.

27. Domain: $-1 \le x \le 1$
Range: $-\pi \le y \le \pi$

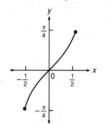

29. Domain; all real x

Range: $2 - \dfrac{\pi}{2} < y < 2 + \dfrac{\pi}{2}$

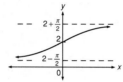

31. Domain: $-\dfrac{1}{2} \le x \le \dfrac{1}{2}$

Range: $-\dfrac{\pi}{4} \le y \le \dfrac{\pi}{4}$

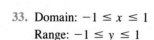

33. Domain: $-1 \le x \le 1$
Range: $-1 \le y \le 1$

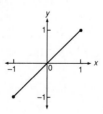

35. $-\dfrac{7}{11}$ **37.** $\dfrac{220}{221}$ **39.** $\dfrac{7}{9}$ **41.** $-\dfrac{3}{2}$ **43.** -2

45. Let $y = \arcsin x$. Then $x = \sin y; -x = -\sin y = \sin(-y); -y = \arcsin(-x); -\arcsin x = \arcsin(-x)$.

47. Using the addition formula for tan gives $\theta = \arctan(.25) = .24$

49. (a) $\sin \theta = x$ $\qquad$ $\csc \theta = \dfrac{1}{x}$ $\qquad$ **(b)** $\sin \theta = \dfrac{x}{\sqrt{1 + x^2}}$ $\qquad$ $\csc \theta = \dfrac{\sqrt{1 + x^2}}{x}$

$\cos \theta = \sqrt{1 - x^2}$ $\qquad$ $\sec \theta = \dfrac{1}{\sqrt{1 - x^2}}$ $\qquad$ $\cos \theta = \dfrac{1}{\sqrt{1 + x^2}}$ $\qquad$ $\sec \theta = \sqrt{1 + x^2}$

$\tan \theta = \dfrac{x}{\sqrt{1 - x^2}}$ $\qquad$ $\cos \theta = \dfrac{\sqrt{1 - x^2}}{x}$ $\qquad$ $\tan \theta = x$ $\qquad$ $\cot \theta = \dfrac{1}{x}$

51. $(f \circ g)(x) = \arcsin(3x + 2); (g \circ f)(x) = 3\arcsin x + 2$ **53.** $g(x) = \arccos x; f(x) = \ln x$

CHAPTER 8 TEST: STANDARD ANSWER (page 445)

1. Amplitude = 2; period = π.

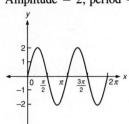

2. (a) Increasing on $\left(-\dfrac{\pi}{2}, \dfrac{\pi}{2}\right)$;

   concave down on $\left(-\dfrac{\pi}{2}, 0\right)$;

   concave up on $\left(0, \dfrac{\pi}{2}\right)$

   (b) Shift the graph of $y = \tan x$ a length of $\dfrac{\pi}{2}$ to the left.

3. Domain: all $x \neq \dfrac{\pi}{2} + k\pi$

   Range: all $y \geq 1$ or $y \leq -1$

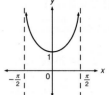

4. $-\frac{1}{4}(\sqrt{6} + \sqrt{2})$ 5. $-\frac{1}{2}\sqrt{2 - \sqrt{3}}$ 6. $\dfrac{2\sqrt{8}}{9} = \dfrac{4\sqrt{2}}{9}$ 7. $\sqrt{3}$

8. $\sin\left(\dfrac{\pi}{4} - \theta\right) = \sin\dfrac{\pi}{4}\cos\theta - \cos\dfrac{\pi}{4}\sin\theta = \dfrac{\sqrt{2}}{2}\cos\theta - \dfrac{\sqrt{2}}{2}\sin\theta = \dfrac{\sqrt{2}}{2}(\cos\theta - \sin\theta)$

9. $\csc 4x = \dfrac{1}{\sin 4x} = \dfrac{1}{2\sin 2x \cos 2x} = \dfrac{1}{4\sin x \cos x \cos 2x} = \dfrac{1}{4}\csc x \sec x \sec 2x$

10. $\sin^2\dfrac{x}{2} = \dfrac{1 - \cos x}{2} = \dfrac{\tan x - \tan x \cos x}{2\tan x} = \dfrac{\tan x - \sin x}{2\tan x}$ 11. $\dfrac{\pi}{6}; \dfrac{5\pi}{6}; \dfrac{3\pi}{2}$

12. $x = \begin{cases} -\dfrac{\pi}{3} + k\pi \\ \dfrac{\pi}{3} + k\pi \end{cases}$, k any integer

13. $a = \frac{1}{2}, b = 3, y = \frac{1}{2}\cos 3x$ 14. $p = \dfrac{\pi}{2}; x = \dfrac{\pi}{4}; x = \dfrac{3\pi}{4}$ 15. $\frac{56}{65}$ 16. $8(2 + \sqrt{3})$

17. (a) Domain: all reals

   Range: $-\dfrac{\pi}{2} < y < \dfrac{\pi}{2}$

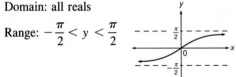

   (b) Symmetric through the origin;

   $y = \pm\dfrac{\pi}{2}$

18. (a) $\dfrac{\pi}{4}$; (b) π; (c) $-\dfrac{\pi}{6}$ 19. (a) .57; (b) 1.76; (c) $-.23$ 20. (a) $\frac{2}{7}\sqrt{10}$; (b) $\frac{1}{2}$; (c) x^2 21. $x = -\frac{3}{8}$

22. $h(x) = 2x$, $g(x) = \csc x$, $f(x) = x^3$ 23. Period = $\dfrac{2\pi}{3}$; amplitude = $\dfrac{1}{2}$; phase shift = $\dfrac{\pi}{6}$

24. Period = π; amplitude = 2; phase shift = $-\dfrac{\pi}{2}$ 25. $b = 18$

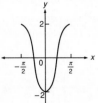

CHAPTER 8 TEST: MULTIPLE CHOICE (page 447)

1. (d)   2. (a)   3. (d)   4. (b)   5. (e)   6. (a)   7. (a)   8. (c)   9. (c)   10. (d)   11. (c)   12. (b)   13. (a)
14. (e)   15. (b)

CHAPTER 9: ADDITIONAL APPLICATIONS OF TRIGONOMETRY

9.1 The Law of Cosines (page 455)

1. $a = \sqrt{93}$ **3.** $\alpha = 30°$ **5.** $a = 6.1$; $\beta = 26.3°$ **7.** $\gamma = 90°$; $\alpha = 67.4°$ **9.** $\beta = 37°$; $\gamma = 9.2°$
11. $c = 11.2$; $\alpha = 125.7°$ **13.** $a = 5.0$; $\beta = 17.3°$ **15.** 151 meters   **17.** 4.8 miles   **19.** 22.6 centimeters
21. 42.7   **23.** 111 feet   **25.** 33.6 miles   **27.** 112.5   **29.** 34.4   **31.** 24   **33.** 25   **35.** 9.1   **37.** 435

9.2 The Law of Sines (page 465)

1. $a = c = 18\sqrt{3}$ **3.** $\gamma = \dfrac{\pi}{6}$ **5.** $\beta = 100°$; $a = 5.1$; $c = 10.0$ **7.** $\gamma = 67.8°$ $a = 5.8$ $c = 6.0$
9. No solution   **11.** One solution   **13.** No solution   **15.** $\beta = 86.7°$, $c = 9.7$ or $\beta = 93.3°$, $c = 8.3$
17. No solution   **19.** $\gamma = 27.6°$ or $\gamma = 152.4°$ **21.** 133.8°   **23.** 1.6 miles   **25.** 454 feet   **27.** 11.7 centimeters
29. 16   **31.** 724 meters   **33.** 26 pounds; 23°   **35.** 42 pounds; 63°   **37.** 197 mph; N 22° E   **39.** 157 pounds
41. 1150 pounds

9.3 Trigonometric Form of Complex Numbers (page 474)

1.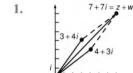

3.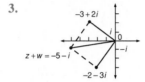

5.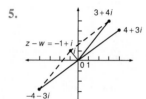

7.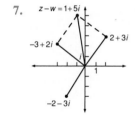

9.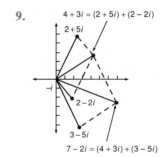

11.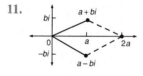

13.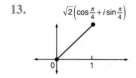

15.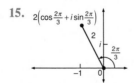

17.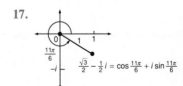

19.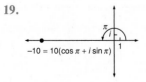

21.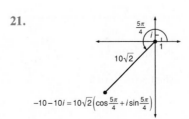

23. $-\dfrac{3}{2} + \dfrac{3\sqrt{3}}{2}i$ 25. $-5i$ 27. $\dfrac{\sqrt{2}}{2} + \dfrac{\sqrt{6}}{2}i$ 29. $2\sqrt{2} - 2\sqrt{2}i$ 31. $.9848 - .1736i$

33. $zw = 5(\cos 150° + i \sin 150°) = -\dfrac{5\sqrt{3}}{2} + \dfrac{5}{2}i$

$\dfrac{z}{w} = \dfrac{1}{20}(\cos 50° + i \sin 50°) = .0321 + .0383i$

$\dfrac{w}{z} = 20(\cos 310° + i \sin 310°) = 12.8558 - 15.3209i$

35. $zw = 8\sqrt{2}(\cos \pi + i \sin \pi) = -8\sqrt{2}$

$\dfrac{z}{w} = \dfrac{\sqrt{2}}{8}\left(\cos \dfrac{3\pi}{2} + i \sin \dfrac{3\pi}{2}\right) = -\dfrac{\sqrt{2}}{8}i$

$\dfrac{w}{z} = 4\sqrt{2}\left(\cos \dfrac{\pi}{2} + i \sin \dfrac{\pi}{2}\right) = 4\sqrt{2}i$

37. $zw = \left[7\left(\cos \dfrac{\pi}{2} + i \sin \dfrac{\pi}{2}\right)\right]\left[2\left(\cos \dfrac{\pi}{6} + i \sin \dfrac{\pi}{6}\right)\right] = 14\left(\cos \dfrac{2\pi}{3} + i \sin \dfrac{2\pi}{3}\right) = -7 + 7\sqrt{3}i$

$\dfrac{z}{w} = \dfrac{7}{2}\left(\cos \dfrac{\pi}{3} + i \sin \dfrac{\pi}{3}\right) = \dfrac{7}{4} + \dfrac{7\sqrt{3}}{4}i$

39. $zw = \left[4\left(\cos \dfrac{2\pi}{3} + i \sin \dfrac{2\pi}{3}\right)\right]\left[4\left(\cos \dfrac{7\pi}{6} + i \sin \dfrac{7\pi}{6}\right)\right] = 16\left(\cos \dfrac{11\pi}{6} + i \sin \dfrac{11\pi}{6}\right) = 8\sqrt{3} - 8i$

$\dfrac{z}{w} = \cos \dfrac{3\pi}{2} + i \sin \dfrac{3\pi}{2} = -i$

41. $zw = \left[\sqrt{2}\left(\cos \dfrac{3\pi}{4} + i \sin \dfrac{3\pi}{4}\right)\right]\left[\sqrt{2}\left(\cos \dfrac{7\pi}{4} + i \sin \dfrac{7\pi}{4}\right)\right] = 2\left(\cos \dfrac{\pi}{2} + i \sin \dfrac{\pi}{2}\right) = 2i$

$\dfrac{z}{w} = \cos \pi + i \sin \pi = -1$

43. $\dfrac{\sqrt{2}}{2}(\cos 90° + i \sin 90°) = \dfrac{\sqrt{2}}{2}i$ 45. $\dfrac{3}{2}(\cos 199° + i \sin 199°) = -1.4183 - .4884i$

47. $|\bar{z}| = |\overline{a + bi}| = |a - bi|$

$\qquad = \sqrt{a^2 + (-b)^2} = \sqrt{a^2 + b^2} = |z|$

49. $\quad |zw| = |(a + bi)(c + di)| = |(ac - bd) + (bc + ad)i|$

$\qquad = \sqrt{(ac - bd)^2 + (bc + ad)^2} = \sqrt{a^2c^2 + b^2d^2 + b^2c^2 + a^2d^2}$

$|z||w| = |a + bi||c + di| = \sqrt{a^2 + b^2}\sqrt{c^2 + d^2}$

$\qquad = \sqrt{(a^2 + b^2)(c^2 + d^2)} = \sqrt{a^2c^2 + b^2d^2 + b^2c^2 + a^2d^2}$

51. Circle: $(x - 1)^2 + y^2 = 4$ 53. Circle: $(x - 2)^2 + (y - 3)^2 = 1$ 55. Ellipse: $\dfrac{x^2}{16} + \dfrac{y^2}{12} = 1$

9.4 De Moivre's Theorem (page 481)

1. $\cos 60° + i \sin 60° = \dfrac{1}{2} + \dfrac{\sqrt{3}}{2}i$ 3. $\cos 320° + i \sin 320° = .7660 - .6428i$

5. $\dfrac{1}{64}\left(\cos \dfrac{3\pi}{4} + i \sin \dfrac{3\pi}{4}\right) = -\dfrac{\sqrt{2}}{128} + \dfrac{\sqrt{2}}{128}i$ 7. $4(\cos 5\pi + i \sin 5\pi) = -4$ 9. $216\left(\cos \dfrac{9\pi}{2} + i \sin \dfrac{9\pi}{2}\right) = 216i$

11. $\cos \dfrac{35\pi}{2} + i \sin \dfrac{35\pi}{2} = -i$ 13. $\cos 10\pi + i \sin 10\pi = 1$

15. $3\left(\cos \dfrac{\pi}{45} + i \sin \dfrac{\pi}{45}\right)$, $3\left(\cos \dfrac{31\pi}{45} + i \sin \dfrac{31\pi}{45}\right)$, $3\left(\cos \dfrac{61\pi}{45} + i \sin \dfrac{61\pi}{45}\right)$

17. $2\left(\cos \dfrac{\pi}{40} + i \sin \dfrac{\pi}{40}\right)$, $2\left(\cos \dfrac{17\pi}{40} + i \sin \dfrac{17\pi}{40}\right)$, $2\left(\cos \dfrac{33\pi}{40} + i \sin \dfrac{33\pi}{40}\right)$, $2\left(\cos \dfrac{49\pi}{40} + i \sin \dfrac{49\pi}{40}\right)$,

$2\left(\cos \dfrac{65\pi}{40} + i \sin \dfrac{65\pi}{40}\right)$

19. 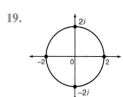   **21.** 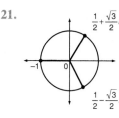   **23.** 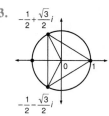   **25.**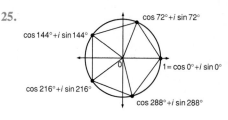

27. $\dfrac{3}{2} + \dfrac{3\sqrt{3}}{2} i$, -3, $\dfrac{3}{2} - \dfrac{3\sqrt{3}}{2} i$ **29.** $2i$, $-\sqrt{3} - i$, $\sqrt{3} - i$

31. $2^{1/6}\left(\cos \dfrac{5\pi}{12} + i \sin \dfrac{5\pi}{12}\right)$, $2^{1/6}\left(\cos \dfrac{13\pi}{12} + i \sin \dfrac{13\pi}{12}\right)$, $2^{1/6}\left(\cos \dfrac{21\pi}{12} + i \sin \dfrac{21\pi}{12}\right)$

33. (a) For $k = 0$, the formula gives $r^{1/n}\left(\cos \dfrac{\theta}{n} + i \sin \dfrac{\theta}{n}\right)$

For $k = n$, the formula gives

$r^{1/n}\left(\cos \dfrac{\theta + 2n\pi}{n} + i \sin \dfrac{\theta + 2n\pi}{n}\right) = r^{1/n}\left[\cos\left(\dfrac{\theta}{n} + 2\pi\right) + i \sin\left(\dfrac{\theta}{n} + 2\pi\right)\right] = r^{1/n}\left(\cos \dfrac{\theta}{n} + i \sin \dfrac{\theta}{n}\right)$

(b) For $k = 1$, the formula gives $r^{1/n}\left(\cos \dfrac{\theta + 2\pi}{n} + i \sin \dfrac{\theta + 2\pi}{n}\right)$

For $k = n + 1$, the formula gives

$r^{1/n}\left(\cos \dfrac{\theta + 2(n + 1)\pi}{n} + i \sin \dfrac{\theta + 2(n + 1)\pi}{n}\right) = r^{1/n}\left[\cos\left(\dfrac{\theta + 2\pi}{n} + 2\pi\right) + i \sin\left(\dfrac{\theta + 2\pi}{n} + 2\pi\right)\right]$

$= r^{1/n}\left(\cos \dfrac{\theta + 2\pi}{n} + i \sin \dfrac{\theta + 2\pi}{n}\right)$

(c) Use k in the formula to get $r^{1/n}\left(\cos \dfrac{\theta + 2k\pi}{n} + i \sin \dfrac{\theta + 2k\pi}{n}\right)$

Use $k + n$ in the formula to get

$r^{1/n}\left(\cos \dfrac{\theta + 2(n + k)\pi}{n} + i \sin \dfrac{\theta + 2(n + k)\pi}{n}\right) = r^{1/n}\left[\cos\left(\dfrac{\theta + 2k\pi}{n} + 2\pi\right) + i \sin\left(\dfrac{\theta + 2k\pi}{n} + 2\pi\right)\right]$

$= r^{1/n}\left(\cos \dfrac{\theta + 2k\pi}{n} + i \sin \dfrac{\theta + 2k\pi}{n}\right)$

35. (a) $[r(\cos \theta + i \sin \theta)]^{-n} = r^{-n}(\cos \theta + i \sin \theta)^{-n} = r^{-n}[(\cos \theta + i \sin \theta)^{-1}]^{n}$

$= r^{-n}[(\cos \theta - i \sin \theta)]^{n}$ (by Exercise 34)

$= r^{-n}[\cos(-\theta) + i \sin(-\theta)]^{n}$

$= r^{-n}[\cos(-n\theta) + i \sin(-n\theta)]$ (since $n > 0$)

$= r^{-n}(\cos n\theta - i \sin n\theta)$

(b) Assuming the rule $z^{0} = 1$, the formula holds for $n = 0$. The formula is given in Section 9.4 for positive integers n. If in part (a) we let $m = -n$, where m is a negative integer, then

$[r(\cos \theta + i \sin \theta)]^{m} = [r(\cos \theta + i \sin \theta)]^{-n} = r^{-n}(\cos n\theta - i \sin n\theta)$ (by part (a))

$= r^{-n}[\cos(-n\theta) + i \sin(-n\theta)]$

$= r^{m}[\cos(m\theta) + i \sin(m\theta)]$

which shows that the formula holds for negative integers.

37. $1 + z + z^{2} = 1 + \cos \dfrac{2\pi}{3} + i \sin \dfrac{2\pi}{3} + \cos \dfrac{4\pi}{3} + i \sin \dfrac{4\pi}{3} = 1 - \dfrac{1}{2} + i\dfrac{\sqrt{3}}{2} - \dfrac{1}{2} - i\dfrac{\sqrt{3}}{2} = 0$

9.5 Polar Coordinates (page 486)

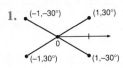
1.

3.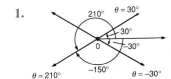
$(-4, 0°)$ 0 $(4, 0°)$

5.
$\left(\frac{1}{2}, \pi\right)$ 0 $\left(-\frac{1}{2}, \pi\right)$
$\left(\frac{1}{2}, -\pi\right)$ $\left(-\frac{1}{2}, -\pi\right)$

7. **(a)** $(1, 405°)$; **(b)** $(1, -315°)$; **(c)** $(-1, 225°)$; **(d)** $(-1, -135°)$

9. **(a)** $\left(3, \frac{19\pi}{6}\right)$; **(b)** $\left(3, -\frac{5\pi}{6}\right)$; **(c)** $\left(-3, \frac{\pi}{6}\right)$; **(d)** $\left(-3, -\frac{11\pi}{6}\right)$ 11. $(0, -7)$ 13. $\left(0, -\frac{3}{2}\right)$

15. $\left(\frac{2\sqrt{2}}{5}, \frac{2\sqrt{2}}{5}\right)$ 17. $(2, 0)$ 19. $(2, \pi)$ 21. $\left(6\sqrt{2}, \frac{5\pi}{4}\right)$ 23. $\left(4, \frac{\pi}{4}\right)$ 25. $\left(2, \frac{11\pi}{6}\right)$

27. $(-7, \pi)$ 29. $\left(-2, \frac{11\pi}{6}\right)$ 31. $\left(-\sqrt{6}, \frac{5\pi}{4}\right)$ 33. $(10, -90°)$

35. $r \cos \theta = 0 \cos \theta = 0 = x$; $r \sin \theta = 0 \sin \theta = 0 = y$

37. Using the polar coordinates $(-r, \theta + \pi)$, $x = -r \cos(\theta + \pi)$ and $y = -r \sin(\theta + \pi)$ since $-r > 0$. Then

$$x = -r(\cos \theta \cos \pi - \sin \theta \sin \pi) = -r(-\cos \theta) = r \cos \theta$$
$$y = -r(\sin \theta \cos \pi + \cos \theta \sin \pi) = -r(-\sin \theta) = r \sin \theta$$

9.6 Graphing Polar Equations (page 495)

1.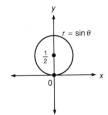

3.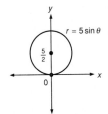

5.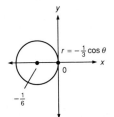

7. $\theta = -6°$ 9. $\theta = \frac{\pi}{6}$

11. $x^2 + (y - \frac{1}{2})^2 = \frac{1}{4}$

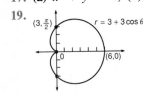

13. $x^2 + (y - \frac{5}{2})^2 = \frac{25}{4}$

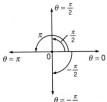

15. $(x + \frac{1}{6})^2 + y^2 = \frac{1}{36}$

17. **(a)** $x^2 + y^2 = 4$; **(b)** $x^2 + y^2 = 4$; **(c)** $x^2 + y^2 = 5$

19.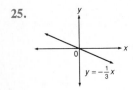

21. $r = 2(1 - \cos \theta)$

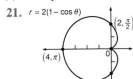

23.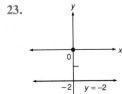

25.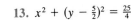

27.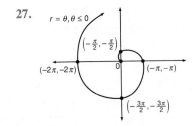

29.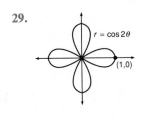

31.
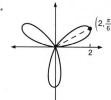

33.
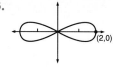

35.
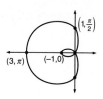

37. $y^2 = 16 - 8x$ (parabola) **39.** $x = -5$ (vertical line) **41.** $(x - 3)^2 + (y - 4)^2 = 25$ (circle)

43. $2x + 3y = 3$ (line) **45.** $x + \sqrt{3}y = 2\sqrt{3}$ (line)

47. Let $P(r, \theta)$ be a point on the line $x = c \neq 0$ and draw OP. From the reference triangle, $\cos \theta = \dfrac{x}{r} = \dfrac{c}{r}$, so that

$r \cos \theta = c$. If $c = 0$, then (r, θ) is on ray $\dfrac{\pi}{2}$ so that $r \cos \dfrac{\pi}{2} = 0$.

49. $\left(\dfrac{3}{2}, \dfrac{\pi}{3}\right), \left(\dfrac{3}{2}, \dfrac{5\pi}{3}\right)$ **51.** $\left(0, \dfrac{\pi}{2}\right), \left(\dfrac{\sqrt{3}}{2}, \dfrac{\pi}{6}\right), \left(-\dfrac{\sqrt{3}}{2}, \dfrac{5\pi}{6}\right)$ **53.** $\left(1, -\dfrac{\pi}{3}\right)$

CHAPTER 9 TEST: STANDARD ANSWER (page 499)

1. (a) $4\sqrt{7}, \dfrac{\pi}{3}$; **(b)** $\sin B = \dfrac{8 \sin 60°}{4\sqrt{7}} = \dfrac{\sqrt{21}}{7}$

2. (a) None; **(b)** two **3.** $60\sqrt{3}$ cm² **4.** $\dfrac{20\sqrt{6}}{3}$ **5.** 105 cm² **6.** 31 cm² **7.** $\dfrac{24}{\sqrt{3}}$ or $8\sqrt{3}$

8. $\alpha = 37.4°, \beta = 118.8°$ **9.** Two solutions: $\beta = 53.1°, \gamma = 96.9°, c = 19.9; \beta = 126.9°, \gamma = 23.1°, c = 7.9$

10. 313 pounds **11.** 15 mph **12.** N 22° W **13.** $-5 + 5i$

14. $z = \dfrac{1}{2}\left(\cos \dfrac{\pi}{6} + i \sin \dfrac{\pi}{6}\right), w = 2\left(\cos \dfrac{2\pi}{3} + i \sin \dfrac{2\pi}{3}\right), \dfrac{z}{w} = \dfrac{1}{4}\left(\cos \dfrac{3\pi}{2} + i \sin \dfrac{3\pi}{2}\right) = -\dfrac{1}{4}i$

15. $\dfrac{1}{32}\left(\cos \dfrac{5\pi}{6} + i \sin \dfrac{5\pi}{6}\right) = -\dfrac{\sqrt{3}}{64} + \dfrac{1}{64}i$ **16.**

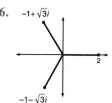

17. $2\left(\cos \dfrac{3\pi}{16} + i \sin \dfrac{3\pi}{16}\right), 2\left(\cos \dfrac{11\pi}{16} + i \sin \dfrac{11\pi}{16}\right), 2\left(\cos \dfrac{19\pi}{16} + i \sin \dfrac{19\pi}{16}\right), 2\left(\cos \dfrac{27\pi}{16} + i \sin \dfrac{27\pi}{16}\right)$

18. (a) $(-3, -300°)$; **(b)** $(3, 240°)$. **19. (a)** $\left(\dfrac{\sqrt{3}}{6}, -\dfrac{1}{6}\right)$; **(b)** $\left(14, \dfrac{3\pi}{4}\right)$. **20.** $r = -6 \cos \theta$

21. $x^2 + (y - 1)^2 = 1$; a circle with radius 1 and center $(0, 1)$ in rectangular coordinates.

22. $\left(x + \dfrac{3}{2}\right)^2 + (y - 2)^2 = \dfrac{25}{4}$; a circle with radius $\dfrac{5}{2}$ and center $\left(-\dfrac{3}{2}, 2\right)$ in rectangular coordinates.

23.

24.
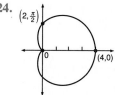

25.
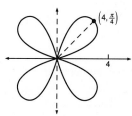

CHAPTER 9 TEST: MULTIPLE CHOICE (page 500)

1. (e) **2. (d)** **3. (b)** **4. (b)** **5. (c)** **6. (a)** **7. (c)** **8. (d)** **9. (c)** **10. (a)** **11. (b)** **12. (d)** **13. (c)**

14. (a) **15. (b)**

CHAPTER 10: LINEAR SYSTEMS, MATRICES, AND DETERMINANTS

10.1 Solving Linear Systems Using Matrices (page 508)

1. $(-4, -1)$ **3.** $(0, 9)$ **5.** $(\frac{1}{8}, \frac{1}{2})$ **7.** $(1, 0)$ **9.** $(2 - \frac{3}{2}c, c)$ for all c, or $(d, \frac{4}{3} - \frac{2}{3}d)$ for all d
11. No solutions **13.** $(5, 2, -1)$ **15.** $(3, 4, 2)$ **17.** No solutions
19. $(\frac{3}{11} + \frac{5}{11}c, -\frac{4}{11} + \frac{8}{11}c, c)$ for all c, or $(d, -\frac{4}{5} + \frac{8}{5}d, -\frac{3}{5} + \frac{11}{5}d)$ for all d, or $(\frac{1}{2} + \frac{5}{8}e, e, \frac{1}{2} + \frac{11}{8}e)$ for all e
21. $(-1, 3, 4, -2)$ **23.** No solutions

10.2 Matrix Algebra (page 515)

1. $\begin{bmatrix} 6 & 1 \\ 0 & 8 \end{bmatrix}$ **3.** Does not exist **5.** $\begin{bmatrix} 3 & -1 & 4 \\ -7 & 4 & 2 \\ -2 & -3 & 5 \end{bmatrix}$ **7.** $[-7]$ **9.** $\begin{bmatrix} 2 & 6 \\ 9 & 10 \end{bmatrix}$ **11.** $\begin{bmatrix} 7 \\ -11 \end{bmatrix}$

13. $\begin{bmatrix} 21 & -11 & 2 \\ 17 & -17 & 19 \end{bmatrix}$ **15.** Does not exist **17.** $\begin{bmatrix} 2 & -5 \\ 4 & 3 \\ 4 & 2 \end{bmatrix}$ **19.** $\begin{bmatrix} 1 & 2 & -15 & 4 \\ 27 & -6 & -15 & 8 \end{bmatrix}$ **21.** $\begin{bmatrix} 0 \\ 0 \\ 0 \end{bmatrix}$ **23.** H

25. $B + C = \begin{bmatrix} 0 + 3 & 6 + 7 \\ 5 - 1 & -4 + (-2) \end{bmatrix} = \begin{bmatrix} 3 & 13 \\ 4 & -6 \end{bmatrix}$; $C + B = \begin{bmatrix} 3 + 0 & 7 + 6 \\ -1 + 5 & -2 + (-4) \end{bmatrix} = \begin{bmatrix} 3 & 13 \\ 4 & -6 \end{bmatrix}$

27. $A + (B + C) = A + \begin{bmatrix} 3 & 13 \\ 4 & -6 \end{bmatrix} = \begin{bmatrix} 5 & 14 \\ 0 & -5 \end{bmatrix}$; $(A + B) + C = \begin{bmatrix} 2 & 7 \\ 1 & -3 \end{bmatrix} + C = \begin{bmatrix} 5 & 14 \\ 0 & -5 \end{bmatrix}$

29. $(A + B)C = \begin{bmatrix} 2 & 7 \\ 1 & -3 \end{bmatrix} C = \begin{bmatrix} -1 & 0 \\ 6 & 13 \end{bmatrix}$; $AC + BC = \begin{bmatrix} 5 & 12 \\ -13 & -30 \end{bmatrix} + \begin{bmatrix} -6 & -12 \\ 19 & 43 \end{bmatrix} = \begin{bmatrix} -1 & 0 \\ 6 & 13 \end{bmatrix}$

31. $(AB)C = \begin{bmatrix} -5 & 4 & 2 \\ 3 & 21 & 33 \\ 8 & 4 & 12 \end{bmatrix} C = \begin{bmatrix} -31 \\ 6 \\ 44 \end{bmatrix}$; $A(BC) = A\begin{bmatrix} -9 \\ 11 \end{bmatrix} = \begin{bmatrix} -31 \\ 6 \\ 44 \end{bmatrix}$

33. (a) $\begin{bmatrix} d & e & f \\ a & b & c \\ g & h & i \end{bmatrix}$ **(b)** $F = \begin{bmatrix} 0 & 0 & 1 \\ 0 & 1 & 0 \\ 1 & 0 & 0 \end{bmatrix}$ **35.** $E = \begin{bmatrix} 1 & 0 & 0 \\ 0 & 1 & 0 \\ 3 & 0 & 1 \end{bmatrix}$ $F = \begin{bmatrix} 1 & 0 & 0 \\ 0 & 1 & 0 \\ 0 & -1 & 1 \end{bmatrix}$

37. (a) $x = -2$, $y = -3$ **(b)** If there were such x, y, then $\begin{bmatrix} 2 & 2 \\ x & x \end{bmatrix} = \begin{bmatrix} 0 & 0 \\ 0 & 0 \end{bmatrix}$, which cannot be true since $2 \neq 0$.

39. (a) $\begin{bmatrix} x^2 & 0 & 0 \\ 0 & y^2 & 0 \\ 0 & 0 & z^2 \end{bmatrix}$ **(b)** $\begin{bmatrix} x^3 & 0 & 0 \\ 0 & y^3 & 0 \\ 0 & 0 & z^3 \end{bmatrix}$ **(c)** $\begin{bmatrix} x^n & 0 & 0 \\ 0 & y^n & 0 \\ 0 & 0 & z^n \end{bmatrix}$

10.3 Solving Linear Systems Using Inverses (page 524)

1. $\begin{bmatrix} 0 & \frac{1}{2} \\ -1 & 2 \end{bmatrix}$ **3.** Does not exist **5.** $\begin{bmatrix} -\frac{4}{7} & -\frac{5}{7} \\ -\frac{3}{7} & -\frac{2}{7} \end{bmatrix}$ **7.** $\begin{bmatrix} -1 & 2 \\ 2 & -1 \end{bmatrix}$ **9.** $\begin{bmatrix} 0 & \frac{1}{b} \\ \frac{1}{a} & 0 \end{bmatrix}$ **11.** $A^{-1} = A$

13. $\begin{bmatrix} -\frac{1}{2} & \frac{3}{4} & \frac{1}{4} \\ -1 & \frac{1}{2} & \frac{1}{2} \\ \frac{3}{4} & -\frac{3}{8} & -\frac{1}{8} \end{bmatrix}$ **15.** Does not exist **17.** $\begin{bmatrix} 1 & 0 & 2 \\ 2 & -1 & 3 \\ 4 & 1 & 8 \end{bmatrix}$ **19.** $\begin{bmatrix} -5 & 0 & \frac{5}{2} & 3 \\ 2 & 0 & -\frac{1}{2} & -1 \\ -\frac{3}{2} & \frac{1}{2} & \frac{3}{4} & 1 \\ 2 & 0 & -1 & -1 \end{bmatrix}$ **21.** $3x + y = 9$
$2x - 2y = 14$

23. $(6, 1)$ **25.** $(18, -21)$ **27.** $(-8, -8, 6)$ **29.** $(-\frac{1}{2}, -\frac{1}{2}, -2)$ **31.** $(13, 7, -6)$ **33.** $(14, -4, 2, -4)$

35. $(AB)^{-1} = \begin{bmatrix} 1 & 2 \\ -4 & 8 \end{bmatrix}^{-1} = \begin{bmatrix} \frac{1}{2} & -\frac{1}{8} \\ \frac{1}{4} & \frac{1}{16} \end{bmatrix}$; $B^{-1}A^{-1} = \begin{bmatrix} 1 & 1 \\ \frac{1}{2} & 0 \end{bmatrix}\begin{bmatrix} \frac{1}{2} & \frac{1}{8} \\ 0 & -\frac{1}{4} \end{bmatrix} = \begin{bmatrix} \frac{1}{2} & -\frac{1}{8} \\ \frac{1}{4} & \frac{1}{16} \end{bmatrix}$

37. (a)
$$AB = I$$
$$C(AB) = CI$$
$$(CA)B = C$$
$$IB = C$$
$$B = C$$

(b) The inverse of a matrix is unique.

39. (a) $(A^2)^{-1} = (AA)^{-1} = A^{-1}A^{-1}$ by Ex. 38(b)
$$= (A^{-1})^2$$

(b) $(A^n)^{-1} = (A^{-1})^n$

10.4 Introduction to Determinants (page 529)

1. 17   **3.** 94   **5.** -60 **7.** 0   **9.** $(-1, 2)$ **11.** $(3, 2)$ **13.** $(1, -1)$ **15.** $(12, 24)$ **17.** $(5, 5)$

19. $(-5, -7)$ **21.** $(\frac{6}{5}, -\frac{3}{10})$ **23.** Inconsistent   **25.** Dependent   **27.** Inconsistent   **29.** 6   **31.** $-\frac{5}{4}$ **33.** $(\frac{2}{3}, -\frac{7}{6})$

35.
$$\begin{vmatrix} a_1 & b_1 \\ a_2 & b_2 \end{vmatrix} = a_1b_2 - a_2b_1 = a_1b_2 - b_1a_2 = \begin{vmatrix} a_1 & a_2 \\ b_1 & b_2 \end{vmatrix}$$

37. Let $b_1 = ka_1$, $b_2 = ka_2$; then $\begin{vmatrix} a_1 & b_1 \\ a_2 & b_2 \end{vmatrix} = \begin{vmatrix} a_1 & ka_1 \\ a_2 & ka_2 \end{vmatrix} = \begin{vmatrix} a_1 & a_2 \\ ka_1 & ka_2 \end{vmatrix}$ (by Exercise 35)
$$= 0 \quad \text{(by Exercise 36)}$$

39.
$$\begin{vmatrix} 27 & 3 \\ 105 & -75 \end{vmatrix} = 3\begin{vmatrix} 9 & 1 \\ 105 & -75 \end{vmatrix} = 45\begin{vmatrix} 9 & 1 \\ 7 & -5 \end{vmatrix} = -2340$$

$$\begin{vmatrix} 27 & 3 \\ 105 & -75 \end{vmatrix} = 3\begin{vmatrix} 9 & 1 \\ 105 & -75 \end{vmatrix} = 9\begin{vmatrix} 3 & 1 \\ 35 & -75 \end{vmatrix} = 45\begin{vmatrix} 3 & 1 \\ 7 & -15 \end{vmatrix} = -2340$$

41.
$$\begin{vmatrix} a_1 + kb_1 & b_1 \\ a_2 + kb_2 & b_2 \end{vmatrix} = \begin{vmatrix} a_1 & b_1 \\ a_2 & b_2 \end{vmatrix} + \begin{vmatrix} kb_1 & b_1 \\ kb_2 & b_2 \end{vmatrix} \quad \text{(by Exercise 40)}$$

$$= \begin{vmatrix} a_1 & b_1 \\ a_2 & b_2 \end{vmatrix} + 0 \quad \text{(by Exercise 37)}$$

$$= \begin{vmatrix} a_1 & b_1 \\ a_2 & b_2 \end{vmatrix}$$

43. (Sample solution)
$$\begin{vmatrix} 12 & -42 \\ -6 & 27 \end{vmatrix} = 6\begin{vmatrix} 2 & -42 \\ -1 & 27 \end{vmatrix} = 12\begin{vmatrix} 1 & -21 \\ -1 & 27 \end{vmatrix} = 12\begin{vmatrix} 1 & -21 \\ 0 & 6 \end{vmatrix} = 12(6) = 72$$
$$\quad\text{Exercise 38} \qquad \text{Exercise 38} \qquad \text{Exercise 41}$$

10.5 Solving Linear Systems Using Third-Order Determinants (page 536)

1. $|A| = -123$ **3.** 35   **5.** -45 **7.** 4   **9.** $x = 2, x = 3$ **11.** $x = -1, x = 3$ (3 is a double root)   **13.** -92

15. 2   **17.** $(-1, 0, 2)$ **19.** $(-1, 7, 2)$ **21.** $(\frac{2}{3}, 1, \frac{1}{2})$

23.
$$\begin{vmatrix} a_1 & b_1 & c_1 \\ a_2 & b_2 & c_2 \\ a_3 & b_3 & c_3 \end{vmatrix} = a_1b_2c_3 + a_2b_3c_1 + a_3b_1c_2 - a_1b_3c_2 - a_2b_1c_3 - a_3b_2c_1 = \begin{vmatrix} a_1 & a_2 & a_3 \\ b_1 & b_2 & b_3 \\ c_1 & c_2 & c_3 \end{vmatrix}$$

25.
$$\begin{vmatrix} c_1 & b_1 & a_1 \\ c_2 & b_2 & a_2 \\ c_3 & b_3 & a_3 \end{vmatrix} = c_1b_2a_3 + c_2b_3a_1 + c_3b_1a_2 - c_1b_3a_2 - c_2b_1a_3 - c_3b_2a_1$$

$$= a_3b_2c_1 + a_1b_3c_2 + a_2b_1c_3 - a_2b_3c_1 - a_3b_1c_2 - a_1b_2c_3$$
$$= a_1b_3c_2 + a_2b_1c_3 + a_3b_2c_1 - a_1b_2c_3 - a_2b_3c_1 - a_3b_1c_2$$
$$= -(a_1b_2c_3 + a_2b_3c_1 + a_3b_1c_2 - a_1b_3c_2 - a_2b_1c_3 - a_3b_2c_1)$$

$$= -\begin{vmatrix} a_1 & b_1 & c_1 \\ a_2 & b_2 & c_2 \\ a_3 & b_3 & c_3 \end{vmatrix}$$

27. $\begin{vmatrix} a_1 + kb_1 & b_1 & c_1 \\ a_2 + kb_2 & b_2 & c_2 \\ a_3 + kb_3 & b_3 & c_3 \end{vmatrix} = (a_1 + kb_1)\begin{vmatrix} b_2 & c_2 \\ b_3 & c_3 \end{vmatrix} - (a_2 + kb_2)\begin{vmatrix} b_1 & c_1 \\ b_3 & c_3 \end{vmatrix} + (a_3 + kb_3)\begin{vmatrix} b_1 & c_1 \\ b_2 & c_2 \end{vmatrix}$

$$= \left(a_1\begin{vmatrix} b_2 & c_2 \\ b_3 & c_3 \end{vmatrix} - a_2\begin{vmatrix} b_1 & c_1 \\ b_3 & c_3 \end{vmatrix} + a_3\begin{vmatrix} b_1 & c_1 \\ b_2 & c_2 \end{vmatrix} \right)$$

$$+ kb_1\begin{vmatrix} b_2 & c_2 \\ b_3 & c_3 \end{vmatrix} - kb_2\begin{vmatrix} b_1 & c_1 \\ b_3 & c_3 \end{vmatrix} + kb_3\begin{vmatrix} b_1 & c_1 \\ b_2 & c_2 \end{vmatrix}$$

$$= \begin{vmatrix} a_1 & b_1 & c_1 \\ a_2 & b_2 & c_2 \\ a_3 & b_3 & c_3 \end{vmatrix} + k(b_1b_2c_3 - b_1b_3c_2 - b_1b_2c_3 + b_2b_3c_1 + b_1b_3c_2 - b_2b_3c_1)$$

$$= \begin{vmatrix} a_1 & b_1 & c_1 \\ a_2 & b_2 & c_2 \\ a_3 & b_3 & c_3 \end{vmatrix}$$

29. 0, since third column is -2 times first column.   **31.** $80 = 20(4)$

33. (Sample Solution)

$\begin{vmatrix} 5 & -4 & 3 \\ -6 & 6 & 2 \\ -7 & 3 & 4 \end{vmatrix} = \begin{vmatrix} 14 & -4 & 3 \\ 0 & 6 & 2 \\ 5 & 3 & 4 \end{vmatrix}$ 3 times third column
added to first column
(Exercise 27)

$$= \begin{vmatrix} 14 & -13 & 3 \\ 0 & 0 & 2 \\ 5 & -9 & 4 \end{vmatrix}$$ -3 times third column
added to second column
(Exercise 27)

$$= -2\begin{vmatrix} 14 & -13 \\ 5 & -9 \end{vmatrix}$$ Expansion by minors along row 2

$$= 2\begin{vmatrix} 14 & 13 \\ 5 & 9 \end{vmatrix}$$ (Exercise 26)

$$= 2(126 - 65)$$
$$= 122$$

35. $2\begin{vmatrix} 0 & 5 & -3 \\ 2 & -4 & 6 \\ 3 & 0 & 1 \end{vmatrix} - (-1)\begin{vmatrix} 1 & 5 & -3 \\ 0 & -4 & 6 \\ -5 & 0 & 1 \end{vmatrix} + 3\begin{vmatrix} 1 & 0 & -3 \\ 0 & 2 & 6 \\ -5 & 3 & 1 \end{vmatrix} - 0\begin{vmatrix} 1 & 0 & 5 \\ 0 & 2 & -4 \\ -5 & 3 & 0 \end{vmatrix} = -144$

37. (1) The property in Exercise 26
(2) y times the second column added to the first (Exercise 27)
(3) z times the third column added to the first (Exercise 27)
(4) Substituting for the given values of d_1, d_2, d_3 in the general system
(5) Divide by D

10.6 Systems of Linear Inequalities (page 542)

1.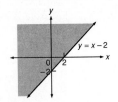

3.

5.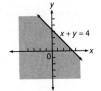

7.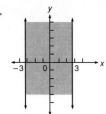

9.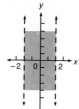

11.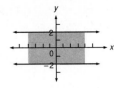

13.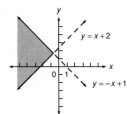

15.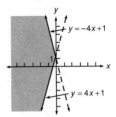

17.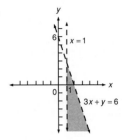

19.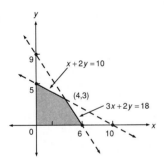

21.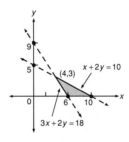

23.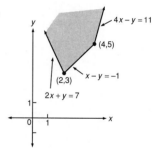

25.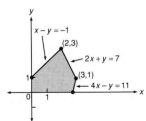

27.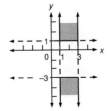

29. $2x + y < 6$
$3x - 4y \leq 12$

31. $y \leq 4$
$x + y \geq 4$
$x - y \leq 8$

33. $y < 4$
$x + y > 4$
$x - y \geq 8$

35. $y \geq 4$
$x + y > 4$
$x - y < 8$

37. $y < 4$
$x + y \leq 4$
$x - y < 8$

1. (a)

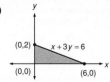

(b) $(0, 0)$, $(6, 0)$, $(0, 2)$ **(c)** Maximum $= 6$; minimum $= 0$ **(d)** Maximum $= 36$, minimum $= 0$
(e) Maximum $= 18$; minimum $= 0$

3. (a)

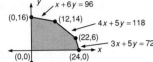

(b) $(0, 0)$; $(24, 0)$; $(22, 6)$; $(12, 14)$; $(0, 16)$ **(c)** Maximum $= 28$; minimum $= 0$
(d) Maximum $= 212$; minimum $= 0$; **(e)** Maximum $= 150$; minimum $= 0$

5. (a)

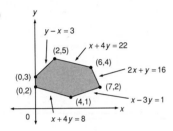

(b) $(0, 2)$, $(4, 1)$, $(7, 2)$, $(6, 4)$, $(2, 5)$, $(0, 3)$ **(c)** Maximum $= 10$; minimum $= 2$
(d) Maximum $= 76$; minimum $= 20$ **(e)** Maximum $= 49$; minimum $= 17$

7. For $p = x + 3y$, maximum $= 36$ at $(0, 12)$; minimum $= 10$ at $(10, 0)$
For $q = 4x + 3y$, maximum $= 40$ at $(10, 0)$; minimum $= 12$ at $(0, 4)$
For $r = \frac{1}{2}x + \frac{1}{4}y$, maximum $= 5$ at $(10, 0)$; minimum $= 1$ at $(0, 4)$

9. For $p = 2x + 2y$, maximum $= 24$ at $(7, 5)$; minimum $= 8$ at $(4, 0)$
For $q = 3x + y$, maximum $= 26$ at $(7, 5)$; minimum $= 9.5$ at $(\frac{3}{2}, 5)$
For $r = x + \frac{1}{5}y$, maximum $= 8$ at $(7, 5)$; minimum $= \frac{5}{2}$ at $(\frac{3}{2}, 5)$

11.

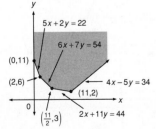

(a) No maximum value; minimum $= 8$ **(b)** No maximum value; minimum $= 63$
(c) No maximum value; minimum $= 10$

13. (a) $x \geq 0$, $y \geq 0$ because a negative number of either model is not possible; $\frac{3}{2}x + y$ is the amount of time that machine M_1 works per day, and $\frac{3}{2}x + y \leq 12$ says that M_1 works at most 12 hours daily. The remaining inequalities are the constraints for machines M_2 and M_3; the explanations are similar, as for M_1.

(b) $p = 5x + 8y$

(c)

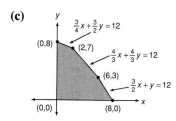

(d) Maximum $= \$66$ when $x = 2$ and $y = 7$ **(e)** Maximum $= \$64$ when $x = 8$ and $y = 0$

15. 400 of model A and 500 of model B. Profit $\$100,000$.

CHAPTER 10 TEST: STANDARD ANSWER (page 553)

1. No solutions **2.** $(4, -1, -6)$

3. $\left(-\frac{2}{3}c + \frac{2}{3}, -\frac{4}{3}c + \frac{7}{3}, c\right)$ for all c, or $\left(d, 1 + 2d, 1 - \frac{3}{2}d\right)$ for all d, or $\left(-\frac{1}{2} + \frac{1}{2}e, e, \frac{7}{4} - \frac{3}{4}e\right)$ for all e.

4. (a) $\begin{bmatrix} -6 & 3 \\ 10 & -5 \end{bmatrix}\begin{bmatrix} x \\ y \end{bmatrix} = \begin{bmatrix} 9 \\ -12 \end{bmatrix}$ **(b)** $\begin{bmatrix} 2 & -1 & 2 \\ 1 & 4 & -3 \\ -4 & 2 & -3 \end{bmatrix}\begin{bmatrix} x \\ y \\ z \end{bmatrix} = \begin{bmatrix} -3 \\ 18 \\ 0 \end{bmatrix}$

5. (a) $\begin{bmatrix} 1 & -1 & 0 \\ 6 & 1 & 4 \\ -6 & -5 & 3 \end{bmatrix}$ **(b)** Does not exist **6.** $\begin{bmatrix} 25 & -14 \\ -3 & 23 \end{bmatrix}$ **7.** $\begin{bmatrix} -1 & 0 & 31 \\ -10 & -4 & 104 \end{bmatrix}$ **8. (a)** -8 **(b)** 51

9. $\begin{bmatrix} -11 & -11 \\ -18 & -18 \end{bmatrix}$ **10.** Does not exist **11.** $\begin{bmatrix} 27 & 0 \\ 13 & 1 \end{bmatrix}$ **12.** $\begin{bmatrix} -2 & 2 & -4 \\ 2 & -2 & 4 \\ 4 & -4 & 8 \end{bmatrix}$ **13. (a)** 4 by 3 **(b)** Does not exist

14. $[-8]$ **15.** $\begin{bmatrix} \frac{1}{4} & \frac{1}{8} \\ \frac{1}{4} & -\frac{3}{8} \end{bmatrix}$ **16.** $(4, -3)$ **17.** $\begin{bmatrix} 0 & -\frac{2}{3} & \frac{1}{3} \\ 0 & 1 & 0 \\ -\frac{1}{2} & -\frac{1}{3} & \frac{1}{6} \end{bmatrix}$ **18.** $(-1, 1, -5)$ **19. (a)** $13ab$ **(b)** -45 **20.** $\left(1, -\frac{11}{2}\right)$

21. $(1, -2, 3)$ **22.**

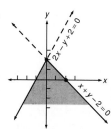

23.

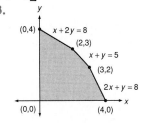

24. 13 **25.** 300 of model A
300 of model B

CHAPTER 10 TEST: MULTIPLE CHOICE (page 554)

1. (d) **2.** (e) **3.** (c) **4.** (a) **5.** (b) **6.** (d) **7.** (d) **8.** (c) **9.** (b) **10.** (a) **11.** (c) **12.** (a) **13.** (e)
14. (b) **15.** (c)

CHAPTER 11: SEQUENCES AND SERIES

11.1 Sequences (page 561)

1. $1, 3, 5, 7, 9$ **3.** $-1, 1, -1, 1, -1$ **5.** $-4, 2, -1, \frac{1}{2}, -\frac{1}{4}$ **7.** $-1, 4, -9, 16$ **9.** $\frac{3}{10}, \frac{3}{100}, \frac{3}{1000}, \frac{3}{10,000}$

11. $\frac{3}{100}, \frac{3}{10,000}, \frac{3}{1,000,000}, \frac{3}{100,000,000}$ **13.** $\frac{1}{2}, \frac{1}{6}, \frac{1}{12}, \frac{1}{20}$ **15.** $64, 36, 16, 4$ **17.** $-2, 1, 4, 7$ **19.** $0, \frac{1}{3}, \frac{1}{2}, \frac{3}{5}$

21. $1, \frac{3}{2}, \frac{16}{9}, \frac{125}{64}$ **23.** $-2, -\frac{3}{2}, -\frac{9}{8}, -\frac{27}{32}$ **25.** $\frac{1}{2}, \frac{1}{2}, \frac{3}{8}, \frac{1}{4}$ **27.** $-\frac{3}{2}, -\frac{5}{6}, -\frac{7}{12}, -\frac{9}{20}$ **29.** $4, 4, 4, 4$ **31.** 122

33. 0.000003 **35.** 12 **37.** -1331 **39.** $4, 6, 8, 10$ **41.** $5, 10, 15, 20, 25; s_n = 5n$

43. $-5, 25, -125, 625, -3125; s_n = (-5)^n$ **45.** $15, 21, 28$ **47.** $12, -4, 4, 0, 2, 1, \frac{3}{2}, \frac{5}{4}$

49. $1, 0, -1, 0, 1, 0, -1, 0$ **51.** $\frac{3}{2}, \frac{15}{8}, \frac{35}{16}, \frac{315}{128}$

11.2 Sums of Finite Sequences (page 565)

1. 45 **3.** 55 **5.** 0.33333 **7.** 510 **9.** 60 **11.** 254 **13.** $\frac{381}{64}$ **15.** 105 **17.** 40 **19.** $\frac{13}{8}$ **21.** 0 **23.** 57

25. 1111.111 **27.** 60 **29.** $\frac{35}{12}$ **31.** 0.010101 **33.** $\sum_{k=1}^{10} 5k$ **35.** $\sum_{k=-3}^{8} 3k$ **37. (a)** $4, 9, 16, 25, 36$; **(b)** n^2

39. $a_n = a_1 + a_{n-1} = a_1 + (a_1 + a_{n-2}) = \cdots = \overbrace{a_1 + a_1 + a_1 + \cdots + a_1}^{n \text{ terms}} = 2 + 2 + 2 + \cdots + 2 = 2n$

41. $\sum_{k=1}^{n} s_k + \sum_{k=1}^{n} t_k = (s_1 + s_2 + \cdots + s_n) + (t_1 + t_2 + \cdots + t_n) = (s_1 + t_1) + (s_2 + t_2) + \cdots + (s_n + t_n)$

$$= \sum_{k=1}^{n} (s_k + t_k)$$

43. $\sum_{k=1}^{n} (s_k + c) = (s_1 + c) + (s_2 + c) + \cdots + (s_n + c) = (s_1 + s_2 + \cdots + s_n) + (c + c + \cdots + c)$

$$= \left(\sum_{k=1}^{n} s_k \right) + nc$$

45. $\sum_{k=1}^{10} \frac{2}{k(k + 2)} = \sum_{k=1}^{10} \left(\frac{1}{k} - \frac{1}{k + 2} \right)$

$$= \left(1 - \frac{1}{3}\right) + \left(\frac{1}{2} - \frac{1}{4}\right) + \left(\frac{1}{3} - \frac{1}{5}\right) + \left(\frac{1}{4} - \frac{1}{6}\right) + \left(\frac{1}{5} - \frac{1}{7}\right) + \left(\frac{1}{6} - \frac{1}{8}\right) + \left(\frac{1}{7} - \frac{1}{9}\right)$$

$$+ \left(\frac{1}{8} - \frac{1}{10}\right) + \left(\frac{1}{9} - \frac{1}{11}\right) + \left(\frac{1}{10} - \frac{1}{12}\right)$$

$$= 1 + \frac{1}{2} - \frac{1}{11} - \frac{1}{12} = \frac{264 + 132 - 24 - 22}{2 \cdot 11 \cdot 12} = \frac{350}{2 \cdot 11 \cdot 12} = \frac{175}{132}$$

11.3 Arithmetic Sequences and Series (page 572)

1. $5, 7, 9; 2n - 1; 400$ **3.** $-10, -16, -22; -6n + 8; -1100$ **5.** $\frac{17}{2}, 9, \frac{19}{2}; \frac{1}{2}n + 7; 245$

7. $-\frac{4}{5}, -\frac{7}{5}, -2; -\frac{3}{5}n + 1; -106$ **9.** $150, 200, 250; 50n; 10,500$ **11.** $30, 50, 70; 20n - 30; 3600$ **13.** 455

15. $\frac{63}{4}$ **17.** 289 **19.** $15,150$ **21.** $94,850$ **23.** $\frac{173,350}{7}$ **25.** $567,500$ **27.** 5824 **29.** $15,300$

31. (a) $10,000$; **(b)** n^2 **33.** -36 **35.** 4620 **37.** $\frac{3577}{4}$ **39.** $\frac{5}{2}n(n + 1)$ **41.** $-\frac{9}{4}$ **43.** $\frac{228}{15} = \frac{76}{5}$ **45.** $\$1183$

47. (a) $\$240, \$220, \$200$;

    (b) monthly loan balance $= 12,000 - 1000(k - 1)$, monthly interest $= [12,000 - 1000(k - 1)](0.02) = 240 - 20k$;

    (c) $\$1560, 13\%$

49. 930 **51.** 9080 **53.** $45,540$ **55.** $a_1 = -7; d = -13$ **57.** $u = \frac{16}{3}; v = \frac{23}{3}$ **59.** 520

11.4 Geometric Sequences and Series (page 579)

1. 16, 32, 64; 2^n **3.** 27, 81, 243; 3^{n-1} **5.** $\frac{1}{9}$, $-\frac{1}{27}$, $\frac{1}{81}$; $-3(-\frac{1}{3})^{n-1}$ **7.** $-125, -625, -3125$; -5^{n-1}
9. $-\frac{16}{9}$, $-\frac{32}{27}$, $-\frac{64}{81}$; $-6(\frac{2}{3})^{n-1}$ **11.** 1000, 100,000, 10,000,000; $\frac{1}{1000}(100)^{n-1}$ **13.** 126 **15.** $-\frac{1330}{81}$ **17.** 1024
19. 3 **21.** $-\frac{1}{2}$ **23.** 1023 **25.** $2^n - 1$ **27.** $\frac{121}{27}$ **29.** $\frac{211}{54}$ **31.** $-\frac{85}{64}$ **33.** $-\frac{3}{21}$ **35.** \$10,737,418
37. 512,000; $1000(2^{n-1})$ **39.** \$3117.79 **41.** **(a)** \$800(1.11)^n; **(b)** \$1350 **43.** \$703.99

45. **(a)** If the volume of the first container is V, then $\frac{1}{2}V + (\frac{1}{2})^2V + (\frac{1}{2})^3V + (\frac{1}{2})^4V + (\frac{1}{2})^5V = \frac{31}{32}V$ is the sum of the volumes of the other five. Since $\frac{31}{32}V < V$, the answer is yes.

(b) $\sum_{k=1}^{5}(\frac{2}{3})^kV = \frac{422}{243}V > V$; therefore, no.

11.5 Infinite Geometric Series (page 587)

1. 4 **3.** $\frac{125}{4}$ **5.** $\frac{2}{3}$ **7.** $\frac{10}{9}$ **9.** $-\frac{16}{7}$ **11.** The numerator at the right should be $a = \frac{1}{4}$, not 1.
13. The denominator at the right should be $1 - (-\frac{1}{3})$ since $r = -\frac{1}{3}$, not $\frac{1}{3}$. **15.** $\frac{3}{2}$ **17.** $\frac{1}{6}$ **19.** 4 **21.** $\frac{20}{9}$
23. No finite sum **25.** $\frac{10}{3}$ **27.** No finite sum **29.** 30 **31.** $\frac{20}{11}$ **33.** 4 **35.** $\frac{4}{21}$ **37.** $\frac{7}{9}$ **39.** $\frac{13}{99}$ **41.** $\frac{13}{990}$

43. 1 **45.** **(a)** $\frac{4}{3}, \frac{4}{9}, \frac{4}{27}, \ldots, \frac{4}{3^n}, \ldots$; **(b)** $\sum_{n=1}^{\infty}\frac{4}{3^n} = \frac{\frac{4}{3}}{1 - \frac{1}{3}} = 2$ **47.** 8 hours

49. The time for the last $\frac{1}{2}$ mile would have to be $\sum_{n=1}^{\infty}\frac{2}{5}(\frac{10}{9})^{n-1}$, which is not a finite sum, since $\frac{10}{9} > 1$.
51. **(a)** $\frac{1}{2}(AC)(CB) = \frac{1}{2}(4)(4) = 8$

(b) $4 + 2 + 1 + \frac{1}{2} + \cdots = \frac{4}{1 - \frac{1}{2}} = 8$

(c) For the odd-numbered triangles:

$$4 + 1 + \frac{1}{4} + \cdots = \frac{4}{1 - \frac{1}{4}} = \frac{16}{3}$$

For the even-numbered triangles:

$$2 + \frac{1}{2} + \frac{1}{8} + \cdots = \frac{2}{1 - \frac{1}{4}} = \frac{8}{3}$$

$$\frac{16}{3} + \frac{8}{3} = 8$$

53. $\frac{1}{2}(9)(3) + \frac{1}{2}(6)(2) + \frac{1}{2}(4)(\frac{4}{3}) + \cdots = \frac{27}{2} + 6 + \frac{8}{3} + \cdots = \frac{\frac{27}{2}}{1 - \frac{4}{9}} = \frac{243}{10}$ **55.** **(a)** \$3000 **(b)** \$2000

11.6 Mathematical Induction (page 594)

For these exercises, S_n represents the given statement where n is an integer ≥ 1 ($n \geq 2$ when appropriate). The second part of each proof begins with the hypothesis S_k, where k is an arbitrary positive integer.

1. Since $1 = \frac{1(1 + 1)}{2}$, S_1 is true. Assume S_k and add $k + 1$ to obtain

$$1 + 2 + 3 + \cdots + k + (k + 1) = \frac{k(k + 1)}{2} + (k + 1) = \frac{k^2 + 3k + 2}{2} = \frac{(k + 1)(k + 2)}{2}$$

$$= \frac{(k + 1)[(k + 1) + 1]}{2}$$

Therefore, S_{k+1} holds. Since S_1 is true and S_k implies $S_{k + 1}$, the principle of mathematical induction makes S_n true for all integers $n \geq 1$. *Note:* The preceding sentence is an appropriate final statement for the remaining proofs. For the sake of brevity, however, it will not be repeated.

3. Since $\sum\limits_{i=1}^{1} 3i = 3 = \dfrac{3(1 + 1)}{2}$, S_1 is true. Assume S_k and add $3(k + 1)$ to obtain the following.

$$\sum_{i=1}^{k+1} 3i = \left(\sum_{i=1}^{k} 3i\right) + 3(k + 1) = \dfrac{3k(k + 1)}{2} + 3(k + 1); \quad \sum_{i=1}^{k+1} 3i = \dfrac{3(k^2 + 3k + 2)}{2} = \dfrac{3(k + 1)(k + 2)}{2} =$$

$\dfrac{3(k + 1)[(k + 1) + 1]}{2}$. Therefore, S_{k+1} holds.

5. Since $\dfrac{5}{3} = \dfrac{1(11 - 1)}{6}$, S_1 is true. Assume S_k and add $-\dfrac{1}{3}(k + 1) + 2$ to obtain $\dfrac{5}{3} + \dfrac{4}{3} + 1 + \cdots +$

$\left(-\dfrac{1}{3}k + 2\right) + \left[-\dfrac{1}{3}(k + 1) + 2\right] = \dfrac{k(11 - k)}{6} + \left[-\dfrac{1}{3}(k + 1) + 2\right] = \dfrac{10 + 9k - k^2}{6} = \dfrac{(k + 1)(10 - k)}{6} =$

$\dfrac{(k + 1)[11 - (k + 1)]}{6}$. Therefore, S_{k+1} holds.

7. Since $\dfrac{1}{1 \cdot 2} = \dfrac{1}{1 + 1}$, S_1 is true. Assume S_k and add $\dfrac{1}{(k + 1)[(k + 1) + 1]}$ to obtain $\dfrac{1}{1 \cdot 2} + \dfrac{1}{2 \cdot 3} + \cdots$

$+ \dfrac{1}{k(k + 1)} + \dfrac{1}{(k + 1)(k + 2)} = \dfrac{k}{k + 1} + \dfrac{1}{(k + 1)(k + 2)} = \dfrac{k^2 + 2k + 1}{(k + 1)(k + 2)} = \dfrac{(k + 1)^2}{(k + 1)(k + 2)} = \dfrac{k + 1}{k + 2} =$

$\dfrac{k + 1}{(k + 1) + 1}$ Therefore, S_{k+1} holds.

9. S_1 is true since $-2 = -1 - (1^2)$. Assume S_k and add $-2(k + 1)$ to obtain

$$-2 - 4 - 6 - \cdots - 2k - 2(k + 1) = -k - k^2 - 2(k + 1) = -(k + 1) - (k^2 + 2k + 1)$$
$$= -(k + 1) - (k + 1)^2$$

Therefore, S_{k+1} holds.

11. S_1 is true since $1 = \frac{5}{3}[1 - (\frac{2}{5})^1]$. Assume S_k and add $(\frac{2}{5})^k$ to obtain

$$1 + \tfrac{2}{5} + \tfrac{4}{25} + \cdots + (\tfrac{2}{5})^{k-1} + (\tfrac{2}{5})^k = \tfrac{5}{3}[1 - (\tfrac{2}{5})^k] + (\tfrac{2}{5})^k$$
$$= \tfrac{5}{3}[1 - (\tfrac{2}{5})^k + \tfrac{3}{5}(\tfrac{2}{5})^k]$$
$$= \tfrac{5}{3}[1 - (\tfrac{2}{5})^k(1 - \tfrac{3}{5})]$$
$$= \tfrac{5}{3}[1 - (\tfrac{2}{5})^k(\tfrac{2}{5})]$$
$$= \tfrac{5}{3}[1 - (\tfrac{2}{5})^{k+1}]$$

Therefore, S_{k+1} holds.

13. S_1 is true since $1^3 = 1 = \dfrac{1^2(1 + 1)^2}{4}$. Assume S_k and add $(k + 1)^3$ to obtain

$$1^3 + 2^3 + 3^3 + \cdots + k^3 + (k + 1)^3 = \dfrac{k^2(k + 1)^2}{4} + (k + 1)^3$$
$$= \dfrac{k^2(k + 1)^2 + 4(k + 1)^3}{4}$$
$$= \dfrac{(k + 1)^2[k^2 + 4k + 4]}{4}$$
$$= \dfrac{(k + 1)^2[k + 2]^2}{4}$$
$$= \dfrac{(k + 1)^2[(k + 1) + 1]^2}{4}$$

Therefore, S_{k+1} holds.

15. S_1 is true since $\displaystyle\sum_{i=1}^{1} ar^{i-1} = a = \frac{a(1-r^1)}{1-r}$. Assume S_k and add $ar^{(k+1)-1}$ to obtain

$$\sum_{i=1}^{k} ar^{i-1} + ar^k = \frac{a(1-r^k)}{1-r} + ar^k$$

$$\sum_{i=1}^{k+1} ar^{i-1} = \frac{a(1-r^k)}{1-r} + ar^k$$

$$= \frac{a(1-r^k) + ar^k - ar^{k+1}}{1-r}$$

$$= \frac{a - ar^{k+1}}{1-r}$$

$$= \frac{a(1-r^{k+1})}{1-r}$$

Therefore, S_{k+1} holds.

17. $2^5 = 32 > 20 = 4 \cdot 5$. So S_5 is true. Assume $2^k > 4k$. Then multiply by 2 to get $2^{k+1} > 8k = 4k + 4k \geq 4k + 20 > 4k + 4 = 4(k+1)$. Therefore $2^{k+1} > 4(k+1)$ and S_{k+1} holds.

19. For $n = 1$, $a^1 = a < 1$ since $0 < a < 1$ (given). Thus S_1 is true. Assume $a^k < 1$. Then, since $a > 0$, $a^{k+1} < a$. But $a < 1$. Therefore $a^{k+1} < 1$ and S_{k+1} holds.

21. S_1 is true since $(ab)^1 = a^1 b^1$. Assume that $(ab)^k = a^k b^k$ and multiply by ab to obtain

$$(ab)^k(ab) = (a^k b^k)ab$$

$$(ab)^{k+1} = (a^k a)(b^k b)$$

$$(ab)^{k+1} = a^{k+1} b^{k+1}$$

Therefore, S_{k+1} holds.

23. S_1 is true since $|a_0 + a_1| \leq |a_0| + |a_1|$. Assume S_k. Then

$$|a_0 + a_1 + \cdots + a_k + a_{k+1}| = |(a_0 + a_1 + \cdots + a_k) + a_{k+1}|$$

$$\leq |a_0 + a_1 + \cdots + a_k| + |a_{k+1}| \qquad \text{(by } S_1\text{)}$$

$$\leq (|a_0| + |a_1| + \cdots + |a_k|) + |a_{k+1}| \qquad \text{(by } S_k\text{)}$$

$$= |a_0| + |a_1| + \cdots + |a_{k+1}|$$

Therefore, S_{k+1} holds.

25. (a) Since $\dfrac{a^2 - b^2}{a - b} = a + b = a^{2-1} + b^{2-1}$, S_2 is true. Assume S_k. Then

$$\frac{a^{k+1} - b^{k+1}}{a - b} = \frac{a^k a - b^k b}{a - b}$$

$$= \frac{a^k a - b^k a + b^k a - b^k b}{a - b}$$

$$= \frac{a(a^k - b^k) + b^k(a - b)}{a - b} = \frac{a(a^k - b^k)}{a - b} + b^k$$

$$= a[a^{k-1} + a^{k-2}b + \cdots + ab^{k-2} + b^{k-1}] + b^k \qquad \text{(by } S_k\text{)}$$

$$= a^k + a^{k-1}b + \cdots + a^2 b^{k-1} + ab^{k-1} + b^k$$

Therefore, S_{k+1} holds.

(b) Since $\dfrac{a^n - b^n}{a - b} = a^{n-1} + a^{n-2}b + \cdots + ab^{n-2} + b^{n-1}$, multiplying by $a - b$ gives

$$a^n - b^n = (a - b)(a^{n-1} + a^{n-2}b + \cdots + ab^{n-2} + b^{n-1})$$

27. (a) 1; 4; 9; 16; 25; **(b)** n^2 **(c)** S_1 is true since $1 = 1^2$. Assume S_k and add $k + (k + 1)$ to obtain

$$1 + 2 + 3 + \cdots + (k - 1) + k + (k + 1) + k + (k - 1) + \cdots + 3 + 2 + 1 = k^2 + k + (k + 1)$$
$$= k^2 + 2k + 1$$
$$= (k + 1)^2$$

Therefore, S_{k+1} holds.

29. For $n = 1$, $7^1 - 1 = 6$ which is divisible by 6. So S_1 holds. Assume S_k. Then $7^k - 1 = 6b$ and $7^k = 6^b + 1$. Multiply by 7 to get $7^{k+1} = 42^b + 7$ and $7^{k+1} - 1 = 42b + 6 = 6(7b + 1)$ which is divisible by 6. Thus S_{k+1} holds.

CHAPTER 11 TEST: STANDARD ANSWER (page 598)

1. $1, -1, -1, -\frac{8}{7}; -\frac{800}{97}$ **2.** $\frac{1}{300}$ **3.** $0, 3, 2, 5$ **4.** $-\frac{1}{3}$ **5.** $-\frac{2}{3}$ **6.** $-\frac{76}{15}$ **7.** 21 **8.** 35

9. $12, -3, \frac{3}{4}; -768(-\frac{1}{4})^{n-1}$ **10.** 8925 **11.** 26 **12.** $13{,}845$ **13.** $\sum\limits_{k=1}^{4} 8\left(\frac{1}{2}\right)^k = \dfrac{4\left(1 - \dfrac{1}{2^4}\right)}{1 - \frac{1}{2}} = \dfrac{15}{2}$ **14.** $15{,}554$

15. $24\left(1 - \dfrac{1}{2^8}\right) = \dfrac{765}{32}$ **16.** $-\frac{1}{2}$ **17.** 18 **18.** No finite sum since $r = \frac{3}{2} > 1$ **19.** $\frac{6}{115}$ **20.** $\frac{4}{11}$ **21.** $\$652.60$

22. 250 **23.** 36 feet   **24.** For $n = 1$, we have $5 = \dfrac{5 \cdot 1(1 + 1)}{2}$. If $5 + 10 + \cdots + 5k = \dfrac{5k(k + 1)}{2}$, then

$$5 + 10 + \cdots + 5k + 5(k + 1) = \dfrac{5k(k + 1)}{2} + 5(k + 1)$$
$$= \dfrac{5k(k + 1) + 10(k + 1)}{2}$$
$$= \dfrac{5(k + 1)(k + 2)}{2}$$
$$= \dfrac{5(k + 1)[(k + 1) + 1]}{2}$$

Thus the statement holds for the $k + 1$ case, and by the principle of mathematical induction the statement is true for all $n \geq 1$.

25. For $n = 1$, $\dfrac{1}{1 \cdot 3} = \dfrac{1}{2 \cdot 1 + 1}$. For $n = k$, assume that

$$\dfrac{1}{1 \cdot 3} + \dfrac{1}{3 \cdot 5} + \cdots + \dfrac{1}{(2k - 1)(2k + 1)} = \dfrac{k}{2k + 1}$$

Add $\dfrac{1}{(2k + 1)(2k + 3)}$ to get

$$\dfrac{1}{1 \cdot 3} + \dfrac{1}{3 \cdot 5} + \cdots + \dfrac{1}{(2k - 1)(2k + 1)} + \dfrac{1}{(2k + 1)(2k + 3)}$$
$$= \dfrac{k}{2k + 1} + \dfrac{1}{(2k + 1)(2k + 3)}$$
$$= \dfrac{k(2k + 3) + 1}{(2k + 1)(2k + 3)} = \dfrac{(2k + 1)(k + 1)}{(2k + 1)(2k + 3)} = \dfrac{k + 1}{2(k + 1) + 1}$$

CHAPTER 11 TEST: MULTIPLE CHOICE (page 599)

1. (b)   **2.** (a)   **3.** (b)   **4.** (a)   **5.** (d)   **6.** (d)   **7.** (c)   **8.** (c)   **9.** (a)   **10.** (d)   **11.** (e)   **12.** (a)
13. (d)   **14.** (b)   **15.** (c)

CHAPTER 12: PERMUTATIONS, COMBINATIONS, AND PROBABILITY

12.1 Permutations (page 607)

1. 7 **3.** 66 **5.** 120 **7.** 4 **9.** $\dfrac{n!}{[n-(n-3)]!} = \dfrac{n!}{3!}$ **11.** 15 **13.** 60 **15.** 12 **17.** 6 **19.** 36 **21.** 504

23. 224 **25.** 224 **27.** (a) 120; (b) 12; (c) 48; (d) 72 **29.** 362,880; 40,320 **31.** 720 **33.** 468,000; 676,000

35. (a) 10 (b) 4

37. 36;

(1, 1)	(2, 1)	(3, 1)	(4, 1)	(5, 1)	(6, 1)
(1, 2)	(2, 2)	(3, 2)	(4, 2)	(5, 2)	(6, 2)
(1, 3)	(2, 3)	(3, 3)	(4, 3)	(5, 3)	(6, 3)
(1, 4)	(2, 4)	(3, 4)	(4, 4)	(5, 4)	(6, 4)
(1, 5)	(2, 5)	(3, 5)	(4, 5)	(5, 5)	(6, 5)
(1, 6)	(2, 6)	(3, 6)	(4, 6)	(5, 6)	(6, 6)

39. (a) 1,000,000,000 (b) 3,628,800 (c) 996,371,200 (d) $10 \cdot 7(9 \cdot 8 \cdot 7 \cdot 6 \cdot 5 \cdot 4) = 4,233,600$

41. (a) 720; (b) 120; (c) 240; **43.** 5 **45.** (a) 6720; (b) 1260; (c) 151,200

12.2 Combinations (page 613)

1. 10 **3.** 1 **5.** 1 **7.** 4060 **9.** (a) 4845 (b) 116,280 **11.** 20 **13.** 3003 **15.** 435 **17.** 46,200

19. (a) n (b) $\dfrac{n(n-1)}{2}$ **21.** 70 **23.** (a) 2^2 (b) 2^3 (c) 2^4 (d) 2^5

25. Each time a subset of r elements is chosen out of n elements there are $n - r$ elements left over. Likewise, when $n - r$ elements are chosen out of n elements there are $n - (n - r) = r$ elements left over. Therefore, there must be the same number of subsets of size r, $\dbinom{n}{r}$, as there are subsets of size $n - r$, $\dbinom{n}{n-r}$.

27. A set of n elements has a total of 2^n subsets of all possible sizes, including the empty set and the set itself.

29. $4\dbinom{13}{5} = 5148$ **31.** $13\dbinom{4}{2}12\dbinom{4}{3} = 3744$ **33.** (a) 205,931,880 (b) 147,094,200 **35.** 180

37. (a) 1008 (b) 2254 **39.** 256 **41.** $210 \times 5! \times 3! = 151,200$

12.3 The Binomial Expansion (page 620)

1. $x^5 + 5x^4 + 10x^3 + 10x^2 + 5x + 1$ **3.** $x^7 + 7x^6 + 21x^5 + 35x^4 + 35x^3 + 21x^2 + 7x + 1$

5. $a^4 - 4a^3b + 6a^2b^2 - 4ab^3 + b^4$ **7.** $243x^5 - 405x^4y + 270x^3y^2 - 90x^2y^3 + 15xy^4 - y^5$

9. $a^{10} + 5a^8 + 10a^6 + 10a^4 + 5a^2 + 1$

11. $1 - 10h + 45h^2 - 120h^3 + 210h^4 - 252h^5 + 210h^6 - 120h^7 + 45h^8 - 10h^9 + h^{10}$

13. $\frac{1}{16} - \frac{1}{2}a + \frac{3}{2}a^2 - 2a^3 + a^4$ **15.** $\dfrac{1}{x^6} - \dfrac{6}{x^3} + 15 - 20x^3 + 15x^6 - 6x^9 + x^{12}$ **17.** $3 + 3h + h^2$

19. $3c^2 + 3ch + h^2$ **21.** $2(5x^4 + 10x^3h + 10x^2h^2 + 5xh^3 + h^4)$

23. $(1 + 1)^{10} = 1 + 10 + 45 + 120 + 210 + 252 + 210 + 120 + 45 + 10 + 1 = 1024$

25. $c^{20} + 20c^{19}h + 190c^{18}h^2 + 1140c^{17}h^3 + 4845c^{16}h^4 + \cdots + 4845c^4h^{16} + 1140c^3h^{17} + 190c^2h^{18} + 20ch^{19} + h^{20}$

27. The nth row of the triangle contains the coefficients in the expansion of $(a + b)^n$.

29. $a^6 + 12a^5 + 60a^4 + 160a^3 + 240a^2 + 192a + 64$ **31.** $81p^4 + 216p^3q + 216p^2q^2 + 96pq^3 + 16q^4$

33.
```
       1   7   21    35    35   21   7    1
     1   8   28    56    70    56   28    8   1
   1   9   36    84   126   126   84   36   9   1
 1  10   45   120   210   252   210  120   45  10   1
```

35. $x^{10} - 10x^9h + 45x^8h^2 - 120x^7h^3 + 210x^6h^4 - 252x^5h^5 + 210x^4h^6 - 120x^3h^7 + 45x^2h^8 - 10xh^9 + h^{10}$

37. $8064a^5b^5$ **39.** $35x^{-5/2}$ **41.** $48,384x^5y^3$ **43.** $-\dfrac{15309}{8}a^5b^5$

45. $(2 + .1)^4 = 2^4 + 4(2^3)(.1) + 6(2^2)(.1)^2 + 4(2)(.1)^3 + (.1)^4 = 19.4481$

47. $(4 - .02)^3 = (4)^3 - 3(4)^2(.02) + 3(4)(.02)^2 - (.02)^3 = 63.044792$

12.4 Probability (page 627)

1. $\frac{1}{8}$ **3.** $\frac{7}{8}$ **5.** $\frac{1}{2}$ **7.** $\frac{1}{2}$ **9.** $\frac{9}{16}$ **11.** $\frac{1}{4}$ **13.** $\frac{3}{8}$ **15.** $\frac{1}{6}$ **17.** $\frac{1}{6}$ **19.** $\frac{1}{9}$ **21.** $\frac{1}{2}$ **23.** $\frac{25}{102}$ **25.** 0   **27.** $\frac{1}{2}$

29. $\frac{1}{2704}$ **31.** $\frac{144}{169}$ **33.** $\frac{1}{169}$ **35.** $\frac{28}{143}$ **37.** $\frac{70}{143}$ **39.** **(a)** $\frac{1}{1024}$ **(b)** $\frac{1}{1024}$ **(c)** $\frac{7}{128}$ **41.** $\frac{77}{102}$

43. **(a)** $\dfrac{48}{\binom{52}{2}} = 0.0000185$ **(b)** $\dfrac{13 \cdot 48}{\binom{52}{2}} = 0.0002401$ **(c)** $\dfrac{4\binom{13}{5}}{\binom{52}{5}} = 0.0019808$ **(d)** $\dfrac{42}{\binom{52}{5}} = \dfrac{1}{649,740}$

45. 1 to 12; 12 to 1   **47.** 7 to 2   **49.** 3 to 5

51. Let $x = P(E)$. Then $P(\text{not } E) = 1 - x$ and we have the odds for $E = \dfrac{a}{b} = \dfrac{x}{1 - x}$. Then

$$a(1 - x) = bx$$
$$a - ax = bx$$
$$a = (a + b)x$$
$$\frac{a}{a + b} = x \quad \text{or} \quad P(E) = \frac{a}{a + b}$$

CHAPTER 12 TEST: STANDARD ANSWER (page 631)

1. **(a)** 720 **(b)** 120   **2.** 120   **3.** **(a)** 900 **(b)** 648   **4.** 455   **5.** 450   **6.** 360   **7.** 420   **8.** 2970

9. 6   **10.** **(a)** 32 **(b)** $\frac{1}{32}$ **(c)** $\frac{1}{32}$ **11.** $\frac{2}{3}$ **12.** **(a)** $\frac{1}{36}$ **(b)** $\frac{35}{36}$ **(c)** $\frac{1}{6}$ **13.** **(a)** $\frac{2}{169}$ **(b)** $\frac{2}{221}$ **14.** $\dfrac{\binom{12}{4} \cdot 40}{\binom{52}{5}} = 0.0076184$

15. $\dfrac{\binom{20}{4} \cdot 15}{\binom{35}{5}} = 0.2238689$ **16.** **(a)** $\frac{1}{16}$ **(b)** $\frac{1}{16}$ **17.** $\frac{2}{11}$ **18.** $\frac{9}{40}$ **19.** $\frac{39}{80}$ **20.** $\frac{7}{27}$

21. $x^5 - 10x^4y + 40x^3y^2 - 80x^2y^3 + 80xy^4 - 32y^5$ **22.** $\dfrac{1}{1024a^{10}} - \dfrac{5b}{256a^9} + \dfrac{45b^2}{256a^8} - \dfrac{15b^3}{16a^7}$

23. $\binom{11}{6}(3a)^5b^6 = 112,266a^5b^6$ **24.** $\binom{16}{8}(2x)^8(-y)^8 = 3,294,720x^8y^8$ **25.** 92.3521

CHAPTER 12 TEST: MULTIPLE CHOICE (page 632)

1. (c)   **2.** (d)   **3.** (c)   **4.** (b)   **5.** (b)   **6.** (e)   **7.** (d)   **8.** (a)   **9.** (b)   **10.** (c)   **11.** (a)   **12.** (d)   **13.** (c)
14. (a)   **15.** (b)

INDEX

GEOMETRIC FORMULAS

Triangle

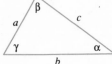

$$\frac{a}{a'} = \frac{b}{b'} = \frac{c}{c'}$$
$$\alpha = \alpha'$$
$$\beta = \beta'$$
$$\gamma = \gamma'$$

Right Triangle

Pythagorean theorem
$a^2 + b^2 = c^2$

Special Triangles

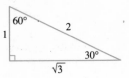

Square

Area $= s^2$
Perimeter $= 4s$

Rectangle

Area $= \ell w$
Perimeter $= 2\ell + 2w$

Parallelogram

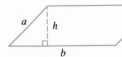

Area $= bh$
Perimeter $= 2a + 2b$

Trapezoid

Area $= \frac{1}{2} h (a + b)$

Circle

Area $= \pi r^2$
Circumference $= 2\pi r$

Circular Sector

Area $= \frac{1}{2} r^2 \theta$
Arc length $s = r\theta$

Sphere

Volume $= \frac{4}{3} \pi r^3$
Surface area $= 4\pi r^2$

Right Circular Cylinder

Volume $= \pi r^2 h$
Lateral surface area $= 2\pi r h$

Right Circular Cone

Volume $= \frac{1}{3} \pi r^2 h$
Lateral surface area $= \pi r \sqrt{r^2 + h^2}$

Rectangular Solid

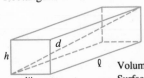

Volume $= \ell w h$
Surface area $= 2\ell w + 2\ell h + 2wh$
Diagonal: $d^2 = \ell^2 + w^2 + h^2$